I0848277

VLSI FABRICATION
PRINCIPLES

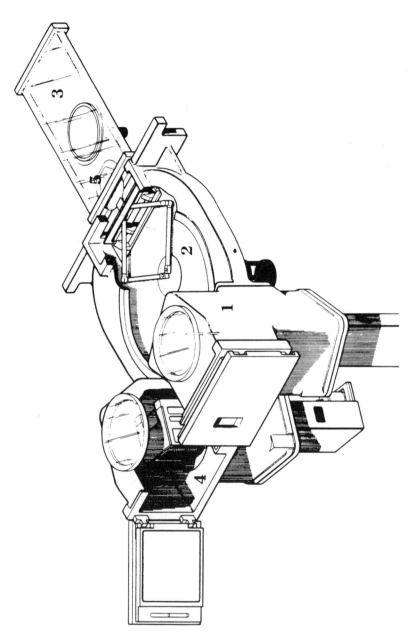

Schematic of a fully automatic, cassette-loaded, epitaxial reactor. *Key:* 1,4: Cassette ports; 2: wafer transfer chamber; 3: process chamber; 5: wafer handling mechanism. *Courtesy of ASM Epitaxy, Phoenix, Arizona.*

VLSI FABRICATION PRINCIPLES
Silicon and Gallium Arsenide

Second Edition

SORAB K. GHANDHI
Rensselaer Polytechnic Institute
Troy, New York

A Wiley-Interscience Publication
JOHN WILEY & SONS, INC.
New York / Chichester / Brisbane / Toronto / Singapore

Library of Congress Cataloging in Publication Data:

Ghandhi, Sorab Khushro, 1928–
    VLSI fabrication principles: silicon and gallium arsenide / Sorab
K. Ghandhi.—Rev. ed. [i.e. 2nd ed.]
        p.   cm.
    Includes bibliographical references and index.
    ISBN 0-471-58005-8 (alk. paper)
    1. Integrated circuits—Very large scale integration. 2. Silicon.
3. Gallium arsenide. I. Title.
TK7874.G473   1994
621.3815′2—dc20   93-40716
                                                CIP

Printed in the United States of America

10 9 8 7 6 5 4 3 2

To my wife, Cecille
and
my sons, Khushro, Rustom, and Behram
for making it all worthwhile

PREFACE TO THE SECOND EDITION

It is now 10 years since the publication of the first edition of this volume. In its Preface, I had noted that "The aim of this book is to provide graduate students and practicing engineers with a body of knowledge of VLSI fabrication principles that will not only *bring* them up-to-date, but also allow them to *stay* up-to-date in the field." This is still the primary focus of the second edition; in fact, it is even more important now because of the proliferation of new techniques and technologies that are being brought to bear on this field.

I have used this book in teaching courses to both graduate students and practicing engineers, and found that the best way to achieve this goal is to emphasize the basic principles governing the direction of developments in this field. Consequently, this approach has been carried over to this edition, as has the overall organization of its chapters. Within each chapter, however, the reader will find that many changes have been made to reflect advances that have occurred in each area.

For example, developments in the areas of Deposited Films and Etching have been very rapid, with a proliferation of new techniques on an almost daily basis. This chapter has been expanded to reflect this growth; moreover, I have attempted to bring some order to this apparently chaotic situation by emphasizing the basic principles underlying these developments. Thus, many new technological developments have been described but always with an emphasis on their underlying physical principles.

In areas such as Diffusion and Epitaxy, for example, our understanding of the basic physical principles has advanced to the point where the technology is on a considerably firmer footing than it was 10 years ago. Here, the material has been completely rewritten to reflect this understanding.

My experience in teaching this material has brought about many changes in the order of its presentation. Often, what appeared logical at first writing did not come across that way in the classroom environment. I have attempted to correct this problem and made many changes, based on my teaching experience since the publication of the first edition. The list of references with each chapter has been greatly expanded and updated to allow the interested reader to probe further into the topics which have been covered. Finally, in response to many requests from colleagues in industry, tables of important data have been collected at the end of each chapter for ready reference.

It is almost impossible to provide a comprehensive list of acknowledgments for a book that is based on more than 40 years of interaction with colleagues in the area of solid state. Certainly, all of those singled out for thanks in the first edition require a redoubling of these thanks, since they have been of continuing help to me. My special thanks go to Dr. Krishna K. Parat for his critical reading of many sections of this book, and to Professor Ishwara B. Bhat for many stimulating discussions, as well as for his help in solving most of the problems that are provided at the end of each chapter. Ms. Priscilla Magilligan deserves my thanks for doing a superb job of typing this manuscript and its interminable revisions, using an arcane computer software package that was far beyond my comprehension.

SORAB K. GHANDHI

Niskayuna, New York
January 1994

PREFACE TO THE FIRST EDITION

This book discusses the basic principles underlying the fabrication of semiconductor devices and integrated circuits. Its emphasis is on processes that are especially useful for Very Large Scale Integration (VLSI) schemes; however, many processes used for discrete devices and medium-density integrated circuits have also been included, since they are the precursors of VLSI technology.

The aim of this book is to provide graduate students and practicing engineers with a body of knowledge of VLSI fabrication principles that will not only *bring* them up-to-date, but also allow them to *stay* up-to-date in this field. This is difficult to do in as fast-moving an area as VLSI. I believe that an effective approach is to concentrate on a broad background in the area, and to emphasize the basic principles governing the direction of developments in this field. As a practicing engineer for 13 years, and a professor and consultant to industry for an additional 19, I believe that this approach will best serve the long-term needs of workers in the area.

My experience as an author has convinced me that the above approach is the correct one for this book. In 1968 I wrote a book on microelectronics* using this approach. It has been a source of deep satisfaction to me that, even after 14 years, it is still considered to be a well-regarded "honest" book, notwithstanding the fact that many new technologies have become firmly established since. The reason for this, I believe, was the book's emphasis on basic principles rather than on the latest technological details.

VLSI Fabrication Principles was written with this idea in mind. Unlike my last volume, however, this book emphasizes fabrication principles. I have

*The Theory and Practice of Microelectronics. Published by John Wiley and Sons, New York.

avoided discussions of the physics of device operation as much as possible, since this topic is already covered in excellent books* in the area. Even so, this new volume is about one and one-half times the length of the previous one, because of the many advances in the field.

This book is about both elemental and compound semiconductors. Silicon is clearly the most widely used material today, and will remain the mainstay of the industry. However, many new materials (gallium arsenide is only one) are under investigation because of their unique capabilities. Gallium arsenide has already advanced technologically to the point where it is being used to make many unique devices and integrated circuits. In terms of both research and advanced development, its exploitation represents the most rapidly growing segment of solid state technology today.

From an academic viewpoint, I believe that a student's valuable time in a university can be put to the best use by acquiring the broadest base of knowledge in the field, rather than by premature specialization in any one specific area of technology. This is particularly true because many of the new problems being uncovered by workers in the latest silicon technology bear a strong resemblance to problems encountered many years ago by workers in gallium arsenide. For example, the anomalous behavior of arsenic diffusion in silicon, which was first explained in 1975, is governed by the same body of mathematics that describes the diffusion of zinc in gallium arsenide (1963). It is my hope, then, that students of this volume will acquire a broad perspective on the subject, and be flexible enough to work in *any* area of semiconductor fabrication technology, as the need arises.

This book emphasizes the basic processes that are involved in integrated circuit fabrication. Each chapter discusses principles common to all semiconductors, with separate sections where necessary to discuss problems and characteristics that are unique to silicon or gallium arsenide. The emphasis is on VLSI techniques; however, in order not to be unduly restrictive, I have also considered techniques that apply to the fabrication of medium and large-scale integrated circuits, as well as of discrete devices.

Chapter 1 describes basic Material Properties, especially those that are important in device processing. This is followed by Chapter 2 on Phase Diagrams, which outlines a valuable tool for investigating the behavior of combinations of two or more materials in intimate contact, when subjected to heat treatments. Chapter 3 describes basic aspects of the Growth of starting materials, and limitations imposed by the different growth technologies.

Chapters 4–6 describe the "anchor" technologies of Diffusion, Epitaxy, and Ion Implantation. Emphasis has been placed on these topics, because they are the key fabrication processes used today. Again, the stress is on the basic principles underlying these processes, rather than on the very latest development

---

*By way of example, Sze's *Physics of Semiconductor Devices*, 2nd edition. Published by John Wiley & Sons, New York (1981).

in each area. Of necessity, some very new techniques have been omitted, since their eventual role in device fabrication is not yet clear.

Chapter 7 discusses Native Oxide Films, which are grown out of the semiconductor. These play a key role in control of the device surface and thus in its long-term stability. Here, the emphasis is on accepted technologies, and on their limitations.

Deposited Films are discussed in Chapter 8. The large variety of available choices prevents a detailed discussion of every combination. Consequently, these films have been grouped on a functional basis—films for protection and masking, films for interconnections, films for ohmic contacts, and films for Schottky devices. I hope that this approach will provide some cohesiveness to this topic.

Chapter 9 outlines Etching and Cleaning processes. Both wet and dry processes are considered, with special emphasis on the latter, since they represent a clear direction for VLSI fabrication.

Chapter 10 outlines the basic principles of Lithographic Processes. The emphasis here is on photolithography and electron beam lithography. However, promising approaches such as X-ray lithography are considered as well. A discussion of the principles underlying the various resist systems is also included.

Chapter 11, on Device and Circuit Fabrication, is a synthesis of all of the preceding chapters, leading to the fabrication of complete microcircuits. No attempt is made to describe the many combinations of process steps that can be used for this purpose. Instead, basic techniques common to all VLSI schemes are first considered. These include isolation, self-alignment, local oxidation, and planarization. This is followed by a detailed discussion of microcircuits based on the metal-oxide-semiconductor, metal-Schottky gate-semiconductor, and bipolar junction transistor devices. An extensive reference list is provide to guide the reader to many of the important variations that are being considered at the present time, but have not been fully evaluated in terms of performance and cost. Finally, a short Appendix provides the necessary mathematical background for the chapter on diffusion.

Wherever possible, problems have been provided, many of which deal with practical situations and are intended to bring out points not covered in detail in the text. In addition, there are extensive references at the end of each chapter. No attempt has been made to use them to give credit to persons who did the original work; rather, their choice has been based on the need to give the reader means for further study.

A book such as this takes many years to write, and springs from many sources of inspiration and encouragement. It would indeed be difficult to single out all of these for acknowledgment. I know, however, that my primary thanks must go to my many graduate students who provide me with much intellectual stimulation and challenge. They have been exposed to this material, in one form or another, over the past five years. Their many penetrating questions have often led to rethinking and reworking the text over these years.

Next, I must acknowledge the encouragement and understanding I have received from Dr. Donald Feucht and John Benner of the Solar Energy Research Institute. Their generous support of funded research in this area has provided a continuing forum for the development of new ideas and new understanding, which have been incorporated into this book. Discussions with many friends in industry, especially those at the Radio Corporation of America and the General Electric Company, have added much to the relevance of this book, and are gratefully acknowledged.

A number of my friends and colleagues have been kind enough to read and comment upon chapters of this book in their areas of specialization. These include Drs. B. Jayant Baliga, Ronald J. Gutmann, Shinji Okazaki, Kenneth Rose, Shambhu K. Shastry, and David S. Yaney.

My wife and family have been very supportive during what must have been a trying period for them. Indeed, their pride in this endeavor has done much to make it all very worthwhile.

Finally, much credit and thanks must go to R. Carla Reep for typing the manuscript and also for editing and checking it from its typed version to the page-proof stage. Her participation has greatly reduced my work, and allowed many revisions to be made in order to bring this manuscript into its final form.

S. K. GHANDHI

Niskayuna, New York
November 1982

CONTENTS

5 EPITAXY   258

VLSI FABRICATION
PRINCIPLES

CHAPTER 1

MATERIAL PROPERTIES

Although many elements and intermetallic compounds exhibit semiconducting properties, silicon is used almost exclusively in the fabrication of semiconductor devices and microcircuits. Of the many reasons for this choice, the most important are the following:

1. Silicon is an elemental semiconductor. Together with germanium, it can be subjected to a large variety of processing steps without the problems of decomposition that are ever present with compound semiconductors. For much the same reason, more is known today about the preparation and properties of extremely pure single crystal germanium and silicon than for any other element in the periodic table.

2. Silicon has a wider energy gap than germanium. Consequently it can be fabricated into microcircuits capable of operation at higher temperatures than their germanium counterparts. At the present time the upper operating ambient temperature for silicon microcircuits is between 125°C and 175°C, which is entirely acceptable for both commercial and military applications.

3. Unlike germanium, silicon lends itself readily to surface passivation treatments. This takes the form of a layer of thermally grown SiO_2 which provides a high degree of protection to the underlying device. Although the fabrication of devices such as metal-oxide-semiconductor (MOS) transistors has emphasized that this oxide falls short of providing perfect control of surface phenomena, it is safe to say that the development of this technique provided a decisive advantage for silicon over germanium as the starting material in microcircuits [1].

As a result of the above, a significant technological base has been established to take advantage of its characteristics. This includes the development of a number of advanced processes for deposition and doping of silicon layers, as well as sophisticated equipment for forming and defining intricate patterns for very-large-scale integration (VLSI).

Although silicon is the workhorse of the semiconductor industry, it is not an optimum choice in every respect. Many compound semiconductors, for example, are direct band gap, whereas the band gap of silicon is indirect. This results in unique properties which allow functions that *cannot* be performed by silicon. These include transferred electron oscillators, lasers and light-emitting devices, and a variety of highly efficient, lightweight, photovoltaic devices for space as well as terrestrial applications.

Of the many compound semiconductors, GaAs has come under the greatest scrutiny, and its technology is the most highly developed. Its low-field electron mobility is larger than that of silicon, so that majority carrier devices in GaAs are faster than in silicon. GaAs has a lower saturation field than silicon, so that GaAs devices have a smaller power-delay product. Finally, GaAs can be made semi-insulating (SI), with a bulk resistivity on the order of 10^9 Ω cm. Devices and circuit interconnections, made in substrates of SI GaAs, have reduced parasitic capacitances which lead to further improvements in speed over silicon integrated circuits. In sum, both discrete components and integrated circuits made in GaAs are faster than those made in silicon [2].

The importance of GaAs as an electronic material is greatly enhanced by the fact that it is closely lattice-matched to aluminum arsenide, which has a much wider energy gap (2.16 eV). Combinations of these materials, the ternary alloys $Al_xGa_{1-x}As$, can thus be grown on GaAs with a wide range of compositions (and energy gaps), and still have a reasonably defect-free interface. The growth of device-quality AlGaAs has allowed the fabrication of practical heterostructure devices, such as highly efficient double heterostructure lasers, high-electron-mobility transistors, and a number of novel bipolar device structures. Comparative studies [3] of silicon emitter coupled logic circuits and AlGaAs/GaAs heterojunction bipolar transistor logic have predicted a propagation delay improvement over silicon by a factor of three or more, for devices with the same feature size.

Recent advances in molecular beam epitaxy and in organometallic vapor-phase epitaxy have allowed the fabrication of structures with multiple layers of GaAs and AlGaAs, controlled to atomic dimensions. This has not only resulted in improvements in conventional devices, but has also opened up the possibility for entirely new device concepts involving superlattices and quantum well structures [4].

In addition to GaAs, other new materials and material systems are under active consideration for device and circuit applications. These include almost all combinations of III–V compounds, and more recently the II–VI compound semiconductors. The technology of these materials is in its infancy; nevertheless, unique application advantages have resulted in a few becoming commercially viable today. By way of example, $Ga_{1-x}In_xAs$ can be grown lattice-matched on InP for $x = 0.53$. The energy gap of this material makes it ideally suited for emitters and detectors with low-loss optical fibers, operating in the 1.3- to 1.5-μm range [5]. A second example is the system CdTe–HgTe, whose binary components are lattice-matched to within 0.28%. This allows the fabrication of $Hg_{1-x}Cd_xTe$ devices which can be optimized over the 0.9- to 16-μm range of wavelengths [6]. The large variety of such material combinations prevents them all from becoming developed to the same level as GaAs. They will not be considered in this book, except in passing. Nonetheless, it is important to recognize that the fabrication principles of GaAs can often serve as a useful starting point for work with other compound semiconductors.

In this chapter some of the properties of silicon and GaAs are considered, especially those which have a bearing on the fabrication processes that follow as well as on the properties of devices made by these fabrication technologies. A short list of these properties is provided in Table 1.1, at the end of this chapter.*

1.1 PHYSICAL PROPERTIES

Together with carbon, germanium, tin, and lead, silicon belongs in group IV of the periodic table. In single-crystal form, it adopts the diamond lattice structure, with each atom covalently bonded to four nearest neighbors. Many of its physical properties result from this strong covalent bonding. In pure form, its lattice constant is 5.43086 Å at 300 K, increasing by +0.02% with doping. The nearest neighbor distance between silicon atoms in the diamond lattice is 2.35163 Å.

Single-crystal GaAs adopts the zincblende structure, with each Ga atom bonded to four nearest-neighbor As atoms, and vice versa. The edge dimension of the unit cell of undoped, stoichiometric GaAs is 5.65325 Å, with a nearest-neighbor distance of 2.44793 Å between Ga and As atoms [7, 8]. Heavily doped GaAs has a slightly larger lattice constant, by about 0.02%. The lattice parameter is reduced by about 0.01% for nonstoichiometric As-rich gallium arsenide, and increased by about this same amount for Ga-rich material. Its limits for stoichiometry are between 49.998 and 50.009 at.% of arsenic.

The presence of dissimilar atoms makes GaAs about 69% covalent, and significantly alters its electronic band structure. Figure 1.1 shows simplified band structures of both silicon and GaAs at 300 K, along two principal directions in the crystal. For both materials the valence band maximum occurs at the zero wave vector point (Γ). With GaAs, the minimum conduction energy is located at the same point, so that it has a direct energy gap, whose value is approximately 1.43 eV at 300 K. With silicon, however, the minimum conduction energy is located near the χ point, so that it has an indirect energy gap of about 1.11 eV at 300 K. Other important intervalley spacings are indicated in Fig. 1.1 as well.

*The comparative properties of GaAs, AlAs, and Al_xGa_{1-x} As are detailed in Ref. 8.

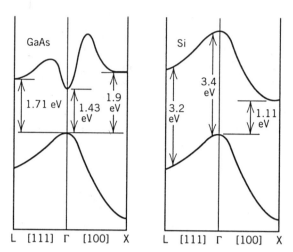

Fig. 1.1 Simplified energy band structure of silicon and GaAs.

The wider energy gap of GaAs results in two important advantages over silicon. First, the intrinsic carrier concentration is about four orders of magnitude lower (10^6 cm^{-3} as compared to 10^{10} cm^{-3} for silicon at 300 K), so that truly SI GaAs is possible whereas SI silicon is not. Next, the barrier height of metals on GaAs is sufficiently large ($\simeq 0.85$ eV) so as to give low-leakage Schottky diode structures with a wide variety of metals. In contrast, most metal–Si combinations (with the notable exception of Pd–Si and Pt–Si) have low barrier heights, around 0.6 eV, and result in correspondingly leaky devices.

The surface properties of silicon and GaAs are strikingly different. Silicon, especially when thermally oxidized, has a significantly lower density of surface states than does GaAs. As a result, its surface can be inverted by the use of a simple metallic gate, placed on this oxide. This has allowed development of the MOS transistor, which is the cornerstone of silicon-based VLSI technology.

In contrast, the fermi level in GaAs is strongly pinned near the midgap, because of the high surface state density. For this reason, an equivalent MOS-based technology is not available in this material. However, the combination of GaAs with AlGaAs to form a heterojunction circumvents this problem, and also allows the fabrication of high-electron-mobility transistors (HEMT) which have a significant speed advantage over silicon devices [3].

A second consequence of fermi-level pinning in GaAs is that the Schottky barrier height is relatively independent of the choice of metal, and of the surface processing conditions. With *n*-type GaAs the barrier height is about 0.85 eV. By the same token, the barrier height on *p*-type GaAs is also relatively fixed, at about 0.6 eV, so that low leakage Schottky devices can only be made on lightly doped material.

As we have pointed out, only platinum and palladium can be used to make satisfactory Schottky diodes on *n*-silicon. Considerable attention must be paid to surface processing conditions in order to fabricate these structures with consistent values of barrier height. This necessitates the deposition of the metal on a silicon substrate which has been *in-situ* cleaned. Sputtering is the preferred deposition process for these metals, since back-sputtering can be used for surface cleaning, prior to the actual deposition.

The effective mass of electrons and holes is inversely proportional to the curvature of the energy bands, and is listed in Table 1.1. Note that the electron effective mass m_n^* for silicon is considerably larger than that for GaAs; the electron mobility is correspondingly lower, typically 1500 cm^2/ V s for silicon as compared to 8500 cm^2/V s for GaAs at 300 K. This accounts in large measure for the speed advantage of GaAs over silicon in majority carrier *n*-type devices [9]. This is because the larger electron mobility results in a higher low field velocity at low electric field strength, as well as reduced parasitic resistance. In addition, the ability to attain a high drift velocity at low electric fields allows GaAs digital circuits to be operated with lower power-speed products than those made in silicon.

The energy band structure of GaAs also exhibits an upper satellite valley, about 0.38 eV removed from the lower [7]. The effective mass of electrons in this valley is considerably larger ($1.2\,m_0$ as compared with $0.067m_0$ for the lower valley), so that the electron mobility is correspondingly reduced. The transfer of electrons from the lower to the upper valley at electric fields above 3.5 kV/cm consequently results in a fall in mobility, and gives GaAs the unique differential mobility character shown in Fig. 1.2, which is not present in silicon. Bulk microwave oscillators, sometimes referred to as *Gunn oscillators* [10, 11], have exploited this property and are in common use today.

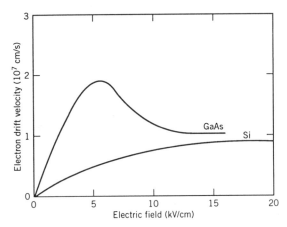

Fig. 1.2 Velocity versus electric field strength characteristics of silicon and GaAs at 300 K. Adapted from McCumber and Chynoweth [10].

The hole effective masses of silicon and GaAs are comparable, as are their mobility values; typically the latter are 500 cm^2/V s and 450 cm^2/V s for lightly doped silicon and GaAs, respectively, at 300 K. Majority carrier devices, based on hole transport, have not been developed in GaAs for this reason. On the other hand, *p*-type silicon (and GaAs) are preferred over *n*-type materials in *minority* carrier devices because the electron diffusion length is larger than the hole diffusion length.

Commercially available silicon is nearly uncompensated. Thus *n*-type material has a negligibly small amount of shallow or deep acceptors unintentionally incorporated in it. Similarly, the concentration of donors which are unintentionally present in *p*-type silicon is negligibly small. As a result, the drift mobility of starting silicon, shown in Fig. 1.3, is essentially that of uncompensated material. A plot of resistivity as a function of carrier concentration is shown in Fig. 1.4.

Although relatively free of compensation, bulk silicon typically has in it a significant amount of inert impurities such as carbon, oxygen, fluorine, sodium and calcium. By way of example, Table 3.1 in Chapter 3 lists the impurity content of a typical silicon crystal, grown by different manufacturing techniques [12].

The purification of GaAs is not as advanced as that of silicon, so that many shallow and deep impurities are unintentionally present in GaAs. These are listed, together with their typical concentrations, in Table 3.2 in Chapter 3 [13].

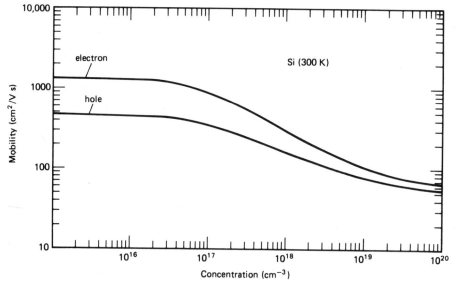

Fig. 1.3 Electron and hole drift mobilities in silicon.

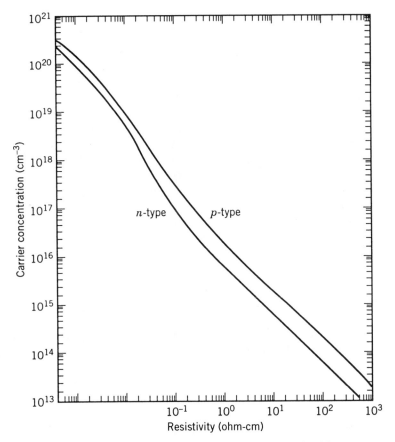

Fig. 1.4 Resistivity versus carrier concentration in silicon.

Some of these impurities form complexes with gallium or arsenic; some are inactive, while yet others are active and of both n- and p-type. In addition, they can be incorporated substitutionally on vacant gallium or arsenic lattice sites, whose concentrations are a function of the temperature and the ambient partial pressures of the gallium and arsenic vapor species. As a result, all GaAs is compensated to some extent; that is, it has in it both ionized acceptors N_A^- and ionized donors N_D^+. The degree of compensation in n-GaAs is often specified by the ratio $(N_D^+ + N_A^-)/n$, where n is the free electron concentration ($\simeq N_D^+ - N_A^-$). An alternative approach is to specify the compensation ratio as N_A^-/N_D^+.

In compensated material, the total number of ionized impurities is larger than the free carrier concentration, so that the mobility is lower than what would be expected if it were uncompensated. The results of theoretical calculations of electron mobility, which include all major scattering mechanisms, are shown in

Fig. 1.5, for drift mobility* as a function of the free electron concentration, at 300 and 77 K, respectively, and for different values of the compensation ratio [14].

Experimentally, the *net* carrier concentration in *n*-type GaAs can be obtained readily by Hall measurements at any given temperature. However, determination of the *total* ionized impurity concentration necessitates (a) an analysis of the temperature dependence of the Hall mobility and (b) fitting these data to the relevant scatter processes [15]. Based on this approximation, a semiempirical relation, which takes into account all scattering mechanisms, has been developed for *n*-type GaAs. This relationship, shown in Fig. 1.6, shows the 77K Hall mobility measured at 5 kG, as a function of $N_D^+ + N_A^-$, and n_{77K}. This curve can be used to estimate the compensation ratio, since n_{77K} is also obtained from the Hall measurements at this temperature.

The above method becomes increasingly inaccurate when the free electron concentration exceeds $2 \times 10^{17}/\text{cm}^3$. Alternative computations, correlated with measurements of the free carrier absorption in addition to the electron mobility, have been made to extend the range to $5 \times 10^{18}/\text{cm}^3$ [16]. However, deviations from band parabolicity, due to either (a) interaction between impurities, (b) many body effects, or (c) inhomogeneous distribution of impurities causing band-gap fluctuation, can cause inaccuracies in this approach. For these reasons, a more direct approach, utilizing quantitative secondary ion mass spectrometry measurements in conjunction with Hall data, has been favored for determination of the total impurity concentration in heavily doped GaAs [17].

A plot of hole mobility in *p*-type GaAs arsenide is shown in Fig. 1.7. Studies of this material [7] have not been extensive, since its primary role has been in injecting p^+ contacts to *n*-type GaAs. Recently, however, it has become more important because of its use in photovoltaic devices, since its minority carrier diffusion length is larger than that for holes in *n*-type material.

The number of hole–electron pairs generated by an electron, as it travels unit distance in a semiconductor, is described by an electron ionization rate, α_n. Similarly, α_p relates to hole–electron pairs created by hole transport in a semiconductor. These ionization rates determine the upper operating voltage of semiconductor devices. Devices such as Impact Avalanche Transit Time (IMPATT) diodes, which are important microwave power sources, are designed to operate in a mode where these effects are exploited. Figure 1.8 shows the ionization rates for electrons and holes in silicon and GaAs. Here, we note that the coefficients for GaAs are similar, while those for silicon are not. It can be shown that IMPATT diodes made of GaAs are more efficient than silicon devices for this reason [18]. Empirical relations for the ionization rates of electrons and holes are given in Table 1.1.

---

*The Hall mobility is approximately equal to 1.93 times the electron drift mobility if the dominant scatter mechanism is due to ionized impurities. The Hall and drift mobilities of holes are approximately equal.

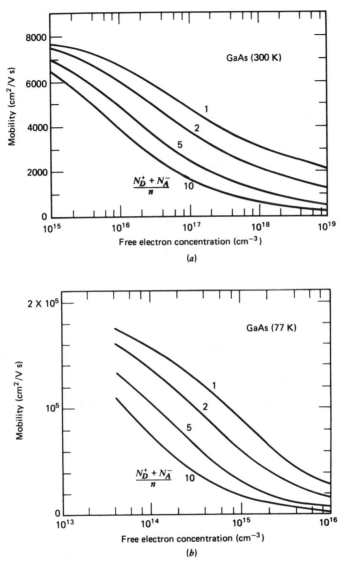

Fig. 1.5 Electron mobility at 77 and 300 K, as a function of $(N_D^+ + N_A^-)/n$. From Rode [14]. Reprinted with permission from *Physical Review*.

The nature of the energy band gaps of silicon and GaAs give rise to uniquely different optical properties, as seen in the absorption characteristics of these materials for radiation above the energy gap [19, 20]. As shown in Fig. 1.9, the variation of the absorption coefficient of silicon is more gradual than that of GaAs, for photon energies beyond their respective band gaps. This is a con-

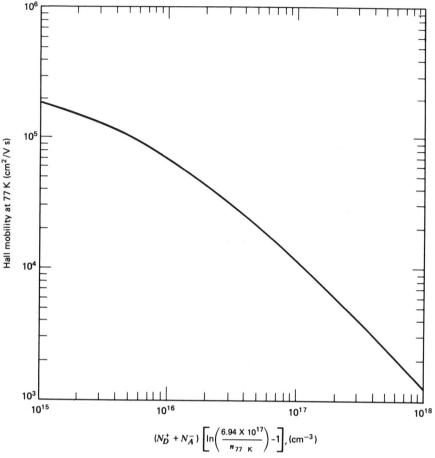

$$(N_D^+ + N_A^-) \left[\ln\left(\frac{6.94 \times 10^{17}}{n_{77 \text{ K}}} \right) - 1 \right], (\text{cm}^{-3})$$

Fig. 1.6 $\mu_{77 \text{K}}$, as a function of $N_D^+ + N_A^-$ and n. Adapted from Stillman and Wolfe [15].

sequence of the indirect band-gap nature of silicon, where phonons must be involved in order to conserve momentum during the recombination process. The absorption edge of GaAs has been studied in much detail, and is related to the free electron and hole concentration in the material. This is illustrated in Fig. 1.10.

The direct gap character of GaAs results in a number of features of importance in optically related applications. First, almost complete absorption of light occurs in an extremely short penetration depth in GaAs as compared to that in silicon—typically 1 μm as compared to 100 μm. As a result, optical devices in GaAs have dimensions which are comparable to those of integrated circuits. The optical integrated circuit is thus an achievable goal in GaAs. Second, GaAs devices are extremely efficient emitters of optical radiation, since the recombination process is accompanied primarily by photon emission. Third,

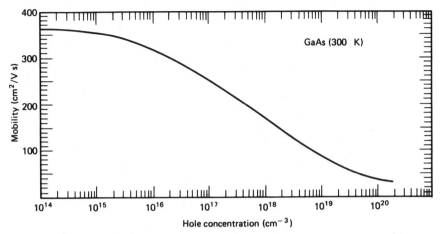

Fig. 1.7 Hole mobility in gallium arsenide. Adapted from Blakemore [7].

GaAs recombination processes are faster than in silicon, because of the absence of phonon participation. For these reasons, high-speed optical devices have become the almost exclusive domain of GaAs and other direct gap compound semiconductors.

Additional properties of silicon and GaAs, which impact directly on the ease with which devices can be manufactured in a factory environment, are given in Table 1.1. By way of example, the critical resolved shear stress is 7.75×10^6 dyn cm^{-2} for GaAs, and 3.61×10^7 dyn cm^{-2} for silicon. As a consequence, GaAs

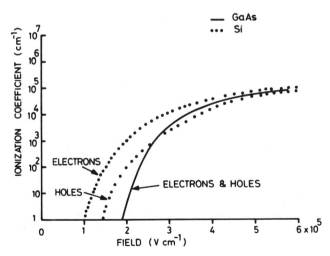

Fig. 1.8 Ionization coefficient for electrons and holes in GaAs and Si at 300 K. From Howes and Morgan [9]. ©1985. Reprinted by permission of John Wiley and Sons.

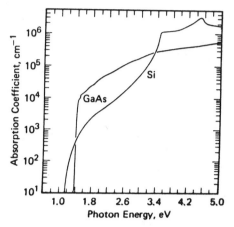

Fig. 1.9 Band-gap absorption of Si and semi-insulating GaAs. From Hovel [19].

slices larger than 75–100 mm in diameter are extremely fragile, and cannot be subjected to heat treatment or physical handling during processing without fear of breakage [21]. Silicon, on the other hand, can be readily handled in slices as large as 200–300 mm in diameter in a manufacturing environment.

The thermal conductivity of silicon is about 2.75 times larger than that of GaAs. As a consequence, the power-handling capacity of a GaAs integrated circuit is severely limited by the thermal resistance of the substrate. Often, these slices must be mechanically or chemically thinned prior to mounting on a header. This is a delicate process at best, with a high chance of wafer breakage.

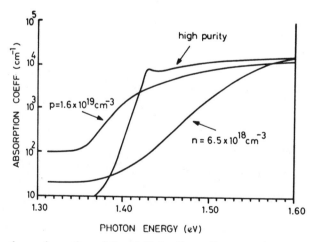

Fig. 1.10 Band-gap absorption of doped GaAs. From Casey and Stern [20]. Reprinted with permission from the *Journal of Applied Physics*.

An important consequence of this poor thermal conductivity of GaAs is that packing densities achievable in silicon microcircuits cannot be matched in circuits built on bulk GaAs substrates. This limitation has given rise to considerable research into the growth of thin single-crystal films of GaAs on silicon substrates [22], in the hope of circumventing the heat dissipation problem. If successful, this approach would also solve the physical handling problems of large GaAs slices.

1.2 CRYSTAL STRUCTURE

Both silicon and GaAs belong to the cubic class of crystals and exhibit the following structures:

1. *Simple Cubic (s.c.).* This is illustrated in Fig. 1.11a. Very few crystals exhibit as simple a structure as this one; an example is polonium, which exhibits this structure over a narrow range of temperatures.

2. *Body-Centered Cubic (b.c.c.).* This is illustrated in Fig. 1.11b. Molybdenum, tantalum, and tungsten exhibit this crystal structure.

3. *Face-Centered Cubic (f.c.c.).* This is illustrated in Fig. 1.11c. The structure is exhibited by a large number of elements, such as copper, gold, nickel, platinum, and silver. (The face-centered atoms are shown different from the corner atoms for illustrative purposes.)

4. *The Zincblende or Sphalerite Structure.* This structure consists of two interpenetrating f.c.c. sublattices, with one atom of the second sublattice located at one-fourth of the distance along a major diagonal of the first sublattice. This configuration is illustrated in Fig. 1.12a and 1.12b, where solid circles belong to the first sublattice and open circles belong to the second. In GaAs, each sublattice contains atoms of only one type (either gallium or arsenic). The *diamond* lattice is a degenerate form of this structure, with identical atoms in each sublattice. Silicon belongs to this class.

The position of the various atoms in the zincblende lattice can be calculated in multiples of the lattice constant a. Thus for the f.c.c. structure the various coordinates (normalized to a) for the corner lattice sites are 0, 0, 0; 0, 0, 1; 0, 1, 0; 0, 1, 1; 1, 0, 0; 1, 0, 1; 1, 1, 0; and 1, 1, 1. Coordinates for the face-centered sites are $\frac{1}{2},\frac{1}{2},0$; $\frac{1}{2},\frac{1}{2},1$; $0,\frac{1}{2},\frac{1}{2}$; $1,\frac{1}{2},\frac{1}{2}$; $\frac{1}{2},0,\frac{1}{2}$; and $\frac{1}{2},1,\frac{1}{2}$, respectively. For the zincblende lattice it is necessary to include the coordinates of the second sublattice, spaced at $\frac{1}{4},\frac{1}{4},\frac{1}{4}$ from those of the first. Within the unit cell, these are $\frac{1}{4},\frac{1}{4},\frac{1}{4}$; $\frac{3}{4},\frac{3}{4},\frac{1}{4}$; $\frac{1}{4},\frac{3}{4},\frac{3}{4}$; and $\frac{3}{4},\frac{1}{4},\frac{3}{4}$, respectively. In Fig. 1.12b these lattice sites are shown different from those of the original sublattice. With reference to Fig. 1.12, the following comments may be made:

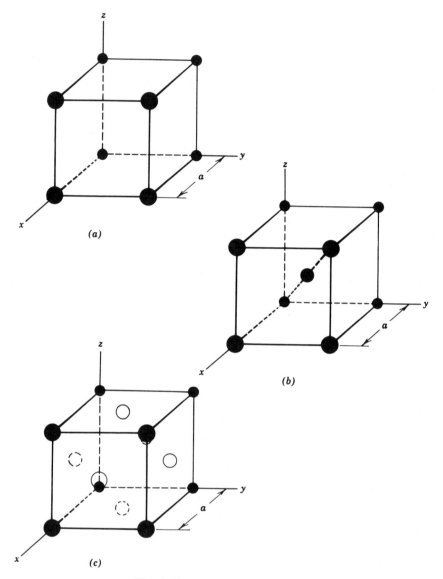

Fig. 1.11 Cubic crystal lattices.

1. The *coordination number* for the zincblende lattice is 4; that is, each atom has four nearest neighbors which belong to a different sublattice. In silicon each atom has four valence electrons which provide covalent bonding with these nearest neighbors. In GaAs, however, each arsenic atom (with five valence electrons) has four neighboring gallium atoms, each of which has three valence electrons. In like manner, each gallium atom (with three valence electrons) has

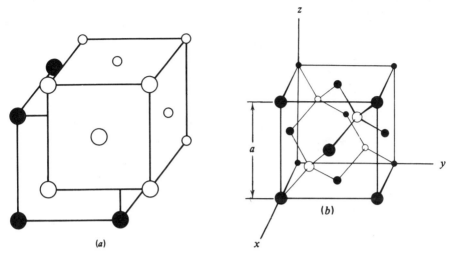

Fig. 1.12 The zincblende lattice.

four neighboring arsenic atoms, each having five valence electrons. Together, these gallium–arsenic pairs enter into bonding, which is about 69% covalent in character, and 31% ionic. Figure 1.13 shows an enlarged picture of a sub-cell with side $a/2$ in order to delineate *tetrahedral* covalent bonds of the type described here.

2. The distance between two neighboring atoms is $(\sqrt{3}/4)a$, where a is the lattice constant. For silicon, the lattice constant is approximately 5.43 Å at room temperature, so that this distance is 2.351 Å. The radius of the silicon atom is thus 1.18 Å, if we assume a "hard sphere" model for atoms. Since each atom is situated within a tetrahedron comprising its four neighbors, this is referred to as the *tetrahedral radius*.

For GaAs, the lattice constant is approximately 5.65 Å at room temperature, so that the distance between neighbors is 2.44 Å. The tetrahedral radii of gallium and arsenic are 1.26 Å and 1.18 Å, respectively; together they add up to 2.44 Å.

3. Using the hard-sphere model, 34% of the silicon lattice and 33.8% of the GaAs lattice is occupied by atoms. Thus these are relatively loosely packed structures as compared to f.c.c. crystals, where the packing density is approximately 74%.

Table 1.2 at the end of this chapter lists the tetrahedral radii* and energy levels of various impurities that are commonly introduced into the silicon lattice

---

*Note that the effective radius of an impurity atom in a zincblende lattice is independent of the chemical components of this lattice. The concept of a constant tetrahedral radius is an empirical but very useful one. This radius, is, however, not the same as the radius of the atom in its own lattice (i.e., its *ionic radius*), since the internal field conditions are quite different for these cases.

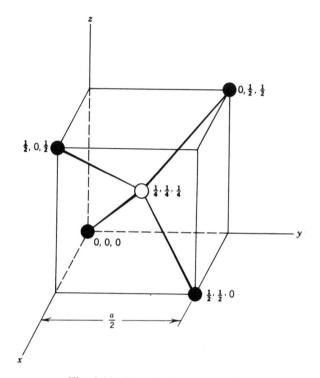

Fig. 1.13   The zincblende subcell.

to control its electronic behavior, or may be present as contaminants. If r_0 is the tetrahedral radius of the silicon atom, the radius of the impurity atom may be written as $r_0(1 \pm \epsilon)$. The quantity ϵ is defined as the *misfit factor*, and is indicative of the degree of strain present in the lattice as a result of introducing this impurity. To a fair approximation, it is also an indication of the amount of dopant which can be incorporated into electronically active sites in the lattice, before the onset of strain induced damage to the crystal lattice.

Impurities used for doping GaAs are listed in Table 1.3 at the end of this chapter. Some of these dopants are incorporated on only one sublattice, corresponding to a single misfit factor. Yet others can substitute on either or on both lattices, so that two misfit factors describe their behavior more appropriately.

1.3 CRYSTAL AXES AND PLANES

Directions in crystals of the cubic class are very conveniently described in terms of Miller notation [23]. Consider, for example, any plane in space, which satisfies the equation

$$\frac{x}{a} + \frac{y}{b} + \frac{z}{c} = 1 \tag{1.1}$$

Here a, b, and c are the intercepts made by the plane at the x, y, and z axes, respectively. Writing h, k, and l as the reciprocals of these intercepts, the plane may be described by

$$hx + ky + lz = 1 \tag{1.2}$$

The Miller indices for such a plane* are written as (hkl). Integral values are usually chosen in multiples of the edge of the unit cell.

Figure 1.14a shows a cubic crystal with some of its important planes indicated. Here the plane $ABCD$ is designated (110), while the plane EDC is designated (111). The (100) and (010) planes are also shown in this figure. Figure 1.14b shows an example of how planes with negative indices may be described. Thus the plane $PQRS$ is defined by $(0, -1, 0)$ and is commonly written as $(0\bar{1}0)$. The plane $RSTU$ is written in like manner as $(1\bar{1}0)$.

The atom configurations in many of the Miller planes in a cubic crystal are identical. Thus the planes (001), (010), (100), $(00\bar{1})$, $(0\bar{1}0)$, and $(\bar{1}00)$ are essentially similar in nature. For convenience they are written as the {001} planes.

Figure 1.14c shows examples of planes with higher indices. Thus the plane $GHKJ$ is denoted by $(1, \frac{1}{2}, 0)$ or preferably by (210). Similarly, the plane HKL is written in Miller notation as (212).

Planes with higher Miller indices may be sketched by extending these principles. They are not, however, often encountered in discussions of the material properties of semiconductors. Often, these planes can be considered as stepped structures, with the steps and the risers consisting of different low-Miller-index planes.

Indices of lattice plane direction (i.e., of the line normal to the lattice plane) are simply the vector components of the direction resolved along the coordinate axes. Thus the (111) plane has a direction written as [111], and so on. This is an extremely convenient feature of the Miller index system for cubic crystals. For this notation the set of direction axes [001], [010], [100], $[00\bar{1}]$, $[0\bar{1}0]$, $[\bar{1}00]$ is written as $\langle 001 \rangle$.

The angle θ included between two planes $(u_1v_1w_1)$ and $(u_2v_2w_2)$ is given by

$$\cos \theta = \frac{u_1u_2 + v_1v_2 + w_1w_2}{\sqrt{(u_1^2 + v_1^2 + w_1^2)(u_2^2 + v_2^2 + w_2^2)}} \tag{1.3}$$

*It should be noted that (hkl) refers to any one of a series of parallel planes in a cubic crystal. This may be seen by a simple shifting of the origin for the references axes.

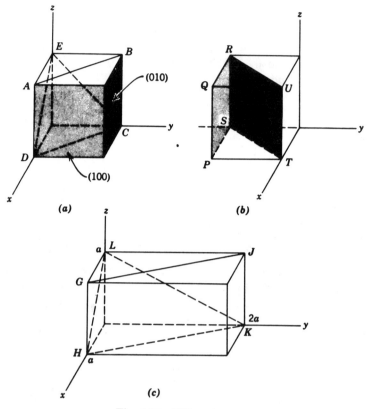

Fig. 1.14 Miller planes.

The line describing the intersection of these planes is $[u \ v \ w]$, where

$$u = v_1 w_2 - v_2 w_1 \tag{1.4a}$$

$$v = w_1 u_2 - w_2 u_1 \tag{1.4b}$$

$$w = u_1 v_2 - u_2 v_1 \tag{1.4c}$$

The separation between two adjacent parallel planes $\{hkl\}$ is given by

$$d = \frac{a}{\sqrt{h^2 + k^2 + l^2}} \tag{1.5}$$

This separation is equal to a for the 100 planes, to $0.707a$ for the $\{110\}$ planes, and to $0.577a$ for the $\{111\}$ planes. Thus the $\{111\}$ planes are the closest spaced among the low-index planes.

1.4 ORIENTATION EFFECTS

Many fabrication processes are orientation-sensitive; that is, they depend on the direction in which the crystal slice is cut. This is to be expected, since many mechanical and electronic properties of the crystal and its surface are also orientation-dependent. Some of the consequences of crystal orientation are now described.

1.4.1 Silicon

For silicon the {111} planes exhibit the smallest separation (3.135 Å). Therefore growth of the crystal along a $\langle 111 \rangle$ direction is the slowest, since it results in the setting down of one atomic layer upon another in closest packed form. Based on packing considerations, $\langle 111 \rangle$ silicon is therefore the easiest to grow. It is thus the least expensive, and is used in many bipolar devices and microcircuits today.

The atom density in the principal planes can be shown to be in the ratio {100} : {110} : {111} = 1 : 1.414 : 1.155. Since atoms in these planes have 2, 1, and 1 dangling bonds respectively, the bond densities are in the ratio 1 : 0.707 : 0.577. Crystal dissolution is related to the density of broken bonds; therefore it is slowest in the $\langle 111 \rangle$ directions, and will delineate these faces. As a consequence, parallel-plane alloyed junctions can only be made on $\langle 111 \rangle$ silicon. Crystal dissolution by chemical etching is also slowest in the $\langle 111 \rangle$ directions, for this reason. Consequently, selective etches will preferentially etch silicon by exposing {111} planes. The use of this technique for cutting V-grooves in a (100)-oriented silicon slice, as well as deep vertical trenches with parallel sides, is outlined in Chapter 9, together with a number of suitable etch formulations. This forms the basis for many important microelectronic processes which are used today, as well as for a variety of micromachining applications [24].

The ultimate tensile strength of silicon (0.35×10^{10} dyn cm^{-2}) is a maximum in the $\langle 111 \rangle$ directions. In addition, the modulus of elasticity in the $\langle 111 \rangle$ directions is higher than in the $\langle 110 \rangle$ or $\langle 100 \rangle$ directions (1.9×10^{12}, 1.7×10^{12}, and 1.3×10^{12} dyn cm^{-2}, respectively). As a result, silicon tends to cleave along the {111} planes [25], since they are normal to the $\langle 111 \rangle$ directions.

Silicon microcircuits, especially those fabricated in large diameter (≥ 100 mm), are generally diced by sawing apart through most of the slice thickness, and then breaking apart by deforming the slice. Smaller-diameter slices (which are about 200 μm thick) are usually separated into individual chips by cleaving. This is done by scribing the surface with a diamond tool into a rectangular pattern, and deforming it until it breaks by the propagation of the scribe cracks through the bulk of the silicon, along the natural cleavage planes.

The {111} cleavage planes within a slice meet the (100) plane of the surface at an angle of 54.74° along the $\langle 110 \rangle$ directions. Thus it is desirable to make scribe lines along the ⟨110⟩ directions, for easy cleaving. In practice, each slice is supplied by the manufacturer with a reference flat ground into it so as to allow the *first* scribe line to be made along an easy cleavage plane. The ⟨110⟩ directions on the (100) surface are mutually at 90° to each other, so that *both* sets of scribe lines can be made along easy cleavage planes as shown in Fig. 1.15. Note, however, that the sides of chips cut apart in this manner will not be at right angles to the plane of the surface.

Separation of (111) silicon into dice presents a special problem, since the {111} cleavage planes intersect the surface on ⟨110⟩ directions which are mutually at 60°, as seen in Fig. 1.15. Thus, only one side of a rectangular chip can be scribed in the ⟨110⟩ direction. As a result, cleaving of the other side results in a jagged, zigzag line, with each individual jag along one of the ⟨110⟩ directions. Chips that are separated by this technique must have considerable spacing provided between them to prevent damage to the microcircuit.

It is difficult to grow silicon in the [110] direction, so that this material is not used for conventional applications. However, (110) silicon plates can be readily cut out of suitably oriented (111) silicon. An unusual property of (110) silicon is that some of the {111} planes intersect its surface at 90°. As a result, this material finds use in situations where it is required to etch deep vertical, parallel-sided grooves in silicon [24].

Other processes which are orientation-dependent are diffusion and oxidation. These are taken up in detail in Chapters 4 and 7, respectively.

The electronic properties of the silicon surface are related to the density of dangling bonds on the surface, as well as to their bond strength. Typically it has been observed that the surface state density for ⟨100⟩ silicon is lower than

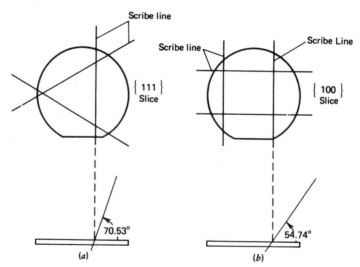

Fig. 1.15 Scribe lines and cleavage planes for {111} and {100} silicon.

that for $\langle 111 \rangle$ silicon by a factor of about 3 [26]. As a result, $\langle 100 \rangle$ silicon is often used for bipolar applications where low $1/f$ (or flicker) noise is required. Furthermore, almost all MOS circuits are built on $\langle 100 \rangle$ silicon today, because its use results in improved control of the threshold voltage. This represents a very large application area, so that the gap between slice costs for these two orientations has been sharply reduced.

1.4.2 Gallium Arsenide

As noted earlier, GaAs comprises two interpenetrating f.c.c. sublattices. One of these is displaced one-quarter of the way along the main diagonal of the other, resulting in asymmetry [27]. This is seen in Fig. 1.16, which shows a schematic view of the gallium arsenide lattice, with the [111] axis (body diagonal) in the plane of the paper. We note that the crystal structure consists of hexagonal rings, stacked in plane of the paper, but with different spacings. Assigning the layers to gallium and arsenic, their succession in the [111] direction is Ga-As—Ga-As—Ga-As—, whereas in the $[\bar{1}\,\bar{1}\,\bar{1}]$ direction it is As-Ga—As-Ga—As-Ga—. In silicon these two directions are identical; with GaAs they can be distinguished from each other, and the [111] axis is a polar axis for this structure. The (111) plane in this figure is referred to as the (111) Ga face, whereas the $(\bar{1}\,\bar{1}\,\bar{1})$ plane is called the (111) As face. These faces are often referred to as (111) A and (111) B, or simply as the A and B faces, respectively.

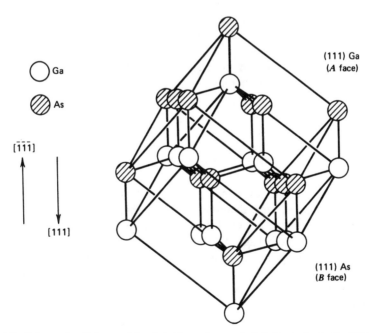

Fig. 1.16 Schematic of the zincblende lattice, observed at right angles to the [111] axis.

Looking down on the crystal from the top, at the (111) Ga face, we have two possibilities: We can get a gallium atom connected by three covalent bonds to arsenic atoms in the next (lower) layer, with one dangling bond. Alternatively, it is possible to have an arsenic atom connected by one covalent bond to a gallium atom in the next (lower) layer, with three dangling bonds. Energetically, the first of these situations is more favored, and the second does not appear to even exist. In fact, crystal growth and etching always seem to occur by the growth or dissolution of such double layers. Looking at the (111) As face from the bottom, the opposite situation is observed, with arsenic atoms connected by three bonds to the next layer of gallium, and so on.

As a result of these bonding configurations, the (111) Ga face has gallium atoms with no free electrons, since all three of their valence electrons are attached to arsenic atoms in the lower layer. The (111) As face, on the other hand, consists of arsenic atoms, each with two free electrons, since only three of their five valence electrons are attached downwards. The (111) As face is thus more electronically active than the (111) Ga face.

This difference in activity is manifested in a number of situations. Thus, etching of the (111) As face occurs very rapidly with a resultant smooth polish. The (111) Ga face, on the other hand, etches very slowly so that all imperfections become delineated, resulting in a rough surface. This is also true for mechanical processes, where it has been observed that the (111) As face is more readily lapped than the (111) Ga face.

At temperatures below 770°C, surface evaporation occurs more rapidly from the (111) As face, suggesting that the surface energy of atoms on this face is higher than for the (111) Ga face. Differences in the evaporation rate are not observed above 800°C, where molecular dissociation, with an activation energy of about 112 kcal/mole (4.86 eV/molecule), dominates the process. Crystal oxidation occurs more readily on the (111) As face than on the (111) Ga face. Again, this is due to the higher electronic activity associated with the (111) As face.

The {111} planes of GaAs consist of alternate layers of group III and group V atoms. These are differently charged, so there is a strong electrostatic attraction between them. As a result, it is difficult to cleave GaAs along {111} planes. The situation for the {110} faces is quite different, since these contain an equal density of gallium and arsenic atoms. Each atom is attached by one bond to an atom in the lower layer; two bonds connect in the surface plane to two nearest neighbors, leaving a fourth dangling bond. Interatomic forces are thus strong within the {110} planes, but weak between adjacent {110} planes. As a result, this is the preferred cleavage plane for GaAs. This situation is exploited in the fabrication of laser diodes made on {110} GaAs. Rectangular chips, with perfectly parallel faces, can be cleaved from this material because the {110} cleavage planes intersect the surface at right angles and along mutually orthogonal directions.

The {100} faces of gallium arsenide consist of either all gallium or all arsenic atoms. In either case, each atom is attached by two bonds to atoms in the lower layer, leaving two free dangling bonds. The properties of this face do not depend on whether the face is made up of gallium or of arsenic atoms.

1.5 POINT DEFECTS

Localized defects of atomic dimensions, which can occur in an otherwise perfect crystal lattice, are called *point defects*. These include vacancies, interstitials, misplaced atoms, dopant impurity atoms deliberately introduced for the purpose of controlling the electronic properties of the semiconductor, and impurity atoms which are inadvertently incorporated as contamination during material growth or processing.

A study of point defects is important because most mechanisms for diffusion and crystal growth are defect-induced. Many defects are introduced during the actual act of device fabrication. Finally, all point defects (chemical or otherwise) alter the electrical properties of the semiconductor in which they are present.

Consider, at first, the silicon lattice. Here the most elementary point defect is the *vacancy*. This is present when, as a result of thermal fluctuations, an atom is removed from its lattice site and moved to the surface of the crystal, which can serve as a sink. Defects of this type are known as *Schottky defects* and are associated with an energy of formation of about 2.6 eV and an energy of migration of about 0.18 eV. Figure 1.17a shows schematically how such a defect occurs in an otherwise regular silicon lattice.

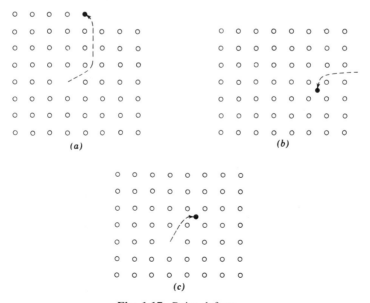

Fig. 1.17 Point defects.

A second elementary point defect that may be present in a crystal lattice is the *interstitial*. Such a defect occurs when an atom becomes located in one of the many interstitial voids within the crystal structure. Figure 1.17*b* shows schematically how this may occur in an otherwise regular crystal lattice. In silicon, which is a loosely packed crystal structure, the energy of formation of this defect is around 1.1 eV.

A vacancy–interstitial pair, or *Frenkel defect*, occurs when an atom leaves its regular site in a crystal and takes up an interstitial position, as shown in Fig. 1.17*c*. This interstitial is usually in the vicinity of the newly formed vacancy, so that the energy of formation of Frenkel defects is comparable to that of interstitial defects—that is, about 1.1 eV.

Figure 1.18 shows the unit cell for the diamond lattice. Within this unit cell are the centers of five interstitial voids, at $\frac{1}{2}, \frac{1}{2}, \frac{1}{2}; \frac{1}{4}, \frac{1}{4}, \frac{1}{4}; \frac{3}{4}, \frac{3}{4}, \frac{1}{4}; \frac{3}{4}, \frac{1}{4}, \frac{3}{4};$ and $\frac{1}{4}, \frac{3}{4}, \frac{3}{4}$. Another three voids, with their centers located at the midpoint of each of the twelve cube edges (each is shared by four unit cells), add up to a total of eight voids per unit cell. Each of these is large enough to contain an atom (again assuming hard spheres), even though there is a constriction in passing from one void to another. From a purely geometric viewpoint, therefore, interstitial defects can be expected to be quite common in silicon.

Various combinations of these defects can also occur. Thus a single vacancy is created by the breaking of four covalent bonds, whereas two vacancies, side by side, require the breaking of only six bonds. Consequently the energy of formation of a divacancy of this type is less than that required to form two separate vacancies. The divacancy is thus commonly encountered. On the other hand, the di-interstitial is much more difficult, if not impossible, to form.

The classification of point defects is somewhat more complex for the GaAs lattice. Here,

1. Schottky defects may exist in the form of either gallium or arsenic vacancies.
2. Either a gallium or an arsenic atom can be interstitially located in a void. As a result, there are two possible types of interstitials.
3. There are two kinds of Frenkel pairs, depending on the type of atom which it is displaced into a void.
4. It is possible for a gallium atom to be located on an arsenic site, or vice versa. These are known as *antisite* or *antistructure* defects. By the same token, the number of combinations of defects which can occur is much larger than for silicon. A detailed study of these possibilities is beyond the scope of this book.

An important type of point defect is created by chemical impurities which are intentionally introduced into the lattice (dopants), or unintentionally in the form of contamination. Impurity atoms that take up their locations at sites ordinarily

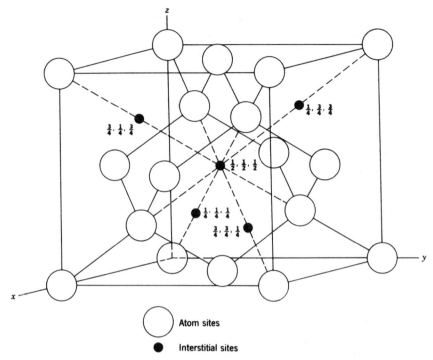

Fig. 1.18 Interstitial sites in the zincblende lattice.

occupied by lattice atoms are referred to as *substitutional impurities*. Alternatively, *interstitial* impurities are located in the many interstitial voids that are
present in the lattice.

Substitutional impurities are usually electronically active and determine the
conductivity type. Interstitial impurities, on the other hand, are usually inactive.
This is not always true, a notable exception being lithium in silicon, which is
interstitial but behaves like a donor.

1.5.1 Thermal Fluctuation Effects

The defect concentration (vacancies, interstitials, Frenkel pairs, etc.) in a semiconductor is caused by thermal fluctuations in the material and by the vapor
pressure of the species surrounding it. With silicon processing, vapor pressure
effects are negligible, since the vapor pressure of silicon is only 10^{-6} torr at
1100°C. It is for this reason that silicon processing can be carried out readily
in an open-tube environment.

The presence of defects in the material changes both the internal energy of
the crystal as well as its entropy. Consequently, their equilibrium concentration
is a function of the energy of formation and of the equilibrium temperature. On

the other hand, the concentration of a chemical defect is primarily a function of the amount available for introduction into the crystal, and of its solid solubility.

The equilibrium concentration of Schottky defects in silicon may be determined on the assumption that all other defects can be neglected [27]. Let:

N = total number of atoms in a crystal of unit volume ($\simeq 5 \times 10^{22}$ cm^{-3} for silicon)

n_s = number of Schottky defects per unit volume

E_s = energy of formation of a Schottky defect—that is, the energy required to move an atom from its lattice site within the crystal to a lattice site on its surface ($\simeq 2.6$ eV).

The number of ways in which a Schottky defect can occur is given by

$$\prescript{N}{}{C}_{n_s} = \frac{N!}{(N - n_s)!n_s!} \qquad (1.6)$$

The entropy associated with this process is

$$S = k \ln (\text{number of ways})$$
$$= k \ln \prescript{N}{}{C}_{n_s} \qquad (1.7)$$

where k is Boltzmann's constant (8.62×10^{-5} eV/K). The internal energy E is given by $n_s E_s$. The change in free energy (neglecting volume changes) is

$$F = E - TS$$
$$= n_s E_s - kT[\ln N! - \ln n_s! - \ln (N - n_s)!] \qquad (1.8)$$

The most probable equilibrium condition is the one where the free energy is a minimum with respect to changes in n_s—that is, the case for

$$\left(\frac{\partial F}{\partial n_s} \right)_{T=\text{const}} = 0 \qquad (1.9)$$

Differentiating Eq. (1.9) and setting to zero, gives

$$E_s = kT \frac{\partial}{\partial n_s}[\ln N! - \ln n_s! - \ln (N - n_s)!] \qquad (1.10)$$

The factorial terms may be simplified by using Stirling's formula for the factorial of a large number. Thus

$$\ln x! \simeq x \ln x - x \tag{1.11}$$

so that Eq. (1.10) reduces to

$$E_s = kT \ln \left(\frac{N - n_s}{n_s} \right) \tag{1.12}$$

or

$$n_s = \frac{N}{1 + e^{-E_s/kT}} \tag{1.13a}$$

$$\simeq N e^{-E_s/kT} \tag{1.13b}$$

The equilibrium concentration of Frenkel defects may be found by an analogous approach. Again, this calculation is made independently of all other defects which are present in silicon. Let:

N = number of atoms in a crystal of unit volume

N' = number of available interstitial sites per unit volume ($= N$)

n_f = number of Frenkel defects (i.e., vacancy–interstitial pairs) per unit volume

E_f = energy of formation of a Frenkel defect ($\simeq 1.1$ eV)

A vacancy can occur in $C^{N}_{n_f}$ ways, and an interstitial can occur in $C^{N'}_{n_f}$ ways. Consequently, a Frenkel defect can occur in $C^{N}_{n_f} C^{N'}_{n_f}$ ways if these events are assumed to be statistically independent.

The entropy associated with this situation is

$$S = k \ln C^{n}_{n_f} C^{N'}_{n_f} \tag{1.14}$$

The internal energy is given by $E = n_f E_f$. The change in free energy is thus

$$F = n_f E_f - kT \ln \mathop{C}_{n_f}^{N} \mathop{C}_{n_f}^{N'} \tag{1.15}$$

As before,

$$\left(\frac{\partial F}{\partial n_f} \right)_{T=\text{const}} = 0 \tag{1.16}$$

in thermal equilibrium. Differentiating Eq. (1.15) and using Stirling's formula gives

$$E_f = kT \ln \left[\frac{(N - n_f)(N' - n_f)}{n_f^2} \right] \tag{1.17a}$$

so that

$$n_f \simeq \sqrt{NN'} e^{-E_f/2kT} = N e^{-E_f/2kT} \tag{1.17b}$$

Concentrations of point defects in excess of the equilibrium value may be obtained by subjecting the semiconductor to nonequilibrium processes. Thus excessively fast cooling (quenching) can result in a supersaturated concentration of these defects. Self-interstitials and vacancies are usually not observed because of their extremely low concentration at processing temperatures. Ion implantation and nuclear radiation damage also result in increasing the defect concentration over its equilibrium value.

1.5.1.1 Charged Defects

The arguments presented earlier have tacitly assumed that defects are neutral. In the case of vacancies, for example, Eq. (1.13) can be used to calculate the concentration of neutral vacancies in silicon as a function of temperature. However, vacancies in silicon have been shown to exist in the charge states V^+, V^-, and V^{2-}. As a result, the total vacancy concentration is more correctly given by

$$[V] = [V^0] + [V^+] + [V^-] + [V^{2-}] \tag{1.18}$$

where concentrations are represented by square brackets.

Charge interactions in a solid can be formally treated in the same way as chemical interactions in solutions [28]. Mass-action relationships can consequently be used to determine the concentration of vacancies in silicon [29]. Degeneracy factors are ignored in the following analysis.

In a semiconductor with arbitrary doping, the neutral vacancy concentration will remain the same throughout its volume, since it is unaffected by any electric field due to the impurity concentration gradients in the material. It follows that the concentration of neutral vacancies in extrinsic material is equal to that in intrinsic material, and is a function of temperature alone. On the other hand, the concentrations of the various charge states of the impurity are related by the position of the fermi level with respect to the band edge. Thus, the total vacancy concentration is a function of the impurity concentration.

The reactions involved in the production of these charged vacancies are

$$V^0 + h^+ \rightleftharpoons V^+ \tag{1.19a}$$

$$V^0 + e^- \rightleftharpoons V^- \tag{1.19b}$$

$$V^0 + 2e^- \rightleftharpoons V^{2-} \tag{1.19c}$$

so that

$$k_1 = \frac{[V^+]}{p[V^0]} = \frac{[V^+]_i}{n_i[V^0]_i} \tag{1.20a}$$

$$k_2 = \frac{V^-}{n[V^0]} = \frac{[V^-]_i}{n_i[V^0]_i} \tag{1.20b}$$

$$k_3 = \frac{[V^{2-}]}{n^2[V^0]} = \frac{[V^{2-}]_i}{n_i^2[V^0]_i} \tag{1.20c}$$

where k_1, k_2, and k_3 are equilibrium constants. Combining the above relations, it follows that

$$[V] = [V^0]_i + [V^+]_i\frac{p}{n_i} + [V^-]_i\frac{n}{n_i} + [V^{2-}]_i\frac{n^2}{n_i^2} \tag{1.21}$$

It is interesting to note that these charged vacancies tend to have a self-compensating effect on the semiconductor, which increases with the doping level. For example, p-type material will have compensating V^+, which is donor-like. Conversely, in n-type material, the vacancies will be V^- or V^{2-}, depending on the position of the fermi level.

1.5.2 Vapor Pressure Effects

The situation with GaAs is quite different from that with silicon. Although this material melts at 1238°C, the surface layers decompose into gallium and arsenic long before this point is reached. The vapor pressures of these individual components are quite different, so there is a preferential loss of the more volatile species (arsenic). If processing is carried out in an evacuated ampoule, this arsenic goes into its volume until it establishes a sufficient partial pressure to prevent further decomposition of the GaAs. Although gallium and arsenic vacancies are generated by thermal fluctuations as well, the vacancy concentration in this material will be dominated by vapor pressure effects so that thermal fluctuation effects can be neglected.

In its vapor phase, arsenic consists of As, As_2, and As_4. All of these species are present, and in equilibrium over the gallium arsenide. Furthermore, some gallium in the form of vapor is also present. The partial pressure of these species is shown in Fig. 1.19 as a function of temperature [30]. Note that all of these curves are double-valued. The upper branch of the arsenic curves and the lower branch of the gallium curves are for conditions over GaAs which is preferentially rich in arsenic. In like manner, the lower branch of the arsenic curves and the upper branch of the gallium curves are for conditions over gallium-rich GaAs.

Under processing conditions, GaAs will usually be gallium-rich, with predominantly As_2 and As_4 in the vapor phase. Note, however, that at temperatures below 637°C, the partial pressures of arsenic and gallium over GaAs are approximately equal. As a consequence, GaAs evaporates congruently below this point, with essentially no gallium or arsenic vacancy generation. In molecular beam epitaxy, thermal cleaning of the GaAs is performed routinely under vacuum at about 600°C for a short period of time, to take advantage of this property (see Chapter 5).

Mass-action relationships can be applied to solids in order to determine the role of vapor pressure in controlling the vacancy concentrations [28]. Consider the decomposition reaction and the vacancy formation reactions of GaAs. The decomposition reaction is

$$GaAs(s) \rightleftharpoons Ga(g) + \tfrac{1}{2}As_2(g) \tag{1.22a}$$

assuming that As_2 is the dominant arsenic species. For this reaction the mass-action law relationship gives

$$k_1 = p_{Ga}\,p_{As_2}^{1/2} \tag{1.22b}$$

where p_{Ga} and p_{As_2} are the pressures of Ga and As_2, respectively. Gallium vacancies are generally considered to exist as V_{Ga}^0 and V_{Ga}^- [31]. The formation reactions for these are

$$Ga(s) \rightleftharpoons V_{Ga}^0 + Ga(g) \tag{1.23a}$$

$$Ga(s) \rightleftharpoons V_{Ga}^- + Ga(g) + h^+ \tag{1.23b}$$

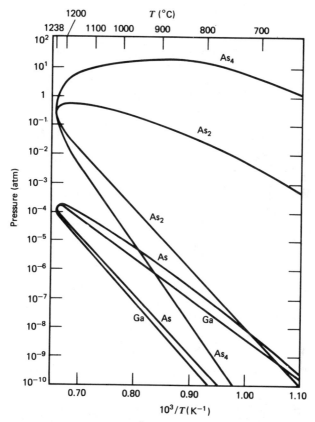

Fig. 1.19 Partial pressures of gallium and arsenic over gallium arsenide, as a function of temperature. From Arthur [30]. Reprinted with permission from the *Journal of Physics and Chemistry of Solids.*

Applying the mass action principle,

$$k_2 = [V_{Ga}^0]p_{Ga} \tag{1.24a}$$

$$k_3 = p[V_{Ga}^-]p_{Ga} \tag{1.24b}$$

Combining these equations with Eq. (1.22b) gives

$$[V_{Ga}^0] = \frac{k_2}{k_1}p_{As_2}^{1/2} \tag{1.25a}$$

$$[V_{Ga}^-] = \frac{k_3}{k_1}\frac{n}{n_i^2}p_{As_2}^{1/2} \tag{1.25b}$$

so that

$$[V_{Ga}] = [V_{Ga}^0] + [V_{Ga}^-]$$
$$= \left[\frac{k_2}{k_1} + \frac{k_3}{k_1} \frac{n}{n_i^2} \right] p_{As_2}^{1/2} \qquad (1.26)$$

Thus, the concentration of gallium vacancies is directly proportional to the square root of the arsenic partial pressure.

Arsenic vacancies are generally considered to exist as V_{As}^0 and V_{As}^+ [31]. Assuming that arsenic is in the form of As_2, the formation reactions are

$$As(s) \rightleftharpoons V_{As}^0 + \tfrac{1}{2} As_2(g) \qquad (1.27a)$$
$$As(s) \rightleftharpoons V_{As}^+ + \tfrac{1}{2} As_2(g) + e^- \qquad (1.27b)$$

Applying the mass action principle as before, we obtain

$$[V_{As}] = [V_{As}^0] + [V_{As}^+]$$
$$= \left[k_5 + k_6 \frac{p}{n_i^2} \right] p_{As_2}^{-(1/2)} \qquad (1.28)$$

where k_5 and k_6 are the equilibrium constants for Eqs. (1.27a) and (1.27b), respectively. Thus, the concentration of arsenic vacancies is inversely proportional to the square root of arsenic pressure. Equations (1.26) and (1.28) can both be simplified when applied to extrinsic GaAs, where one of the terms can be neglected.

Additionally, the product of the uncharged and charged vacancy concentrations, $[V_{Ga}^0][V_{As}^0]$ and $[V_{Ga}^-][V_{As}^+]$, is a function of temperature alone and is independent of the arsenic partial pressure. This interesting relationship is similar to the np product in a material, which is also a function of temperature, and not of doping.

Calculations for the arsenic and gallium vacancy concentrations in GaAs, under equilibrium pressure conditions, have given

$$[V_{Ga}] = 3.33 \times 10^{18} e^{-0.4/kT} \qquad (1.29a)$$
$$[V_{As}] = 2.22 \times 10^{20} e^{-0.7/kT} \qquad (1.29b)$$

These values must be considered as very approximate. Experimental data at 1100°C indicate an uncertainty in $[V_{As}]$ of less than a factor of 10 [32]. No experimental data are available for the gallium vacancy concentration.

1.5.3 Chemical Point Defects

As mentioned earlier, point defects can also take the form of impurities which are present in the semiconductor lattice, either introduced deliberately for doping purposes, or inadvertently as contamination during processing. They are also present as both dopants and contaminants in as-grown bulk material.

Chemical point defects of these types are located in substitutional as well as interstitial sites. As a general rule, substitutional impurities are electronically active whereas many contaminants are interstitial in behavior and are electronically inactive. Among impurities which are electronically active, donors and acceptors which have energy levels within $3kT$ of the conduction and valence band edges, respectively, are referred to as *shallow* impurities, and are used for doping purposes. They are fully ionized at room temperatures. Impurities which have one or more energy levels outside of this range are referred to as *deep*. Their presence in the semiconductor reduces its minority carrier lifetime, and strenuous attempts are made to avoid their incorporation during microcircuit fabrication. In some cases, however, they are deliberately introduced for the purpose of reducing lifetime in high-speed devices.

Silicon belongs to group IV of the periodic table; impurities from groups III and V are substitutional in nature and are electronically active in it. A group V atom has an excess valence electron which does not enter into covalent bonding with its four silicon neighbor atoms. Consequently, this electron is loosely bound, and is free to participate in the conduction process. Group V atoms thus behave as *n-type* impurities.

An estimate of the binding energy of this electron to its atom may be made by noting that a group V impurity can be represented by a nucleus with a single orbiting electron; that is, it is hydrogen-like in character. Elementary atomic theory shows that the energy levels of the hydrogen atom are given by

$$E = -\frac{m_0 q^4}{8a^2 h^2 \epsilon_0^2} \qquad (1.30a)$$

where ϵ_0 is the permittivity of free space, h is Planck's constant, q is the magnitude of the electron charge, and a takes on the values $1, 2, \ldots$. An equivalent situation can be considered for a crystal by replacing the mass of the electron with its effective mass (m_n^*) and the permittivity of free space by that of the crystal, so that

$$E = -\frac{m_n^* q^4}{8a^2 h^2 (\epsilon \epsilon_0)^2} \qquad (1.30b)$$

where ϵ is the relative permittivity of the crystal.

The energy required to remove an electron from the ground state of the hydrogen atom is 13.6 eV. This is known as the *first ionization potential*. The comparable value in a crystal would thus be

$$E_{ion} = \frac{13.6}{\epsilon^2} \frac{m_n^*}{m_0} \qquad (1.31)$$

Assuming an effective mass ratio* of about 0.6 and a permittivity of 11.8 for silicon gives an ionization energy of about 0.06 eV. This is relatively shallow, so that these impurities are almost fully ionized at room temperature. A group III impurity has a hole (the absence of an electron) which is loosely bound to it. Using analogous reasoning, we arrive at a very similar figure for its ionization energy.

Table 1.2 at the end of this chapter lists group III and V impurities in silicon, together with their ionization energies. All are seen to follow the arguments presented above, with the exception of indium, which is generally considered to be a deep impurity in silicon.

The above reasoning can also be applied to the behavior of impurities in GaAs, which is a group III-V compound semiconductor. Here, impurities belonging to group VI will usually be incorporated substitutionally, and on the arsenic sublattice. Consequently, each impurity atom of this type contributes one loosely bound electron for conduction purposes; that is, it is n-type. In like manner, impurities from group II are also substitutional, but are incorporated on the gallium sublattice where they result in p-type conduction.

As with silicon, these impurities are shallow, and their ionization energies can be estimated by using the hydrogen model. The effective mass of holes in GaAs is $m_p^* \simeq 0.5m_0$, so that the ionization energy for p-type impurities is approximately 0.045 eV. On the other hand, the effective electron mass under low-field conditions is much smaller ($m_n^* \simeq 0.067m_0$) so that the ionization energy for electrons is only 0.005 eV. As a result, n-type GaAs is degenerate at most useful concentration levels. Typically these n-type impurities are fully ionized, even at liquid nitrogen temperatures (77 K).

Impurities from group IV (carbon, germanium, silicon, tin, lead) are usually incorporated substitutionally into GaAs, partly on each sublattice, depending on the relative vacancy concentrations and site occupation probabilities [33]. Group IV impurities will be n-type on the gallium sublattice, but p-type on the arsenic sublattice. The net free carrier concentration is thus less than the impurity concentration and is either n- or p-type, depending upon the conditions under which these impurities are incorporated.

The site occupation probability for impurity incorporation is strongly related to the conditions under which the material is grown. Thus, silicon incorporates almost entirely as n-type during vapor-phase growth, which is carried out under highly nonequilibrium conditions. On the other hand, it can incorporate

---

*This is roughly the average of the conductivity effective mass ratio and the density-of-states effective mass ratio.

either n- or p-type during liquid-phase epitaxy, which is carried out under conditions approaching thermal equilibrium. This situation is used to advantage in the growth of successive n- and p-layers for the fabrication of light-emitting diodes, using silicon as the dopant in both cases. Table 1.3 lists shallow impurities in GaAs together with their ionization energies and conductivity types.

In principle, a shallow impurity may exhibit more than one energy level for each of its charge states. Only one impurity level, however, is normally observed within the energy gap for shallow donors and acceptors. Additional levels, corresponding to the second and higher ionization potentials, have also been observed at low temperatures [34].

It should be emphasized that energy levels given in Tables 1.2 and 1.3 are not precise, but are reasonably accurate for moderate doping concentrations. With heavier doping, these impurity levels broaden into bands. In addition, the impurity atoms come closer together (their average distance varies inversely as the cube root of the doping concentration), with a resulting decrease in their potential energy. As a result, the activation energy of the impurity, measured as the minimum energy difference between the impurity level and the appropriate band edge, falls. Experimentally the activation energy can be fitted to

$$E_0(N) = E_0(0) - \alpha N^{1/3} \tag{1.32}$$

where $E_0(N)$ is the activation energy at a doping level of N impurity atoms cm^{-3}, and $E_0(0)$ is the activation energy at low doping levels. For boron and phosphorus in silicon, $E_0(0)$ values are 0.08 eV and 0.054 eV, respectively, and α takes on the value of 4.3×10^{-8} eV cm [35]. Thus boron-doped silicon and phosphorus-doped silicon are degenerate $[E_0(N) = 0$ eV$]$ at doping levels in excess of 6.44×10^{18} and 1.98×10^{18} cm^{-3}, respectively. Similar considerations apply to acceptors in GaAs. As mentioned earlier, donors in GaAs are degenerate at most useful concentration levels.

The hydrogen model is an extremely elementary one. Indeed it is remarkable that it can predict the energy levels of shallow impurities with any degree of accuracy, but it cannot be extended beyond this point. Thus it cannot explain the ionization energy of indium (which belongs to group III) in silicon, or of impurities from other groups in the periodic table.

Doping with elements other than those from groups III to V of the periodic table often gives rise to a complex energy-level structure in silicon. In general these impurities exhibit more than one energy level, often of more than one type (i.e., both donor and acceptor levels). Furthermore, these levels are usually found quite deep in the forbidden gap. The effective mass theory described above does not apply to these impurities because their electronic wave functions are more highly localized than for shallow impurities [36]. The deliberate introduction of deep levels is sometimes done in order to reduce minority carrier lifetime in high-speed silicon microcircuits.

A relatively simple model can be used to illustrate the general character of deep levels created by the introduction of these impurities into silicon [29]. Consider, for example, what happens when a monovalent impurity is introduced into a substitutional site. Such an atom, in the neutral state, has only one attached electron which provides covalent bonding with its neighboring silicon lattice atoms. When additional electrons are attached to it, it is successively transferred to a more and more negatively charged state. Each additional electron gives rise to a possible new energy level. Since these electrons are attached successively to more negatively charged atoms, it is probable that the value of the associated energy level will continually increase in sequence until it goes beyond the edge of the conduction band. At this point the atom will lose this electron to the conduction band, resulting in no further identifiable energy levels. Finally, a monovalent impurity atom may lose an electron and be promoted to a positive (donor) charge state. Based on the above arguments, it is reasonable to postulate that the energy-level structure of a monovalent impurity atom may consist of as many as one donor level and three acceptor levels, progressively spaced in order of increasing negative charge. In most instances, only a few of these levels are identifiable within the energy gap. A notable exception is gold in germanium, which exhibits all four energy levels.

Similar considerations can be applied to the behavior of deep impurities in GaAs. Here, their energy levels will be progressively spaced in order of increasing negative charge, but will depend also on whether the impurity occupies a vacant gallium or arsenic site. Again, only a few of these energy levels are seen for each specific impurity in practice.

Associated with each charge state is a concentration, so that the solid solubility of the impurity will depend upon the position of the fermi level in the semiconductor. Consider, for example, the situation with gold in silicon, which is known to adopt the charge states Au^-, Au^0, and Au^+. For this impurity,

$$Au^0 + e \rightleftharpoons Au^- \tag{1.33a}$$

$$Au^0 + h \rightleftharpoons Au^+ \tag{1.33b}$$

Applying the principle of mass balance, and writing k_1 and k_2 as the equilibrium constants, gives

$$k_1 = \frac{[Au^0]n}{[Au^-]} \tag{1.34a}$$

$$k_2 = \frac{[Au^0]p}{[Au^+]} \tag{1.34b}$$

Moreover,

$$k_1 = \frac{[Au^0]_i n_i}{[Au^-]_i} \qquad (1.35a)$$

$$k_2 = \frac{Au^0]_i n_i}{[Au^+]_i} \qquad (1.35b)$$

where the subscript i refers to concentrations in intrinsic silicon.

A typical device, in which there are multiple regions of different doping and impurity type, will have a spatial distribution of electric field. However, the equilibrium distribution of Au^0 will not be affected by this electric field. Rather, it will be set by the diffusion of gold into silicon from an external source. Hence,

$$[Au^0] = [Au^0]_i \qquad (1.36)$$

Consequently, assuming an unlimited source, we have

$$[Au] = [Au^0] + [Au^-] + [Au^+] \qquad (1.37a)$$

$$= [Au^0]_i + [Au^-]_i \frac{n}{n_i} + [Au^+]_i \frac{p}{n_i} \qquad (1.37b)$$

From Eqs. (1.37a) and (1.37b), it is seen that the solubility of an impurity with multiple charge states is a strong function of the background doping level of the semiconductor.

A further complication can arise with some impurities, which incorporate in a more complex manner than described here. By way of example, experimental studies of gold in silicon have shown that about 90% incorporates in active substitutional sites, while the rest is neutral and interstitial. With platinum, about 90% is in active, substitutional sites; the remaining 10% is also active, but its exact lattice configuration is as yet unidentified. One possibility is that it forms active platinum–silicon complexes in silicon. With nickel in silicon, as much as 99.9% is inactive and in interstitial sites. Finally, we note that the fraction of an impurity which is active is also a function of the doping concentration. This is because the introduction of a dopant at high concentrations is usually accompanied by the generation of strain in the lattice, caused by misfit.

Impurities in both silicon and GaAs are sometimes incorporated in the lattice in the form of electronically active complexes. For GaAs, many impurities exhibit energy levels which have been identified as being associated with impurity–gallium and impurity–arsenic pairs. It is also possible for an impurity to combine with a vacancy, to form an electronically active defect. One example of this type of defect is a deep acceptor in GaAs, which results from the combination of a gallium vacancy and a shallow donor.

The ionization energies of some deep-lying impurities in silicon are also listed in Table 1.2. Many of these are present unintentionally in the starting material. Table 1.3 shows this information for GaAs. A number of specific impurities are now considered, together with their characteristics.

1.5.3.1 Impurities in Silicon

Gold. Gold is of special importance in silicon technology, where it is commonly used for minority carrier lifetime reduction in high-speed digital circuits [37]. It is also used for the control of lifetime in semiconductor power devices. It has a solid solubility of 2×10^{17} atoms cm^{-3}, and can be incorporated in large concentrations into silicon without the formation of any complexes. It exhibits both a donor level at $E_c - 0.76$ eV, and an acceptor level at $E_v + 0.57$ eV, depending on the particular charge state in which it is incorporated. It is commonly referred to as an *amphoteric impurity* because of this property.

Platinum. Platinum is increasingly used as a substitute for gold in silicon [38]. About 90% of it is incorporated into substitutional sites, where it exhibits an acceptor energy level at $E_v + 0.92$ eV, and a donor level at $E_c - 0.85$ eV. The remaining platinum is believed to be in the form of an electronically active complex, and behaves as an acceptor at $E_v + 0.42$ eV. Its capture cross section is extremely large, and accounts for most of the lifetime reduction properties of this dopant.

The energy levels of platinum in silicon are highly asymmetric with respect to their location in the energy gap, and they result in very different lifetime characteristics from those obtained with gold in silicon. In particular, the space charge generation lifetime in platinum-doped devices at room temperature is many hundred times longer than the low-level lifetime. Thus platinum doping can reduce low-level lifetime in the neutral regions of a diode without a comparable increase in its leakage current. In contrast, a direct consequence of the symmetrical location of the gold acceptor level is that leakage current in p–n diodes is inversely related to the lifetime in the neutral regions.

Oxygen. Silicon is commonly grown in silica crucibles, so that a large amount of this impurity is usually present. Depending on the crystal growth technique, this ranges from 10^{16} to over 10^{18} cm^{-3}. The dissolved oxygen is mostly in interstitial sites, and can be identified by a characteristic absorption peak at 1107 cm^{-1} ($\simeq 9.1\ \mu$m). About 0.1% of this oxygen can be converted into an active Si–O complex during heat treatment [39] at low temperatures (400–500°C). This complex is donor-like and has been tentatively identified as SiO_4. It disappears with processing at higher temperatures and can be eliminated by a 650°C anneal for as little as 20 min, followed by quenching. However, extended heat treatment at this temperature can result in the formation of a new donor, especially in crystals with a high carbon content.

The concentration of dissolved oxygen in silicon is given [40] by

$$O_i = 5.5 \times 10^{20} e^{-0.89/kT} \tag{1.38}$$

This is far above the solubility limit at integrated circuit processing temperatures. As a result, during device processing, most of it combines with silicon, and precipitates in the form of hexagonal platelets of SiO_2, typically 1 μm across and 1 μm apart [41]. Although inert, their presence distorts the potential lines in the depletion layer of p–n junctions. This results in premature breakdown and "soft spots."

Carbon. Carbon has a high solid solubility and is often incorporated in silicon in the 10^{16}- to 10^{18}-cm^{-3} range, during its chemical purification process. Its presence is indicated by an infrared absorption line at 603 cm^{-3} ($\approx 16.7\,\mu$m). It is electrically inactive, and forms silicon–carbon complexes in the form of microprecipitates. Again, its presence in high concentrations leads to premature breakdown of p–n junctions. A number of metal–carbon complexes have been reported in the literature as well [42].

In addition to those listed in Table 1.2, a number of other deep levels have been reported in the literature, but there is little agreement concerning their positions and concentrations. Many have been attributed to oxygen complexes with other impurities such as copper, cobalt, and nickel.

1.5.3.2 Impurities in Gallium Arsenide

Many impurities, both shallow and deep, are present in the form of complexes with gallium or arsenic. Both active and inactive complexes have been identified in GaAs, and little is known of the manner in which they are incorporated into the lattice. A number of specific impurities are now considered briefly, together with some of their characteristics:

Selenium and Tellurium. Both are n-type and located on the arsenic sublattice. At high concentrations they tend to form compounds with gallium (Ga_2Se_3 and Ga_2Te_3) which are inactive. Evidence of complexes with gallium vacancies have also been observed.

Tin. Although belonging to group IV, tin is almost always n-type in gallium arsenide which is grown by chemical vapor transport processes (see Chapter 5). Tin is located on both types of lattice sites, with an increasing fraction incorporated on arsenic sites at high concentration. Thus, its electron concentration varies sublinearly with the tin concentration. Tin-doped GaAs, grown by liquid-phase epitaxy, exhibits a shallow donor level.

Rare Earths. Interest in doping with rare earths comes about because they exhibit extremely narrow emission lines. This is true, regardless of the semiconductor in which they are introduced. Erbium is of commercial importance as a dopant in GaAs because its level corresponds to 1.54 μm, the wavelength at which silica fibers have minimum optical transmission loss [43]. Its solid solubility in GaAs is around 3×10^{17} cm^{-3}.

Silicon. This is also a group IV element, and is incorporated on both sublattices. It is of special importance because its relative incorporation can be readily controlled in liquid-phase epitaxy to give either *p*-type GaAs by low-temperature processing, or *n*-type GaAs by processing at high temperatures (see Section 2.2.5). It is used extensively as a dopant in (a) vapor phase epitaxy, (b) molecular beam epitaxy, and (c) ion implantation, where it behaves *n*-type.

Silicon is present in all GaAs as a contaminant. Often it is in the original materials from which it was made. Processing in silica vessels is also an important contributory factor. The background concentration of undoped GaAs is thus critically dependent on how this contaminant is incorporated into the lattice. Both growth temperature and arsenic overpressure play an important role here, since they determine the relative vacancy concentrations, and hence the incorporation of this impurity. In addition, silicon–oxygen complexes, with the silicon on gallium sites and the oxygen as an interstitial, have been identified [44] as the cause for a number of doping anomalies in GaAs.

Carbon. This is also present as a contaminant in GaAs. Although belonging to group IV, it has been observed as a shallow acceptor and as a deep donor, at concentration levels as high as 8×10^{20} cm^{-3}. It incorporates readily during growth by organometallic vapor-phase epitaxy [45], where it exhibits exclusively *p*-type behavior.

Copper. Copper is a deep, triple acceptor [46] in GaAs. It has a high solid solubility (6×10^{18} cm^{-3}) and moves extremely rapidly, even at relatively low processing temperatures (300–400°C). It is often present as a contaminant, and is very effective in reducing the diffusion length of *n*-type GaAs.

Chromium and Iron. Chromium behaves as a single acceptor, with an impurity level that is extremely close to the center of the energy gap. It is used during crystal growth to intentionally counterdope *n*-type GaAs to make it semi-insulating (SI), with a resistivity of as high as 10^9 Ω-cm. This makes it of great importance technologically, since it permits the possibility of using GaAs as an insulating substrate on which active layers of doped GaAs can be grown. The incorporation of chromium in GaAs during crystal growth is covered in detail in Chapter 3.

Chromium-doped, SI GaAs has been extensively studied in recent years. It has been found [47] that only some slices of this material retain their SI properties upon heat treatment, while others do not. This behavior is related to impurities which are initially present in the material, in addition to the chromium. Stable SI GaAs requires a low background concentration of these impurities; this, in turn, permits the use of low concentrations of chromium for counterdoping purposes.

Iron can also be used for making SI GaAs. However, because it has no midgap energy levels, the highest resistivity obtained with this impurity is $4 \times 10^4 \, \Omega$-cm. Its solid solubility is in excess of $10^{17} \, cm^{-3}$.

Oxygen. Oxygen can best be described as a problem contaminant. It exhibits two levels in the energy gap, one at $E_c - 0.14$ eV and one at $E_c - 0.57$ to 0.75 eV, and has a solid solubility in excess of $10^{17} \, cm^{-3}$ [48]. Its presence in *n*-type GaAs results in raising the resistivity of material, until SI behavior is achieved with a resistivity on the order of $10^8 \, \Omega$-cm. This is somewhat surprising, since the fermi level in this material is well above the oxygen donor level. A generally accepted theory for this behavior is based on the fact that undoped, *n*-type GaAs contains silicon as the primary donor contaminant. Incorporated oxygen combines with this silicon to form inactive silicon–oxygen complexes, thus tying up this donor so as to shift the material toward *p*-type. This in turn allows the remaining oxygen to become ionized and moves the fermi level toward the center of the gap. Oxygen-doped, SI GaAs has a relatively high mobility ($\simeq 4000 \, cm^2/V$ s), which is not what would be expected in a highly compensated material. This would lend strength to the above arguments.

Oxygen cannot be used for making SI GaAs, since it is highly mobile at processing temperatures. Its real importance lies in the restrictions it places on the thermal processing of GaAs in an open-air ambient. Such processing poses the ever-present risk of converting the material inadvertently to high resistivity. This is perhaps the primary reason why many of the simple thermal processes available for silicon are not possible with GaAs.

Finally, it is worth noting that GaAs can have in it many inactive impurities which have only a slight effect on the mobility or the free carrier concentration. Thus high-purity GaAs, with a free electron concentration of 10^{13}–$10^{14} \, cm^{-3}$, can contain as much as 10^{16}–10^{17} atoms cm^{-3} of carbon and oxygen, in addition to such impurities as aluminum, calcium, potassium, nitrogen, strontium, and tantalum in the $\leq 10^{16}$-cm^{-3} range. These impurities are generally interstitial, or form complexes which are electronically inactive.

1.5.4 Contamination Control

Chemical point defects, in the form of undesired impurities, can be incorporated inadvertently during every microcircuit fabrication step. As a result, the avoidance of these impurities is of prime consideration, and is a continuous part of the fabrication process. The use of ultrapure, "semiconductor grade" starting chemicals and substrates, and the conduct of processes under clean room conditions, is aimed at achieving this goal. Nevertheless, impurities do get incorporated during this process. As noted above, these usually behave as deep levels, and contribute to phonon-assisted recombination in both GaAs and silicon. In GaAs, the minority carrier lifetime is dominated by radiative processes, and is extremely short, typically in the 10- to 100-ns range. The presence of these impurities reduces it further, and affects the current collection efficiency of detectors and solar cells. It also deteriorates the performance of light-emitting diodes and lasers, since recombination via these impurities is usually nonradiative.

Silicon is an indirect gap semiconductor, in which the lifetime is large, often in the 500 μs range. Here, the presence of deep impurities significantly reduces the minority carrier lifetime and degrades the performance of devices such as bipolar transistors and solar cells. It also reduces the refresh time of storage diodes used in dynamic random access memories.

The metal-oxide-semiconductor (MOS) transistor is a majority carrier device, so that its transport properties are not affected by the presence of impurities. However, the leakage current of the source and drain regions of these devices increases with the concentration of these deep-lying impurities. This is a serious problem in VLSI applications, since it increases the standby power dissipation in these devices and limits the number of devices which can be operated on a single chip. As a result, special attention must be taken to minimize the concentration of these impurities.

As mentioned earlier, almost every chemical impurity outside of groups III and V of the periodic table exhibits one or more deep impurity levels in silicon. Some are present as contaminants in as-grown silicon. Yet others are introduced into silicon as trace impurities in the chemical solutions and transport gases. Some impurities are present in large quantities in the air, in glassware, in diffusion tube liners, and in the quartz diffusion tubes themselves. The amount of these impurities that is incorporated into the silicon depends on the processing time and temperature involved and on the diffusion-induced stress. Thus contamination problems are most severe with deep-diffused, heavily doped structures.

Almost all deep impurities diffuse primarily by an interstitial mechanism and move rapidly through the slice at processing temperatures. Their diffusivity is typically 5–6 orders of magnitude higher than that of substitutional impurities such as boron or phosphorus. During processing, they tend to condense around dislocations so as to form metallic precipitates, which can cause distortion of the potential lines if they are located in the depletion layer of a p–n junction, leading to localized regions of high electric field through which excess leakage current flows. Thus a second consequence of deep impurities is that they "decorate" dislocations, and give rise to "soft" spots and "soft" breakdown characteristics with excess reverse currents at voltages below their avalanche breakdown value. The elimination or reduction of these precipitates leads to improvement in the sharpness of the reverse characteristic, as well as to improvement in the minority carrier lifetime of the material.

The most straightforward approach for lifetime improvement is to use clean processing, to minimize deterioration of the original lifetime of the as-grown silicon. Sealed tube diffusions, for example, are almost exclusively employed for the formation of deep *p*-diffusions in silicon high voltage power devices. Diffusions of this type invariably result in junctions with long lifetimes and hard reverse characteristics.

The deliberate manipulation of clusters and deep impurities *away* from active regions presents a powerful alternative technique for their reduction by a process known as *damage gettering*. Here, the strategy is to create a region of damage far from the active device; upon subsequent heat treatment, rapidly moving impurities segregate in this region in order to relieve the strain created by this damage.

Sandblasting or otherwise abrading the back surface has been used with both silicon and GaAs [49]. This results in the formation of an almost infinite supply of vacancies. During high-temperature processing, these vacancies act as sinks for fast-moving deep impurities that diffuse throughout the slice.

Controlled amounts of damage can also be introduced by the use of deposited films, such as layers of silicon nitride [50], on the back surface of the wafer. Here the high interfacial stress created by the deposition process results in the formation of a dislocation network, toward which it is desired that deep impurities migrate. Polycrystalline silicon can be used for the same purpose, and has the advantage of being extremely pure compared to other deposited materials.

Ion implantation of heavy neutral species such as argon has also been explored as an alternative technique for producing the desired surface damage [51, 52]. A disadvantage of this method is that the damage anneals out with subsequent high-temperature processing, so that its benefits are relatively short-lived.

Impurities of special concern in silicon processing are copper, iron, and gold [53]. All diffuse rapidly through the silicon lattice, and have reasonably large solid solubilities and capture cross sections. In contrast, other impurities tend to form inactive complexes and to have extremely low values of solid solubility.

Both copper and iron diffuse by a rapid interstitial–substitutional mechanism, but ultimately take up substitutional sites by combining with vacancies. Moreover, since their substitutional solid solubility is lower than their interstitial value, they freeze out upon cooling and tend to condense around dislocations. Once there, however, they may either remain in the slice or leave by out-diffusion. The vapor pressure of copper is 15–20 times higher than that of iron. Consequently, nearly all this impurity leaves by out-diffusion. On the other hand, there is little out-diffusion with iron; typically, about 33% segregates in this layer of dislocations, the rest being retained in the bulk.

Gold moves primarily by an interstitial–substitutional mechanism and also takes up substitutional sites in the lattice by combining with vacancies. Unlike copper and iron, the solid solubility of substitutional gold is much higher than that of the interstitial species; thus it remains in solution upon cooling. Furthermore, since gold does not form compounds with silicon, it does not segregate by this process.

Layers of borosilicate (BSG) and phosphosilicate (PSG) glasses, in contact with the silicon at elevated temperature, are extremely effective in removing deep impurities from silicon. A theory that has been proposed to explain the effectiveness of this gettering process is that heat treatment of the sample results in creation of a diffusion-induced stress well ahead of the diffusing front. In addition, it produces a viscous layer during heat treatment, in which rapidly moving impurities can be immobilized.

During heat treatment, these lifetime killers (copper, iron, and gold) all move rapidly by interstitial motion, and are trapped to a certain extent in the glassy layer or under it. PSG is generally found to be more effective than BSG for the removal of gold in silicon. Its superiority has been attributed to the enhanced solid solubility of substitutional gold in silicon in the presence of phosphorus. On the other hand, BSG results in boron diffusion into silicon, which getters gold by the formation of compounds at moderate to heavy contamination levels. However, it is less effective for dilute gold concentrations.

Intrinsic gettering is also used for reducing deep-lying impurities in silicon. Here, the basic idea is to take advantage of the large amount of dissolved oxygen ($\approx 10^{18}/cm^3$) which is present in crucible-grown material. During prolonged heat treatment at high temperature (1100–1200°C), this oxygen diffuses away from the surface region and towards the central core of the slice. A subsequent nucleation step at 650–800°C results in the reaction of this oxygen to form platelet-like clusters of SiO_2. Each of these steps is typically 2–8 h in duration.

The large volume change associated with this process builds up a strain field [54], which serves as a sink for metallic impurities. A relatively clean *denuded* zone of silicon is left in the surface regions, in which devices are subsequently fabricated [55]. Thus, intrinsic gettering is accomplished by means of a long, high-temperature treatment to remove oxygen from the surface regions, followed by a lengthy low-temperature step to create SiO_2 clusters and their associated strain. More complex denuding procedures, consisting of multistep thermal cycles, have also been proposed for this purpose.

For successful intrinsic gettering, it is necessary that the starting silicon have a relatively high oxygen concentration, typically in excess of $1 \times 10^{18}/cm^3$. On the other hand, excessively large concentrations can result in severe warpage of the silicon slices during this process, because of the volume change involved in forming SiO_2. Thus, control of the oxygen concentration during crystal growth, over the entire length of the boule, is essential for the successful implementation of this method.

The effect of subsequent processes (which may include damage gettering) must also be considered. For example, interstitial oxygen tends to precipitate during oxidation and diffusion processes, which are conducted at 950–1150°C. The depth of the denuded zone is also a function of the resistivity of the silicon, and of the components of the microcircuit which is fabricated in it. Consequently, the gettering procedure must be optimized for each situation.

Yet another approach for lifetime improvement is known as *metallic gettering*. Here, a metal, such as nickel, is evaporated on the back surface of the slice which is subsequently heat-treated [56, 57]. Impurities such as gold tend to move preferentially into this layer because of its higher solubility, resulting in an improvement in the lifetime. Metallic getters have not been found to be very effective; in addition, they often cause gross damage to the surface on which they are deposited, and thus can only be used on the back surface of the slice.

Special diffusion and substrate cleaning procedures are used for reducing the minority carrier lifetime in silicon and GaAs microcircuits. These will be discussed in detail in Chapters 4 and 9, respectively.

1.6 DISLOCATIONS

A dislocation is a one-dimensional array of point defects in an otherwise perfect crystal, and results in a geometric fault in the lattice. It occurs when the crystal is subjected to stresses in excess of the elastic limit—for example, during the cool-down phase of its growth from a melt. Although the nature of dislocations is quite complex, they are usually composed of combinations of two basic types: the screw dislocation and the edge dislocation. A simple cubic lattice is considered in the following sections. The diamond lattice is considerably more complex; the general properties of dislocation types, however, are very similar to those of the cubic lattice.

1.6.1 Screw Dislocations

Figure 1.20 shows the manner in which a regular crystal lattice may be subjected to shear stresses in order to establish a screw dislocation. For this crystal, consider the plane *ABCD*, which is one of its regular lattice planes, and let the two halves of the crystal on either side of this plane be subjected to shearing forces that are sufficiently large to cause them to be separated by one atomic spacing. The line of the screw dislocation so formed is *AD*, since this marks the boundary in the plane *ECBF* which divides the perfect crystal from the imperfect.

A rough estimate can be made of the strain energy associated with a screw dislocation by considering a cylinder of material, of length l, with axis *AD* and inner and out radii R_i and R_o, respectively, as shown in Fig. 1.21 [58]. It is assumed that the crystal behaves as an elastic solid within the cylinder defined by these radii. Let

b = amount of shear present in a shell of radius r and thickness dr

μ = shear modulus

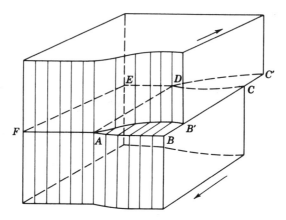

Fig. 1.20 Screw dislocation.

Then the elastic shear strain is given by $b/2\pi r$. The elastic energy of the shell, dE_μ, is

$$dE_\mu = \tfrac{1}{2}\mu \,(\text{shear strain})^2 dV \qquad (1.39)$$

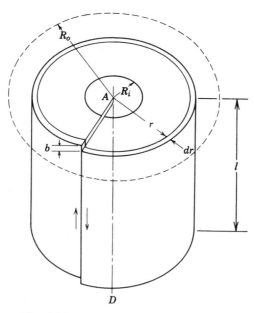

Fig. 1.21 Details of a screw dislocation.

where dV is the volume of the shell. Hence

$$dE_\mu = \left(\frac{\mu b^2 l}{4\pi} \right) \left(\frac{dr}{r} \right) \tag{1.40}$$

so that

$$E_\mu = \left(\frac{\mu b^2 l}{4\pi} \right) \ln \left(\frac{R_o}{R_i} \right) \tag{1.41}$$

If this is the only dislocation in an infinite volume of material, then $R_o = \infty$ and the energy associated with it is infinite. In practice, however, crystals usually contain many dislocations, randomly distributed. As a result, their strain fields are also randomly distributed and cancel each other at distances approximately equal to the mean distance between them. In typical crystals, R_o is about 10^5 atom spacings. The inner radius limit R_i is set by the fact that a region of atomic dimensions can no longer be considered as an elastic continuum, and the theory of elasticity ceases to hold. As a consequence, it is reasonable to eliminate the inner 4–5 atoms from consideration.

Practical values of the ratio R_o/R_i are usually taken around 10^4. Using this value, the strain energy for a screw dislocation in silicon may be calculated as about 10–19 eV/atom length. (By way of comparison, values for aluminum and diamond are 3.1 and 29 eV, respectively.)

1.6.2 Edge Dislocations

An edge dislocation is shown in Fig. 1.22. Here an extra half-plane of atoms, *ABCD*, is present in the otherwise regular lattice, with most of the distortion concentrated around the line *AD*. A dislocation of this type is created by applying a shearing force along the face of the crystal, parallel to a major crystallographic plane. When this force exceeds that required for elastic deformation, the upper half of the crystal moves by a slip mechanism. The plane along which slip occurs is commonly referred to as a *slip plane*.

The strain energy associated with an edge dislocation is given [59] by

$$E_\mu = \left[\frac{\mu b^2 l}{4\pi(1 - \nu)} \right] \ln \left(\frac{R_o}{R_i} \right) \tag{1.42}$$

where ν is Poisson's ratio ($\simeq 0.3$ for both silicon and GaAs). Its magnitude is thus approximately 50% larger than that for a screw dislocation.

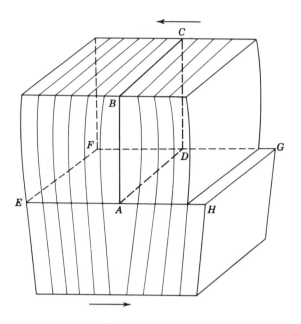

Fig. 1.22 Edge dislocation.

In view of the large energies of formation for both basic dislocation types, it must be concluded that their equilibrium concentration is negligible. They *cannot* be created by purely thermal means, since the thermal energy associated with processing temperatures as high as 1500 K is under 0.15 eV ($\approx 3kT$). Rather, mechanical forces, induced during thermal processing, are the driving forces behind the generation of these dislocations.

1.6.3 Dislocation Movement and Multiplication

Figure 1.23 indicates the manner in which an edge dislocation may move completely through a crystal. The mechanism for such a movement is called *slip*. It is a characteristic of the slip mechanism that it results in movement along planes of high atomic density where opposing forces are at a minimum.

The displacement of a screw dislocation also takes place along a slip plane. In Fig. 1.20 this slip plane is given by *ABCD*. The end result of such a displacement is identical to the movement of an edge dislocation, even though the strain pattern is different.

In addition to slip, *climb* is an alternative method by which a dislocation can move in a crystal. For an edge dislocation, such as that shown in Fig. 1.22, climb of the plane *ABCD* takes place at right angles to the slip plane *EFGH*. Figure 1.24 shows that this may occur as the result of the movement of either substitutional or interstitial atoms out of the plane *ABCD*. Alternatively, climb

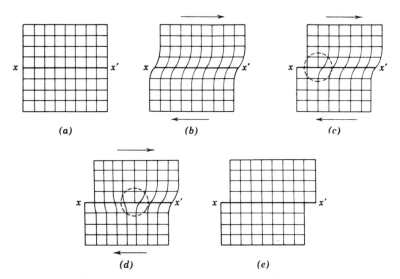

Fig. 1.23 Crystal movement along a slip plane.

may also occur by atoms moving into the plane. Intuitively, it is seen that the energy of formation associated with such a process is on the same order of magnitude as that for the energy of migration (≈ 0.18 eV) of a point defect. In fact, it is somewhat less, since the migration of these atoms is aided by the stress field surrounding the dislocation.

Climb in a screw dislocation occurs by a complex motion. Here the screw dislocation line twists itself into a helix, which can then climb. The actual movement of dislocations in a crystal is made up of combinations of these and other types of movements.

The energy of movement of a dislocation has been shown to be 0.15 eV/atom spacing for silicon and somewhat less for GaAs. This is the energy barrier that must be overcome in order for a dislocation to move in a crystal. A comparison with the energy of formation of a dislocation (about $10-19$ eV) shows that it is

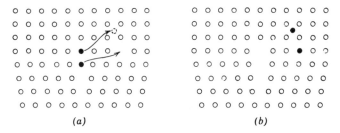

Fig. 1.24 Climb of an edge dislocation.

extremely easy to induce dislocation *motion* in a crystal by thermal means, even though it is almost impossible to *create* a dislocation in this manner. Thus one of the more important problems of crystal growth and device processing is to avoid (or minimize) the formation of dislocations in the first place. Alternatively, if such dislocations are unavoidable, they can sometimes be relieved by annealing. During crystal growth, techniques are available for inducing these dislocations to grow out of the crystal, leaving behind a relatively dislocation-free lattice.

There is considerable evidence indicating that dislocation multiplication occurs in a crystal in addition to dislocation movement. Examination of deformed crystals has shown that this is indeed the case, and various mechanisms have been suggested for this multiplication. Figure 1.25 shows a model for the Frank–Read mechanism by which this can occur [60]. Consider a dislocation, as shown in Fig. 1.25a. Under the application of a force F, the dislocation tends to expand along its length by climb. If, however, it is pinned at xx', possibly by the presence of some obstruction such as an oxygen or metallic cluster, it will tend to bow out of its slip plane, as shown in Fig. 1.25b. In doing so, it becomes longer and requires a greater force to maintain its new radius. A critical condition is reached at which the dislocation line becomes semicircular. For a force in excess of that required for this condition, the dislocation becomes unstable and progresses as shown in Fig. 1.25c and d. Eventually it returns to its original form by the collapse of the cusp, as shown in Fig. 1.25e, leaving an expanding loop in addition to it. The process now repeats itself, resulting in multiple dislocation loops from a single dislocation source of this type. A

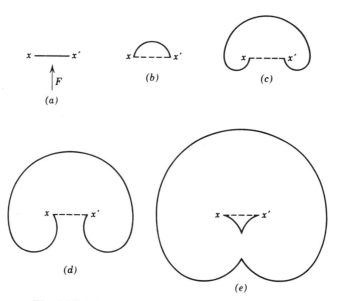

Fig. 1.25 Mechanism for dislocation multiplication.

conclusion to be drawn from the above is that the presence of insulating and metallic clusters in the semiconductor should be avoided, since they can lead to dislocation multiplication in the presence of thermally induced stress. On the other hand, the presence of clusters in a material can also serve to harden it, as is the case for SiO_2 in silicon. As a result, a balance must be struck between these extremes.

1.6.4 Process-Induced Dislocations

As-purchased single-crystal silicon has a dislocation content which is below the measurement limit, and can be essentially considered to be dislocation-free. GaAs is somewhat poorer because of its combination of relatively low critical resolved shear stress (7.75×10^6 dyn cm^{-2} as compared to 3.61×10^7 dyn cm^{-2} for silicon) and is typically available with 5000 dislocations cm^{-2} at the present time. In both these materials, the energy of formation of dislocations is so large that it is extremely difficult to form them by purely thermal means. However, many conventional processing steps, especially those involving high impurity concentrations, can lead to the formation of dislocations. These must form closed loops, or else terminate on the crystal surface where the microcircuit is fabricated. This, in turn, can affect the performance of the circuit. A few of the common causes for dislocation formation are now considered, but the list is by no means comprehensive.

The diffusion of an impurity will introduce stress in the crystal lattice, whose magnitude is related to the misfit factor and to the impurity concentration. Both lattice contraction or dilation can occur, depending on whether the tetrahedral radius of the impurity is smaller or larger than that of the semiconductor, respectively. Diffusion-induced dislocations will occur if this stress exceeds the elastic limit of the semiconductor. In silicon technology, boron presents a serious problem because of its combination of large misfit factor ($\epsilon = 0.254$) and its high solid solubility (2.5×10^{20} cm^{-3} at 1100°C). Diffusion at this temperature creates a stress which is seven times larger than the elastic limit of silicon ($\simeq 1 \times 10^9$ dyn cm^{-2}) and results in damaged material, with a high dislocation content.

Ion implantation, where substrates are subjected to high energy bombardment, is another source of these dislocations. Their complete removal can only be accomplished by subsequent annealing at temperatures within a few hundred degrees of the melting point of the semiconductor. Annealing at lower temperatures results in varying degrees of effectiveness, depending on the nature and extent of ion damage.

Rapid removal of a slice from a furnace causes its edges to cool faster than the center. This results in both radial and tangential stresses which are proportional to the temperature gradient, and dislocations are generated when this stress exceeds the critical resolved shear stress (CRSS) of the semiconductor. This problem is avoided by slow heating and cooling of the semiconductor during thermal processing. It is of special concern during rapid thermal processing.

An important process in silicon technology is the growth of an oxide on the substrate by means of its thermal oxidation (see Chapter 7). Conversion of the silicon to silicon dioxide in this manner results in a volume change by a factor of 2.23. This fact, combined with the difference in coefficients of thermal expansion of these materials, often results in the formation of dislocations in the silicon during the cool-down process. Moreover, it can cause cracking of the oxide film and thus a loss of its masking properties. In practice, the oxide film thickness is kept as small as possible in order to avoid this problem.

The deposition of thin films of materials such as silicon dioxide and silicon nitride, which are often used for masking purposes, also results in strain in the underlying conductor. The actual amount of strain and its sign (tensile versus compressive) is related to the deposition parameters. This is especially true for silicon nitride films which are frequently grown by the reaction of silane and ammonia in a hydrogen plasma. Here, both system pressure and reactant composition can affect the magnitude and sign of the residual strain in the film.

Highly localized stress can be introduced in both grown and deposited films when windows are cut in them to delineate regions for selective processes such as regrowth, diffusion, and ion implantation. This stress occurs in the corners of the windows, and results from stress relief over the rest of the region. Here too, the stress generally increases with film thickness. Thus, it is prudent practice to make masking films as thin as possible.

Thin metal films, deposited at room temperature by vacuum evaporation or sputtering, often have a built-in stress. The magnitude and sign of this stress depends upon the nature of the film growth process, and on the material itself. Subsequent processing can further alter the stress at the interface. For example, ductile aluminum, if excessively heat-treated in contact with silicon, forms an extremely strong alloy which places considerable stress on the underlying semiconductor during device fabrication, and with subsequent thermal cycling during device operation.

The above are only a few of the processes which can result in the formation of dislocations in the semiconductor. In all cases, an understanding of the problem can generally indicate the direction for its partial or total relief.

Dislocations can also be generated during device *operation*. During electron–hole recombination, for example, one of the methods by which excess energy is released is by its transference to the lattice in the form of phonon vibrations. This excess energy can often increase the rate of defect reactions. For example, in GaAs lasers and light-emitting diodes, dislocations are a source of nonradiative recombination. Rapid degradation of these devices has been traced to the enhanced motion of these dislocations by a climb process, when they are operated at high optical power densities [61]. Additionally, this rapid movement leads to the development of networks of dislocations, which originate preferentially at hetero-interfaces. Many studies of these *dark line defects* have shown an increase in the dislocation velocity in the presence of optical irradiation. Moreover, it has been established that they develop rapidly in material with a high initial density of defects [62]. Thus, solution of this problem lies in the use of high-quality starting substrates, and processing techniques which do not result in the production of dislocations.

1.6.5 Two-Dimensional Defects

Two-dimensional defects can be created at the boundary separating an error in the stacking sequence in a crystal. Defects of this type are known as *stacking faults*. Usually, these faults are decorated by fast-moving impurities which cluster around them. This leads to enhanced leakage current if the fault is in the vicinity of a *p–n* junction.

Twinning is a gross form of two-dimensional defect that may occur in a crystal. Its presence is usually indicative of material that has a high dislocation content and is not suited for device or microcircuit fabrication. Experimental evidence shows that excessive twinning is encountered if the material is physically restricted during its growth from a melt. Thus crucible-grown materials are highly prone to this defect, since it is difficult to prevent sticking of the semiconductor to the walls of the crucible.

Twinning occurs when one portion of a crystal lattice takes up an orientation with respect to another, the two parts being in intimate contact over their bounding surfaces. This bounding surface is called the *twinning plane*. Figure 1.26 shows a two-dimensional representation of twinned and untwinned parts of a crystal. For this case, atoms along xx' are common to both twinned and untwinned sections, and the twinning plane is sometimes referred to as the *composition plane*.

Finally, the surface of a semiconductor represents an important two-dimensional defect, since it is a discontinuity in an otherwise reasonably periodic lattice. This is true for an ideal, atomically clean surface with dangling bonds, as well as for the practical case where surface bonds are terminated by an interfacial layer or by contaminants. The electronic properties of this region have a strong influence on microcircuits which are essentially surface-oriented in character.

1.7 ELECTRONIC PROPERTIES OF DEFECTS

The deliberate insertion of chemical defects into the semiconductor lattice is the basis for the fabrication and control of electrical properties of semiconductor devices and microcircuits. This subject has been discussed in Section 1.5.3. However, attention must also be paid to the electronic behavior of defects that are intrinsic to the crystal and to complexes of these defects with impurities.

1.7.1 Point Defects

In silicon the presence of a vacancy in a crystal results in four unsatisfied bonds which would ordinarily be used to bind the atom to its tetrahedral neighbors. Thus a vacancy tends to be acceptor-like in behavior. The addition of each electron to this vacancy results in successively higher values of energy level because of the large mutual electrostatic repulsion present between them. It is highly improbable, however, that such a vacancy will exhibit as many as four energy levels within the band gap. One acceptor level at $E_v + 0.54$ eV and a second acceptor level at $E_v + 1.0$ eV have been positively identified in silicon by a number of workers. A donor level at $E_c - 0.98$ to 1.06 eV has also been identified and is considered to be due to a distorted bond configuration.

In like manner, an interstitial has four valence electrons that are not involved in covalent binding with other lattice atoms and which may be lost to the conduction band. As a result, it should exhibit donor-like behavior, and one or more levels within the energy gap. A singly ionized donor level, at about $E_c - 0.71$ eV, and a singly ionized acceptor level, at about $E_v + 0.62$ eV, have been identified here.

Many complex vacancy–interstitial combinations are also electronically active in silicon. For example, electron irradiation in the 1- to 2-MeV range gives rise to four energy levels at $E_v + 0.27$ eV, $E_c - 0.23$ eV, $E_c - 0.17$ eV, and $E_c - 0.41$ eV [63]. Annealing for 36 h at 300°C alters the defect structure, with the last two levels converting to one level at $E_c - 0.36$ eV.

The energy levels associated with vacancies and interstitials in silicon are deep. They serve as localized centers for minority carrier recombination, and result in a fall in the lifetime. In general there is an inverse relationship between the concentration of these levels and the minority carrier lifetime. This has been verified by numerous experiments on material with as-grown defects as well as with induced defects (by electron and nuclear radiation, and by plastic deformation techniques). In fact, the deliberate introduction of deep levels into silicon by electron irradiation provides a promising technique for controlling lifetime in high-power semiconductor devices.

Intrinsic defects in GaAs include both arsenic and gallium vacancies, their concentrations being determined by the overpressure of arsenic during processing. Many workers have considered these defects to be neutral [64]. However, it has been established that, in addition to being neutral, arsenic vacancies behave as deep donors, whereas gallium vacancies exhibit deep acceptor-like behavior [65].

Many other intrinsic defects are also observed in GaAs, with their nature and concentration being a function of the manner in which the material is grown [31]. The majority of these tend to thermally anneal out by 250°C. A notable exception is the antisite defect, As_{Ga}, which is stable to 950°C. Some of these defects are listed in Table 1.4 at the end of this chapter.

Intrinsic defects also occur in the form of complexes with other intrinsic defects, as well as with impurities. Yet other defects take the form of impurity–impurity pairs, which result from the presence of residual impurities in the GaAs.

An important defect–impurity complex is the gallium vacancy–shallow donor pair (V_{Ga}^--donor), which behaves as a deep acceptor [66]. It occurs with all group IV and group VI shallow donors, and is identifiable by a broad photoluminescence signal centered around $E_c - 1.2$ eV [67]. It is believed that the gallium vacancy is an ionized acceptor, and that the donor atom is bound to it by coulombic forces in this complex. One consequence of the presence of this center is that the free carrier concentration in *n*-type GaAs is generally less than the dopant concentration, so that the material is compensated [68].

The energy levels of many of these complexes are also listed in Table 1.4. Often, these levels have been identified by photoluminescence, and their impurity type is unknown.

An important defect in GaAs, commonly called EL2, is present in material which is grown from an arsenic-rich melt, and is considered to be an $As_{Ga}-V_{As}$ complex. This defect is donor-like in character and is located at $E_c - 0.76$ eV; that is, in the middle of the energy gap [69]. It is highly stable, and can withstand thermal processing at temperatures above 900°C. It has electron and hole capture cross sections of 1.5×10^{-13} cm^2 and 2×10^{-18} cm^2, respectively, that is, it is an electron trap.

EL2 is of great technological importance, because it can convert *p*-type GaAs to semi-insulating material, and is widely used for this purpose because of its thermal stability. Its properties and manner of incorporation are described in Chapter 3.

1.7.2 Dislocations

Many of the electronic properties of dislocations are similar to those of an ensemble of point defects. Thus there is considerable evidence to show that dislocations in silicon are acceptor-like, because of the presence of dangling (or unfilled) bonds at the edge of the half-plane comprising the dislocation.

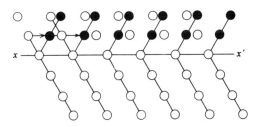

Fig. 1.26 A twinned structure.

However, it is possible to obtain a diffusional rearrangement of the atoms which reduces the formation of dangling bonds of this type [70]. This would explain the absence of strong acceptor-like properties for these dislocations.

The behavior of dislocations in GaAs is not firmly established. As noted earlier, they serve as recombination centers, and greatly lower the internal quantum efficiency in light-emitting diodes and lasers. Their presence results in accelerated degradation of laser devices.

Dislocations take the form of a line of defects, with an associated distortion of the energy-band structure in their vicinity. Thus they behave as anisotropic scattering centers for carriers because of the extended space charge region surrounding them. Furthermore, this distortion of the energy-band structure leads to the formation of trapping sites for free holes and electrons, and a concurrent reduction in minority carrier lifetime, in proportion to their concentration.

The most important property of a dislocation is that it interacts with chemical and other point defects in its neighborhood. This interaction exists between the localized disturbance, due to impurity atoms, and the strain field in the vicinity of the dislocations. Its extent is directly proportional to the misfit between the impurity and the lattice atom. Thus the presence of a dislocation is usually associated with an enhanced rate of impurity diffusion in its neighborhood, leading to the formation of diffusion pipes. Often the presence of the dislocation results in the segregation of metallic impurities in its vicinity. Taken together, they lead to such problems as excessive leakage and premature breakdown in semiconductor junctions made on this material.

1.7.3 Two-Dimensional Defects

The twinning plane in a crystal is a two-dimensional region with a large concentration of broken, or unsatisfied, bonds. This region is known as a *grain boundary*, and the actual number of such broken bonds is related to the angle of the grain boundary. This is shown in Fig. 1.27 for a symmetric grain boundary with an angle θ. If d is the atomic spacing, the distance between broken bonds is given by

$$D = \frac{d}{2 \sin (\theta/2)} \qquad (1.43)$$

If $\theta \leq 5°$, this distance is on the order of ≥ 11 lattice spacings. This is comparable to the lattice spacing for dopant atoms with a concentration of 3.3×10^{19} cm^{-3}. As a result, a crystal having a *low-angle grain boundary* of this type can still be considered as a coherent structure. The term *lineage* is used for grain boundaries where $\theta \leq 1°$. Here any loss in coherency due to this type of grain boundary can be ignored.

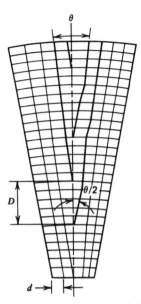

Fig. 1.27   Model for a low-angle grain boundary.

Isolated stacking faults in silicon have been shown to exhibit localized states at $E_c - 0.1$ eV and at $E_c - 0.3$ eV. Usually, these faults are decorated with fast-moving impurities, whose effect dominates the electrical behavior of the device.

From purely quantum-mechanical considerations, it can be shown that a freshly cleaved semiconductor surface exhibits a number of deep-lying levels throughout the energy gap [71]. These are due to the presence of unsaturated covalent bonds, and are acceptor-like in character. In a practical surface, many of these bonds are terminated with a partial coherence of the lattice. This leads to a reduction in the density of deep levels, and alters their distribution in the energy gap [26]. The detailed nature of these surface states is covered in Chapter 7.

TABLE 1.1 Properties of Silicon and Gallium Arsenide at 300 K

Property	Silicon	Gallium Arsenide
Crystal Structure	Diamond	Zincblende
Lattice constant (Å)	5.43086	4.65325 ± 0.00002
Distance between neighboring atoms (Å)	2.35163	2.44793 ± 0.00002
Atoms or molecules (cm^{-3})	4.99441×10^{22}	2.21×10^{22}
Atomic or molecular weight	28.0855	144.642 (69.72 for Ga, 74.922 for As)
Density (g cm^{-3})	2.328	5.3174
Dielectric constant	11.7	$12.4 \ (1 + 1.2 \times 10^{-4}T)$ (12.85 at 300 K)
Refractive index ($\lambda = 2\ \mu m$)	3.45	3.317
Infrared refractive index	—	$3.255 \ (1 + 4.5 \times 10^{-5}T)$ (3.299 at 300 K)
Melting point (°C)	1412	1238
Thermal coefficient of expansion (K^{-1})	2.33×10^{-6} at 300 K 3.0×10^{-6} at 400 K 4.0×10^{-6} at 650 K 5.0×10^{-6} at 900 K 4.5×10^{-6} at 1100 K	$5.69 \times 10^{-6} + 1.5 \times 10^{9}T$ (6.05×10^{-6}) at 300 K

Intrinsic carrier concentration (cm^{-3})	$3.1 \times 10^{16} T^{3/2} \exp\left(-\dfrac{1.206}{2kT}\right)$ $(1.25 \times 10^{10}$ at 300°C)	2.25×10^6 at 300 K 8.78×10^{11} at 500 K 3.02×10^{14} at 700 K 1.01×10^{16} at 900 K 6.5×10^{16} at 1050 K
Effective density of states		
Valence band (cm^{-3})	1.83×10^{19}	9.6×10^{18}
Conduction band (cm^{-3})	3.22×10^{19}	4.2×10^{17}
Effective mass electrons	$0.3m_0$(near χ)	$0.063m_0(\Gamma)$ $0.35m_0$(near χ)
Holes		
Heavy	$0.5m_0(\Gamma)$	$0.5m_0(\Gamma)$
Light	$0.16m_0(\Gamma)$	$0.076m_0(\Gamma)$
Split-off		$0.155m_0(\Gamma)$
Electron mobility (cm^2/Vs)	1350	8500
Hole mobility (cm^2/Vs)	490	400
Thermal conductivity (W/cm°C)	1.5	0.58
Specific heat (J/g°C)	0.7	0.325

TABLE 1.1 Properties of Silicon and Gallium Arsenide at 300 K (Continued)

Property	Silicon	Gallium Arsenide
Thermal diffusivity, $\dfrac{\text{thermal conductivity}}{\text{density} \times \text{specific heat}}$ (cm^2/s)	0.9	0.27
Energy gap (eV)	$1.17 - \dfrac{4.73 \times 10^{-4} T^2}{T + 636}$ (1.21 eV at 300 K)	$1.519 - \dfrac{5.405 \times 10^{-4} T^2}{T + 204}$ (1.422 eV at 300 K)
Peak electron drift velocity (cm/s)	8×10^7	2.1×10^7
Threshold field (kV/cm)	—	3.3
Ionization rate for electrons (cm^{-1})	$1.6 \times 10^6 \exp\left\{-\left(\dfrac{1.65 \times 10^6}{\mathscr{E}}\right)\right\}$	$1.18 \times 10^5 \exp\left\{-\left(\dfrac{5.55 \times 10^5}{\mathscr{E}}\right)^2\right\}$
Ionization rate for holes (cm^{-1})	$5.5 \times 10^5 \exp\left\{-\left(\dfrac{1.65 \times 10^6}{\mathscr{E}}\right)\right\}$	$1.18 \times 10^5 \exp\left\{-\left(\dfrac{5.55 \times 10^5}{\mathscr{E}}\right)^2\right\}$
Young's modulus (dyn cm^{-2})	1.9×10^{12}	7.466×10^{11}
Upper yield point (dyn cm^{-2})	4.5×10^8 at 850°C	5×10^8 at 500°C
Critical resolved shear stress (dyn cm^{-2})	3.61×10^7	7.75×10^6

TABLE 1.2   Properties of Impurities in Silicon

Dopant	Type[a]	Tetrahedral Radius (Å)	Misfit Factor	Acceptor Level, Distance from Valence Band (eV)	Donor Level, Distance from Conduction Band (eV)
As	n	1.18	0	—	0.049
P	n	1.1	0.068	—	0.044
Sb	n	1.36	0.153	—	0.039
Al	p	1.26	0.068	0.057	—
B	p	0.88	0.254	0.045	—
Ga	p	1.26	0.068	0.065	—
In	d	1.44	0.22	0.16	—
Ag	d	1.52	0.29	0.89	0.79
Au	d	1.5	0.272	0.57	0.76
Cu	d	1.28		0.24, 0.37, 0.52	—
Mo	d			0.3	—
Ni	d	1.24		0.21, 0.76	—
O	d			—	0.16
Pt	d			0.42, 0.92	0.85
Ti	d			0.26	—
Zn	d			0.31, 0.56	—
Fe	d	1.26			

[a] d = deep impurity.

TABLE 1.3 Properties of Impurities in Gallium Arsenide[a]

Dopant	Type[b]	Tetrahedral Radius (Å)	Misfit Factor on As Site	Misfit Factor on Ga Site	Acceptor Level, Distance from Valence Band (eV)	Donor Level, Distance from Conduction Band (eV)
S	n	1.04	0.119	—	—	0.0061
Se	n	1.14	0.034	—	—	0.0059
Te	n	1.32	0.119	—	—	0.0058
Sn	n	1.4	0.186	0.111	—	0.006
Be	p	1.06	—	0.159	0.028	—
Cd	p	1.48	—	0.175	0.035	—
Li	p				0.023, 0.05	
Mg	p	1.4	—	0.111	0.028	—
Zn	p	1.31	—	0.04	0.031	—
C	n/p	0.77	0.347	0.389	0.026	0.006
Ge	n/p	0.22	0.034	0.032	0.04	0.0061
Si	n/p	1.18	0.0	0.063	0.035	0.0058
Ag	d	1.52	0.228	0.206	0.11	
Au	d	1.5	0.271	0.19	0.09	
Ca	d				0.16	
Co	d				0.16, 0.56	—
Cr	d				0.79	
Cu	d				0.14, 0.24, 0.44	0.16
Fe	d				0.38, 0.52	
Mn	d				0.9	
Ni	d				0.35, 0.42	—
V	d				1.27	0.65–0.75

[a]See Refs. 16 and 22.
[b]d = deep impurity.

TABLE 1.4   Point Defects in Gallium Arsenide

Defect	Distance from E_c (in eV)	Distance from E_v (in eV)
V_{As}	0.045	
As_i	0.5	
V_{As}–As_i	0.3	0.06
As_{Ga}–V_{As}	0.76	
	0.35	
Ga_{As}		0.077
		0.23
V_{Ga}–C_{Ga}	1.17	
V_{Ga}–Si_{Ga}	1.18	
V_{Ga}–Se_{As}	1.22	
V_{Ga}–Te_{As}	1.21	
V_{Ga}–S_{As}	1.20	
Cu_{Ga}–Se_{As}	1.30	
Cu_{Ga}–Te_{As}	1.32	
Cu_{Ga}–S_{As}	1.32	

REFERENCES

1. M. M. Atalla, E. Tannenbaum, and E. J. Scheibner, Stabilization of Silicon Surfaces by Thermally Grown Oxides, *Bell Syst. Tech. , J.* **38**, 749 (1959).

2. B. M. Welch, Advances in GaAs LSI/VLSI Processing Technology, *Solid State Technol.* (Feb. 1980), p. 95.

3. P. Ashburn, A. A. Rezazadeh, E.-F. Chor, and A. Brunnschweiler, Comparison of Silicon Bipolar and GaAlAs/GaAs Heterojunction Bipolar Technologies Using a Propagation Delay Expression, *IEEE J. Solid State Circuits* **24**, 512 (1989).

4. R. K. Richardson and A. C. Beers, Eds., Applications of Multiquantum Wells, Selective Doping and Superlattices, in *Semiconductors and Semimetals*, Vol. 24, Academic Press, New York, 1985.

5. Y. Suematsu, Long Wavelength Optical Fiber Communication, *Proc. IEEE* **71**, 692 (1981).

6. R. K. Willardson and A. C. Beers, Eds., Mercury Cadmium Telluride, in *Semiconductors and Semimetals*, Vol. 18, Academic Press, New York, 1981.

7. J. S. Blakemore, Semiconducting and Other Major Properties of Gallium Arsenide, *J. Appl. Phys.* **53**, R123 (1985).

8. S. Adachi, GaAs, AlAs and $Al_x Ga_{1-x}As$: Material Parameters for Use in Device and Research Applications, *J. Appl. Phys.* **58**, R1 (1985).

9. M. J. Howes and D. V. Morgan, Eds., *Gallium Arsenide Materials, Devices and Circuits*, John Wiley and Sons, New York, 1985.

10. D. E. McCumber and A. G. Chynoweth, Theory of Negative Conductance Amplification and of Gunn Instabilities in Two-Valley Semiconductors, *IEEE Trans. Electron Dev.* **ED-13**, 4 (1966).

11. C. Hilsum, Transferred Electron Amplifiers and Oscillators, *Proc. IRE* **50**, 185 (1962).

12. L. D. Crossman and J. A. Baker, Polysilicon Technology, in *Semiconductor Silicon*, H. R. Huff and E. Sirtl, Eds., The Electrochemical Society, Princeton, NJ, 1977, p. 18.

13. J. Krauskopf, J. D. Meyer, B. Wiedemann, M. Waldschmidt, K. Bethge, G. Wolf, and W. Schütz, Impurity Analysis of Gallium Arsenide, in *5th Conference on Semi-Insulating III–V Materials, Malmo, Sweden*, G. Grossmann and L. Lebedo, Eds., Adam Hilger, Philadelphia, PA, 1988, p. 165.

14. D. L. Rode, Low Field Electron Transport, in *Semiconductors and Semimetals*, Vol. 10, R. K. Willardson and A. C. Beer, Eds., Academic Press, New York, 1975.

15. G. E. Stillman and C. M. Wolfe, Electrical Characterization of Epitaxial Layers, *Thin Solid Films* **31**, 69 (1976).

16. W. Walukiewicz, L. Lagowski, L. Jastrzebski, M. Lichtensteiger, H. C. Gatos, Electron Mobility and Free-Carrier Absorption in GaAs: Determination of the Compensation Ratio, *J. Appl. Phys.* **50**, 899 (1979).

17. R. Venkatasubramanian, S. K. Ghandhi, and T. F. Kuech, Compensation Mechanisms in n^+-GaAs Doped with Sulfur, *J. Cryst. Growth* **97**, 827 (1989).

18. G. Salmer and J. Pribetich, Theoretical and Experimental Study of GaAs Impatt Oscillator Efficiency, *J. Appl. Phys.* **44**, 314 (1973).

19. H. Hovel, *Semiconductors and Semimetals, Vol. 11: Solar Cells*, R. K. Willardson and A. C. Beer, Eds., Academic Press, New York, 1975.

20. H. C. Casey and F. Stern, Concentration Dependent Absorption and Spontaneous Emission of Heavily Doped GaAs, *J. Appl. Phys.* **47**, 631 (1976).

21. R. E. Williams, *Modern Gallium Arsenide Processing Methods*, Artech House, Boston, MA, 1985.

22. S. J. Pearton, S. M. Vernon, C. R. Abernathy, K. T. Short, R. Caruso, M. Stavola, J. M. Gibson, V. E. Haven, A. E. White, and D. C. Jacobson, Characteristics of GaAs Layers Grown Directly on Si Substrates by Metalorganic Vapor Phase Deposition, *J. Appl. Phys.* **62**, 862 (1987).

23. C. Kittel, *Introduction to Solid State Physics*, John Wiley and Sons, New York, 1966.

24. J. D. Meindl and K. D. Wise, Eds., Special Issue on Solid State Sensors, Actuators and Interface Electronics, *IEEE Trans. Electron. Dev.* **ED-26**, 1861 (1979).

25. K. E. Bean and P. S. Gleim, The Influence of Crystal Orientation on Silicon Semiconductor Processing, *Proc. IEEE* **57**, 1469 (1969).

26. E. H. Nicollian and J. R. Brews, *MOS Physics and Technology*, John Wiley and Sons, New York, 1982.

27. R. G. Rhodes, *Imperfections and Active Centers in Semiconductors*, Pergamon, New York, 1964.

28. H. Reiss, C. S. Fuller, and F. J. Morin, Chemical Interactions Among Defects in Germanium and Silicon, *Bell Syst. Tech. J.* **35**, 535 (1956).

29. W. Shockley and J. T. Last, Statistics of the Charge Distribution for a Localized Flaw in a Semiconductor, *Phys. Rev.* **107**(2), 392 (1957).

30. J. R. Arthur, Vapor Pressures and Phase Equilibria in the Ga–As System, *J. Phys. Chem. Solids* **28**, 2257 (1967).

31. J. C. Bourgoin, H. J. von Bardeleben, and D. Stievenard, Native Defects in Gallium Arsenide, *J. Appl. Phys.* **64**, R65 (1988).

32. H. R. Potts and G. L. Pearson, Annealing and Arsenic Overpressure Experiments on Defects in Gallium Arsenide, *J. Appl. Phys.* **37**, 2098 (1966).

33. A. G. Milnes, *Deep Impurities in Semiconductors*, John Wiley and Sons, New York, 1973.

34. W. Kohn, Shallow Impurity States in Silicon and Germanium, *Solid State Phys.* **5**, 257 (1957).

35. G. L. Pearson and J. Bardeen, Electrical Properties of Pure Silicon and Silicon Alloys Containing Boron and Phosphorus, *Phys. Rev.* **75**, 865 (1949).

36. M. Lannoo and J. Bourgoin, *Point Defects in Semiconductors. I. Theoretical Aspects*, Springer-Verlag, New York, 1981.

37. W. M. Bullis, Properties of Gold in Silicon, *Solid State Electron.* **9**, 143 (1966).

38. K. P. Lisiak and A. G. Milnes, Platinum as a Lifetime Control Deep Impurity in Silicon, *J. Appl. Phys.* **46**, 5229 (1975).

39. S. Kishino, Y. Matsushita, M. Kanamori and T. Iizuka, Thermally Induced Microdefects in Czochralski-Grown Silicon: Nucleation and Growth Behavior, *Jpn. J. Appl. Phys.* **21**, 1 (1982).

40. W. Wijaranakula, Solubility of Interstitial Oxygen in Silicon, *Appl. Phys. Lett.* **59**, 1185 (1991).

41. A. Kanamori, Annealing Behavior of the Oxygen Donor in Silicon, *Appl. Phys. Lett.* **34**, 287 (1979).

42. T. Nozaki, Concentration and Behavior of Carbon in Semiconductor Silicon, *J. Electrochem. Soc.* **117**, 1566 (1970).

43. S. Pomrenke, H. Ennen, and Y. Haydi, Photoluminescence Optimization and Characterization of the Rare Earth Element Erbium Implanted in GaAs, InP and GaP, *Appl. Phys.* **59**, 601 (1986).

44. M. E. Weiner and A. S. Jordan, Analysis of Doping Anomalies in GaAs by Means of a Silicon-Oxygen Complex Model, *J. Appl. Phys.* **43**, 1767 (1972).

45. T. F. Kuech and E. Veuhoff, Mechanism of Carbon Doping in MOCVD GaAs, *J. Cryst. Growth* **68**, 148 (1984).

46. R. N. Hall and J. H. Racette, Diffusion and Solubility of Copper in Extrinsic Germanium, Silicon and Gallium Arsenide, *J. Appl. Phys.* **35**, 379 (1964).

47. D. C. Look, The Electrical Characterization of Semi-insulating GaAs, *J. Appl. Phys.* **48**, 5141 (1977).

48. M. Skowronski, S. T. Neild, and R. E. Kremer, Location of Energy Levels of Oxygen–Vacancy Complex in GaAs, *Appl. Phys. Lett.* **57**, 902 (1990).

49. F. C. Wang and M. Bujatti, The Effect of Substrate Gettering on GaAs MESFET Performance, *IEEE Trans. Electron. Dev.* **ED-32**, 2839 (1985).

50. P. M. Petroff, G. A. Rozgonyi, and T. T. Sheng, Elimination of Process-Induced Stacking Faults by Preoxidation Gettering of Si Wafers, *J. Electrochem. Soc.* **123**, 565 (1976).

51. T. W. Sigmon, L. Csepregi, and J. W. Mayer, Ion Implantation Gettering of Gold in Silicon, *J. Electrochem. Soc.* **123**, 1116 (1976).

52. C. O. Bosler, J. P. Donnelly, W. T. Lindley, and R. A. Reynolds, Impurity Gettering in Semi-insulating Gallium Arsenide Using Ion Implantation Damage, *Appl. Phys. Lett.* **29**, 698 (1976).

53. S. K. Ghandhi, *Semiconductor Power Devices*, John Wiley and Sons, New York, 1977.

54. D. Huber and J. Reffle, Precipitation Process Design for Denuded Zone Formation in CZ–Silicon Wafers, *Solid State Technol.*, (Aug. 1983), p. 137.

55. T. Y. Tan, E. E. Gardner, and W. K. Tice, Intrinsic Gettering by Oxide Precipitate Induced Dislocations in Czochralski Si, *Appl. Phys. Lett.* **30**, 175 (1977).

56. S. Kishino, K. Nagasawa, and T. Iisuka, Heavy Metal Gettering by an Intrinsic Gettering Technique Using Microdefects in Czochralski-Grown Silicon Wafers, *Jpn. J. Appl. Phys.* **19**, L455 (1980).

57. T. J. Magee, J. Peng, J. D. Hong, W. Katz, and C. A. Evans, Back Surface Gettering of Au in GaAs, *Phys. Status. Solidi.* A **55**, 161 (1979).

58. D. Hull and D. J. Bacon, *Introduction to Dislocations*, 3rd edition, Pergamon Press, New York, 1989.

59. H. F. Matare, *Defect Electronics in Semiconductors*, Wiley-Interscience, New York, 1971.

60. F. C. Frank and W. T. Read, Jr., Multiplication Processes for Slow Moving Dislocations, *Phys. Rev.* **79**, 722 (1950).

61. P. W. Hutchinson and P. S. Dobson, Defect Structures of Degraded GaAlAs–GaAs Double Heterojunction Lasers, *Philos. Mag* **32**, 745 (1975).

62. P. M. Petroff, Point Defects and Dislocation Climb in III–V Compound Semiconductors, *J. Phys. Colloq. Suppl.* **40**, 201 (1979).

63. A. O. Evwaraye and B. J. Baliga, The Dominant Recombination Centers in Electron-Irradiated Semiconductor Devices, *J. Electrochem. Soc.* **124**, 913 (1977).

64. H. C. Casey, Jr., Diffusion in the III–V Compound Semiconductors, in *Atomic Diffusion in Semiconductors*, D. Shaw, Ed., Plenum, New York, 1973, p. 351.

65. S. Y. Chiang and G. L. Pearson, Properties of Vacancy Defects in GaAs Single Crystals, *J. Appl. Phys.* **45**, 2986 (1975).

66. D. T. J. Hurle, Revised Calculation of Point Defect Equilibria and Non-stoichiometry in Gallium Arsenide, *J. Phys. Chem. Solids* **40**, 613 (1979).

67. E. W. Williams, Evidence for Self-Activated Luminescence in GaAs: The Gallium Vacancy–Donor Center, *Phys. Rev.* **158**, 922 (1968).

68. J. Nishizawa, H. Otsuka, S. Yamakoshi, and K. Ishida, Nonstoichiometry of Te-Doped GaAs, *Jpn. J. Appl. Phys.* **13**, 46 (1974).

69. S. Makram-Ebeid, P. Langdale, and G. M. Martin, Nature of EL2: The Main Native Midgap Electron Trap in VPE and Bulk GaAs, in *Semi-insulating III–V Materials*, D. C. Look and J. S. Blakemore, Eds., Shiva Publishing Ltd., England, 1984, p. 184.

70. J. Hornstra, Dislocations in the Diamond Lattice, *J. Phys. Chem. Solids* **5**, 129 (1958).

71. A. Many, Y. Goldstein, and N. B. Grover, *Semiconductor Surfaces*, North-Holland, Amsterdam, 1965.

PROBLEMS

1. Compute the atom density for the principal planes of GaAs in terms of the cube edge. Identify these atoms and their positions by means of a sketch. Calculate the bond density for these planes.

2. A (100) surface, misoriented 2.29° towards the nearest [110], can be characterized as a series of steps, with risers of height equal to one-half the cube edge. Calculate the step length and the Miller index of this plane.

3. We wish to chemically delineate a window in a slice, with vertical walls. Beginning with (110) silicon, and using an anisotropic etch which exposes

{111} planes, indicate all the directions of the edges of this oxide cut, and also all the planes delineated by this etch. Sketch a plan view of the window.

A flat is provided on the slice to aid in making the first cut. What is the plane of this flat? What is the direction of the line made by the intersection of this flat with the (110) surface?

4. It is possible for a tetrahedral stacking fault to be initiated during the growth of an epitaxial layer on (111) silicon. This fault usually begins at the substrate–layer interface, and propagates along {111} planes until it reaches the surface.

   Sketch the outline of a stacking fault within a unit cell, and identify its various {111} planes. Determine the thickness of the epitaxial layer in terms of the length of one side of the stacking fault, as it penetrates to the surface. It is only necessary to show the unit cell outline.

5. Stacking faults also arise during epitaxial growth in (100) silicon. Again, these propagate along {111} planes. Repeat Problem 4 for this situation. Illustrate by means of the unit cell outline.

6. When InAs is epitaxially grown on GaAs, a 2-μm region is formed in which the film accommodates to the substrate lattice. What is the defect concentration in this region, assuming that it is uniform, and that growth is on the (100) plane. Note that this will give you a pessimistic number, because effects such as lattice distortion or strain relief have not been considered. Assume that $a = 6.058$ Å for indium arsenide.

7. What is the equilibrium concentration of Schottky defects in silicon at 300 and 1500 K. Repeat for Frenkel defects and compare the numbers. Comment on the results.

8. Gold is introduced into n-type silicon. Show that it will behave p-type. Repeat for p-silicon. Here, show that the gold will behave n-type. In both cases, assume that the gold concentration is well below the background concentration.

9. A piece of n-GaAs has a total donor concentration of 5×10^{16} cm^{-3}. Plot the 77-K mobility as a function of the compensation ratio $(N_D^+ + N_A^-)/n$.

CHAPTER 2

PHASE DIAGRAMS AND SOLID SOLUBILITY

A number of different materials are used in the fabrication of semiconductor devices and microcircuits. As a consequence, combinations of two or more of these are often encountered at various points within the structure. By way of example, ohmic contacts to silicon and GaAs are made by the alloying of various metals to the semiconductor. Occasionally, combinations of materials are inadvertently formed during heat treatment or during the storage of devices at elevated temperature. The most notorious of these, formed at the interface between gold leads and aluminum-bonding pads, results in the so-called *purple plague*,* which sets an upper limit to the temperature at which many silicon microcircuits can be stored.

In this chapter, the behavior of these combinations is described by means of phase diagrams [1–4]. This behavior is important since it often determines the nature and choice of the fabrication process. In addition, it provides a clue as to the problems that may arise when certain combinations of materials are used together.

A *phase* is defined as a state in which a material may exist, which is characterized by a set of intensive properties, such as composition, electrical polarization, color, and so on. If these phases are presented for equilibrium conditions, the resulting diagram is called an *equilibrium diagram*. Since equilibrium conditions are attained at rates that are much slower than the freezing rate, most diagrams involving one or more solid phases are usually called *phase diagrams* and represent quasi-equilibrium conditions.

*This topic is treated in Section 2.2.6.

69

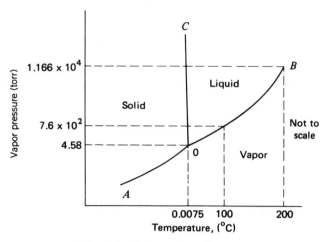

Fig. 2.1 Unitary phase diagram.

2.1   UNITARY DIAGRAMS

These are diagrams showing the phase change in a single element, usually as a function of temperature and pressure. They also apply to compounds* that undergo no chemical change over the range for which the diagram is constructed.

In its simplest form, the unitary diagram consists of three lines that intersect at a common point, thus delineating three areas on a two-dimensional plot. Figure 2.1 shows such a diagram for water, and illustrates the three phases in which it can exist. The common point, referred to as a *triple point*, is invariant for the system and defines the temperature and pressure at which solid, liquid, and gaseous phases are all in equilibrium with one another. *OA*, *OB*, and *OC* are univariant lines. Water can coexist in two phases for any pressure–temperature combination represented by these lines. Phase changes, which are represented by crossing these boundaries, occur at sharply defined temperatures, because the heat evolved (or absorbed) during this transition is the latent heat. This characteristic is used in the experimental construction of phase diagrams.

2.2   BINARY DIAGRAMS

These are phase diagrams showing the relationship between two components as a function of temperature. The second variable, pressure, is usually set at 1 atm. In this way a relatively complex three-dimensional representation is avoided.

---

*The term *component* is often used to denote elements and compounds of this type interchangeably.

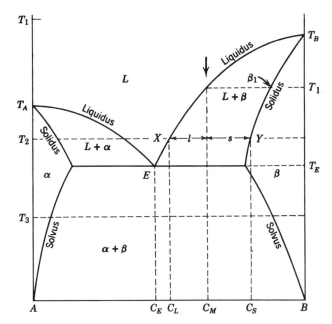

Fig. 2.2 Binary eutectic phase diagram.

Figure 2.2 shows one of the many different types of binary diagrams that are encountered in practice. Here, the abscissa represents various compositions of two components A and B, usually specified in atom percent, mole fraction or weight percent. Each end represents a pure component, which may be an element or a compound.

2.2.1 The Lever Rule

At any temperature, the equilibrium composition of the two single phases L and β, which coexist to make up a two-phase region, may be determined as follows (see Figure 2.2): Consider a melt of initial composition C_M (the mole fraction of B in the melt). Let this melt be cooled from some temperature T_1 to a temperature T_2, corresponding to a point in the two-phase region. Let

$$N_L = \text{number of moles of liquid } (L, \text{ for this example}) \text{ at this temperature}$$
$$N_S = \text{number of moles of solid (in the } \beta \text{ phase, for this example)}$$
$$C_L, C_S = \text{composition of the liquid and solid respectively (mole fraction of } B)$$

Then $N_L C_L$ and $N_S C_S$ are the number of moles of B in the liquid and solid, respectively. But the total number of moles of B is $(N_L + N_S)C_M$. Hence by the conservation of matter

$$\frac{N_S}{N_L} = \frac{C_M - C_L}{C_S - C_M} = \frac{l}{s} \tag{2.1}$$

where l and s are the length of the two lines shown in Fig. 2.2. This is known as the *lever rule* and can be used to analyze compositional changes during the freezing of a crystal from the melt. The line XY is commonly referred to as a *tie line*.

2.2.2 The Phase Rule

The correct interpretation of phase diagrams is greatly helped by knowledge of the *phase rule*. This rule, which is based on thermodynamic considerations, states that for any system in thermal equilibrium, the sum of the number of phases (P) and the number of degrees of freedom (F) is related to the number of components (C) by

$$P + F = C + 2 \tag{2.2}$$

Here, the degrees of freedom are the number of variables which can be independently changed while still preserving a specific phase. For example, $P+F = 3$ for a single-component diagram of the type shown in Fig. 2.1. For water in its liquid phase $(P = 1)$ we have 2 degrees of freedom; that is, either pressure or temperature (or both) can be changed independently of each other, and still maintain water in this liquid phase. Along OB, however, $P = 2$ (liquid and vapor) so that we have 1 degree of freedom. Now, either pressure or temperature (but not both) can be independently varied while still preserving this two-phase coexistence. At O, $P = 3$ and $F = 0$; that is, there is a unique temperature–pressure combination where water coexists in all three phases.

For a binary phase diagram such as that shown in Fig. 2.2, $F+P = C+2 = 4$. If we confine our discussions to systems at atmospheric pressure, then $F = 3 - P$, since 1 degree of freedom is now preassigned. Furthermore, we note that composition is assigned a single degree of freedom, since X_A (the fraction of A) and X_B (the fraction of B) are related by $X_A + X_B = 1$.

In Fig. 2.2, regions L, α and β are each one-phase regions, so that $F = 2$ (*both* temperature *and* composition can be varied independently of each other). In the two-phase regions $L+\alpha, L+\beta$, and $\alpha+\beta$, we have $F = 1$. These are univariant regions; at any given temperature, for example, the compositions of both the liquid and solid phase are fixed by the lever rule. Thus, *either* temperature *or* composition can be varied independently in this region, but not both. At the point E, there are three phases, L, α and β, so that $P = 3$. Hence, there are 0 degrees of freedom; E is an invariant point, whose temperature and composition are *both* fixed.

As a result of the phase rule, we note that a binary diagram can have only regions of one- and two-phase. (Degenerated three-phase regions, represented by a point such as E, are also permitted.) Furthermore, two-phase regions cannot be next to each other. It follows that, in traversing a binary phase diagram from one side to another at any constant temperature, all single-phase regions are separated by two-phase regions as the composition is varied. By way of example, the phases that exist at T_2 are $\alpha, L + \alpha, L, L + \beta$, and β in succession. On the other hand, the phases that are present at T_3 are $\alpha, \alpha + \beta$, and β, in succession.

Various types of binary phase diagrams are encountered in practice, depending on the components involved and their degree of miscibility in the solid and liquid state. Some of these are now described.

2.2.3 Isomorphous Diagrams

The isomorphous diagram is characteristic of components that are completely soluble in each other. It has been empirically found that phase diagrams of this type are restricted to binary systems in which the components are within 15% of each other in atomic radius, have the same valence and crystal structure, and have no appreciable difference in electronegativity. As a consequence, only a few binary systems belong to this class; some examples are Cu–Ni, Ag–Pd, and Au–Pt. Limited solubility of components is by far the more common occurrence in binary systems.

Silicon and germanium are very similar in structure and atomic properties. As expected, they are completely miscible in both the liquid and solid phase, and are characterized by the isomorphous phase diagram of Fig. 2.3. Here, α represents the full range of *solutions* of germanium and silicon; a single-crystal, diamond lattice material can be grown for any composition. In this lattice, however, atom sites will be randomly occupied by silicon and germanium, while preserving a specific overall composition ratio. It must be emphasized that this material is quite different to a compound, such as GaAs, where lattice sites are specifically assigned to either gallium or arsenic.

Mixtures of III/V compounds, such as GaAs and GaP, are also isomorphous in behavior, as seen in the GaAs–GaP phase diagram of Fig. 2.4 [5]. Solutions of this type are known as *mixed compound semiconductors, ternary compounds,* or *ternary alloys.* The ternary compound $GaAs_yP_{1-y}$ consists of a solution of the binary compounds GaAs and GaP, in a $y/(1 - y)$ ratio. Here, all group III sites are occupied by gallium atoms, whereas group V sites comprise randomly distributed arsenic and phosphorus atoms in this ratio.

The lattice parameter of ternary compounds is related linearly to that of its binary components (Vegard's law). For GaAs, $a = 5.6532$ Å whereas $a = 5.660$ Å for AlAs. Thus, there is a 0.12% mismatch between these extremes. As a result, all $Al_xGa_{1-x}As$ compositions are closely matched to each other, so that a single continuous crystalline structure can be grown with successive layers of different composition. In addition, bulk GaAs can be used as a substrate for this purpose. The AlAs–GaAs system is technologically very important for

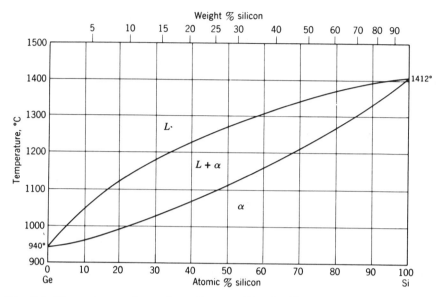

Fig. 2.3 Binary isomorphous phase diagram: The Ge–Si System. from Hansen and Anderko, *Constitution of Binary Alloys*, McGraw-Hill, New York (1958) [2]. Used with permission of the McGraw-Hill Book Company.

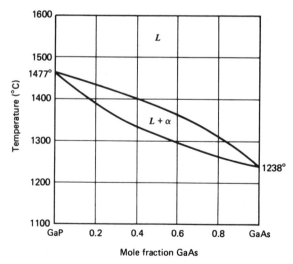

Fig. 2.4 The GaAs–GaP system. From Antypas [5]. Reprinted with permission of the publisher, The Electrochemical Society, Inc.

this reason, and most modern GaAs-based semiconductor devices utilize the electronic properties of heterostructures which exploit this combination.

The energy gap of most ternary alloy semiconductors also varies between the two extremes associated with its binary components. Here, the relationship is only approximately linear. For $Al_xGa_{1-x}As$ [6], the direct energy gap (Γ point) is given by

$$E_\Gamma(x) = 1.424 + 1.087x + 0.438x^2 \qquad (2.3)$$

The indirect gap (χ point) for this system also varies monotonically, with

$$E_\chi(x) = 1.905 + 0.10x + 0.16x^2 \qquad (2.4)$$

Combining these equations, it is seen that $Al_xGa_{1-x}As$ is direct gap for $x \le 0.43$, but indirect gap at higher x values. This property is exploited in the use of indirect gap $Al_xGa_{1-x}As$ for optically transparent windows in GaAs detectors. Moreover, the surface recombination velocity of GaAs is greatly reduced when its surface dangling bonds are terminated by AlGaAs, which is grown as an extension of its crystal lattice. These advantages, taken together, have been exploited in the fabrication of high-efficiency GaAs solar cells and lasers.

The ternary alloy $GaAs_{0.6}P_{0.4}$ is a direct gap material which is used in red light emitting diodes. The lattice parameter of GaP is 5.4512 Å, so that $GaAs_{0.6}P_{0.4}$ is mismatched to gallium arsenide by 1.4%. As a consequence, device quality material cannot be grown on it directly. In practice, a 40-μm-thick buffer layer, graded from $x = 0$ at the GaAs substrate to $x = 0.4$ at its top, is grown between the substrate and the active layer in order to relieve the misfit dislocations which would arise if the active layer were grown directly on the GaAs substrate.

More complex solutions of compound semiconductors can also be made. For example, *quaternary alloys* with compositions corresponding to $III_x^A III_{1-x}^B V_y^C V_{1-y}^D$, $III_x^A III_y^B III_{1-x-y}^C V^D$, and $III^A V_x^B V_y^C V_{1-x-y}^D$ are also possible, and are used in some device applications. A discussion of these is beyond the scope of this book.

2.2.4 Eutectic Diagrams

Eutectic diagrams result when the addition of either component to a melt lowers its overall freezing point, as shown in Fig. 2.2. Here the freezing point of the molten mixture has a minimum value T_E, below T_A and T_B. This minimum value is known as the *eutectic temperature*, and the corresponding mixture C_E is called the *eutectic composition*. Eutectic systems usually occur when two components are completely miscible in the liquid phase but are only partly soluble in the solid phase. Most semiconductor systems fall into this class.

Referring to Fig. 2.2, consider the cooling of a melt of initial composition C_M. As the temperature is reduced below T_4, a solid of composition β_1 first freezes out. With falling temperature, the liquid composition moves along the liquidus line, becoming richer in A, until the eutectic temperature T_E is reached. Simultaneously, β-phase material of varying composition freezes out, as the solid composition moves down the solidus line. At T_E, the melt is of eutectic composition C_E, and freezes isothermally to form the $(\alpha + \beta)$ phase. Thus the final solid consists of clumps of β-phase material in an $(\alpha + \beta)$-phase mixture of eutectic composition. A solid of this type has a tendency to break along the boundaries of the β-phase clumps, if subjected to mechanical strain. A starting melt of eutectic composition, on the other hand, solidifies into a $(\alpha + \beta)$ phase without any clumping, and is thus considerably stronger.

Figure 2.5 shows the lead–tin system which exhibits this type of characteristic. Here the eutectic point has a 38% lead–62% tin composition by weight and a eutectic temperature of 183°C. On the other hand, the freezing points of pure lead and pure tin are 327°C and 232°C, respectively. Lead–tin combinations, of eutectic composition, are used routinely as solders in electronic circuit wiring.

Variations of the eutectic diagram of Fig. 2.2 are often seen in practice,

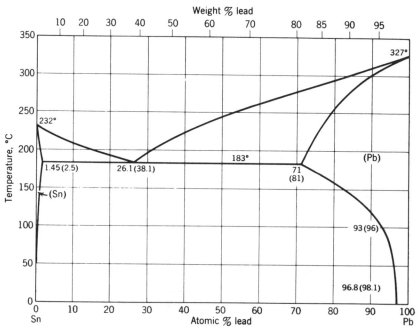

Fig. 2.5 The lead–tin system. From Hansen and Anderko, *Constitution of Binary Alloys*, McGraw-Hill, New York (1958) [2]. Used with permission of the McGraw-Hill Book Company.

depending on the nature of the terminal solid solutions. Thus in the Pb–Sn system the terminal solid solubility (atom fraction) of tin in lead is significant (29%), as is that of lead in tin (1.45%). In the Al–Si system, on the other hand, the terminal solid solubility of silicon in aluminum is 1.59%, while that of aluminum in silicon is so small (<0.1%) that the β phase cannot be indicated in this diagram. Figure 2.6 shows the phase diagram for this system, which is

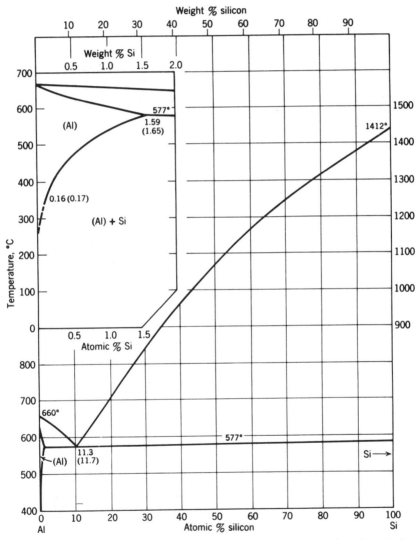

Fig. 2.6 The aluminum–silicon system. From Hansen and Anderko, *Constitution of Binary Alloys*, McGraw-Hill, New York (1958) [2]. Used with permission of the McGraw-Hill Book Company.

of importance in the fabrication of ohmic contacts* to *p*-type and degenerate *n*-type silicon.

Aluminum is sometimes used for the fabrication of Al–Si Schottky diodes in high-speed digital logic circuits. Here, subsequent processing can alter the interface and cause changes in diode characteristics. Thus, the phase diagram gives vital information about potential problems which can arise during the fabrication of circuits incorporating these devices.

The Au–Si system is shown in Fig. 2.7. Here both terminal solid solubilities are too small to be indicated on the diagram. The sharply depressed eutectic temperature of this combination leads to its widespread use in the die bonding of discrete silicon devices and microcircuits to substrates. An alternative combination, which provides a better wetting action and has a lower eutectic temperature, is gold and germanium, whose phase diagram is shown in Fig. 2.8. In this case, the resulting bond is a ternary Au–Ge–Si alloy. Gold alloys formed with these systems are extremely hard and strong, and cannot be used in power devices which are subject to significant thermal cycling. Lead–tin solders are more suitable for these applications because of their plastic behavior.

Gold–germanium alloys are extensively used in GaAs technology for making ohmic contacts to *n*-GaAs. Although a very small Ge fraction is necessary for this purpose, the eutectic composition is invariably used because of its depressed melting point.

Occasionally, binary systems may exhibit an eutectic composition that is very close to one of the components. Figure 2.9 shows the Ga–Ge system which belongs to this class. Ge–Ga alloys are created by the heat treatment of Au–Ge eutectic with GaAs, during contact formation. Thus, this figure can be used as an indicator of problems during this process. The more detailed Ga–As–Ge ternary phase diagram (Fig. 2.26) can be used for this purpose, as well.

2.2.5 Compound Formation

In many complex systems, one or more intermediate compounds may be formed at specific temperatures and compositions. In contrast to mixtures, these intermediate compounds are ordered on an atomic scale. They usually occur within an extremely narrow compositional range, corresponding to very small departures from stoichiometry. Consequently, they are indicated on the phase diagram in the form of discrete vertical lines, as shown for the Ga–As system of Fig. 2.10 at the 50% Ga–50% As composition. Since these lines appear to represent a change from one phase directly into another, without any apparent alteration in composition, they are sometimes called *congruent transformations*.

The formation of a congruently melted compound in this manner effectively isolates the system on each side of it. Thus, this figure can be considered as

---

*This topic is covered in Chapter 8.

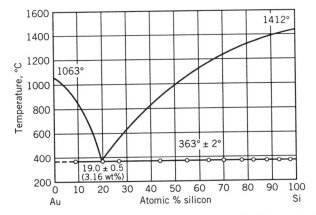

Fig. 2.7 The gold–silicon System. From Massalski [1]. Used with permission of ASM International.

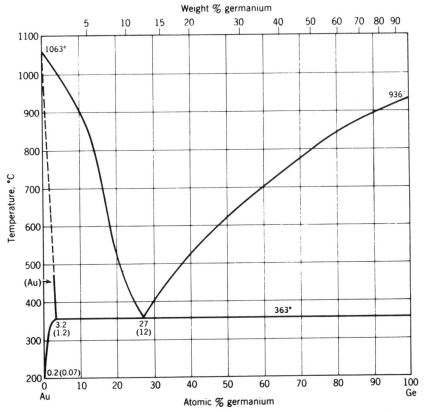

Fig. 2.8 The gold-germanium system. From Hansen and Anderko, *Constitution of Binary Alloys*, McGraw-Hill, New York (1958) [2]. Used with permission of the McGraw-Hill Book Company.

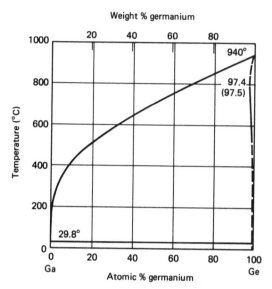

Fig. 2.9 The germanium-gallium system. From Hansen and Anderko, *Constitution of Binary Alloys*, McGraw-Hill, New York (1958) [2]. Used with permission of the McGraw-Hill Book Company.

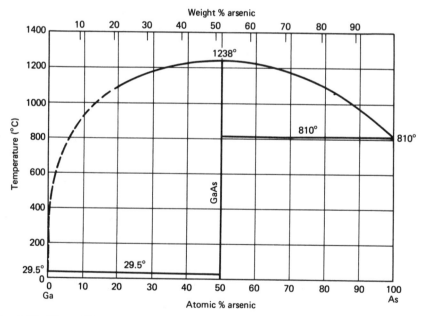

Fig. 2.10 The gallium–arsenic system. From Hansen and Anderko, *Constitution of Binary Alloys*, McGraw-Hill, New York (1958) [2]. Used with permission of the McGraw-Hill Book Company.

separate Ga–GaAs and GaAs–As systems, each of which is seen to be eutectic in nature. In this manner it is possible to break up relatively complex phase diagrams into a number of simpler ones.

A vertical line, drawn to indicate a congruently melted compound, is somewhat an oversimplification. In general, congruent melting will appear as a region of finite width if a sufficiently expanded compositional scale is used. In materials such as GaAs, the presence of large concentrations of vacancies and interstitials of both Ga and As (with their associated different free energies) results in melting at a point slightly removed from that of the ideal 1 : 1 stoichiometric composition. A greatly expanded phase diagram [7] around this region is sketched in Fig. 2.11. We note from this diagram that the solidification of GaAs can occur in a region of excess gallium or arsenic, resulting in the formation of excess arsenic or gallium vacancies respectively. This represents the stoichiometry range of GaAs, which is between 49.998 and 50.009 at.% of As.

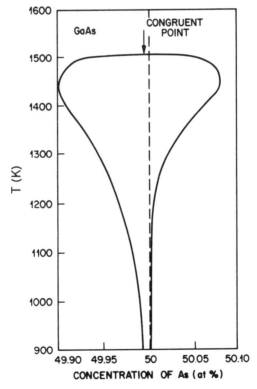

Fig. 2.11 Details of the GaAs phase diagram. From Swaminathan and Macrander. ©1991 [6, p. 45] Prentice-Hall, Englewood Cliffs, New Jersey. Reprinted with permission.

As mentioned in Chapter 1, many of these excess vacancies can become occupied by impurity atoms which are either intentionally or unintentionally present during crystal growth, with a resultant shift in the conductivity. Furthermore, the amount of shift in conductivity, as well as its direction (i.e., towards *p*- or *n*-type), will be determined by the temperature at which solidification occurs from the melt. This has important consequences in crystal growth and epitaxy, as shown in Chapters 3 and 5.

2.2.6 Peritectic and Other Reactions

A peritectic reaction is an important pathway by which an intermediate compound can be formed in a binary system. By way of example, Fig. 2.12 shows the arsenic–silicon system, in which the compound $SiAs_2$ is formed by a peritectic reaction, between SiAs and a liquid. Note that both these compounds are depicted by vertical lines, since they occur within a narrow compositional range. A eutectic point also exists for the SiAs–As system, corresponding to a 96% As–4% Si composition by weight.

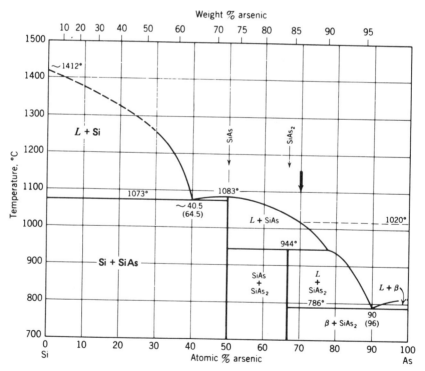

Fig. 2.12 The arsenic-silicon system. From Hansen and Anderko, *Constitution of Binary Alloys*, McGraw-Hill, New York (1958) [2]. Used with permission of the McGraw-Hill Book Company.

Consider what happens if a mixture of 86% As–14% Si by weight (indicated by an arrow in this figure) is cooled from some high temperature. At and below 1020°C, solid SiAs is precipitated from the melt, which becomes arsenic-rich until the temperature reaches 944°C. At this point, the liquid composition is 90% As–10% Si by weight. Further cooling results in a thermal arrest period at this temperature, during which time the solid SiAs combines with some of this excess liquid to form an $(L + SiAs_2)$ phase. At 786°C and lower, both the $SiAs_2$ and the β phase precipitate from the remainder of this excess liquid to form a solid $(\beta + SiAs_2)$ phase as shown.

Cooling through a peritectic temperature causes the formation of a nonequilibrium structure by the process of "surrounding." Here the $SiAs_2$ is formed by the surrounding of a solid SiAs core by the liquid. Under normal cooling conditions, this $SiAs_2$ layer creates a barrier to the diffusion of liquid to the SiAs, resulting in a reaction that proceeds at an ever-decreasing rate. Thus a peritectic reaction is usually accompanied by the formation of relatively large (micron-size) precipitates.* This is also true if the starting melt was of the exact composition as the peritectic compound—for example, a melt of Si:As in a 1:2 atomic ratio in the example shown in Fig. 2.12. Here, too, cooling the melt to 944°C results in the precipitation of solid SiAs and a liquid of 90% As–10% Si by weight. Under thermal equilibrium conditions, this liquid would react completely with the SiAs to form the $SiAs_2$. In a practical situation, under nonequilibrium conditions, this reaction would slow down with the formation of a $SiAs_2$ skin surrounding each SiAs core, so that excess liquid would remain. This would eventually cool to form a solid $(\beta + SiAs_2)$ phase as before.

Peritectic phases usually occur as part of more complex phase diagrams. Figure 2.13, for example, shows the phase diagram of the aluminum–gold system, which is of importance in evaluating the bonding of gold leads to aluminum contact pads on microcircuits. Of the many possible intermetallic compounds that are indicated here, the most significant are $AuAl_2$ (a dark purple, strongly bonding, highly conductive compound) and Au_2Al (a tan-colored, brittle, poorly conducting compound). This latter compound** has a melting point of 624°C. Nevertheless, significant compound formation occurs at 300°C and higher. This is a serious cause for lead-attachment failure in microcircuits which are stored for period of times at these high temperatures [8].

Experiments have shown that the formation of Au_2Al is enhanced by the presence of silicon. Although the ternary Al–Au–Si system has not been studied in detail, it can be expected that the effect of silicon addition to the Al–Au system produces simple eutectic lowering of its phase diagram [8]. Thus the formation of these compounds may be expected to be accelerated in the presence of the silicon. This has been observed experimentally.

---

*The anomalous behavior of arsenic doping at high concentrations can be explained by the formation of As–Si clusters by this reaction.

**For many years, the tan Au_2Al was not noticed in the presence of the purple $AuAl_2$. Thus the term *purple plague* was wrongly given to this phenomenon, and still persists in the literature.

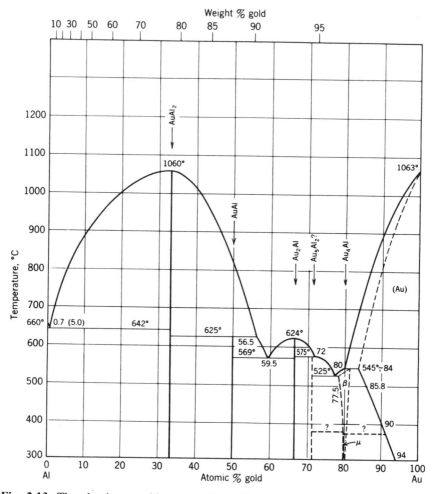

Fig. 2.13 The aluminum–gold system. From Hansen and Anderko, *Constitution of Binary Alloys*, McGraw-Hill, New York (1958) [2]. Used with permission of the McGraw-Hill Book Company.

Figure 2.14 shows the platinum–silicon system, and indicates the phases involved in the formation of a PtSi Schottky barrier layer on a silicon surface. Typically, a layer of platinum, 500 Å thick, is deposited on the silicon which is then heated to about 600°C for 15 min. This suffices to complete* the reaction and form PtSi, even though the melting point of this compound is 1229°C. This

---

*It should be emphasized that phase diagrams provide no information about the kinetics of a reaction.

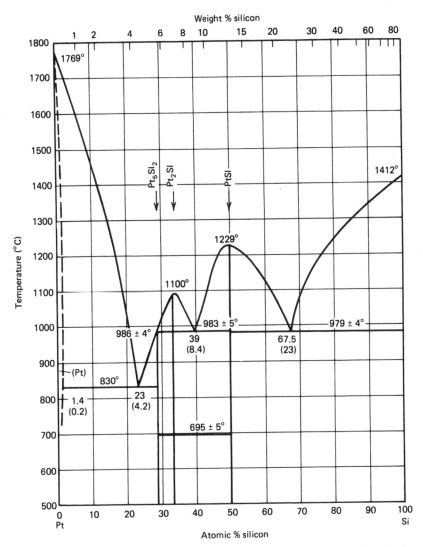

Fig. 2.14 The platinum–silicon system. From Hansen and Anderko, *Constitution of Binary Alloys*, McGraw-Hill, New York (1958) [2]. Used with permission of the McGraw-Hill Book Company.

PtSi layer serves as a base which is followed by successive layers of titanium, platinum, and gold. This complex metallization scheme forms the basis of the beam lead system [9], which is described in Chapter 8.

A common contacting scheme for *n*-type GaAs consists of depositing a layer of germanium, followed by a layer of gold to which leads can be readily bonded. Phase diagrams of the Ge–As system, shown in Fig. 2.15, together with those for the Ge–Ga system (Fig. 2.9) and the Au–Ga system of Fig. 2.16, are useful in evaluating potential problems here. The phase diagram for the ternary Ga–As–Au system is shown in Fig. 2.25. The Ga–As–Ge system is shown in Fig. 2.26.

In addition to eutectic and peritectic behavior, a number of other less common reactions may occur in binary systems. Thus a single phase can cool to form two phases in one of three possible ways, as follows:

(a) Monotectic: $L_1 \xrightarrow{\text{cooling}} \alpha + L_2$

(b) Eutectic: $L \xrightarrow{\text{cooling}} \alpha + \beta$

(c) Eutectoid: $\gamma \xrightarrow{\text{cooling}} \alpha + \beta$

Here, L, L_1, and L_2 represent liquid phases and α, β, and γ represent solid phases, including compounds.

In addition, three possibilities occur when two phases react to form a third, different phase, as follows:

(a) Syntectic: $L_1 + L_2 \xrightarrow{\text{cooling}} \beta$

(b) Peritectic: $L + \alpha \xrightarrow{\text{cooling}} \beta$

(c) Peritectoid: $\alpha + \gamma \xrightarrow{\text{cooling}} \beta$

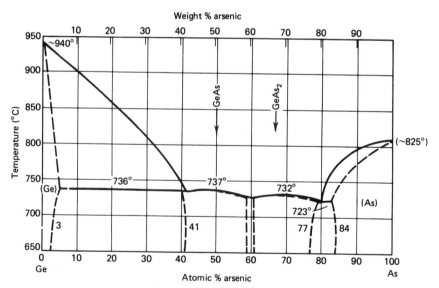

Fig. 2.15 The germanium-arsenic system. From Hansen and Anderko, *Constitution of Binary Alloys*, McGraw-Hill, New York (1958) [2]. Used with permission of the McGraw-Hill Book Company.

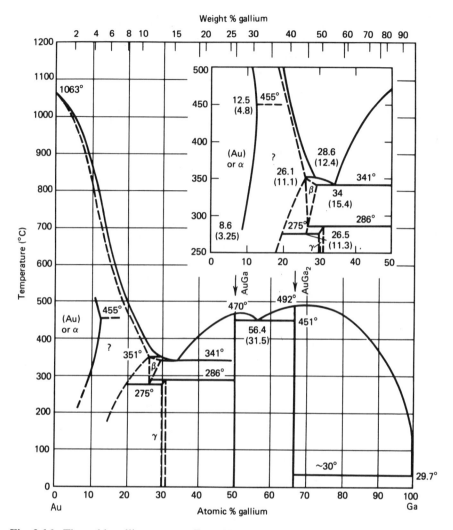

Fig. 2.16 The gold–gallium system. From Hansen and Anderko, *Constitution of Binary Alloys*, McGraw-Hill, New York (1958) [2]. Used with permission of the McGraw-Hill Book Company.

Examples of some of these reactions are shown in the phase diagrams of Figs. 2.12 to 2.16.

2.2.7 Phase Diagrams for Oxide Systems

Phase diagrams involving oxide systems are especially important in both silicon and GaAs technology. Thus, boron doping of silicon is carried out from a

B_2O_3–SiO_2 source which is deposited on the silicon slice prior to diffusion. The phase diagram of Fig. 2.17 is relevant to this process. Although of a relatively straightforward eutectic type, a number of structural phase change regions are seen here, as the SiO_2 goes through its cristobalite and tridymite phases [10].

Figure 2.18 shows the phase diagram for the P_2O_5–SiO_2 system. Phosphosilicate glass (PSG) has considerable importance in device technology [11], since it is used for phosphorus doping of silicon, for gettering to improve lifetime in silicon devices, as a passivation layer over silicon microcircuits, as a diffusion mask for GaAs, and as a cap for open tube diffusions in GaAs [12].

2.3 SOLID SOLUBILITY

It has been noted that many binary systems exist in which the terminal solid solubility of one component in the other is extremely small. This is usually the case for donor and acceptor impurities in silicon. As a consequence, it is necessary to expand greatly the scale of the phase diagram in order to show this important region. Figure 2.19 shows this part of the phase diagram, and is typical of most impurities in silicon. The solid solubility of these impurities is seen to increase with temperature, reach a peak value, and fall off rapidly as the melting point of the silicon is approached. This is commonly referred

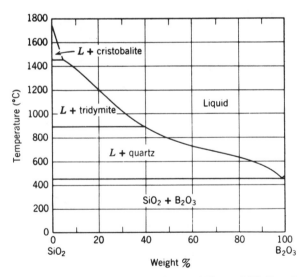

Fig. 2.17 The B_2O_3–SiO_2 system. From Rockett and Foster [10]. Reprinted with permission from the American Ceramic Society.

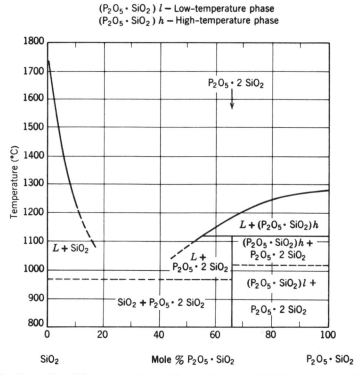

Fig. 2.18 The P_2O_5–SiO_2 system. From Tien and Hummel [11]. Reprinted with permission from the American Ceramic Society.

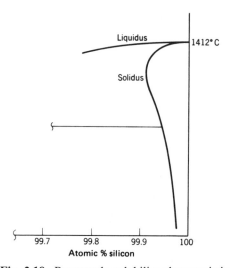

Fig. 2.19 Retrograde solubility characteristic.

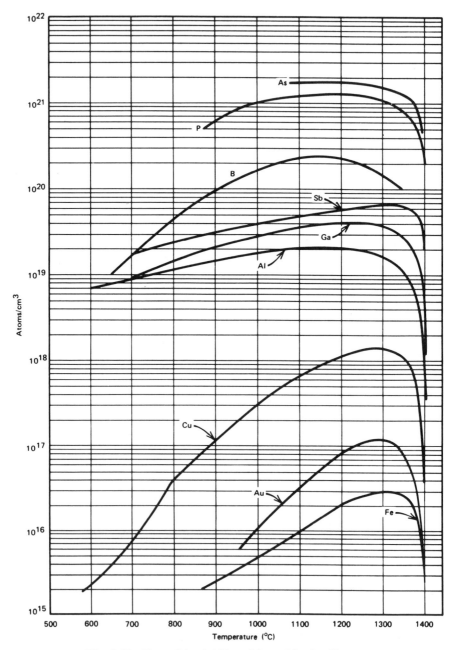

Fig. 2.20 The solid solubility of impurities in silicon.

to as a *retrograde solid-solubility characteristic*. Figure 2.20 shows values for various commonly used dopants in undoped silicon [13, 14]. Note that nearly all of these elements indicate a retrograde behavior in their solubility characteristics. Finally, it must be emphasized (see Chapter 1) that the solid solubility of dopants exhibiting multiple charge states is a function of the background concentration—that is, a function of the fermi level.

2.4 TERNARY DIAGRAMS

Ternary phase diagrams are required for systems with three components. They are of special use in the study of the behavior of III–V semiconductors such as GaAs, in the presence of an additional component. Very few systems of this type have been studied in detail. However, even fragmentary information, which is available for many such systems, is useful in evaluating the potential problems which can arise during device processing.

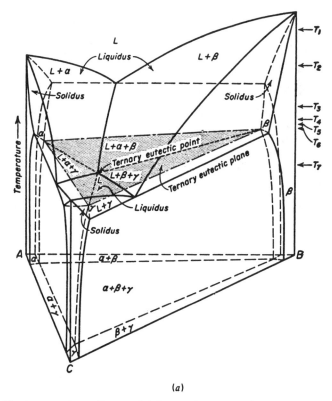

(a)

Fig. 2.21a The ternary phase diagram (a) Isometric construction. From Rhines, *Phase Diagrams for Metallurgy* McGraw-Hill, New York (1956) [15]. Used with permission of the McGraw-Hill Book Company.

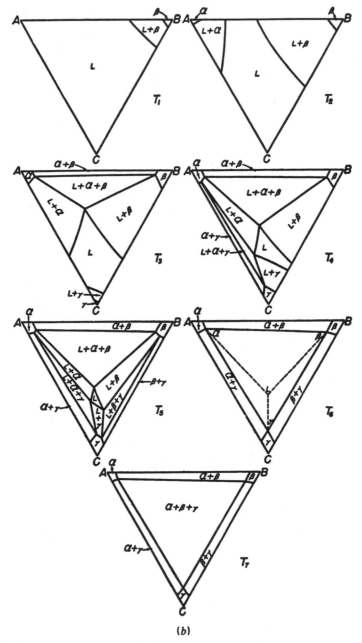

(b)

Fig. 2.21b The ternary phase diagram. (b) Isothermal sections. From Rhines, *Phase Diagrams for Metallurgy* McGraw-Hill, New York (1956) [15]. Used with permission of the McGraw-Hill Book Company.

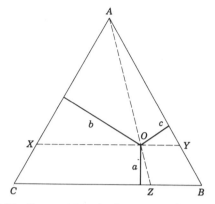

Fig. 2.22 Compositions in the ternary phase diagram.

A ternary system, where all alloys solidify eutectically, is shown in Fig. 2.21a and illustrates many of the characteristics of these diagrams [15]. We note that this is a three-dimensional plot, and is rather complex to construct, even for such a simple situation. The base of this diagram is an equilateral triangle with sides of unit length, as shown in Fig. 2.22, with each side corresponding to a binary composition, usually on an atom fraction basis. Any line parallel to a side of this triangle represents compositions in which the fraction of the component in the opposite vertex is held constant (for example, all compositions with 25% A are represented by the line XY in Fig. 2.22). Any line through a vertex represents all compositions with a fixed ratio of the components at the other two vertices. Thus, compositions where $B : C = 3 : 1$ are represented by the line AZ. The composition at O is thus 25% A, 56.25% B, and 18.75% C.

Another interesting property of the equilateral triangle is seen by dropping perpendiculars to the three sides from the point O. Let the length of these perpendiculars be a, b, c drawn as shown to the sides opposite the vertices A, B, C respectively. Then the composition represented by the point O is given by $A : B : C = a : b : c$. Furthermore, $a + b + c = \sin 60°$. For this example, $a = 0.25 \sin 60°, b = 0.5625 \sin 60°$, and $c = 0.1875 \sin 60°$, so that the composition defined by O is 25% A, 56.25% B, and 18.75% C as before.

2.4.1 Isothermal Sections

Ternary phase diagrams of the type shown in Fig. 2.21a are usually drawn in simplified form as a series of isothermal sections taken at different temperatures of interest [16]. Sections of this type are shown in Fig. 2.21b and correspond to the temperatures indicated in Fig. 2.21a. Very few such systems have been completely determined. Nevertheless, many sections can be drawn qualitatively from a knowledge of the behavior of the individual binary systems $A - B, B - C$, and $C - A$.

A study of isothermal sections of this type shows the following:

1. All two-phase regions are enclosed by four boundaries. Two opposite boundaries are adjacent to three-phase regions, or may be boundaries of the phase diagram. The other two opposite boundaries separate this region from one-phase regions. In addition, contact is made with two-phase regions at each vertex.

2. All three-phase regions are triangular, bounded by two-phase regions on three sides. In addition, they are in contact with one-phase regions at each vertex.

2.4.2   Congruently Melting Compounds

These are always present in the Ga–As–impurity system. Examples are binary compounds such as GaAs or Zn_3As_2 (in the Ga–As–Zn system). The presence of congruently melting compounds of this type serves to divide the ternary system into subsidiary eutectic systems.

2.4.3   Degrees of Freedom

In ternary systems, $F + P = 3 + 2 = 5$. Since 1 degree of freedom is usually assigned to pressure, $F + P = 4$ under ordinary circumstances. In addition, we note that two compositions must be assigned in order to determine the third, since $X_A + X_B + X_C = 1$.

For any isothermal section, the temperature is also preassigned, so that $F = 3 - P$. In a one-phase region as shown in Fig. 2.23 (such as L, α, β, γ)

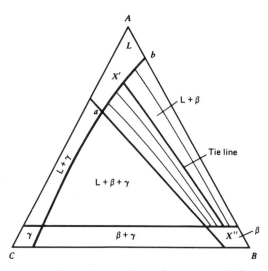

Fig. 2.23 Diagram illustrating regions of one, two and three phases.

there are 2 degrees of freedom. Both X_A and X_B (and hence X_C) can be varied independently over this region. For a two-phase region ($L + \alpha, L + \beta, L + \gamma, \alpha + \beta$, etc.), only 1 degree of freedom is possible. Thus, only one of the components can be varied. This is illustrated in Fig. 2.23. Here, a given value of X_A results in a liquid of composition X' along the boundary ab. This fixes, via a tie line, the composition of the solid solution β at X'' so that the system is fully defined.

In ternary diagrams, these tie lines may be straight or curved, and must be determined experimentally. In general, however, their positions are usually estimated by noting that they vary gradually from one boundary to the other, without crossing each other. Furthermore, they must run between two one-phase regions. A series of such tie lines is shown in this figure.

Figure 2.23 also shows a three-phase region, $L + \beta + \gamma$. This is an invariant region, with no degrees of freedom. For this temperature, the composition of the liquid and solid phases (L, β, and γ) is fixed at the values given by the vertices of the triangle bounding this region. Thus, the partial pressure of these three components is fixed by temperature alone. Compositions of this type are particularly useful as diffusion sources for this reason. The application of ternary considerations to diffusion will be considered in Chapter 4.

2.4.4 Some Ternary Systems of Interest

It is customary to show these systems in the form of a contour map, looking down on the composition triangle, with isotherms as contour lines. Also indicated on these diagrams are unique features (eutectics, peritectics, compounds, etc.) which aid in constructing specific isothermal projections as required. The valleys of the liquidus are also projected on to the composition triangle, with arrows indicating decreasing temperature.

The diagrams for GaAs, with Ag, Au, Ge, S, Sn, Te, or Zn as the third component are shown in Figs. 2.24 to 2.30, respectively [16–18]. Most of these diagrams are incomplete, with many lines estimated from the appropriate binary diagrams. Nevertheless, they are useful for assessing problems which might arise during the growth or alloying of GaAs [19], and also for evaluating conditions during diffusion in GaAs [20]. This latter topic is taken up in Chapter 4, where a series of isothermal sections are shown for the Ga–As–Zn and Ga–As–S systems.

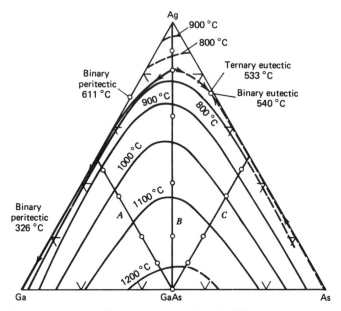

Fig. 2.24   The ternary system Ga–As–Ag. From Panish [18]. Reprinted with permission of the publisher, The Electrochemical Society, Inc.

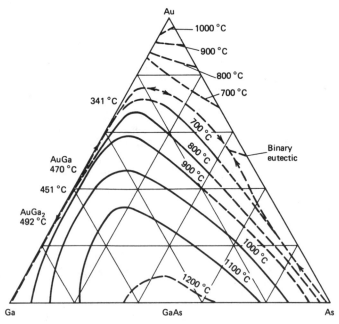

Fig. 2.25   The ternary system Ga–As–Au. From Panish [18]. Reprinted with permission of the publisher, The Electrochemical Society, Inc.

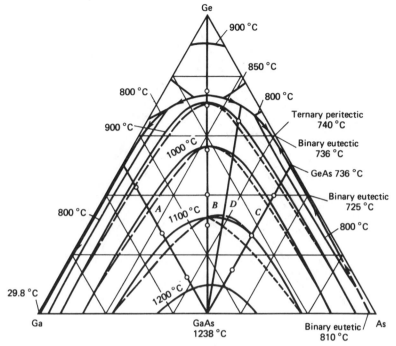

Fig. 2.26 The ternary system Ga–As–Ge. From Panish [19]. Reprinted with permission from the *Journal of Less-Common Metals*.

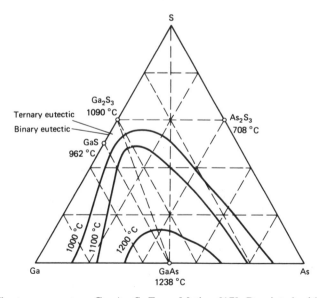

Fig. 2.27 The ternary system Ga–As–S. From Matino [17]. Reprinted with kind permission from Pergamon Press, Ltd., Oxford, UK.

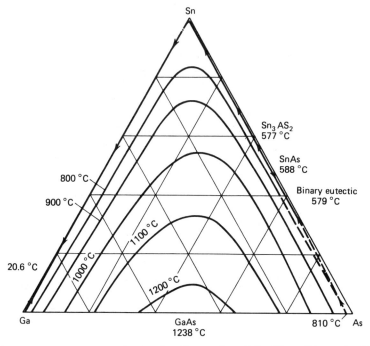

Fig. 2.28 The ternary system Ga–As–Sn. From Alper, Ed., *Phase Diagrams*, Vol. III (1970) [16]. Reprinted with permission of Academic Press, Inc.

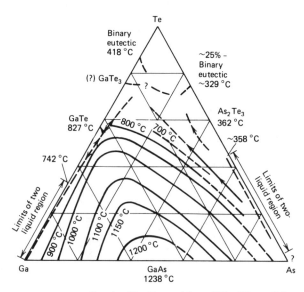

Fig. 2.29 The ternary system Ga–As–Te. From Alper, Ed., *Phase Diagrams*, Vol. III (1970) [16]. Reprinted with permission of Academic Press, Inc.

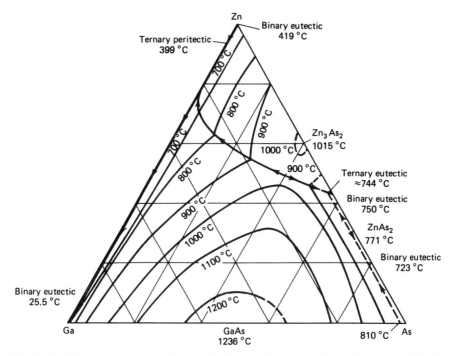

Fig. 2.30 The ternary system Ga–As–Zn. From Alper, Ed., *Phase Diagrams*, Vol. III (1970) [16]. Reprinted with permission of Academic Press, Inc.

REFERENCES

A large collection of phase diagrams for binary alloys may be found in Ref. 1 as well as in Ref. 2, with its more up-to-date supplement [3]. A compendium of phase diagrams for refractory oxides may be found in Ref. 4. Phase diagrams for ternary systems of interest to GaAs processing technology are scattered throughout the technical literature.

1. T. B. Massalski, Ed., *Binary Alloy Phase Diagrams*, 2nd edition, ASM International, New York, 1990.

2. M. Hansen and K. Anderko, *Constitution of Binary Alloys*, McGraw-Hill, New York, 1958.

3. R. P. Elliot, *Constitution of Binary Alloys* (A supplement to Ref. 1), McGraw-Hill, New York, 1965.

4. E. M. Levin, C. R. Robbins and H. F. McMurdie, *Phase Diagrams for Ceramists*, American Ceramics Society, Columbus, OH (1964).

5. G. A. Antypas, The GaP–GaAs Ternary Phase Diagram, *J. Electrochem. Soc.* **117**, 700 (1970).

6. V. Swaminathan and A. T. Macrander, *Materials Aspects of GaAs and InP-Based Structures*, Prentice-Hall, Englewood Cliffs, NJ, 1991.

7. C. D. Thurmond, Phase Equilibria in the GaAs and the GaP Systems, *J. Phys. Chem. Solids* **26**, 785 (1965).

8. B. Selikson and T. Longo, A Study of Purple Plague and Its Role in Integrated Circuits, *Proc. IEEE* **52**, 1638 (1964).

9. J. H. Forster and J. B. Singleton, Beam-Lead Sealed Junction in Integrated Circuits, *Bell Lab Record* **44**, 312 (1966).

10. T. J. Rockett and W. R. Foster, Phase Relations in the System Boron Oxide–Silica, *J. Am. Ceram. Soc.* **48**, 75 (1965).

11. T. Y. Tien and R. A. Hummel, The System SiO_2–P_2O_5, *J. Am. Ceram. Soc.* **45**, 422 (1962).

12. B. J. Baliga and S. K. Ghandhi, PSG Masks for Diffusion in Gallium Arsenide, *IEEE Trans. Electron Dev.* **ED-19**, 761 (1972).

13. F. A. Trumbore, Solid Solubility of Impurity Elements in Germanium and Silicon, *Bell Syst. Tech. J.* **39**, 205 (1960).

14. G. L. Vick and K. M. Whittle, Solid Solubility and Diffusion Coefficients of Boron in Silicon, *J. Electrochem. Soc.* **116**, 1142 (1969).

15. F. N. Rhines, *Phase Diagrams in Metallurgy*, McGraw-Hill, New York, 1956.

16. H. C. Yeh, Interpretation of Phase Diagrams, in *Phase Diagrams*, Vol. 1, A. M. Alper, Ed., Academic Press, New York, 1970.

17. H. Matino, Reproducible Sulfur Diffusions in GaAs, *Solid State Electron* **17**, 35 (1974).

18. M. B. Panish, Ternary Condensed Phase Systems of Gallium and Arsenic with Group 1B Elements, *J. Electrochem. Soc.* **114**, 516 (1967).

19. M. B. Panish, The Gallium-Arsenic-Tin and Gallium-Arsenic-Germanium Ternary Systems, *J. Less-Common Metals*, **10**, 416 (1966).

20. H. C. Casey, Jr. and M. B. Panish, Reproducible Diffusion of Zinc into GaAs: Applications of the Ternary Phase Diagram and the Diffusion and Solubility Analyses, *Trans. Met. Soc. AIME* **242**, 406 (1968).

PROBLEMS

1. You are provided with a phase diagram of the type shown in Fig. 2.2, where the abscissa is a linear scale based on wt. %. Develop the lever rule for this situation.

2. The lattice parameters of GaAs, InAs and InP are 5.6532 Å, 6.0583 Å and 5.8687 Å respectively. Determine the value of x for which a $Ga_xIn_{1-x}As$ layer is exactly lattice matched to an InP substrate. Indicate this composition in Fig. 2.22.

Assuming that the energy gap is given by

$$E(x) = 0.356 + 0.7x + 0.4x^2$$

calculate the wavelength corresponding to the lattice matched condition.

3. Assuming Vegard's law, show that the lattice parameter of the quarternary alloy $Ga_xIn_{1-x}As_yP_{1-y}$ is given by:

$$a(Ga_xIn_{1-x}As_yP_{1-y}) = xy[a(GaAs)] + x(1-y)[a(GaP)]$$
$$+ y(1-x)[a(InAs)] + (1-x)(1-y)[a(InP)]$$

Can you suggest a reason why anyone would want to use a quaternary alloy of this type in preference to a ternary?

4. The lattice parameters of GaAs, InAs, GaP and InP are 5.6532 Å, 6.0583 Å, 5.4512 Å and 5.8687 Å respectively. Determine the composition of GaInAsP which can be grown lattice matched to InP substrates, for phosphorus fractions of 0.15 and 0.5. Assuming Vegard's law, determine the (approximate) energy gap for these compositions. Assume that the energy gaps for GaAs, InAs, GaP, and InP are 1.42 eV, 0.36 eV, 2.26 eV and 1.35 eV, respectively.

5. From a study of Fig. 2.12, can you determine a simple criterion for differentiating a peritectic compound from a congruent transformation?

CHAPTER 3

CRYSTAL GROWTH AND DOPING

Demands placed on the quality of starting materials have become more severe as device dimensions have shrunk and packing densities have increased. Today, the requirements for very-large-scale integration (VLSI) circuits necessitate large-diameter crystals, virtually free of dislocations and variations in radial and axial resistivity. These goals have not been achieved; however, progress in this area has been very rapid. This chapter describes common techniques for growing single-crystal silicon and GaAs of a quality suitable for commercial device and microcircuit fabrication. Doping of the starting materials in the melt is also considered, followed by a discussion of the properties of material grown by these methods.

Semi-insulating (SI) GaAs is of high commercial importance, since it is the starting material for integrated circuits made with this semiconductor. Details of its methods of growth and the properties of the resulting materials will also be covered in this chapter. Relevant tables are collected at the end of this chapter.

3.1 STARTING MATERIALS

Polycrystalline silicon is used as the starting material from which device-quality, single-crystal silicon is grown. Silicon is commercially synthesized by heating silica, which is widely present in the earth's surface, and carbon, in an electric furnace with a reducing gas ambient. The resulting material is only 95–97% pure. Further purification is usually carried out by converting it to trichlorosilane and chemically processing this compound until it is 99.99999% (seven nines) pure. Semiconductor-grade polycrystalline silicon is made by its thermal decomposition in a hydrogen atmosphere, at a temperature of around 1100°C. This is used as the starting material for the manufacture of single-crystal silicon. The impurity content of polycrystalline silicon, produced by this process, as well as that of single-crystal silicon which is made from it by melting and recrystallization, is listed in Table 3.1 [1]. Additional impurities become incorporated in this material during crystal growth, as well as during the various processing sequences which are used to form the final microcircuit.

Gallium, which is a relatively scarce material, is found in nature as a trace element in germanite ore, from which it can be purified. Arsenic, on the other hand, is quite abundant. It is found in ores in the form of sulfides and arsenates, but most commonly in arsenopyrite from which it can be extracted by heating. Both gallium and arsenic, upon subsequent purification to a seven nines (99.99999%) level, are used as the starting materials for the synthesis of polycrystalline GaAs. The formation of GaAs, with near-perfect stoichiometry, is complicated by the highly different vapor pressures of its individual components. Moreover, the formation reaction is exothermic, and proceeds with considerable violence.

GaAs must be synthesized in an overpressure of arsenic in order to avoid simultaneous decomposition into its separate constituents. The appropriate arsenic pressure over molten GaAs is seen from Fig. 1.19 to be about 1.0 atm. This pressure can be established by conducting the reaction in a sealed tube maintained slightly above 1238°C, with a precisely measured quantity of arsenic. A more practical approach is to provide what is essentially an infinite supply of arsenic which is maintained at a lower temperature. For this situation, the pressure in the tube is approximately equal to the equilibrium vapor pressure of the volatile component (arsenic) at the lower temperature. Values for these vapor pressures are displayed in a pressure–temperature (P–T) diagram [2], as shown in Fig. 3.1. Here the solid curve shows the pressure of gaseous arsenic over GaAs, which is necessary to prevent decomposition of the compound, as a function of the temperature at which the compound is maintained. The dashed curve in this figure shows the pressure of gaseous arsenic over solid arsenic for this same temperature range, as obtained from standard vapor pressure tables. From these curves it is seen that the required arsenic overpressure to avoid decomposition can be achieved by keeping the elemental arsenic at about 600–620°C, while the GaAs is maintained at around 1240–1250°C.

Figure 3.2 shows an evacuated quartz sealed-tube configuration which can be used for the synthesis of GaAs, in a furnace with two temperature zones. The boats in this figure are usually made of quartz, and sometimes of graphite, because of its closer thermal expansion match to GaAs. Increasingly, however, pyrolytic boron nitride (PBN) boats are used to meet the purity requirements of modern devices and integrated circuits.

In operation, one boat is loaded with a charge of pure gallium, with the arsenic kept in a separate boat. A plug of quartz wool is often placed between these boats to act as a diffuser. Next, the tube is evacuated, sealed, and brought

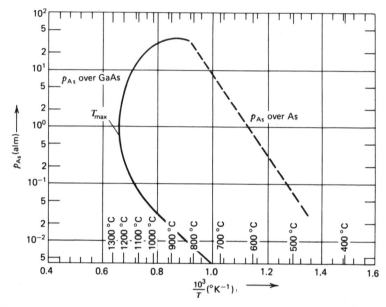

Fig. 3.1 *P–T* projection of the system Ga–As. From Boomgaard and Scholl [2]. With permission from *Philips Research Reports.*

up to system temperature, as shown in Fig. 3.2. This results in the transport of arsenic vapor to the gallium, with its conversion into GaAs in a slow, controlled manner. Such a process typically takes many hours to accomplish, for a starting charge of a few kilograms of gallium.

A pressure–composition (*P–x*) diagram can also be drawn by combining the *P–T* diagram with the phase diagram (*T–x*) for the gallium–arsenic system (Figs. 2.10 and 2.11). This diagram is shown in Fig. 3.3 and illustrates the necessity for maintaining extremely close control of the arsenic overpressure in order to avoid departures from stoichiometry in the resulting GaAs. This is

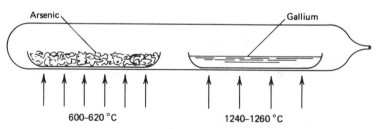

Fig. 3.2 Sealed-tube system for gallium arsenide synthesis.

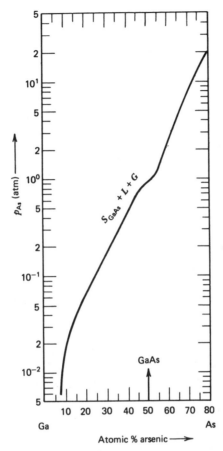

Fig. 3.3 *P–x* diagram for gallium arsenide. From Boomgaard and Scholl [2]. With permission from *Philips Research Reports*.

usually done by selecting the precise arsenic temperature by a trial-and-error process.

Polycrystalline GaAs, formed in this manner, is often used as the starting material for single-crystal growth. It is also possible to convert gallium and arsenic directly into single-crystal GaAs by a process known as *in situ*, or *direct*, compounding. This is carried out by reacting these elements in a pressurized vessel, at a temperature of around 700°C, which is well below the melting point of 1238°C. Alternatively, the reaction can be carried out at high temperatures in a low-pressure vessel, provided this is done is a very slow, controlled manner. These processes will be described in Section 3.2.1.2.

The impurity content of typical undoped GaAs, as measured [3] by spark source mass spectrometry, is listed in Table 3.2 at the end of this chapter.

3.2   GROWTH FROM THE MELT

Single crystals can be grown by controlled freezing of a melt in a boat or ampoule. This approach has the disadvantage of possible adhesion of the freezing melt to the walls. Contact of this type can initiate the formation of dislocations, and special care must be taken to minimize this effect.

Alternative approaches, such as the Czochralski (CZ) and float zone (FZ) techniques, allow the crystal to be grown with a free surface and thereby avoid this problem. Silicon is almost universally grown by these methods, and in particular by the CZ technique. This technique is increasingly used for GaAs as well; however, boat-grown material has some important advantages, and is used in a number of applications.

3.2.1   The Czochralski Technique

Figure 3.4 shows the schematic of a CZ system which can be used for both silicon [4] and GaAs [5] crystal growth. Here, the melt is contained in a crucible and kept in a molten condition by heating. A seed crystal, suitably oriented, is suspended over the crucible in a chuck. For growth, the seed is inserted into the melt until its end is molten. It is now slowly withdrawn, resulting in a single crystal which grows by progressive freezing at the liquid–solid interface. A pull rate of about 50–100 mm/hr is typical for both silicon and GaAs. Provisions are also made to rotate the crystal, and sometimes the crucible as well, during the pulling operation. A series of annular heat shields are provided between the growth region and the walls of the reactor, in order to control radial thermal gradients during the solidification process.

For silicon the entire assembly is enclosed within an envelope which is water-cooled and flushed with an inert gas. With GaAs, on the other hand, it is important to prevent decomposition of the melt during crystal growth by maintaining an overpressure of about 1.0 atm of arsenic. In principle, this can be done by maintaining the entire chamber at high temperature (600–620°C), to prevent arsenic condensation on its walls. In practice, however, this is extremely difficult to accomplish because of the highly corrosive nature of arsenic at these temperatures. Modified approaches, to be described later, are used for this purpose.

Most modern crystal pullers of the CZ type are resistively heated, and use replaceable crucible liners. Direct-current heating is generally used to prevent induced movement in the melt by eddy currents. In some systems, three-phase a.c. has been used to set up contra-rotation of the melt in order to control the growth conditions.

Carbon contamination occurs in both silicon and GaAs CZ systems. Generally, this comes about because of the reaction of moisture with hot carbon components and fittings, such as susceptors, crucible supports, and heaters, which are within the growth chamber. This produces carbon monoxide, which is sub-

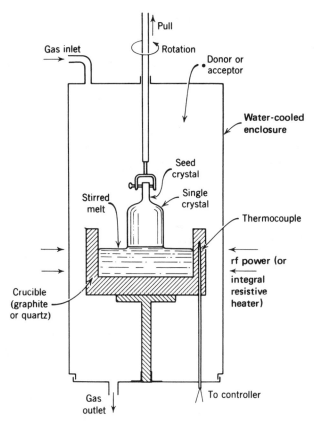

Fig. 3.4 Czochralski crystal-growing apparatus.

sequently incorporated into the melt. Some carbon is also present in the starting materials, as a contaminant.

Fused quartz crucibles are used for silicon growth. Considerable dissolution of this vessel takes place by the corrosive action of the molten silicon, resulting in the formation of SiO [6]. A large amount of this material is incorporated as dissolved oxygen in the crystal, with some evaporation of SiO from the melt into the ambient. In addition, impurities such as iron and carbon are incorporated from the silica crucible, which is usually 99.6% pure. The incorporation of these materials increases with spin rate. As a result, silicon crystals are rotated very slowly ($\simeq$1–10 rpm) during CZ growth.

Fused silica cannot be used for the growth of GaAs since silicon is a shallow donor and oxygen is a deep donor (see Section 1.5.1.3). Here, materials such as graphite, graphite coated with pyrolytic graphite, alumina, and PBN have been used; of these, PBN has been found to give a two-decade reduction in impurity incorporation over other materials [7], and is the universal choice today.

GaAs, grown in these crucibles, is weakly *p*-type because of residual carbon incorporation.

An inert gas ambient is used in CZ crystal growth, sometimes at slightly reduced pressure, to promote the evaporation of contaminants. In silicon crystal pullers, provision is sometimes made for operation at 2–50 torr to promote the evaporative loss of SiO, and thus reduce the amount of dissolved oxygen in the crystal. Both argon and nitrogen gas have been used for GaAs growth. Recent work [8] has shown that the use of krypton gas results in material of higher purity than is obtained with either of these gases. It has been proposed that this improvement comes about because krypton has the lowest heat transfer coefficient of these gases, and thus reduces temperature gradients in the grown crystal.

It is necessary that the latent heat of fusion be removed from the crystal-melt system during the growth process. This heat is lost by radiation from the crystal surface and by thermal conduction along the crystal axis. A detailed treatment of crystal growth, based on the interaction of these terms, is beyond the scope of this book. However, we can arrive at a number of important conclusions based on some simplifying assumptions. For example, since growth is accomplished by withdrawal of a solidified crystal, it follows that the maximum pull rate is limited by the rate at which the crystal can freeze (i.e., by the axial thermal gradient in the solid) near the growing interface. Analysis has shown that this thermal gradient is inversely proportional to the square root of the crystal diameter [9]. This result has also been experimentally verified [10].

If H_i is the rate of heat input to the system, and H_o is the rate of heat loss, then the difference $H_o - H_i$ is largely accounted for by the latent heat of crystallization, L. Writing ρ as the density and A as the cross section of the growing crystal, respectively, the heat balance equation based on this assumption is

$$H_o - H_i = AL\rho(dx/dt) \qquad (3.1)$$

where dx/dt is the pull rate. From this equation, it can be seen that the crystal diameter can be controlled by adjusting the heat input and/or the pull rate. In a practical situation, this type of control is used in the early phases of growth to expand the seed to the required crystal diameter, and also during growth to maintain this diameter within process tolerances. Once a steady-state diameter is achieved, a feedback control system is necessary to hold the temperature within ±0.5°C, even though thermal conditions are continually changing as the melt becomes depleted. Usually, the diameter of the growing crystal is optically measured by a remote video camera, or a sensitive load cell is used to measure variations in its weight. Diameter adjustments can be made during crystal growth, using the information derived in this manner. With GaAs, diameter control can be achieved by means of a disc of Si_3N_4, with a hole of the appropriate size. This disc, known as a *coracle*, floats on the molten liquid and defines the resulting crystal dimension. This technique has been found to be suitable for growth in the $\langle 111 \rangle$ directions, but causes problems with twinning during $\langle 100 \rangle$ growth. Both techniques allow diameter control to within ±2%.

Thermal stresses are created during cool down of the crystal as it is pulled in the CZ apparatus. In GaAs, these stresses can cause the crystal to shatter during subsequent slicing into wafers, or in subsequent processing. A long, slow anneal at around 950°C in an evacuated, sealed ampoule is necessary after growth to relieve these stresses and to homogenize defects in the material. This also improves the uniformity of electrical parameters such as resistivity and mobility over the entire crystal. Similar heat treatments are often given to silicon as well, although its surface fracture strength is about three times that of GaAs. In this case, their primary function is to homogenize the dissolved oxygen in the crystal.

3.2.1.1 Liquid Encapsulation

Czochralski growth of GaAs is seriously hampered by the need for maintaining an arsenic overpressure of about 1.0 atm during growth. This can be done by keeping the entire chamber hot, at about 600–620°C. Unfortunately, arsenic is highly reactive at these temperatures, thus greatly restricting the choice of materials for this purpose. The liquid encapsulated Czochralski (LEC) technique has been developed to avoid these problems, and consists of using a cap layer of an inert liquid to cover the melt [11, 12]. This cap prevents decomposition of the GaAs as long as the pressure on its surface is 1.0 atm or more. This pressure can be readily provided by means of an inert gas such as helium, argon, or nitrogen.

Requirements on this cap are that it float on the GaAs surface without mixing, be chemically stable, and have a low vapor pressure at the melting point of GaAs. In addition, it should be optically transparent so that the crystal can be viewed during growth, and impervious to the diffusion of arsenic (which is the volatile component). Although success has been obtained with materials such as $BaCl_2$ and $CaCl_2$, B_2O_3 is most commonly used for this purpose. B_2O_3 has a density of 1.5 g cm^{-3} as compared to 5.71 g cm^{-3} for GaAs. Moreover, its vapor pressure at the melting point of GaAs is below 0.1 torr, although it melts at 450°C. Cap layer thicknesses of B_2O_3 range from 5 to 10 mm; some work [13] has indicated that the use of thicker cap layers (19–27 mm) reduces the etch pit density by a factor of 5.

The LEC technique is found to result in almost no oxygen contamination of the GaAs. Typically, the incorporated boron content is in the 10^{15} to 10^{16}-cm^{-3} range. Furthermore, it has been observed that impurities initially present in the B_2O_3 cap remain preferentially in this layer, rather than becoming incorporated into the GaAs. The residual contamination in this material is usually carbon, which is *p*-type.

The presence of the B_2O_3 cap modifies thermal conditions in the freezing crystal, since the heat transfer coefficients of this layer are 10–20 times larger than that of the gas ambient. This introduces large thermal gradients as the freezing crystal emerges from the cap. One approach to circumventing this problem is to use a thick cap layer, and to limit the growth to a short crystal which is grown entirely *under* the cap layer. This approach [14], known as the liquid encapsulated Kryopoulos (LEK) technique, results in crystals with a lower dislocation content. However, it has not received significant attention because of its higher cost as compared to the LEC method.

3.2.1.2   *Direct Compounding*

There is considerable economic advantage to combining, in a single system, (a) the synthesis of GaAs from its elements and (b) the growth of the single-crystal material. Moreover, many of the contaminants in GaAs are incorporated during the compound formation operation described in Section 3.1. Thus the avoidance of this step, by *in situ* compounding in a CZ system, is highly desirable.

The vapor pressure of arsenic (over arsenic) at the melting point of GaAs is about 60 atm. Initially, this has prevented direct compounding. However, the availability of crystal pullers which can operate at as high as 150 atm, in combination with the liquid encapsulation technique, makes this process feasible.

Here, high-purity gallium and arsenic, in near-stoichiometric proportion, are placed in a PBN crucible liner within a heated graphite container [15]. Vacuum-baked B_2O_3, with a controlled amount of water added, is placed on its surface and the system pumped down to remove volatile oxides. The system is now pressurized to 3 atm of nitrogen and heated to 450°C, so that the B_2O_3 melts. Next the pressure is raised to about 60 atm and the crucible heated slowly to 700°C, at which point the reaction of gallium with arsenic takes place. Once complete, the pressure is reduced to about 2 atm and the crucible temperature raised so as to be slightly above the melting point of GaAs (1238°C). A single crystal is grown by inserting a seed crystal in the melt, and pulling in the manner described earlier. High-purity material can be grown by this approach [16].

The high-pressure LEC technique (HPLEC) described here is routinely used in the manufacture of GaAs crystals up to 75–100 mm in diameter. Methods have also been developed for compounding GaAs at low pressures (LPLEC) using an injection system, by which arsenic can be added to a melt of gallium in a slow, controlled fashion [17].

Perhaps the greatest advantage of the LPLEC method is a commercial one, in that it allows the use of low-pressure (1–5 atm) crystal pullers of the type commercially used for the growth of silicon, with only minor modifications to improve the thermal shielding [18]. These machines have benefited from many years of operational experience, so that their use should result in rapid reductions in the cost of single-crystal GaAs slices.

3.2.2   Gradient Freeze Techniques

Both silicon and GaAs can be grown by gradient freeze methods; in fact, early efforts [19, 20] concentrated on this approach for growing single-crystal materials. Even today, this technique is used to produce material with the lowest dislocation densities. In one approach, the polycrystalline starting material is loaded into a long, horizontal boat, and melted. The melt is now cooled from one end, which is usually necked down in order to restrict nucleation to a (hopefully) single event during freezing. This allows a single grain to propagate at the liquid–solid interface. Eventually the melt is frozen to take up this crystalline structure. A boat that is suitable for crystal growth by this technique has a seed crystal placed in the narrow end in order to establish a specific crystallographic orientation.

An alternative approach is to use a vertical ampoule, which is necked down at the bottom end in which a single-crystal seed is placed. Again, the melt is progressively cooled from this end until the single-crystal growth propagates over the entire volume of the semiconductor.

Gradient freeze techniques can be implemented by moving the boat or ampoule in an appropriate temperature gradient, by moving the furnace as shown in Fig. 3.5 [21]. Yet another approach is to use a furnace which has a large number of narrow heat zones [22], which can be separately computer-controlled in order to dynamically alter the temperature gradient profile during crystal growth.

The range of techniques outlined here can be described generically as gradient freeze (GF) methods. These include horizontal gradient freeze (HGF) and vertical gradient freeze (VGF), depending on the orientation of the semiconductor material during crystal growth. Silicon is rarely grown by GF techniques.

In the HGF process, the GaAs boat, which is typically 50–100 cm long and 100–50 mm wide, is sealed in a long quartz ampoule, with a suitable amount of high-purity (99.99999%) arsenic placed at one end. A loose plug of quartz wool is placed between the boat and the arsenic source. The arsenic is held at a temperature of $617°C \pm 1°C$ throughout the growth run so as to maintain an As overpressure of about 1 atm in the system. Typical values for the temperature gradient are $10°C/cm$, with a heater travel of 15–20 mm/hr.

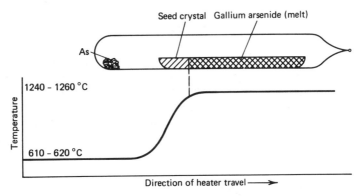

Fig. 3.5 Gradient freeze system.

During a typical growth run, an initial temperature profile is set up so that the GaAs charge is in a molten state, at 1241°C, and the arsenic is at 617°C. Next, the temperature profile is retracted slightly (towards the left in Fig. 3.5) in order to back up onto the seed. This is done by moving the furnace or by controlling its heater windings, and serves to establish a molten interface between the seed and the charge. Next, the profile is moved forward slowly (to the right in Fig. 3.5) to allow the solidification of single-crystal GaAs to occur. Simultaneously, the temperature of the solid GaAs is lowered to 1229°C, and held at this value across the entire solidified length in order to minimize thermal gradients in the grown crystal. A similar approach is used in the case of VGF growth.

The boat or ampoule used in these systems must fulfill two requirements. First, it must be of high purity to prevent crystal contamination. Next, it should not touch the growing crystal at any point, since twinning can be nucleated at points of adhesion between it and the freezing melt. Carbon-coated quartz, vitreous carbon, and lately PBN are in use today. PBN is the material of choice because its purity level is considerably higher than that of quartz or vitreous carbon. Often, these are provided with a rough, sandblasted finish to minimize the possibility of contact with the melt. A woven quartz pad is sometimes placed in the container, with the melt floating on its surface by surface tension.

Crystal growth of GaAs by the GF technique is usually carried out in the $\langle 111 \rangle$ direction which is the preferred direction for low defect growth, although the $\langle 110 \rangle$ orientation is sometimes used. Most device applications require (100) material, as pointed out in Section 1.4.2. Consequently, the D-shaped ingot resulting from the HGF process must be sawed at an angle to its growth axis (54.74° for $\langle 111 \rangle$), or along the axis if the crystal is grown on in the $\langle 110 \rangle$ direction. Any variation in doping along the crystal growth axis will thus show up as a variation across the resulting slice. Crystals grown in this manner are sometimes sold "as is," or are cut to a circular profile before sale.

Although HGF material is grown in a hot wall furnace, vertical temperature gradients of about ±0.5°C are still present [23], due to gravitational effects which create convection currents in the melt. These, as well as axial gradients (of about ±0.1°C), can be minimized in modern furnaces which have multiple quadrant as well as axial heat zones. Using these systems, it is possible to grow crystals, 5–7.5 cm wide, with dislocation densities under $5000/cm^2$.

Liquid encapsulation can also be used with the VGF method [24], as shown schematically in Fig. 3.6. Here, a PBN ampoule, which is open at the top, serves to confine the crystal. This crucible is tapered at the bottom in order to hold the seed crystal. The ampoule is held in a pressurized furnace with multiple heater windings which are computer-controlled to adjust the temperature gradient during crystallization. As before, special care must be taken to prevent sticking of the melt to the crucible walls at any point, to avoid dislocation formation.

Both radial and axial thermal gradients in VGF systems are lower than in HGF systems. Consequently, the resulting GaAs has a lower dislocation content. Here, 5 to 7.5-cm-diameter crystals have been grown, with dislocation densities

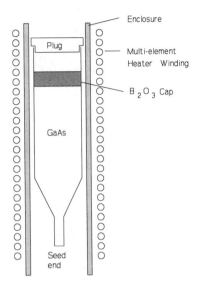

Fig. 3.6 Vertical gradient freeze apparatus.

as low as 100–3000/cm^2. Again, the $\langle 111 \rangle$ growth direction is favored, so that (100) material is obtained by slicing the boule at 54.74°C to its growth axis.

3.3 CONSIDERATIONS FOR PROPER CRYSTAL GROWTH

The problems of growing large area silicon and GaAs crystals are essentially similar, except that they differ in degree. The critical resolved shear stress of silicon is about four to five times that of GaAs. At the same time, the stacking fault energy is about 50% larger, so that the tendency for dislocation formation is much lower. On the other hand, industry requirements are for silicon crystals whose diameter is larger than that of GaAs by a factor of 2–3. With GaAs, small departures from stoichiometry can greatly alter its mechanical strength, with a resultant increase in the defect density in the grown crystal. This problem does not arise with silicon, which is an elemental semiconductor. In this section we describe the various factors which must be taken into account in order to grow single crystals which are relatively free from defects. Many of the considerations outlined here apply to both CZ and GF growth techniques. However, the emphasis will be on the CZ approach because of its greater commercial importance.

3.3.1 The Role of Point Defects

In Chapter 1, it was shown that the concentration of point defects in silicon and GaAs is an exponential function of the crystal temperature. The equilibrium concentration of these defects is thus a maximum at the crystal solidification temperature, and falls as the crystal cools. Condensation of these defects during this post-solidification phase of crystal growth results in their supersaturation, giving rise to the formation of dislocations. This is particularly true in the neighborhood of 950°C, where these materials undergo a plastic to elastic transition. Some quenching of defects can occur if high pull rates are used; however, this results in excessive thermal gradients.

Oxygen is an important point defect in CZ silicon which is conventionally grown in silica crucibles. Here, the combined effects of crystal rotation and thermal convection increase (a) the corrosive action of the crucible wall and (b) the dissolution of oxygen into the silicon melt. Some of this oxygen is lost by evaporation from the melt surface, in the form of SiO. Consequently, the amount of incorporated oxygen is a function of the ratio of the cross-section areas of the crystal and the crucible. With the trend towards large crystal diameters, this ratio has steadily increased because a larger fraction of the available space in the crucible is taken up by the growing boule. At the present time, the amount of dissolved oxygen can range from 10^{16} to 10^{18} cm^{-3}, depending on the spin rate. This oxygen is mostly in interstitial form, dispersed throughout the crystal. Its solid solubility is given [25] by

$$O_i = 5.5 \times 10^{20} e^{-0.89/kT}, \text{ atoms cm}^{-3}. \tag{3.2}$$

Upon cool-down, the oxygen becomes supersaturated, and some of it freezes out in the form of clumps and platelets of silicon dioxide. The volume expansion associated with this process produces strain in the crystal, which provides a gettering action for metallic impurities in the bulk silicon. It should be noted that a moderate amount of precipitated oxygen will strengthen the silicon, because of its ability to pin dislocations. Excessive amounts, however, can produce dislocations throughout the crystal.

Dissolved oxygen in silicon can affect device performance adversely; consequently, once the crystal is cut into slices, a number of different thermal process cycles are used in order to remove this oxygen from the surface regions, where devices and microcircuits will be fabricated [26]. The formation of these denuded regions is accompanied by additional precipitate formation within the interior of the slice. If excessive, it can result in dislocations which propagate to the surface region. Thus, a balance must be struck in order to obtain this gettering effect and still avoid excessive dislocation formation [27].

Carbon in silicon, in the 10^{16}–10^{17} atoms cm^{-3} range, is incorporated from hot graphite components of the crystal puller. It reacts with the quartz crucible walls, and increases their dissolution rate. Although inactive, it acts as sites for the precipitation of microdefects in the silicon. The carbon level can be greatly reduced by coating all graphite components with PBN.

With GaAs, vacancy formation is a strong function of stoichiometry in the melt. Moreover, both gallium and arsenic vacancies participate in the dislocation formation process. Gallium vacancies are the more mobile of these two and generally initiate this process; however, gallium vacancy migration can also produce arsenic vacancies, so that these generation mechanisms are interrelated.

The actual coalescence of these vacancies, which occurs during cool-down at around 1100 K, depends upon their charge state, and is more rapid if they are uncharged. Thus, the position of the fermi level (i.e., the doping concentration) during this post-solidification phase plays an important role in the defect structure of the material. If the GaAs is extrinsic and p-type at these temperatures, the gallium vacancies (which are deep acceptors) will be neutral during the cool-down process. Thus, they can coalesce readily to produce voids and dislocations. In n-type material, on the other hand, these vacancies are ionized so that the tendency for coalescence is suppressed. This has been illustrated in a study of HGF grown p- and n-GaAs with different doping concentrations, where the position of the fermi level (at 1100 K) was seen to track the dislocation density [28]. It is also observed in general practice, where commercially available n-type material is consistently superior to p-type. The importance of crystal stoichiometry was also demonstrated in experiments with HGF material using various arsenic source temperatures to control the arsenic overpressure. Here, it has been shown that optimum growth required the arsenic to be held at a temperature of $617 \pm 1°C$; an increase of dislocation density by more than two decades was observed for arsenic source temperature variations of $\pm 4°C$.

As noted earlier, PBN crucibles are used because they result in crystals with a minimum of contamination. Boron, carbon, and silicon are present as contaminants in the GaAs when grown in PBN vessels, by the LEC process. A small, controlled amount of water in the B_2O_3, around 100 ppm, has been found to reduce the silicon contamination, and is routinely used for this purpose. It has been proposed that this water converts the dissolved silicon into SiO_2, which is subsequently scavenged by the B_2O_3 cap. The resulting GaAs is thus weakly p-type, since both carbon and boron are acceptors.*

Carbon is an important extrinsic point defect in gallium arsenide which is grown by the LEC process. Gaseous oxides of carbon, rather than carbon itself, have been shown to be the source of this impurity, which is p-type. It can be largely suppressed by the use of PBN coatings on all graphite components which are in the hot zone of the crystal puller, including the heaters. By this approach, in addition to the use of PBN crucible liners, it is possible to keep the carbon concentration below 1×10^{15} cm^{-3}.

3.3.2 Thermal Gradients in the Crystal

Since crystal growth takes place by the controlled freezing of a part of the melt at any give time, both radial and axial temperature gradients are inherent to this process. Gradient freeze techniques are carried out in an environment in which there are almost no radial thermal gradients; in addition, new furnace designs

---

*Boron incorporates as a deep double acceptor on the As sublattice.

described earlier have allowed the axial thermal gradients to be minimized. The situation is quite different during growth by the CZ process, where the latent heat of fusion, liberated during the solidification process, is lost by convection and radiation from the surfaces of the growing crystal. With silicon, the thermal conductivity is high, and there is an almost linear temperature gradient from the solid–liquid interface to the neck of the boule [29]. In LEC growth of GaAs, however, the situation is further complicated by the presence of the B_2O_3 cap layer whose thermal transfer coefficients are very different from those of the neutral gas ambient. Crystal growth by these processes is thus accompanied by large thermal gradients, both axially as well as radially. In general, these give rise to stress in the crystal. Eventually, this stress is released by the generation of a network of glide plane dislocations known as *slip*.

Detailed calculations and computer simulations have been made of stress during crystal growth [30, 31], and are beyond the scope of this book. However, an understanding of this problem can be achieved by means of a simple model of a crystal cooled primarily by radial heat flow. During this cooling process, the outer regions of the crystal (near its circumference) are cooler than the core, and hence shrink. This process results in a tangential stress component, σ_θ, which is largest at the circumference, and falls as the center is approached. The cooling process also sets up a radial compressive stress, σ_r, which is highest at the crystal center, and falls to zero at the surface. The maximum resolved shear stress is then given [32] by τ_s, where

$$\tau_s = 0.5|\sigma_r - \sigma_\theta| \tag{3.3}$$

This shear stress is largest at the center and at the edge, but falls off to a low value in the mid-region. Slip will be created in a growing crystal if the resolved shear stress exceeds some critical value which is a function of the upper yield point. This results in glide of the {111} planes, which intersect the surface in the ⟨110⟩ directions.

The incidence of slip is more serious with the growth of GaAs than with that of silicon, for three reasons. First, the thermal conductivity of GaAs is about one-third that of silicon (see Table 1.1) so that thermal gradients are larger. Next, the critical resolved shear stress (CRSS) for GaAs is about one-fourth to one-fifth that for silicon. Finally, the presence of the cap layer results in large thermal gradients in GaAs, because of its large thermal transfer coefficients. This last problem does not arise in silicon growth, where the cap layer is not used.

Figure 3.7 shows calculated contours of constant dislocation density for a (001) GaAs crystal grown in the [100] direction [30]. These create a W-shape of dislocations along a radius. Impurities tend to preferentially segregate in these regions, resulting in a characteristic M-shaped pattern (i.e., an inverse W-shape). A similar pattern of dislocations has been observed for silicon crystals grown by the CZ process [33].

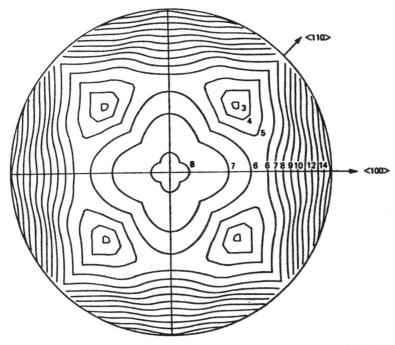

Fig. 3.7 Contours of constant dislocation density for a (100) GaAs crystal. From Jordan et al. [30]. ©1980. AT&T. All rights reserved. Reprinted with permission.

Elaborate heat shielding is used to reduce the problems of radial thermal gradients in crystal growth apparatus. In GaAs systems, additional radiation shields are sometimes used on top of the crucible to reduce axial thermal gradients, with the crystal being pulled through a hole in the shield.

3.3.3 Turbulences in the Melt

Crystal growth by the CZ process takes place under conditions which promote considerable turbulence in the melt. Here, the material at the crucible walls is hotter than at the center, because of direct contact with the heat source. Buoyancy forces cause the melt to move upwards at the walls, and downwards near the axis of the system. The resulting vortex flow produces thermal fluctuations in the melt, and in the solidifying crystal as well, leading to nonuniform growth behavior. Both silicon and GaAs growth can be affected by these instabilities.

Thermal convection effects of this type can be greatly reduced in both silicon and GaAs melts by CZ growth in a magnetic field [34, 35]. Both axial and transverse fields, up to 4000 gauss, have been used for this purpose and found to effectively suppress thermal fluctuations to as low as ±0.1°C. This melt stabilization smooths out axial resistivity fluctuations in the crystal. Moreover, it increases the boundary layer thickness [36] and hence the effective distribution coefficient, thus allowing improved control of doping along the boule length.

In silicon technology, reduction of the melt turbulence also reduces the reactivity of the melt with the quartz crucible wall, and hence the amount of dissolved oxygen in the crystal, typically by a factor of 2–5. This also reduces variations in oxygen content along the length of the boule.

3.3.4 Pull and Spin Rate

In general, the liquid–solid interface is not flat during CZ growth. Mechanical forces, which can give rise to dislocations, are thus unevenly distributed. This in turn can result in a radial pattern of defects, and also of impurities which segregate preferentially around them. These problems can be minimized if the liquid–solid interface is kept as flat as possible, and at right angles to the direction of crystal growth. The flatness of the interface can be enhanced by increasing the spin rate. Unfortunately, this also increases the corrosive effect of the melt on the crucible walls during the growth of silicon in a quartz crucible.

Silicon crystals are often grown with a meniscus that is slightly Ω-shaped. This has been experimentally found to result in a greatly increased allowable pull rate while still giving crystals with acceptable quality. With GaAs, however, a curved interface results in large radial temperature differences because of its relatively low thermal conductivity. This can set up a condition where stresses in the crystal can become excessive. A relatively flat interface is usually maintained for this reason.

3.3.5 Crystal Orientation

Crystal growth proceeds by the successive addition of layers of atomic planes at the liquid–solid interface. Since the {111} planes are the most closely packed, growth is easiest in the ⟨111⟩ directions for both silicon and GaAs. It is for this reason that {111} material is commonly used in bipolar silicon device and integrated circuit technology. On the other hand, {100} is required for silicon-based low-threshold MOS microcircuits and in low-noise operational amplifiers, so that considerable effort has been expended along these lines. Crystal growth in the ⟨100⟩ direction is highly desirable for GaAs, which cleaves readily along the {110} planes, resulting in rectangular dice with parallel faces.

Considerable care must be taken during ⟨100⟩ growth, in order to avoid any accidental thermal or mechanical shock, which can result in the crystal changing its orientation by twinning, with subsequent growth in a different direction [9]. Based on geometric and energy considerations, this will usually occur with a common {111} plane. If crystal growth is in the ⟨111⟩ directions, the twin plane will be at an angle of 70.54° to the growth plane, so it will work its way out of the ingot as growth proceeds. This new growth can be shown to be along one of the ⟨511⟩ directions. In practice, a (111) crystal which has twinned will almost invariably twin again and proceed as a $(\bar{1}\,\bar{1}\,\bar{1})$. Thus growth in the $\langle 111 \rangle$ direction tends to be self-correcting.

A crystal growing in the $\langle 100 \rangle$ direction can also twin along $\{111\}$ planes, which are at $54.74°$ to the growth plane, with the twinned section growth in one of the $\langle 221 \rangle$ directions. Although the twin plane will again work itself out of the crystal, further growth will usually continue along the $\langle 221 \rangle$ directions, provided that no further twinning occurs. With GaAs such ingots can eventually be cut in the $\langle 100 \rangle$ direction, resulting in noncircular slices. The state of the silicon technology is sufficiently advanced so that this material is rejected.

Crystals growing in the $\langle 110 \rangle$ directions can also twin along $\{111\}$ planes, with further growth in some of the $\langle 411 \rangle$ or $\langle 110 \rangle$ directions. Some of these $\{111\}$ twin planes are at right angles to the growth plane, and thus are propagated throughout the entire length of the crystal. Consequently, it is extremely difficult to grow $\{110\}$ material that is free from twinning.

It has been noted in Chapter 1 that the energy of formation of a dislocation is on the order of 10–20 eV [37], while its energy of propagation is lower by a factor of 100. Consequently, the strategy of good crystal growth is to begin with a very small, accurately aligned, dislocation-free crystal as the seed [38]. In CZ growth of GaAs, this seed crystal is typically 3–5 mm in diameter. After insertion into the melt, the seed is slowly withdrawn. Heat and pull rate are varied to neck this down to about 1 to 2-mm diameter. This increases the chances for the remaining dislocations to grow *out* of the freezing crystal. After about 1–2 cm is grown, the system parameters are altered so as to slowly expand the crystal diameter in a conical taper, until the required dimension is achieved, at which point the majority of the crystal is grown.

Silicon crystals are considerably larger in both diameter and weight, with melts of 40 kg or more. Consequently, somewhat larger seeds (6 to 12-mm diameter) are used, with necking to 2 to 4-mm diameter before eventual expansion to the final boule diameter.

3.3.6 Crystal Hardening Techniques

The generation of thermally induced glide dislocations can be reduced if the crystal is strengthened, so as to increase its critical resolved shear stress. In silicon growth, for example, this occurs to some extent by the incorporation of oxygen due to the corrosive action of the SiO_2 crucible walls. In GaAs, the intentional doping with group III and V impurities has been used for this purpose [39]. Of these *isoelectronic* impurities, indium has been found to be the most suitable, in concentrations of about 5×10^{20} cm^{-3}, without any evidence of impurity segregation. Using this approach, 100-mm-diameter gallium arsenide crystals have been grown with less than 300 dislocations/cm^2.

The mechanism of solution hardening is not fully understood at the present time. It has been proposed that it is caused by the strain field created by the lattice misfit due to the impurity, which inhibits the propagation of dislocations, once they are formed. As indicated in Chapter 1, the energy of movement of dislocations is much lower than their energy of formation, which may account for the effectiveness of this technique.

Isoelectronic doping is not used commercially for a number of reasons. First, although the material is almost dislocation-free, it is extremely brittle so that it is hard to manufacture into slices, and tends to fracture during device fabrication. Moreover, indium tends to outdiffuse from these slices during heat treatment. Next, the addition of indium to the GaAs increases its lattice parameter, so that any subsequent growth on its surface occurs under conditions of lattice mismatch. This sets an upper limit to layer thickness, before the onset of misfit dislocations at the interface. Finally, it was shown that boule annealing of GaAs provided crystals which met the requirements for threshold voltage uniformity of field effect transistor devices across a slice, without the need for solid solution hardening. Thus, the need for this complication does not exist at the present time.

3.4 DOPING IN THE MELT

This is usually done by adding a known mass of dopant to the melt in order to obtain the desired composition in the crystal. Raw dopants are not used since the amounts to be added are unmanageably small; moreover, their physical characteristics are often quite different from those of the melt (e.g., element phosphorus cannot be added to molten silicon without disastrous results). It is common practice to add the dopant in the form of a highly doped powder of crushed semiconductor, of about 0.01 Ω-cm resistivity. In this manner both these problems are avoided.

Boron and phosphorus are the most common dopants for p- and n-type silicon, respectively. Antimony and arsenic are useful for n^+ substrates because they are relatively slow diffusers compared to phosphorus. The variety of selections for GaAs is much larger. Thus zinc and cadmium are commonly used for p-type GaAs; tellurium, selenium and silicon are used for n-type material. Chromium is used for SI GaAs. Increasingly, however, this material is grown in undoped form, as discussed in Section 1.5.1.3.

The addition of impurities to the melt, accompanied by stirring, results in a doped liquid from which the crystal is grown. In general, the concentration of the solute will be quite different in the solid and liquid phases of a crystal because of energy considerations. Consider a crystal at any given point during its growth, where C_S and C_L are the concentration of solute (atoms/cm^3) in the solid and the liquid phases, respectively, in the immediate vicinity of the interface. Then, a distribution coefficient k may be defined, where

$$k = \frac{C_S}{C_L} \tag{3.4}$$

Table 3.3 at the end of this chapter gives values of k for some impurities in silicon [40, 41]. To an approximation, these values are independent of concentration. It is seen that, except for oxygen, k is normally less than unity, and ranges from as high as 0.8 for boron to as low as 2.0×10^{-3} for aluminum. To a good approximation, the value of k is equal to 10 times the maximum solid solubility, expressed as an atom fraction X_0. This empirical relation holds for both deep and shallow impurities in silicon.

Table 3.4 shows k values for some impurities in GaAs [42]. As with silicon, we note that the segregation coefficients for shallow impurities are, in general, larger than for those exhibiting deep levels. Here, however, values of k in excess of unity have been measured for a number of impurities. Moreover, wide ranges of k values have been observed, depending on the growth technique [43].

If k is less than unity, excess solute is thrown off at the interface between the melt and the crystal. Consequently, the melt becomes increasingly solute-rich as crystal growth progresses, resulting in a crystal of varying composition. Analogous considerations apply for the situation where k is greater than unity. In either event, it is possible to calculate the axial dopant incorporation for different growing conditions.

3.4.1 Rapid-Stirring Conditions

Assume that rapid stirring is involved during very slow growth of the crystal. Then the solute in the immediate vicinity of the freezing interface is immediately removed [44] into the melt. In addition, assume that there is no diffusion of the solute within the crystal during the growth process. This is reasonable, since the diffusion coefficient of impurities in the melt ($\approx 5 \times 10^{-4}$ cm^2/s for most situations) is about six orders of magnitude larger than in the solid crystal [45]. Let:

V_M = initial volume of the melt

C_M = initial concentration of solute in the melt (atom cm^{-3})

At a specified point in the growth process, when a crystal of volume V has been grown, let:

C_L = concentration of solute in liquid (atom cm^{-3})

C_S = concentration of solute in crystal (atom cm^{-3})

S = amount of solute in melt

Consider an element of the crystal of volume dV. During its freezing, the amount of the solute lost from the melt is $C_S dV$. Thus

$$-dS = C_S \, dV \qquad (3.5)$$

At this point the volume of the melt is $V_M - V$, and

$$C_L = \frac{S}{(V_M - V)} \tag{3.6}$$

Combining these equations and substituting $C_S/C_L = k$, we obtain

$$\frac{dS}{S} = -k\frac{dV}{V_M - V} \tag{3.7}$$

The initial amount of the solute is $C_M V_M$. Consequently, Eq. (3.7) may be integrated as

$$\int_{C_M V_M}^{S} \frac{dS}{S} = -k \int_{0}^{V} \frac{dV}{V_M - V} \tag{3.8}$$

Solving and combining with Eq. (3.6) gives

$$C_S = kC_M \left(1 - \frac{V}{V_M} \right)^{k-1} \tag{3.9a}$$

$$= kC_M(1 - X)^{k-1} \tag{3.9b}$$

where X is the fraction of the melt that is solidified.

Figure 3.8 illustrates the crystal composition, described by this expression, where the range of values of k is below unity. It is seen that, as crystal growth progresses, the composition continually increases from an initial value of kC_M. In addition, considerably more uniform crystal composition is obtained as values of k approach unity.

Figure 3.8 also shows that the incorporation of impurities with low values of k is extremely difficult. This is especially fortuitous, since many deep-lying impurities, which are lifetime killers, are selectively rejected during the crystal growth process. Of these, the most important are copper, gold, and iron, in both silicon and in GaAs.

Although $k > 1$ for oxygen in silicon, its incorporation is largely controlled by changes in the amount of dissolved oxygen in the melt during the growth process. Thus, the area of the crucible in contact with the silicon falls with increasing X, as do thermal convection effects. Consequently, the oxygen concentration falls with an increase in the fraction that is solidified. In the initial stages of crystal growth, it approaches 2.5×10^{18} cm^{-3}, which is the solid solubility at the melt temperature. It goes without saying that Eq. (3.9) does not apply to this situation.

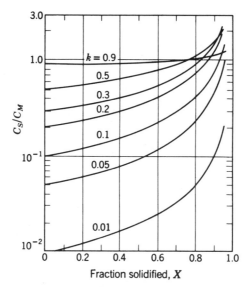

Fig. 3.8 Crystal composition during growth. Adapted from Bridges et al. [44].

The process of freezing from a doped melt may also be described in terms of a phase diagram. By way of example, Fig. 3.9 shows a greatly expanded version of the solid-solution region for a semiconductor B which is doped with A. Starting with a melt of composition C_M (atom cm^{-3} of the solute in the melt), the crystal begins to freeze out initially at a composition kC_M, where k is the ratio of the slopes of the liquidus and solidus curves (assumed straight lines). With further cooling, the crystal composition moves along the solidus line until the composition of the solid solution is C_M. During this freezing, the melt composition is initially C_M and moves along the liquidus curve until its composition becomes C_M/k. It should be noted that k is the distribution coefficient described in Section 3.4 and is given by the ratio of the slopes of the liquidus and solidus curves in the vicinity of the melt composition.

The validity of Eq. (3.9) breaks down as X, the fraction of the melt solidified, approaches unity. In this region, the melt becomes excessively rich in the solute, and the crystal composition is determined by its solid-solubility characteristic. Eventually the solid growth becomes polycrystalline.

3.4.2 Partial-Stirring Conditions

For realistic growth and stirring conditions, the rejection rate of the solute atoms at the interface is higher than the rate at which they can be transported into the melt. Consequently, the solute concentration at the interface will exceed the concentration in the melt. This results in a crystal with a doping concentration in excess of that obtained for the case of full stirring.

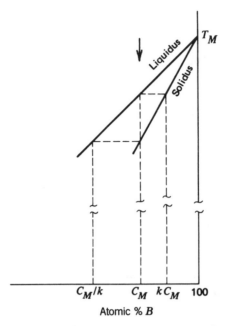

Fig. 3.9 Phase diagram for crystal doping.

The crystal composition may be determined by postulating the presence of a thin stagnant boundary layer of liquid immediately adjacent to the liquid–solid interface, through which solute atoms flow by diffusion alone. Equilibrium conditions prevail beyond this layer. The boundary layer thickness, δ, is given [46] by

$$\delta = 1.6D^{1/3}\nu^{1/6}/(2\pi n)^{-1/2} \text{ cm} \tag{3.10}$$

where ν is the kinematic viscosity of the melt, n is the rotational rate, and D is the diffusivity of the impurity in the melt. As a result of this boundary layer, the concentration of the melt at the interface exceeds the equilibrium concentration.

Figure 3.10 shows the liquid–solid interface and the solute concentration beyond this interface, for the case where $k < 1$. With reference to this figure, an effective distribution coefficient may now be defined for partial stirring conditions, such that $k_e = C_S/C_L$. This coefficient is larger than k ($= C_S/C'_L$) and depends on the growth parameters of the crystal. Let

$R =$ growth rate for crystal (i.e., rate of movement of the liquid–solid interface).

$D =$ diffusivity for solute atoms in liquid

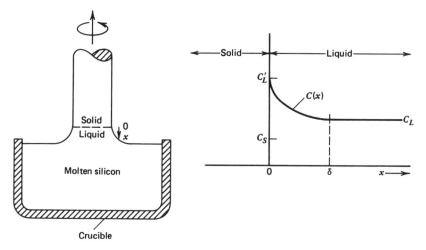

Fig. 3.10 Partial-stirring conditions.

The equation governing the diffusion of solute atoms in the layer may now be written. Noting that the amount of solute rejected from the solid is equal to that gained by the liquid, the stationary distribution is given by

$$D\frac{d^2C}{dx^2} + R\frac{dC}{dx} = 0 \tag{3.11}$$

so that

$$C = Ae^{-Rx/D} + B \tag{3.12}$$

and

$$\frac{dC}{dX} = -\frac{AR}{D}e^{-Rx/D} \tag{3.13}$$

where A and B are constants of integration.

The first boundary condition is that $C = C'_L$, at $x = 0$. The second boundary condition is obtained by noting that the sum of the impurity fluxes at a boundary must be zero. Again assuming that diffusion of solute atoms in the solid is negligible compared with diffusion in the liquid, this condition may be written as

$$D\left(\frac{dC}{dx}\right)_{x=0} + (C'_L - C_S)R = 0 \tag{3.14}$$

so that

$$\left(\frac{dC}{dx}\right)_{x=0} = -\frac{R}{D}(C'_L - C_S) \tag{3.15}$$

Substituting these boundary values into Eq. (3.12) and noting that $C = C_L$ at $x = \delta$, we obtain

$$\frac{C_L - C_S}{C'_L - C_S} = e^{-R\delta/D} \tag{3.16}$$

Substituting for k and k_e, and rearranging terms, gives

$$k_e = \frac{k}{k + (1 - k)e^{-R\delta/D}} \tag{3.17}$$

Finally, the crystal composition for partial stirring may be derived from the results for complete stirring by substituting k_e for k in Eq. (3.9b), so that

$$C_S = k_e C_M (1 - X)^{k_e - 1} \tag{3.18}$$

Values of k_e thus approach unity for large values of the normalized growth parameter $R\delta/D$. Uniform crystal composition is thus approximated with a high pull rate and/or a low spin rate (since δ falls with increasing spin rate). In practice, these growth parameters are set by considerations outlined in Section 3.3. Thus the pull and spin rates are optimized to grow crystals of the desired diameter with low dislocation concentration.

A number of approaches are used to obtain the slice-to-slice dopant uniformity which is necessary for commercial microcircuit fabrication. These include the use of small sections from the grown crystal, the growth of ingots of small length, and growth with a magnetic field to reduce thermal fluctuations. An additional approach is to program the growth parameters to obtain uniform composition over a large fraction of the crystal growth. Success has also been achieved by adding doped silicon feedstock, in molten form and at a controlled rate, during the continuous growth of large crystals.

3.4.3 Radial Doping Variations

During growth, heat is lost to the freezing crystal by conduction along its axis, and radially towards its surface, which is exposed to a purging gas flow. The radial temperature gradient depends on both the growth rate and the surface cooling rate, and can be as high as $40°C$ in the immediate vicinity of the liquid–solid interface. This can produce variations in the radial doping concentration because of changes in k with temperature.

Radial variations of doping concentration can also come about because the boundary layer thickness in the immediate vicinity of a moving surface is inversely proportional to the square root of its linear velocity [46]. Thus, δ is large at the axis of the growing crystal, where the effective linear motion is slow, and falls off as we approach the perimeter, where the motion is rapid. As a result, there is a radial variation of k_e due to this effect.

The effect distribution coefficient also varies with growth rate. This gives rise to an important difference in the doping behavior of semiconductors that are grown in the $\langle 111 \rangle$ and $\langle 100 \rangle$ directions [47]. Consider a growing crystal whose solid–liquid interface is convex. Noting that growth occurs by the successive addition of $\{111\}$ layers, a microscopic view of the growth interface can be drawn as a series of steps and risers, as in Fig. 3.11a. From this figure, it is seen that the movement of the steps for $\{111\}$ growth is directly proportional to the cotangent of θ, where θ is the slope of the growth surface. Thus, the growth front velocity increases rapidly as the crystal center is approached. This results in a rise in k_e, and hence in the dopant incorporation, which takes on a concave character. The situation is quite different with $\{100\}$ growth, where the velocity of the growing surface is relatively independent of the shape of the interface, as in Fig. 3.11b.

These considerations are especially important for silicon, which is normally grown with a curved interface, because of the relatively high growth rate. Here, radial resistivity variations for $\{111\}$ growth are large for dopants which have low segregation coefficients, such as phosphorus ($k = 0.35$), but not for boron ($k = 0.8$). Antimony, with a segregation coefficient of 0.023, presents an especially severe problem for this reason [48].

GaAs is normally grown with a flat interface, so that some of these considerations do not apply. Here, however, radial resistivity variations are dominated by the W-shaped dislocation pattern described earlier, and take on an M-shaped

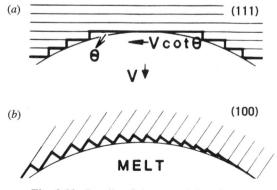

Fig. 3.11 Details of the growth interface.

character. Additionally, native defects such as EL2 are found to segregate preferentially in the same manner. This problem is minimized by careful attention to heat shielding in the crystal pulling apparatus in order to minimize thermal gradients.

3.4.4 Constitutional Supercooling

The presence of dopants in a melt greatly reduces the rate at which a crystal can be grown without the loss of crystallinity. Consider, again, the process of crystal growth in a partial-stirring situation where $k_e < 1$. Here the growing crystal results in a pileup of impurity at the liquid–solid interface. The amount of this pileup is a function of the melt volume and the degree of stirring. In addition, it increases with higher growth rate. Too great a pileup can lead to supersaturation in the boundary layer, which can cause spurious nucleation and polycrystalline growth [37]. Such a condition is known as *constitutional supercooling*.

The growing conditions under which constitutional supercooling can occur, or can be avoided, are illustrated in Fig. 3.12. Here, Fig. 3.12*b* is a plot of the equilibrium melt temperatures associated with the compositions shown in Fig. 3.12*a*, on the assumption that an increase in doping concentration results in a linear fall in the melting point (see Fig. 3.9). Thus the curved regions of both figures are exponential in character, for the steady-state growth conditions shown.

If it is assumed that heat transfer from the crystal interface to the (cooler) melt region is by conduction, the actual temperature in the liquid will be a linear function of distance, as illustrated for one specific growth rate by the dashed line *XA*. Here, supersaturation of the melt will occur for a distance *a* beyond the interface, with the resulting polycrystalline growth. The line *XB* shows the temperature in the liquid for a second, slower growth rate. Here no region of supercooling is present, and single-crystalline growth is the result.

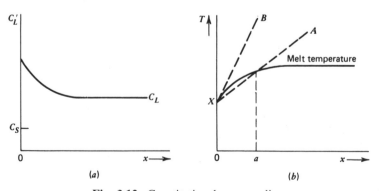

Fig. 3.12 Constitutional supercooling.

From the above it is seen that the growth rate of a doped crystal is considerably lower than if the crystal were undoped. Moreover, this difference is even larger for dopants (and stirring conditions) with lower values of k_e.

3.5 SEMI-INSULATING GALLIUM ARSENIDE

The thrust of GaAs single-crystal growth is in the commercial development of semi-insulating (SI) material. The technological importance of this material is based on the fact that devices made in it by direct ion implantation are self-isolating, so that it is ideally suited to integrated circuit fabrication. Moreover, field effect transistors made in this material have a much lower drain capacitance than those made into doped substrates, and are thus faster.

Two basic requirements must be met to exploit the unique properties of SI GaAs. First, its bulk resistivity must be larger than some high value, typically $10^6 \, \Omega$-cm, over the entire substrate. Next, it must preserve this minimum resistivity value during all subsequent processing. In particular, this resistivity must be maintained after the material is subjected to a post-implant anneal, which is commonly carried out at 850–950°C, for periods of 30–60 min.

Undoped GaAs is semi-insulating, but it is not possible to achieve this property by purification alone. Moreover, the compensation of residual impurities by shallow impurities of opposite type cannot be achieved with sufficient precision for this purpose. However, doped GaAs can be made semi-insulating if counterdoped appropriately with a deep impurity of opposite conductivity type, whose impurity level is close to the middle of the bandgap.

Consider n-type GaAs with a carrier concentration N_d, which is counterdoped with a midgap acceptor of concentration N_a. Solution of the equilibrium equations for free carriers in this material shows that the electron concentration falls with increasing acceptor concentration, becomes zero when $N_a \simeq 2N_d$, and becomes slightly p-type with a further increase of deep acceptors. A similar reasoning applies to p-type GaAs which is counterdoped with a deep donor.

In principle, this approach can be used regardless of the initial doping concentration, as long as a sufficient amount of the deep level impurity can be used to counter-dope it. However, changes in either the impurity or the residual background can result in variations of the resistivity; in some cases, thermal conversion of the semiconductor impurity type can occur. These effects are minimal if the actual concentrations of the impurity and of the counterdopant are low. Therefore, the best strategy is to grow GaAs with the highest purity level possible, and counterdope it with the smallest amount of deep level that is necessary to make it semi-insulating.

We have shown that GaAs grown in quartz vessels is usually n-type, because of silicon incorporation from the walls. Material grown in PBN crucibles, with a B_2O_3 cap layer, is usually p-type because of both carbon and boron incorporation. Thus, both p- and n-type deep levels can be used as counterdopants in order to obtain SI material.

Midgap impurities which can make GaAs semi-insulating are oxygen, chromium, and the native defect EL2. The discovery of the SI property of oxygen-doped GaAs was first made during the thermal processing of this material under open air conditions, and SI material was prepared using oxygen doping [49, 50]. As indicated in Chapter 1, however, its use as a dopant for making SI GaAs is unfortunately impractical since it is highly mobile at temperatures above 650°C, which are required for devices processes.

3.5.1 EL2-Doped GaAs

Recent years have seen great improvement in the quality of LEC-grown GaAs. This is because of the many advances which have been made in eliminating residual contaminants in the starting materials. Moreover, the use of PBN crucibles has greatly reduced the incorporation of n-type contaminants such as silicon into the melt. The incorporation of p-type impurities comes about due to the inclusion of boron from the B_2O_3 cap layers, which is incorporated as a deep double acceptor on the As sublattice. Carbon is also present in bulk GaAs material, and is incorporated as a shallow acceptor on the As sublattice. Although the use of a small but controlled amount of water in B_2O_3 reduces the incorporation of these impurities, the net chemical impurities in high-purity GaAs are p-type with a concentration in the 5×10^{14} to 10^{15} cm^{-3} range [51, 52]. As a result, the native defect EL2, which is a deep donor, can be used to make this material semi-insulating. This defect is located at about $E_c -0.75$ eV in the energy gap, with a temperature dependence given by

$$(E_c - E_T) = 0.73 - 3.5 \times 10^{-4} \frac{T^2}{T + 204} \tag{3.19}$$

EL2 is a doubly ionized native donor defect in GaAs, and is stable at temperatures up to 950°C. It is well established that the concentration of this defect is related to the stoichiometry of the melt from which the bulk GaAs is grown, and increases with the arsenic concentration [53]. EL2 has been identified in SI material grown from melts with > 0.47 As atom fraction, and concentrations of 2×10^{16} cm^{-3} are found in material grown from melts with 0.51 As atom fraction [54]. The concentration of this defect also varies in epitaxial gallium arsenide where it increases with the As overpressure.

Studies have confirmed that EL2 involves the native defect As_{Ga}, which is related to the arsenic overpressure, and is probably created during the postgrowth cooling of the crystal [55]. However, the precise atomic model for EL2 has not been established at the present time. One model is based on the theory that the formation of As_{Ga}, by the movement of an arsenic atom onto a gallium site, leaves behind an arsenic vacancy in its proximity. Thus, the EL2 defect is an (As_{Ga}–V_{As}) complex. A one-to-one correlation has been observed between EL2 and a shallow donor 26 meV below the conduction band [56]. It has been proposed that an additional defect center, involving the arsenic vacancy V_{As} on a neighboring site, gives rise to this shallow donor level.

An alternative model has been proposed, based on the characterization of EL2 as an As_{Ga}–As_i complex for this purpose [57]. Other research has suggested that an oxygen-related complex is also involved, indicated by the presence of an electron trap at $E_c - 0.65$ eV.

We have noted that the concentration of EL2 in GaAs is primarily related to the stoichiometry of the melts from which it is grown, and that small departures in the As/Ga ratio are required for this purpose [58]. In addition, the EL2 concentration falls off rapidly as the material is doped with shallow donors such as sulfur, silicon, and tellurium, in concentrations above 10^{17} cm^{-3} [59]. This behavior can be explained if we assume that the antisite defect is created by the migration of a vacancy to a nearest neighbor:

$$As_{As}^0 + V_{Ga}^- \rightleftharpoons As_{Ga}^{2+} + V_{As}^+ + 4e^- \tag{3.20}$$

so that

$$[As_{Ga}^{2+}] = k \frac{[V_{Ga}^-]}{[V_{As}^+]} n^{-4} \tag{3.21}$$

It is possible for undoped SI GaAs to convert to p-type during high-temperature processing in hydrogen gas, during the post-implant anneal step. This effect is seen in the first 2- to 3-μm surface layer, in which the integrated circuit is fabricated. It has been associated with the indiffusion of atomic hydrogen, which results in the formation of stable As–H bonds. These bonds deactivate the EL2 centers with a net fall in their active concentration. LEC-grown GaAs has a residual p-background, so that the SI material eventually converts to p-type. The chance of thermal conversion is thus reduced if the residual acceptors incorporated during crystal growth are kept to a minimum.

The EL2 defect can also be created, and altered, by a wide variety of processing conditions. Ion implantation has been shown to produce small concentrations of this defect. Post-implantation anneals also change the EL2 concentration in GaAs, depending on capping conditions as well as on the time and temperature associated with the anneal process.

3.5.2 Chromium-Doped GaAs

Chromium introduces a deep acceptor whose level is very close to the midgap [60], with a temperature dependence given by

$$E_c - E_T = 0.759 - 2.4 \times 10^{-4} \frac{T^2}{T + 204} \tag{3.22}$$

It can be intentionally incorporated to a concentration of several 10^{17} cm^{-3}, where it can serve as either an electron or a hole trap in GaAs, depending on the position of the fermi level prior to its introduction. Thus, it can be used with silica boat-grown material (which is usually n-type) as well as with CZ material (which is usually p-type) to produce SI GaAs. Since its energy level is extremely close to the midgap, its use results in material with a resistivity in excess of 10^9 Ω cm. It is somewhat mobile at high processing temperatures, so that out-diffusion effects are observed with this dopant.

Historically, large amounts of chromium ($\approx 10^{17}$ cm^{-3}) were required for this purpose because of the high concentration of residual impurities in GaAs. This led to many problems during heat treatment of Cr-doped material, where its out-diffusion resulted in loss of SI character [61]. More recently, however, the ability to grow high-purity material has required lower ($\leq 10^{16}$ cm^{-3}) concentrations of this dopant, so that Cr-doped material is now generally free from thermal conversion problems.

We have noted that the native defect EL2 is present in all GaAs, regardless of its preparative method. Consequently, it is present in Cr-doped material as well. The compensation mechanism of Cr-doped material is thus somewhat complex; both n- and p-type SI material can be obtained, depending on whether the Cr concentration is low ($< 10^{16}$ cm^{-3}) or high ($> 10^{17}$ cm^{-3}), respectively.

Impurities such as iron and vanadium have also been investigated, with the hope of finding a deep level which is less thermally mobile than chromium. This work has not been pursued to any extent, however.

3.6 PROPERTIES OF MELT-GROWN CRYSTALS

Single-crystal silicon material, which is used for both linear and digital devices and integrated circuits, is made by the CZ technique. GaAs, grown by HGF methods and more recently by VGF methods, is used for optical emitter and detector applications where low dislocation count is required. Czochralski-grown GaAs, in combination with liquid encapsulation and *in situ* compounding, is the preferred choice for volume production of SI material for integrated circuits. Moreover, circular slices which result by this technique are amenable to automatic manufacturing processes which are already available for silicon.

The state of silicon technology is highly advanced, whereas that of GaAs is primitive in comparison. At the present time, silicon ingots of 40- to 80-kg weight and 125 to 150-mm diameter are routinely grown, whereas the comparable numbers for GaAs are 1- to 2-kg weight and 100-mm diameter. In addition, most achievable specifications for silicon are considerably better (and tighter) than those for GaAs.

Slices cut from single-crystal material area usually evaluated on the basis of the following properties:

Dislocation Content. It is possible to obtain 200-mm-diameter CZ silicon that is entirely free from dislocations, to the point where they cannot be measured. In practice, however, the dislocation content of acceptable material ranges from zero to as high as 10^2 cm^{-2}. In addition, crystals which freeze along a curved interface show corresponding radial patterns of dislocation concentration.

The dislocation content of 75-mm-diameter CZ GaAs crystals is generally 10^3–10^4 cm^{-2}. Values for gradient freeze material are a factor of 10 lower [62]. The growth of dislocation-free CZ material has also been reported for small-diameter (15 mm) GaAs crystals.

Resistivity. Values as high as 100 Ω cm can be achieved with silicon. Here the upper limit is set by residual contamination from the high-purity quartz crucible in which the material is grown. Typically, starting silicon for bipolar microcircuits is specified at 3–10 Ω cm p-type (with many manufacturers setting no upper limit), whereas 5–10 Ω cm n- or p-type is used in MOS circuits. The starting material for discrete devices is typically 0.01–0.001 Ω cm of both p- and n-types. Arsenic, antimony, and phosphorus are all used to dope silicon n-type. Boron is used for p-type material.

Melt-grown GaAs is available in heavily doped form. Tellurium, selenium, and silicon are common n-type dopants, in the 10^{17}-to 10^{18}-cm^{-3} range, whereas cadmium and zinc are common p-type dopants in the 10^{16}- to 10^{18}-cm^{-3} range.

Chromium-doped SI GaAs is available with a starting resistivity of 10^8–10^9 Ω cm. Thermal conversion is not considered to be a problem, because the amount of chromium required for this purpose has been reduced to the low 10^{16}-cm^{-3} level [61]. EL2-doped SI GaAs is routinely available in the 10^6- to 10^7 Ω-cm range. EL2 is relatively stable up to 950°C; high temperature hydrogen plasma processing sometimes results in its conversion to p-type because of the formation of As–H bonds.

Radial Resistivity. Variations in the impurity incorporation are related to the shape of the liquid–solid interface during crystal growth, to the crystal orientation, and to the defect structure. The resulting resistivity variation, as expected, is more severe with high-resistivity material. Thus, radial resistivity variations may run as high as ±30 – 50% on high-resistivity silicon slices.

Radial resistivity variations are also a function of the distribution coefficient of the dopant, as pointed out in Section 3.4.3. Typically, for doping concentrations in the 10^{15} Ω-cm^{-3} range, these variations are about ±7.5% for boron doping and ±15% for phosphorus doping for CZ silicon. In integrated circuit processing, a "substrate-adjust implant" is often used to correct for this variation. For bipolar silicon microcircuits, a radial resistivity variation of ±5–10% is the maximum that can be tolerated.

A ±25% variation of radial resistivity is typical in SI GaAs. This is caused by the W-shaped pattern associated with dislocations and with EL2. As long as the minimum resistivity is sufficiently high, this does not present a problem, since the starting material is used as a substrate for the active layer, which is epitaxially grown or ion-implanted. Radial resistivity variations in heavily doped GaAs are typically ±5%.

Oxygen Content. The presence of oxygen in silicon is due to the corrosive action of the melt on the walls of the silicon crucible. Oxygen concentrations of 10^{16}–10^{18} atoms cm^{-3} can be obtained, with the higher values corresponding to higher spin rates. Most of the oxygen is present in the form of clumps and platelets, while some is in electronically active, interstitial sites. Some oxygen is desirable because it pins dislocations, strengthens the material, and getters metallic impurities. On the other hand, a high oxygen content is usually accompanied by a large dislocation concentration. As a consequence, it is necessary to minimize variations in oxygen concentration along the length of the boule. At the present time, oxygen concentrations of 1–1.4 $\times 10^{18}$ cm^{-3}, with a variation of 5% along the boule length, are required for VLSI applications.

The behavior of oxygen in GaAs has been described in Section 1.5.1.3. Its presence is undesirable; fortunately, it can be avoided during crystal growth, since GaAs reacts only slightly with the walls of the silica container. In addition, the use of PBN-lined, graphite crucibles has almost completely eliminated this problem.

3.7 SOLUTION GROWTH

The basic techniques of crystal growth by freezing from a melt have the disadvantage that this melt must be cooled from a relatively high temperature (1412°C for silicon and 1238°C for gallium arsenide). An alternative approach, which is an extension of this technique, is to grow the crystal from a solution of the desired material in a solvent. The underlying principle can be described by means of the simple phase diagram of Fig. 3.13 for the semiconductor B (silicon or GaAs) and its solvent A. The solidus region (β) has been greatly enlarged in this figure, for illustrative purposes. Let T_A and T_B be the melting points of A and B respectively. Then, the semiconductor B can be grown (in β-form) by freezing a solution of composition C_M, which is liquid at a temperature that is much lower than T_B. Tin is a convenient solvent for silicon since it is inert in this semiconductor, although both gallium and aluminum can be used, if p^+-type silicon is desired. Gallium is suitable for GaAs, since it is one of its constituent elements. Tin is used when n^+-type GaAs is required.

Single crystals of B can be grown in this manner so that it is, in theory, a low-temperature alternative to melt growth. However, growth rates for this technique must be kept extremely small, in order to avoid constitutional supercooling. Consequently, it is restricted to the growth of thin epitaxial films where the low growth rate becomes an advantage, since it allows control of the layer

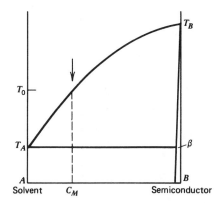

Fig. 3.13 Phase diagram for solution growth.

thickness. This technique, referred to as *liquid phase epitaxy*, has been used extensively with GaAs, and is considered in some detail in Chapter 5. Its use with silicon has been limited to special situations [63].

3.8 ZONE PROCESSES

Zone processes can be used to grow material with higher purity than is normally obtained by CZ or gradient freeze techniques [64]. They require a bar of cast starting material and hence are more expensive than melt growth techniques. Their use is thus restricted to situations where this increased cost is justified.

In silicon technology, a bar of this semiconductor is held in a vertical position, and rotated during the operation. A small zone of the crystal is kept molten by means of a r.f. heater which is moved so that it can traverse the length of the bar. Alternatively, the system can be moved while the r.f. heater is held stationary. A seed crystal is provided at the starting point where the molten zone is initiated and arranged so that its end is just molten. As the zone traverses the bar, single-crystal silicon freezes at its retreating end and grows as an extension of the seed crystal. Figure 3.14 shows a schematic of the necessary equipment for the growth of float zone (FZ) silicon. As with the CZ process, the bar is enclosed in a cooled silica envelope in which an inert atmosphere is maintained.

Zone processes have also been used for GaAs growth. This technique, combined with liquid encapsulation, is potentially capable of producing large diameter ingots, but has met with only partial success at the present time [65]. Serious problems associated with instabilities of the molten zone due to temperature gradients, and corresponding variations in the melt composition, have limited the growth of this material to 38 mm in diameter.

The primary advantage of FZ silicon over CZ material is that its oxygen content can be lowered by a factor of 10–50. For many applications, this is actually

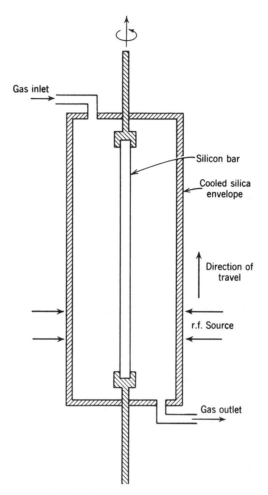

Fig. 3.14 Float zone apparatus.

a disadvantage, since dissolved oxygen plays an important role in strengthening the silicon, and also in the gettering of impurities. On the other hand, the low oxygen concentration results in silicon which is thus relatively free of oxygen clumps. In addition, the absence of a crucible allows the growth of material with less dissolved impurities in it. As a consequence, FZ silicon is used extensively for semiconductor devices which must support reverse voltages in excess of 750–1000 V for these reasons [66]. Recently, it has received attention for use in VLSI schemes where the requirement for clump-free material is considerably more stringent than can be met by the CZ method.

The problems of maintaining a large-diameter molten zone, which is mechanically stable, are quite formidable. Consequently, the FZ apparatus has been modified by the inclusion of a "needle-eye" r.f. heater, which allows growth of a large-diameter crystal, with a molten zone of much smaller dimensions. Crystals as large as 125 mm in diameter have been grown using this technique. FZ silicon is commercially available at the present time in diameters up to 100 mm.

Float zone processes have been developed to take advantage of techniques used for minimizing dislocations in CZ growth. Here, too, a small-diameter seed crystal is used. Growth is initiated by necking this down until dislocations can work out of the crystal. Next, growth parameters are altered so as to bring the crystal to its full diameter, while still maintaining a small molten zone. A schematic of this approach is shown in Fig. 3.15 [67]. Here, mechanical stabilizers are used to prevent the crystal from toppling over during the growth process.

Zone processes are primarily used for refining silicon, which is subsequently doped. These processes are now described.

3.8.1   Zone Refining

In this process, the charge is usually in the form of a cast silicon rod, of doping concentration C_M (atom fraction), with a seed crystal juxtaposed at one end. The molten zone is initiated at this end, and passes along the bar. Figure 3.16 shows an elementary model for this process. Here, let

$L =$ length of the molten zone, at a distance x along the bar,
$S =$ amount of solute present in the molten zone at any given time
$A =$ cross-section area of the bar

As the zone advances by dx, the amount of solute added to it at its advancing end is $C_M A \, dx$. The amount of solute removed from it at the retreating end is $(k_e S/L) \, dx$, where k_e is the effective distribution coefficient. Thus,

$$dS = C_M A \, dx - \frac{k_e S}{L} dx \qquad (3.23)$$

This equation is subject to the boundary value

$$S = C_M AL \quad \text{at } x = 0 \qquad (3.24)$$

Solving, gives

$$S = \frac{C_M AL}{k_e} [1 - (1 - k_e)e^{-k_e x/L}] \qquad (3.25)$$

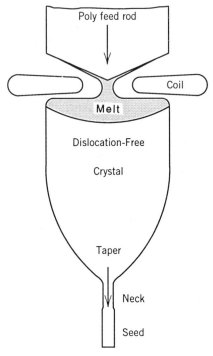

Fig. 3.15 Zone refining with a "needle-eye" r.f. heater. From Kramer [67]. Reprinted with permission from *Solid State Technology*.

But C_S, the concentration of the solute in the crystal at the retreating end, is given by

$$C_S = \frac{k_e S}{AL} \qquad (3.26)$$

so that

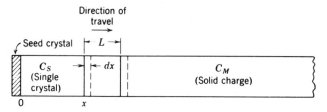

Fig. 3.16 Model for zone refining.

$$C_S = C_M[1 - (1 - k_e) e^{-k_e x/L}] \tag{3.27}$$

Figure 3.17 shows the doping concentration as a function of distance, for a relatively high value of k_e (=0.1) and for a single pass of the zone along the length of the bar. Note that this process is ideally suited to crystal refinement. This is seen by the relative compositions obtained in a series of successive passes of the zone, also shown in this figure. This is often accomplished by a zone reversal process. In this method the direction of the zone is reversed after it has traversed the bar. Thus deviations in composition resulting from the first pass are compensated by the reverse pass. This results in very uniform compositions, except for the two end zones where all the remaining impurities eventually freeze out.

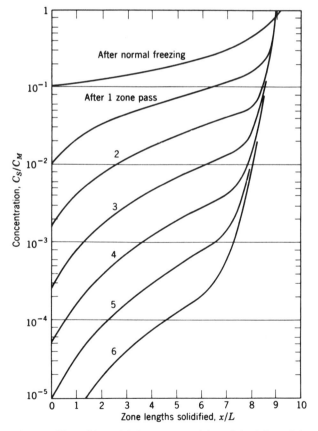

Fig. 3.17 Doping profiles with multiple passes and $k_e = 0.1$. Adapted from Bridges et al. [44].

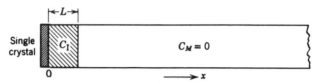

Fig. 3.18 Model for zone doping.

3.8.2 Doping by Zone Leveling

Zone leveling is a technique by which a zone-refined bar can be doped. Consider a bar of undoped material with a molten zone of length L. Initially, let this zone contain a charge of doping concentration C_I. A seed crystal is also juxtaposed, as illustrated in Fig. 3.18. Using the same notation as that of Section 3.8.1, we may rewrite Eq. (3.23) as

$$dS = -\frac{k_e S}{L} dx \tag{3.28}$$

since $C_M = 0$. At $x = 0$,

$$S = C_I A \rho L \tag{3.29}$$

Solving Eq. (3.28) and inserting this boundary value gives

$$S = C_I A L e^{-k_e x/L} \tag{3.30}$$

But

$$C_S = \frac{k_e S}{AL} \tag{3.31}$$

hence

$$C_S = k_e C_I e^{-k_e x/L} \tag{3.32}$$

From this equation, it is seen that, for low values of k_e, very uniformly doped crystals may be obtained in a single pass.

3.8.3 Neutron Transmutation Doping

A major problem with FZ material is the presence of radial microresistivity variations. These take the form of large, abrupt resistivity changes of $\pm 7\%$ to $\pm 15\%$, which occur over intervals $10-100$ μm wide and can be detected by high-resolution, spreading-resistance probe techniques. These variations are generally considered to be caused by a misalignment of the rotational axis of the crystal and the thermal axis of symmetry of the crystal growth apparatus. They can be reduced, but not eliminated, by careful mechanical design of the system.

Microresistivity variations can be virtually eliminated by a technique known as *neutron transmutation doping* (NTD) [68]. This approach is based on the fact that bombardment of silicon with thermal neutrons gives rise to the formation of finite amounts of phosphorus, which is an *n*-type dopant. Specifically, the ^{30}Si isotope form occurs as a component of native silicon, with a concentration of about 3%. Neutron bombardment causes this ^{30}Si to change to ^{31}Si, liberating gamma rays; simultaneously, the ^{31}Si undergoes a transmutation to phosphorus (^{31}P) with the liberation of beta rays, having a half-life of 2.62 hr. As a consequence, the technique is limited to *n*-type silicon.

A secondary process results in the conversion of ^{31}P to ^{32}P, which transmutes to ^{32}S. This reaction has a half-life of 14.3 days and is present in significant quantities only for heavy doping. Typically, neutron doping down to $5-10 \, \Omega$-cm can be accomplished with a cool-down period of $3-4$ days, and is completely safe by international standards.

The penetration of thermal neutrons in silicon falls off exponentially with distance, and has a characteristic length of about 19 cm. Variations in resistivity are thus minimized by rotating the crystal while it is being irradiated. In this manner, a ±0.5% resistivity variation can be obtained across silicon crystals of 100-mm diameter [69]. Thus, starting with slices of FZ silicon having an average resistivity well in excess of what is required, it is possible to obtain uniformly doped slices, free from both macro- and microresistivity variations.

Annealing for as little as 2 minutes at 750°C has been found to be sufficient to stabilize the neutron-doped silicon against resistivity changes [70, 71]. Radiation damage of crystals does not present a problem, since normal crystal processing during device fabrication suffices to anneal out all defects created in this manner.

The NDT process is used extensively for high-voltage semiconductor devices at the present time. However, this technique has also been used to control the resistivity of *n*-type epitaxial layers that are required for bipolar VLSI circuits.

3.9 PROPERTIES OF ZONE-PROCESSED CRYSTALS

Because of its two step nature (i.e., from the melt to cast rod to single crystal), the zone process is more expensive than the CZ process. Only silicon is available in zone-processed form at the present time. NTD doping is used extensively in conjunction with this process.

The low oxygen content of FZ material results in high-temperature properties which are inferior to those of CZ material. Thus, dislocation generation during the post-solidification step is a relatively severe problem with this material.

Dislocation Content. Typically, values range from 10^3 to 10^4 dislocations/cm^2. This is because the region near the retreating solid–liquid interface (where the crystal is grown) is highly stressed owing to the weight of the molten zone. New techniques, using single-turn "needle-eye" r.f. heaters, results in a reduction in the dislocation content by a factor of $10-100$.

Resistivity. The residual impurity concentration of FZ material can be kept to below 2×10^{12} cm^{-3}, since no outside matter is added during growth and the chamber can be made quite free from contamination. In fact, it is possible to conduct the process in a vacuum and purify the crystal of volatile impurities such as zinc. Finally, repeated processing is possible by successive passes and reversals of the molten zone. As a consequence, silicon with resistivities in excess of 15,000 Ω-cm for *n*-type material (and 45,000 Ω-cm for *p*-type material) can be obtained by this process.

Radial Resistivity. Variations of this parameter are comparable to those obtained with CZ crystals in the same resistivity range. More serious, however, is the presence of microresistivity variations. The use of neutron doping can reduce variations in radial resistivity to under $\pm 1\%$ for 5- to 10 Ω-cm silicon, which is useful as a starting material for MOS microcircuits. Note, however, that this process is unique to *n*-type doping.

Oxygen Content. Float zone silicon can be grown with 1–2 decades less oxygen than CZ material. Although primarily used for high-voltage devices, this material, in combination with neutron doping, is an attractive candidate for MOS-based VLSI circuits.

TABLE 3.1   Residual Impurities in Silicon [1]

Element	Concentration (cm^{-3})			
	Polycrystal	CZ	Magnetic CZ	FZ
Group III	$< 1 \times 10^{12}$	$5 \times 10^{13} - 10^{14}$	5×10^{12}	1×10^{12}
Group V	$< 1 \times 10^{13}$			2×10^{12}
Heavy metals	$< 0.5 \times 10^{13}$			$< 10^{12}$
Oxygen	$< 10^{15}$	10^{18}	10^{17}	$< 10^{15}$
Carbon	$< 10^{15}$	$< 10^{16}$	$< 10^{16}$	$< 10^{15}$
Others, each	$< 5 \times 10^{10}$			

TABLE 3.2   Residual Impurities in Gallium Arsenide [3]

Element	Concentration (cm^{-3})	Element (cm^{-3})	Concentration (cm^{-3})
B	4.42×10^{15}	Cl	1.33×10^{15}
C	3.1×10^{16}	Ca	3.54×10^{16}
N	4.42×10^{15}	Cr	3.09×10^{14}
O	2.21×10^{16}	Mn	2.21×10^{14}
F	8.84×10^{15}	Fe	8.84×10^{14}
Mg	8.84×10^{14}	Ni	1.77×10^{15}
Al	8.84×10^{14}	Co	1.33×10^{14}
Si	8.84×10^{14}	Cu	4.42×10^{14}
P	4.42×10^{15}	Zn	2.21×10^{15}
S	4.42×10^{14}	Cd	8.84×10^{14}

TABLE 3.3   Distribution Coefficients of Impurities in Silicon [40, 41]

Dopant	Type[a]	Atomic Weight	Distribution Coefficient
As	n	74.92	0.3
P	n	30.97	0.35
Sb	n	121.75	0.02
Al	p	26.98	2×10^{-3}
B	p	10.81	0.72
Ga	p	69.72	7.2×10^{-3}
In	d	114.82	3.6×10^{-4}
Ag	d	107.87	1.7×10^{-5}
Au	d	196.97	2.25×10^{-5}
Co	d	58.93	2×10^{-5}
Cr	d	52.00	1.1×10^{-5}
Cu	d	64.54	8×10^{-4}
Fe	d	55.85	6.4×10^{-6}
Mn	d	54.94	1.5×10^{-5}
Ni	d	58.71	1.3×10^{-4}
O	d	16.00	1.25
Pd	d	106.50	5×10^{-5}
Sn	Neutral	118.69	3.2×10^{-2}
Ti	d	47.90	2×10^{-6}
V	d	50.94	4×10^{-6}

[a]d = deep lying.

TABLE 3.4   Distribution Coefficients of Impurities in Gallium Arsenide [42]

Dopant	Type	Atomic Weight	Distribution Coefficient
S	*n*	32.06	0.3–0.5
Se	*n*	78.96	0.1–0.3
Te	*n*	127.6	0.03–0.059
Sn	*n/p*	118.69	0.08
C	*n/p*	12.01	0.2–0.8
Ge	*n/p*	72.59	0.01
Si	*n/p*	28.09	0.14–2
Be	*p*	9.01	3
Mg	*p*	24.31	0.1
Zn	*p*	65.37	0.4–1.9
Cr	SI	52.00	5.7×10^{-4}
Fe	SI	55.85	1×10^{-3}

REFERENCES

1. L. D. Crossman and J. A. Baker, *Semiconductor Silicon*, H. R. Huff and E. Sirtl, eds., The Electrochemical Society, Princeton, NJ, 1977, p. 18.

2. J. van den Boomgaard and K. Scholl, The P–T–x Phase Diagrams of the Systems In–As, Ga–As, and InP, *Philips Res. Rep.* **12**, 127 (1957).

3. J. Krauskopf, J. D. Meyer, B. Wiedemann, M. Waldschmidt, K. Bethge, G. Wolf, and W. Schültze, Impurity Analysis of Gallium Arsenide, in *5th Conference on Semi-insulating III–V Materials, Malmo, Sweden, 1988*, G. Grossman and L. Ledebo, Eds., Adam-Hilger, New York, 1988, p. 165.

4. G. K. Teal and J. B. Little, Growth of Germanium Single Crystals, *Phys. Rev.* **78**(5), 647 (1950).

5. R. N. Thomas, H. M. Hobgood, P. S. Ravishankar, and T. T. Braggins, Melt Growth of Large Diameter Semiconductors: Part 1, *Solid State Technol.* (March 1990) p. 83.

6. H. Hirata and K. Hoshikawa, The Dissolution Rate of Silica in Molten Silicon, *Jpn. J. Appl. Phys.* **19**, 1573 (1980).

7. J. S. C. Chang, D. E. Mcall, and E. H. Wong, Quartz vs. PBN—The Effect of Crucible Type on Undoped LEC GaAs, *IEEE Trans. Electron Dev.* **ED-32**, 1124 (1985).

8. H. Emori, K. Tereshima, F. Orito, T. Kikuta, and T. Fukuda, Effect of Ambient Gas on LEC Undoped Semi-insulating GaAs, in *Semi-Insulating III–VI Compounds*, D. C. Look and J. S. Blakemore, Eds., Shiva Publishing Ltd., England, 1984, p. 111.

9. W. R. Runyan, *Silicon Semiconductor Technology*, McGraw-Hill, New York, 1965.

10. S. N. Rea, Czochralski Silicon Pull Rate Limits, *J. Cryst. Growth* **54**, 267 (1981).

11. E. P. A. Metz, R. C. Miller, and J. Mazelski, A Technique for Pulling Crystals of Volatile Materials, *J. Appl. Phys.* **33**, 2016 (1962).

12. J. B. Mullin, B. W. Straughan, and W. S. Brickell, Liquid Encapsulation Techniques: The Use of an Inert Liquid in Suppressing Dissociation During the Melt Growth of InAs and GaAs Crystals, *J. Phys. Chem. Solids* **26**, 782 (1965).

13. R. N. Thomas, H. M. Hobgood, G. W. Eldridge, D. L. Barrett, T. T. Braggins, B. L. Ta, and S. K. Wang, High Purity LEC Growth and Direct Implantation of Gallium Arsenide for Monolithic Microwave Circuits in *Semiconductors and Semimetals*, Vol. 20, R. K. Willardson and A. C. Beer, Eds., Academic Press, New York, 1984, p. 1.

14. G. Jacob, A Novel Crystal Growth Method for GaAs: The Liquid Encapsulated Kryopoulos Method, *J. Cryst. Growth* **58**, 455 (1982).

15. T. R. AuCoin, R. L. Ross, M. J. Wade, and R. O. Savage, Liquid Encapsulated Compounding and Czochralski Growth of Semi-insulating Gallium Arsenide, *Sol. State Technol.* (Jan. 1979) p. 59.

16. H. M. Hobgood, G. E. Eldridge, D. L. Barrett, and R. N. Thomas, High Purity Semi-insulating GaAs for Monolithic Microwave Integrated Circuits, *IEEE Trans. Electron Dev.* **ED-28**, 140 (1981).

17. C. A. Stolte, Ion Implantation and Materials for GaAs Integrated Circuits in *Semiconductors and Semimetals*, Vol. 20, R. K. Willardson and A. G. Beer, Eds., Academic Press, New York, 1984, p. 89.

18. W. M. Duncan, G. H. Westphal, and J. B. Sherer, A Direct Comparison of LEC GaAs Grown Using Low- and High-Pressure Techniques, *IEEE Electron Dev. Lett.* **EDL-4**, 199 (1983).

19. C. H. L. Goodman, Ed., *Crystal Growth: Theory and Techniques*, Vol. 1, Plenum, New York, 1974.

20. N. B. Hannay, Ed., *Semiconductors*, Reinhold, New York, 1959.

21. T. R. AuCoin and R. O. Savage, in *Gallium Arsenide Technology*, Vol. I, D. K. Ferry, Ed., Howard W. Sams and Co., Indianapolis, IN, 1985, p. 47.

22. 3D Precision Heat Systems, The Mellen Company Inc., New Hampshire, USA.

23. P. D. Greene, Growth of GaAs Ingots with High Free Electron Concentrations, *J. Cryst. Growth* **50**, 612 (1980).

24. W. A. Gault, E. M. Monberg, and J. E. Clemens, A Novel Application of the Vertical Gradient Freeze Method to the Growth of High Quality III–V Crystals, *J. Crys. Growth* **74**, 491 (1986).

25. W. Wijaranakula, Solubility of Interstitial Oxygen in Silicon, *Appl. Phys. Lett.* **59**, 1185 (1991).

26. R. A. Craven and H. W. Korb, Internal Gettering in Silicon, *Solid State Technol.* (July 1981) p. 55.

27. T. Aoshima, Y. Kosaka, and A. Yoshinaka, Soft Intrinsic Gettering for VLSI Process, in *Semiconductor Silicon*, H. R. Huff, K. G. Barraclough, J. -I. Chikawa, Eds., The Electrochem. Society, Pennington, NJ, 1990, p. 724.

28. J. Lagowski, H. C. Gatos, T. Aoyama, and D. G. Lin, On the Dislocation Density in Melt-Grown GaAs, in *Semi-insulating III–V Materials*, D. C. Look and J. S. Blakemore, Eds., Shiva Publishing Ltd., England, 1984, p. 60.

29. A. J. R. de Kock and W. M. van de Wijgert, The Effect of Doping on the Formation of Swirl Defects in Dislocation-Free Czochralski-Grown Silicon, *J. Cryst. Growth* **49**, 718 (1980).

30. A. S. Jordan, R. Caruso, and A. R. Von Neida, A Thermoelastic Analysis of Dislocation Generation in Pulled GaAs Crystals, *Bell Syst. Tech. J.* **59**, 593 (1980).

31. M. Duseaux, Temperature Profile and Thermal Stress Calculations in GaAs Crystals Growing from the Melt, *J. Cryst. Growth* **61**, 576 (1983).

32. S. P. Timoshenko and J. N. Goodier, *The Theory of Elasticity*, McGraw-Hill, New York, 1970.

33. J. C. Brice, *Crystal Growth Processes*, John Wiley and Sons, New York, 1986.

34. T. Suzuki, N. Isawa, Y. Okubo, and K. Hoshi, in *Semiconductor Silicon*, H. R. Huff, R. J. Kreigler, and Y. Takeishi, Eds., The Electrochemical Society, Inc., Pennington, NJ, 1981, p. 90.

35. K. Tereshima, T. Katsumata, F. Orito, T. Kikuta, and T. Fukuda, Electrical Resistivity of Undoped GaAs Single Crystal Grown by Magnetic Field Applied LEC Techniques, *Jpn. J. Appl. Phys.* **22**, 401 (1983).

36. D. T. J. Hurle and R. W. J. Series, Effective Distribution Coefficients in Magnetic Czochralski Growth, *J. Cryst. Growth* **73**, 1 (1985).

37. R. G. Rhodes, *Imperfections and Active Centres in Semiconductors*, Pergamon, New York, 1964.

38. W. C. Dash, Silicon Crystals Free of Dislocations, *J. Appl. Phys.* **29**, 736 (1958).

39. H. M. Hobgood, R. N. Thomas, D. L. Barret, G. W. Eldridge, M. M. Sopira, and M. C. Driver, Large Diameter Low Dislocation In-Doped GaAs: Growth, Characterization and Implications for FET Fabrication, in *Semi-insulating GaAs*, D. C. Look and J. S. Blakemore, Eds., Shiva Publishing Ltd., England, 1984, p. 149.

40. F. A. Trumbore, Solid Solubilities of Impurity Elements in Germanium and Silicon, *Bell Syst. Technol. J.* **39**, 205 (1960).

41. R. H. Hopkins and A. Rohatgi, Impurity Effects in Silicon High Efficiency Solar Cells, *J. Cryst. Growth* **75**, 67 (1986).

42. A. G. Milnes, *Deep Impurities in Semiconductors*, John Wiley and Sons, New York, 1973.

43. V. Swaminathan and A. T. Macrander, *Materials Aspects of GaAs and InP Based Structures*, Prentice-Hall, Englewood Cliffs, NJ, 1991.

44. H. E. Bridges et al., *Transistor Technology*, Vol. 1, Van Nostrand, New York, 1958.

45. H. Kodera, Diffusion Coefficients of Impurities in Silicon Melts, *Jpn. J. Appl. Phys.* **2**, 212 (1963).

46. J. A. Burton, R. C. Prim, and W. P. Slichter, The Distribution of Solute in Crystals Grown from the Melt, Part I. Theoretical, *J. Chem. Phys.* **21**, 1987 (1953).

47. T. Abe, K. Kikuchi, S. Shirai, and S. Muraoka, Impurities in Silicon Crystals—A Current View, in *Semiconductor Silicon*, H. R. Huff, R. J. Kreigler, and Y. Takeishi, Eds., The Electrochemical Society, Inc., Pennington, NJ, 1981, p. 54.

48. K. M. Kim, Interface Morphological Instability in Czochralski Silicon Crystal Growth from Heavily Doped Sb-Melt, *J. Electrochem. Soc.* **126**, 875 (1979).

49. J. W. Whelan and G. H. Wheatley, The Preparation and Properties of Gallium Arsenide Single Crystals, *J. Phys. Chem. Sol.* **6**, 169 (1958).

50. N. G. Ainsley, S. E. Blum, and J. F. Woods, On the Preparation of High Purity Gallium Arsenide, *J. Appl. Phys.* **33**, 2392 (1962).

51. J. R. Oliver, R. D. Fairman, R. T. Chen, and P. W. Yu, Undoped Semi-insulating LEC GaAs: A Model and a Mechanism, *Electron Lett.* **17**, 839 (1981).

52. A. T. Hunter, H. Kimura, J. B. Baukus, H. V. Winston, and O. J. Marsh, Carbon in Semi-Insulating Liquid Encapsulated Czochralski GaAs, *Appl. Phys. Lett.* **44**, 74 (1984).

53. S. Makram-Ebeid, P. Langdale, and G. M. Martin, Nature of EL2: The Main Native Midgap Electron Trap in VPE and Bulk GaAs, in *Semi-insulating III–V Materials*, D. C. Look and J. S. Blakemore, Eds., Shiva Publishing Ltd., England, 1984, p. 184.

54. R. Elliot, R. T. Chen, S. G. Greenbaum, and R. J. Wagner, Identification of As_{Ga} Antisite Defects in LEC GaAs, in *Semi-insulating III–V Compounds*, D. C. Look and J. S. Blakemore, Eds., Shiva Publishing Ltd., England, 1984, p. 239.

55. A. Goltzene, B. Meyer, and C. Schwab, Electron Paramagnetic Resonance Determination of the Generation Rate of As Antisites in Fast Neutron Irradiated GaAs, *J. Appl. Phys.* **54**, 3117 (1983).

56. J. Lagowski, M. Kaminska, J. M. Parsey, H. C. Gatos, and W. Walukiewicz, Microscopic Model of the EL2 Level in GaAs, *Inst. Phys. Conf. Ser.* **65**, 41 (1983).

57. B. K. Meyer, D. M. Hoffman, J. R. Niklas, and J. M. Spaeth, Arsenic Antisite Defect As_{Ga} and EL2 in GaAs, *Phys. Rev.* **B36**, 1332 (1987).

58. D. E. Holmes, R. T. Chen, K. R. Elliot, C. G. Kirkpatrick, and P. W. Yu, Compensation Mechanism in Liquid Encapsulated Czochralski GaAs, *IEEE Trans. Electron. Dev.* **29**, 1045 (1982).

59. J. Lagowski, H. C. Gatos, J. M. Parsey, K. Wada, M. Kaminska, and W. Walukiewicz, Origin of the 0.82 Electron Trap in GaAs and Its Annihilation by Shallow Donors, *J. Appl. Phys.* **342** (1982).

60. G. M. Martin, J. P. Farges, G. Jacob, J. P. Hallais, and G. Poiblaud, Compensation Mechanism in GaAs, *J. Appl. Phys.* **51**, 2840 (1980).

61. P. F. Lindquist, A Model Relating Electrical Properties and Impurity Concentrations in Semi-Insulating GaAs, *J. Appl. Phys.* **48**, 1262 (1977).

62. J. E. Clemans and J. H. Conway, Vertical Gradient Freeze Growth of 75 mm Diameter Semi-Insulating GaAs, in 5th Conference on Semi-insulating III–V Materials, Malmo, Sweden, 1988, G. Grossmann and L. Ledebo, Eds., Adam Hilger, New York, 1988, p. 423.

63. B. J. Baliga, Ed., *Epitaxial Silicon Technology*, Academic Press, New York, 1986.

64. W. G. Pfann, *Zone Melting*, John Wiley and Sons, New York, 1958.

65. E. S. Johnson, Liquid Encapsulated Floating Zone Melting of GaAs, *J. Cryst. Growth* **30**, 249 (1975).

66. S. K. Ghandhi, *Semiconductor Power Devices*, John Wiley and Sons, New York, 1977.

67. H. G. Kramer, Float-Zoning of Semiconductor Silicon: A Perspective, *Solid State Technol.* (Jan. 1983) p. 137.

68. H. A. Herrmann and H. Herzer, Doping of Silicon by Neutron Radiation, *J. Electrochem. Soc.* **122**, 1568 (1975).

69. J. M. Meese, Ed., *Neutron Transmutation Doping in Semiconductors*, Plenum, New York, 1979.

70. H. M. Janus and O. Malmros, Application of Thermal Neutron Irradiation for Large Scale Production of Homogeneous Phosphorus Doping of Float Zone Silicon, *IEEE Trans. Electron Dev.* **ED-23**, 797 (1976).

71. A. Senes and G. Sifre, Stabilization of Transmutation Doped Silicon, in *Semiconductor Silicon, 1977*, H. R. Huff and E. Sirtl, Eds., The Electrochemical Society, Princeton, NJ, 1977, p. 135.

PROBLEMS

1. Diffusions into GaAs are often carried out in a sealed ampoule with an arsenic overpressure to prevent decomposition. Junction depths for diffusions in GaAs are a function of partial pressures, set up in this ampoule.

   Zinc diffusions are to be carried out at 800°C in an ampoule having a volume of 100 ml. The arsenic overpressure is to be 0.2 atm (arsenic exists almost entirely as As_2 at 800°C). Calculate the amount of arsenic that is to be placed in the ampoule. What would happen if you omitted the arsenic?

2. It is desired to grow gold-doped and boron-doped crystals by the CZ method, such that the concentration of dopant in each case is 10^{17} atoms/cm^3 when one-half of the crystal is grown. Assuming that growth is undertaken under rapid-stirring conditions, compare the relative amounts of gold and boron that should be added to a pure silicon melt on an atom ratio basis. Comment on the feasibility of this approach to doping.

3. An antimony-doped crystal is required to have a resistivity of 1.0 Ω cm when one half of the crystal is grown. Assuming that a 1000-g pure silicon charge is used, what is the amount of 0.01-Ω-cm antimony-doped silicon that must be added. Note that the mass of one atom of hydrogen is 1.6×10^{-24} g. The electron mobility should be assumed to be 150 cm^2/V s.

4. A CZ furnace is loaded with 60 kg of silicon, to which 10 mg of boron is added. Determine the doping concentration throughout the length of the boule, assuming rapid stirring conditions.

5. A small silicon bar is doped with both boron and antimony. One end of the bar is inserted into a furnace and melted. The bar is now withdrawn. Sketch the doping profile that results from the regrowth process and calculate the base width of the resulting transistor. Assume equilibrium values for k, and that the original antimony/boron atom ratio was 33.

6. Neglecting the effects of impurity distribution in the final zone, show that the limiting value of the solute concentration in a zone-refined crystal of length l is given by:

$$C_S = Ae^{Bx}$$

where

$$A = \frac{C_M Bl}{e^{Bl} - 1}$$

$$k = \frac{BL}{e^{BL} - 1}$$

Here L is the length of the molten zone, and k is the distribution coefficient.

CHAPTER 4

DIFFUSION

Diffusion is a relatively straightforward process by which impurities may be introduced into selected regions of a semiconductor, for the purpose of altering its electronic properties [1, 2]. Both single and multiple diffusion steps can be used for this purpose. The laws governing the diffusion of impurities [3] were established many years before the development of semiconductors. It is only since reasonably defect-free materials of high purity have become available, however, that the method has become generally accepted for semiconductor device fabrication. At the present time, diffusion is a basic process step in the fabrication of both discrete devices and microcircuits.

In silicon technology, diffusion allows the formation of (a) sources and drains for metal-oxide-semiconductor (MOS) devices and (b) the active regions of bipolar transistors. It is extensively used because it is ideally adapted to batch processes where many slices are handled in a single operation. Unlike alternative technologies such as ion implantation, it does not produce crystal damage; thus high-quality junctions, with a minimum leakage current, can be made easily by this method.

Diffusions are also used in GaAs technology, but not as extensively as for silicon. One reason is that many advanced GaAs structures use juxtaposed layers of different materials, such as AlGaAs, and require epitaxial processes. Additionally, the behavior of diffusion in GaAs is more complex, and less well understood than in silicon. As a result, diffusion can only be controlled precisely in special situations. Thus, it is not the general-purpose, flexible process which it is with silicon technology.

There are, however, a number of unique application areas for diffusion in GaAs. Of these, perhaps the most interesting lies in its use as a technique for disordering superlattices. Here, it has been shown [4] that the diffusion of impurities can disorder a superlattice, converting it into an alloy of the same average composition. This discovery has important applications in the isolation of devices in optical integrated circuits, and has greatly increased our interest and understanding of diffusion processes in GaAs and other compound semi-conductors.

A number of simplifying assumptions can be made in the development of diffusion theory as it applies to semiconductor device fabrication. Thus the single-crystal nature of the solid in which diffusion takes place often allows the effects of grain boundaries to be ignored. Moreover, since a very small number of impurities are involved, dimensional changes during the diffusion process are not considered. Finally, the almost exclusive use of parallel-plane device and circuit structures has allowed the initial development of this tool, and our understanding of the basic physics of diffusion, by relatively simple mathematical methods. More recently, computer-aided methods have extended the application of these physical principles to situations where two-dimensional (and sometimes three-dimensional) effects are important.

All tables of diffusion parameters are provided at the end of this chapter for easy reference purposes.

4.1 THE NATURE OF DIFFUSION

Diffusion describes the process by which atoms move in a crystal lattice. Although this includes self-diffusion phenomena, our goal is to study the dif-fusion of impurity atoms that are introduced into the lattice for the purpose of altering its electronic properties. In addition to concentration gradient and temperature, crystal structure and defect concentration play an important part in this process.

The wandering of an impurity in a lattice takes place in a series of random jumps [5]. These jumps occur in all three dimensions, and a flux of diffusing species results if there is a concentration gradient. The mechanisms by which jumps can take place are now outlined.

Interstitial Diffusion. This is illustrated in Fig. 4.1, where an impurity atom moves through the crystal lattice by jumping from one interstitial site to the next. Interstitial diffusion requires that jump motion occur from one interstitial site to another adjacent interstitial site. This process is relatively fast, because of the large number of vacant sites of this type in a semiconductor. Impurities such as sodium and lithium move in silicon by this mechanism.

Substitutional Diffusion. Here (see Fig. 4.2) an impurity atom wanders through the crystal by jumping from one lattice site to the next, thus substituting for the original host atom. However, it is necessary that this adjacent site be vacant; that is, vacancies must be present to allow substitutional

Fig. 4.1 Diffusion by the interstitial mechanism.

diffusion to occur. These are provided by the mechanisms discussed in Section 1.5.1. Since the equilibrium concentration of vacancies is quite low, it is reasonable to expect substitutional diffusion to occur at a much slower rate than interstitial diffusion. Many dopants used in silicon microcircuit fabrication are substitutional diffusers.

Divacancies are also present in the semiconductor, so that diffusion can be accomplished by movement into these sites. As indicated in Section 1.5, their energy of formation is somewhat lower than that for isolated vacancies. The diffusion of *n*-type impurities in GaAs has been ascribed to this type of movement.

Interstitial-Substitutional Diffusion. In this case, impurity atoms occupy substitutional as well as interstitial sites. However, they only move at a significant rate when in interstitial sites (by interstitial diffusion). The *dissociative* mechanism, by which a substitutional impurity atom can become an interstitial, leaving behind a vacancy, can be the controlling factor for this process. As a result, the effective diffusivity is a function of the dissociation rate, and depends on both impurity concentration and crystal quality. Figure 4.3 illustrates this mechanism in schematic form. Copper and nickel move in silicon by this mechanism, as do zinc, cadmium, and copper in GaAs.

An alternative pathway for interstitial diffusion is the *kick-out* mechanism illustrated in Fig. 4.4. Here, a rapidly moving interstitial diffuser can move into

Fig. 4.2 Diffusion by the substitutional mechanism.

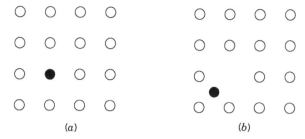

Fig. 4.3 Interstitial–substitutional diffusion by the dissociative mechanism.

a substitutional site by displacing an atom which is already in place, resulting in the formation of a self-interstitial. The behavior of both gold and platinum in silicon has been described by this process.

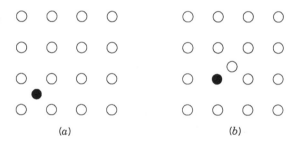

Fig. 4.4 Interstitial–substitutional diffusion by the kick-out mechanism.

Interstitialcy Diffusion. This is a modified version of substitutional diffusion. Here, as shown in Fig. 4.5, interstitial host atoms (self-interstitials) can be annihilated by pushing substitutionally located impurity atoms into

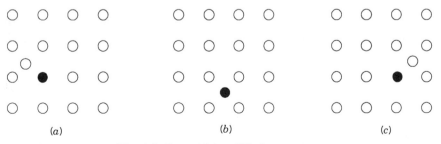

Fig. 4.5 Interstitialcy diffusion process.

interstitial sites. These impurities can now diffuse to adjacent substitutional sites and create new self-interstitials. Thus, the interstitial position of the diffusing impurity atom is purely a transition state, in moving from one substitutional site to another. This increases the concentration of vacant substitutional sites over its equilibrium value, so that interstitialcy diffusion is somewhat faster than substitutional diffusion. All substitutional diffusers move, in part, by this mechanism. The diffusion of boron and phosphorus in silicon is dominated by this process.

Interchange Diffusion. This occurs when two or more atoms diffuse by an interchange process. Such a process is known as a *direct interchange* process when it involves an impurity and host atom, and as a *cooperative interchange* when a larger number of host atoms is involved. The probability of occurrence of interchange diffusion effects is extremely low.

Grain Boundary Diffusion. Diffusion also occurs by movement along dislocations and grain boundaries. This process is anisotropic, being much faster in directions parallel to the dislocation core or to the grain boundary edge than at right angles to it. Furthermore, atom movement is two or more orders of magnitude faster than that obtained by the lattice diffusion processes outlined above. Enhanced diffusion along dislocations often leads to emitter–collector shorts in bipolar devices [6], by the formation of "diffusion pipes".

Combination Effects. Combinations of these mechanisms may occur within a crystal. For example, a certain fraction of impurity atoms may diffuse substitutionally and the rest interstitially, resulting in a two-stream process. With many substitutional diffusers, it is necessary to assume that diffusion proceeds in part by a substitutional mechanism, and partly by an interstitialcy mechanism.

Finally, it should be emphasized that diffusion describes the movement of impurities, and not their ultimate destination. Thus, it is possible that some of the diffusing species may end in substitutional sites, with others in interstitial sites.

4.2   DIFFUSION IN A CONCENTRATION GRADIENT

The jumping of impurities in a semiconductor is thermally activated, and random in character. However, directed motion occurs, by diffusion, in the presence of a concentration gradient. The net flux of diffusing species is now determined for this situation, for impurities which diffuse substitutionally in the zincblende lattice. However, the arguments have broad general application to all types of diffusion mechanisms.

Consider a crystal of cross section A, divided up by a series of parallel planes at right angles to the [100] axis. (Analysis of impurity motion in other directions follows along similar lines of reasoning.) Let the spacing between planes be

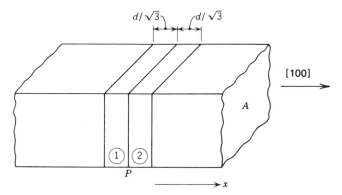

Fig. 4.6 Diffusion due to a concentration gradient.

$d/\sqrt{3}$, where d is the distance between tetrahedral sites. Figure 4.6 depicts this arrangement, with the crystal divided into layers 1, 2, and so on, of this width. Thus, each layer contains atoms (impurity and host) whose tetrahedral neighbors are in adjacent layers on either side; substitutional diffusion takes place by jump motion between tetrahedral sites. A single jump has projections of length $d/\sqrt{3}$ along the [100] direction.

Let n_1 and n_2 be the number of atoms in layers 1 and 2, respectively, and let N_1 and N_2 be the volume concentrations. Then

$$N_1 = \frac{\sqrt{3}n_1}{Ad} \tag{4.1a}$$

$$N_2 = \frac{\sqrt{3}n_2}{Ad} \tag{4.1b}$$

For diffusion in this lattice, atoms in any one layer must jump into neighboring layers. Each atom has four such neighbors, two in the layer to the left and two in the layer to the right. Let ν be the jump frequency* for these atoms. Then, in time $1/\nu$, half of the moving atoms jump right while the other half jump left, on the average.

The net flow of atoms across the plane P in a direction x is given by

$$\frac{\Delta n}{\Delta t} = \frac{(n_1 - n_2)/2}{1/\nu}$$

$$= \frac{\nu}{2}\frac{Ad}{\sqrt{3}}(N_1 - N_2) \tag{4.2}$$

*The calculation of this parameter is outlined in Section 4.2.3.

But

$$\frac{\Delta N}{\Delta x} = \frac{N_2 - N_1}{d/\sqrt{3}}$$ (4.3)

so that

$$\frac{\Delta n}{\Delta t} = -A\frac{\nu d^2}{6}\frac{\Delta N}{\Delta x}$$ (4.4)

Writing the flux density j as the time rate of change of the number of impurities per unit area, Eq. (4.4) reduces to

$$j = -\frac{\nu d^2}{6}\frac{\Delta N}{\Delta x}$$ (4.5)

By defining a *diffusivity* D such that

$$D = \frac{\nu d^2}{6}$$ (4.6)

we may rewrite Eq. (4.5) in partial derivative form as

$$j = -D\frac{\partial N}{\partial x}$$ (4.7)

where j is the flux density (in atoms cm^{-2}), D the diffusion coefficient (in cm^2 s^{-1}), N the volume concentration (in atoms cm^{-3}), and x the distance (in cm). This is Fick's first law. It states that, under diffusion conditions, the flux density is directly proportional to the concentration gradient.

Similar arguments apply to interstitial diffusers, except that jumps between interstitial voids must be considered. Since each void also has four nearest neighbors, the same line of reasoning applies. However, the jump frequency is much larger for this situation.

4.2.1 Field-Aided Motion

In addition to diffusion, ionized impurities can move by drift in the presence of an electric field. Assume that the random scattering of these impurities can be represented by a restraining force that is directly proportional to the drift velocity. Then, the equation of motion for an impurity atom of valence Z may be written as

$$F = Zq\mathscr{E} = m^* \frac{dv}{dt} + \alpha v \tag{4.8}$$

where F is the force on the impurity ion, Zq its charge, $\mathscr{E}$ the electric field, v the average velocity, m^* the effective mass, and α a factor of proportionality. Note that the sign of the charge on the impurity ion determines the direction of this force. Thus a positive force is exerted on a positively charged impurity in a positive $\mathscr{E}$ field. Its magnitude can be significant at high impurity concentrations.

In steady state a drift velocity v_d is attained, such that

$$v_d = \frac{Zq\mathscr{E}}{\alpha} = \mu\mathscr{E} \tag{4.9}$$

where μ is defined as the mobility of the impurity ion in cm^2/V s.

Assume that the movement of impurities due to drift and diffusion are uncorrelated events. Then, in the presence of an electric field, the net flow of atoms across the plane P in the direction x (see Fig. 4.6) is given by modifying Eq. (4.2) so that

$$\frac{\Delta n}{\Delta t} = \frac{v}{2} \frac{Ad}{\sqrt{3}}(N_1 - N_2) + \mu N \mathscr{E} A \tag{4.10}$$

Thus the flux density is given by

$$\begin{aligned} j &= -\frac{vd^2}{6} \frac{\Delta N}{\Delta x} + \mu N \mathscr{E} \\ &= -D \frac{\partial N}{\partial x} + \mu N \mathscr{E} \end{aligned} \tag{4.11}$$

in partial derivative form.

An electric field can be internally generated during the diffusion of substitutional impurities at high doping levels, if the semiconductor remains extrinsic at diffusion temperatures. That this often occurs in practical situations is seen from Fig. 4.7, which displays the intrinsic carrier concentration for silicon and GaAs as a function of temperature.

The effect of this field is now described for the diffusion of an n-type impurity [7]. Here the impurity is ionized, with an electron concentration n. These electrons have a much higher diffusion rate than the impurity atoms and tend to outrun them, creating a space charge which establishes an electric field. The

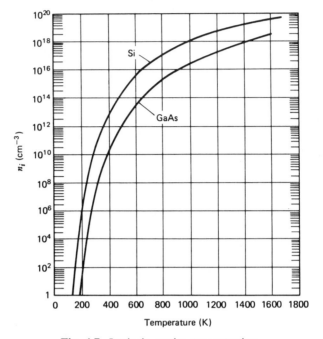

Fig. 4.7 Intrinsic carrier concentration.

direction of the $\mathscr{E}$ field due to the space charge associated with the positively charged impurity ions and the mobile electrons aids the drift of these ions in the direction of the diffusing flux, and enhances the rate of impurity movement* through the lattice.

By analogy with Eq. (4.11), the steady-state electron flux is given by

$$j_n = 0 = -D_n \frac{\partial n}{\partial x} - \mu_n n \mathscr{E} \tag{4.12}$$

Substituting the Einstein relation,

$$D = (kT/q)\mu \tag{4.13a}$$

gives

$$\mathscr{E} = -\frac{kT}{q} \frac{1}{n} \frac{\partial n}{\partial x} \tag{4.13b}$$

*A similar conclusion can be reached for p-type impurities.

Substituting into Eq. (4.11),

$$j = -D\left(1 + \frac{N}{n}\frac{\partial n}{\partial N}\right)\frac{\partial N}{\partial x} \tag{4.14}$$

so that the impurity moves with an effective diffusivity D_{eff}, where

$$D_{eff} = D\left(1 + \frac{N}{n}\frac{\partial n}{\partial N}\right) \tag{4.15}$$

The quantity in parentheses can thus be considered as a *field enhancement factor*. For an n-type impurity we have

$$\frac{n}{n_i} = \frac{N}{2n_i} + \left[\left(\frac{N}{2n_i}\right)^2 + 1\right]^{1/2} \tag{4.16}$$

If h is the field enhancement factor, then

$$h = \left(1 + \frac{N}{n}\frac{\partial n}{\partial N}\right) = 1 + \frac{N}{2n_i}\left[1 + \left(\frac{N}{2n_i}\right)^2\right]^{-1/2} \tag{4.17}$$

Thus the effect of this field enhancement factor is to (at most) double the magnitude of the diffusivity under field-free conditions. Consequently, this mechanism cannot account for the large increase in diffusivity with doping concentration that has been experimentally observed for substitutional diffusers. Here interaction with charged defects can play a dominant role, as will be seen in Section 4.2.3.1.

4.2.2 Interstitial Diffusion

Interstitial diffusion involves impurity jumps via interstitial voids. These voids are arranged tetrahedrally in the zincblende lattice (see Fig. 1.12). Although some are occupied by point defects, their equilibrium concentration is low, even at normal diffusion temperatures (700–1100°C). Consequently, nearly all interstitial sites are available for receiving impurity atoms as they wander through the lattice. Calculation of the diffusivity under these conditions is a relatively straightforward process.

Each tetrahedral void can readily accommodate an interstitial atom. However, there is an energy barrier which must be surmounted in order for interstitially located impurity atoms to jump from one void to the next. This barrier for interstitial diffusers is periodic in nature, as shown in Fig. 4.8. The jump frequency, ν, is the frequency with which thermal energy fluctuations occur with sufficiently large magnitude to overcome this potential barrier. Let:

E_m = activation energy of impurity migration, in eV

T = temperature of the lattice, in K

ν_0 = frequency of lattice vibrations, about $10^{13}-10^{14}$/s at typical diffusion temperatures. This is the frequency with which atoms strike the potential barrier depicted in Fig. 4.8.

Assuming a Boltzmann energy distribution, the probability that an atom has an energy in excess of E_m is given by $e^{-E_m/kT}$. Atoms in the zincblende lattice can jump from one interstitial site to the next in four different ways. If these jumps are uncorrelated, and all sites are vacant, the frequency of jumping is given by

$$\nu = 4\nu_0 e^{-E_m/kT} \tag{4.18}$$

Typical values for E_m are about 0.6–1.2 eV. Thus an interstitial impurity atom will jump from one void to another at a rate of about once every minute at room temperature. The rate is considerably higher at typical diffusion temperatures (700–1100°C).

Interstitially dissolved oxygen is an exception to this general behavior [8]. Here, $E_m \simeq 2.44$ eV, so that the jump rate is considerably lower than that for other interstitials. This has been attributed to the strong chemical binding between oxygen and silicon.

Substituting Eq. (4.6) into Eq. (4.18), the diffusivity of an interstitial impurity is given by

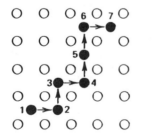 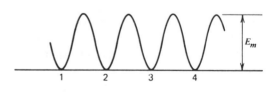

Fig. 4.8 Interstitial diffusion by jumping.

$$D = \frac{\nu d^2}{6} = \frac{4\nu_0 d^2}{6} e^{-E_m/kT}$$
$$= D_0 \, e^{-E_m/kT} \tag{4.19}$$

Both D_0 and the exponential term are functions of temperature. However, the variation of D with temperature is largely controlled by changes in the exponential factor. As a result, D_0 is usually considered to be constant over a range of a few hundred degrees of diffusion temperature, and is referred to as the *diffusion constant* in cm^2 s^{-1}. Equations of this type are usually written in the general form

$$D = D_0 \, e^{-E_d/kT} \tag{4.20}$$

where D is the diffusivity, D_0 is the diffusion constant, and E_d is the activation energy of diffusion.

4.2.3 Substitutional Diffusion

The jumping of substitutional diffusers requires first that there be available vacant sites into which they can move. These vacant sites are Schottky defects, with an energy of formation given by E_s (≈ 2.6 eV for silicon). Thus the atom fraction of such defects in a crystal is $e^{-E_s/kT}$. The charge state of these defects, which is related to the impurity concentration, must also be included in the calculation of diffusivity.

Consider, first, the situation when the charge states of the defects can be ignored. The substitutional jump process, shown in Fig. 4.2, requires the breaking of certain covalent bonds, and the making of yet others. If the potential barrier associated with impurity migration by this process is E_n, then the probability of atoms having a thermal energy in excess of this value is $e^{-E_n/kT}$. Typical values for E_n range from 0.6 to 1.2 eV. Finally, each lattice site has four tetrahedrally situated neighbors, so that each jump can be made in one of four different ways. Again assuming that these jumps are uncorrelated,* we have

$$\nu = 4\nu_0 \, e^{-E_s/kT} e^{-E_n/kT}$$
$$= 4\nu_0 \, e^{-(E_n+E_s)/kT} \tag{4.21}$$

Values of $E_n + E_s$ ($= E_d$), predicted from this simple theory, are found to be slightly larger than those actually observed. This is because the binding energy between an impurity atom and its neighboring atom is less than that between two adjacent atoms in the lattice. Consequently, the energy of formation of a vacancy next to an impurity atom is less than that of forming a vacancy at

---

*A correlation factor of 0.5 is sometimes used in these computations [1].

any other site. This is borne out by the experimental fact that values of E_d range from 3 to 4 eV for substitutional impurities in silicon, while the activation energy of self-diffusion in silicon is 5.13 eV. From Eq. (4.21), the jump rate of a substitutional impurity at room temperature is about once every 10^{45} years, as compared to the jump rate of about once every minute for interstitial diffusers.

The diffusivity for a substitutional impurity is thus

$$D = \frac{4\nu_0 d^2}{6} e^{-E_d/kT}$$
$$= D_0 e^{-E_d/kT} \tag{4.22}$$

As before, D_0 is usually considered to be constant over a range of a few hundred degrees of diffusion temperature, and the variation of D is largely controlled by changes in the exponential factor.

4.2.3.1 Interaction with Charged Defects

The theory of substitutional diffusion is based on the interaction of the impurity ion with vacancies in the zincblende lattice. The charge states of these vacancies can strongly influence the nature of this interaction, and hence the effective diffusivity.

It was shown in Chapter 1 that any "flaw" in a semiconductor lattice can usually be represented by a series of energy levels, in the sequence $E^+, E^-, E^{2-}, E^{3-}, \ldots$, in addition to being neutral [9]. In general, all deep levels, as well as vacancies, exhibit one or more of these charge states, and in this sequence. Let us assume that vacancies can be represented in this way, corresponding to $V^+, V^-, V^{2-}, V^{-3}, \ldots$, in addition to V^0.

These vacancies will interact with the diffusing impurity ions in different ways. Consequently, the jump statistics, and hence the diffusivity and activation energy associated with each impurity–vacancy combination, will be different. However, if it is assumed that no correlation exists between the separate jump processes, the fluxes associated with each process can be added, and the effective diffusivity will be given by the sum of these terms taken over their respective contributions.

Consider the dilute case, where an extremely small amount of an impurity A diffuses in an undoped semiconductor lattice. Since the semiconductor is intrinsic, this diffusion can be characterized by intrinsic diffusivities $D_i^0, D_i^+, D_i^-, D_i^{2-}, D_i^{3-}, \ldots$, where these terms are associated with $A–V^0, A–V^+, A–V^-, A–V^{2-}, A–V^{3-}, \ldots$ pair interactions, respectively. It follows that the overall intrinsic diffusivity is given by*

---

*The right-hand side of this equation is sometimes multiplied by 0.5, which is the correlation factor for tracer self-diffusion in the zincblende lattice.

$$D_i = D_i^0 + D_i^+ + D_i^- + D_i^{2-} + D_i^{3-} + \ldots \tag{4.23}$$

Diffusion under extrinsic conditions results in a displacement of the fermi level, and a change in the concentration of the various defect species. Consequently, the diffusivity for this condition is given [10] by

$$D = D_i^0 \frac{[V^0]}{[V^0]_i} + D_i^+ \frac{[V^+]}{[V^+]_i} + D_i^- \frac{[V^-]}{[V^-]_i} + D_i^{2-} \frac{[V^{2-}]}{[V^{2-}]_i}$$
$$+ D_i^{3-} \frac{[V^{3-}]}{[V^{3-}]_i} + \ldots \tag{4.24}$$

where $[V^0], [V^+], [V^-], [V^{2-}], [V^{3-}], \ldots$ refer to the concentration of these species under extrinsic conditions, and $[V^0]_i, [V^+]_i, [V^-]_i, [V^{2-}]_i, [V^{3-}]_i, \ldots$ pertain to intrinsic conditions.

The occupation statistics for vacancies must be determined in order to apply this equation. To do so it must be recognized that, during impurity diffusion, the neutral vacancy concentration will remain the same throughout the semiconductor, since it is unaffected by the electric field due to the impurity concentration gradient, and is thus independent of the position of the fermi level with respect to the band edge. It follows that the extrinsic concentration of neutral vacancies will be equal to the concentration in intrinsic material, which is a function of temperature alone. On the other hand, the concentration of the various ionized states is determined by the position of the fermi level, so that the total defect concentration changes with shifts in its position.

The ionized vacancy ratios can be determined using the law of mass action. Degeneracy factors are ignored in this analysis. For the V^+ case we have

$$V^0 + h^+ \rightleftharpoons V^+ \tag{4.25}$$

so that

$$k_1 = \frac{[V^+]}{p[V^0]} \tag{4.26}$$

Under intrinsic conditions, $p = n = n_i$, so that

$$k_1 = \frac{[V^+]_i}{n_i[V^0]_i} \tag{4.27}$$

Combining these equations, and noting that $[V^0] = [V^0]_i$, gives

$$\frac{[V^+]}{[V^+]_i} = \frac{p}{n_i} \tag{4.28}$$

In like manner, the reaction for a vacancy in the $-r$ charge state is given by

$$V^0 + re^- \rightleftharpoons V^{-r} \tag{4.29}$$

so that

$$k_2 = \frac{[V^{-r}]}{n^r[V^0]} = \frac{[V^{-r}]_i}{n_i^r[V^0]_i} \tag{4.30}$$

and

$$\frac{[V^{-r}]}{[V^{-r}]_i} = \left(\frac{n}{n_i}\right)^r \tag{4.31}$$

Thus, diffusion in an extrinsic semiconductor is characterized by

$$D = D_i^0 + D_i^+\left(\frac{p}{n_i}\right) + D_i^-\left(\frac{n}{n_i}\right) + D_i^{2-}\left(\frac{n}{n_i}\right)^2 + D_i^{3-}\left(\frac{n}{n_i}\right)^3 + \ldots \tag{4.32}$$

This equation can be further modified by a field enhancement factor, h, which takes into consideration the internally generated electric field term (see Section 4.2.1). Thus the effective extrinsic diffusivity is given by

$$D_{\text{eff}} \simeq h\left[D_i^0 + D_i^+\left(\frac{p}{n_i}\right) + D_i^-\left(\frac{n}{n_i}\right) + D_i^{2-}\left(\frac{n}{n_i}\right)^2 + D_i^{3-}\left(\frac{n}{n_i}\right)^3 + \ldots\right] \tag{4.33}$$

Not all of these terms apply to every situation, so that only relevant terms must be considered.

Vacancies in silicon exhibit the charge states V^+, V^-, and V^{2-}, in addition to being neutral. Consequently, Eq. (4.33) reduces to

$$D_{\text{eff}} = h\left[D_i^0 + D_i^+\left(\frac{p}{n_i}\right) + D_i^-\left(\frac{n}{n_i}\right) + D_i^{2-}\left(\frac{n}{n_i}\right)^2\right] \tag{4.34}$$

Figure 4.9*a* shows [11] the energy-band diagram for vacancies in silicon. From this figure it is seen that only $A-V^+$ interactions are present during the diffusion of *p*-type impurities, since the fermi level is close to the valence-band edge. By the same reasoning, the diffusion of *n*-type impurities is dominated by interactions with the negatively charged species. Thus Eq. (4.34) is further simplified when applied to specific diffusion situations.

The arguments presented here can also be extended to include interaction of the impurity with the charge states of the interstitials [12]. Localized energy states [13] associated with silicon self-interstitials are shown in Fig. 4.9*b*. Their incorporation leads to an additional series of diffusivity terms associated with $A-I^0$, $A-I^+$, and $A-I^-$ which take the same form as those of Eq. (4.34). These terms would add to the diffusivity, if jump processes are assumed to be statistically independent.

The situation with GaAs is considerably more complex because of the possibility of defect interactions with the vacancies and interstitials of both gallium and arsenic. Thus, a more general form of Eq. (4.33), involving four sets of terms, is required for a complete characterization of the diffusivity. Some computations, based on the assumption of arsenic vacancies V_{As}^0 and V_{As}^+, gallium vacancies V_{Ga}^0 and V_{Ga}^-, and gallium di-vacancies $V_{Ga_2}^0$ and $V_{Ga_2}^-$ have also been made [14] in order to determine the concentration of these defects.

Considerable impetus towards sorting out the relative importance of various point defect species has been provided by the discovery [15, 16] that the diffusion of impurities into AlGaAs/GaAs superlattices results in their rapid disordering. Here, doping apparently enhances the interdiffusion of aluminum and gallium in order to bring this about. In addition, since gallium diffuses more

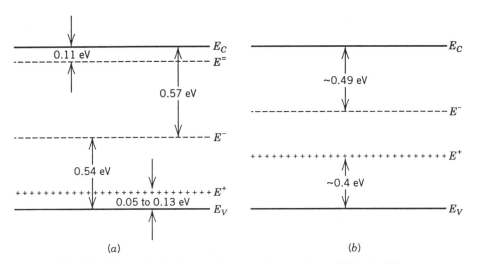

(a)   (b)

Fig. 4.9 Energy-band diagram for vacancies and interstitials in silicon.

slowly than aluminum, it has been proposed that gallium atom movement, associated with the interaction of charged gallium vacancies (or interstitials), is the primary diffusion mechanism. From these studies, the diffusivity of gallium in intrinsic GaAs has been obtained as

$$D_{Ga} = 2.9 \times 10^8 e^{-6/kT} \, \text{cm}^2 \, \text{s}^{-1} \tag{4.35}$$

Studies of gallium diffusion in n-GaAs have shown a cubic dependence of the diffusivity on doping. Moreover, the disordering rate of GaAs/AlAs superlattices is found to increase with the arsenic overpressure. Taken together, this indicates that the most likely point defect species involved in this process is V_{Ga}^{3-}. Yet other studies have suggested that Ga–V_{Ga}^{-} is the active point defect involved in this process.

A similar approach with p-GaAs has indicated that gallium diffusion probably involves point defect interaction with I_{Ga}^{r+}. The magnitude of the charge state has not been determined at the present time.

4.2.4 Interstitialcy Diffusion

We have noted that substitutional diffusion requires the presence of thermally activated vacancies in order to occur. Thus the fraction of such vacancies is an important factor in determining the jump rate for this process. Interstitialcy diffusion also requires the presence of vacant sites to which the impurity can move. Here, however, these sites can be created by the process of ejecting lattice atoms to form self-interstitials, in addition to thermal activation. Thus, the fraction of vacant sites is larger than that for substitutional diffusion.

Analysis for the jump rate of impurity atoms follows along the lines of substitutional diffusers. In this case the atom fraction of vacancies is $e^{-E_s'/kT}$, where E_s' is somewhat lower than the energy of formation of Schottky defects. From Eq. (4.21), it follows that

$$\nu = 4\nu_0 e^{-(E_n + E_s')/kT} \tag{4.36}$$

so that

$$D = D_0 e^{-(E_n + E_s')/kT} \tag{4.37}$$

As a result, the diffusivity is somewhat larger than that of substitutional diffusers.

The actual value of E_s' is about 1 eV lower than the energy of formation of a Schottky defect, and is a function of the local position of the fermi level, and of conditions under which the diffusion is carried out. If diffusion takes place in an oxidizing ambient, the value of E_s' can be even lower, since the oxidation of silicon promotes vacancies in the near-surface region by the creation of self-interstitials.

Almost all substitutional diffusers in silicon move partially by the interstitialcy mechanism. The diffusion of boron and phosphorus is dominated by this process.

4.2.5 Interstitial–Substitutional Diffusion

Many impurities are incorporated both substitutionally and interstitially in the host lattice, and diffuse by the combined motion of these two species. For such situations, let N_s and N_i be the substitutional and interstitial solubilities, respectively, at the diffusion temperature. Usually the solubility of the substitutional component is higher* than that of the interstitial component by about an order of magnitude. On the other hand, the rates of movement of these components through the lattice are different by many orders of magnitude, with the impurity comparatively immobile when in substitutional sites, but free to move rapidly when it is interstitial. Assuming *independent* motion of these two types of diffusers, the impurity will spend $N_i/(N_i + N_s)$ of its time in an interstitial state, and $N_s/(N_i + N_s)$ in a substitutional state. It follows that the effective jump frequency ν_{eff} is given by

$$\nu_{\text{eff}} = \frac{\nu_s N_s}{N_i + N_s} + \frac{\nu_i N_i}{N_i + N_s} \tag{4.38}$$

where ν_s and ν_i are the substitutional and interstitial jump frequencies, respectively. Typically, the second term is much larger than the first.

Atom movement by two *independent* streams rarely occurs in silicon or GaAs. Rather, the two streams are *interdependent*, and may be created by the possibility of dissociation of a substitutional atom into an interstitial atom and a vacancy, as in Fig. 4.3. If A is the impurity atom, and V is the vacancy produced by this process, the reaction can be written as

$$A_s \rightleftharpoons A_i + V \tag{4.39}$$

where A_s and A_i represent the substitutional and interstitial impurity atom, respectively, and V is the vacancy created by this process. It follows that the diffusion is controlled by the rate of production of the interstitial species and vacancies by this *dissociative mechanism*.

Equation (4.39) is driven from left to right in regions of high impurity concentration, and from right to left in regions of low concentration. In view of the fact that the interstitial species is a fast diffuser whereas the substitutional one

---

*The behavior of copper in silicon is an exception to this rule.

is (by comparison) immobile, the rate of diffusion will be enhanced in regions of high concentration, and retarded in regions of low concentration.

If the crystal contains many defects, the vacancy concentration readily attains its equilibrium value, so that the supply of the interstitial specie is the rate limiter. Since the impurity spends $N_i/(N_i + N_s)$ of its time in an interstitial state,

$$\nu_{\text{eff}} \simeq \frac{\nu_i N_i}{N_i + N_s} \tag{4.40}$$

If local equilibrium prevails,

$$\nu_{\text{eff}} \simeq \frac{\nu_i N_i^{\text{eq}}}{N_i^{\text{eq}} + N_s^{\text{eq}}} \simeq \frac{\nu_i N_i^{\text{eq}}}{N_s^{\text{eq}}} \tag{4.41a}$$

since $N_s^{\text{eq}} > N_i^{\text{eq}}$. Thus, the jump frequency is somewhat less than that for the interstitial. The effective diffusivity is given by

$$D_{\text{eff}} \simeq D_i \frac{N_i^{\text{eq}}}{N_s^{\text{eq}}} \tag{4.41b}$$

where D_i is the interstitial diffusivity. The ratio $N_i^{\text{eq}}/N_s^{\text{eq}}$ can be determined if the charge states of the individual species are known.

For low-dislocation material, however, the formation of substitutional atoms requires the generation of new vacancies which must come from the surface of the semiconductor—that is, by a relatively slow process. Thus, the diffusion of these vacancies is the rate limiter. If N_v is their concentration, then the fraction of substitutional impurity sites which are vacant is $N_v/(N_v + N_s)$, so that the effective jump frequency becomes

$$\nu_{\text{eff}} = \frac{\nu_v N_v}{N_v + N_s} \tag{4.42}$$

where ν_v is the jump frequency associated with vacancies. If local equilibrium prevails, this relation can be further simplified

$$\nu_{\text{eff}} \simeq \frac{\nu_v N_v^{\text{eq}}}{N_v^{\text{eq}} + N_s^{\text{eq}}} \simeq \frac{\nu_v N_v^{\text{eq}}}{N_s^{\text{eq}}} \tag{4.43a}$$

since $N_s^{eq} \gg N_v^{eq}$. It follows that

$$D_{\text{eff}} \simeq D_v \frac{N_v^{eq}}{N_s^{eq}} \qquad (4.43b)$$

The *kick-out mechanism*, illustrated in Fig. 4.4, is an alternative means of providing interdependent streams of substitutional and interstitial impurity atoms. Here,

$$A_i \rightleftharpoons A_s + I \qquad (4.44)$$

where I represents self-interstitials which are created during this process and are removed by diffusion to the surface. In this case also, Eq. (4.44) is driven from left to right in regions of low concentration, and from right to left in regions of high concentration. Again, the end result is that the diffusivity is enhanced in regions of high concentration, and retarded in regions of low concentration.

Two limiting cases are also possible when the kick-out mechanism is operative. In material with many defects, the self-interstitial concentration rapidly attains its equilibrium value, so that indiffusion of the impurity is the rate limiter. Since the impurity spends $N_i/(N_i + N_s)$ of its time in the interstitial state, its jump rate is given by

$$\nu_{\text{eff}} = \frac{\nu_i N_i}{N_i + N_s} \simeq \frac{\nu_i N_i}{N_s} \qquad (4.45a)$$

and the effective diffusivity, under local equilibrium conditions, is

$$D_{\text{eff}} \simeq \frac{D_i N_i^{eq}}{N_s^{eq}} \qquad (4.45b)$$

If the material has only a few defects, however, then the rate limiter is the out-diffusion of self-interstitials to the surface. The fraction of self-interstitials involved in this process is $N_I/(N_I + N_s)$, so that the corresponding jump frequency is given by

$$\nu_{\text{eff}} = \frac{\nu_I N_I}{N_I + N_s} \simeq \frac{\nu_I N_I}{N_s} \qquad (4.46a)$$

where the subscript I refers to self-interstitials. The corresponding diffusivity is thus given by

$$D_{\text{eff}} \simeq \frac{D_I N_I}{N_s} \qquad (4.46b)$$

Measurements of self-diffusion in silicon, made by radio-tracer methods, have established that the jump rates for both interstitials and vacancies are comparable to those for substitutional diffusers.

Applying the law of mass action to Eq. (4.44) gives

$$K_1 = \frac{[A_s][I]}{[A_i]} = \frac{[A_s]^{eq}[I]^{eq}}{[A_i]^{eq}} \tag{4.47}$$

so that

$$\frac{N_s N_I}{N_i} = \frac{N_s^{eq} N_I^{eq}}{N_i^{eq}} \tag{4.48}$$

Combining with Eq. (4.46) and noting that concentration of interstitial impurities is relatively unchanged from its equilibrium value gives

$$D_{\text{eff}} \simeq \frac{D_I N_I^{eq} N_s^{eq}}{N_s^2} \tag{4.49}$$

Radioactive tracer techniques are commonly used for measuring the diffusivities of vacancies and self-interstitials in silicon. One set of values [17] obtained by these methods gives

$$D_v = 10^3 e^{-2.0/kT} \tag{4.50a}$$
$$D_I = 0.1 e^{-0.4/kT} \tag{4.50b}$$

Studies of gold diffusion in silicon [18], which proceeds by the kick-out mechanism, have given

$$D_I N_I^{eq} = 9.14 \times 10^6 e^{-4.84/kT} \tag{4.51}$$

for heavily dislocated material.

The physical interpretation of D_v in GaAs is less clear, since each gallium atom has arsenic as its nearest neighbor, and vice versa. Thus substitutional motion for these atoms would necessitate a large number of anti-structure defects, which are not observed in practice. Other mechanisms, such as interstitial–substitutional diffusion, have been proposed for self-diffusion in GaAs, but the situation has not been resolved at the present time. From annealing studies of GaAs [19], the diffusivity of gallium and arsenic vacancies is given by

$$D(V_{Ga}) = 2.1 \times 10^{-3} e^{-2.1/kT} \tag{4.52a}$$
$$D(V_{As}) = 7.9 \times 10^{3} e^{-0.4/kT} \tag{4.52b}$$

The role of self-interstitials in GaAs is even less clear, and has only been considered in a few situations. Many of the experimentally observed results can be interpreted without involving their presence.

In summary, the characteristics of interstitial–substitutional diffusion, by either the dissociative or the kick-out mechanism, are (a) a jump rate that is large in materials with a high defect concentration, and (b) a jump rate that is small in materials with a low defect concentration. Impurity movement by this process has many of the characteristics of interstitial diffusion. These impurities are sometimes included in the general class of interstitial diffusers.

4.3 THE DIFFUSION EQUATION

Fick's second law may be derived by applying considerations of continuity to Eq. (4.7). Consider the flow of particles in a crystal of cross section A, between planes P_1 and P_2 separated by dx, as shown in Fig. 4.10. The rate of accumulation of particles in the region between planes is $A(\partial N/\partial t) dx$. This can also be written as the difference between the fluxes flowing into and out of the region. The flux entering the region at P_1 is Aj, and the flux leaving the region at P_2 is $A(j + dj)$. The net flux entering the region is thus $-A\,dj$. Hence

$$A\frac{\partial N}{\partial t}dx = -A\,dj \tag{4.53}$$

But $dj = (\partial j/\partial x)\,dx$. Hence, from Eq. (4.7),

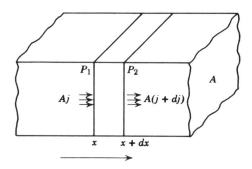

Fig. 4.10 Diffusing flux.

$$\frac{\partial N}{\partial t} = \frac{\partial}{\partial x}\left(D\frac{\partial N}{\partial x}\right) \tag{4.54}$$

This expression is in a more convenient form than Eq. (4.7) since it involves only measurable quantities such as volume concentration, diffusion depth, and diffusion time. It can be used to obtain the impurity distribution for a variety of diffusion conditions.

4.3.1   The $D =$ Constant Case

The diffusion of shallow impurities in silicon, at relatively low concentrations (n or $p < n_i$ at the diffusion temperature), can be approximated by assuming that the diffusivity is not a function of the impurity concentration. Making this simplification, Fick's second law reduces to

$$\frac{\partial N}{\partial t} = D\frac{\partial^2 N}{\partial x^2} \tag{4.55}$$

where N is the volume concentration (atoms cm^{-3}), D the diffusivity (cm^2 s^{-1}), x the distance (cm), and t the diffusion time (s). In the presence of an electric field, this equation is modified to

$$\frac{\partial N}{\partial t} = D\frac{\partial^2 N}{\partial x^2} - \mu\mathscr{E}\frac{\partial N}{\partial x} \tag{4.56}$$

Equation (4.55) may be solved for a number of situations which arise in practice [20–24]. Computations have also been made by several authors for a variety of additional diffusion conditions. Some of these may be found in the Appendix.

4.3.1.1   *Diffusion from an Unlimited Source*

This situation occurs when a wafer is exposed to what is, for all practical purposes, an infinite amount of the impurity during the diffusion period. These conditions result in a surface concentration of N_0. Here the impurity concentration at any given distance and time is $N(x,t)$, such that

$$N(x,t) = N_0\,\mathrm{erfc}\,\frac{x}{2\sqrt{Dt}} \tag{4.57}$$

where N_0 is the impurity concentration at the surface (atoms cm^{-3}), D the diffusivity at the specific diffusion temperature, (cm^2 s^{-1}), x the penetration depth (cm), and t the diffusion time (s). The solution of this equation is provided in the Appendix.

Figure 4.11 shows a sketch of this concentration for various diffusion times. Its most significant characteristic is that the surface concentration is constant whereas the diffusion depth increases with time. A normalized plot of Eq. (4.57) is shown in Fig. 4.12. A diffusion of this type is referred to as a *complementary error function* (erfc) diffusion.

The erfc diffusion is performed by exposing the slice to an unlimited supply of the impurity during the entire process. At low levels the surface concentration is given by the value that is in equilibrium with the surrounding gas. Usually, the source concentration is very large, so that N_0 is ultimately set by the solid solubility of the impurity for that specific diffusion temperature. The emitter diffusion of bipolar silicon transistors, source/drain regions in MOS devices, and most diffusions in GaAs are usually of this type. Here a high surface concentration is desired, necessitating the use of relatively high diffusion temperatures.

Since the diffusion equation is linear, superposition may be used to calculate the effect of a finite background. For example, a *p–n* junction can be fabricated by diffusing a *p*-type impurity into an *n*-type substrate of background concentration N_C, as shown in Fig. 4.11. The effective doping concentration at any point is given by the difference of acceptor and donor concentrations, whichever is larger. If the *p*-type impurity profile is given by

$$N(x,t) = N_0 \operatorname{erfc} \frac{x}{2\sqrt{Dt}} \qquad (4.58)$$

the junction is located at x_j, such that

$$N_0 \operatorname{erfc} \frac{x_j}{2\sqrt{Dt}} = N_C \qquad ((4.59)$$

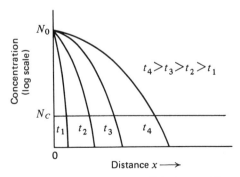

Fig. 4.11 Constant-source diffusion profiles.

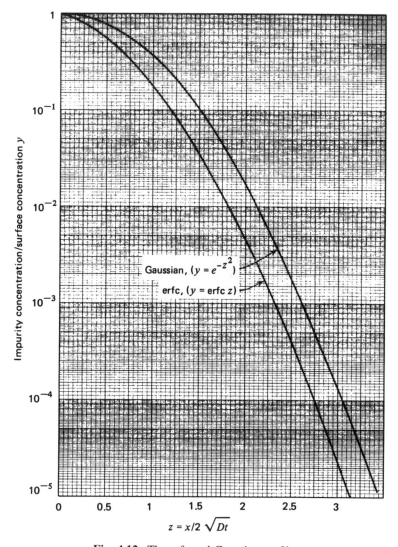

Fig. 4.12 The erfc and Gaussian profiles.

4.3.1.2 Diffusion from a Limited Source

Here a finite quantity of the diffusing matter is first placed on the wafer. Diffusion proceeds from this limited source, and it is assumed that all of this matter is consumed during the process. The resulting impurity concentration is given by (see Appendix)

$$N(x, t) = \frac{Q_0}{\sqrt{\pi D t}} e^{-(x/2\sqrt{Dt})^2} \tag{4.60}$$

where Q_0 is the amount of material placed on the surface prior to diffusion (atoms cm^{-2}), D the diffusivity (cm^2 s^{-1}), x the diffusion distance (cm), and t the diffusion time (s). A diffusion of this type is known as a *gaussian diffusion*.

Figure 4.13 shows a diffusion profile for various diffusion times. The effect of a background concentration N_C in establishing the position of the junction is also shown here. A normalized plot of Eq. (4.60) is shown in Fig. 4.12.

A comparison of Figs. 4.11 and 4.13 shows that the significant difference between these two types of diffusion lies in the fact that the surface concentration of the gaussian diffusion falls with increasing diffusion times, whereas that of the erfc diffusion does not. Otherwise they are essentially similar; as seen from Fig. 4.12, they can be approximated by exponential profiles at concentration levels that are two or more orders of magnitude below the surface concentration.

Limited source diffusion is ideally suited for those cases where a relatively low value of surface concentration is required in conjunction with a high diffusion depth.* The base diffusion of a bipolar silicon transistor presents this type of situation. Surface concentrations encountered here are lower than those needed for the emitter diffusion, whereas the diffusion depth is higher.

*Although the low surface concentration can also be obtained by diffusion from an unlimited source at low diffusion temperature, the amount of time required for the desired penetration depth would be prohibitive.

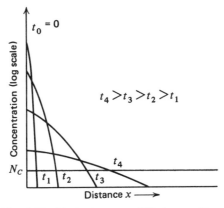

Fig. 4.13 Limited-source diffusion profiles.

Figure 4.14 shows the doping profile of a bipolar silicon transistor, formed by successive gaussian and erfc diffusions into a wafer with a collector background concentration of 3×10^{16} atoms cm^{-3}. Typical parameters for a modern high-speed digital transistor are also indicated.

The quantity of matter required for a gaussian diffusion is considerably less than a monolayer. Thus special techniques must be used for its accurate placement on the slice. A direct approach is to use ion implantation methods, which are described in Chapter 6. An indirect approach, which can also be used, is to conduct a constant-source diffusion at a low temperature ($\approx 900°$C) for a short time. By this process, known as *predeposition*, a small quantity of the impurity is transported into an extremely thin layer *in* the silicon slice. Since the penetration depth is negligibly small, this provides a suitable approximation to placing the matter *on* the surface. The actual amount of matter transported during the predeposition phase is given by (see the Appendix)

$$Q_0 = 2N_{01} \left(\frac{D_1 t_1}{\pi} \right)^{1/2} \tag{4.61}$$

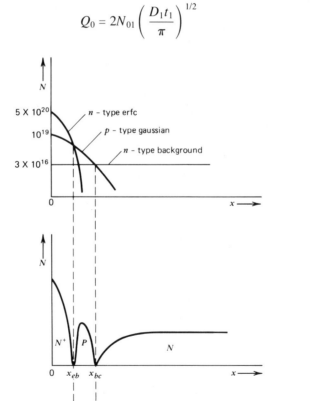

Fig. 4.14 Bipolar transistor formation by successive diffusions.

where Q_0 is the amount of material entering the silicon during predeposition (atoms cm^{-2}), N_{01} the surface concentration at the predeposition temperature (atoms cm^{-3}), D_1 the diffusivity at the predeposition temperature ($cm^2 s^{-1}$), and t_1 the predeposition time (s).

The external source of impurity is now removed from the surface of the slice, which is then subjected to a high-temperature drive-in phase for an appropriate time and temperature. The final impurity concentration is given by

$$N(x, t_1, t_2) = \frac{2N_{01}}{\pi} \left(\frac{D_1 t_1}{D_2 t_2} \right)^{1/2} e^{-(x/2\sqrt{D_2 t_2})^2} \tag{4.62}$$

where the subscript 2 refers to drive-in parameters, and the schedule is arranged such that

$$\sqrt{D_1 t_1} \ll \sqrt{D_2 t_2} \tag{4.63}$$

Thus, Eq. (4.62) assumes an extremely short predeposition phase compared with the drive-in phase.

4.3.1.3 The Two-Step Diffusion

The two-step diffusion represents a practical implementation of the last process. Because it has so many advantages over alternative methods, it is the most common diffusion process today for silicon devices and microcircuits.

This process is initiated by first conducting a constant-source diffusion at a low temperature for a short time. The impurity supply is shut off and the drive-in phase initiated in an oxidizing ambient. This results in the formation of a surface oxide on the silicon, preventing the further indiffusion of impurities which are present *on* the slice and inhibiting the out-diffusion of impurities which were already predeposited *in* the slice. Although some of the impurity is consumed during the oxidation step, its final concentration is related to the predeposition and drive-in parameters. If the subscripts 1 and 2 are used to denote predeposition and drive-in, as before, an erfc diffusion results when $D_1 t_1 \gg D_2 t_2$. Conversely, a gaussian profile results if $D_1 t_1 \ll D_2 t_2$.

If neither of these inequalities holds, the results are modified to give the final impurity distribution [25] as

$$N(x, t_1, t_2) = \frac{2 N_{01}}{\pi} \int_0^U \frac{e^{-\beta(1+U^2)}}{1 + U^2} dU \tag{4.64}$$

where

$$U = \left(\frac{D_1 t_2}{D_2 t_2} \right)^{1/2} \tag{4.65}$$

and

$$\beta = \left(\frac{x}{2\sqrt{D_1 t_1 + D_2 t_2}} \right)^2 \tag{4.66}$$

In addition, the final surface concentration is given by

$$N_{02} = \frac{2 N_{01}}{\pi} \tan^{-1} U \tag{4.67}$$

Consequently,

$$N(x, t_1, t_2) = \frac{N_{02}}{\tan^{-1} U} \int_0^U \frac{e^{-\beta(1+U^2)}}{1 + U^2} dU \tag{4.68}$$

Table 4.1 gives values of the integral in Eq. (4.68), and is provided at the end of this chapter. In addition, Fig. 4.15 shows normalized concentration profiles for two-step diffusions with various values of U, as well as the limiting cases of erfc and gaussian diffusions. Consumption of the impurities during the oxidation step is usually included in computer simulations of this process.

In summary, the two-step process can be used to approximate an erfc diffusion profile without the accumulation of large amounts of impurity on the surface of the silicon. Thus it avoids the possibility of surface damage. A slow oxidation process,* using dry oxygen, is used for this purpose.

It can also be used to approximate a gaussian diffusion while avoiding the necessity of exposing the surface of the slice at any time. A fast oxidation process, using steam, is required in this situation. Junctions are always formed beneath the protective layers and show excellent breakdown characteristics. Finally, the glass may be used as a mask for the next diffusion step.

4.3.1.4 Successive Diffusions

It is often required to calculate the total effect of diffusion during a series of temperature cycles. In a bipolar transistor, for example, the emitter-diffusion step takes place after the base drive-in. Thus the impurities in the base are

*Oxidation processes are covered in detail in Chapter 7.

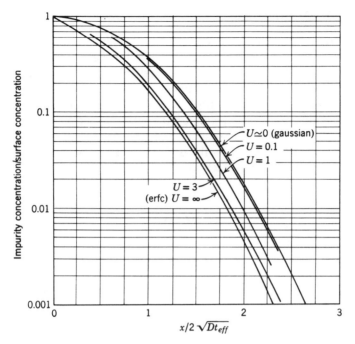

Fig. 4.15 Profiles for the two-step diffusion process.

subjected to one set of time and temperature values during the base drive-in phase and to a second set during the emitter-diffusion phase. If a gold-diffusion step follows, the base is subjected to yet another time and temperature cycle. To compute the total effect of these cycles, it is necessary to obtain an effective Dt product for this region. The effective Dt product is given by

$$(Dt)_{\text{eff}} = \sum D_1 t_1 + D_2 t_2 + D_3 t_3 + \cdots \qquad (4.69)$$

where $t_1, t_2, t_3, \ldots$ are the different diffusion times, and $D_1, D_2, D_3, \ldots$ are the appropriate diffusivities in effect during these times. For the specific example given here, t_1, t_2 and t_3 are the base-, emitter-, and gold-diffusion times, respectively; D_1, D_2, and D_3 are the diffusivities for the base region impurity at the base-, emitter-, and gold-diffusion temperatures, respectively.

Cooperative diffusion effects can also occur, especially in situations where a high-concentration diffusion is performed. Often this results in enhancing the atom movement of earlier diffusions. These effects will be considered for specific dopants in Section 4.4.

4.3.2 The $D = f(N)$ Case

The majority of diffusion situations are described by a concentration-dependent diffusivity. Calculations of the doping profile for these cases are not available in closed form. However, computer-derived solutions have been obtained for some important situations. These computations have been made for an unlimited source diffusion, so that the surface concentration and diffusivity are constant for any specific diffusion temperature. All of these result in profiles which differ from those given by the erfc function, for which D is independent of N.

Consider the situation where the diffusivity can be written directly as some power of the concentration, such as

$$D = KN^r \tag{4.70}$$

This equation can be rewritten in normalized form as

$$D = D_{sur} \left(\frac{N}{N_{sur}} \right)^r \tag{4.71}$$

Here N_{sur} is the surface concentration and D_{sur} is the diffusivity at this concentration. For these cases it can be shown (see Section A.3 in the Appendix) that Fick's second law can be rewritten as an ordinary differential equation, and solved by numerical computation techniques. These solutions, for a constant-source diffusion, are shown in Fig. 4.16 for $r = 1, 2$, and 3, together with the solution for $D = $ constant [26]. Also shown is the profile for $r = -2$, which is relevant to interstitial–substitutional diffusion by the kick-out mechanism. Tabulated values for these distributions are given in Table A.2.

From this figure it is seen that, for $r = 1, 2$, and 3, doping profiles are extremely steep at low concentrations ($\ll N_{sur}$). Consequently, highly abrupt junctions are formed when diffusions are made into a background of opposite impurity type. Furthermore, the steepness of the doping profile results in a junction depth which is given by the value for which $N \simeq 0$; that is, it is almost independent of the background concentration.

From Table A.2, the junction depth for these types of diffusions is given by

$$x_j = 1.616 \, (D_{sur}t)^{1/2}, \qquad D \propto N \tag{4.72}$$

$$x_j = 1.092 \, (D_{sur}t)^{1/2}, \qquad D \propto N^2 \tag{4.73}$$

$$x_j = 0.872 \, (D_{sur}t)^{1/2}, \qquad D \propto N^3 \tag{4.74}$$

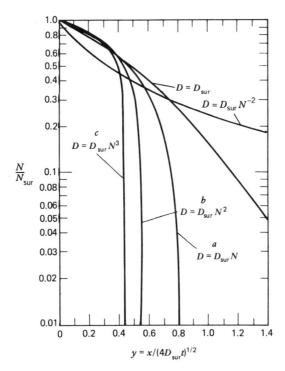

Fig. 4.16 Normalized diffusion profiles for concentration dependent diffusions. From Weisberg and Blanc [26]. Reprinted with permission of the *American Physical Society.*

Analytical solutions are available for limited source diffusions with dopants which exhibit concentration-dependent behavior [27]. Details of these solutions will not be considered here.

Finally, it is often necessary to characterize concentration-dependent diffusional behavior by a series of power law terms of the type described here. For these cases the junction depth occurs at some intermediate point. A number of computer-aided process simulation programs [28, 29] can be used to provide numerical solutions of these equations. These programs provide two- and three-dimensional simulations, and can include complex effects such as impurity consumption during the oxidation phase. A discussion of these programs, which are routinely used in manufacturing, is beyond the scope of this text.

4.3.3   Lateral Diffusion Effects

In practice, diffusion in microcircuits is always carried out through windows cut in a mask that is placed on the slice. The one-dimensional diffusion equation represents a satisfactory means of describing this process, except at the edge of the mask window. For this situation, the dopant source provides impurities which diffuse at right angles to the semiconductor surface as well as parallel to it (i.e., laterally).

Contours of constant doping concentration which illustrate lateral diffusion effects are shown in Fig. 4.17a for unlimited-source diffusing conditions [30], on the assumption of a concentration-independent diffusivity. These contours are, in effect, a map of the location of the junctions created by diffusing into various background concentrations. It is seen from this figure that the lateral penetration is about 75–85% of the penetration in the vertical direction for concentrations that are two or more orders of magnitude below the surface concentration.

Figure 4.17b shows the contours of constant doping concentration for limited-source diffusions, again assuming a concentration-independent diffusivity. Here too, the penetration ratio (lateral to vertical) is about 75–85% for background concentrations that are two or more orders of magnitude below the surface concentration. Note that the depletion of this source results (for some cases) in actually terminating the contours within the window. This is of no consequence in practice, since a junction is usually formed by doping into a background that is considerably below the surface concentration.

Computer-aided solutions have been obtained for the case where the diffusivity is concentration-dependent [31]. These are of special interest for shallow diffusions with high surface concentration, of the type encountered in many very-large-scale integration (VLSI) applications. Here it has been shown that, in addition to the more abrupt doping profile in both directions, the lateral junction penetration is greatly reduced. Typically the ratio of lateral to vertical penetration for these diffusion conditions ranges from 65% to 70%.

In some VLSI situations the width of a diffusion window may become so narrow that it can be treated as a line source. An unlimited source diffusion through such a slot is approximately cylindrical in nature for this limiting case, and is represented by a gaussian profile in the radial direction. For the constant diffusivity case, it can be shown that (see Appendix)

$$N(r, t) = \frac{\delta}{2\pi D t} e^{-r^2/4Dt} \tag{4.75}$$

where r is the radius and δ is the number of impurities per unit length of the slot. For a narrow slot of width W,

$$\delta \simeq Q_0 W \tag{4.76}$$

where Q_0 is the dopant density on the silicon surface, in atoms cm^{-2}. Analysis of this equation shows that junctions formed in this manner will have a smaller depth than those made through a wide slot, which approximates a parallel-plane source.

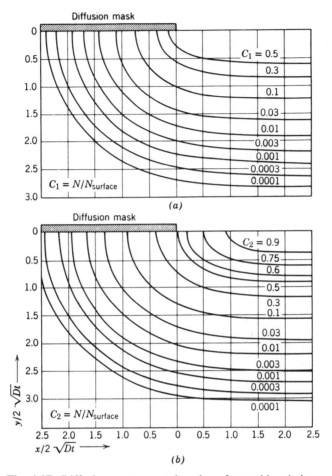

Fig. 4.17 Diffusion contours at the edge of an oxide window.

4.4 IMPURITY BEHAVIOR: SILICON

The movement of impurities belonging to groups III and V of the periodic table is caused by a combination of substitutional and interstitialcy diffusion. Their motion is thus strongly influenced by the concentration and the charge state of lattice point defects. They include the p-type impurities aluminum, boron, gallium, and indium, as well as the n-type impurities antimony, arsenic, and phosphorus. With the exception of indium, all exhibit shallow impurity levels in silicon. They are technologically important since they are used in junction formation in discrete devices and integrated circuits.

Impurities which move interstitially in silicon belong to groups I and VIII of the periodic table. These include alkali metals such as lithium, potassium, and sodium, and gases such as argon, helium, and hydrogen. They occupy interstitial sites in silicon and are usually electronically inactive. A notable exception is lithium, which is *n*-type. These impurities are of little interest in microcircuit fabrication technology, and are not considered further.

Most transition elements (cobalt, copper, gold, iron, nickel, platinum, and silver, for example) diffuse by an interstitial–substitutional mechanism, and end up in both types of sites, the difference in solubility between these sites being substantial. The dissociation of the substitutional into an interstitial and a vacancy is the mechanism involved in most cases. On the other hand, the kick-out mechanism defines the diffusional movement of gold and platinum in silicon.

4.4.1 Shallow Impurities

Figure 4.18 shows Arrhenius plots of the diffusivity D as a function of temperature, for shallow impurities in silicon [32]. Plots of this type can be used to obtain values of an apparent activation energy, E_d, listed in Table 4.2, and are of an average nature, independent of concentration. They are obtained by diffusing the impurities into silicon wafers of known (but opposite type) background concentration and by making measurements of the *p–n* structures thus formed. This experimental technique essentially duplicates the fabrication process for *p–n* junctions, so that the results are reasonably accurate for many junction-formation situations. From Table 4.2, we note that the apparent activation energy of arsenic and antimony is about 4 eV, which is to be expected for substitutional diffusers in silicon [8]. On the other hand, the corresponding values for phosphorus, and all the *p*-type impurities, are significantly lower (about 3.5 eV), indicating the presence of an interstitialcy diffusion component.

The data of Table 4.2 become increasingly poor when used to establish doping profiles for shallow junctions with high surface concentration. Table 4.3, which lists values of the intrinsic diffusivity and activation energy for commonly used impurities, with their separate $A - V$ contributions, is more useful in these situations.

A closed-form solution of the diffusion equation, using the data of Table 4.2 for the effective diffusivity, is not possible because the dopant and carrier concentration change as a function of penetration. As a result, the concentration dependence of the diffusivity varies with distance. An additional complicating factor is that bandgap narrowing effects become important at high carrier concentrations [33], and result in altering the intrinsic carrier concentration at these levels. This is especially true for arsenic and phosphorus, which are commonly used in situations where heavy doping is required. Values of n_{ie}, the intrinsic electron concentration for solid-solubility-limited diffusions with these impurities, are shown in Fig. 4.19.

Computer simulations have been made to handle all of these complexities in a manufacturing environment. However, our understanding of the behavior of

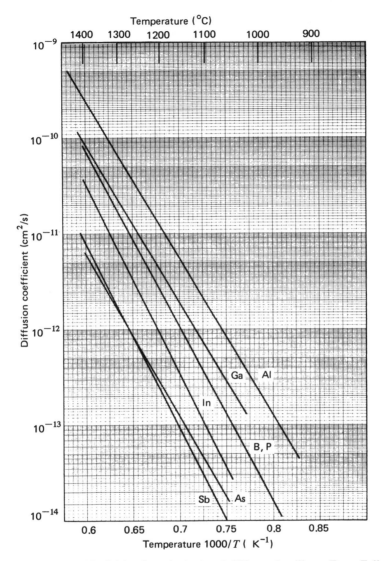

Fig. 4.18 Average diffusivities for substitutional diffusers in silicon. From Fuller and Ditzenberger [32]. With permission from the *Journal of Applied Physics*.

substitutional diffusers can be greatly enhanced by making a number of appropriate simplifications. Details for individual impurities are now considered.

4.4.1.1 Arsenic in Silicon

Arsenic has a low misfit factor ($\epsilon \simeq 0$), and can be incorporated to a high concentration without causing strain in the lattice. This fact, coupled with its

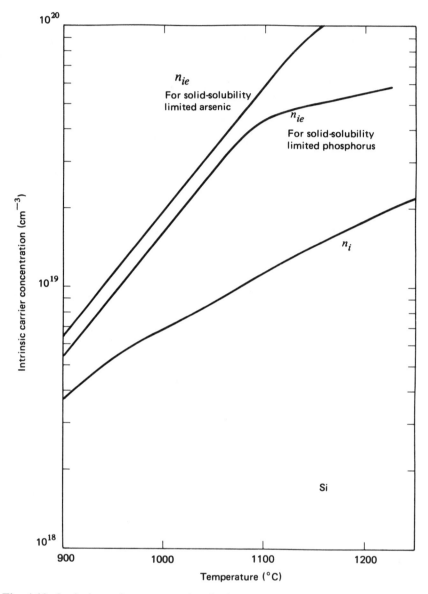

Fig. 4.19 Intrinsic carrier concentration for heavy doping conditions. From Colclaser [35].

low diffusivity and its abrupt doping profile, makes it ideally suited for shallow diffused structures. Thus its use as a dopant in VLSI applications is of extreme importance.

Anomalous behavior, noted by a number of workers, is seen in a fall of the conductivity of arsenic-diffused (1100–1200°C) layers which are subsequently heat treated in the 500–900°C range. The conductivity of these layers can be fully recovered by reprocessing at high temperatures (1100–1200°C), so that this behavior is not due to loss of arsenic by outdiffusion from the silicon surface. Although not understood at the present time, possible reasons for this effect are that it is due to the formation of As–As clusters or to the formation of $SiAs_2$, both of which are electronically inactive. The solid solubility of arsenic, as well as its electrically active component, is shown in Fig. 4.20 as a function of temperature [33].

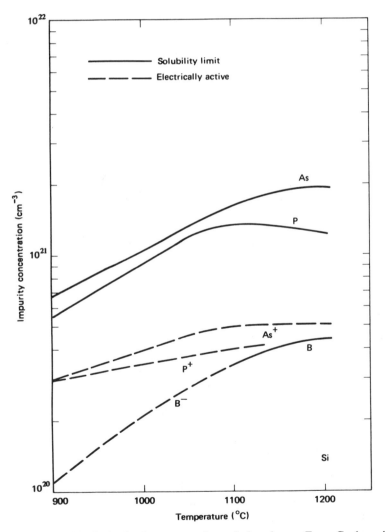

Fig. 4.20 Solubility limits for boron, arsenic, and phosphorus. From Coclaser [35].

The concentration-dependent diffusivity of arsenic in silicon is described by Eq. (4.34), in which the interaction of A–V^0 and A–V^- must be considered, so that

$$D = h\left[D_i^0 + D_i^-\left(\frac{n}{n_{ie}}\right)\right] \qquad (4.77)$$

where n_{ie} is the intrinsic electron concentration for solid-solubility-limited diffusions at high concentration. The field enhancement factor can be approximated by 2 for these diffusions. Assuming that the electron concentration is given by the doping concentration N, and noting that $N \gg n_{ie}$, this equation can be rewritten as

$$D \simeq 2D_i^-\left(\frac{N}{n_{ie}}\right) \qquad (4.78)$$

This is of the same form as Eq. (4.70), with $r = 1$, and can be normalized to

$$D_{sur} = 2D_i^-\left(\frac{N_{sur}}{n_{ie}}\right) \qquad (4.79)$$

so that the junction depth is given by

$$x_j = 2.29\left(\frac{N_{sur}}{n_{ie}}\right)^{1/2}(D_i^- t)^{1/2} \qquad (4.80)$$

where (see Table 4.2)

$$D_i^- = 22.9e^{-4.1/kT} \qquad (4.81)$$

A closed-form solution to Eq. (4.78) can also be written in polynomial form [34] as

$$N = N_{sur}(1 - 0.87Y - 0.45Y^2) \qquad (4.82)$$

where

$$Y = \frac{x}{2\sqrt{D_{sur}t}} \qquad (4.83)$$

and results in the same junction depth. This equation can be readily manipulated to determine additional properties of arsenic-diffused layers in silicon [34].

We have noted that arsenic diffusion, while primarily substitutional in nature, involves a small interstitialcy component as well. Thus, slight enhancement of

D_{eff} is observed during diffusion in an oxidizing ambient [35]. This effect is significantly larger for both boron and phosphorus diffusion under oxidizing conditions.

4.4.1.2 Boron in Silicon

Boron diffusions are often made with a low surface concentration, so that a constant diffusivity can be used to characterize them. In some situations, however, high concentrations are required. Here the interaction of $A-V^0$ and $A-V^+$ must be considered in Eq. (4.34), so that

$$D = h\left[D_i^0 + D_i^+\left(\frac{p}{n_i}\right)\right] \tag{4.84}$$

For low surface concentration diffusions, the field enhancement factor can usually be neglected. Qualitatively, however, the boron profile can be expected to be more abrupt than the erfc (i.e., the D = constant) case, the degree of abruptness increasing with higher surface concentrations. Figure 4.21 shows computer simulations [8] for boron diffusion into silicon with varying surface concentration. Also shown is the dilute case, where $p/n_i \ll 1$, which results in an erfc function.

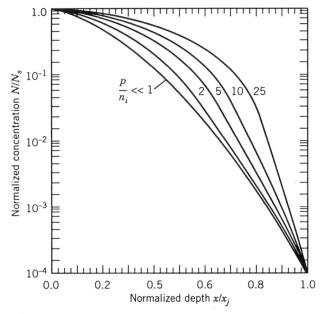

Fig. 4.21 Normalized diffusion profiles for boron in silicon at different doping concentrations.

Empirical data on a large number of high-concentration ($N_{sur} \geq 2 \times 10^{19}$ cm^{-3}) boron diffusions made into background material that is doped to less than 10^{18} cm^{-3} have shown that the junction depth can be approximated [36] by

$$x_j \simeq 2.45 \left(\frac{N_{sur}}{n_i} \right)^{1/2} (Dt)^{1/2} \tag{4.85}$$

where

$$D = 3.17 e^{-3.59/kT} \tag{4.86}$$

For these conditions, the doping profile can be approximated by

$$N \simeq N_{sur}(1 - Y^{2/3}) \tag{4.87}$$

where

$$Y = \left(\frac{x^2}{6D_{sur}t} \right)^{3/2} \tag{4.88}$$

A plot of this profile, for a boron diffusion with high surface concentration, is shown in Fig. 4.22. Also shown for comparison are the profile for arsenic in silicon, and the erfc function which holds for the constant diffusivity case. Note that the boron profile is intermediate to these two.

The enhanced diffusion of boron in oxidizing ambients has been observed during the fabrication of isolation wells in low-threshold MOS circuits, which are usually made on (100) silicon. It can be understood by noting that the oxidation rate for silicon is orientation-dependent. Incomplete oxidation of silicon during diffusion in an oxidizing ambient generates silicon interstitials in excess of their equilibrium value. This, in turn, enhances the diffusivity of boron which moves by the interstitialcy mechanism [37, 38], and also makes it orientation-dependent. It is important to note that the presence of an oxide that is already on the surface of the silicon does not affect the diffusivity of boron, since this does not lead to the formation of silicon interstitials.

4.3.1.3 Phosphorus in Silicon

Phosphorus has a high solid solubility in silicon, and has been extensively used for the fabrication of n^+-regions in MOS and bipolar microcircuits. Its misfit factor is small ($\simeq 0.068$), and an active carrier concentration of 3×10^{20} cm^{-3} can be achieved in practice. Its solid solubility limit is close to 10^{21} cm^{-3}. At high concentrations, the total and electronic concentrations are related [39]

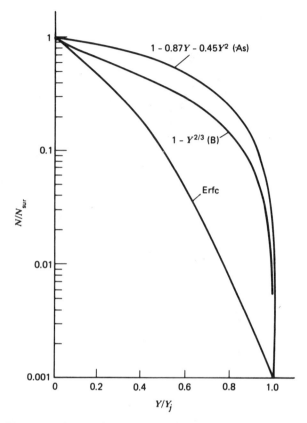

Fig. 4.22 Normalized diffusion profiles for boron and arsenic in silicon.

by

$$N_{tot} = n + 2.04 \times 10^{-41} n^3 \qquad (4.89)$$

for values of N_{tot} from 10^{19} to 10^{21} cm^{-3}. Figure 4.20 shows the electrically active component of phosphorus as a function of temperature.

The anomalous behavior of shallow phosphorus diffusions in silicon is illustrated by the doping profile of Fig. 4.23, which displays both the total and the electronically active phosphorus concentrations. Essential features of the electronically active phosphorus profile are an initial flat-top region, a kink at around 10^{20} cm^{-3}, followed by a rapidly falling concentration with distance, and a tail region beyond this point. The tail region represents an enhancement of the diffusivity. Figure 4.24 shows the effective diffusivity of phosphorus in this region, and illustrates the magnitude of this problem.

Many theories have been advanced to explain this anomalous behavior and provide a quantitative fit to the experimental profiles. One of these [39], based

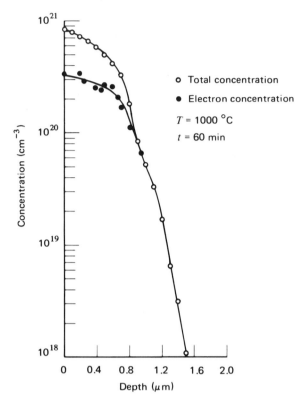

Fig. 4.23 High-concentration phosphorus diffusion profile. From Fair and Tsai [39]. Reprinted with permission of the publisher, The Electrochemical Society, Inc.

on interaction of the phosphorus ion with neutral and charged vacancies, is extensively used to model phosphorus diffusion in silicon. It does not, however, take into consideration the role of interstitials, in spite of the fact that there is ample evidence of phosphorus diffusion by an interstitialcy mechanism [40].

A complete theory that describes the diffusional behavior of phosphorus in a physically consistent manner is not available at the present time. However, it has been proposed that the in-diffusion of phosphorus produces a large supersaturation of self-interstitials, which diffuse into the silicon as well as towards the silicon surface (where they attain their equilibrium value). In the near-surface region, we can expect the role of self-interstitials to be small. Thus, phosphorus–vacancy interaction is the controlling process in this initial, high-concentration region. Here, the dominant term in Eq. (4.34) is associated with $A-V^{2-}$ interactions. Furthermore, because there is essentially no change in concentration in this region, the enhancement factor is unity, and the diffusivity is given by

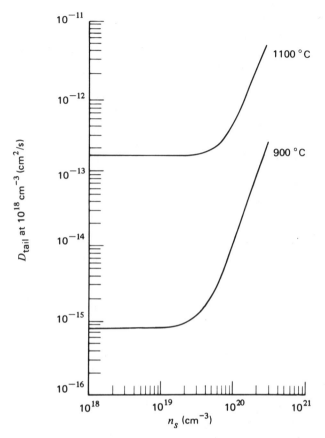

Fig. 4.24 Phosphorus diffusivity in the tail region. From Fair and Tsai [39]. Reprinted with permission of the publisher, The Electrochemical Society, Inc.

$$D_i \simeq D_i^{2-} \left(\frac{n}{n_i} \right)^2 \tag{4.90}$$

where

$$D_i^{2-} = 44.2 e^{-4.37/kT} \tag{4.91}$$

The doping profile for this concentration dependent diffusivity is extremely abrupt, as shown in curve *b* in Fig. 4.16.

The self-interstitial concentration builds up as we move further from the surface so that enhanced diffusion, by the interstitialcy mechanism, dominates the behavior in the tail region. Experimental studies have correlated the extent of the tail with the amount of supersaturation of interstitials. Diffusion techniques which reduce the amount of supersaturation are also found to decrease the prominence of this tail region [41].

In summary, there is a growing consensus that phosphorus diffuses in silicon via a two-stream process, with (a) vacancy-dominated interactions in the initial, high-concentration region and (b) interstitialcy motion in the tail region. At the present time, there is no complete theory to describe its physical behavior, although its characteristics have been modeled by a number of different approaches.

4.4.2   Deep-Lying Impurities

Table 4.4 shows the diffusion constant and the apparent activation energy of diffusion for a number of deep-lying impurities in silicon. It must be emphasized that these data are in the form of average values and estimates. Reliable values are difficult to obtain for the following reasons:

1. These impurities usually diffuse by an interstitial–substitutional mechanism, at about five to six orders of magnitude faster than substitutional diffusers. As a result, considerable error can occur from out-diffusion effects when the specimens are cooled to room temperature. Attempts at rapid quenching usually result in the generation of a large number of crystal defects and obscure the interpretation of data.

2. Their movement is described by a diffusivity which is a function of both concentration and temperature. Thus the data in Table 4.4 are an attempt to cast their behavior in a simplified form, which only holds for substitutional diffusers.

3. On freezing, most deep-lying impurities end up in both electronically active and electronically inactive sites. The fractions of each type differ widely from element to element. Thus about 90% of the gold resides in active sites, while the corresponding figure for nickel is only 0.1%. The analysis of the results is complex, since analytical techniques such as resistivity measurements only provide information on the part that is electronically active. On the other hand, secondary ion mass spectrometry techniques result in information on the entire impurity content.

4. The electronically active part of all these dopants exhibits one or more deep levels. Thus their average diffusion parameters cannot be measured by p–n junction techniques, as for substitutional dopants.

5. The interaction energy between an interstitial–substitutional diffuser and the strain field associated with a dislocation tends to favor clustering in the neighborhood of this dislocation. Thus the diffusion process is dominated by the defect nature of the crystal, and the experimental data are difficult to interpret meaningfully. In addition, many of these impurities form compounds with silicon over certain ranges of diffusion temperature. These tend to segregate

in clusters in the silicon lattice and are often electronically inactive. Material doped with these impurities is usually sensitive to heat treatments.

As a result of the above, the data of Table 4.4 must be considered as highly tentative.

Most impurities exhibit one or more deep-lying levels in silicon. Their presence reduces the minority carrier lifetime in devices, and increases leakage currents in p–n junctions. Thus they are important contaminants, and considerable effort is taken to avoid their inadvertent incorporation during microcircuit processing. Copper, gold, and iron are the most important of these contaminants, largely because they have relatively high solid solubilities in silicon.

4.4.2.1 Gold in Silicon

The intentional introduction of gold in high-speed saturated logic circuits reduces the minority carrier lifetime and constitutes an important fabrication step. Its energy levels are shown in Fig. 4.25a; Table 4.5 lists its capture rate constants [42, 43].

Because of its relatively high solid solubility in silicon (see Fig. 2.20), gold incorporation can be controlled over a wide range of values. Moreover, it does not form any compounds with silicon (as seen from the phase diagram of the gold–silicon system in Fig. 2.7). Thus its behavior is free from anomalous effects that may be caused by the formation or decomposition of these compounds. Such effects are commonly encountered with many deep-lying impurities.

The diffusion of gold in silicon follows that of interstitial–substitutional diffusers which move by the kick-out process. Consequently, its diffusivity is a strong function of the concentration of interstitials, in addition to the doping level and the temperature. Upon diffusion, approximately 90% of the

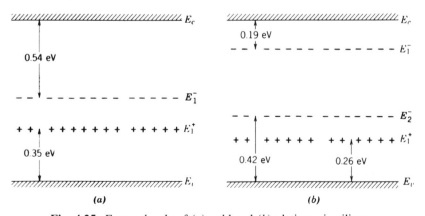

Fig. 4.25 Energy levels of (a) gold and (b) platinum in silicon.

gold freezes out into substitutional sites, where it is electronically active. The remaining gold is interstitially located in the silicon lattice, and is electronically inactive.

Gold in silicon adopts the charge states Au^0, Au^+, and Au^-, whose relative concentrations depend on the position of the fermi level, which is established by the background doping concentration into which it is introduced. In n-type silicon, the gold is primarily in its Au^- state, and is formed by

$$Au^0 + e \rightleftharpoons Au^- \qquad (4.92)$$

Applying the principle of mass action, and noting that $[Au^0]$ is unaffected by the concentration gradient, it has been shown in Section 1.5.3 that

$$[Au] = [Au^0]_i + [Au^-]_i \frac{n}{n_i} \qquad (4.93)$$

$$\simeq [Au^-]_i \frac{N}{n_i} \qquad (4.94)$$

where N is the background concentration. Thus, the solid solubility of gold in n-type silicon is greatly enhanced [44] at high background doping levels. The effectiveness of gold doping is thus related to the doping concentration of the material in which it is introduced.

Gold behaves as a donor in p-type silicon, where it can be shown that

$$[Au^+] \simeq [Au^+_i] \frac{N}{n_i} \qquad (4.95)$$

where N is the background (p-type) concentration. Here, however, the enhanced solid solubility is not as great as in the case of n-type silicon, because of the relative position of the energy levels of gold.

Gold diffusion is generally the last step in the fabrication of high-speed, digital microcircuits. Thus diffusion takes place into a wafer in which various doped regions are already present, each of different defect concentration, depending on the type and amount of impurity in it. In actual practice, no attempt is made to establish a specific doping profile, and the wafer is doped by maintaining it an an elevated temperature for what is, for all practical purposes, an infinitely long time (15–30 min). Thus varying amounts of gold are present in the entire microcircuit. The diffusion temperature is altered to adjust the gold concentration in the silicon. However, the actual concentration is also a function of such variables as the manner in which the wafer is cooled to room temperature. The choice of the actual diffusion temperature is determined by trial and error, with a desired end effect in mind.

4.4.2.2 Platinum in Silicon

Intentional doping with platinum has also been investigated, as a substitute for gold in silicon [45, 46]. Like gold, it has a relatively high solid solubility, and diffuses predominantly by an interstitial-substitutional process involving the kick-out mechanism. About 90% of the diffused platinum ends up in substitutional sites and is electronically active, with a donor level E_1^+ and an acceptor level E_1^- as shown in Fig. 4.25b. Unlike gold, however, the remainder* is also active, and exhibits an acceptor level E_2^-. This is the dominant level for lifetime control. The capture rate constants of platinum are listed in Table 4.5.

The impurity levels of platinum in silicon are highly asymmetric, whereas one gold level (the acceptor) is almost in the center of the gap. As a result, platinum doping can be used to reduce the lifetime in silicon [46] without a proportional increase in the leakage current, as is the case with gold. This is especially important for high-voltage devices, where leakage currents must be kept very low in order to minimize off-state power dissipation.

4.5 IMPURITY BEHAVIOR: GALLIUM ARSENIDE

The preferred method for junction formation in GaAs is by ion implantation, followed by heat treatment. An alternative approach, which is widely used, is epitaxial growth. As a result, extensive studies of the fundamental mechanisms for the diffusion of impurities in GaAs have not been made in recent time. Nevertheless, there is a wealth of experimental data (some conflicting in nature) from which the basic diffusion characteristics can be enunciated. Only one impurity, zinc, has been studied in detail, and many of the results have been extrapolated to other impurities.

Radioactive tracer analysis of the diffusion of gallium and arsenic in GaAs gives the following values for the diffusion coefficients:

$$D_{Ga} = 0.1e^{-3.2/kT} \text{cm}^2 \text{s}^{-1} \tag{4.96a}$$
$$D_{As} = 0.7e^{-5.6/kT} \text{cm}^2 \text{s}^{-1} \tag{4.96b}$$

These very different rates indicate that self-diffusion of gallium and arsenic occurs along separate sublattices, with each atom moving on its own sublattice. This argument has been extended to impurities in GaAs as well. Typically, impurities from group II (beryllium, cadmium, magnesium, mercury, and zinc) are shallow and p-type, and move along the gallium sublattice. This cannot be accomplished by substitutional diffusion alone, since each gallium site has four arsenic sites as its nearest neighbors. It is generally accepted that these impurities move by an interstitial–substitutional mechanism.

---

*A distorted substitutional site has been proposed for this species.

The characteristics of interstitial–substitutional diffusers in GaAs are very similar to those in silicon. Thus, the substitutional solid solubility is much larger than the interstitial; the interstitial diffusivity is orders of magnitude larger than the substitutional, and dominates the diffusion process; and the diffusion is enhanced in regions of high concentration but is retarded in regions of low concentration.

Impurities from group VI (selenium, sulfur, and tellurium) are shallow and n-type. Here a divacancy mechanism has been proposed for this type of movement. This approximates substitutional behavior, so that these impurities are extremely slow diffusers. Impurities from group IV (carbon, germanium, silicon, and tin) can be either n- or p-type, depending on the sublattice on which they are preferentially located. It is assumed that they move on both sublattices, but no attempts have been made to propose a detailed diffusion model for them. These have also been found to be extremely slow diffusers. Finally, most other impurities are deep in nature, often exhibiting multiple levels. No information is available on their mechanism of diffusion. In spite of these many uncertainties, it is common practice to cast the diffusion data in the Arrhenius form, $D = D_0 e^{-E_d/kT}$. Values for D_0 and E_d are given in Table 4.6 for various impurities [2].

4.5.1   Zinc in Gallium Arsenide

Zinc is an important interstitial–substitutional diffuser in GaAs, and has been well studied for this reason. It is commonly used for diffusions with a high surface concentration ($\geq 10^{20}$ cm^{-3}). For these conditions, the vacancy concentration maintains its equilibrium value. Zinc movement is by the dissociative mechanism, so that its effective diffusivity is given by Eq. (4.40) as

$$D_{\text{eff}} = D_i \frac{N_i}{N_s} \tag{4.97}$$

where D_i is the interstitial diffusivity, and the superscript eq is removed for simplicity.

The nature of the dissociative reaction determines the ratio of N_i/N_s in this equation. Here, it is generally accepted that substitutional zinc is a singly ionized acceptor, whereas interstitial zinc is a singly ionized donor. For this situation, the dissociation reaction can be written as

$$\text{Zn}_s^- \rightleftharpoons V_{\text{Ga}}^- + \text{Zn}_i^+ + e^- \tag{4.98}$$

where Zn_s^- and Zn_i^+ represent substitutional and interstitial zinc, respectively, and V_{Ga}^- is the gallium vacancy. Applying the mass-action principle gives

$$k_3 = \frac{[V_{Ga}^-]N_i n}{N_s} \tag{4.99}$$

But

$$V_{Ga}^- \rightleftharpoons V_{Ga}^0 + e \tag{4.100}$$

so that

$$k_4 = \frac{[V_{Ga}^0]n}{[V_{Ga}^-]} \tag{4.101}$$

Combining Eqs. (4.99) and (4.101), and noting that the hole concentration is due to the substitutional zinc, we obtain

$$k_3 k_4 = \frac{[V_{Ga}^0]N_i n_i^4}{N_s^3} \tag{4.102}$$

Substituting into Eq. (4.97), it follows that

$$D_{eff} \simeq D_i \left\{ \frac{k_3 k_4}{[V_{Ga}^0]n_i^4} \right\} N_s^2 \tag{4.103a}$$

$$\simeq D_i \left\{ \frac{k_3 k_4}{[V_{Ga}^0]_i n_i^4} \right\} N_s^2 \tag{4.103b}$$

since the concentration of neutral flaws is independent of the localized doping concentration.

For any specific diffusion temperature, the term in the brackets is fixed. Consequently, Eq. (4.103b) can be rewritten in the normalized form of Eq. (4.71), where $r = 2$. Diffusion profiles of zinc in GaAs closely follow curve b in Fig. 4.16, which represents this case, and result in a junction depth of 1.092 $(D_{sur}t)^{1/2}$.

At extremely high surface concentrations, in excess of 10^{20} cm^{-3}, workers have observed the presence of a kink which modifies the shape of curve b in Fig. 4.16. This phenomenon is not fully understood at the present time. However, a number of theories have been advanced to explain this behavior. Thus, it has been proposed that zinc diffusion occurs partially by the kick-out mechanism [47]. A second theory is based on the proposition that the Zn_i donor has multiple ionization states [48]. More complex theories, which include the generation of interstitial $(Zn-V_{Ga}^-)$ pairs [49], have also been advanced to explain this anomalous behavior.

The surface concentration can be determined by an analysis of the incorporation reaction for zinc vapor into GaAs, in the form of substitutional zinc. Assume a diffusion situation during which the vapor pressures of zinc and arsenic are held at p_{Zn} and p_{As_2}, respectively. The incorporation reaction is given as

$$Zn(g) + V_{Ga}^- \rightleftharpoons Zn_s^- \tag{4.104}$$

so that

$$k_5 = \frac{N_s}{p_{Zn}[V_{Ga}^-]} \tag{4.105}$$

where k_5 is an equilibrium constant. Combining with Eq. (4.101), and noting that the hole concentration is due to substitutional zinc, gives

$$\frac{k_5}{k_4} = \frac{N_s^2}{p_{Zn}n_i^2[V_{Ga}^0]} \tag{4.106}$$

However, from Eq. (1.25a) we obtain

$$[V_{Ga}^0] = \frac{k_2}{k_1}p_{As_2}^{1/2} \tag{4.107}$$

Combining with Eq. (4.106) gives

$$N_s^2 = \frac{k_1 k_5 n_i^2}{k_2 k_4} p_{Zn} p_{As_2}^{1/2} \tag{4.108a}$$

$$N_s = k p_{Zn}^{1/2} p_{As_2}^{1/4} \tag{4.108b}$$

Thus the surface concentration is proportional to the square root of the zinc partial pressure, and is a very weak function of the arsenic pressure. In practice, the square-root dependence is seen to hold for doping concentrations below 10^{19} cm^{-3}. Above this point, however, a slight departure from this law is observed. This can be explained by noting that the above analysis holds for low-concentration diffusions, where the mass-action principle applies. However, it has been shown [50] that equations of the same form can be developed by introducing the concept of a hole activity coefficient γ_p linking the Boltzmann and Fermi statistics that apply to the low- and high-concentration cases, respectively. This activity coefficient is unity at low concentrations, but falls rapidly to about 0.5 at hole concentrations above 3×10^{19} cm^{-3} (for diffusions made at 700°C), and remains constant beyond this point.

The effect of zinc and arsenic overpressures on the junction depth can also be determined. Combining Eqs. (4.103a) and (4.106) gives

$$D_{\text{eff}} = \frac{k_3 k_5}{n_i^2} p_{\text{Zn}} D_i \qquad (4.109a)$$

$$= K^2 p_{\text{Zn}} D_i \qquad (4.109b)$$

From a solution of the concentration-dependent diffusion equation, it follows that the junction depth is given by

$$x_j = 1.092 K p_{\text{Zn}}^{1/2} (D_i t)^{1/2} \qquad (4.110a)$$

The effect of surface concentration on the junction depth can be obtained by combining this equation with Eq. (4.108b). Thus

$$x_j = 1.092 (K/k)^{1/2} p_{\text{As}_2}^{-1/4} N_s (D_i t)^{1/2} \qquad (4.110b)$$

Thus the junction depth is relatively independent of the arsenic pressure. This has been observed in a comparison of diffusions made with a Ga–As–Zn alloy source to those made with an elemental zinc source [51]. The zinc source provided an increase in p_{Zn} by a factor of approximately 3, and an increase in the arsenic pressure by a factor of approximately 10^8. However, junction depths were larger by a factor of only 10 with this source.

4.5.2 Silicon in Gallium Arsenide

The importance of silicon as an n-type dopant in GaAs is due to the fact that silicon implants can be annealed with relative ease. The diffusional behavior of silicon during the annealing process has been studied in epitaxial GaAs [52]. Using the formalism of Eq. (4.34), it has been shown that the diffusivity can be approximated by

$$D_{\text{eff}} = D_i^{2-} \left(\frac{n}{n_i} \right)^2 \qquad (4.111a)$$

$$\simeq 60.1 e^{-3.9/kT} \left(\frac{n}{n_i} \right)^2 \text{cm}^2 \text{ s}^{-1} \qquad (4.111b)$$

Interaction with V_{Ga}^{2-} is thus the primary diffusional mechanism which has been proposed. Other workers, using studies of AlGaAs/GaAs superlattice disorder [16], have proposed that point defect interactions with V_{Ga}^{3-} are dominant. The discrepancy between these results may be due to complicating factors associated with the motion of aluminum in the latter case.

4.6   DIFFUSION SYSTEMS

Many aspects of diffusion technology are common to both silicon and GaAs. This section deals with these aspects. The special problems of these materials are considered at a later point.

The basic requirements of any diffusion system are that a means be provided for bringing the diffusing impurity in contact with a suitably prepared slice, and maintained there for a specified time and temperature. Within the broad framework of these two requirements, the following additional features are also desired:

1. The diffusion process should not result in damage to the surface of the slice. This is an extremely important requirement, because the entire microcircuit is fabricated within the first few micrometers below this surface.

2. After diffusion, material residing on the surface of the slice should be capable of easy removal, where desired. This is a necessary requirement if subsequent processing, such as contacting or additional diffusions, is needed.

3. The system should give reproducible results from one diffusion run to the next and from slice to slice within a single run.

4. The system should be capable of processing a number of slices at a time; that is, it should be capable of batch processing.

Diffusion systems may be of either sealed- or open-tube type. In the sealed system the slices and dopant are enclosed in a clean, evacuated quartz tube prior to heat treatment. After diffusion the slices are removed by breaking the tube. Systems of this type can be easily maintained free from contamination. In general, their use for silicon diffusions has been restricted to the laboratory because of the inconvenience of sealing and unsealing these tubes. Arsenic diffusions are sometimes made by this technique (see Section 4.7.3). Gallium diffusions, for high-voltage power devices, are also made by this technique [46].

Sealed-tube systems operate by thermal evaporation of the dopant source, transport in the gas phase, adsorption on the surface of the semiconductor and the tube walls, and eventual diffusion of the dopant into the slices. As a result, surface conditions can influence the outcome of the diffusion by altering the sticking probability [53], which determines the amount of material exchange. The residual gas pressure in the ampoule can also affect the diffusion process. Pressures in excess of 10 torr can cause problems with dopant transport, presumably by covering the surface with residual contamination. On the other hand, pressures below 0.01 torr can result in a mean free path for dopant atoms which exceeds the capsule dimensions, and lead to shadowing effects. A pressure of around 1 torr has been shown to be a satisfactory compromise.

Sealed tube diffusions for silicon devices are usually carried out at temperatures in the 1200–1225°C range. Here, pressures as high as 200–250 torr are used, to prevent undue mechanical strain in the ampoule at the diffusion temperature.

Surface concentrations obtained in sealed-tube diffusion can be made to approach the solid-solubility limit for the impurity at the diffusion temperature, provided that the dopant surface area is large compared to the rest of the system. This is achieved by using a granular dopant source. In addition, the diffusion time must be sufficient so that both slice and wall coverage attain steady-state conditions. Thus long, deep diffusions are favored for this process.

The open-tube method is favored for practical diffusion systems. Here slices are placed in a clean, horizontal diffusion tube made of high-purity fused quartz. A separate diffusion furnace, diffusion tube, and slice carrier are reserved for each impurity because they are contaminated with it during the process. Insertion is done (either by hand or mechanically) from one end of the tube, whereas the other end is used for the flow of gases or impurities in vapor form. This method is capable of handling a number of slices at one time and is considerably more convenient than the sealed-tube system.

Vertical furnaces represent a recent development in diffusion systems. These have the advantage that no contact is made between the slice carrier (and slices) and the tube walls at any time, thus minimizing dust problems. Moreover, they occupy less space on the factory floor than their horizontal counterparts, and can be equipped with cassette slice carriers for automatic loading and unloading.

The process of inserting slices into a diffusion tube, as well as their removal, results in radial thermal gradients. These can cause crystal damage in the form of dislocations and slip [54], with a resulting fall in the minority carrier lifetime. In extreme situations, loss of planarity results, causing the slice to take on a "potato-chip" shape.* These problems have become increasingly severe with the trend toward the use of large-diameter slices. One approach to alleviating them is to insert and remove the slice carrier in a slow, controlled manner, using a mechanical puller. A pull rate on the order of 10 cm/min is typical. A disadvantage of this approach is that wafers at the front end of the slice carrier are diffused for a longer time than those at the rear. A better approach is to insert the slice carrier into a relatively cool furnace, and bring it up to the diffusion temperature at a linear rate. Modern diffusion furnaces use a programmed controller to perform this ramp-up function, as well as ramp-down at the end of the diffusion. Typical ramp rates are 3–10°C/min for high-temperature diffusions.

Consider a furnace whose temperature is ramped down from T_0 with a linear ramp at rate C. Then it can be shown (see Section A.2.1) that the ramp-down

---

*On the other hand, slow cooling results in an improved lifetime in heavily doped regions, since it prevents precipitation in the diffused layer.

process is equivalent to an extra diffusion time of $kT_0^2/(CE_d)$ at the initial diffusion temperature, where E_d is the activation energy of diffusion and k is Boltzmann's constant. This is usually a small fraction of the total diffusion time. The problem of ramp-up can be handled in an analogous manner.

Diffusion furnaces are usually operated at temperatures ranging from 600° to 1200°C, and are equipped with feedback controls to maintain a central flat zone with a ±0.5°C tolerance. The length of this zone varies from 10 cm for laboratory systems to 100 cm for industrial units. In large systems, as many as five heater zones are used to control the temperature profile.

4.6.1 Choice of Source

Diffusion proceeds by heat treatment of the semiconductor with which the impurity species is in contact. This serves to drive in the impurity to the desired depth and surface concentration, as described earlier. The end goal is to achieve doping uniformity across the surface area, consistency from the front to the rear of multiple slices, and repeatability from run to run. One approach is to bring a large amount of the impurity, or an impurity-bearing compound, in contact with the slice by means of vapor-phase transport.

Solid source techniques can be used for this purpose. Figure 4.26 shows a sketch of the diffusion system in which a platinum boat is used to hold a solid source of the dopant species upstream from the carrier with the semiconductor wafers. In operation, the carrier gas transports vapors from this source and deposits them on the semiconductor slices. Source shutoff is usually accomplished by moving the dopant source to a colder region of the furnace.

The success of this technique depends critically on the vapor pressure of the source. In some cases, this necessitates a two-temperature furnace, with the source maintained at a lower temperature than is used for the diffusions.

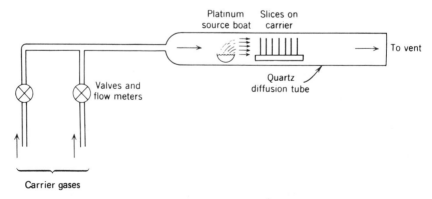

Fig. 4.26 Solid-source diffusion system.

Often, however, the source boat and the slices can be maintained at the same temperature, avoiding the need for a two-zone furnace.

Liquid sources can also be used. Here a carrier gas is bubbled through the liquid which is transported in vapor form to the surface of the slices. It is common practice to saturate this gas with the vapor so that the concentration is relatively independent of gas flow. The surface concentration is thus set by the temperature of the bubbler and of the diffusion system (see Fig. 4.27). Additional lines are also provided for other gases in which the diffusion is performed, and for flushing out the system after use.

Liquid-source systems are extremely convenient since the doping process can be readily initiated (or terminated) by control of the gas through the bubbler. In addition, the amount of dopant transported to the slices is relatively easy to control in these systems, by adjustment of the bubbler temperature. Finally, a number of halogenic dopant compounds are available as liquids. Use of these sources greatly reduces heavy metal contamination in diffusion systems.

Gaseous sources are even more convenient than the liquid ones. Again, it is common practice to use an excess dopant gas concentration, so that these systems are relatively insensitive to the gas-flow rate. Figure 4.28 shows the schematic for a typical diffusion system using a gaseous dopant source. Here provision is made for an ambient carrier gas in which the diffusion takes place. In addition, a chemical trap is often incorporated to dispose of unreacted dopant gases, which are often highly dangerous.

All vapor transport methods rely on the fact that the surface concentration of the incorporated dopant is solid-solubility-limited, so that it is relatively insensitive to the vapor pressure of the reactant species. Still, massive depletion of the source reactant is possible as it travels down the diffusion tube. Additional

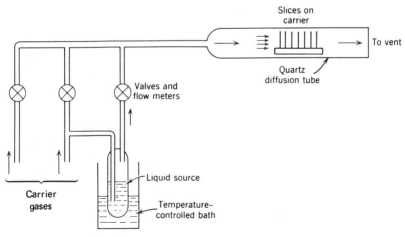

Fig. 4.27 Liquid-source diffusion system.

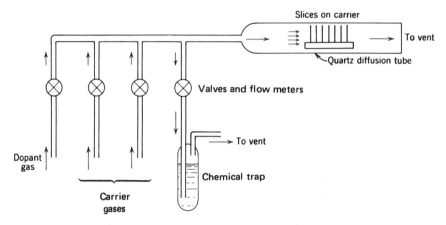

Fig. 4.28 Gaseous-source diffusion system.

depletion occurs during transport of the reactant vapor between slices. Thus, it is difficult to maintain doping uniformity over all the slices during a diffusion run, and even over the surface area of each individual slice. Both these problems are exacerbated with the present trend towards large slice diameters, and the use of a large number of slices in a single run, which necessitates close spacings between individual slices.

Spin-on techniques can be used for obtaining uniform source coverage over a slice. Here, a dilute mixture of a dopant compound in an organosilane is used as the source chemical [55]. The semiconductor slice is held in a vacuum chuck and rotated at high speed (2500–5000 rpm). A drop of the source chemical is next applied to form a thin layer ($\approx$5000 Å) across the slice by means of centrifugal force. With appropriate attention to proper viscosity control, relatively uniform layers of dopant can be obtained in this manner. Upon heat treatment above 200°C, the organosilanes decompose into SiO_2, which serves as an inert matrix for the dopant.

Spin-on techniques are versatile, and almost any dopant for silicon or GaAs can be applied in this manner, by the addition of a few drops of source chemical to an undoped organosilane. A side benefit of this approach is that the SiO_2 layer serves as a cap during the diffusion process. This is especially advantageous with GaAs, since it inhibits its decomposition during diffusion. The degree of control provided by these sources is poor by modern microcircuit standards, so that they are only used in high-concentration diffusions, where dopant control is not critical.

Planar dopant sources are another approach for avoiding this problem. These can be directly placed on the semiconductor slice by chemical vapor deposition techniques, which are described in Chapter 8. They have the disadvantage of requiring an extra processing step; however, their use avoids the necessity for uniform transport of dopant in vapor form to the slice. As a result, they can be used in VLSI applications, where tight diffusion control over large-diameter slices is important. Additional flexibility is provided by their use since they can be patterned to allow selective diffusions on the semiconductor slice. This technique also provides a protective cap layer to be deposited in place after the dopant source, and is especially important for diffusions in GaAs for this reason.

Very dilute dopant concentrations can be obtained with solid sources by adjusting the dopant-to-binder ratio. Diffusion from these sources results in a surface concentration which is controlled by the concentration of the dopant in the oxide and not by its solid-solubility limit in the semiconductor [56]. Thus, they can be used for limited source diffusions of the type described in Section 4.3.1.2.

In silicon technology, solid planar sources, consisting of dopant species and a binder, and fabricated in the form of discs, can be used to provide vapor transport over large diameter slices [57]. These discs, of the same diameter as the silicon slices, are stacked in the arrangement shown in Fig. 4.29, with each dopant slice acting as a source for two adjacent microcircuit slices. Typically 2- to 3-mm spacing is provided between each source and silicon slice. Uniform vapor transport occurs, provided that the spacing between the dopant and the semiconductor slices is held constant. Impurities in these sources are held to less than 1 ppm each for a wide range of heavy metals, so that they are adequate for microcircuit fabrication.

An important advantage of these sources is that they require extremely simple gas handling systems, and present no environmental problems with effluent handling. Typically, all that is needed is a flow of a few liters of nitrogen (or argon) per minute, to avoid back-diffusion effects. In some systems, oxygen (and/or water) is also used for controlling the extent of vapor transport.

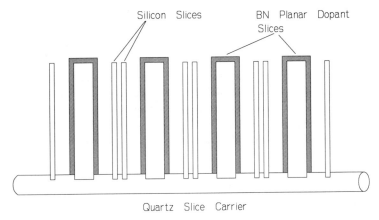

Fig. 4.29 Planar source arrangement.

Solid sources are available for common dopants (As, B, P) which are used for silicon technology. With each, careful attention must be paid to proper handling of the dopant discs in order to obtain run-to-run reproducibility. Often, pretreatment of the discs is necessary in order to freshen their surface prior to diffusions. Detailed procedures are provided by the manufacturer of each product in order that this can be accomplished.

Ion implantation is an alternative method for obtaining uniform impurity diffusion over large area slices. This is a technique by which impurity atoms, traveling at high energy, are made to impinge on the semiconductor and are thus deposited in it. Details of this important process are treated in Chapter 6. This approach to semiconductor placement of the dopant on (or rather, in) the semiconductor has a number of significant characteristics. First, extremely high-purity sources are available, since *in-situ* mass spectrometric techniques are used to separate them. Next, impurity deposition is done by scanning the ion beam, so that it can be made uniform over large diameter slices. Finally, *in-situ* dopant monitoring techniques are available, so that a dopant accuracy of 1% is readily achievable (as compared to 5–10%, which is the best that can be achieved by the approaches mentioned earlier). Moreover, this control is available over many decades of dopant concentration, so that both limited and unlimited source diffusions can be made by this technique.

A disadvantage of this process is that it results in considerable damage to the semiconductor. This damage is relatively easy to remove for light ion doses, if they are followed by a high-temperature diffusion step. Conversely, heavily doped shallow regions, which require high ion doses and relatively low subsequent diffusion temperatures, are difficult to anneal by this means. In spite of these problems, ion implantation is the technique of choice for placement of the dopant on the slice prior to the actual diffusion step, in both silicon and GaAs technology.

4.6.1.1 Halogenic Dopant Sources

The prevention of unwanted impurities, as well as their removal, plays an important part in any semiconductor process. The general aspects of this problem have been covered in Section 1.5.4. Here, we shall address issues relevant to the diffusion step. Contamination control is especially important with open-tube diffusions, where the slices are subjected to high-temperature processing, and impurities can enter from the surroundings by back-diffusion. It goes without saying that all gases and dopant sources should be of semiconductor grade. In addition, chemical cleaning of the semiconductor is of special importance, and is discussed in detail in Chapter 9.

The diffusion furnace itself is a major source of impurity contamination. Commercial diffusion tubes, made of fused quartz, are typically 96–97% pure; thus they must be cleaned upon installation, and also regularly during service. Typically this is accomplished by flowing anhydrous hydrogen chloride gas through them for 15–30 min at diffusion temperatures. This tends to leach out impurities which are removed by conversion to their more volatile halides.

The firebricks and heater elements in a diffusion furnace are a major source of alkali ion and copper contamination. In addition, since they are hotter than the semiconductor slices, there is a thermal gradient which assists the transport of these impurities into the diffusion tube, by rapid movement through the quartz walls. The use of a high-purity, high-density mullite liner (made of mixtures of alumina and zirconia) is common practice since this serves as a barrier to the transport of alkali ions. Liners of this type are especially important in MOS-based fabrication processes, where alkali ions contamination critically affects the threshold stability of active devices.

Ultra-pure silicon carbide liners can also be used, and will block the transport of both sodium and copper. Although considerably more expensive than mullite liners, they have great potential for reducing furnace-associated contamination in diffusion systems.

The unwanted incorporation of heavy metallic impurities (such as iron, copper, gold, etc.) results in a fall in minority carrier lifetime, and in an increase in the leakage current in p–n junctions. Thus their removal is important for bipolar as well as field effect devices. One approach here is to use dopant sources which are halogenic compounds. With these the halogen is liberated during diffusion, and reacts with any metallic impurities in the incoming gas, as well as with impurities within the semiconductor that reach its surface during their rapid movement at high temperatures. This reaction converts them to their more volatile halides, which leave the system by incorporation into the gas stream. The use of a halogenic dopant source thus effectively getters the semiconductor. These sources must be used with care, however, since excessive use can result in local dissolution of the semiconductor and cause its pitting.

Halogenic source chemicals are available for most dopants used in diffusion technology. They are especially useful during later steps in the manufacturing process, where they serve to "clean up" the contamination incorporated during the early phases. An important disadvantage of ion implantation is that halogenic scavanging of impurities is not possible with this approach.

4.7 DIFFUSION SYSTEMS FOR SILICON

Diffusion systems for silicon are of the open-tube type, with quartz being the most common tube material. Tubes of semiconductor grade silicon are sometimes used for this purpose. Although considerably higher in initial cost, they are extremely clean, and block the movement of contaminants such as sodium through their walls. In addition, their useful life is very long compared to quartz tubes since they do not devitrify during use and are impervious to thermal shock.

The *p*-type impurities are aluminum, boron, gallium, and indium. Figure 2.20 indicates the maximum solid solubilities that are attainable using these impurities. Of these, the only practical choice is boron, since indium is actually a moderately deep-lying impurity in silicon, with an acceptor level at 0.16 eV above the valence band. Furthermore, neither aluminum or gallium can be masked by silicon dioxide. Although other forms of masking are possible, they have not received much commercial use in microcircuit technology.

The *n*-type impurities are antimony, arsenic, and phosphorus. None of these exhibits any undesirable characteristic that precludes their use in microcircuit fabrication. As seen from Fig. 2.20, they are all highly soluble in silicon; consequently, the choice of impurity depends on the specific application for which each is uniquely suited. Thus,

1. The diffusivity of phosphorus is comparable to that of boron, and is about ten times larger than that of arsenic or antimony. It is used in the majority of diffusion applications, because it is both uneconomical and undesirable to operate diffusion systems for longer periods of time than necessary.

2. The low diffusivity of arsenic and antimony makes these impurities ideal for use in early phases of device fabrication, since they are relatively less affected by subsequent fabrication steps. They are also more suitable for the fabrication of shallow junctions, since they do not exhibit the anomalous behavior associated with phosphorus diffusion (see Section 4.4.1.3).

3. Arsenic can provide a considerably higher useful surface concentration than antimony, because of its excellent fit to the silicon lattice. Most arsenic sources are highly poisonous, and require elaborate precautions for safe handling. In addition, surface depletion of arsenic by evaporation is often a serious problem during diffusion. Antimony is somewhat easier to use than arsenic, because of its lower vapor pressure. It is occasionally chosen on the mistaken assumption that it is less dangerous. Ion implantation is extensively used for arsenic diffusions, since it avoids these problems.

Specific aspects of both *p*- and *n*-type diffusion processes are now considered.

4.7.1 Boron

Boron has an intrinsic diffusivity of about 10^{-12} cm^2 s^{-1} at 1200°C. It has a high solid solubility and can be diffused with a surface concentration as large as 4×10^{20} atoms cm^{-3}. Its tetrahedral radius is 0.88 Å, corresponding to a misfit factor $\epsilon = 0.254$. As a result, the presence of large amounts of boron in the silicon lattice is accompanied by strain-induced defects which lead to considerable crystal damage. This sets an upper limit of about 5×10^{19} atoms cm^{-3} to the impurity concentration that can actually be achieved in practical structures, with the rest being electronically inactive.

Boron diffusion is accomplished by means of a surface reaction between boron trioxide (B_2O_3) and the silicon slice, given by

$$2B_2O_3 + 3Si \rightleftharpoons 4B + 3SiO_2 \qquad (4.112)$$

Diffusion proceeds from elemental boron, produced in this manner. Excessive amounts of B_2O_3 can lead to the formation of silicides and other compounds of boron on the silicon surface. This *boron skin* causes a dark brown stain, is electrically insulating in character, and often results in device failure due to open contacts.

Once formed, boron skin is difficult to remove in any acid. However, wet oxidation of the silicon can be used to convert this layer into a borosilicate glass which can then be dissolved in hydrofluoric acid. At any rate, it is far easier to avoid this skin by carrying out diffusions in a weakly oxidizing atmosphere (3–10% oxygen by volume), which promotes the formation of some SiO_2.

Spin-on sources for boron diffusion may involve the use of mixtures of carborane and an alkylsiloxane. An initial bake out before diffusion results in their decomposition into B_2O_3 and SiO_2, respectively.

Disks of boron nitride are commonly used as planar dopant sources for boron diffusion [58]. Typically, they must be preoxidized at 750–1100°C for about 30 min in order to form a thin skin of B_2O_3 on their surface, which serves as the diffusion source. This preoxidation step must be done on a regular schedule, depending on the frequency of usage, if consistent results are to be maintained.

There are a number of limitations to the use of solid sources of this type. Thus, the rate of transfer of B_2O_3 to the silicon slice is very sensitive to traces of oxygen or water vapor in the diffusion furnace. This fact has been exploited in an approach where a small amount of water vapor is briefly injected into the furnace ambient [59]. This transforms some of the B_2O_3 on the BN wafer into boric acid, HBO_2, whose vapor pressure is several orders of magnitude higher than that of B_2O_3. Thus, the amount of dopant vapor which is transported can be controlled by the injection period of the water vapor.

Diffusions from boron nitride sources generally result in high surface concentrations. Their doping profiles follow closely the shape described in Section 4.4.1.2 [60], rather than the erfc function. With these sources, the B_2O_3 (or HBO_2) on the silicon slice is not diluted with SiO_2, so that boron skin can be a problem. One approach to preventing this is to use slices consisting of mixtures of B_2O_3 with materials such as BaO, MgO, Al_2O_3, and SiO_2 [61]. All of these have very high negative free energies of formation as compared to SiO_2, and are not reduced to their elements during diffusion. Different concentrations of B_2O_3 are available, so that it is possible to obtain sheet resistances from 1 to 200 Ω/ square by direct use of these sources. With proper source preparation and storage, it is possible to obtain a doping uniformity of $\pm 2\%$ across the slice, $\pm 3\%$ along the length of the boat, and $\pm 4\%$ from run to run by this approach.

Various liquid sources may be used in the system shown in Fig. 4.27. In each case a preliminary reaction results in the deposition of B_2O_3 on the silicon slices. For *trimethylborate (TMB)*, the preliminary oxidizing reaction is

$$2(CH_3O)_3B + 9O_2 \xrightarrow{900°C} B_2O_3 + 6CO_2 + 9H_2O \qquad (4.113)$$

The vapor pressure of TMB is usually maintained at 0°C since it is extremely volatile at room temperatures.

Boron tribromide is a halogenic source, and can be used for gettering of metallic impurities during diffusion. Here,

$$4BBr_3 + 3O_2 \rightarrow 2B_2O_3 + 6Br_2 \qquad (4.114)$$

There is a tendency for pitting to occur if excessive vapor concentrations are used or if the carrier gas is not sufficiently oxidizing in nature. Since bromine is a reaction product, it is essential that provision be made for venting the system.

The preliminary reaction with *boron trichloride* is

$$4BCl_3 + 3O_2 \rightarrow 2B_2O_3 + 6Cl_2 \qquad (4.115)$$

Again, system venting is necessary, as well as care to avoid halogen pitting effects at high concentrations.

It is considerably more difficult to obtain uniform diffusions from one end of the slice carrier to the other with BCl_3 than with BBr_3. This is because the oxidizing reaction of BCl_3 takes a long time (as much as 100 s in some situations), whereas that for BBr_3 is relatively short (on the order of 5 s for similar operating parameters). Consequently, all the BBr_3 is rapidly converted to B_2O_3, and is available for use as a diffusion source, before delivery over the entire length of the slice carrier. With BCl_3, on the other hand, much of the B_2O_3 is not available to slices at the leading end of the carrier because of its slow rate of oxidation [62].

The BCl_3 reaction is greatly accelerated in the presence of water vapor. It is customary to provide this by introducing a small amount of hydrogen into the furnace during the predeposition phase, together with the oxygen flow.

Gaseous sources are used in the configuration of Fig. 4.28. Again, B_2O_3 is formed as a primary reaction product. *Diborane* is a highly poisonous, explosive gas which can be used for this purpose. It is used in a 99.9% argon dilution by volume, which is reasonably convenient to handle. The preliminary oxidizing reaction for the system is

$$B_2H_6 + 3O_2 \xrightarrow{300°C} B_2O_3 + 3H_2O \qquad (4.116)$$

Since only water is released from this reaction, the method is not prone to the pitting effects experienced with BBr_3. Venting the system, as well as the installation of an input trap with a weak hydrochloric acid solution, is essential in order to capture any unreacted diborane during this process.

Higher surface concentrations can be obtained, especially for low-temperature (800-900°C) predeposition, by using carbon dioxide gas [63] instead of oxygen. Here

$$B_2H_6 + 6CO_2 \rightarrow B_2O_3 + 6CO + 3H_2O \tag{4.117}$$

This is because less SiO_2 is formed in this process, since carbon dioxide is a weaker oxidizing agent than oxygen.

4.7.2 Phosphorus

Phosphorus has a diffusivity which is comparable to that of boron. Its misfit factor is considerably lower (0.068 as compared to 0.254), so that an active carrier concentration of 3×10^{20} cm^{-3} can be achieved in practical structures. Its solid solubility limit is close to 10^{21} cm^{-3}.

The most commonly used compound that participates in phosphorus diffusion is P_2O_5. At diffusion temperatures, the surface reaction from which diffusion proceeds is given by

$$2P_2O_5 + 5Si \rightleftharpoons 4P + 5SiO_2 \tag{4.118}$$

Phosphorus-bearing compounds, used for this purpose, ultimately result in the transport of P_2O_5 to the surface of the silicon slice. This readily combines with SiO_2 to form a phosphosilicate glass (PSG) that is viscous at diffusion temperatures. The phase diagram for the SiO_2–P_2O_5 system is presented in Fig. 2.18 and shows those regions that are relevant to the diffusion process.

Spin-on techniques have been used successfully for phosphorus diffusions. Organic compounds of phosphorus, such as triphenylphosphate $[(C_6H_5O)_3PO]$, are used in a suitable organosilane mixture. Phosphorus diffusion can also be conducted in systems of the type used for boron. A brief description of these follows.

Solid-Source Systems. It is difficult to obtain close control of surface concentration with the use of a P_2O_5 source, because of its hygroscopic nature. Other sources, such as ammonium dihydrogen phosphate (NH_4HPO_4) and ammonium mono-hydrogen phosphate $[(NH_4)_2HPO_4]$, do not suffer from this problem and can be used at source temperatures from 450°C to 900°C. Again the preliminary reaction results in the delivery of P_2O_5 to the silicon slices.

Planar solid sources can also be used. One such source, consisting of 25% of silicon pyrophosphate in an inert binder [64], directly converts to a P_2O_5–SiO_2

glass at diffusion temperatures, so that P_2O_5 is released as a vapor from its surface. Alternative sources [65], containing $Al(PO_3)_3$ and LnP_5O_{14}, are also available in the form of hot pressed discs for this purpose.

Liquid-Source Systems. A commonly used liquid source is phosphorus oxychloride ($POCl_3$), which is maintained in a bubbler at $0-40°C$. An oxidizing gas mixture is used, resulting in the formation of P_2O_5 at some point in the system before the diffusion zone. The preliminary reaction is

$$4POCl_3 + 3O_2 \rightarrow 2P_2O_5 + 6Cl_2 \qquad (4.119)$$

The presence of oxygen aids in preventing halogen pitting effects, which are noticeable only at high surface concentrations. $POCl_3$ systems are free from moisture problems and allow a high degree of control of surface concentration by adjusting the bubbler temperature. An alternative liquid source is phosphorus tribromide (PBr_3). This dopant source has somewhat superior gettering properties to $POCl_3$ and is being increasingly used for this reason.

The use of a halogenic dopant source for the final diffusion in a microcircuit process is highly advantageous since it getters contaminants from all previous high-temperature steps. Both $POCl_3$ and PBr_3 are commonly preferred over other phosphorus sources for use at this point during microcircuit fabrication.

Gaseous Systems. The most commonly used gas is phosphine, which is both highly toxic and explosive. It is relatively convenient to handle in dilute form (with 99.9% N_2 or Ar). A slightly oxidizing carrier gas is used, the preliminary reaction being

$$2PH_3 + 4O_2 \xrightarrow{400°C} P_2O_5 + 3H_2O \qquad (4.120)$$

As with other reactions, P_2O_5 is delivered to the silicon slices. Adequate venting is necessary to prevent the release of trace amounts of unreacted PH_3. In addition, an acidic $CuCl$ trap must be used in the inlet gas lines to remove any unused gas, for reasons of safety. Characteristics of the system are similar to those of the diborane system.

4.7.3 Arsenic

Arsenic has a diffusivity which is about one-tenth that of boron or phosphorus. Consequently, it is used in situations where it is important that the dopant be relatively immobile with subsequent processing. It is also desirable for shallow diffusions, where its abrupt doping profile and low diffusivity make control of the junction depth more precise. Finally, arsenic movement is primarily substitutional in nature, with a small interstitialcy component. Thus, it is relatively free from anomalous diffusion tails of the type observed with phosphorus at high doping levels. Consequently, it is the preferred dopant for n-type regions in VLSI circuits.

The tetrahedral radius of arsenic is identical to that of silicon, so that it can be introduced in large concentration without causing lattice strain. Its maximum solid solubility in silicon is 2×10^{21} cm^{-3}, and active electron concentrations as high as 5×10^{20} cm^{-3} are achievable with this dopant. Diffusion can be carried out directly from the reaction of arsenic trioxide (As_2O_3) and silicon. At the diffusion temperature [66],

$$2As_2O_3 + 3Si \rightleftharpoons 3SiO_2 + 4As \qquad (4.121)$$

A solid As_2O_3 source, held at 150–250°C, can be used for this purpose. In actual operation, vapors of the oxide are transported into the diffusion zone by means of a carrier gas of almost pure nitrogen, since arsenic is readily masked by the presence of any oxide on the silicon slices. Surface concentrations obtained by this technique generally do not exceed 2×10^{19} cm^{-3}, because of rapid depletion by evaporation from the silicon surface.

Spin-on dopants for arsenic consist of a number of different arsenosiloxanes in an appropriate organic binder. Upon-bake out at 250°C, these react to form an arsenosilicate glass which acts as the source of this dopant. Higher doping concentration can be achieved by this method, since the glass serves as an evaporation barrier to the arsenic during diffusion.

Planar solid sources for arsenic consist of aluminum arsenate ($AlAsO_4$) embedded in an inert binder [67]. At diffusion temperature (925–1000°C)

$$2AlAsO_4 \rightleftharpoons Al_2O_3 + As_2O_3 + O_2 \qquad (4.122)$$

The incorporation of arsenic from this source is extremely sensitive to the amount of oxygen present in the diffusion system. As a result, it is difficult to control in practical situations.

Sealed-tube technology has also been used successfully for arsenic [68]. Here silicon slices are stacked within a quartz tube together with the arsenic source, which consists of doped silicon powder having an arsenic concentration of about 3%. A short, high-temperature bake is used to remove traces of water vapor from the tube, at which point it is evacuated and sealed. After diffusion the tube is cut open. This approach allows surface concentrations of as high as 10^{21} cm^{-3} to be achieved because of the complete absence of source evaporation, and of oxygen. In addition, uniformity from slice to slice is better than 2% because of the closed nature of the system. Finally, a large number of wafers can be diffused at a single time, since there are no problems of dopant depletion from one end of the carrier to the other.

Gaseous systems, using arsine (AsH_3), have been successfully operated. They are more convenient than systems using As_2O_3, but do not result in surface concentrations above 3×10^{19} cm^{-3}.

Ion implantation is by far the most successful technique [69] for the predeposition of arsenic before drive-in, and has displaced all of the above methods. A high-energy implant serves to deposit this dopant deep into the semiconductor surface, and thus minimizes its depletion during the drive-in step. It is usually carried out through a SiO_2 layer, which serves as a barrier to the loss of arsenic during the drive-in step. Active concentrations in excess of 3 to 5 $\times 10^{20}$ cm^{-3} are readily achieved by this method.

4.7.4 Antimony

Antimony can be used as an alternative to arsenic because of its comparable diffusion coefficient. Its misfit factor in silicon is 0.153, and its electronically active surface concentration is limited to about 5×10^{19} atoms cm^{-3}. However, it has a lower vapor pressure than arsenic, so that surface concentrations close to this value can be achieved in open-tube systems.

Ion implantation is commonly used for antimony predeposition. An alternative approach is to use the trioxide Sb_2O_3, which is a solid source. A two-zone diffusion furnace is required, with the Sb_2O_3 maintained at 600–650°C. Some systems have also been operated with the tetroxide Sb_2O_4 at a source temperature of about 900°C. As with arsenic, care must be taken to operate these systems in an almost oxygen-free ambient, because antimony is relatively easily masked by small traces of SiO_2 on the silicon surface.

Liquid-source systems have also been used with antimony pentachloride (Sb_3Cl_5) in a bubbler arrangement, with a carrier gas of oxygen. This results in the delivery of antimony trioxide to the silicon surface, from which the diffusion takes place. Stibine (SbH_3) cannot be used in gaseous systems since it is unstable.

Antimony diffusion is almost purely substitutional in nature, with no interstitialcy component. Thus, it is completely free from anomalous effects, and is superior to arsenic in this regard. It is occasionally used in VLSI applications for this reason.

4.7.5 Gold and Platinum

The reasons for the choice of gold over other impurities exhibiting deep-lying levels have already been outlined in Section 4.4.2. Since gold is an extremely rapid diffuser (with an effective diffusion coefficient about five orders of magnitude larger than that of boron or phosphorus), it is usually the last impurity to be introduced into the microcircuit.

Gold diffusion is usually carried out from the element, which is vacuum evaporated on the silicon in a layer about 100 Å thick. Diffusion proceeds from a liquid gold–silicon alloy, and results in damage to the silicon surface. For this reason, it is always performed from the back of the wafer, remote from the actual microcircuit.

Diffusion is normally carried out in the temperature range of 800–1050°C. The concentration is controlled by means of this temperature, and the time (about 10–15 min) is more than enough to cause the gold to diffuse through the entire microcircuit. The precise diffusion temperature is selected on a cut-and-try basis to obtain the desired end result. In-processing monitoring of the collector–emitter voltage of a test transistor, driven into saturation, is often used to determine the process time.

Gold-doped silicon slices must be removed rapidly from the diffusion furnace and brought to room temperature in a manner that is repeatable from run to run. The repeatability of this quenching cycle is important in ensuring consistent results because out-diffusion effects are significant with fast diffusers. Moreover, gold diffusion is sensitive to the defect character of the silicon and to the doping concentration, and hence to the nature of the device or circuit in which it is introduced. As a result, gold doping is perhaps the most difficult process to control in semiconductor manufacture.

Platinum diffusions are primarily used in semiconductor power devices. The behavior of this dopant in silicon is poorly understood. As a result, the usual approach for diffusion is to use techniques that have worked with gold [46]. Unlike gold, however, results with platinum diffusion from the element are highly variable, so that spin-on dopants are more commonly used. Typically, these consist of platinum chloride in an alkyloxysilane.

4.8 SPECIAL PROBLEMS IN SILICON DIFFUSION

This section discusses problems and anomalies encountered during silicon diffusion. These have become increasingly important with the trend toward large-scale integration, and the resultant use of shallow diffusions with high surface concentration. A few of these problems have been analyzed by closed-form solutions, while computer-aided models [70] have been used extensively for the others.

4.8.1 Redistribution During Oxide Growth

In microcircuit fabrication, diffusions are made through windows cut in a mask on the surface of the silicon. The most convenient mask is one of SiO_2, grown out of the slice itself by subjecting it to an oxidizing gas at elevated temperatures. This can be done in a separate step after a diffusion. Alternatively, the two-step diffusion process may be used to grow it simultaneously with the drive-in phase. This latter technique is preferred, because it not only provides immediate protection for the silicon surface, but also saves an additional masking step. In addition, it minimizes the time the slice is maintained at elevated temperatures and thus reduces the undesired movement of already diffused impurities.

Common to these processes is the fact that some of the impurity-doped silicon is consumed to form an oxide. This effect has its parallel in the freezing of a doped crystal from a melt, as described in Chapter 3, and results in a redistribution of impurities. The extent of this redistribution is a function of the rate at which the silicon is consumed and of the relative diffusivities and solid solubilities of the impurities in silicon and SiO_2.

The ratio of the equilibrium concentration of the impurity in silicon to its equilibrium concentration in the oxide is given by the distribution coefficient for the impurity in the $Si–SiO_2$ system. Experimentally derived values for this distribution coefficient m are about 20 for gallium and 10 for phosphorus, antimony, and arsenic. Values of m for boron are a function of crystal orientation [71], and are given by

$$m_{(100)} \simeq 33\, e^{-0.52/kT} \tag{4.123a}$$

$$m_{(111)} \simeq 20\, e^{-0.52/kT} \tag{4.123b}$$

where T is the diffusion temperature in Kelvin. Typically, m ranges from 0.15 to 0.3 for {100} silicon.

For the case in which $m < 1$, the growing oxide takes up the impurity from the silicon. Thus for boron diffusion, impurity depletion occurs on the silicon side of the interface. The extent of this depletion also depends on the rate at which the boron diffuses through the SiO_2 layer to its surface and escapes into the gaseous ambient. The diffusion rate of boron in SiO_2 is orders of magnitude below that in silicon; hence boron depletion is dominated by redistribution effects.

For the case in which $m > 1$, the growing oxide rejects the impurity. If the diffusion through the oxide is slow, as with phosphorus, this results in a pileup of the impurity at the silicon surface. If, however, the impurity diffuses rapidly through the oxide (as is the case with gallium), the net result can still be an impurity depletion.

Finally, even if $m = 1$, the process of oxidation results in impurity depletion at the silicon surface. This is because the silicon roughly doubles in volume on forming silicon dioxide. Thus some diffusion will occur from the highly doped silicon into the relatively lightly doped oxide.

Figure 4.30 shows the impurity equilibrium concentrations that may occur in uniformly doped silicon in the presence of a growing oxide surface. For practical diffusions the problem is more complicated since the dopant concentration is not uniform. In addition, the impurity is diffusing both into the silicon and into the growing oxide at the same time.

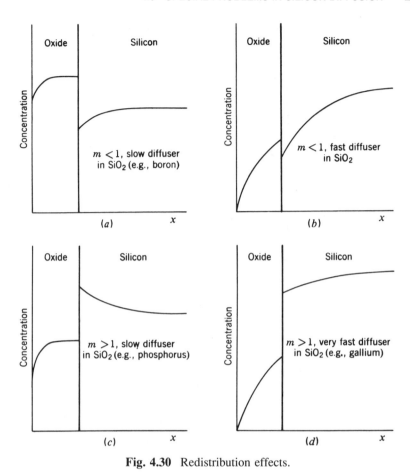

Fig. 4.30 Redistribution effects.

Figure 4.31 shows the results of calculations [72] for boron drive-in with simultaneous oxidation, and $m = 0.1$. These data are for typical drive-in situations which follow a predeposition* with a Dt product of 3.6×10^{-12} cm^2. In addition, it is assumed that the diffusivity of boron in silicon and SiO$_2$ is 0.6×10^{-12} and 1.02×10^{-13} cm^2 s^{-1}, respectively, at the drive-in temperature. As seen from this figure, redistribution effects cannot be ignored for boron. In a typical practical situation involving a 45-min drive-in at 1200°C (15 min in dry oxygen followed by 30 min in wet oxygen), the surface concentration is approximately 50% of the value that would be obtained if redistribution effects were absent.

In practice, the pileup effect at the silicon surface is relatively small in phosphorus diffusion and may be ignored. This is because of the fortuitous cancel-

---

*A closed form solution for the ion-implanted case is also available [73]

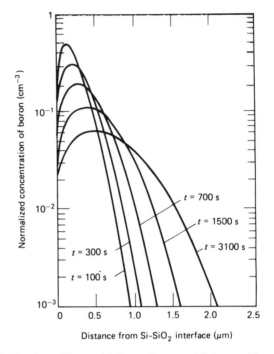

Fig. 4.31 Redistribution effects with boron in an oxidizing ambient. From Kato and Nishi [72]. Reprinted with permission from the *Japanese Journal of Applied Physics*.

lation of the redistribution and diffusion effects under typical processing conditions.

4.8.2 Diffusion During Oxide Growth

Surface oxidation of silicon also results in the enhanced diffusion of most group III and group V dopants, with the notable exception of antimony. It has been proposed that this enhancement is due to the creation of self-interstitials during the oxidation process. In effect, this provides an interstitialcy diffusion component to what is normally considered as a substitutional diffusion process. It is most significant with boron in silicon, which moves predominantly by the interstitialcy mechanism.

Boron diffusion is isotropic when carried out in a nitrogen ambient. Its diffusivity increases significantly in an oxidizing ambient, and becomes a function of crystal orientation [37, 38]. It is highest for boron diffusions performed in (100) silicon and somewhat lower for diffusions in (111) silicon. Curves of $D_i = D_i^0 + D_i^+$ are shown in Fig. 4.32, together with the curve for D_i in a nonoxidizing ambient, and illustrate this effect. It can be explained by noting that the boron atom moves by pushing one of its nearest silicon neighbors into an adja-

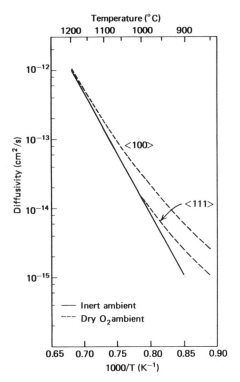

Fig. 4.32 Boron diffusivity in an oxidizing ambient. From Antoniadis et al. [38]. Reprinted with permission of the publisher, The Electrochemical Society, Inc.

cent interstitial site, and taking up the substitutional site which has been vacated in this manner. At any given diffusion temperature, there exists an equilibrium concentration of silicon interstitials. Interaction with these is implicitly included in the formulation for diffusion in a nonoxidizing ambient. The situation is quite different when diffusion and oxidation occur simultaneously. Incomplete oxidation at the Si–SiO$_2$ interface results in the formation of silicon interstitials in excess of the equilibrium value, at concentrations which are dependent on the oxide growth rate and the crystal orientation. Since boron diffusion is due to the interstitialcy mechanism, it follows that an increased concentration of silicon interstitials results in an enhanced diffusion rate. Moreover, this rate will further increase if diffusion is carried out under more rapid oxidizing conditions [74].

All impurities whose diffusional behavior has an interstitialcy component will exhibit enhanced diffusion under oxidizing conditions. This effect is the greatest in boron because diffusion is entirely by this mechanism. Strong enhanced diffusion effects are also observed with phosphorus, for this reason. The effect is relatively small with arsenic. Antimony is a purely substitutional diffuser, and does not exhibit this characteristic.

Self-interstitials which are generated during oxidation can reduce the vacancy concentration according to the reaction

$$I + V \rightleftharpoons 0 \qquad (4.124)$$

under some conditions. This can lead to a retardation of impurity diffusion. Studies of antimony behavior in silicon have shown diffusion retardation under long, high-temperature oxidation conditions, but enhancement for shorter times [75]. From their data, the lifetime associated with the I–V annihilation process is about 1 hr at 1100°C.

4.8.3 Cooperative Diffusion Effects

The cooperative diffusion effect was first observed in bipolar transistor fabrication, where it resulted in enhanced diffusion under the emitter, as shown in Fig. 4.33. It is sometimes referred to as the *emitter push effect* for this reason. Experimentally, it has been noted that the parameter δ in this figure varies approximately inversely with the unenhanced base diffusion depth [76], ranging from 0.2 to 0.4 μm for values of x_{bc} from 2 to 0.5 μm, respectively. Thus it is particularly important in shallow structures of the type used in VLSI applications, and has been the subject of much study.

This enhanced diffusion effect can be attributed to two mechanisms. The first of these can be studied by noting that the flux of base impurities consists of both a concentration-dependent and a field-dependent term, as given by Eq. (4.11). In the diffusion of a single species, this field-dependent term is self-produced, and aids the movement of the impurity. The emitter diffusion, however, sets up its own electric field, which acts against the base concentration gradient. If sufficiently large, diffusion "up" the gradient is possible, and can even result in a fall in the base doping profile. This simple model is illustrated in Fig.

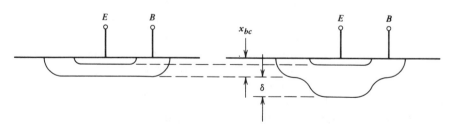

Fig. 4.33 Enhanced diffusion under the emitter.

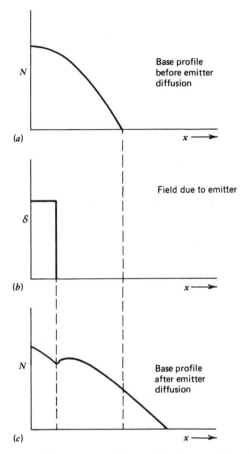

Fig. 4.34 Model for base profile resulting from emitter diffusion.

4.34. Here Fig. 4.34a shows the initial base doping profile, Fig. 4.34b shows the electric field produced by a high-concentration emitter diffusion, and Fig. 4.34c shows the base doping profile after emitter diffusion.

This model alone does not predict the large base diffusion enhancement which has been shown to extend well below the emitter. In addition, the enhancement occurs in all directions. A second mechanism, which can account for these factors, is that enhanced diffusion occurs because of self-interstitials generated during the emitter diffusion, which can migrate a considerable distance. This has been shown in Section 4.4.1.3 to produce a tail region in high-concentration phosphorus diffusions, where the dominant process is the creation of self-interstitials in the bulk of the silicon. These effects are shown in Fig. 4.35 for a transistor with a phosphorus-doped emitter and a boron-doped base.

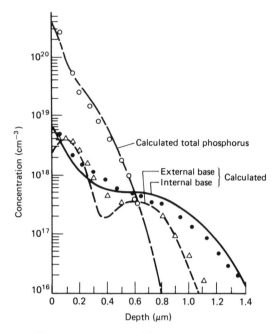

Fig. 4.35 Doping profiles for a transistor with a phosphorus-doped emitter and a boron-doped base. From Fair [11]. Reprinted with permission of the publisher, The Electrochemical Society, Inc.

Also shown are the results of calculations which take both these mechanisms into account [11].

A dip in the base doping profile is also observed when arsenic is used for the emitter diffusion. However, no extended push effect is seen in this case. This is because the diffusion of arsenic into silicon has a very small interstitialcy component. Finally, the emitter push effect is not observed with antimony, whose movement is purely substitutional in character.

4.9 DIFFUSION SYSTEMS FOR GALLIUM ARSENIDE

Diffusion systems for GaAs are based on the same considerations as those for silicon. Here, too, the primary requirement is to bring an impurity in contact with the slice, and maintain it there for a fixed time and at an elevated temperature. With GaAs, however, this can result in a preferential loss of arsenic, and deterioration of the surface morphology. Consequently, techniques must be used to prevent this occurrence. These include diffusion in sealed ampoules with a sufficient overpressure of arsenic, diffusion in semiclosed "leaky boxes," and diffusion in open-tube furnaces with cap layers of refractory materials such as silica, phosphosilicate glass, or silicon nitride. Much of the early work in this area has been done with sealed ampoules, with a known weight of arsenic (or an arsenic-bearing compound) to establish a suitable arsenic overpressure. This technique, which provides a closed thermodynamic situation, is ideal for the studies of basic diffusion mechanisms, and has resulted in much of our understanding of the physical processes underlying diffusion in GaAs. From a practical point of view, however, its use involves the necessity of sealing and breaking open these ampoules, so that there is much interest in alternative methods which are more convenient.

A second problem is in the choice and amount of dopant source. Excessive amounts of elemental sources often result in surface damage to the GaAs because of the formation of reaction products. This problem is worse for some dopants than for others. It has usually been minimized by using dilute concentrations of the dopants, often as alloys or compounds of gallium or arsenic ($ZnAs_2$, GaS, and Zn–Ga alloys, for example). Tellurium and selenium sources are particularly bad, because of their ability to form chalcogenide glasses upon reaction with GaAs.

A great variety of early diffusion studies resulted in a wide scatter in diffusion data [76, 77], by as much as one to two orders of magnitude. Since then, significant advances in the understanding and control of diffusion in GaAs have been made, once attention was focused on the role of vacancies during this process, and on the importance of ternary considerations [51] in these systems. A number of authors have shown that reproducible diffusions can be made in sealed-tube systems [26, 78, 79], if attention is paid to these considerations.

The emphasis of diffusion technology for GaAs has shifted away from sealed ampoule techniques to "leaky box" methods, which involve systems which are only partially confined. These have achieved excellent uniformity and reproducibility and are used in many practical situations.

More recently the use of doped oxide sources with cap layers has resulted in diffusion processes [80, 81] which can be conducted in an open-tube environment. In some cases, the degree of reproducibility has exceeded that which is conventionally obtained in silicon technology. Thus the prognosis for this approach is very promising. Even so, the work with GaAs diffusion has been limited to studies of only a few dopants.

Ion implantation is an important technique for incorporating impurities into GaAs in a controlled manner. It is widely used for channel doping in field effect devices for this reason. Silicon is the most commonly used dopant for this purpose because it can be more readily annealed than other, heavier dopants. Conventional annealing necessitates the use of a cap layer of PSG or Si_3N_4. With RTA, however, proximity annealing can be used, and is favored because of its convenience.

Finally, it should be noted that no truly flexible diffusion process, capable of a wide range of surface concentrations and junction depths, is available for GaAs. Thus, all of the approaches which follow have been developed with special-purpose applications in mind.

4.9.1 Ternary Considerations

During diffusion, the impurity and the GaAs surface with which it is in contact must reach an equilibrium composition, which determines the partial pressure of the three components and hence the vacancy concentrations. These quantities are crucial to establishing the conditions for the diffusion process. In a ternary system, at any given temperature and pressure, it follows from the phase rule that $F = 3 - P$, where P is the number of condensed phases. This rule can be used to predict what happens when GaAs is placed in contact with the vapor during diffusion.

The behavior of zinc in GaAs is considered as a case in point [51], since it has been well-studied. Here Fig. 4.36a shows an isothermal section at 1000°C for this system (see Fig. 2.30). From this figure we note that diffusion sources represented by compositions in the region A are liquid. Sources with compositions in regions B and C involve a liquid in equilibrium with a solid. Diffusions with either of these types of sources are undesirable since they result in liquid formation on both the source and the GaAs, thus destroying the surface quality flatness of the semiconductor.

Figure 4.36b shows an isothermal cut at 775°C. Here, a number of regions have three condensed phases and thus no degrees of freedom. Consequently, the arsenic and zinc vapor pressures are fixed by the source composition and temperature, and are independent of the weight of the source or the volume of the ampoule. Note, however, that one condensed phase is a liquid in each case,

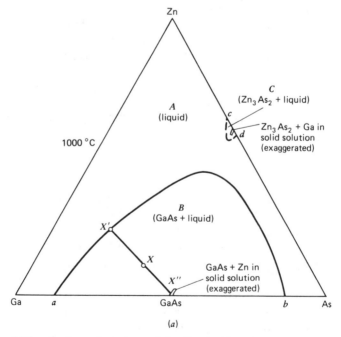

Fig. 4.36a Isothermal sections of the Ga–As–Zn ternary phase diagram.

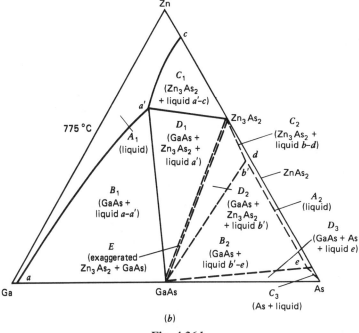

Fig. 4.36*b*

so that the use of sources in these regions, and at this diffusion temperature, will also lead to poor surfaces.

Figure 4.36*c* shows an isothermal cut at 700°C. Compositions in the regions labeled B_1, D_1, G, and G_1 are all invariant ($F = 0$). However, the G and G_1 regions are bounded by only solid phases. Diffusion is most desirable with sources in these regions since they lead to undamaged surfaces, in addition to being independent of the amount of the source or the volume of the ampoule. Here it should be noted that compositions in G_1 will result in higher arsenic vapor pressures than compositions in G.

Some workers have used diffusion sources whose compositions fall on the lines connecting GaAs and Zn_3As_2 or $ZnAs_2$. These have no liquid phase, but are bounded by two solid phases so that $F = 1$. Consequently, control of the source weight and ampoule volume is necessary for reproducible diffusions.

4.9.2 Sealed-Tube Diffusions

The sealed tube system provides a controlled environment for studies of diffusion in GaAs, and has been used extensively as a vehicle for this purpose. Its use for commercial diffusion applications is limited by the inconvenience and cost of sealing quartz ampoules, and of their subsequent cutting apart after the process is complete. Both *p*- and *n*-type diffusions have been made into GaAs

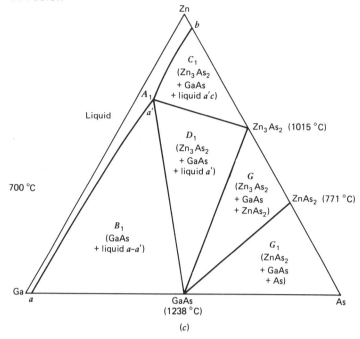

Fig. 4.36 c

by this technique. Most of the work with p-type diffusions has been confined to the use of zinc, although some work has also been done with other group II dopants.

Sealed-tube systems for zinc require both arsenic and zinc sources. Elemental sources can be used; more commonly, however, compounds of zinc and arsenic, such as Zn_3As_2 or alloys of Zn–Ga–As, are employed to obtain control of the arsenic and zinc pressures.

Sealed-tube zinc diffusions have also been made in GaAs whose surface was coated with a layer of sputtered silica. Here a 6500-Å layer of SiO_2 was used to protect the GaAs surface from dissociation at diffusion temperatures, as well as to prevent surface damage due to alloying with the zinc. A zinc–gallium alloy served to establish the partial pressure of zinc; no additional arsenic was included in the ampoule.

Results of these experiments showed a reduction in surface concentration to about 25% of what was obtained on uncoated wafers which were diffused at the same time. Thus the SiO_2 is highly transparent to the zinc. In addition, the SiO_2 prevented loss of arsenic from the GaAs during heat treatment. Consequently, the surface finish of the coated wafers was protected by this process.

Group VI impurities in GaAs are sulfur, selenium, and tellurium, and are n-type. Both selenium and tellurium lead to the formation of chalcogenide glasses upon reaction with GaAs, so little work has been done with these. Even the nature of their diffusion mechanism has not been established at the present time.

All of these impurities exhibit extremely low values of diffusivity, which are comparable to those obtained for substitutional movement. A divacancy model has been proposed to account for this behavior. In contrast, p-type impurity motion is more closely approximated by interstitial diffusion, with diffusivity values that are five to six decades larger and an abrupt doping profile.

Some sealed-tube work has been done using sulfur as the dopant. Here it has been proposed [82] that the dominant vapor species is Ga_2S and that the sulfur vapor density is directly proportional to the partial pressure of this compound. The incorporation reaction for sulfur from Ga_2S is given by

$$Ga_2S(g) + V_{As} \rightleftharpoons S_{As}^+ + 2Ga(g) + e^- \qquad (4.125)$$

Application of the mass-action principle results in a surface concentration which is linearly proportional to p_{Ga_2S}, and thus to the sulfur vapor pressure. This has been observed to be the case.*

It has been experimentally observed that the diffusivity of sulfur increases as the square root of the arsenic partial pressure. Again, the divacancy mechanism has been invoked as a possible explanation. However, the reasons for this behavior are not firmly established at the present time.

Reproducible sulfur diffusions have been made [79] using mixtures of GaS and arsenic at temperatures of 860°C and 910°C. The phase diagram of the Ga–As–S ternary system is shown in Fig. 2.27. The appropriate isothermal section of this phase diagram is shown in Fig. 4.37. For this work, compositions

---

*The molecular species of sulfur are S_2, S_6, and S_8. Direct incorporation of any of these into the gallium arsenide would result in a surface concentration that is proportional to the $\frac{1}{2}, \frac{1}{6}$ and $\frac{1}{8}$ power of the sulfur vapor density, respectively.

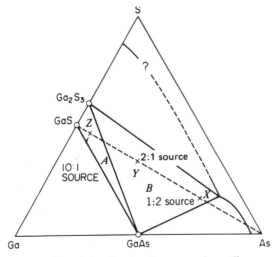

Fig. 4.37 Isothermal section of the Ga–As–S ternary phase diagram at 910°C. From Matino [79]. Reprinted with permission from Pergamon Press, Ltd, Oxford, UK.

were selected to fall into the invariant regions A and B of Fig. 4.37, so that the diffusion was independent of the weight of the dopant or the volume of the ampoule. These sources were made up of mixtures of GaS and arsenic in $1:2, 2:1$, and $10:1$ atomic ratios, respectively; they are labeled X, Y, and Z in Fig. 4.37.

Curves of junction depth versus the square root of time are shown in Fig. 4.38 for these sources, and for diffusions at 860°C. These are seen to be straight lines, so that values of D can be extracted from them. Values of D at 860°C were 10^{-12} cm^2 s^{-1} for the $1:2$ and $2:1$ sources, respectively, and 2×10^{-13} cm^2 s^{-1} for the $10:1$ source. This is to be expected, since the arsenic pressure of the $10:1$ source is considerably lower than that for the other sources.

Although tin is a group IV impurity, it behaves as a donor when diffused into GaAs. Sealed-tube diffusions, using elemental tin as a vapor source, have been made in this manner. More recently, however, a doped oxide source, formed by the simultaneous oxidation of tetramethyltin and silane,* has been used for this purpose. This source prevents loss of arsenic from the semiconductor during diffusion, and thus protects the surface, even when no arsenic overpressure is established in the ampoule. Sealed-tube diffusions [83], made at 1000°C, resulted in a junction depth which was linearly proportional to the square root of the diffusion time; a diffusivity of 1.8×10^{-13} cm^2 s^{-1} was obtained at this temperature.

*Details of source preparation are provided in Chapter 8.

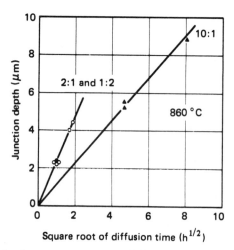

Fig. 4.38 Junction depth versus square root of time for sulfur diffusions. From Matino [79]. Reprinted with permission from Pergamon Press, Ltd., Oxford, UK.

4.9.3 Leaky Box Diffusions

In this approach, the GaAs slice, the dopant source, and a source which provides an overpressure of arsenic are all placed in a chamber of restricted volume, within an open diffusion tube. This chamber is not tightly closed, so that pressure cannot build up within it. Additionally, the diffusion tube is flushed with flowing nitrogen gas, and suitably vented.

Leaky chambers of varying degrees of sophistication can be used for this purpose. The simplest [84] consists of a graphite box with a loose-fitting graphite lid. Next in complexity [85] is a stoppered quartz chamber with a thermocouple well, which is capable of holding a few slices at a time.

Because of their relatively small volume, the dopant, the arsenic source and the gallium arsenide are all maintained at the same temperature. This limits the choice of species which can be used in this method. Moreover, care must be taken to prevent surface reactions which can result in destroying the GaAs morphology during diffusion.

Conventional diffusion tubes, capable of holding many slices, have also been used [86] for this purpose. One advantage of this approach is that it is possible to hold the slices at a slightly higher temperature than the dopant source. This helps prevent condensation of excess dopant species on them during the diffusion process, and thus minimizes surface morphology problems.

The same considerations which apply to sealed tube diffusions hold for the leaky-box process as well. As before, the dopant source is usually the element, an alloy of the element or a compound. In the case of zinc diffusion, for example, elemental zinc, Ga–Zn alloys, and Zn_3As_2 have all been used successfully.

In leaky-box diffusions, the amount of dopant cannot be controlled to any degree. As a result, most diffusions made by this method can be treated as if made from an unlimited source. Diffusion profiles which result are generally of the type described in Section 4.3.2, with junction depths which are proportional to the square root of time.

Pieces of GaAs or InAs can be used to provide the arsenic overpressure for these diffusions. InAs provides a higher overpressure, since its decomposition temperature is lower than that of GaAs. However, its use results in the incorporation of indium as a diffusing species which may be undesirable. In some cases, as in the use of diffusion to form a highly doped region for ohmic contact purposes, it is advantageous since the diffused GaInAs region has a slightly lower bandgap than the GaAs.

Sulfur diffusions into undoped SI GaAs have also been made [87] by the leaky-box technique, using a mixture of GaAs and Ga_2S_3 as the source. Doping profiles obtained by this technique were of the erfc type. Using these profiles, the diffusivity of sulfur in undoped SI gallium arsenide, over the 750–900°C temperature range, was

$$D \simeq 0.27e^{-3.0/kT} \, \text{cm}^2 \, \text{s}^{-1} \qquad (4.126)$$

and the surface concentration was

$$N_{sur} = 4.42 \times 10^{23} e^{-1.32/kT} \, cm^{-3} \tag{4.127}$$

The leaky-box technique has also been used for the diffusion of zinc in $Al_xGa_{1-x}As$ [88, 89]. Diffusion behavior was found to be quite complex, with the diffusivity initially falling as x was increased from 0 to 0.07. Subsequent increase in the aluminum fraction resulted in an increase in the diffusivity. The argument presented for this eventual increase is that higher values of aluminum fraction go hand in hand with an increase in the melting point, and hence in the formation energy of the group III vacancy [90]. This, in turn, would increase the amount of time spent by zinc in its interstitial state during diffusion, and hence its jump rate.

4.9.4 Open-Tube Diffusions

Two techniques have been used for making open-tube p-type diffusions with zinc. The first of these requires that diffusions be carried out with a gas such as arsine to provide the necessary arsenic overpressure. Diffusions of this type, using dimethylzinc (DMZn) as the dopant source, have been made in GaAs [91].

Although this approach avoids the necessity of using sealed ampoules, it has many disadvantages when compared to its silicon counterpart. Thus both arsine and DMZn are highly dangerous, and are only partially consumed during the diffusion process. In addition, decomposition of the arsine gas results in a large amount of arsenic deposits on the walls, as well as arsenic dust in the diffusion tube. Consequently, elaborate effluent handling precautions must be taken with systems of this type. An alternative approach, which consists of using doped oxides with cap layers, is considerably more convenient. Diffusions using this approach can be carried out in an inert gas ambient, and are as convenient as diffusions in silicon.

Both n- and p-type dopants are available in liquid form, and usually consist of proprietary formulations of organosilanes, mixed with a small amount of dopant. They can be applied by a spin-on technique; after drying, this coating is heated to about 450°C in order to remove volatile components, and to convert the organosilane to silicon dioxide. Diffusion is carried out directly in an open tube furnace, using this dopant source as a cap. In some cases, a second cap, of an undoped organosilane, is spun-on to provide an additional protective layer. Diffusion behavior is typical of that obtained with an infinite source.

A more elegant approach is to use sources which can be applied by chemical vapor deposition (CVD) methods. In p-type diffusions of this sort, a $ZnO-SiO_2$ source was formed on the GaAs slice by the simultaneous oxidation of DMZn and SiH_4 [81] at 300–400°C. This represents an excess of zinc, so that neither the layer composition nor its thickness was important. A cover layer of phosphosilicate glass (PSG), formed by the oxidation of PH_3 and SiH_4, was next deposited at 350°C. Diffusions were made in an open-tube furnace with a small flow of nitrogen gas or a 50% N_2–50% H_2 gas mixture. Undamaged surfaces were obtained after both cap and doped oxide layers were removed in dilute HF.

Figure 4.39 shows that the junction depth varied linearly with the square root of time over a wide range of diffusion temperatures [92]. It is also seen that the diffusion rate was extremely rapid, as compared to that for n-type impurities. By way of example, it is seen from this figure that $D_{sur} = 1.6 \times 10^{-12}$ cm^2/s at 600°C. It is interesting to note that approximately the same value of diffusivity was obtained with sulfur diffusions at 860°C, and with tin diffusions at 1010°C.

Open-tube tin diffusions have been made using a spin-on SnO_2–SiO_2 source covered with a CVD layer of SiO_2 [80], but results with this technique were highly variable. It is probable that the spin-on source film induced considerable stress in the GaAs at the diffusion temperature and thus affected the diffusion process. These problems have been avoided [93, 94] by growing a CVD layer of SnO_2–SiO_2 by the simultaneous oxidation of tetramethyltin and silane gas. In addition, the use of a PSG cap greatly reduced the interface stress during diffusion, because of its better thermal expansion match to the GaAs. Diffusions

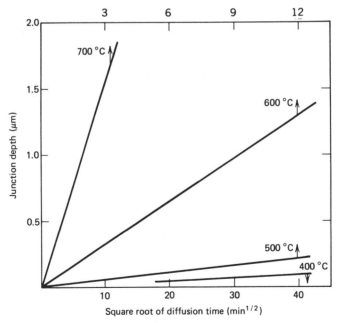

Fig. 4.39 Junction depth versus square root of time for open-tube zinc diffusions. From Field and Ghandhi [92]. Reprinted with permission of the publisher, The Electrochemical Society, Inc.

made at 800°C for 30 min resulted in a 10^{17}-cm^{-3} surface concentration with a junction depth of 0.1 μm, and with ±10% reproducibility. Figure 4.40 shows the variation of junction depth as a function of time for these diffusions.

It should be noted that control of the doped oxide thickness and composition were necessary for this application, since the surface concentration was well below the equilibrium solubility limit for this diffusion temperature. However, high-surface-concentration (10^{18}–10^{19} cm^{-3}) diffusions, which are solid-solubility-limited, were readily achieved without careful attention to this factor.

4.9.5 Masked Diffusions

Masked diffusions can be made in GaAs, using deposited films* of either silica, PSG, or Si_3N_4 as mask materials. An anodically grown oxide of GaAs has also been used for some low-temperature diffusions. GaAs films can be used as diffusion masks for $Al_xGa_{1-x}As$, especially for high x values.

All of these materials have both advantages and disadvantages, so that the ideal mask is not available. Deposited silica films are certainly the worst. They do not inhibit the decomposition of GaAs at elevated temperatures, since they permit the rapid diffusion of gallium through them. They are relatively transparent to both zinc and tin, so thick layers ($\geq 0.5\,\mu$m) must be used for masking purposes. Finally, they are very poorly matched to GaAs in their thermal expan-

---

*Techniques for the growth of these films are described in Chapter 8.

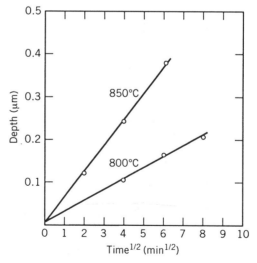

Fig. 4.40 Junction depth versus square root of time for open-tube tin diffusions. From Ledakovitch et al. [93]. Reprinted with kind permission from Pergamon Press, Ltd., Oxford, UK.

sion characteristics, so that thick films often crack during thermal processing, or damage the semiconductor.

Thermal expansion problems can be greatly reduced [94] by the use of a PSG cap. Moreover, this effectively blocks the transport of both zinc and tin, so that thinner films (0.1 μm) can be used for masking. Its blocking characteristics for gallium are not established at the present time. However, indications are that it is considerably better than silica in this respect.

Silicon nitride is an almost impermeable barrier to zinc, tin, and gallium. However, it has a bad thermal expansion match to GaAs, and must be used in extremely thin layers (300–500 Å) to avoid problems during processing. Its deposition technology is considerably more difficult than that of either silica or PSG.

A number of workers have noted extremely large lateral diffusion effects (lateral to vertical ratios of as much as 50 : 1) with masked diffusions, and have developed techniques for reducing them to manageable proportions (2 : 1). Yet others have reported that these effects are not present with some dopants, but are with others. No definitive work has been done in this area, but some general observations can be made at the present time.

1. These effects are confined to the first 50–100 Å of the GaAs surface.

2. They are sensitive to interfacial stress between the mask and the semiconductor [95]. Close matching of the respective thermal expansion coefficients greatly reduces the lateral diffusion [96].

3. Lateral diffusion effects are also sensitive to interface reactions between the mask and the GaAs surface. Such reactions are most severe with silica and least with silicon nitride. PSG is somewhat intermediate; lateral diffusion effects are reduced as the phosphorus content of the glass is increased.

4. Lateral diffusion effects are worst when there is a large available reservoir of the dopant, since this provides the driving force for the diffusion. Thus ion implantation techniques [97-99] are effective in minimizing this problem. Here, beryllium and silicon are the p and n dopants of choice since they are relatively light ions, and cause minimal damage in GaAs.

In summary, then, there are no magic dopants that work and others that do not. Rather, the technique by which the masked diffusion is made appears to be the dominant factor in the extent of this effect.

4.10 EVALUATION TECHNIQUES FOR DIFFUSED LAYERS

Many techniques are available for evaluating the properties of a diffused layer [100]. One approach is to diffuse the sample with a radioactive isotope and evaluate the radiation intensity upon removal of successive layers by chemical etching [101]. This allows the detailed nature of the diffusion profile to be obtained for the entire impurity content, under the assumptions that the isotope atoms disperse uniformly among the normal atoms of the diffusing impurity and that the diffusivity of the isotope is the same as that of the impurity. This technique is both inaccurate and time-consuming, and has been superseded by the use of secondary ion mass spectroscopy [102]. Here, depth profiling of the semiconductor is achieved by ion beam etching of successive layers of material. Accuracy of depth profile is in the 10- to 100-Å range, using this method.

Details of these techniques are beyond the scope of this book. However, knowledge of the impurity profile permits the use of Boltzmann–Matano techniques (see Section A.4.1 in the Appendix) to extract the value of the diffusion coefficient on a point-by-point basis, as a function of the concentration. Their primary disadvantage is that they are not suited for monitoring diffusion processes during microcircuit fabrication. In addition, they provide information on the entire impurity content and not on the electronically active part with which we are concerned.

The properties of the diffused layer are readily determined by measurements of masked patterns on *p-n* junctions formed by diffusion into a slice of known background concentration, but opposite impurity type. This method gives results which bear directly on the end goal of the diffusion process, and it is commonly used. Details of this evaluation technique now follow.

4.10.1   Junction Depth

To determine the junction depth, a small chip of the diffused slice is lapped on an angle so as to expose the actual junction. This angle is on the order of $0.5°$ to $1°$, so that the junction region is visually magnified. Next the junction is delineated by means of a selective etch. A number of such etches are in use, a common one consisting of an extremely dilute solution of $CuSO_4$ in dilute HF. In operation the acid serves to dissolve surface oxides; the $CuSO_4$ selectively plates the region with copper so as to delineate the junction. Additionally, the sample is strongly illuminated in order to cause the junction to be forward-biased during this process.

An alternative technique consists of grinding a cylindrical groove into the wafer. Again, this visually magnifies the junction region, which is delineated by means of a selective etch. The grooving technique is commonly used in a manufacturing environment, because of its greater convenience and ease of execution.

Junction-depth measurements based on geometric considerations can lead to considerable error if the lapping angle or groove radius is not precisely known. Consequently, an interferometric method is preferred, with the upper surface of the chip serving as a reference plane. An optical flat is placed on this chip, as shown in Fig. 4.41, and is vertically illuminated by collimated monochromatic light. The resulting fringe pattern gives a direct measure of the vertical depth in wavelengths of the illuminating source. For a sodium vapor

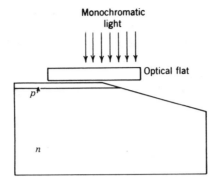

Fig. 4.41   Junction-depth measurement.

lamp whose spectral radiation is concentrated at 4895.93 and 5889.96 Å (the D_1 and D_2 lines, respectively), the distance between fringes is approximately 0.29 μm.

Multiple-beam interferometers [103] are also commonly employed. Their use results in sharply delineating the fringes, so that they are more accurate. Electron-beam-induced current (EBIC) methods can also be used in conjunction with a scanning electron microscope for junction-depth measurements [104].

4.10.2   Sheet Resistance

It is not possible to determine the specific resistivity of a diffused layer, because it is inhomogeneous. For a layer of this type, a *sheet resistance* is more appropriate. Consider a rectangular layer of diffused material of length l, width w, and thickness t. If its resistance is measured across the faces of width w and thickness t, then

$$R = \frac{\rho(t)}{t}\frac{l}{w} \tag{4.125}$$

where $\rho(t)$ is the specific resistivity of the material (in Ω-cm) and varies with depth. This equation may be more conveniently rewritten in the form

$$R = R_s\frac{l}{w} \tag{4.126}$$

where R_s is defined as the sheet resistance of the layer (in Ω). If the layer takes the form of a square sheet, its resistance is given by R_s, regardless of its actual dimensions. Hence the sheet resistance is often specified in "Ω per square."

In microcircuit fabrication, the sheet resistance of a diffused layer can be directly measured if it is made in the patterned shape shown in Fig. 4.42. Here, as shown for a *p*-channel MOS transistor, a *p*-type source/drain diffusion is made into an *n*-type substrate. After this diffu-break sion is conducted, a constant current is applied across the points *AB*, and the voltage developed across *CD* is read with the aid of a high-impedance voltmeter. With reference to this figure,

$$R_s = \frac{V}{I}\frac{w}{l} \qquad (4.127)$$

Since w/l is known for a specific test pattern, the sheet resistance is directly found.

Figure 4.43 shows the manner in which the sheet resistance of the emitter diffusion (*n*-type) is determined for a bipolar *n–p–n* transistor. Here, an initial *p*-type region is formed during the base-diffusion step, and the *n*-type emitter diffusion is made within it. The sheet resistance of this layer is measured as described for the base diffusion.

Patterns of these types are usually made as part of a test chip; one or more of these are placed on each slice, and used for diagnostic purposes. Often these test patterns are located in the dead space between microcircuit chips. More complex patterns (multi-stage ring counters, for example), are also located in this space, and used for functional testing purposes.

4.10.3 Surface Concentration

The surface concentration can be determined from the sheet resistance and the junction depth. For a diffusion made into a background concentration N_B, the

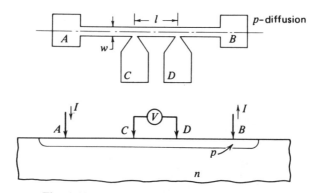

Fig. 4.42 Source/drain resistance test pattern.

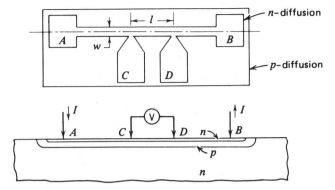

Fig. 4.43 Emitter sheet resistance test pattern.

sheet resistance of the diffused layer is given by R_s, where

$$R_s = \left[\int_0^{x_j} q\mu(x)[N(x) - N_B]\, dx \right]^{-1} \tag{4.128}$$

Here $\mu(x)$ is the mobility for the majority carrier in the diffused layer and is a function of the concentration at any specific depth. This equation can be solved to determine the surface concentration, provided that the doping profile is known, as well as the functional dependence of majority carrier mobility on the doping. A number of special cases are now described.

4.10.4 Diffused Layers in Silicon

Both high- and low-concentration diffusions are used in silicon microcircuit fabrication technology. Layers with low surface concentration can usually be fitted to the erfc or the gaussian diffusion profiles shown in Fig. 4.12. High-concentration diffusions can also be approximated by the erfc profile provided that they are deep.

Computer-derived solutions have been obtained [105] for both erfc and gaussian layers in silicon, for different values of background concentration. These are displayed in Figs. 4.44 and 4.45, and interrelate the surface concentration, the sheet resistance, and the junction depth for a wide variety of background concentrations.

The resistivity of base layers under an emitter can also be determined by integrating the $q\mu N$ product over the base layer width. Computer-derived curves [106] of this type are also available for this purpose.

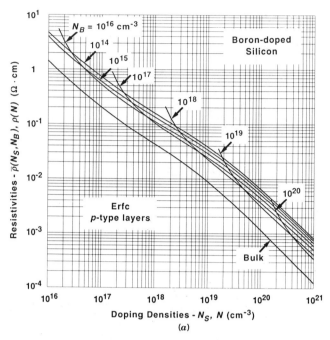

Fig. 4.44 *a* Average resistivity of boron-doped erfc *p*-layers in silicon. The resistivity of boron-doped bulk silicon is also shown. From Bulucea [105]. Reprinted with kind permission from Pergamon Press, Ltd., Oxford, UK.

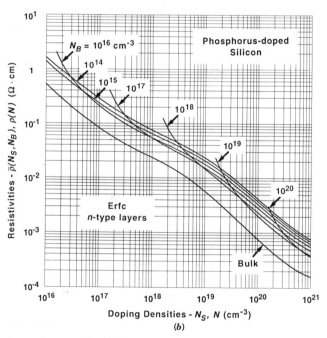

Fig. 4.44 *b* Average resistivity of phosphorus-doped erfc *n*-layers in silicon. The resistivity of phosphorus-doped bulk silicon is also shown. From Bulucea [105]. Reprinted with kind permission from Pergamon Press, Ltd., Oxford, UK.

240

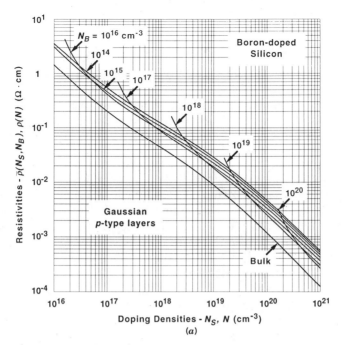

Fig. 4.45 a Average resistivity of boron-doped Gaussian *p*-layers in silicon. The resistivity of boron-doped bulk silicon is also shown. From Bulucea [105]. Reprinted with kind permission from Pergamon Press, Ltd., Oxford, UK.

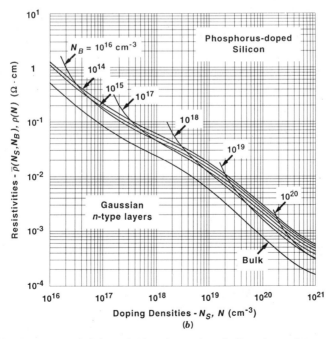

Fig. 4.45 b Average resistivity of phosphorus-doped Gaussian *n*-layers in silicon. The resistivity of phosphorus-doped bulk silicon is also shown. From Bulucea [105]. Reprinted with kind permission from Pergamon Press, Ltd., Oxford, UK.

4.10.4.1 High-Concentration Diffused Layers

Many diffusions that are used in VLSI fabrication technology are made with a high surface concentration. The doping profiles for these diffusions are described in Section 4.4. Most profiles are abrupt, so that the junction depth is relatively independent of the background concentration. Profile parameters for these diffusions are now described.

Boron. Here an average hole mobility (55 cm^2 V^{-1} s^{-1}) can be assigned to layers with high surface concentration, resulting in straightforward relationships between R_s, x_j, and N_{sur}. These are obtained [36] if the background concentration is assumed to be less than 1% of the surface value. These relationships are compiled in Table 4.7.

Arsenic. The extremely steep doping profile for this impurity allows an average value of mobility to be assigned to this layer, and interrelations of x_j, R_s, and N_{sur} have been developed [107, 108] based on this assumption. The polynomial approximation of the doping profile (see Section 4.5.2.2), as well as the assumption that the junction depth is independent of background concentration, were used to derive the relationships listed in Table 4.8.

Phosphorus. The anomalous doping profile for high-concentration phosphorus diffusions has been described in Section 4.4.1.3. A systematic analysis of its detailed features has permitted the necessary interrelations between R_s, x_j, and N to be extracted [109]. These results are now presented without proof.

1. The surface electron concentration n_s and the surface impurity concentration N_{tot} are related by

$$N_{tot} = n_s + 2.04 \times 10^{-41} n_s^3 \qquad (4.129)$$

2. The relationship of n_s and the total number of diffused impurities per unit area, Q_{tot} is shown in Fig. 4.46. Also shown is the graph for Q_{el}, the electronically active component of Q_{tot}. Figure 4.47 shows the fraction Q_{el}/Q_{tot} as a function of n_s.

3. The sheet resistance is related to the junction depth and to Q_{el} by

$$R_s = [q(75Q_{el} + 1.8 \times 10^{20} x_j)]^{-1} \qquad (4.130)$$

Profile parameters can be obtained by use of the above data. However, the approximate nature of these relationships must be stressed.

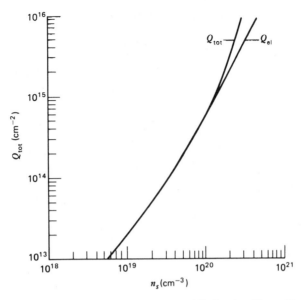

Fig. 4.46 n_s versus Q_{el} and Q_{tot} for phosphorus diffusion in silicon. From Fair [109]. Reprinted with permission of the publisher, The Electrochemical Society, Inc.

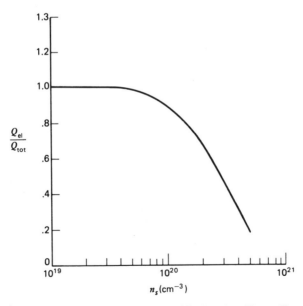

Fig. 4.47 n_s versus Q_{el}/Q_{tot} for phosphorus diffusion in silicon. From Fair [109]. Reprinted with permission of the publisher, The Electrochemical Society, Inc.

4.10.5 Diffused Layers in Gallium Arsenide

The behavior of *n*-type dopants in GaAs is similar to that of substitutional diffusers, and is characterized by an erfc doping profile. This is especially true for low-concentration ($<10^{19}$ cm^{-3}) diffusions, which are commonly required with these dopants. Figure 4.48 shows computer-derived curves which relate the profile parameters for these junctions, for substrates with varying background concentration [110].

p-type dopants, such as zinc, are characterized by a steep doping profile, and relatively high surface concentrations ($\geq 10^{20}$ cm^{-3}). Figure 4.49 shows curves relating N_{sur} and x_j with the average conductivity of zinc-diffused layers in GaAs [111]. These curves are based on sheet resistance measurements of these layers for a wide variety of diffusion conditions, using elemental zinc as well as alloy sources in a sealed-tube system.

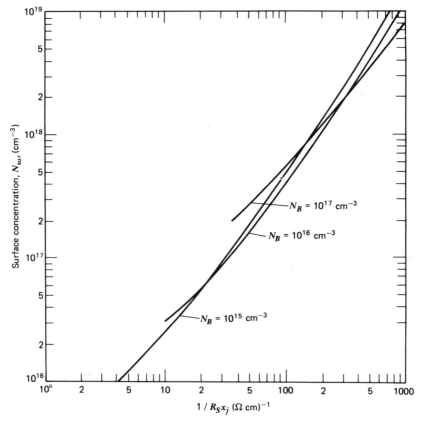

Fig. 4.48 *n*-type erfc diffusions into gallium arsenide. From Baliga [110]. Reprinted with permission of the publisher, Pergamon Press, Ltd.

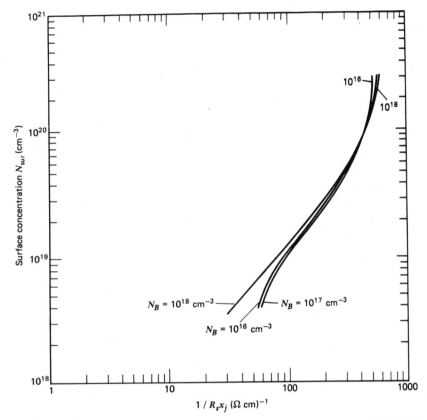

Fig. 4.49 p-type zinc diffusions into gallium arsenide. Adapted from Kendall [111].

TABLE 4.1   Values of the Smith Integral [25]

					β							
U	0.1	0.2	0.3	0.4	0.5	0.6	0.7	0.8	0.9	1.0	1.1	1.2
0.1	0.09015	0.08155	0.07376	0.06672	0.06035	0.05459	0.04938	0.04467	0.04040	0.03655	0.03306	0.02990
0.2	0.17838	0.16119	0.14566	0.13162	0.11894	0.10748	0.09713	0.08777	0.07931	0.07167	0.06477	0.05853
0.3	0.26295	0.23723	0.21403	0.19310	0.17422	0.15719	0.14182	0.12795	0.11545	0.10416	0.09398	0.08479
0.4	0.34254	0.30837	0.27761	0.24993	0.22501	0.20259	0.18240	0.16422	0.14786	0.13314	0.11988	0.10794
0.5	0.41626	0.37374	0.33557	0.30132	0.27058	0.24299	0.21822	0.19599	0.17603	0.15812	0.14203	0.12759
0.6	0.48366	0.43290	0.38751	0.34692	0.31062	0.27814	0.24908	0.22308	0.19982	0.17900	0.16036	0.14368
0.7	0.54464	0.48580	0.43340	0.38673	0.34515	0.30809	0.27505	0.24562	0.21937	0.19596	0.17508	0.15645
0.8	0.59940	0.53264	0.47347	0.42100	0.37447	0.33317	0.29652	0.26398	0.23508	0.20940	0.18657	0.16628
0.9	0.64829	0.57380	0.50812	0.45017	0.39903	0.35385	0.31393	0.27864	0.24742	0.21979	0.19532	0.17365
1.0	0.69176	0.60975	0.53784	0.47475	0.41935	0.37066	0.32783	0.29013	0.25693	0.22765	0.20183	0.17903
1.1	0.73033	0.64100	0.56318	0.49529	0.43600	0.38415	0.33877	0.29900	0.26411	0.23348	0.20655	0.18286
1.2	0.76448	0.66808	0.58465	0.51232	0.44950	0.39486	0.34726	0.30574	0.26946	0.23772	0.20991	0.18553
1.3	0.79470	0.69148	0.60276	0.52634	0.46035	0.40327	0.35377	0.31078	0.27336	0.24074	0.21225	0.18734
1.4	0.82144	0.71164	0.61797	0.53781	0.46901	0.40979	0.35870	0.31449	0.27616	0.24286	0.21385	0.18855
1.5	0.84509	0.72899	0.63069	0.54714	0.47586	0.41482	0.36238	0.31720	0.27815	0.24431	0.21492	0.18933
1.6	0.86601	0.74388	0.64130	0.55469	0.48123	0.41865	0.36511	0.31914	0.27953	0.24530	0.21562	0.18983
1.7	0.88454	0.75666	0.65010	0.56076	0.48542	0.42153	0.36710	0.32051	0.28048	0.24595	0.21607	0.19014
1.8	0.90095	0.76759	0.65739	0.56562	0.48865	0.42369	0.36854	0.32147	0.28112	0.24638	0.21636	0.19033
1.9	0.91549	0.77693	0.66340	0.56948	0.49114	0.42529	0.36956	0.32213	0.28154	0.24665	0.21653	0.19045
2.0	0.92838	0.78491	0.66833	0.57254	0.49303	0.42646	0.37029	0.32258	0.28182	0.24682	0.21664	0.19051
2.5	0.97404	0.81009	0.68228	0.58029	0.49735	0.42887	0.37165	0.32335	0.28225	0.24707	0.21678	0.19059
3.0	0.99920	0.82094	0.68698	0.58234	0.49825	0.42928	0.37183	0.32343	0.28229	0.24708	0.21679	0.19059
∞	1.02834	0.82795	0.68892	0.58291	0.49843	0.42933	0.37184	0.32343	0.28229	0.24709	0.21679	0.19059

					β							
U	1.3	1.4	1.5	1.6	1.7	1.8	1.9	2.0	2.5	3.0	4.0	5.0
0.1	0.02705	0.02446	0.02213	0.02002	0.01811	0.01638	0.01481	0.01340	0.00811	0.00491	0.00180	0.00066
0.2	0.05289	0.04779	0.04319	0.03903	0.03527	0.03187	0.02880	0.02603	0.01568	0.00945	0.00343	0.00125
0.3	0.07651	0.06903	0.06228	0.05620	0.05071	0.04575	0.04128	0.03725	0.02228	0.01333	0.00477	0.00171
0.4	0.09720	0.08752	0.07881	0.07097	0.06391	0.05756	0.05183	0.04668	0.02766	0.01640	0.00577	0.00204
0.5	0.11462	0.10297	0.09251	0.08312	0.07468	0.06711	0.06030	0.05419	0.03178	0.01866	0.00645	0.00224
0.6	0.12875	0.11538	0.10340	0.09268	0.08308	0.07448	0.06678	0.05988	0.03475	0.02021	0.00688	0.00236
0.7	0.13982	0.12499	0.11174	0.09992	0.08936	0.07993	0.07150	0.06398	0.03677	0.02120	0.00712	0.00242
0.8	0.14824	0.13219	0.11790	0.10519	0.09387	0.08379	0.07481	0.06680	0.03806	0.02180	0.00724	0.00244
0.9	0.15444	0.13741	0.12230	0.10889	0.09699	0.08642	0.07702	0.06867	0.03885	0.02213	0.00730	0.00245
1.0	0.15889	0.14109	0.12535	0.11141	0.09907	0.08814	0.07844	0.06985	0.03931	0.02231	0.00733	0.00246
1.1	0.16200	0.14361	0.12739	0.11307	0.10041	0.08923	0.07933	0.07056	0.03956	0.02240	0.00734	0.00246
1.2	0.16411	0.14529	0.12872	0.11412	0.10125	0.08989	0.07985	0.07098	0.03969	0.02244	0.00735	0.00246
1.3	0.16552	0.14638	0.12956	0.11478	0.10176	0.09028	0.08016	0.07122	0.03976	0.02246	0.00735	0.00246
1.4	0.16643	0.14706	0.13008	0.11517	0.10205	0.09051	0.08033	0.07134	0.03979	0.02247	0.00735	0.00246
1.5	0.16700	0.14749	0.13039	0.11540	0.10222	0.09063	0.08042	0.07141	0.03980	0.02247	0.00735	0.00246
1.6	0.16736	0.14774	0.13057	0.11552	0.10231	0.09070	0.08046	0.07144	0.03981	0.02247	0.00735	0.00246
1.7	0.16757	0.14789	0.13067	0.11559	0.10236	0.09073	0.08049	0.07146	0.03981	0.02247	0.00735	0.00246
1.8	0.16770	0.14797	0.13073	0.11563	0.10239	0.09075	0.08050	0.07147	0.03982	0.02247	0.00735	0.00246
1.9	0.16777	0.14802	0.13076	0.11565	0.10240	0.09075	0.08050	0.07147	0.03982	0.02247	0.00735	0.00246
2.0	0.16781	0.14804	0.13078	0.11566	0.10240	0.09076	0.08051	0.07147	0.03982	0.02247	0.00735	0.00246
2.5	0.16786	0.14807	0.13079	0.11567	0.10241	0.09076	0.08051	0.07147	—	—	—	—
3.0	0.16786	0.14807	0.13079	0.11567	0.10241	0.09076	0.08051	0.07147	—	—	—	—
∞	0.16786	0.14807	0.13079	0.11567	0.10241	0.09076	0.08051	0.07147	0.03982	0.02247	0.00735	0.00246

TABLE 4.2   Apparent Activation Energy of Substitutional Diffusers in Silicon

Impurity	E_d (eV)
B	3.51–3.67
Al	3.36
Ga	3.75
In	3.6
P	3.51–3.67
As	4.05–4.34
Sb	3.89–4.05

TABLE 4.3   Intrinsic Diffusivities and Activation Energies of Substitutional and Self-Diffusers in Silicon[a] [11]

		P	As	Sb	B	Al	Ga	Si
D_i^0	D_0	3.85	0.066	0.214	0.037	1.385	0.374	0.015
	E_0	3.66	3.44	3.65	3.46	3.41	3.39	3.89
D_i^+	D_0	—	—	—	0.76	2480	28.5	1180
	E_0	—	—	—	3.46	4.20	3.92	5.09
D_i^-	D_0	4.44	22.9	13	—	—	—	16
	E_0	4.0	4.1	4.0	—	—	—	4.54
D_i^{2-}	D_0	44.2	—	—	—	—	—	10
	E_0	4.37	—	—	—	—	—	5.1

[a]D_0 in cm^2 s^{-1}; E_0 in eV.

TABLE 4.4 Diffusion Constant and Apparent Activation Energy of Diffusion for Deep-Lying Impurities in Silicon [2]

Impurity	D_0 (cm^2 $s-1$)	E_d (eV)
Ag	2×10^{-3}	1.6
Au	1.1×10^{-3}	1.12
Co	0.16	1.12
Cu	4×10^{-2}	1.0
Fe	6.2×10^{-3}	0.87
Ni	1.3×10^{-2}	1.4
O	0.21	2.44
S	0.92	2.2
Zn	0.1	1.4

TABLE 4.5 Capture Rate Constants for Gold and Platinum in Silicon [42, 45]

	Capture Rate Constants (cm^3 s^{-1})	
	Au	Pt
Electron capture at E_1^+	6.3×10^{-8}	2.2×10^{-8}
Hole capture at E_1^+	2.4×10^{-8}	1.2×10^{-9}
Electron capture at E_1^-	1.65×10^{-9}	2.5×10^{-9}
Hole capture at E_1^-	1.15×10^{-7}	1.5×10^{-7}
Electron capture at E_2^-	—	3.2×10^{-7}
Hole capture at E_2^-	—	2.7×10^{-5}

TABLE 4.6   Diffusion Constant and Apparent Activation Energy of Diffusion for Impurities in Gallium Arsenide [2]

Impurity	D_0 $(cm^2\ s^{-1})$	E_d (eV)
Au	2.9×10^1	2.64
Be	7.3×10^{-6}	1.2
Cd		2.43
Cr	4.3×10^3	3.4
Cu	3×10^{-2}	0.53
Li	5.3×10^{-1}	1.0
Mg	2.6×10^{-2}	2.7
Mn	6.5×10^{-1}	2.49–2.75
O	2×10^{-3}	1.1
S	1.85×10^{-2}	2.6
Se	3.0×10^3	4.16
Sn	3.8×10^{-2}	2.7
Zn		2.49
Hg	$D = 5 \times 10^{-14}$ @1000°C	
Te	$D = 10^{-13}$ @1000°C	
	$D = 2 \times 10^{-12}$ @1100°C	

TABLE 4.7   Properties of High-Concentration Boron Diffusions in Silicon [36]

	Diffused from a Constant-Source Concentration	Implanted with a Dose of Q_0 ions/cm^2 and Diffused for t s
x_j (cm)	$2.45\left(N_{sur}\dfrac{D_i t}{n_i}\right)^{1/2}$	$Q_0(0.4N_{sur})^{-1}$
N_{sur} (cm^{-3})	$2.78 \times 10^{17} (R_s x_j)^{-1}$	$0.53\left(Q_0^2\dfrac{n_i}{D_i t}\right)^{1/3}$
D_i (cm^2 s^{-1})	$3.17e^{-3.59/kT}$	

TABLE 4.8   Properties of High-Concentration Arsenic Diffusions in Silicon [108, 109]

	Diffused from a Constant-Source Concentration	Implanted with a Dose of Q_0 ions/cm^2 and Diffused for t s
x_j (cm)	$2.29 \left(N_{sur} \dfrac{D_i^- t}{n_{ie}} \right)^{1/2}$	$2 \left(Q_0 \dfrac{D_i^- t}{n_{ie}} \right)^{1/3}$
$R_s (\Omega)$	$1.56 \times 10^{17} (N_{sur} x_j)^{-1}$	$\dfrac{1.7 \times 10^{10}}{Q^{7/9}} \left(\dfrac{n_{ie}}{D_i^- t} \right)^{1/9}$
N_{sur} (cm^{-3})	N_{sur}	$Q_0 (0.45 x_j)^{-1}$
D_i (cm^2 s^{-1})		$6.26 \times 10^{15} (R_s x_j)^{3/2}$
	$22.9 e^{-4.1/kT}$	

REFERENCES

1. B. Tuck, *Introduction to Diffusion in Semiconductors*, IEE Monograph Series 16, Peter Peregrinus Ltd., England, 1974.

2. D. Shaw, Ed., *Atomic Diffusion in Semiconductors*, Plenum, New York, 1973.

3. W. Jost, *Diffusion in Solids, Liquids and Gases*, Academic Press, New York, 1962.

4. W. D. Laidig, N. Holonyak, Jr., M. D. Camras, K. Hess, J. J. Coleman, P. D. Dapkus, and J. Bardeen, Disorder of an AlAs-GaAs superlattice by impurity diffusion, *Appl. Phys. Lett.* **38**, 776 (1981).

5. N. B. Hannay, Ed., *Semiconductors*, Reinhold, New York, 1960.

6. F. Barson, Emitter–Collector Shorts in Bipolar Devices, *IEEE J. Solid State Circuits* **SC-11**, 505 (1976).

7. S. M. Hu and S. Schmit, Interactions in Sequential Diffusion Processes in Semiconductors, *J. Appl. Phys.* **39**, 4272 (1968).

8. P. M. Fahey, P. B. Griffin, and J. D. Plummer, Point Defects and Dopant Diffusion in Silicon, *Rev. Modern Phys.* **61**, 289 (1989).

9. W. Shockley and J. T. Last, Statistics of the Charge Distribution for a Localized Flaw in a Semiconductor, *Phys. Rev.* **107**(2), 392 (1957).

10. D. Shaw, Self- and Impurity-Diffusion in Ge and Si, *Phys. Status. Solidi. B* **72**, 11 (1975).

11. R. B. Fair, Recent Advances in Implantation and Diffusion Modelling for the Design and Process Control of Bipolar ICs, in *Semiconductor Silicon*, 1977, H. R. Huff and E. Sirtl, Eds., The Electrochemical Society, Princeton, NJ, 1978, p. 968.

12. V. Gösele, Diffusion in Semiconductors, in *Microelectronic Materials and Processes*, R. A. Levy, Ed., Kluwer Academic Publishers, Boston, 1989, p. 583.

13. H. Lefevre, Trap Centers of Self-Interstitials in Silicon, *Appl. Phys. Lett.* **22**, 15 (1980).

14. Logan and Hurle, Calculation of Point Defect Concentrations and Non-stoichiometry in GaAs, *J. Phys. Chem. Solids*, **32**, 1739 (1971).

15. D. G. Deppe and N. Holonyak, Jr., Atom Diffusion and Impurity-Induced Layer Disordering in Quantum Well III–V Semiconducting Heterostructures, *J. Appl. Phys.* **64**, R93 (1988).

16. T. Y. Tan and V. Gösele, Self-Diffusion in GaAs, *Appl. Phys. Lett.* **52**, 1240 (1988).

17. T. Y. Tan and U. Gösele, Point Defects, Diffusion Processes and Swirl Defect Formation in Silicon, *Appl. Phys.* **A37**, 1 (1985).

18. N. A. Stolwijk, B. Schuster and J. Holzl, Diffusion of Gold in Silicon, Studied by Neutron-Activation Analysis and Spreading-Resistivity Measurements, *Appl. Phys.* **A33**, 133 (1984).

19. S. Y. Chiang and G. L. Pearson, Properties of Vacancy Defects in GaAs Single Crystals, *J. Appl. Phys.* **46**, 2986 (1975).

20. F. M. Smits and R. C. Miller, Rate Limitation at the Surface for Impurity Diffusion in Semiconductors, *Phys. Rev.* **104**, 1242 (1956).

21. W. R. Rice, Diffusion of Impurities During Epitaxy, *Proc. IEEE* **52**, 284 (1964).

22. T. I. Kucher, The Problem of Diffusion in an Evaporating Solid Medium, *Sov. Phys. Solid State* **3**, 401 (1961).

23. R. L. Batdorf and F. M. Smits, Diffusion of Impurities into Evaporating Silicon, *J. Appl. Phys.* **30**, 259 (1959).

24. R. B. Allen, H. Bernstein, and A. D. Kurtz, Effect of Oxide Layers on the Diffusion of Phosphorus into Silicon, *J. Appl. Phys.* **31**, 334 (1960).

25. R. C. T. Smith, Conduction of Heat in the Semi-infinite Solid with a Short Table of an Important Integral, *Aust. J. Phys.* **6**, 129 (1953).

26. L. R. Weisberg and J. Blanc, Diffusion with Interstitial–Substitutional Equilibrium. Zinc in GaAs, *Phys. Rev.* **131**, 1548 (1963).

27. M. Ghezzo and G. Gildenblat, Analytical Solutions for High Concentration Shallow Arsenic Profiles in Silicon, *Solid State Electron* **28**, 909 (1985).

28. D. A. Antoniadis, S. E. Hansen, and R. W. Dutton, SUPREME II—A Program for IC Process Modeling and Simulation, *Stanford Electron Lab Technical Report*, No. 5019-2, Palo Alto, CA, 1978.

29. W. Fichtner, Process Simulation, in *Microelectronic Materials and Processes*, R. A. Levy, Ed., Kluwer Academic Publishers, Boston, 1989, p. 775.

30. D. P. Kennedy and R. R. O'Brien, Analysis of the Impurity Atom Distribution Near the Diffusion Mask for a Planar *p–n* Junction, *IBM J. Res. Dev.* **9**, 179, 1965.

31. D. D. Warner and C. L. Wilson, Two-Dimensional Concentration Dependent Diffusion, *Bell Syst. Tech. J.* **59**, 1 (1980).

32. C. S. Fuller and J. A. Ditzenberger, Diffusion of Donor and Acceptor Elements in Silicon, *J. Appl. Phys.* **27**, 544 (1956).

33. D. A. Antoniadis, A. M. Lin, and R. W. Dutton, Oxidation-Enhanced Diffusion of Arsenic and Phosphorus in Near-Intrinsic ⟨100⟩ Silicon, *Appl. Phys. Lett.* **33**, 1030 (1978).

34. R. B. Fair and J. C. C. Tsai, The Diffusion of Ion-Implanted Arsenic in Silicon, *J. Electrochem. Soc.* **122**, 1689 (1975).

35. R. A. Colclaser, *Microelectronics: Processing and Device Design*, John Wiley & Sons, New York, 1980.

36. R. B. Fair, Boron Diffusion in Silicon: Concentration and Orientation Dependence, Background Effects, and Profile Estimation, *J. Electrochem. Soc.* **122**, 800 (1975).

37. G. Masetti, S. Solmi, and G. Soncini, Temperature Dependence of Boron Diffusion in (111), (110), and (100) Silicon, *Solid State Electron.* **19**, 545 (1976).

38. D. A. Antoniadis, A. G. Gonzalez, and R. W. Dutton, Boron in Near Intrinsic ⟨100⟩ and ⟨111⟩ Silicon Under Inert and Oxidizing Ambients—Diffusion and Segregation, *J. Electrochem. Soc.* **125**, 813 (1978).

39. R. B. Fair and J. C. C. Tsai, A Quantitative Model of Phosphorus in Silicon and the Emitter Dip Effect, *J. Electrochem. Soc.* **124**, 1107 (1977).

40. V. Gösele and H. Strunk, High Temperature Diffusion of Phosphorus and Boron in Silicon via Vacancies or Self-Interstitials, *Appl. Phys.* **20**, 265 (1979).

41. G. Masetti, P. Nobili and S. Solmi, Profiles of Phosphorus Predeposited in Silicon and Carrier Concentration in Equilibrium with SiP Precipitates, in *Semiconductor Silicon, 1977*, H. R. Huff and E. Sirtl, Eds., The Electrochemical Society, Princeton, NJ, 1978, p. 648.

42. J. M. Fairfield and B. V. Gokhale, Gold as a Recombination Center in Silicon, *Solid State Electron.* **8**, 685 (1965).

43. W. M. Bullis, Properties of Gold in Silicon, *Solid State Electron.* **9**, 143 (1966).

44. S. F. Cagnina, Enhanced Gold Solubility Effect in Heavily *n*-Type Silicon, *J. Electrochem. Soc.* **116**, 498 (1969).

45. K. P. Lisiak and A. G. Milnes, Platinum as a Lifetime Control Deep Impurity in Silicon, *J. Appl. Phys.* **46**, 5229 (1975).

46. M. D. Miller, Differences Between Platinum and Gold-Doped Silicon Power Devices, *IEEE Trans. Electron Dev.* **ED-23**, 1279 (1976).

47. U. Gösele and F. Morehead, Diffusion of Zinc in Gallium Arsenide: A New Model, *J. Appl. Phys.* **52**, 4617 (1981).

48. S. Reynolds, W. Vook, and J. F. Gibbons, Open Tube Zinc Diffusion in GaAs Using Diethylzinc and Trimethylarsenic, *J. Appl. Phys.* **63**, 1052 (1988).

49. K. B. Kahen, Model for the Diffusion of Zinc in Gallium Arsenide, *Appl. Phys. Lett.* **55**, 2117 (1989).

50. H. C. Casey, Jr., M. B. Panish, and L. L. Chang, Dependence of the Diffusion Coefficient on the Fermi Level: Zinc in Gallium Arsenide, *Phys. Rev.* **162**, 660 (1967).

51. H. C. Casey, Jr., and M. B. Panish, Reproducible Diffusion of Zinc into GaAs: Application of the Ternary Phase Diagram and the Diffusion and Solubility Analyses, *Trans. Met. Soc. AIME* **242**, 406 (1968).

52. J. J. Murray, M. D. Deal, E. L. Allen, D. A. Stevenson, and N. Nozaki, Modeling Silicon Diffusion in GaAs Using Well Defined Silicon Doped Molecular Beam Epitaxy Structures, *J. Electrochem. Soc.* **139**, 2037 (1992).

53. A. Kostka, R. Gereth and K. Kreuzer, A Physical and Mathematical Approach to Mass Transport in Capsule Diffusion Processes, *J. Electrochem. Soc.* **120**, 971 (1973).

54. K. Mokuya and I. Matsuba, Prediction of Defect Onset Conditions in Heat Cycling Based on a Thermoelastic Wafer Model, *IEEE Trans. Electron Dev.* **36**, 319 (1989).

55. J. A. Becker, Silicon Wafer Processing by Application of Spun-On Doped and Undoped Silica Layers, *Solid State Electron.* **17**, 87 (1974).

56. R. N. Ghoshtagore, Model of Doped-Oxide-Source Diffusion in Silicon, *Solid State Electron.* **17**, 1065 (1974).

57. N. Goldsmith, J. Olmstead, and J. Scott, Jr., Boron Nitride as a Diffusion Source for Silicon, *RCA Rev.* **28**, 344 (1967).

58. Technical Literature, Carborundum Corp., Graphite Products Div., P. O. Box 577, Niagara Falls, NY 14302, 1975.

59. D. Rupprecht and J. Stach, Oxidized Boron Wafers as an *In-Situ* Boron Dopant for Silicon Diffusions, *J. Electrochem. Soc.* **120**, 9 (1973).

60. J. Stach and A. Turley, Anomalous Boron Diffusion in Silicon from Planar Boron Nitride Sources, *J. Electrochem. Soc.* **131**, 722 (1974).

61. BoronPlus®High Purity Planar Dopant Sources, Owens-Illinois Technical Products, Toledo, OH.

62. V. Geiss and E. Fröschle, Mass Spectrometric Investigation of the Reaction Velocities of BCl_3 and BBr_3 with Oxygen and Water Vapor in a Diffusion Furnace, *J. Electrochem. Soc.* **123**, 133 (1976).

63. H. Nakamura, Diborane–Carbon Dioxide System for Boron Diffusion into Silicon at Low Temperature, *Jpn. J. Appl. Phys.* **11**, 761 (1972).

64. N. Jones, D. M. Metz, J. Stach, and R. E. Tressler, A Solid Planar Source for Phosphorus Diffusion, *J. Electrochem. Soc.* **123**, 1565 (1976).

65. PhosPlus®High Purity Planar Dopant Sources, Owens-Illinois Inc., Toledo, OH.

66. R. B. Fair, High Concentration Arsenic Diffusion in Silicon from a Doped Oxide Source, *J. Electrochem. Soc.* **119**, 1389 (1972).

67. J. R. Dalcin, M. C. Rogers, J. L. Tworek, and E. R. Wojtanik, Arsenic Solid Source Doping: The Deposition Process, *Solid State Technol.* (Aug. 1988) p. 123.

68. J. S. Sandhu and J. L. Reuter, Arsenic Source Vapor Pressure Kinetics and Capsule Diffusion, *IBM J. Res. Dev.* (Nov. 1971) p. 464.

69. R. B. Fair and J. C. C. Tsai, The Diffusion of Ion-Implanted Arsenic in Silicon, *J. Electrochem. Soc.* **122**, 1677 (1975).

70. D. A. Antoniadis and R. W. Dutton, Models for Computer Simulation of Complete IC Fabrication Processes, *IEEE J. Solid State Circuits* **SC-14**, 412 (1979).

71. J. W. Colby and L. E. Katz, Boron Segregation at the Si–SiO_2 Interface as a Function of Temperature and Orientation, *J. Electrochem. Soc.* **123**, 409 (1976).

72. T. Kato and Y. Nishi, Redistribution of Diffused Boron in Silicon by Thermal Oxidation, *Jpn. J. Appl. Phys.* **3**, 377 (1964).

73. C. P. Wu, E. C. Douglas, and C. W. Mueller, Redistribution of Ion-Implanted Impurities in Silicon During Diffusion in Oxidizing Ambients, *IEEE Trans. Electron Dev.* **ED-23**, 1095 (1976).

74. A. M-R. Lin, D. A. Antoniadis, and R. W. Dutton, The Oxidation Rate Dependence of Oxidation-Enhanced Diffusion of Boron and Phosphorus in Silicon, *J. Electrochem. Soc.* **128**, 1141 (1981).

75. D. A. Antoniadis and I. Moskowitz, Diffusion of Substitutional Impurities in Silicon at Short Oxidation Times: An Insight Into Point Defect Kinetics, *J. Appl. Phys.* **53**, 6788 (1982).

76. C. L. Jones and A. F. W. Willoughby, Studies of the Push-Out Effect in Silicon, Pt. I, *J. Electrochem. Soc.* **122**, 1531 (1975); Pt. II, *J. Electrochem. Soc.* **123**, 1531 (1976).

76. L. R. Weisberg, Diffusion in Gallium Arsenide, *Trans. Met. Soc. AIME* **230**, 291 (1964).

77. F. A. Cunnell and C. H. Gooch, Diffusion of Zinc in Gallium Arsenide, *J. Phys. Chem. Solids* **15**, 127 (1960).

78. K. K. Shih, High Surface Concentration Zn Diffusion in GaAs, *J. Electrochem. Soc.* **123**, 1737 (1976).

79. H. Matino, Reproducible Sulfur Diffusion into GaAs, *Solid State Electron.* **17**, 35 (1974).

80. Y. I. Nissim, J. F. Gibbons, C. A. Evans, Jr., V. R. Deline, and J. C. Norberg, Thermal Diffusion of Tin in GaAs from a Spin-On SnO_2/SiO_2 Source, *Appl. Phys. Lett.* **37**, 89 (1980).

81. S. K. Ghandhi and R. J. Field, Precisely Controlled Shallow p^+-Diffusions in GaAs, *Appl. Phys. Lett.* **38**, 267 (1981).

82. A. B. Y. Young and G. L. Pearson, Diffusion of Sulfur in Gallium Phosphide and Gallium Arsenide, *J. Phys. Chem. Solids* **31**, 517 (1970).

83. B. J. Baliga and S. K. Ghandhi, Planar Diffusions in Gallium Arsenide from Tin-Doped Oxides, *J. Electrochem. Soc.* **126**, 135 (1979).

84. M. Oren and A. N. M. Masum Choudhury, Zn Diffusion in GaAs Obtained by a Simple Open-Tube Technique, *J. Appl. Phys.* **60**, 3379 (1986).

85. A. J. Springthorpe and M. N. Svilans, Low Temperature Zinc Diffusion in GaAs, InP and GaInAs Using the Box-Diffusion Technique, in *Gallium Arsenide and Related Compounds, International Physics Conference Series*, No. 65 (1983) G. E. Stillman, Ed., Institute of Physics, Bristol, U.K., 1983, p. 288.

86. R. J. Roedel, J. Edwards, A. Righter, P. Holm, and H. Erhaya, *J. Electrochem. Soc.* **131**, 1726 (1984).

87. F. C. Prince, M. Oren, and M. Lam, High Mobility GaAs Layers Obtained by Open Tube Sulfur Diffusion, *Appl. Phys. Lett.* **48**(8), 546 (1986).

88. S. K. Ageno, R. J. Roedel, N. Mellen, and J. S. Escher, Diffusion of Zinc Into $Ga_{1-x}Al_xAs$, *Appl. Phys. Lett.* **47**, 1193 (1985).

89. A. N. M. Masum Choudhury, M. Oren, M. A. Rothman, and S. K. Shastry, A Simple Open-Tube Zn-Diffusion Technique for GaAs and AlGaAs, *J. Electrochem. Soc.* **134**, 2631 (1986).

90. Y-R Yuan, K. Eda, G. A. Vawter, and J. L. Merz, Open Tube Diffusion of Zinc Into AlGaAs and GaAs, *J. Appl. Phys.* **54**, 6044 (1983).

91. M. Dohsen, J. Kasahara, Y. Kato, and N. Watanabe, GaAs J-Fet, Formed by Localized Zn Diffusion, *IEEE Trans. Electron Dev. Lett.* **EDL-2**, 157 (1981).

92. R. J. Field and S. K. Ghandhi, An Open Tube Method for Diffusion of Zinc into GaAs, *J. Electrochem. Soc.* **129**, 1567 (1982).

93. P. J. Ledakovitch, R. Agarwal, and S. K. Ghandhi, Fabrication and Characteristics of *n*-Layers into Galium Arsenide by an Open-Tube Diffusion Process, *Solid State Electron.* **30**, 681 (1987).

94. B. J. Baliga and S. K. Ghandhi, PSG Masks for Diffusion in Gallium Arsenide, *IEEE Trans. Electron Dev.* **ED-19**, 761 (1972).

95. R. J. Roedel, H. H. Erkaya, and J. L. Edwards, Sputtered Dielectric Films for GaAs Diffusion Masks, *J. Electron. Mater.* **17**, 243 (1988).

96. B. J. Baliga and S. K. Ghandhi, Lateral Diffusion of Zinc and Tin in Gallium Arsenide, *IEEE Trans. Electron Dev.* **ED-21**, 410 (1974).

97. M. J. Helix, K. V. Vaidyanathan, and B. G. Streetman, Properties of Be-Implanted Planar GaAs p–n Junctions, *IEEE J. Solid State Circuits* **SC-13**, 426 (1978).

98. K. Asai, Y. Iishi, and K. Kurumada, P-Column Gate FET, *IEEE Electron Dev. Lett.* **EDL-1**, 83 (1980).

99. K. Kurumada, K. Asai, and Y. Iishi, P-Type Anode Structure for GaAs TEDs on Semi-insulating Substrates, *IEEE Electron Dev. Lett.* **EDL-1**, 167 (1980).

100. P. F. Kane and G. B. Larrabee, *Characterization of Semiconductor Materials*, McGraw-Hill, New York, 1970.

101. B. I. Boltaks, *Diffusion in Semiconductors*, Academic, New York, 1963.

102. J. I. Goldstein and H. Yakowitz, Eds., *Practical Scanning Electron Microscopy and Ion Microprobe Analysis*, Plenum, New York, 1975.

103. S. Tolansky, *Multiple-Beam Interferometry of Surfaces and Films*, Clarendon Press, Oxford, 1948.

104. P. E. Russell, SEM Based Characterization Techniques for Semiconductor Technology, *Proc. SPIE* **452**, 183 (1983).

105. C. Bulucea, Recalculation of Irvin's Resistivity Curves for Diffused Layers in Silicon using Updated Resistivity Data, *Solid State Electron.* **36**, 489 (1993).

106. J. C. Irvin, Resistivity of Bulk Silicon and Diffused Layers in Silicon, *Bell Syst. Tech. J.* **41**, 387 (1962).

107. R. B. Fair, Profile Estimation of High Concentration Diffusions in Silicon, *J. Appl. Phys.* **43**, 1278 (1972).

108. R. B. Fair and J. C. C. Tsai, Profile Parameters of Implanted-Diffused Arsenic Layers in Silicon, *J. Electrochem. Soc.* **123**, 583 (1976).

109. R. B. Fair, Analysis of Phosphorus-Diffused Layers in Silicon, *J. Electrochem. Soc.* **125**, 323 (1978).

110. B. J. Baliga, Conductivity of Complementary Error Function n-Type Diffused Layers in Gallium Arsenide, *Solid State Electron.* **20**, 321 (1977).

111. D. L. Kendall, Diffusion, in *Semiconductors and Semimetals*, Vol. 4, R. K. Willardson and A. C. Beer, Eds., Academic Press, New York, 1968, p. 163.

PROBLEMS

1. A diffusion furnace operating at $1000°C$ has a $\pm 1°C$ tolerance. What is the corresponding tolerance on the diffusion depth. Assume an erfc diffusion, and an apparent activation energy of 3.69 eV for the dopant.

2. A p-diffusion is made into an n-region such that $N_{sur} = 1000 N_B$. Show that the junction depth is proportional to $(Dt)^{1/2}$ and determine the factor of proportionality. A constant-source diffusion may be assumed.

3. A diffusion furnace is ramped up (from $600°C$) for 15 min, held at $100°C$ for 30 min, and ramped down to $600°C$ in 15 min. Calculate the effective diffusion time, assuming phosphorus in silicon.

4. The following diffusions are performed into n-type silicon with background concentration of 3×10^{16} atoms cm^{-3}: (a) a 10-min constant-source boron predeposition at 900°C, followed by a 45-min drive-in at 1100°C; (b) a 30-min constant-source phosphorus diffusion at 1100°C. Determine the surface concentrations and junction depths that result from this process. Sketch the doping profiles on semilog paper. Assume that $D = 2 \times 10^{-15}$ cm^2 s^{-1} for boron at 900°C and 3×10^{-13} cm^2 s^{-1} for phosphorus at 1100°C.

5. It is required to make a base diffusion in silicon with a surface concentration of 4×10^{18} cm^{-3}, into a background of 3×10^{16} cm^{-3}, with a junction depth of 1.0 micron. Prove that this cannot be done with an unlimited source diffusion.

6. Show that the logarithm of the sheet resistance after a predeposition is (approximately) a linear function of reciprocal temperature. What is the activation energy associated with this Arrhenius plot?

CHAPTER 5

EPITAXY

Once an ingot is grown, it is cut into slices by a diamond saw. These slices, after rough polishing with an abrasive grit, are chemi-mechanically polished to remove surface damage. These are the starting materials of which semiconductor devices and microcircuits are made. For many applications, the slice serves merely as a mechanical support on which is first grown one or more upper layers of material of the appropriate resistivity and conductivity type, in which the microcircuit is fabricated. This process of growth, by which an amount of material is set down upon a crystalline substrate while the overall single-crystal structure is still preserved, is known as *epitaxy*.

In silicon technology, metal-oxide-semiconductor (MOS)-based microcircuits are often made directly in the bulk slice. Increasingly, however, very-large-scale integration (VLSI) applications require that these circuits be fabricated in epitaxial layers on heavily doped substrates, in order to minimize latchup when they are powered, or exposed to ionizing radiation. Many GaAs devices and circuits are fabricated by direct ion implantation into semi-insulating substrates. However, epitaxial layers are required for modern electro-optical devices and integrated circuits.

In its simplest form, epitaxy involves the extension of the substrate lattice by the overgrowth of a layer of identical material (e.g., Si on Si or GaAs on GaAs). This process is known as *autoepitaxy* (or *homoepitaxy*). No problems of compatibility or mismatch occur because of the similarity (orientation, chemical properties, lattice parameters, crystal structure, etc.) of the epitaxial layer and the substrate on which it is grown.

At the other extreme, the term epitaxy can apply to any two materials of different crystalline structure and orientation. The growth of crystalline (001)

258

Si (diamond lattice) on ($1\bar{1}02$) sapphire (hexagonal) is an example of this type of *heteroepitaxy*. Yet another example is the epitaxy of (001) GaAs on (001) Si, which requires the growth of a polar semiconductor on a nonpolar substrate; additionally, there is a 4.5% lattice mismatch between these materials, so that the growth process involves a considerable amount of strain.

Ideal autoepitaxy is rarely encountered in practice, since the epitaxial layer usually has a different doping concentration and/or impurity type from its substrate. Thus, a small amount of strain is associated with the presence of impurity atoms with different misfit factor. The growth of AlGaAs on GaAs represents also a small departure from autoepitaxy, since the lattice parameter of AlAs is about 0.13% larger than that of GaAs. As a result, there will be strain associated with the growth of the mixed III–V compound $Al_xGa_{1-x}As$ on GaAs, its amount depending on the magnitude of the aluminum fraction, x, and on the layer thickness. However, the mismatch in this system is so small that it can be considered as autoepitaxy for typical layer thicknesses which are used in practice.

A number of important problems must be surmounted in heteroepitaxy. First, the substrate must be physically and chemically inert to the growth environment, and capable of being prepared with a damage-free surface. Next, there must be chemical compatibility between the materials, as evidenced by their phase diagrams, in order to avoid compound formation, and/or massive dissolution of one layer by the other. In addition, the layer and the substrate should be closely matched in thermal expansion characteristics, to prevent the formation of excess stress upon cooling to the ambient. This stress can lead to the formation of dislocations at the interface, or even breaking of the structure. Finally, the layer and the substrate should be closely matched in their lattice parameters. Interestingly, this last requirement is not as serious an impediment to epitaxy as originally predicted, since considerable accommodation can take place during the initial stages of growth. In view of these difficulties, it is no surprise that the art of heteroepitaxy is still in its infancy. However, a number of interesting device applications require the exploitation of this approach [1], so that it is the subject of intensive research at the present time.

Many of the problems associated with heteroepitaxy are absent during autoepitaxial growth. In fact, this approach generally results in layers of better electrical quality than that of the starting melt-grown material, since it is often conducted under more highly controlled conditions, and at temperatures well below the melting point.

Epitaxial layers (often abbreviated as *epilayers*) in silicon microcircuits are typically about 0.5–5 μm thick for high-speed digital circuit applications, and 10–20 μm thick for linear circuit applications. Layers as thick as 50–100 μm are often encountered in silicon power devices. A similar range is used with GaAs, from 0.1 μm for field effect transistors to as thick as 100 μm for some transferred electron device structures.

There are a number of ways in which epitaxial growth can be achieved, the most direct being the physical transport of material to a heated substrate.

This can be accomplished by vacuum evaporation or by *molecular beam epitaxy* (MBE). These techniques are identical in principle, but critically different in their operational details. Specifically, delivery rates for the material to the substrate are many orders of magnitude slower for MBE, as are the corresponding growth rates. Of necessity, MBE systems must operate at much higher vacuum levels (three to four decades lower pressure than vacuum evaporators) to avoid background contamination. An important feature of MBE is that this high-vacuum environment allows the use of sophisticated diagnostic tools, as well as *in-situ* processes such as thermal or ion beam cleaning of substrates, prior to epitaxy. As a result, MBE is a very promising field and is the subject of much research [2]. This is especially true for the area of III–V compounds and mixed III–V compounds where the precise control of multiple impurities, combined with low-temperature deposition processes, is mandatory.

A second approach, *vapor-phase epitaxy* (VPE), is based on the transport of the epilayer constituents (Si, Ga, As, dopants, etc.) in the form of one or more volatile compounds to the substrate, where they react to form the epitaxial layer [3]. The principles underlying this approach are inherently more complex than MBE, since they involve additional considerations associated with chemical reactions. For example, radical species produced during VPE can often be used to provide an inherent substrate cleaning action, so that requirements for substrate preparation are not as stringent as for MBE. Additionally, VPE can be carried out at atmospheric pressure, avoiding the need for ultra-high-vacuum (10^{-10} torr) technology, and greatly reducing the system complexity. High-quality material can be grown by this technique, and thus it is widely used at the present time for both silicon and GaAs.

Chemical vapor deposition (CVD) is a more generic form of VPE, since it is not limited to single crystal growth. It is thus broadly applicable to the growth of a wide variety of deposited films, which are the subject of Chapter 8. The term CVD is often used (incorrectly) to describe VPE processes.

A third approach, *liquid-phase epitaxy* (LPE), involves the growth of epitaxial layers on crystalline substrates by direct precipitation from the liquid phase [4]. It is thus an application of the solution growth technique described in Chapter 3; here, the slow growth rate is an advantage, since thin layers are required in epitaxy.

LPE is a relatively simple process, and requires a small investment in equipment. In addition, it is usually carried out in liquid indium or gallium, both of which are excellent scavengers of deep-lying impurities. As a result, LPE material is characterized by its exceptionally long minority carrier lifetime. In fact, it is only recently that material of comparable lifetime has been grown by any alternative method [5].

The major disadvantage of LPE is its poor surface morphology and the inability to grow extremely thin layers with abrupt interfaces. It is widely used in applications such as light-emitting diodes, where these requirements are not necessary.

This chapter will outline those principles which are common to all types of epitaxy. This will be followed by the specifics of important processes for both silicon and GaAs. Tables of relevance to epitaxy will be provided at the end of the chapter, for ready reference.

5.1   GENERAL CONSIDERATIONS

In this section we describe some aspects of epitaxial growth which are common to all of the approaches mentioned. This will be followed by a discussion of their specifics.

In all epitaxial growth techniques, the constituent material eventually arrives at the heated substrate. In some cases, it is in elemental or molecular form (such as Si, Ga, As_2, As_4, etc.) or in the form of radical species (Ga–CH_3, Ga–Cl, As–H, etc.). The incident flux of these species is given by

$$j = \frac{\beta P_s}{(2\pi mkT)^{1/2}} \tag{5.1}$$

where P_s is the vapor pressure of the reactant species at the substrate, m is the molecular weight, j is the number of atoms cm^{-2} s^{-1}, and β is a proportionality factor. These species move by surface diffusion until attachment, or are lost from the surface by desorption. The desorption energy, in turn, is a function of the crystal surface to which they attach; species adsorbed on a low energy face (with a low binding energy) will have a high probability of desorption, and vice versa. In general, the closest-packed planes have the lowest surface energies, and hence the lowest growth rates. This is observed in silicon epitaxy, where growth is about 40% slower in the $\langle 111 \rangle$ direction than in the $\langle 100 \rangle$ direction. With GaAs, however, the noncentrosymmetric nature of the crystal tends to modify this conclusion somewhat.

Adsorbed species move rapidly by surface diffusion until they finally come to rest, to form the growing layer. Here, attachment is favored at sites where the species can connect by multiple bonds.

5.1.1   Nucleation and Growth

Nucleation is a process by which molecules of the reactants combine to form isolated ensembles which attach to points on the surface. Thus, it is an essential prerequisite to film growth. The simplest nucleation process, known as *homogeneous nucleation*, is one that can come about by direct condensation out of the gas phase, in the absence of other objects such as surfaces. Although rarely occurring in practice, this serves as the starting point for further analysis [6].

If P_∞ is the equilibrium vapor pressure over the solid phase, then the solid will neither condense nor evaporate if exposed to this pressure. If, however, the pressure is higher (i.e., the vapor is supersaturated), conditions for nucleation may prevail.* If P_0 is the actual vapor pressure, the free energy difference per atom between the vapor and the solid is given by

$$G_v = kT \int_{P_0}^{P_\infty} \frac{dP}{P} = -kT \ln \frac{P_0}{P_\infty} \tag{5.2}$$

If ΔG_v is the change in energy per unit volume, then

$$\Delta G_v = -nkT \ln \frac{P_0}{P_\infty} \tag{5.3}$$

where n is the number of atoms per unit volume of the semiconductor.

During nucleation, a number of small embryos are formed. These coalesce and grow** in size as the conversion proceeds from the vapor phase to the solid phase. There is a release of free energy (ΔG_v per unit volume) during this conversion process. At the same time, however, the formation of a surface between these phases involves an increase of free energy (γ per unit area of surface).

The total free energy change for an embryo (i.e., the activation energy required for its homogeneous nucleation) is thus given by

$$\Delta G_{\text{homo}} = (\tfrac{4}{3}\pi r^3)\Delta G_v + (4\pi r^2)\gamma \tag{5.4}$$

Here spherical embryos, of radius r, have been assumed. Note that γ is always positive, but ΔG_v is negative for any spontaneous reaction. From this equation, it is seen that the change in free energy is maximized for embryos of critical radius r_{crit}, where

$$r_{\text{crit}} = -\frac{2\gamma}{\Delta G_v} \tag{5.5}$$

Moreover, the free energy change leading to the formation of these embryos is given by

$$\Delta G_{\text{crit}} = \frac{16\pi\gamma^3}{3\Delta G_v^2} \tag{5.6}$$

Embryos below this size will lower their free energy change by becoming smaller, whereas those that are larger than r_{crit} will lower their free energy by

---

*Note, however, that growth can proceed at sub-critical supersaturations.
**The possibility of nucleus shrinkage has been treated elsewhere [7], and is not considered here.

growing. Note that large values of P_0/P_∞ result in an increase in the value of ΔG_v, thus reducing the critical radius for nucleation as well as the critical change in free energy.

It is possible to calculate the concentration of critical nuclei by Boltzmann statistics, if a quasi-equilibrium situation is assumed. Thus if n_0 is the number of particles of the new solid phase, we may write the number of critical nuclei as

$$n_{\text{crit}} = n_0\, e^{-\Delta G_{\text{crit}}/kT} \tag{5.7}$$

If τ is the average lifetime for critical nuclei, the nucleation rate R_n is given by

$$R_n = \frac{n_0}{\tau} e^{-\Delta G_{\text{crit}}/kT} \tag{5.8}$$

In most situations, nucleation takes place directly on a substrate, rather than in the gas phase. Such a process is known as *heterogeneous nucleation*. Here the presence of the substrate greatly alters the situation because of the possibility of "wetting" its surface. In addition, it leads to better crystal quality since the layer is constituted directly on the surface of the slice in a two-dimensional manner.

Elementary arguments for wetting, based on surface-free energy considerations, are now outlined, with reference to Fig. 5.1. Let γ_{NS}, γ_{NV} and γ_{SV} be the surface energies associated with nucleus–surface, nucleus–vapor, and surface–vapor, respectively. Again assuming spherical nuclei, three situations are possible. No wetting will occur if γ_{NS} is larger than $\gamma_{NV} + \gamma_{SV}$, since this process would require an increase in the net free energy. On the other hand, complete wetting occurs if γ_{SV} is larger than $\gamma_{NS} + \gamma_{NV}$, since the spreading out of the nucleus now lowers the net free energy. The intermediate situation of partial wetting occurs when force equilibrium is established—that is, when

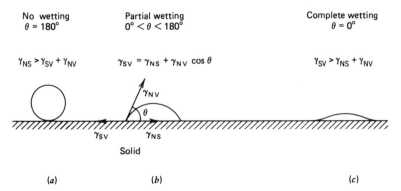

Fig. 5.1 Wetting characteristics of surfaces.

$$\gamma_{SV} = \gamma_{NS} + \gamma_{NV} \cos \theta \qquad (5.9)$$

where θ is the contact angle (Fig. 5.1b).

Writing the respective surface areas as S_{NS} and S_{NV}, the surface energy of this nucleus is given by $\gamma_{NS}S_{NS} + \gamma_{NV}S_{NV} - \gamma_{SV}S_{NS}$. Using this expression and Eq. (5.4), it can be shown that the critical energy for heterogeneous nucleation, ΔG_{het} is related to the value for homogeneous nucleation by

$$\Delta G_{het} = \Delta G_{homo} \left[\frac{(2 + \cos \theta)(1 - \cos \theta)^2}{4} \right] \qquad (5.10)$$

Thus as long as wetting takes place ($\theta \simeq 0°$), the presence of the substrate greatly reduces the activation energy for nucleation. As a consequence, the majority of reactions involving phase change nucleate heterogeneously.

Equation (5.10) illustrates the importance of surface preparation on epitaxial growth. Thus the presence of any cavities or imperfections increases the surface area on a local basis. This results in a large decrease in the activation energy of nucleation at this point.

The equation describing the rate of heterogeneous nucleation is of the same form as Eq. (5.8), except that a smaller critical energy is involved. For both situations, therefore, the nucleation rate increases very rapidly as the saturation ratio P_0/P_∞ increases. As a consequence, careful attention must be paid to the partial pressures of the reactants, the temperature, and the gas flow conditions in the reactor, if high-quality epitaxial layers are desired. Note that ΔG_v approaches zero as the growth temperature approaches the melting point; the nucleation rate falls rapidly with increasing temperature, leading to improved crystal quality.

The above arguments are of special importance to heteroepitaxy, where growth is often three-dimensional in character and proceeds by the expansion and eventual coalescence of isolated islands. This produces defects at their contact surfaces, such as threading dislocations and stacking faults, which consequently propagate through the epitaxial layer. Three-dimensional growth of this type is undesirable.

Complete wetting usually occurs in autoepitaxy, provided that it is carried out on a clean surface, which is free from residues. Here, growth can be considered as the successive addition of material on the surface of a crystalline substrate, i.e., it is two-dimensional in character. For the purpose of illustration, assume [8] the simple cubic crystal structure shown in Fig. 5.2, with growth occurring by the addition of cubes. Each cube represents one added atom (*adatom*) and also one unit cell. Cubes are bonded to their nearest neighbors at a common face, with a binding energy of ϕ_1/face; bonding also occurs with second nearest neighbors at a common edge, with a binding energy of ϕ_2/edge.

Let us consider a few of the many possible bonding configurations for this ideal situation. Epitaxial growth can occur by the addition of atoms at different

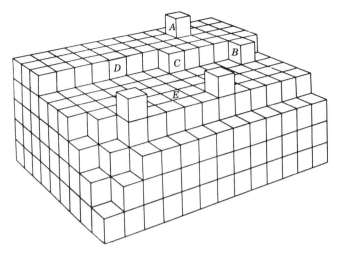

Fig. 5.2 Atomic sites in a simple cubic crystal. Adapted from Pamplin [8].

sites, as labelled in Fig. 5.2. The binding energies associated with this process are as follows.

   Case A: Adsorption on a surface, with a binding energy of $\phi_1 + 4\phi_2$.
   Case B: Adsorption at a step, with a binding energy of $2\phi_1 + 6\phi_2$.
   Case C: Adsorption at a corner, $3\phi_1 + 6\phi_2$.
   Case D: Incorporation into a step, $4\phi_1 + 8\phi_2$.

Finally, incorporation into the volume of the crystal is associated with a binding energy of $6\phi_1 + 12\phi_2$.

   The above progression, with increasing binding energies, allows the postulation of a relatively simple model for epitaxial growth. Atoms impinge on the crystal surface, where they move by diffusional* processes, until they are incorporated into the growing crystal or desorbed from its surface. The desorption rate is highest for atoms with the lowest binding energy; consequently, the probability of desorption of a surface atom is large, unless it can migrate to a *kink* site, as illustrated in Case B or Case C. Crystal growth thus proceeds by the preferential incorporation of atoms at these kink sites.

   It is an oversimplification to assume that crystal growth in this manner proceeds by the successive addition of monolayers. Layer-by-layer growth can only be achieved under extremely slow growth conditions. In practice, growth proceeds on many atomic layers at the same time, for various reasons. For example, defects on the growing layer give rise to steps where additional planes can grow;

---

*To an approximation, the surface diffusion length is on the order of $e^{0.45(E_1-E_2)/2kT}$, where E_1 and E_2 are the activation energies of desorption and diffusion, respectively.

adsorption of clusters of atoms on a crystal plane can initiate growth on multiple planes at the same time.

It is common practice to carry out epitaxial growth on substrates which are cut slightly off axis, by 2–4°. This introduces a series of steps in the growing surface, increasing the number of kink sites and thus the probability that impinging atoms will reach these sites before being desorbed. This allows crystalline perfection to be achieved at higher growth rates than are obtained with growth on precisely oriented substrates.

It is also desirable to conduct epitaxial growth at elevated temperatures, because species which impinge on the substrate move by surface diffusional processes which are thermally activated. Typically, the activation energy of surface diffusion is 25–50% of the bulk diffusion energy, so that surface diffusion rates are five to eight decades faster than bulk diffusion, at the growth temperature. On the other hand, low growth temperatures are required to minimize interdiffusional effects. Thus, the actual growth temperature is a compromise in any practical situation.

The preferred direction for easy epitaxial growth is based on a number of considerations. One factor of importance is the spacing between planes. The distance between (111) planes is 57.5% of the distance between (100) planes. Thus, from packing considerations alone, the stacking of (111) planes is more readily accomplished, favoring epitaxy in this direction. Many silicon-based devices and microcircuits are grown on (111) material for this reason.

There are special considerations associated with the growth of compound semiconductors, such as GaAs, which result from their nonpolar nature. Epitaxial growth by VPE or MBE is relatively constrained; that is, the orientation of the growing interface more or less follows that of the substrate. At the elementary level, it can be looked upon as growth on a layer-by-layer basis. Consider the (001) surface $x–x'$ of GaAs, made up of Ga atoms [9]. Each is bonded downwards into the crystal to two As atoms (see Fig. 5.3a), leaving two unsatisfied bonds. Growth of the next layer $y–y'$ causes the original layer to make bonds upwards to two additional As atoms, again leaving two unsatisfied bonds. As a result, the process of epitaxial growth leaves the number of surface dangling bonds unchanged. It follows that (100) growth presents little significant barrier to nucleation, since there is only a small change in surface energy on the addition of each layer. (For nonpolar semiconductors such as silicon, there would be *no* change in surface energy during growth in this manner.) This ideal situation is based on the assumption that growth results from the alternate deposition of species containing Ga and As; in practice, both impinge on the surface at all times. This results in a rough growth surface on the atomic scale—that is, one which has at all times both Ga and As surfaces.

The situation is very different for growth on a (111) surface. Here (see Fig. 5.3b) the crystal consists of pairs of closely spaced planes (double layers), separated from each other by a relatively wide spacing. In the double layer, each Ga atom in the $x–x'$ plane bonds to three As neighbors in the $y–y'$ plane. The fourth bond, normal to the growth plane, is made upwards to an As atom during

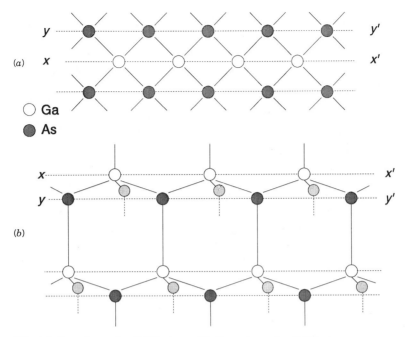

Fig. 5.3 (a) Growth on the (100) plane. (b) Growth on the (111) plane.

growth of the next double layer. Thus, there is a significant change in surface energy as growth progresses, resulting in a barrier to nucleation in this growth direction. However, since {111} planes are the closest spaced, growth on these surfaces tends to have better morphology than growth on (100) surfaces.

5.1.2 Doping

Doping during epitaxy is accomplished by introducing the impurity to the sub-strate, along with the species which contribute to growth. The surface concentration of this impurity is maintained dynamically (i.e., an adsorption– desorption equilibrium exists), and is linearly proportional to the dopant partial pressure.

Low-vapor-pressure dopant species will desorb rapidly with increasing growth temperature, so that their incorporation will fall under these circumstances. Examples of this behavior are arsenic and phosphorus in silicon, and sulfur and zinc in GaAs. On the other hand, the incorporation of less volatile species (boron in silicon, and silicon in GaAs, for example) is relatively independent of the growth temperature.

If the dopant is a slow diffuser, its incorporation is similar to that of the host material, and proceeds by attachment at a kink site, so that it gets trapped upon arrival of the next layer of growth. It follows that the dopant incorporation will be a linear function of the surface concentration of adsorbed dopant—that is, linearly proportional to the partial pressure of the incoming dopant species.

The actual amount will be considerably lower than the surface concentration. This is because the incorporation probability is determined by the capability of the impurity to form a chemical bond to the atoms arriving in the layer above. Therefore, since the reaction has a finite activation energy, E_{act}, only a fraction of adsorbed impurities can get incorporated by this mechanism. The ratio of impurities adsorbed on the surface to the number incorporated in the monolayer below is given by $e^{-E_{act}/kT}$, and is considerably less than unity.

An alternative mechanism for dopant incorporation, by in-diffusion, is also possible. Thus, if the impurity is capable of making a diffusion jump into the layer below, *before* the next monolayer grows over it, it will become incorporated into the epitaxial layer. This will also result in uniform doping, through the layer, since it is a continuous process.

Dopant incorporation by in-diffusion will occur when the impurity diffusivity is larger than a_0 times the growth rate, where a_0 is the lattice parameter. High doping efficiency is characteristic of this process, since all of the surface-adsorbed species will incorporate into the growing layer. However, the doping behavior now depends on the diffusivity of the impurity, which is concentration-dependent. It can be shown that, for low concentrations, dopant incorporation is linearly proportional to the partial pressure of the incoming species [10]. At higher concentrations, however, it shows a square root dependence on the partial pressure, as long as the doping is nondegenerate.

In summary, doping by the trapping mode results in a low incorporation efficiency, but is linear with the partial pressure of the impurity. All dopants in silicon, and *n*-type dopants in GaAs, usually exhibit this behavior. On the other hand, doping by a diffusive process has a high incorporation efficiency, but is dominated by a square root dependence on the dopant partial pressure. Most *p*-type dopants in GaAs are rapid diffusers, and exhibit this behavior.

Doping incorporation generally saturates at high concentrations for a number of reasons. Perhaps the most important is the excessive strain caused by the change in lattice parameter of the epilayer with respect to the substrate. Impurity segregation, in the form of clustering, has also been observed with many dopants at these concentrations.

5.1.3   Dislocations

In Chapter 3, it was indicated that threading dislocations of many types are present in all bulk grown semiconductors due to thermal and mechanical stresses acting on the crystal during growth and/or cooling. These dislocations terminate on available free surfaces, and many will therefore intersect the growth surface of a substrate crystal. Typical threading dislocation densities in high-quality bulk crystals range from 1–10 cm^{-2} for silicon to 10^3–10^4 cm^{-2} for GaAs.

In autoepitaxial growth, dislocations present in substrate materials are typically replicated in the epitaxial layer [11]. In some situations, threading dislocations decorated with precipitates or microloops, can multiply. The end result is the propagation of many dislocations through the epitaxial layer for each dislocation in the substrate.

Epitaxial layers grown on mismatched substrates often have threading dislocation densities which are much larger than those for autoepitaxial growth [12]. Here, in addition to replication of substrate defects, new threading dislocations are generated by the relaxation of mismatch strain during epitaxial growth. This process is now considered.

When a material is grown on a substrate to which it is *not* lattice-matched, initial growth often occurs with perfect registration on an atomic scale, if the mismatch is small and the crystals are of similar structure. This is referred to as *pseudomorphic* growth. The epitaxial layer is strained to match the lattice spacings of the substrate, in the plane of the interface. This in-plane strain also results in a Poisson strain perpendicular to the interface. Pseudomorphic growth may be maintained up to some critical thickness h_c, at which misfit dislocations are introduced at the interface by one or more mechanisms. Examples of such mechanisms include the bending over of threading dislocations and the nucleation of half-loops at the crystal surfaces. The critical thickness is a function of the crystal elastic constants, the mismatch, and also the lateral size of the island (for three-dimensional nucleation) or the lateral size of the deposition area (for selective area deposition). For (100) epitaxy on (100) substrates, this thickness has been calculated on the basis of a force balance model [13], and is given by

$$h_c \simeq \frac{b(1 - \nu/4)}{4\pi|f|(1 + \nu)} \left[\ln \frac{h_c}{b} + 1 \right] \qquad (5.11)$$

Here, f is the lattice mismatch between the epitaxial layer and the substrate, and is given by $(a_e - a_s)/a_e$; ν is the Poisson's ratio (0.265 for Si and 0.312 for GaAs), and b is the length of the Burgers vector [$= a_0/\sqrt{2}$, for growth on (100) substrates].

This criterion leads to values of h_c as low as 10 Å in highly mismatched systems such as GaAs:Si. However, such values must be viewed with caution, because the continuum theory of elasticity fails at dimensions on the order of the atomic spacing. In addition, other mechanisms for lattice relaxation may become active in highly mismatched films, so that this criterion is not always met in practice.

Problems due to misfit dislocations have been studied for boron-doped layers in silicon autoepitaxy, since this impurity has a large misfit factor ($\epsilon = 0.254$). Its presence in silicon results in a reduction in the lattice parameter, so that the critical layer thickness for coherent epitaxy falls with increasing boron concentration. Experiments with 5-μm thick boron-doped epitaxial layers have shown [14] that the strain in these films increases up to a doping concentration of about 10^{19} cm^{-3}. At this point, the layer thickness exceeds its critical value, and the film relieves its strain (as seen in Fig. 5.4), by the formation of misfit dislocations.

5.1.4 Thermally Induced Strain

Lattice mismatch strain is often nearly relaxed at the growth temperature; however, thermal strain results during cool-down and cannot be removed by dislocation motion. The reason for this is that dislocation glide velocities are thermally activated, and may fall by as much as one decade for every 25°C reduction in temperature.

Consider a heteroepitaxial layer, grown at temperature T_g with in-plane strain $\epsilon_{\parallel}(T_g)$, which is cooled to room temperature. If no relaxation occurs during cool-down, then the thermal strain ϵ_{th} associated with this process is

$$\epsilon_{th} = \int_{T_g}^{T_r} [\alpha_s(T) - \alpha_e(T)] \, dT \tag{5.12}$$

where α_s and α_e are the thermal coefficients of expansion (TCE) of the substrate and epitaxial layer, respectively, and T_r is the room temperature. If the TCEs are

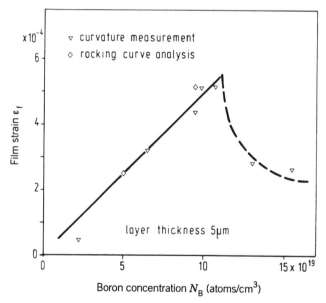

Fig. 5.4 Film strain as a function of boron concentration. From Herzog et al. [14]. Reprinted with permission of the publisher, The Electrochemical Society, Inc.

reasonably constant with temperature, we may approximate the thermal strain by

$$\epsilon_{th} \simeq (\alpha_e - \alpha_s)(T_g - T_r) \tag{5.13}$$

For this situation, the thermal strain is tensile (positive) if $\alpha_e > \alpha_s$, and compressive (negative) if $\alpha_e < \alpha_s$.

The room temperature residual strain measured in the film will be the sum of the growth temperature strain and the thermal strain (again, assuming that there is no relaxation during cool-down), so that

$$\epsilon_{\parallel}(T_r) = \epsilon_{\parallel}(T_g) + \epsilon_{th}. \tag{5.14}$$

Furthermore, if the strain at the growth temperature is given by the equilibrium relationship, and if no relaxation occurs during cool-down, then the room temperature strain for a zincblende semiconductor with 60° dislocations can be shown to be given [12] by

$$\epsilon_{\parallel}(T_r) \approx \frac{f}{|f|} \frac{b(1 - \nu/4)}{4\pi h(1 + \nu)} \left[\ln\left(\frac{h}{b}\right) + 1 \right] + (\alpha_e - \alpha_s)(T_g - T_r) \tag{5.15}$$

where h is the film thickness.

Heteroepitaxial films in tensile stress often relieve their strain energy by dislocation generation. In the extreme case, relief is provided by cracking if the film is too thick. This situation occurs during the epitaxy of GaAs on Si, and limits the film thickness to about 4 μm.

Films in compression will not crack because cracking would increase their strain energy. However, they may separate from the substrate, but this is expected to occur at greater thicknesses than cracking, because the process increases the surface area. The cracking thickness of these films is roughly twice that of comparable films which have residual tensile strain.

5.1.4.1 Slip

During epitaxial growth, substrates are usually brought up to temperature by placement on heated platforms, called *susceptors*. Stress can be created during this process. Consider, for example, a perfectly flat substrate which is placed on a uniformly heated susceptor [15]. The top face of this substrate will be slightly cooler than the lower face, since it loses heat by radiation. As a result, the substrate will bow, because of the differences in thermal expansion of these surfaces. This, in turn, produces a loss of contact at its edges, which become progressively cooler. Additionally, edge radiation effects add to this cooling process, thus increasing the amount of curvature.

As a consequence, there will be a radial distribution of temperature, $T(r)$, with the substrate becoming progressively cooler at its circumference. This results in radial and tangential stress components σ_r and σ_θ, respectively. A stress component which is normal to the face is also present, but it can be ignored; that is, the substrate is essentially two-dimensional in character. The resulting maximum shear stress at any point is given by

$$\tau_{s,\max} = \tfrac{1}{2}|\sigma_r - \sigma_\theta| \qquad (5.16)$$

Dislocations will be generated if this shear stress exceeds the critical resolved shear stress (CRSS) of the material. These dislocations will propagate along well-ordered {111} glide planes, and intersect the surface in ⟨110⟩ directions, to result in *slip* lines. These can be readily observed by appropriate chemical etching, and sometimes by visual inspection. During subsequent device processing, impurity atom movement is enhanced whenever a diffusion crosses a slip line. This leads to leaky junctions, and, in some cases, to shorts between adjacent diffused regions.

For the situation described here, it can be shown [16] that

$$\frac{\sigma_r}{\alpha E} = \frac{1}{r^2}\int_0^r rT(r)\,dr - \frac{1}{R^2}\int_0^R rT(r)\,dr \qquad (5.17\text{a})$$

$$\frac{\sigma_\theta}{\alpha E} = T(r) - \frac{1}{r^2}\int_0^r rT(r)\,dr - \frac{1}{R^2}\int_0^R rT(r)\,dr \qquad (5.17\text{b})$$

where R is the radius of the slice, E is Young's modulus, and α is the TCE. Precise determination of $T(r)$ is difficult, since it depends on the initial curvature of the substrate, the nonuniformity of the susceptor temperature, and the spatial nonlinearity of substrate radiation. However, assuming a parabolic profile, with temperature falling off at the perimeter, it can be shown that the maximum shear stress increases as we go from the center towards the edge of the slice. The magnitude of this stress tends to vary approximately as the square of the slice diameter. Thus, problems due to slip become increasingly severe with the trend to larger slices.

Experimental studies have shown the presence of additional slip at the center of the slice [17]. This has been attributed to entrapment of a hot central region of the slice within a colder annular zone. As a result, slip patterns in ⟨100⟩ material, which is placed on a heated susceptor, take the form shown in Fig. 5.5.

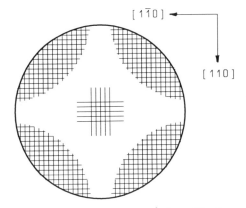

Fig. 5.5 Idealized slip pattern in (100) slices.

5.2 MOLECULAR BEAM EPITAXY

Molecular beam epitaxy (MBE) involves the direct physical transport of the material to be grown, or its components, to a heated substrate [18]. In essence, therefore, it is similar to vacuum evaporation, and growth is achieved by directing atomic or molecular beams in a well-controlled, ultra-high-vacuum system.

Both silicon and GaAs can be grown by MBE; however, the interest in this technology is heavily centered around advancements in GaAs to the point where it is a commercially viable technology. Interest in silicon and Si–Ge alloys for high-frequency devices is growing [19, 20], and much of this topic is being undertaken by MBE. However, detailed discussions of this work are beyond the scope of this book.

Gallium and arsenic are used as source materials for GaAs. Additionally, aluminum is used for the growth of AlGaAs layers. Elemental silicon and germanium are used in the growth of silicon and Si–Ge alloys. All of these materials are available in "seven nines" (99.99999%) purity at the present time.

The beams of material to be transported are usually generated by thermal evaporation from crucibles known as *effusion cells*, which are shuttered in order to initiate and terminate the flux of the evaporant species. A series of effusion cells, each with a separate shutter, are set up so that their flux is directed to the substrate. In the case of silicon MBE, however, electron beam heating is commonly used to produce a significant flux of source material, and thermal evaporation is used for germanium.

The substrate on which the wafer is mounted usually consists of a heated molybdenum block, which can be rotated during growth at a few rpm, for increased layer uniformity. Sample mounting is done by means of indium or gallium on its back face or by means of mechanical fasteners.

Figure 5.6 shows a simple schematic illustrating the basic principle of MBE. Details of a typical MBE growth chamber are shown [21] in Fig. 5.7. This chamber is backed up by a substrate preparation chamber and a sample introduction chamber, each of which is separately pumped. Chambers are interconnected by high-vacuum gate valves, and designed so that the sample and its holder can be transferred from one chamber to another without breaking vacuum.

The sample introduction chamber is the only part of an MBE system which is opened to the atmosphere, for the purpose of loading the sample—typically, a single substrate. Recently, cassette loading has been incorporated to facilitate the handling of many slices at a time. This chamber is capable of rapid pump-down, at which point a gate valve allows movement of the sample into the preparation chamber. Here, provision is made to clean the sample surface, either by heating or by argon ion bombardment. *In-situ* monitoring equipment is also provided to check its surface condition before moving into the growth chamber.

The final surface preparation is carried out in the growth chamber, prior to epitaxial growth. Both diagnostic and growth monitoring equipments are available in this chamber. In many systems, two or more such chambers are available for the growth of widely different materials. An example is a system in which group III–V compounds can be grown in one chamber, and group IV materials in another. The various components of an MBE system are now considered, together with some of their design considerations.

5.2.1 The Growth Chamber

A characteristic feature of MBE is the beam nature of particle flow from the effusion cell to the substrate. In order for this to be achieved, it is impor-

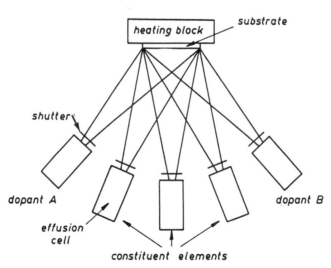

Fig. 5.6 Basic principle of MBE.

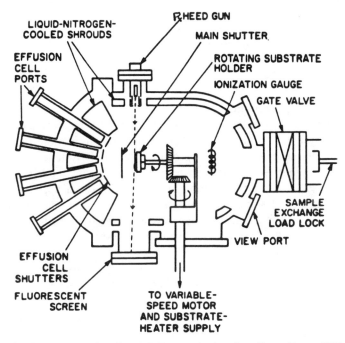

Fig. 5.7 Schematic of an MBE growth chamber. From Tsang [21].

tant that the mean free path of the evaporant be long compared to the distance from the substrate, which is typically 5–30 cm. An approximate computation of this parameter can be made if we assume that all particles are at rest, except one particle which is moving with a velocity c; that is, it travels $c\,dt$ in time dt.

If all particles have the same diameter σ, the traveling particle collides with all particles whose centers are separated by a distance $\sigma(= \sigma/2 + \sigma/2)$. Thus, each particle can be considered to have a collision cross-sectional area of $\pi\sigma^2$. The collision volume swept by this particle is $\pi\sigma^2 c\,dt$.

If n is the number of particles in a volume V, then the collision frequency is $(n/V)\pi\sigma^2 c\,dt$. Thus, the average mean free path is given by

$$\lambda = \left(\frac{n}{V}\pi\sigma^2 \right)^{-1} \tag{5.18}$$

A more rigorous treatment, which takes the relative motion of all particles into account, gives

$$\lambda = \left(\frac{n}{V} \pi \sigma^2 \sqrt{2} \right)^{-1} \tag{5.19a}$$

$$= \frac{kT}{\sqrt{2} \pi \sigma^2 P} \tag{5.19b}$$

where P, the partial pressure, is given by nkT/V.

Typical values for σ range from 2 to 5 Å, so that the mean free path is about 10^3 cm at a system pressure of 10^{-5} torr. This pressure is more than adequate for establishing the beam nature of growth in an MBE system.

A more stringent requirement on base pressure is obtained if we treat the residual gas as a contaminant during layer growth. To an approximation, the incorporation rates of source and contaminant are proportional to their partial pressures. If we define a "clean" layer as one in which the contamination fraction is $\leq 10^{-5}$, then the base pressure should be correspondingly lower by this factor; that is, it should be in the 10^{-10} torr range. MBE systems are generally designed to maintain a base pressure of this value, for this reason. This necessitates the use of a stainless steel chamber with metal O-ring seals, so that it can be baked out during pump down. In addition, the chamber is load-locked, so that it is only opened for maintenance purposes. During actual operation, a low system pressure is maintained by means of an internal shroud, which is liquid-nitrogen cooled and surrounds the entire growth chamber.

In addition to the above, all pumping is oil-free, involving liquid nitrogen sorption pumps and titanium sublimation ion pumps, located at strategic points on each chamber in the system. Turbomolecular pumps are also used in many MBE systems.

5.2.2 Sources for Growth and Doping

Source beams are obtained by thermal evaporation from high purity elements, which are placed in crucibles known as *effusion cells*. A separate effusion cell is used to provide each element needed for the growth of an epitaxial layer, and also for its doping. These cells must be designed to provide a uniform flux density at the substrate surface. Typically, cells are made of pyrolytic boron nitride with tantalum heat shields, and are resistively heated with feedback temperature control to within $\pm 0.1°C$.

The classical treatment of evaporation from a surface into a vacuum, based on the kinetic theory of gases, can be used to show that the evaporation rate from a surface of area A_e is equal to the impingement rate at the equilibrium pressure [22]. Thus

$$\frac{dN_e}{dt} = \frac{A_e P}{(2\pi kTm)^{1/2}} \tag{5.20}$$

where P is the equilibrium pressure at the effusion cell temperature T, and

m is the mass of the evaporant. This equation can be written in terms of the molecular weight of the species, M, as

$$\frac{dN_e}{dt} = \frac{A_e P}{(2\pi k T M / N_A)^{1/2}} \tag{5.21}$$

where N_A is Avogadro's number ($= 6.022 \times 10^{23}$ mole^{-1}). Solving, the effusion rate is given by

$$\frac{dN_e}{dt} = 3.51 \times 10^{22} \frac{P A_e}{(MT)^{1/2}} \text{ molecules s}^{-1} \tag{5.22}$$

where P is the pressure in torr. Note that the equilibrium pressure dependence is exponential with temperature. Typically, a $\pm 0.1\%$ variation in cell temperature is needed to maintain an effusion cell pressure variation within $\pm 1.0\%$.

The flux of species impinging on the substrate is related to (a) its distance l from the evaporating surface and (b) the angle θ between the beam axis and the normal to the substrate plane. To an approximation,

$$j = \frac{\cos \theta}{\pi l^2} \frac{dN_e}{dt} \tag{5.23a}$$

$$= 1.117 \times 10^{22} \frac{P A_e \cos \theta}{l^2 (MT)^{1/2}} \text{ molecules cm}^{-2} \text{ s}^{-1} \tag{5.23b}$$

Equation (5.23b) assumes an effusion cell in which the evaporation source is at its mouth; that is, it is full. In practice, the level is below this mouth, and falls as the source gets depleted. As a result, the spatial flux distribution pattern (and hence the spatial uniformity across the epilayer) will vary as a function of time [23]. This problem can be reduced by maintaining a large cell-to-substrate spacing. Additionally, effusion cells often use a tapered configuration, in the form of a truncated cone, for this reason.

Typically, crucibles have a 1-cm^2 evaporation surface, and are positioned at about 5–20 cm from the substrate. The temperature of the cell is controlled so that the pressure of the effusing species is in the 10^{-3}- to 10^{-2}-torr range [24], resulting in the delivery of about 10^{15}–10^{16} molecules cm^{-2} s^{-1} to the substrate. Assuming a sticking coefficient of unity, this corresponds to a growth rate of about one monolayer per second, which is typically used for GaAs MBE. Cell temperatures for gallium and aluminum are around 1000°C and 1175°C, respectively. A somewhat higher flux of arsenic is required, to prevent preferential desorption of this species from the growing surface. In practice, the As/Ga flux ratio is a function of both the growth temperature and the growth rate. Thus, the arsenic pressure, as set by the temperature of the effusion cell, varies widely for different growth conditions.

The flux from an effusion cell is controlled by means of a pneumatically operated shutter at its mouth. When closed, this shutter reflects heat back to the surface of the source; on opening, this reflected heat is lost, resulting in a slow cooling of the furnace, and a transient in the beam flux, which can last several minutes. The beam flux can vary by as much as 25–50% during this time period, since the equilibrium pressure of the source is an exponential function of temperature. This can present a problem in the growth of structures which require abrupt changes in composition or in doping. A second problem, associated with the thermal mass of the cell, is that it is difficult to grow structures that need compositions which are graded over short distances (graded index lasers, for example). In some cases, two effusion cells, set at different temperatures, have been used for this purpose.

Silicon growth is usually carried out in a separate water-cooled chamber where an electron beam (e-beam) is used to melt a silicon charge which is held in a water-cooled crucible. Control of the e-beam is used instead of a shutter with this source. Although this avoids problems of beam current drift due to the shutter, the silicon flux is now a function of the e-beam parameters, and is extremely difficult to control.

Both p- and n-type dopants can be used, with each dopant in a separate effusion cell. The basic requirements [25] of a suitable dopant source for MBE are that it have a sufficiently low vapor pressure to be used in this cell, that it can provide controlled doping over a wide concentration range, and that it incorporate into GaAs with little or no compensation.

Of the many impurities which have been investigated for this purpose, the most commonly used are silicon and beryllium (n- and p-type respectively). Table 5.1 at the end of this chapter lists their relevant characteristics. Volatile species, such as sulfur and selenium, can also be used, by means of PbS and PbSe sources, and result in uncompensated material. However, their incorporation is critically dependent on the flux ratio and the substrate temperature, because of their volatile character.

Undoped GaAs is usually p-type in MBE due to residual carbon which can be incorporated from many components in the system. Careful attention to details of system design and operation have reduced this to the 10^{13}- to 10^{14}-cm^{-3} range, so that it can be ignored in most practical situations.

5.2.3 Substrate Holders

A molybdenum block is commonly used for this purpose in GaAs systems, because of its high thermal conductivity. The substrate is glued to this block with indium, and sometimes with gallium. Both these materials have low melting points (159°C and 29°C, respectively), so that the substrate is held during growth by surface tension. Both have low vapor pressures at the GaAs growth temperature ($\simeq 600$°C), and thus do not present any problems due to evaporation. Provision is made for both heating and rotating this block at 1–5 rpm, and for monitoring its temperature by means of a thermocouple or pyrometer.

Systems for both silicon and GaAs are increasingly built using pressure clamps for holding the substrate to the molybdenum block. Here, care must be taken to not use a rigid mount, in order to avoid excessive stress in the substrate at the growth temperature. A disadvantage of clamping arrangements of this type is that substrate heating occurs primarily by radiation from the block, because of the lack of an intimate physical contact. As a result, its temperature can be more than 50–100°C below that of the thermocouple in the block. Direct surface temperature monitoring, by means of an infrared pyrometer, is preferred for this reason.

5.2.4 Flux and Growth Rate Monitors

An ionization gauge is commonly used to directly measure the effusion flux. This is usually of the hot filament type, and is operated without a protective envelope. Fouling of its filament can result from continued exposure to the effusing species, so that provision is made to move the gauge out of the flux beam path when a direct measurement is not required.

The deposition rate can also be monitored at the substrate by means of a growth rate monitor. This unit is essentially a crystal controlled oscillator, with one electrode of the crystal placed in the molecular beam. Growth on this electrode alters the oscillation frequency, which is directly translated into a growth rate, for any specific atomic/molecular weight of the growing species. Contamination of the electrode by growth deposits can present a problem here, especially for elements such as gallium, which react with the (usually) gold electrode to form many intermetallic compounds.

5.2.5 Requirements for Good Growth

In MBE, low-energy atoms or molecules arrive at the substrate where they move around without any chemical reaction taking place. As a consequence, the success of this process is critically dependent on the cleanliness of the substrate at the atomic level. Two approaches are used to attack this problem. First, after implementing [26] elaborate wet cleaning procedures, *in-situ* cleaning is accomplished by bombarding the substrate with an inert gas such as argon to remove adsorbed surface contaminants, or by thermal etching. Ion sputtering is necessary to remove any carbon which might be present on the substrate, and must be followed by annealing in an arsenic flux to restore both crystallinity and stoichiometry of the surface region. Thermal cleaning at 500°C will remove surface oxides. Further cleaning is usually carried out at around 640°C, at which point GaAs evaporates congruently. This step can be done in the preparation chamber, or in the growth chamber itself. Next, some form of diagnostic is provided by which it can be determined if a suitable degree of substrate cleanliness has been achieved prior to the initiation of growth.

A commonly used diagnostic is reflective high electron energy diffraction (RHEED, sometimes written as HEED) and consists of a high-energy beam of electrons, about 0.1 mm dia. and in the 10- to 50-keV range, which impinges on the substrate at a grazing angle of 1–2° and reflects onto a fluorescent screen. The reflection pattern on this screen provides information about the substrate surface reconstruction, and is used for determining the start of the growth process [18].

In this technique, the substrate surface acts as a two-dimensional grating, which diffracts the electron beam in the form of a series of streaks, as shown in Fig. 5.8. The beam path is orthogonal to the flux beams associated with MBE growth, and penetrates the growth surface by a few atomic layers. Thus, this tool is essentially a surface diagnostic tool.

Let d be the spacing between atom planes, and θ the angle through which the beam is diffracted. Then, for constructive interference,

$$\lambda = 2d \sin \theta \tag{5.24}$$

where λ is the wavelength of the incident beam in Å, and is related to the beam voltage (V) by

$$\lambda = \frac{12.247}{\sqrt{V(1 + 10^{-6} V)}} \tag{5.25}$$

Finally, if L is the distance between the substrate and the fluorescent screen,

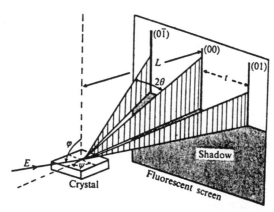

Fig. 5.8 Schematic illustrating the formation of a RHEED pattern.

and $D/2$ is the distance between the diffraction streaks, then

$$d = \frac{2\lambda L}{D} \tag{5.26}$$

Figure 5.9 shows the relationship of the various quantities described above.

The RHEED technique provides a qualitative picture of the crystal surface, and is especially useful in the early stages of crystal growth. A pattern with sharply defined streaks is indicative of a flat crystal surface. Rough surfaces take on the appearance of spots, since the electron beam can now penetrate many layers under the crystal surface. This results in strong multiple scattering effects and local resonances which alter the apparent width of the diffracted streaks and reduce their sharpness. Some penetration of the beam occurs, even under ideal circumstances. As a result, although extremely simple in principle, RHEED patterns are quite difficult to interpret in practice.

It has been estimated that a surface contamination of greater than one tenth of a monolayer is sufficient to prevent the growth of single-crystal material by MBE. Thus, use of the RHEED technique is absolutely essential for the successful growth of high-quality material. More recently, variations in the intensity of RHEED oscillations have been used for characterizing the growth rate and the alloy composition, during the growth of ternary compounds such as AlGaAs. Details of this technique are beyond the scope of this book.

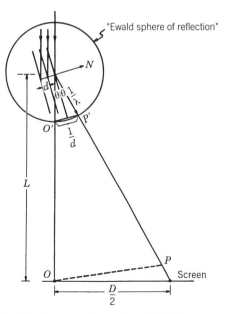

Fig. 5.9 Interpretation of the RHEED pattern.

The ultra-high-vacuum environment of an MBE system is also ideal for the inclusion of a number of additional diagnostic equipments, such as Auger electron spectroscopy, mass spectroscopy, and x-ray photoelectron spectroscopy. Diagnostics of this type are usually incorporated in the sample preparation chamber to avoid contamination from the growth environment.

The growth rate in MBE is proportional to the arrival rate of the flux at the substrate. Precise control of this flux is difficult, for the reasons given in Section 5.2.2. This is of little consequence with binary compounds such as GaAs, where small variations in growth rate are tolerable. On the other hand, the precise control of the alloy composition of ternary compounds is extremely difficult. One such situation is encountered in the growth of $In_xGa_{1-x}As$, where the value of x must be held at precisely 0.53 to achieve lattice match with the InP substrate on which it is normally grown. Composition control to tight tolerances required for materials of this type necessitates close control of the level of material in the effusion cell, or the direct monitoring (and control) of flux by an ion gauge [27].

Elemental arsenic and gallium are used for the growth of GaAs by MBE. At typical growth temperatures around 650°C, the vapor pressure of gallium over gallium is comparable to that of gallium over GaAs, on the order of 10^{-7} torr (see Fig. 1.19). The sticking coefficient of gallium can thus be taken as unity; that is, all gallium atoms arriving at the substrate will remain there and become incorporated into the growing layer. Arsenic, on the other hand, is desorbed since its vapor pressure (over arsenic) is greater than 1000 torr at this temperature, unless it combines with the gallium to produce GaAs. (The desorption energy of arsenic is around 0.4 eV, whereas the bond strength of Ga–As is around 50 eV.)

As a result, a simple MBE growth model for GaAs is that the gallium atoms arrive in an excess of arsenic flux, and that all arsenic which is not incorporated as GaAs is desorbed. This model is complicated by the fact that arsenic vapor is not monatomic, but is in the form of As_2 or As_4, depending on the temperature of the effusion cell [28].

At temperatures below 600 K, the growth of GaAs from As_2 requires the arrival and adsorption of this species on a Ga-stabilized surface. Here, it can desorb as As_2 (or as As_4 by subsequent association), or become chemically bonded as part of the growing layer. Growth from an As_4 source occurs in similar manner, except that it requires an additional second-order surface reaction, by which the tetramer loses two arsenic atoms, with the remaining two attaching to the Ga-stabilized surface as part of the growing layer.

At higher temperatures (800–1000 K), at which growth is normally carried out in order to obtain high-quality material, this process is further complicated by the fact that considerable arsenic is lost by desorption. In extreme cases, this can result in the formation of free gallium on the surface. As a result, the As/Ga flux ratio that is used is critically dependent on both the surface temperature and the absolute magnitude of the fluxes, if high-quality GaAs epilayers are to be grown.

5.2.5.1 Surface Morphology

The use of MBE for the growth of GaAs integrated circuits is limited by surface defects, which prevents the fabrication of large area chips. The most common of these is the "oval defect," typically $10–20$ μm long and elliptic in character, with a $2:1$ aspect ratio. This defect protrudes $1–2$ μm from the surface, and inhibits the successful implementation of fine line lithography.

There are two sources for these defects. The first is particles on the substrate surface, caused by inadequate cleaning procedures. The second is the formation of gallium droplets during the growth process, as evidenced by a Ga-rich core in each defect.

Techniques for reducing the incidence of these oval defects involve the use of scrupulous cleaning procedures and careful control of the gallium flux. Moreover, the use of an arsenic cracker, to form As_2 as the gaseous species, has been found to reduce the oval defect density of MBE-grown GaAs.

5.2.6 Gas Source MBE

Many of the disadvantages of MBE result from the necessity for using elemental solid sources, which require opening the high-vacuum chamber on a regular basis, for the purpose of replenishing them. This can be avoided by the use of gaseous sources for the various elements [29]. These take the form of volatile organometallic alkyls for group III elements such as gallium, aluminum and indium, described in Section 5.5. Gaseous hydrides, such as AsH_3 and PH_3, are used as sources for the group V elements. Typically, they are thermally cracked upon entry into the growth chamber.

An immediate advantage of using these gaseous sources is that group III alkyls can be premixed before entry, so that precise control can be maintained over the composition of ternary alloys such as $Ga_xIn_{1-x}As$ and $Al_xGa_{1-x}As$. Moreover, rapid grading of the alloy composition can be achieved by adjustment of the vapor pressure of the corresponding alkyls. Finally, beam fading effects are not present when shutters are operated in these systems.

The use of gas sources increases the complexity of the MBE growth process, so that this approach is by no means an ideal replacement for conventional MBE. Moreover, pumping systems are greatly complicated by the need for handling high throughputs of chemicals, some of which are in a partially polymerized state. This can result in rapid fouling of titanium sublimation pumps, so that turbomolecular pumping is more common in these systems.

5.3 VAPOR-PHASE EPITAXY

In *vapor-phase epitaxy* (VPE), which is a subset of chemical vapor deposition (CVD), transport of the element(s) [or of their compound(s)] occurs in the form of volatile species which flow toward the substrate. As indicated earlier, these species may be adsorbed (or chemisorbed) on the substrate surface where they react to form the element(s) of which the layer is composed; alternatively, one or more species may thermally convert to this form before being adsorbed on the surfaces. The surface temperature is high, and the species are free to move by surface diffusion, until eventual incorporation, at kink* sites, into the growth layer. This process continues as the epitaxial film builds up.

Growth can also occur by the direct formation of stable nuclei in the gas phase [30]. This process is not common; however, once initiated, these nuclei provide a surface on which heterogeneous nucleation can occur, with the formation of large particles in the gas stream. This is highly undesirable since it leads to defects in the growing layer, and eventually to polycrystalline growth. An additional problem is the creation of dust in the reactor, which further destroys layer quality.

Desorption of reactants and reaction products into the gas stream also takes place. Most of these are carried off to the exhaust; however, some reenter into the chemical processes outlined here.

5.3.1 Source Chemicals

Reactants that are useful for VPE must be available in high purity, either as gases or as liquids with a high vapor pressure. Hydrogen gas is commonly used as a vapor transport medium and also as a diluent gas in both silicon and GaAs systems, where it makes up 90–95% of the total flow through the reaction chamber. Commercial-grade gas, followed by filters for oxygen and water removal, is often used for silicon systems. Increasingly, however, hydrogen is purified (to 99.9999% purity) by diffusing low-cost, commercial-grade hydrogen through a heated palladium–silver membrane.

Hydrogen is also bubbled through liquid sources to provide the appropriate vapor pressures of these reactants. This requires control of the bubbler temperature, which sets its vapor pressure, as well as the inlet gas flow and the exhaust pressure. It is commonly assumed that the carrier gas gets saturated** with the reactant vapor in passing through the liquid in a bubbler.

Solid reactants can also be used. These are transported by preheating to the vapor phase, or by *in-situ* conversion to a volatile species before transport to the substrate. In some cases, their vapor pressure is sufficiently high so that they can be used by directly "bubbling" the carrier gas through them. Run-to-run consistency is difficult to maintain with these sources.

A number of the chlorides of silicon are reasonably stable and can be readily transported in the vapor phase. Of these, silicon tetrachloride ($SiCl_4$) is often used, for historical reasons. Trichlorosilane ($SiHCl_3$) and dichlorosilane (SiH_2Cl_2) are favored because they allow epitaxial growth at lower tempera-

---

*Note that GaAs has both gallium and arsenic kink sites. Silicon, on the other hand, has only one type because of its elemental nature.

**See Problem 3 at the end of this chapter.

tures than the $SiCl_4$ reaction. An alternative material, silane gas (SiH_4), is used in halogen-free systems, and has unique features because of this fact.

Both arsenic and gallium sources are required for the growth of GaAs. Here, arsenic trichloride ($AsCl_3$) is a liquid source, while arsine (AsH_3) is a commonly used gaseous source, which can be used as a dilute (10–20%) mixture in hydrogen gas. Pure arsine, which is a liquid at room temperature, is often used as an alternative. It is safer to use than dilute arsine gas because its cylinder pressure (220 psi) is about a decade lower; moreover, it requires replacement less frequently, thus reducing the hazards associated with tank changes of this highly toxic material.

Sources for gallium are the organometallic compounds trimethylgallium [$(CH_3)_3Ga$] and triethylgallium [$(C_2H_5)_3Ga$], commonly abbreviated as TMGa and TEGa, respectively. Both are liquids at room temperature. Elemental gallium is also used, but it must be separately reacted with a halogen such as chlorine or HCl to convert it into a volatile species. Undoped GaAs itself can also be used as the source material. However, it also requires the use of transport agents, such as halogens, to provide the vapor phase sources that are required for VPE.

Many combinations of sources are possible for GaAs. The Ga–$AsCl_3$ combination is unique since both of these starting materials are available in seven nines (99.99999%) purity or higher. The TMGa–AsH_3 system is also unique because of its flexibility and its complete absence of halogens. The purity of these chemicals is somewhat lower (99.999% or better); however, manufacturing techniques have resulted in the almost complete elimination of active impurities, so that GaAs can now be grown routinely in the 10^{14}-cm^{-3} range.

Source chemicals which are used in epitaxial growth tend to be pyrophoric, so considerable care must be exercised in handling them. Arsine gas is a special problem, since it is highly toxic. A number of less toxic alternatives have been introduced in recent years to replace this hydride [31]. Of these, tertiarybutylarsenic, $(CH_3)_3CAsH_2$, is most commonly used today.

5.3.2 Steady-State Growth

Consider the situation at one specific point along the length of the susceptor, where a single reactant contributes to growth of the epilayer. If this reactant is delivered in a gas stream at a concentration N_g, resulting in a concentration of N_0 at the substrate, then the flux of species arriving at this point is given to a first order by

$$j = h(N_g - N_0) \tag{5.27}$$

where h is the gas-phase mass transfer coefficient. This relation is sometimes referred to as *Henry's law*.

This equation is directly applicable to silicon, where a chlorosilane or silane contributes to layer growth. For GaAs, growth is usually carried out with an excess arsenic overpressure, so that the gallium species determines the growth rate. Here, N_g and N_0 refer to the gallium species, since it is the rate limiter for this reaction.

Reaction proceeds upon arrival at the substrate. Since extremely dilute concentration of source chemicals are used, this reaction can be assumed to be linear, so that the flux of reaction products is given by

$$j = kN_0 \tag{5.28}$$

where k is the surface reaction rate constant. Typically,

$$k = k_0 e^{-E_a/RT} \tag{5.29}$$

where E_a is the activation energy of the process, on the order of 25–100 kcal/mole* for most surface-activated processes.

Steady-state growth requires that the fluxes described by Eqs. (5.27) and (5.28) be equal. Combining these equations and writing n as the number of atoms (or molecules) of grown material in a unit volume ($n \simeq 5 \times 10^{22}$ atoms cm^{-3} for silicon, and 2.22×10^{22} molecules cm^{-3} for GaAs), the growth rate of the epitaxial layer is given by

$$\frac{dx}{dt} = \frac{j}{n} = \frac{N_g}{n}\left(\frac{hk}{h+k}\right) \tag{5.30}$$

At low temperatures, $k \ll h$, so that the growth rate is given by

$$\frac{dx}{dt} = \frac{kN_g}{n} \tag{5.31}$$

Now the growth is *reaction-rate-limited*, and proceeds under kinetic control. At high temperatures, $h \gg k$, so that the growth rate is given by

$$\frac{dx}{dt} = \frac{hN_g}{n} \tag{5.32}$$

and the reaction is said to be *mass-transfer-limited*.

Many situations arise where the overall system behavior is not easily definable into two separate regions of this type. Thus caution must be used in interpreting growth data in terms of the physical initiating process. In practice, most epitaxial reactors are operated in the mass flow control region. An advantage is

---

*1 kcal/mole = 0.0434 eV/molecule.

that the growth rate is relatively unaffected by minor variations on the substrate surface, or by the substrate orientation.

Open-tube epitaxy is generally not carried out under thermal equilibrium conditions. Nevertheless, thermodynamics provides some indication of the nature of the growth reactions as a function of temperature. Consider first the case where this reaction is *endothermic* (i.e., involves a positive heat of reaction) and where growth is favored with increasing temperature. Here, growth in the kinetic control region will fall off with decreasing temperature. With increasing temperature, however, mass transfer will limit the growth rate so that it will be essentially constant. Figure 5.10 *a* illustrates a growth characteristic of this type, which is typical for all chlorinated silicon systems described in this chapter. Note that the mass-transfer region exhibits a small falloff with decreasing temperature, on the order of 3–8 kcal/mole. This is because the gas diffusion constant itself varies slightly with temperature, as mentioned earlier.

Figure 5.10 *b* illustrates the growth characteristic of systems where the formation reaction is *exothermic*. Here the reaction rate limits growth at low temperatures, as before. At elevated temperatures, even though the system is usually in the mass-transfer-limited region, the reaction itself is the rate limiter, and causes a decrease in the growth rate with increasing temperature. GaAs systems of the type described in Sections 5.5.3 and 5.5.4 (the halide and hydride processes) exhibit this characteristic.

The formation reaction for the organometallic process, described in Section 5.5.5, is one of pyrolysis and is essentially nonreversible, with growth being favored on hot surfaces. Consequently, the growth characteristic is similar to that for endothermic systems, as illustrated in Fig. 5.10 *a*. Growth occurs in the kinetic regime at low temperatures, and eventually becomes mass transfer limited at high temperatures.

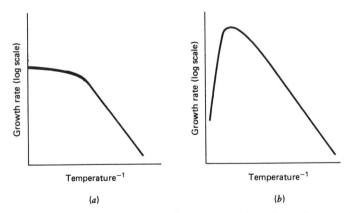

Fig. 5.10 Growth rate versus temperature for a system whose formation reaction is (*a*) endothermic and (*b*) exothermic.

5.3.3 Flow Considerations

Uniformity of growth across slices in a reactor is a prime requirement for the commercial exploitation of epitaxy. Much emphasis, both theoretical and experimental, has been placed on solving this problem, which is essentially fluid dynamic in nature. Here, a major simplification can be made by recognizing that reactors operate under laminar flow conditions, which can be characterized by the smooth flow of one layer over another.

This is readily seen by a study of gas flow in pipes [32]. This flow can be characterized by a dimensionless, empirical Reynolds number N_R, given by

$$N_R = \frac{dv\rho}{\mu} \tag{5.33}$$

where d is the pipe diameter, v the velocity along the pipe, μ the absolute viscosity, and ρ the fluid density. The viscosity of hydrogen, which is commonly used as a carrier gas, is about 200×10^{-6} g cm^{-1} s^{-1} at 700°C and increases to about 250×10^{-6} g cm^{-1} s^{-1} at 1200°C. Its density is approximately 2.5×10^{-5} g cm^{-3} at 700°C and 1.65×10^{-5} g cm^{-3} at 1200°C.

Experimentally, values of N_R between 2000 and 3000 represents the transition between laminar and turbulent flow. In a typical reactor, N_R is usually two decades below the critical value, so that flow is laminar in character, once the gases have traveled some distance in the reaction chamber.

Consider the idealized reactor shown in Fig. 5.11a, with a recessed susceptor which does not disturb the fluid flow conditions. Substrates are placed on this susceptor, which is heated. Only one reactant, of concentration N_g, is assumed to contribute to the growth of material. Consider also that *all* the reactant, arriving at the susceptor, is involved in the growth process, and none is desorbed; that is, N_g is zero at the substrate. Figure 5.11b shows the reactant concentration as a function of y, for a number of different points along the reactor. Here, x_0^- represents all points where $x \le 0$, and x_0^+ represents a point which is slightly beyond the leading edge of the susceptor.

Under mass transport limited conditions, the flux of species arriving at the substrate ($x = 0$) is proportional to the growth rate, and is given by

$$j = -D \left. \frac{\partial N}{\partial y} \right|_{y=0} \tag{5.34}$$

Using this equation, a study of Fig. 5.11b shows no growth at x_0^-, an extremely high growth rate at x_0^+, and a falling growth rate with distance along the susceptor. The anomaly of infinite growth rate at $x = 0$ can be resolved by noting that mass transport conditions do not prevail at extremely high growth rates, so that kinetic conditions are the rate limiter. The effective growth rate takes the form shown in Fig. 5.11c.

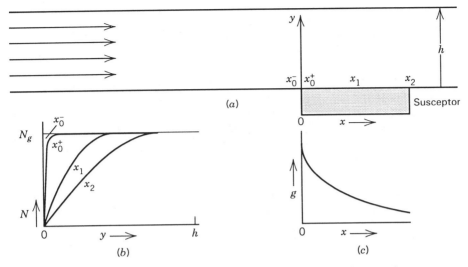

Fig. 5.11 Concentration of reactants in a horizontal reactor, and the growth rate.

The growth rate can be determined if we assume that the gas velocity, v, is constant along the length of the susceptor. Then, under steady-state conditions, the continuity equation can be written in two-dimensional form as

$$0 = \frac{\partial N}{\partial t} = D\frac{\partial^2 N}{\partial x^2} + D\frac{\partial^2 N}{\partial y^2} - v\frac{\partial N}{\partial x} \tag{5.35}$$

Making the further simplification that diffusion can be neglected in the flow direction gives

$$0 = D\frac{\partial^2 N}{\partial y^2} - v\frac{\partial N}{\partial x} \tag{5.36}$$

The appropriate boundary conditions are:

$$N = N_g \qquad \text{at } x = 0, 0 < y < h \tag{5.37a}$$

$$N = 0 \qquad \text{at } 0 < x, y = 0 \tag{5.37b}$$

$$\frac{\partial N}{\partial y} = 0 \qquad \text{at } 0 < x, y = h \tag{5.37c}$$

where N_g is the inlet concentration and h is the reaction chamber height above the susceptor. Finally, the growth rate is determined from Eq. (5.34). Solving Eq. (5.36) subject to the above boundary conditions [33] gives

$$j = \frac{-2DN_g}{h} \sum_{r=0}^{\infty} \exp\left(\frac{-\pi^2 Dx(2r+1)^2}{4vh^2} \right) \tag{5.38}$$

The reactant is delivered to the substrate by transport in the gas stream, with a finite velocity. This velocity must be zero along the substrate, because of friction. This is a reasonable assumption, since otherwise there would be an infinitely large velocity gradient at this boundary. It follows that there is a region, next to the substrate, over which the gas velocity is extremely low. This region can be visualized as a *velocity boundary layer* through which the reactant species must diffuse in order to reach the semiconductor surface. It is convenient to cast Eq. (5.38) in an analogous manner. Combining with Eq. (5.27), the boundary layer thickness which results in the same flux distribution is given by

$$\delta_D(x) = \frac{h}{2} \left[\sum_{r=0}^{\infty} \exp\left(-\frac{\pi^2 Dx(2r+1)^2}{4vh^2} \right) \right]^{-1} \tag{5.39}$$

It is important to recognize that $\delta_D(x)$ is actually a *diffusion boundary layer* in this case.*

Equation (5.39) can be greatly simplified if we impose the restriction that $(h^2 v/\pi D) > x$. For this case, we obtain [34]

$$j \simeq -N_g \sqrt{\frac{Dv}{\pi x}} \tag{5.40}$$

so that the diffusion boundary layer thickness is given by

$$\delta_D(x) \simeq \sqrt{\frac{\pi Dx}{v}} \qquad \text{for} \quad \delta_D(x) \le h \tag{5.41}$$

The simple analysis provided here ignores the effect of temperature and velocity variations in the reactor. Still, the boundary layer concept is useful because it provides the basis for a physical understanding of reactor behavior. Thus, it is seen that growth rate uniformity can be greatly improved if $\delta_D(x)$ is made more uniform with distance. This can be accomplished in many ways. For example, the susceptor can be tilted so that the reactor cross section is reduced

---

*It should be emphasized that the velocity boundary layer is distinctly different from the diffusion boundary layer. It is incorrect to use them interchangeably.

as the reactants move down stream with increasing velocity. Additionally, the total gas flow rate, as well as the system pressure, can be changed to alter the gas velocity. Techniques of this type are routinely used to "fine tune" reactors before they are put into operation. Yet other techniques, such as modifications to the physical configuration of the reactor, can also be made. However, changes of this type involve a major investment in time and effort and are difficult to implement on a cut-and-try basis.

Studies [35] based on variations of the boundary layer arguments presented above have been made with varying degrees of success. Most of these treatments have required a number of experience-based guesses of temperature and velocity profiles in the reactor, as well as the use of adjustment parameters to fit the experimental data.

Analytical approaches, relieved of these approximations and adjustments, are best handled by computation methods [36–38]. Typically, their starting point is the partial differential equations which relate continuity, conservation of momentum (Navier–Stokes equations), conservation of energy, and conservation of mass of diffusing species. These equations are solved simultaneously, with the appropriate boundary conditions which define the reactor and susceptor geometries. Computer simulation allows these conditions to be tested for any given reactor configuration.

The importance of computational methods results from (a) the remarkable accuracy with which they can predict reactor behavior and (b) their ability to evaluate the effects of major reactor design changes, both of which are not possible on an experimental basis. Thus, they have been used to successfully design novel reactor configurations with high-speed rotating susceptors [39], and also close-spaced vertical reactor systems [40] with distributed inlet nozzles.

A detailed study of these models is beyond the scope of this text. However, many computer-based studies have provided fresh insight into the behavior of reactors in practical situations. For example, the inclusion of effects due to gravity have resulted in significant improvements to reactor performance, and are described here.

Silicon epitaxy is usually carried out in the 1100°C range, with "cold" wall temperatures of around 600–700°C. For GaAs, growth is usually in the 700°C range, with wall temperatures around 200–400°C. Thus, there are generally large temperature gradients over the susceptor and in its vicinity. These gradients lead to significant natural convection, due to gravitational effects.

Natural convection causes the cold gas entering the reactor to fall while that being warmed by the susceptor tends to rise. This causes velocity streamlines to take the shape of a "U" as the gas traverses the reactor, and creates a low-pressure region in front of the susceptor, near the roof of the reactor. If the gas velocity is low, and the reactor height sufficiently large, this low-pressure region creates a backward pressure gradient which induces back flow. Hence, the "U"-shaped streamlines will close into recirculating ones. In both cases, the streamlines are squeezed down towards the susceptor and the effective cross section for flow is reduced. Figure 5.12a shows computed streamlines for an

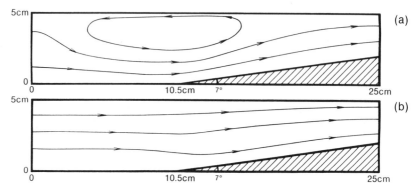

Fig. 5.12 Streamlines in a horizontal reactor: (*a*) atmospheric pressure, (*b*) 76 torr. Adapted from Chinoy et al. [37].

atmospheric pressure reactor operated at 700°C with a gas flow rate of 5 slm, and illustrates this effect [37].

Recirculation effects are detrimental to good epitaxy since they return unreacted products to the susceptor and increase the probability of gas-phase nucleation, with a loss in layer quality. Moreover, the resulting inability to swiftly move the reactants out of the reaction chamber is detrimental in device applications where abrupt layer changes are required [41].

For a constant gas input, the average gas velocity will be increased proportionately as the reactor pressure is reduced. This has the effect of reducing streamline curvature and eliminating recirculation, as seen in Fig. 5.12*b*, and illustrates an important advantage of carrying out the epitaxial growth process at reduced pressures.

Recirculation effects can be eliminated in atmospheric pressure reactors by sufficiently increasing the total gas flow and keeping the reactant flow rate constant. However, this decreases its mole fraction because of dilution effects, and hence the growth rate decreases by a comparable amount. An alternative approach is to increase flow velocity by reducing the height of the reactor. This results in an increased wall temperature which may lead to dusting problems associated with wall reactions. A significant advantage of computational methods is that all of these approaches can be tested, either separately or together, before implementation of the reactor design.

5.3.4 System Design Aspects

Vapor-phase epitaxy involves bringing measured quantities of the reactant and carrier gases to the substrate, reaction to produce the epitaxial growth, and eventually removal of reaction products from the susceptor. As a result, epitaxial reactors can be logically divided into three sections:

1. The gas delivery system, which provides the species in accurately metered form
2. The reaction chamber, where these species are transported to the substrate on which they impinge and subsequently react
3. The effluent handling system

Gas-handling systems are typically all-welded in construction, and utilize stainless steel plumbing and valving for handling of the reactants. All valves are pneumatically operated for positive action and are of the bellows-sealed variety, to prevent system contamination due to leakage from outside the system.

Liquid reactants are commonly obtained in high-purity stainless steel or quartz vessels, and are provided with a valving configuration which allows delivery of their vapor in a manner that is free from transients.

Gaseous reactants are usually provided in dilute form with ultrapure hydrogen as the diluent. Extremely dilute concentrations, in the parts per million (ppm) range, are used for doping purposes, but higher concentrations (10–30%) are used for materials growth. As mentioned earlier, some of the reactant gases are also supplied in liquid form at 100% concentration.

Gas flow may be measured by a flow gauge consisting of a vertical, tapered tube with a spherical ball inside it. The position of the ball in this tube is related to both dynamic and viscous forces which operate against the force of gravity. Consequently, the flow is not known with any degree of precision. Typical values of control which can be achieved range from ±10% for large flow rates to ±30% at low flows.

A widely accepted technique for gas flow monitoring is the mass flow controller (MFC). Here the gas flows through a heated tube, causing a change in temperature across its ends, which is directly related to its thermal capacity. Electronic sensing, together with feedback operation of a control valve, make these units insensitive to changes in inlet or exhaust pressure, and provide a measure of the absolute flow rate with a precision of ±0.5%. An important advantage of MFCs is that they can be adjusted by electronic means, and thus allow the use of computer control of the process. Thus, they are used in all but the simplest reactor systems.

Most VPE systems are operated in the mass flow control regime, so that considerable care must be taken to ensure precise control of the flux of reactant species at the slice. This requires the use of closed-loop MFCs. In addition, tight temperature control of the bubbler, to ±0.1°C or less, is necessary for systems in which the reactant species is a liquid. Normally the bubbler is designed so that the carrier gas is fully saturated with the vapor under flow conditions.

Premixing of gases must be done upon (or just before) entry into the reaction chamber, where laminar flow conditions generally prevail. Here careful attention must be paid to fluid flow considerations in order to minimize turbulences. Often the edge of the substrate is shaped, as seen in Fig. 5.12, so as to present a smooth profile to the incoming gases. Substrates are also recessed in the slice carrier for the same reason.

During epitaxial growth, switching transients in the gas flow are minimized by maintaining a continuous flow, and diverting it through (or around) the reaction chamber in order to initiate (or terminate) growth. A differential pressure transducer, in conjunction with a control valve, is used to equalize the pressure in these two paths, in order to reduce pressure disturbances during these transitions. Pressure-balancing systems are routinely used when abrupt transitions are required, as in thin-layer devices and superlattices.

Three types of reactors are in common use, the simplest having its reaction chamber in the form of a horizontal quartz tube. In this reactor the flow of gases is parallel to the surface of the wafers, as shown in Fig. 5.13a. The falloff in growth rate with distance can be greatly reduced by confining the gas flow in a reaction chamber of rectangular cross section, and by tilting the slice carrier by 3–10°, so that the gas velocity is forced to increase as it travels down the chamber [35].

A vertical reactor, in which the gas flows at right angles to the surface of the slices, is shown in Fig. 5.13b. Here, the slice holder is rotated during growth, resulting in improved uniformity. A disadvantage of stagnation point reactors of this type is that convection effects result in increased turbulence, and grossly alter the laminar flow conditions over the substrate. These effects can be avoided by the use of high-speed rotation (1000–2000 rpm), combined with operation at low pressure [42].

The barrel reactor, shown in Fig. 5.13c, is suitable for high-volume production. Here slices are held (by gravity) in niches along the slightly sloping vertical wall of a large cylindrical slice carrier. Thus it is essentially an expanded adaptation of the horizontal reactor of Fig. 5.13a, since the gas flow is parallel to the slices.* Here, too, the tilt of the slice carriers allows an increase of the gas velocity to compensate for boundary layer and reactant depletion effects.

The nature of the formation reaction is a determining factor in the manner in which the substrates are heated. Thus, cold wall reactors are used for endothermic and pyrolytic processes, whereas hot wall systems are used in the case where the formation reaction is exothermic. Cold wall reactors are usually heated by infrared lamps or by radio frequency, although internal resistive elements are used in some cases. These reactors have the advantage over hot wall systems in that thermal gradients are directed from the slice to the walls. As a result, the movement of particulate matter is away from the slice, rather than toward it.

The growth temperature, and its control, are of secondary importance in epitaxial reactors, which usually operate in the mass-transfer-limited regime. A ±2°C tolerance is normally considered to be adequate for applications requiring undoped material, as well as for doped material if the dopant species is fully

---

*Note that in this context, the terms *horizontal* and *vertical* do not refer to the physical position of the semiconductor slices, but to the relative direction of the gas flow and the axes of the slices.

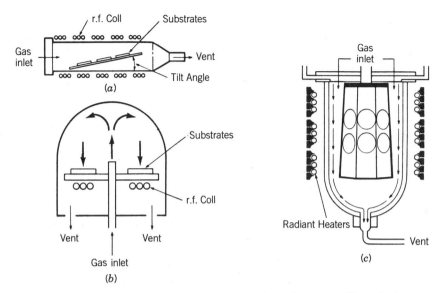

Fig. 5.13 Epitaxial reactor configurations: (*a*) horizontal reactor; (*b*) vertical reactor; (*c*) barrel reactor.

cracked at the growth temperature. Measurement is done by optical or infrared pyrometry; thermocouples, imbedded in the susceptor, are used in low-temperature systems (<900°C).

Provision is often made for vacuum pump-down of the reactor, especially in GaAs systems which are extremely sensitive to atmospheric leaks. This is useful for leak checking, and also greatly aids in flushing and backfilling the reactor prior to start-up. Since hydrogen is the most commonly used carrier gas, care must be taken to ensure hazard-free operation. Finally, provision must be made for adequate venting of the various lines and of the exit port. Gaseous effluents are usually thoroughly scrubbed, and sometimes burned before being exhausted to the atmosphere.

Increasingly, modern epitaxial reactors are designed to operate at low pressures, in the 20- to 200-torr range [43]. In addition to the advantages mentioned earlier, this also reduces hydrogen adsorption on the semiconductor surfaces and allows high-quality layers to be grown at temperatures that are 50–100°C lower than for atmospheric pressure systems [44]. Other advantages are that the boundary layer width also increases at reduced pressures (for the same gas velocity). This causes a significant reduction of autodoping effects, to be described in Section 5.4.4.

Low-pressure operation is not without its problems, however. Thus the thermal resistance between slices and susceptor is increased, so that there is often a 100–150°C difference in their corresponding temperatures. In addition, these systems are considerably more complex than their atmospheric pressure counterparts. This is primarily due to the fact that all pumping and gas pressure control equipment must be capable of operating in a continuous flow of corrosive chemicals, and maintenance problems are especially severe for these systems. Nevertheless, they are proving to be cost-effective and are becoming increasingly popular.

5.4 VPE PROCESSES FOR SILICON

The VPE growth of silicon is highly advanced, with many large automated systems in commercial use. Historically, $SiCl_4$ was used in these systems and its growth characteristics are well understood. Increasingly, systems using $SiHCl_3$ and SiH_2Cl_2 are favored, since they can be operated at somewhat lower temperatures. All of these chlorosilanes are colorless, corrosive liquids which pyrolyze in air to form HCl and silicon hydroxides.

Silane (SiH_4) is also used in silicon epitaxy. It is a colorless, highly pyrophoric gas, which burns on exposure to air. Epitaxial processes using SiH_4, although unique, are difficult to implement, and are found only in situations requiring the growth of very thin epitaxial layers.

The physical and thermodynamic properties of these compounds have been studied extensively [45, 46]. Their vapor pressure versus temperature characteristics are shown in Fig. 5.14.

5.4.1 Chemistry of Growth

The processes involved in the hydrogen reduction of all the chlorosilanes are qualitatively similar in nature. Equilibrium calculations, as well as mass spectrometric studies [47], have shown that, at temperatures above 800°C, all result in the formation of $SiCl_2$ in great quantity in the vapor phase, in addition to H_2 and HCl. Thus the primary effect of introducing any of these reactants, in addition to hydrogen, is to form a mixture of $SiCl_2$ and HCl in the hot zone. Typically this reduction reaction is almost complete when the gas temperature exceeds 1000°C. Some SiH_2Cl_2, $SiHCl_3$, and $SiCl_4$ are also formed, but their partial pressures fall with increased operating temperatures.

Upon arrival at the silicon slice, which is maintained at about 1050–1200°C, the $SiCl_2$ is adsorbed on its surface. The surface reaction is given by

$$2SiCl_2 \rightleftharpoons Si + SiCl_4 \qquad (5.42)$$

For this reaction to occur, two molecules of $SiCl_2$ must be involved in the presence of a host body (the substrate). Thus the reaction is *surface-catalyzed* and is a *heterogeneous* one. $SiCl_4$, which is a reaction product of Eq. (5.42), is desorbed into the gas stream, where it can again undergo reaction.

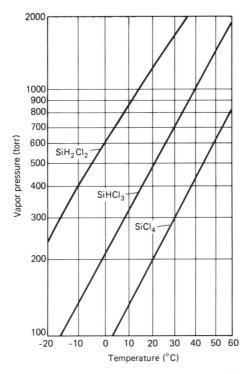

Fig. 5.14 Vapor pressure versus temperature for reactants used in silicon epitaxy.

The overall reaction may be written as

$$SiCl_4 + 2H_2 \rightleftharpoons Si + 4HCl \qquad (5.43)$$

It is important to note that this is a reversible reaction, the end product of which is either the growth of a silicon layer or the etching of the substrate. The exact nature of this reaction is a function of the mole fraction of $SiCl_4$ in H_2. Experimental data for one particular $SiCl_4$ system are shown in Fig. 5.15 and illustrate this effect, with the occurrence of etching at concentrations in excess of 0.28 mole fraction of $SiCl_4$.

Although all of these reactions are qualitatively similar, there are a number of differences which are important in epitaxy. Thus, $SiCl_4$ is a stable, strongly covalent bonded compound with a heat of formation, ΔH_f, of -153.2 kcal/mole at 25°C. The successive replacement of each chlorine atom by a hydrogen makes it increasingly unstable. For example, the heats of formation of $SiHCl_3$ and SiH_2Cl_2 are -112.1 and -75.0 kcal/mole, respectively. As a result, silicon epitaxy can be carried out at temperatures which are successively lower by 50°C, as we progress from $SiCl_4$ to SiH_2Cl_2 [48]. On the other hand, $SiCl_4$

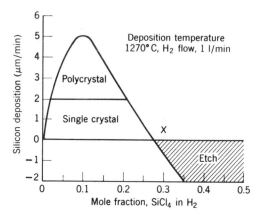

Fig. 5.15 Deposition rate for the $SiCl_4$ process.

has the advantage that it is the most stable chlorosilane and is also quite insensitive to the presence of small amounts of oxidizing species. $SiCl_4$ systems are relatively trouble-free in operation, because of this reason. As expected, the difficulty of system operation increases as we go from $SiCl_4$ to SiH_2Cl_2.

Another difference is the efficiency of the reaction—that is, the ratio of the amount of deposit to the amount of silicon in the incoming reactant. Here SiH_2Cl_2 and $SiHCl_3$ have the highest efficiency, and $SiCl_4$ the lowest. SiH_2Cl_2 is the most attractive for the production of epitaxial silicon. $SiHCl_3$ is often used as the starting material in the production of bulk semiconductor-grade silicon, because it is less expensive than SiH_2Cl_2.

The heat of formation of SiH_4 is +7.8 kcal/mole. It is potentially unstable, although there is a kinetic barrier to its autopyrolysis. The SiH_4 reaction proceeds by its adsorption on the surface, resulting in the formation of SiH_2, as follows:

$$SiH_4(ads) \rightleftharpoons SiH_2(ads) + H_2(ads) \tag{5.44}$$

This species breaks down rapidly at temperatures above 800°C, to form silicon which becomes incorporated into the growing layer. The overall reaction is

$$SiH_4 \overset{500°C}{\rightleftharpoons} Si + H_2 \tag{5.45}$$

Here, the absence of HCl gas as an end product avoids the necessity of elaborate venting. An additional advantage is that it can be carried out at even lower temperatures (50°C lower) because of the relative instability of silane.

There are two problems inherent to the silane process. The first comes about because some degree of homogeneous, gas-phase nucleation is possible with this reaction [30], so that the silane process can result in poor morphology or even in polycrystalline growth.* This problem can be reduced by operating the reactor at significantly higher gas velocities than are common for chloride systems, and by careful control of the deposition temperature. Second, silane is more sensitive to oxidation than any of the chlorosilanes. As a result, silane systems must be extremely free of oxidizing impurities in order to avoid (a) the formation of silica dust and (b) the consequent deterioration of layer quality. Because of these problems, they are not commonly used at the present time. Increasingly, however, the development of high-frequency structures has emphasized the need for thin silicon epitaxial layers which must be grown on heavily doped substrates and with an abrupt doping profile. The use of the silane process is almost mandated for these situations.

5.4.2 Epitaxial Systems and Processes

Silicon epitaxy is carried out in reaction chambers of the type shown in Fig. 5.13. The horizontal reactor is the most commonly used system, with larger systems tending toward the barrel design. Recently, new horizontal reactor designs [49], capable of growing one slice at a time, have been developed. These machines are fully computer-controlled, with cassette loading and automatic slice transport, and are ideally suited for commercial applications. One such design is illustrated in the frontispiece of this book.

In chloride epitaxy, the formation reactions are endothermic, so that the extent of deposit increases with temperature [50]. An induction-heated system, with essentially cold walls, can be used here, with the deposits confined to the slices and the susceptor. Infrared lamp heating is increasingly used, because it is considerably simpler and more cost-effective than induction heating. A significant advantage of this type of heater is that it can be designed to reduce temperature gradients in the slices, and thus lessen the problems of slip during epitaxy.

Slices are placed on a graphite susceptor which is coated with pyrolytic graphite or silicon carbide to reduce contamination. They are usually recessed in the susceptor in order to minimize turbulences in the reactant flow stream. A series of vertically mounted susceptors is used in the barrel reactor system.

Growth temperatures in $SiCl_4$ systems can be as high as 1250°C. Typically, however, growth is carried out in the 1150–1200°C range with this reactant. Systems using $SiHCl_3$ and SiH_2Cl_2 are commonly operated in the 1050–1150°C range, whereas silane systems are usually operated around 1000°C or lower.

Initially, the growth rate of the epitaxial layer increases linearly with the mole fraction of the reactant. However, the quality of the layer becomes progres-

---

*Advantage is taken of this characteristic for the growth of polycrystalline silicon films (see Chapter 8). Here silane is the most commonly used reactant source. Typically, this reaction is conducted at low temperatures ($\approx$650°C) with reasonably fast growth rates (1 μm/min) to promote polycrystalline growth.

sively worse, and eventually polycrystalline. A further increase of the reactant flux leads to the etching condition, as shown in Fig. 5.15 for a $SiCl_4$ system. In practice, a growth rate of about 2 μm/min results in reasonably acceptable single-crystal layers.

Impurity doping is done at the same time as epitaxial growth, using volatile dopant species. Gaseous species such as diborane (B_2H_6), phosphine (PH_3), and arsine (AsH_3) are commonly used for this purpose [51, 52]. Gas mixtures, in the 10- to 100-ppm range, with ultrapure hydrogen as the diluent, are used in order to allow reasonable control of flow rates for the desired doping concentration. Accurate control of the gas flow rate is critical to successful doping, since the resistivity of the grown layer is a function of the dopant concentration in the gas phase.

Dopant incorporation is essentially linear with the partial pressure of the dopant gas, up to concentrations as high as 10^{18} cm^{-3}. At higher concentrations, however, there is a gradual falloff in the dopant incorporation with partial pressure, and an eventual saturation in the doping level. This results from the increased strain in the layers due to misfit, and also to clustering effects. Misfit strain is most apparent with boron doping (see Fig. 5.4), and results in deteriorated crystal quality above 10^{19} cm^{-3}, although saturation occurs at 10^{20} cm^{-3}. Arsenic, because of its low misfit factor ($\epsilon \simeq 0$), does not have this problem. However, arsenic clusters form above 10^{20} cm^{-3}, with a resulting deterioration in crystal quality.

The amount of dopant which is incorporated in the epitaxial layer is also related to the growth temperature. As expected, the incorporation of a non-volatile impurity, such as boron, increases with growth temperature. On the other hand, the incorporation of volatile species (arsenic and phosphorus) falls. Doping incorporation comes about by trapping of the dopant species in the growing layer.

In addition to the above, the growth rate of the layer is slightly altered by dopant incorporation. Enhanced growth occurs with diborane, whereas phosphorus and arsenic doping tend to retard the growth. This behavior has been ascribed to catalytic effects due to the dopant, and is not well understood at the present time.

5.4.3 *In Situ* Etching Before Growth

Perhaps the most important condition for good epitaxial growth is the use of a completely damage-free, oxide-free surface. Silicon slices, obtained from the manufacturer, have been mechanically lapped and chemically polished to remove, as much as possible, the regions of surface damage caused by slicing of the ingot. After thorough degreasing and cleaning, these slices are inserted into the reactor and flushed for a few minutes in dry hydrogen at a temperature of about 1150–1200°C. This serves to reduce all traces of native oxide that may have been formed while transferring the slices into the reactor.

After the hydrogen cleanup step, anhydrous HCl gas, diluted with H_2, is introduced into the system. Etching proceeds by the conversion of surface silicon to volatile $SiCl_4$. A 1–5% mole concentration of HCl is typical, and provides an etch rate of about 0.5–2 μm/min. About 1–2 μm of silicon are removed in this process, resulting in a high-quality, optically flat finish. Excessive HCl concentration in the etching gas leads to halogen pitting of the silicon slice, and must be avoided.

The success of etching, and of the subsequent epitaxial growth, is a critical function of the quality of the etchant gas. Traces of water, nitrogen, or hydrocarbons can lead to the formation of oxides, nitrides, and carbides of silicon, respectively; these, in turn, serve as sites for the initiation of defects in the epitaxial layer. For this reason, the HCl etch step is sometimes omitted, and replaced by an additional bake period (3–5 min) in hydrogen, which results in thermal cleaning of the silicon surface.

Sulfur hexafluoride is sometimes used as an etchant gas [53], and results in a polishing etch at temperatures as low as 1060°C. The etching reaction for this gas is

$$4Si(s) + 6SF_6(g) \rightarrow SiS_2(g) + 3SiF_4(g) \tag{5.46}$$

This reaction is an irreversible one, whose end products are volatile at etching temperatures.

5.4.4 Impurity Redistribution During Growth

Redistribution of impurities occurs during epitaxial growth, resulting in departures from the ideal, abruptly discontinuous, doping profile. When epitaxy is carried out on a uniformly doped substrate, this effect is observed in the growth direction (i.e., the "vertical" direction) and is referred to as *autodoping*.

The process involved in autodoping can be described as follows. During epitaxy, both silicon and dopant are removed from the substrate by thermal evaporation (as is the case for growth from SiH_4), or by evaporation and chemical etching (which occur during growth from the chlorosilanes). Furthermore, this removal may or may not involve preferential leaching of either the silicon or the dopant. Thus thermal evaporation would tend to be preferential, whereas etching may not. Note that the sources of silicon and dopant may be the substrate itself (front and back surface, as well as the sides) or the susceptor which is silicon coated, either before or during the growth process. However, the reactor walls are not a source of evaporant, since silicon epitaxy is carried out in cold wall systems.

Both the silicon and the dopant diffuse into the boundary layer, and thus modify the composition of the incoming gas stream. This, in turn, changes the composition of the deposited material from that which would be present if there were no reverse flux of these constituents. As a result of this redistribution process, the doping concentration of an epitaxial layer varies during its growth until a steady-state situation is eventually reached for layers of sufficient thickness. With thin epitaxial layers, however, it is possible to obtain a continually varying impurity concentration over the entire thickness as a result of this mechanism.

A relatively simple analysis can be made of this process if it is assumed that the only loss of silicon and dopant (by thermal evaporation or chemical reaction) is from the front surface of the silicon substrates. Here we expect the effects of autodoping to be greatest during the initial growth, and diminish as the layer builds up.

Assume that an undoped silicon layer is to be grown on a substrate with a concentration N_S. For this situation it can be shown [54] that the impurity distribution in the epitaxially grown layer is given by

$$N_E(x) = N_S e^{-x/\phi} \tag{5.47}$$

where x is the distance from the interface (in cm), $N_E(x)$ is the impurity concentration (in cm^{-3}) at x cm from the interface, and ϕ is the growth constant (in cm). The growth constant ϕ is a function of the dopant, the reaction, the reactor system, and the process. For example, arsenic is more readily evaporable by thermal means than is boron or phosphorus; ϕ is larger for chlorosilane reactions than for the silane process; ϕ is related to the thickness of the boundary layer and is smaller for large values of layer thickness. Typical values of ϕ range from 1×10^{-4} to 2×10^{-4} cm. Thus, autodoping effects can be observed readily in layers of 1- to 2-μm thickness.

Solution of Eq. (5.47) shows that doping in the epitaxial layer falls off as the substrate becomes covered with the less doped crystalline silicon, until eventually an undoped layer is achieved. In practice, however, this doping limits to some finite value, which represents the minimum background concentration for the reactor (typically $10^{14}-10^{15}$ cm^{-3}). This is because of the effects of continuing evaporation from other surfaces of the substrates (especially from the back face), as well as from the susceptor. Measures for reducing the background doping level include precoating of the entire susceptor, as well as the back surface of the substrates, with silicon nitride or polycrystalline silicon. A reduction by a factor of 10 or more in background concentration can be achieved in this way.

Finally, it must be recognized that diffusional transport of evaporants into the gas stream also occurs during the pregrowth period, when the substrates are maintained at growth temperature in a hydrogen carrier gas ambient, *before* the introduction of reactant species. Thus thermal evaporation *prior* to actual epitaxial growth also contributes to redistribution effects *during* initial growth of the layer.

The elementary theory outlined above can be extended to the situation of epitaxial growth from a doped gas stream onto an undoped substrate. Now,

$$N_E(x) = N_{E_\infty}(1 - e^{-x/\phi}) \tag{5.48}$$

where N_{E_∞} is the doping concentration (in cm^{-3}), if steady-state conditions are reached during the epitaxial growth (i.e., for an infinitely thick layer). In actual situations, where a doped layer is to be grown on a doped substrate, the final impurity distribution is given by the superposition of these two cases. Note that the value of the growth constant, which applies to this equation, is not the same as that in Eq. (5.47), since ϕ depends on the impurity which is involved in the redistribution process.

Figure 5.16a shows the application of this theory to the growth of a lightly doped layer on a heavily doped substrate of the same impurity type, and is representative of the situation which occurs during epitaxial growth over a buried layer in a microcircuit. It is assumed that the substrate concentration is N_S, and that the concentration in the epitaxial layer, if made infinitely thick, is N_{E_∞}. Here, the ideal step of impurity concentration is not realized. In addition, it is possible that the final steady-state value of N_{E_∞} may not be attained for sufficiently thin epitaxial layers. Figure 5.16b shows what can happen if the substrate is heavily doped, or if the growth constant associated with the substrate impurity is small. Here it is possible to obtain a dip in the epitaxial layer doping concentration, close to the substrate–epitaxial layer interface.

Figure 5.16c shows the situation when a heavily doped layer is epitaxially grown on a lightly doped substrate of opposite impurity type. This is commonly encountered in the fabrication of bipolar microcircuits, which require an n-type epitaxial layer on a lightly doped p-type substrate. Here, redistribution causes a shift in the position of the junction delineated by the two layers, known as *junction lag*, and reduces the layer thickness.

Equation (5.47) indicates that autodoping effects can be reduced by lowering the surface concentration of the buried layer in the case of Fig. 5.16a. One approach for doing this is to deep-diffuse the buried layer prior to epitaxy, thus preserving its high conductivity while lowering N_S. A second approach is to maintain the substrate at an elevated temperature for a period of time, just before the initiation of epitaxial growth [55]. This pre-epitaxial bake step results in the outdiffusion of dopant into the boundary layer, thus reducing the surface concentration, which falls as the pre-epitaxial temperature and/or time are increased.

Measurements have shown that, although the surface concentration is reduced for all common dopants by this method, autodoping effects are only reduced in the case of arsenic and antimony [56]. In fact, they actually become *more* pronounced with phosphorus and boron doped substrates. Thus, the simple model which shows that lower surface concentration results in reduced autodoping is only partially correct.

The arguments presented above are based on the premise that all reactants are present as monoatomic species in the gas phase. A more complete theory, which includes chemical processes in the gas phase, has been proposed to explain this anomaly [57]. Thermodynamic computations for gaseous compounds produced during chloride-based epitaxy have shown that the stability of monoatomic compounds is enhanced for boron and phosphorus, at elevated temperature, but not

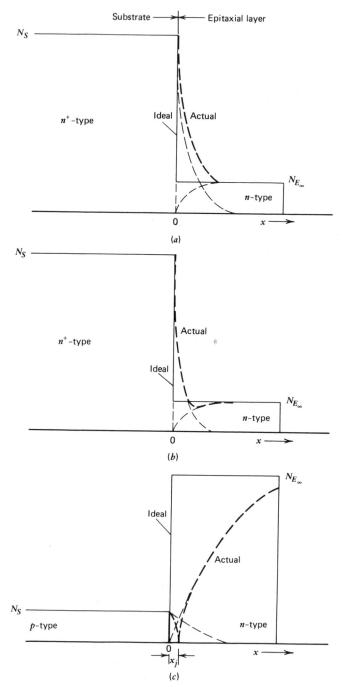

Fig. 5.16 Effect of autodoping on the impurity profile.

for arsenic and antimony. This would tend to explain the observed autodoping behavior, at least in a qualitative manner.

During epitaxy, some of the dopant in the boundary layer is lost by diffusion into the flow gas stream. This process is enhanced by growth at reduced pressure, since the gaseous diffusion constant varies inversely with system pressure. Consequently, autodoping effects should be lowered by reduced pressure epitaxial growth. This is indeed the case for both arsenic- and antimony-doped substrates, but the opposite is true for boron and phosphorus. Chemical studies [57] have shown that the stability of the monoatomic species of boron and phosphorus are enhanced by reduction of the system pressure, but not for arsenic and antimony. Thus, they can be used to explain the characteristics of autodoping at low pressure.

Autodoping effects also depend on the reactant which is used for epitaxial growth. Figure 5.17 shows the results for the growth of an essentially undoped epilayer on a heavily arsenic-doped buried layer, using SiH_4, SiH_2Cl_2, and $SiCl_4$ at 1050°C, with a growth rate of 0.25 μm/min. Here, redistribution effects are seen to be reduced with increasing chlorine content in the silicon species. The reasons for this behavior are not fully understood, but are probably related to the increased formation of arsenic chlorides in the progression from SiH_4 to $SiCl_4$.

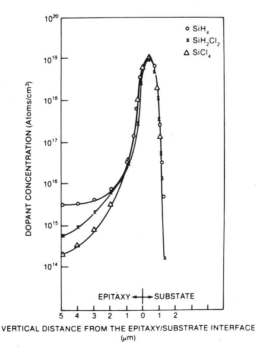

Fig. 5.17 Autodoping effects for different reactant species. From Chang [56]. Reprinted with permission of the publisher, The Electrochemical Society, Inc.

Departure from an abrupt doping profile can also result because of diffusion during epitaxial growth. Here, if the film growth rate g is much higher than the diffusion rate, (i.e., $g \gg \sqrt{Dt}$), the epitaxial film can be considered to be infinitely thick compared to the extent of the region affected by solid-state diffusion. Hence diffusion can be considered as proceeding into an infinite solid. This situation often prevails in silicon epitaxy. For example, a growth rate of 0.25 μm/min, and a diffusivity of 10^{-12} cm^{-2} at the growth temperature, corresponds to a characteristic diffusion length ($\sqrt{Dt}$) of 0.34 μm for a layer which is 5 μm thick. Thus, diffusion effects in this case will extend, at most, to 1 μm (about three diffusion lengths) into the epilayer. This is seen in the initial behavior of the autodoping characteristics of Fig. 5.17, which is similar for all the reactant species for the first micron of epilayer.

More significant, however, are diffusion effects during subsequent thermal processing, which can be of many hours' duration. They can be evaluated by the aid of the effective diffusivity concept, described in Chapter 4. Using this approach, the combined effects of autodoping and interdiffusion can satisfactorily describe dopant behavior in layers which are $\geq 4\,\mu$m in thickness, of the type commonly used in bipolar transistor logic (TTL) and in linear integrated circuits. Increasingly, however, the need for faster devices has necessitated the use of structures with thin ($\leq 2\,\mu$m) epitaxial layers. These are usually grown by the SiH$_2$Cl$_2$ or SiH$_4$ process because of the lower temperatures involved. Arsenic-doped buried layers are commonly used, since this impurity is a slow diffuser in silicon.

Detailed studies of thin epitaxial layers require that both autodoping and diffusion effects be considered. This includes consideration of the following processes: diffusion of the impurity from the substrate surface, both into its bulk and into the growing epitaxial layer; diffusion of evaporants (silicon and impurity) from the surface of the epitaxial layer into the gas stream; diffusion of evaporants from other surfaces into the gas stream; reverse diffusion of reactants from the gas stream to the silicon surface, where growth proceeds; and chemical processes in the gas phase.

Computer-aided studies of the coupled diffusion equations which describe many of these processes have been made [58–60]. These allow the simulation of changes in the process in order that critical steps can be exposed and modified. At the present time, however, they necessitate the use of many adjustment parameters; moreover, autodoping results are extremely machine-sensitive. Consequently, the experimental approach is almost universally adopted during epitaxial growth.

5.4.4.1 Lateral Autodoping

Redistribution effects outlined in the previous section come about because the silicon substrate behaves as a large, uniformly doped surface area from which evaporants leave as if from an infinite plane. However, an additional complexity arises during epitaxial growth over substrates in which there are a number of selectively doped regions, as is the case with the fabrication of bipolar microcircuits. Here it is customary to begin by diffusing a pattern of small n^+-doped layers into a slice of lightly doped, p-type material (3–10 Ω cm). This serves as the substrate for the growth of the n-epitaxial layer in which subsequent process steps are carried out. During epitaxy, each n^+-buried layer acts [61] as a two-dimensional source from which evaporant diffuses onto the surrounding medium.* This results in a redistribution of impurities over the adjoining lightly doped substrate region (*lateral autodoping*), in addition to conventional autodoping over the buried layer. In extreme situations, a slice can serve as a dopant source for other slices that are downstream from it.

The problem of lateral autodoping is especially serious in high-speed bipolar integrated circuits, which require the growth of very thin, lightly doped epitaxial layers, on substrates with heavily n^+-buried layers. Its effect is illustrated in Fig. 5.18 for the growth of an epitaxial layer by the $SiCl_4$ process, on slices with a number of arsenic-doped buried layer regions [62]. In Fig. 5.18a the doping profile is shown directly over the buried layer region. The profile at a point 0.1 mm away from this region, and over the lightly doped substrate (Fig. 5.18b), shows a significant amount of lateral doping for this experimental situation.

Lateral autodoping occurs primarily by the diffusive transport of evaporants *within* the stagnant boundary layer. There is a small velocity gradient in this

*The situation is analogous to that of diffusion through a window in an oxide, described in Chapter 4.

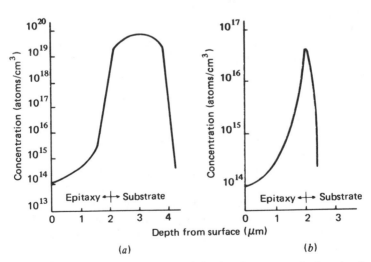

Fig. 5.18 Vertical and lateral autodoping behavior for an arsenic-doped substrate in the $SiCl_4$ process. From Srinivasan [62]. Reprinted with permission of the publisher, The Electrochemical Society, Inc.

region; consequently, substrate regions which are downstream are more heavily doped than those that are upstream. The extent of this effect may be characterized by the maximum value of doping which occurs in a region that is some fixed distance away from the buried layer. It has been experimentally shown that this value falls with increasing pre-epitaxial bake time and temperature, as is the case for vertical redistribution effects.

The physical processes of lateral autodoping are thus the same as those for vertical autodoping. As a result, the effects are similar, as are methods for their reduction. Specifically, lateral autodoping effects are reduced for arsenic and antimony by increasing the time and temperature of the pre-epitaxial bake step, and by growth at reduced pressures. However, the opposite is true with boron and phosphorus. Finally, lateral autodoping with arsenic buried layers is reduced as the chlorine content of the source reactants is increased.

5.4.5 Selective Epitaxy

Epitaxial growth approximately follows the surface contour of the substrate and completely covers it. There are a number of advanced device concepts, however, where epitaxy is required only in selected regions. Thus, selective epitaxy can be used for complementary bipolar microcircuits which require both n- and p-type buried layers. It is used for lateral and vertical device isolation in high-density CMOS structures, as well as in situations where single crystal and polycrystalline silicon have to be deposited simultaneously [63], over different regions of the microcircuit.

One approach to selective epitaxy consists of photolithographically opening windows in an oxide layer over the slice, and filling these with epitaxial material. Typically, growth of this type is carried out under low supersaturation conditions in order to control the crystal morphology at the edge of the window. Often, HCl gas is added to SiH_2Cl_2 in order to etch the material which grows on the oxide at a faster rate than on the silicon [64]. (Surface nucleation effects are probably significant in this process [65].) Additionally, the selectivity of the $HCl + SiH_2Cl_2$ system can be further enhanced by growth at reduced pressures, to enhance the differences in desorption rates for these reactants on silicon and silicon dioxide surfaces, respectively. Recent work has shown that surface treatments of the oxidized substrate can also have a significant effect on the conditions for obtaining selective growth [66].

Selective epitaxy can also be achieved on substrates without masking, if reactor conditions are such that the growth is nonuniform in a prescribed manner. One approach is to operate the reactor in the kinetic regime, where the growth rate is sensitive to both temperature and crystallographic orientation. A second approach is based on the thermodynamics of reversible processes, such as those involving chloride transport. Here growth is generally carried out at high reactant concentrations corresponding to a very low growth rate (point × in Fig. 5.15). Alternatively, the addition of HCl gas can be used to simulate this reactor condition.

One example of this approach is a situation where deep grooves, cut in a silicon surface, can be refilled without masking [67]. For this situation it is possible to adjust the reactant concentration so that the growth rate on the surface is zero. The temperature within the groove is higher, however, so that conditions for growth exist here. As a result, growth proceeds until the grooves are filled, and then ceases.

5.4.5.1 *Pattern Shift*

In chloride-based silicon epitaxy, the growth rate is affected by the surface-free energy of the substrate. This is true to some extent, even though the growth is usually carried out in the mass transport limited region. In general, therefore, epitaxial growth will not precisely follow the contour of the substrate, since a nonplanar slice exposes a number of different surfaces. This presents a problem in the fabrication of bipolar transistor microcircuits, where a buried layer diffusion is made into windows in the slice prior to the growth of the active layer. This leaves an optically sharp pattern of indentations in the substrate surface, due to the oxidation and subsequent photolithographic process. It is important that this pattern be carried through to the surface of the epitaxial layer, since it is used for the alignment of succeeding masks. Typically, however, it suffers from a variety of spatial distortions and displacements, which are referred to as *pattern distortion* and *pattern shift*, respectively, depending on whether there is a change in the pattern size, or only a spatial displacement. In extreme cases, complete obliteration of the pattern, known as *washout*, can occur. Although these problems are somewhat lessened with thin epitaxial layers, the requirements on such layers is comparatively more tight, since they are used in high-density applications.

Reactor conditions during growth, as well as crystal orientation, affect the extent of these problems. Experimental studies have been made with (111) silicon on which layers were grown by the $SiCl_4$ process [68]. Here it has been shown that the amount of pattern shift is reduced at lower growth rates, and/or with increasing growth temperatures. In addition, silicon that is misoriented by 2–4° from the [111] direction, towards the nearest (011) plane, shows a reduction in pattern shift over precisely oriented material. However, misorientation towards the (001) plane results in diffuse boundaries, and eventually washout. For (100) silicon, on the other hand, best results are obtained when the slice is precisely oriented [69]. Here, however, a misorientation of as little as 0.5° can lead to a significant displacement of the pattern.

Conservative design practice makes allowance for an angular pattern shift of 45° due to epitaxial growth. Thus a substrate pattern, after epitaxy, is assumed to shift a maximum of x_0 in all directions, where x_0 is the thickness of the epitaxial layer. Successive masks must be designed to take this uncertainty in the position of the buried layer pattern into account.

Both pattern shift and washout effects are less in systems where the chlorine-to-silicon ratio is lower. Thus the use of SiH_2Cl_2 reduces these problems as compared to $SiCl_4$. The SiH_4 process, which is chlorine-free, essentially eliminates this problem. However, pattern distortion effects become very large if there is any substrate misorientation [70].

5.4.6 Crystal Defects

A number of different types of crystal defects may be present in epitaxial silicon. Some of these are crystallographic in nature and result in a loss of periodicity in the crystal lattice. Some are thermally induced as a consequence of the epitaxial growth procedure. Yet others result from improper handling or cleaning procedures.

5.4.6.1 Stacking Faults

Stacking faults are an important type of crystallographic defect that occurs in epitaxial silicon films. They can be observed by interference contrast microscopy and appear as equilateral triangles and lines on the surface of layers grown on {111} substrates. The orientation of the sides of these stacking faults is in the ⟨110⟩ directions. In addition, all the lines and the sides of these triangles are usually of equal length. Stacking faults on {100} substrates are generally square in shape, with their sides oriented along the ⟨110⟩ directions.

The growth of stacking faults can be described by considering epitaxy as the ordered deposition of atomic planes, one at a time, with nucleation centers for a fresh plane being formed after the last plane has been completed. If, however, there is a small region which is mismatched with respect to the substrate, the regularity of the layers is disrupted. On disruption, the layers continue to grow in this new sequence within the fault. This is shown in Fig. 5.19 for {111} layers in a face-centered cubic (f.c.c.) lattice. For this case the stacking sequence of atomic planes is *abcabc*..., whereas that of the diamond lattice is *aa'bb'cc'aa'bb'*.... Thus the growth mechanisms are essentially similar, although somewhat easier to visualize for the f.c.c. lattice. Here the defect propagates from layer to layer, retaining its shape and increasing in size with each successive atomic plane. The various faces of the defect all fall upon {111} planes, resulting in the tetrahedron of Fig. 5.20a. From geometrical considerations, $h = \sqrt{\frac{2}{3}}\, s$, where h is the height of the stacking fault, and s is the length of one side of the triangle on the surface.

Stacking faults in (100) silicon also propagate along {111} planes. These intersect the surface plane in the ⟨110⟩ directions, which are mutually perpendicular. Thus the fault takes the form of a pyramid with a square base, as shown in Fig. 5.20b, with its apex at the point where the disruption originates. Here, $h = \sqrt{\frac{1}{2}}\, s$, where h and s are defined as before.

Stacking faults generally originate from the substrate surface, where inhomogeneities such as dislocation loops and precipitated impurities act as sites [71]. Surface measurements of these faults are often used to check the thick-

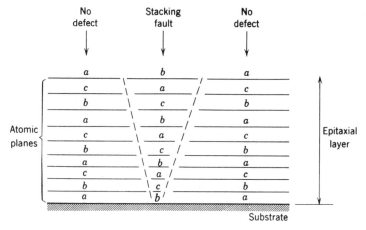

Fig. 5.19 Formation of a stacking fault in (111) material.

ness of the epitaxial layer. One of the primary causes of stacking faults is the presence of traces of oxide on the surface of the crystal. These result in steps which are not equal to an integral number of atomic plane spacings. Lattice periodicity is maintained during epitaxial growth by the initiation of stacking faults that start at the edge of the oxide and grow into the depositing layer. Experimental verification of this theory has been noted in the fact that the concentration of stacking faults is critically dependent on the effectiveness of the hydrogen pre-bake step.*

*Note that the HCl etch step does not remove SiO_2.

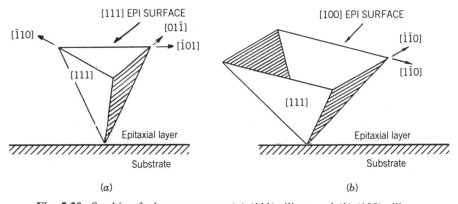

Fig. 5.20 Stacking fault geometry on (a) (111) silicon and (b) (100) silicon.

Scanning electron microscope studies have been made to investigate the electronic activity of stacking faults, that is, their ability to act as recombination centers [72]. Results are inconclusive; however, much evidence of activity has been noted at the corners of the triangles (or squares) formed where the defect penetrates through the surface. This is probably the result of impurity segregation at these points of stress concentration. Thus it is generally considered that stacking faults, taken by themselves, do not significantly alter the electronic properties of the epitaxial layer. They give rise, however, to imhomogeneities in the diffusion of impurities and act as nucleation centers for metal precipitates. Impurity atom movement is enhanced wherever a diffusion front crosses a slip line. As a result, junctions which are built in faulted regions often exhibit soft reverse-breakdown characteristics, with breakdown initiated by the formation of microplasmas at these faults.

The elimination of stacking faults is especially important as devices become smaller and more densely packed in a microcircuit. Basic techniques for their reduction are those that result in the elimination of surface defects on the substrate, and dusting during epitaxial growth. In addition, misorientation of the crystal surface by 1–3° exposes a large number of layers to which atoms can attach during epitaxy. This allows a faster growth rate for the same layer quality, without a corresponding increase in stacking fault density. Thus techniques for the reduction of pattern shift in (111) silicon are also effective in reducing stacking faults for this orientation.

5.4.6.2 Slip

The mechanism of slip was outlined in Section 5.1.3.1, where it was shown to occur when the maximum shear stress in the slice exceeds the critical resolved shear stress. For Czochralski (CZ) silicon, slip occurs at about 1 kg mm^{-2}. Float zone material, on the other hand, is considerably more susceptible to slip, even though it is inherently cleaner than CZ silicon. This comes about because of the higher density of oxygen precipitates in CZ silicon, which pin the dislocations and inhibit their propagation.

Slip can be reduced, or prevented, by reducing the radial temperature gradient during epitaxy. This can be accomplished by growth at reduced temperature, and SiH_2Cl_2 is preferred over $SiCl_4$ for this reason.

Shaped susceptors of different types have been used with radio-frequency (r.f.)-heated systems, with some success [73]. The basic principle underlying these schemes is to reduce the r.f. coupling (and hence the local temperature) in the central region of the depression in which the slice is placed. This results in a reduction in the radial temperature gradient, and reduces the tendency towards slip during epitaxy.

Major improvements in slip reduction (and even elimination) in silicon epitaxy have come about by the use of infrared lamp heating. This provides both front- and back-face heating, since much of the radiation, upon passing through the silicon, is absorbed on the graphite susceptor on which the slices are positioned.

5.4.6.3 Contamination-Induced Defects

Many gross defects can occur during the growth of an epitaxial layer. One of these, the *tripyramidal defect*, is the result of crystal growth by twinning, and is initiated by carbon contamination (in the form of β-SiC microprecipitates). In (111) silicon, it appears as a rosette-shaped cluster of three pyramids with hexagonal bases oriented in the $\langle 110 \rangle$ directions. Other gross defects are caused by leaks in the system, the presence of foreign matter during epitaxial growth, and surface scratches caused by pregrowth handling. Their presence can usually be detected by optical means and often leads to rejection of the slice. Defects of this type are described here very briefly.

Substrate scratches act as sites for the initiation of stacking faults and result in lines of such faults on the surface of the layer.

Pits, voids, and *spikes* are caused by small particles of silicon or silicon dioxide in the reactor during the epitaxial growth.

Haze takes the form of a cloudy appearance and results from a leaky system or from improper cleaning and solvent removal procedures prior to epitaxial growth.

Orange-peel appearance is sometimes caused by preferential etching during the *in-situ* step, prior to layer growth.

All of these defects are indicative of poor fabrication technique and can be corrected by taking suitable precautions.

5.5 VPE PROCESSES FOR GALLIUM ARSENIDE

Epitaxial growth of GaAs is sometimes carried out in sealed-tube systems, because of the ease of maintaining controlled vapor pressures and equilibrium conditions by this method. In this section, however, we confine ourselves to open-tube processes which are commonly used because of their convenience. As with all CVD processes, they involve vapor-phase transport of the active species, followed by adsorption and by a growth reaction at the substrate.

5.5.1 General Considerations

GaAs decomposes into gallium and arsenic upon evaporation, so that its direct transport in the vapor phase is not possible. Transport of its separate components must therefore be considered. Elemental arsenic has a sufficiently high vapor pressure so that it can be transported by direct sublimation. However, the use of arsenic compounds which are in the form of volatile liquids or gases is more common, because of convenience. Of these, the chloride ($AsCl_3$) is the most stable liquid source, and is available in "seven nines" purity (99.99999%).

The hydride, arsine, is used in a 5–50% dilution in ultrapure hydrogen gas, or in undiluted form as a liquid. It is a somewhat less pure source of arsenic (99.999%), with silicon as its primary contaminant. Arsine is highly toxic in character, so that a number of alternatives have been developed as possible replacements [31]. Of these, tertiarybutylarsenic (TBAs) is the most commonly used. It is available in high purity, as a volatile liquid which can be transported by means of a bubbler arrangement. It is about ten times less toxic than arsine gas.

Gallium is also available in "seven nines" purity. Its vapor pressure is negligible at growth temperatures, so that it must be transported in the form of a volatile compound. Gallium chloride (GaCl) is commonly used for this purpose, and must be formed by a secondary reaction at some point upstream from the substrates.* An alternative approach [74] is the use of an organometallic compound such as trimethylgallium (TMGa). TMGa is a liquid which can be transported in vapor form by means of a bubbler arrangement, and is available in sufficient purity (better than 99.999%) to be used for GaAs epitaxy. Triethylgallium (TEGa) is also used as an alternative organometallic source for gallium.

The organometallic compound trimethylaluminum, $(CH_3)_3Al$, is a commonly used liquid source for the growth of AlGaAs. TMAl is highly sensitive to oxidation, so that extremely leak-tight systems are necessary for its successful use.

All of these sources are highly toxic and/or pyrophoric in nature, so that extreme care must be taken in handling them. Their vapor pressure versus temperature characteristics, shown in Fig. 5.21, can be related by

$$\log_{10} P = a - \frac{b}{T} \tag{5.49}$$

where P is the vapor pressure (in torr) and T is the absolute temperature. The constants a and b are listed in Table 5.2 at the end of this chapter.

Although reactor systems for GaAs are very similar to those used for silicon, there are some significant differences which are now considered. Perhaps the most important difference stems from the fact that oxygen is a deep-lying impurity in GaAs. As a result, these systems must be extremely leak-tight for satisfactory operation. Sometimes the reaction chamber, which must be opened for loading purposes, is placed in a glove box which is flushed with an inert gas to prevent the accidental introduction of this impurity. Load-lock systems, which allow the reaction chamber to be sealed to the atmosphere at all times, even during the loading and unloading process, are increasingly popular for this reason.** In addition, elaborate pump-down and back-fill techniques are often used for leak checking and flushing prior to growth.

*Note that GaCl is only stable at elevated temperatures.
**Systems of the type shown in the frontispiece, with cassette loading, are also suitable for GaAs epitaxy.

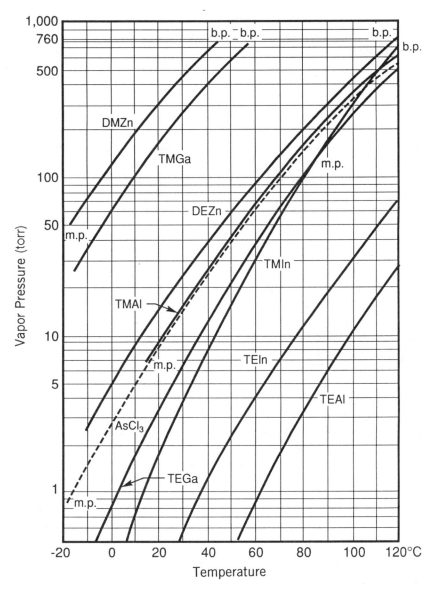

Fig. 5.21 Vapor pressure versus temperature for reactants used in gallium arsenide epitaxy.

The purity requirements for hydrogen, which is used as a carrier gas, are considerably higher than for silicon systems. Palladium-purified hydrogen, with "six nines" purity, is mandatory for this application. In contrast, many silicon systems use commercial-grade hydrogen, followed by filters for oxygen and water removal.

There are two basic approaches for the epitaxy of GaAs: the *halide/hydride* processes and the *organometallic* process, which will be described in detail in the following sections. The formation reaction for the halide/hydride processes is exothermic in character, so that a hot wall reactor is used for this purpose. Typically, its configuration is of the type shown in Fig. 5.13a, except that the reaction chamber is resistance-heated, with at least two temperature zones: one for the source and one for the deposition.

The organometallic process, on the other hand, is pyrolytic in character and requires the use of a cold wall reactor. As a consequence, reaction chambers are similar to those for silicon (see Fig. 5.13a–c), and are usually heated by r.f. or by infrared lamps. The cold wall minimizes premature reaction of the constituent gases as well as the formation of wall deposits.

Requirements for GaAs epitaxy are usually more stringent than those for silicon. Most active devices require layer thicknesses of 0.1–0.2 μm. Modern quantum structures require multiple layers as little as 20 Å in thickness, with abrupt composition and/or impurity type and doping changes from layer to layer. Thus, the control of epitaxial layers must be extremely tight [41, 75]. In addition, there is often the complication that reactants are unstable at temperatures well below the growth temperature. In some instances, competitive reactions can also occur. As a result, most GaAs systems process one to five wafers at a time, and operate with relatively high gas velocities. In almost all systems, provision is made for slice rotation to improve uniformity. Some of these problems can be alleviated by system operation at reduced pressure, and this is the general trend for these reactors.

Epitaxial GaAs is usually grown on (100) substrates, which are misoriented 2–3° towards the nearest [110] direction. This increases the density of kink sites, resulting in improved morphology for the epitaxial layer. The use of (100) substrates allows the fabrication of rectangular chips whose edges are {110} planes, which are preferred for easy cleavage. This orientation is mandatory for laser fabrication, where the parallelism of opposite faces is essential for device operation.

Growth in the ⟨111⟩ direction is usually carried out on its arsenic face. This is because the (111)As face is more electronically active than the (111)Ga face, and thus can be readily etched to a smooth polish. Differences in growth rates on these faces are discussed in Section 5.5.7.

5.5.2 Growth Strategy

A common strategy for good growth is applicable to all GaAs VPE systems. Thus from stoichiometric considerations it is desirable to grow GaAs in the 600–850°C temperature range. Growth at the high end of this range results in excessive gallium vacancy formation. On the other hand, reactants are inefficiently cracked at low temperatures, and this can limit film quality. In extreme cases, the reduced mobility of surface species creates a tendency for films to become polycrystalline.

There must always be sufficient arsenic overpressure to prevent thermal decomposition of the substrate and of the growing film. Moreover, the arsenic-to-gallium mole ratio must be controlled to prevent the presence of excess gallium, which can otherwise lead to the formation of droplets, hillocks, and whiskers on the growing surface. Excess arsenic, on the other hand, remains in vapor form (As_2, As_4, or as an arsenic radical) and thus a zero sticking coefficient.

Substrates must be extremely clean and free from mechanical damage prior to growth. An *in situ* etch step is highly desirable for this purpose. Finally, it is desirable that the growth conditions be mass-transport-limited; operation in the kinetic control regime is orientation-dependent, and tends to accentuate surface defects, resulting in poor crystal morphology.

5.5.3 The Halide Process

In this process [76, 77], transport of the gallium is accomplished by means of the halide, $AsCl_3$. A two-zone, resistance-heated, hot wall reactor is used, since the formation of GaAs by this process involves a negative heat of reaction; that is, it is exothermic. The gallium is held in a graphite crucible in the source zone, as shown in Fig. 5.22. Growth occurs downstream in the deposition zone where the substrates are laid on a quartz or graphite carrier. During growth, part of the hydrogen is sent through an $AsCl_3$ bubbler which is maintained in the 0–20°C range. Both hydrogen and the $AsCl_3$ vapor enter the system, where they react upstream from the gallium source. Here

$$AsCl_3 + \tfrac{3}{2}H_2 \overset{425°C}{\rightleftharpoons} \tfrac{1}{4}As_4 + 3HCl \qquad (5.50)$$

These reaction products flow over the gallium source boat, which is typically held at 800–850°C. GaAs is formed as a crust on the surface of this gallium source:

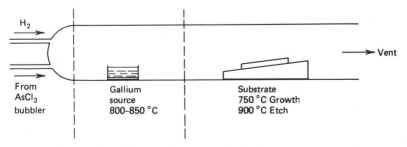

Fig. 5.22 Schematic for the halide process.

$$Ga + \tfrac{1}{4} As_4 \rightleftharpoons GaAs \tag{5.51}$$

Simultaneously, the HCl gas resulting from the first reaction serves to transfer gallium to the substrate in the form of GaCl, as follows:

$$GaAs + HCl \rightleftharpoons GaCl + \tfrac{1}{2} H_2 + \tfrac{1}{4} As_4 \tag{5.52}$$

From thermodynamic data [78] it can be shown that this reaction is driven from left to right at temperatures above 800°C, that is, it is a decomposition reaction for the GaAs. There is some evidence of the formation of both $GaCl_2$ and $GaCl_3$ during this process as well. However, their effect on the growth is of second order.

The substrate is held at a lower temperature, typically 750°C. Here the same reaction occurs, except that it is now driven in the opposite direction so that it is a formation reaction for GaAs. The overall effect is to transport GaAs from the source boat, where it is formed, to the substrate on which it is epitaxially grown. Growth takes place by a surface-catalyzed (heterogeneous) reaction so that excellent crystal quality can be obtained. Typical growth rates of 0.2–0.5 μm/min are maintained in systems of this type.

The growth versus temperature characteristic of this system takes the form of Fig. 5.10b because of the exothermic nature of the reaction. Here the growth rate falls with decreasing temperature due to reaction rate limitations—that is, kinetic control. The growth rate also falls with *increasing* temperature because of the exothermic reaction, even if the reaction is mass-transfer-limited in this region (as is usually the case). However, as is characteristic for mass-transfer control, the growth rate is not orientation-dependent [79] in this temperature range, as seen in Fig. 5.23. Note that growth in the low-temperature kinetic control region is highly sensitive to crystal orientation. At high temperatures, in the mass-transfer-limited region, rates for all orientations become similar.

Figure 5.24 shows the growth rate of GaAs for a large number of different crystal orientations. Here, it is seen that growth on the arsenic-rich surfaces is slower than on gallium-rich surfaces. Thus the addition of gallium atoms appears to be more difficult than arsenic on surfaces of the same geometry and packing. This has been explained by postulating [80] that the addition of a gallium atom requires attachment of GaCl to an arsenic site to form the intermediate compound AsGaCl. This is followed by the subsequent detachment of the chlorine, and its desorption from the surface. The addition of arsenic atoms to gallium sites, on the other hand, does not involve this rate-limiting step, since it occurs by direct adsorption on the surface.

There are two important practical considerations in the successful implementation of the halide process. First, the gallium source must be fully saturated*

---

*The solubility of arsenic in gallium is 4% (atomic) at 850°C.

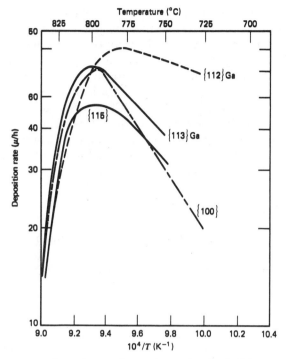

Fig. 5.23 Temperature dependence of growth rate for the halide process. From Shaw [79]. Reprinted with permission of the publisher, The Electrochemical Society, Inc.

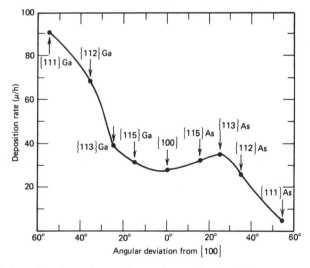

Fig. 5.24 Orientation dependence of growth rate for the halide process. From Shaw [79]. Reprinted with permission of the publisher, The Electrochemical Society, Inc.

with arsenic before consistent operation can be obtained. As a result, it is customary to saturate this source *before* the actual deposition run is made. Typically, this is done by passing the AsCl$_3$ vapor over the source for as much as 24 hours, until a crust of GaAs is formed on its surface. Next, the substrate should be *in situ* etched prior to epitaxial growth by this process, by raising its temperature above 800°C, so as to drive the reaction of Eq. (5.52) from left to right. An alternative approach is to inject HCl or AsCl$_3$ into the system beyond the source boat, but before the deposition zone. This technique is considerably more convenient, since temperature control of a hot wall furnace (with a large thermal mass) is avoided. Here, HCl is a potential source of contamination and AsCl$_3$ should be used if high-purity layers are desired.

5.5.4 The Hydride Process

This process differs from the last in that the fluxes of gallium and arsenic species are formed independently. This allows greater control in the vapor phase, and hence a wider control of the deposition parameters [81]. The flux of arsenic is formed by the decomposition of AsH$_3$, as follows:

$$AsH_3 \overset{400°C}{\rightleftharpoons} \tfrac{1}{4} As_4 + \tfrac{3}{2} H_2 \tag{5.53}$$

The flux of gallium species is formed by passing HCl gas over a heated gallium source, resulting in GaCl

$$Ga + HCl \rightleftharpoons GaCl + \tfrac{1}{2} H_2 \tag{5.54}$$

Finally, both these fluxes arrive at the substrate, where epitaxial growth occurs according to the following reaction:

$$GaCl + \tfrac{1}{4} As_4 + \tfrac{1}{2} H_2 \rightleftharpoons GaAs + HCl \tag{5.55}$$

This is the same reaction as that for the halide process—that is, Eq. (5.52). Thus both systems have many similarities, and require almost identical reactor conditions. Again, the reaction is exothermic, so that a hot wall reactor is used. Typically, the gallium source is held at 800–850°C, and the substrate is maintained at 650–725°C. HCl gas enters the system and passes over the gallium source, which must be presaturated with GaAs. The arsine gas is injected downstream from the gallium source, as shown in Fig. 5.25, and provides the arsenic flux. *In-situ* etching is done by increasing the HCl flow beyond the gallium source.

The growth characteristics of this system, as well as its orientation dependence, are very similar to those that are observed with the halide process. Again,

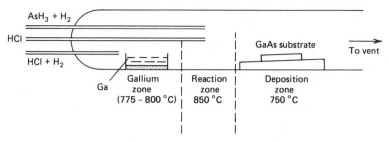

Fig. 5.25 Schematic for the hydride process.

the reaction is surface-catalyzed so that excellent-quality layers are obtained, in the growth range of 0.2–0.5 μm/min.

In the halide system, the primary process variable is the mole fraction of $AsCl_3$, which is varied by altering the carrier gas flow through the bubbler. As a result, the arsenic-to-chlorine ratio of this system is fixed at 0.33. In the hydride process, however, this ratio can also be varied over a wide range, giving an additional degree of freedom. Thus the system is considerably more flexible, and is the preferred approach for most situations where carrier concentrations in excess of 10^{14} cm^{-3} are acceptable. Recent work [82] with these systems, operating at low pressure (30 torr), has resulted in the growth of high-quality material, with a 77-K mobility value of 130,000 cm^2/V s. Growth rates as high as 5 μm/min have been achieved at a substrate temperature of 700°C.

5.5.5 The Organometallic Process

An important limitation of both halide and hydride processes is that they cannot be extended to the growth of AlGaAs by the simultaneous growth of GaAs and AlAs. This is due to the fact that the growth of AlAs occurs around 1100°C

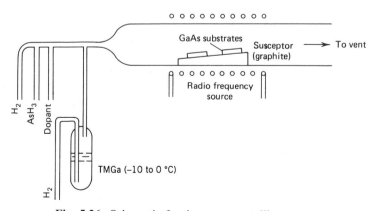

Fig. 5.26 Schematic for the organometallic process.

by these approaches, at which point the GaAs reaction is one of etching. This problem is avoided in the organometallic process [74, 83]. Here, vapor transport of gallium is achieved by use of TMGa, which is transported by a bubbler arrangement maintained in the $-10°C$ to $0°C$ range. TMAl is used to provide a volatile aluminum species, by the same method. Thus the flux of gallium and aluminum reactants can be delivered conveniently and independently, without the necessity of introducing a halogen (see Fig. 5.25). Epitaxial growth by compounds of this type is referred to as *organometallic vapor phase epitaxy* (OMVPE).

In the OMVPE of GaAs, the flux of arsenic is provided by arsine gas, with hydrogen as the carrier. The arsine partial pressure that is used is significantly higher than that of the TMGa, so that the growth rate is limited by transport of the alkyl species to the substrate surface. The overall reaction leading to the growth of GaAs is primarily

$$(CH_3)_3Ga + AsH_3 \rightarrow GaAs + 3CH_4 \qquad (5.56)$$

although ethane and higher hydrocarbons are also products of this reaction [84].

This reaction is pyrolytic in nature, so that cold wall reactors are used to avoid deposition on surfaces other than the substrate. These are similar to the type used for silicon epitaxy, as illustrated in Fig. 5.13. Here the temperature is usually monitored by means of a thermocouple embedded in the susceptor, and approximates the temperature of the substrate. In some instances, the susceptor is heated by infrared lamps placed outside the reaction chamber. Resistance-heated substrates can also be used; however, the heater element must be adequately shielded to prevent system contamination.

OMVPE systems are usually operated at pressures between 0.1 and 1.0 atm. Low pressure operation allows the elimination of circulation effects, and also the rapid removal of reactants; it is desirable when highly abrupt transitions are required.

HCl gas is often provided for reactor and susceptor cleaning, which is necessary from time to time. Typically, arrangements are made to temporarily heat the reactor walls to 800–900°C, and cleaning is accomplished by passing dilute HCl gas ($\simeq 1\%$ in H_2) through the reactor with the susceptor in place. This is followed by a bake-out phase, to remove all adsorbed chlorine from the susceptor surface. Problems with using HCl stem from the fact that it is highly corrosive, and can attack stainless steel plumbing and fittings. This necessitates the use of monel hardware for this part of the reactor plumbing, as well as the isolation of HCl lines from the rest of the system.

OMVPE growth is carried out in an excess of the arsenic species, with the gallium species serving the role of the rate limiter. In the mass transport region, the growth is linearly proportional to the flux of TMGa and is relatively independent of the substrate orientation or the partial pressure of the column V species. Moreover, the growth rate [85] is constant over a wide range of temperatures, as seen in Fig. 5.27. Here, we note that the growth is weakly depen-

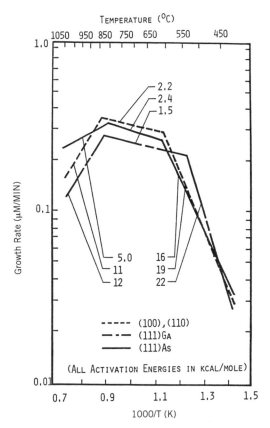

Fig. 5.27 Orientation dependence of growth rate in the organometallic process. From Reep and Ghandhi [85]. Reprinted with permission of the publisher, The Electrochemical Society, Inc.

dent on the deposition temperature between 600°C and 850°C. The apparent activation energy ranges from 1.5 to 2.4 kcal/mole, as is expected for diffusion-controlled, mass-transport-limited growth. Additionally, the growth rate is relatively insensitive to the substrate orientation in this temperature region.

Growth at low temperatures occurs in the kinetically controlled regime with activation energies from 16 to 22 kcal/mole, depending on the substrate orientation. Growth also falls off at high temperatures. A number of processes can contribute to this falloff, including pre-reactions in the gas phase, as well as desorption effects at the substrate. Here, crystal quality is generally poor, so that this regime has not been studied in detail.

The decomposition of TMGa involves dissociation energies of 59.5, 35.4, and 77.5 kcal/mole for the first, second, and third $GaCH_3$ bonds, respectively [86]. Moreover, the rate constant for the dissociation of the first $GaCH_3$ bond is considerably higher than that of the second bond, even though its activation energy is larger. Thus, the rate limiter for the thermal decomposition of TMGa is in all probability the breaking of the last $GaCH_3$ bond.

The decomposition of AsH_3 suggests [87] that the rate limiter is the removal of the first hydrogen atom with an activation energy of 23.2 kcal/mole, followed by chemisorption on the substrate as AsH_2, and with subsequent hydrogen release being more rapid. In a later study, AsH_3 was found to decompose over GaAs at temperatures below 900 K by a surface-catalyzed reaction [88], to yield AsH_2 and AsH.

Based on these observations, a probable growth model is that $GaCH_3$ and AsH are the primary reactant species which participate in the growth process. The reaction of these species, which results in the formation of GaAs, is

$$GaCH_3 + AsH \rightarrow (CH_3Ga\text{---}AsH) \rightarrow GaAs(s) + CH_4(g) \qquad (5.57)$$

with removal of the CH_3 by the presence of active hydrogen.

The incorporation of carbon in GaAs provides much evidence in support of this model [89]. The attachment of $GaCH_3$ to a gallium kink site favors the incorporation of CH_3 (and thus carbon) into an arsenic site, where it behaves as a shallow acceptor upon release of the hydrogen. The amount of carbon that is actually incorporated will depend on the residence time of the adsorbed $GaCH_3$; this in turn depends on the availability of the hydrogen radical from the AsH. Thus, increasing the AsH_3 overpressure results in a dramatic reduction of the incorporated carbon. Typically, values of the V/III ratio that are used for the growth of GaAs from TMGa and AsH_3 range from 16 to 200 for this reason. This corresponds to an arsine partial pressure which is many orders of magnitude larger than the amount required to prevent decomposition of the substrate at the growth temperature.

The growth of GaAs from either elemental arsenic, or an arsenic alkyl such as $(CH_3)_3As$, results in a large increase in the carbon incorporation. This emphasizes the need for the hydrogen radical, which is required for the methyl elimination proposed earlier. In its absence, the breaking of the Ga—CH_3 bond is delayed, with greatly increased carbon incorporation in the GaAs. Finally, the introduction of methane gas into the reactor does not increase the carbon incorporation in GaAs. This indicates that the carbon comes from the CH_3 component of the $GaCH_3$, and not from methane gas evolved in the reaction. Moreover, replacement of the H_2 carrier gas by another diluent such as N_2 or He does not affect the carbon incorporation, implying that the hydrogen required to complete the formation reaction must be in the form of the active radical (i.e., from AsH).

The pyrolysis of triethylgallium (TEGa) has been shown [84] to occur as

$$(C_2H_5)_3Ga \rightarrow GaH\text{---}(C_2H_5)_2 + C_2H_4 \qquad (5.58)$$

This process, referred to as β-hydride elimination, results in the formation of an adsorbed species GaH—$(C_2H_5)_2$, whose ethyl groups can be rapidly removed, without the need for AsH. No detectable carbon has been measured in GaAs grown by the TEGa-AsH$_3$ reaction, and TEGa is referred to as a "clean burner" for this reason.

Among the available organometallic sources of arsenic, TBAs is also a clean burner. This fact, combined with its relatively high vapor pressure, makes it an attractive replacement for arsine gas in OMVPE systems [31]. TBAs has the added advantage of more easy pyrolysis than AsH$_3$, allowing the growth of GaAs to be carried out at 50–100°C lower temperatures.

Some work has also been done [90, 91] with "mixed" systems, using dopants such as diethylgallium chloride, $(C_2H_5)_2$GaCl. Recently, it has been shown that highly selective epitaxy can be carried out using this alkyl [92].

5.5.6 Impurities and Doping

From the applications viewpoint, n-type GaAs is by far the most interesting, and there is a need for growing this material over a wide carrier concentration range. In contrast, p-type material is primarily required in heavily doped regions for p^+–n junction formation, and also for making ohmic contact to lightly doped p-layers. In both cases, however, the ideal approach for obtaining doped epitaxial layers is to grow otherwise-pure, undoped material, and to introduce controlled amounts of specific impurities during the growth process. These dopants must be available in highly purified form, and must be suitable for vapor transport.

The limits of purity of undoped material depends, first, on the degree of purification of the starting source chemicals. Here, zinc is a p-type contaminant in arsenic and gallium compounds, and is present in the starting ores from which As and Ga are refined. It is a shallow acceptor in GaAs. New purification techniques have reduced this contaminant to a minimum ($\approx$0.1 to 0.2 ppm) in recent years.

Silicon and germanium are residual impurities in VPE growth, and often result from contamination associated with the quartz or stainless steel vessels in which these sources are purified and stored [93]. Although amphoteric in behavior, silicon dopes VPE–grown GaAs n-type. This is because vapor–solid equilibrium determines the concentration of silicon incorporated in lattice vacancies. For VPE, silicon incorporation into gallium sites is $\approx 1.6 \times 10^3$ times that into arsenic sites.

Carbon is the most important contaminant of OMVPE-grown GaAs, since it is an inherent component of the metal alkyl that is used for epitaxy. It incorporates into arsenic sites in GaAs by the entrapment of the adsorbed CH$_3$ radical in the growing semiconductor, and behaves as a shallow acceptor. The use of AsH$_3$ has been shown to assist in its removal by conversion of the radical to a stable hydrocarbon. Another approach is to utilize alkyl sources such as TEGa, which are clean burners. Combinations of these approaches are often used to minimize the carbon incorporated in OMVPE-grown semiconductors. It is not surprising, therefore, that extremely high mobilities (210,000 cm^2/V s at 77 K) have been achieved in GaAs grown from TEGa and AsH$_3$ [94].

Impurity incorporation is also related to the specific nature of the contaminant. For example, a nonvolatile contaminant in a source may not even be transported to the reactor as a vapor. Additionally, a volatile, but stable, contaminant may not break down in the growth environment, and may leave the reactor in its vapor phase. This is fortuitously true for many of the organometallic compounds of copper and iron, which exhibit deep levels in GaAs. With chloride transport systems, however, impurities are usually in the form of volatile halides, which break down relatively easily in the reactor environment. As a result, lifetimes in OMVPE-grown GaAs are significantly higher than those obtained for chloride processes, or for MBE [5].

Since both *p*- and *n*-type impurities are present in starting materials for OMVPE, the residual impurity type and carrier concentration are a function of the V/III ratios which are used. A high V/III ratio minimizes carbon incorporation, resulting in the growth of *n*-type material due to residual silicon. On the other hand, carbon incorporation dominates in material grown with a low V/III ratio, resulting in the growth of *p*-type material.

In the chloride process, which is essentially free of carbon, as-grown GaAs is invariably *n*-type, because of silicon contamination. The reaction of chlorine with the silicon forms compounds which are relatively stable at growth temperatures, so that this impurity becomes unavailable for doping the layer. As a result, the doping concentration rapidly falls [77, 95] as the partial pressure of the chlorine-bearing species is increased.

Extrinsic doping is accomplished by the introduction of group II, IV, and VI impurities during epitaxial growth. Group II impurities incorporate on the Ga sublattice, resulting in *p*-type behavior. Conversely, group VI impurities are *n*-type and incorporate on the As sublattice. Group IV impurities are amphoteric in character, and can incorporate on either the Ga or As sublattice, where they behave as *n*- or *p*-type, respectively. In general, however, site selection is heavily favored towards only one of these locations, depending on the particular impurity. Thus silicon incorporates as an *n*-type impurity in VPE-grown GaAs, whereas carbon is invariably *p*-type.

Volatile dopant sources must be used, so that they can be transported into the reactor along with the species required for GaAs growth. During growth, these are adsorbed on the surface, and are incorporated into the semiconductor. The concentration of adsorbed species is proportional to its flux, as well as to the adsorption and desorption rates, which are widely different for different dopant species. As a result, doping is usually a function of the growth temperature. For low-vapor-pressure dopants (Be, Si, and Sn), the kinetics of breakdown is the determining factor, so that the doping efficiency increases with temperature. With high-vapor-pressure dopants (Cd, S, Se, Te, and Zn), however, desorption processes dominate and the doping efficiency falls with temperature. At one time or another, almost all of the impurities mentioned above have been used for extrinsic doping. However, the choice has narrowed to those dopants which are readily available in highly pure form, and can be used conveniently. Uniformity of doping over a substrate is also an important factor in determining the choice of dopant for any practical situation.

Zinc is the most commonly used p-type dopant, and incorporates by a diffusion process. Zinc sources are the alkyls dimethylzinc and diethylzinc, both of which are high-vapor-pressure liquids. Often, they are used in gaseous form, diluted to a few percent in hydrogen, so that they can be metered accurately by means of mass flow controllers. Doping to 10^{20} cm^{-3} can be readily achieved with these sources. Zinc has a high vapor pressure, so that its doping behavior is dominated by surface desorption effects. As a result, its incorporation efficiency falls with growth temperature.

A serious practical problem with zinc sources is that they tend to coat the lines through which they pass. This can lead to memory effects, and create problems in situations where a lightly doped n-type layer must be grown after a p-type layer. In addition, zinc is a rapid diffuser in GaAs [96], because of its interstitial–substitutional behavior. Abrupt changes in p-type doping, of the type required in heterojunction bipolar transistors (HBTs), are impossible to achieve with this dopant. This has resulted in a search for other dopants which are slower-moving. Of these, carbon is the most satisfactory.

Heavy carbon doping, in the 10^{20}-cm^{-3} range, can be obtained by using trimethylarsenic (TMAs) instead of arsine during the growth of GaAs and AlGaAs by OMVPE. Since the use of AsH$_3$ suppresses carbon incorporation, control of this impurity can be achieved by a mixture of TMAs and AsH$_3$ during growth. An alternative approach is to use carbon tetrachloride as a dopant source [97]. Carbon is an extremely slow diffuser in GaAs [98], with a diffusivity of under 10^{-16} cm^2 s^{-1} at 920°C, so that it is ideal for applications where abrupt doping profiles are required.

The n-type dopants belonging to group VI are sulfur, tellurium, and selenium. Hydrogen sulfide gas, diluted in hydrogen, is a suitable source for sulfur, and can be used for controlled doping to 3×10^{18} cm^{-3} in both halide transport and OMVPE systems. Desorption processes dominate its incorporation into GaAs, so that its doping efficiency falls with growth temperature. It has no memory effects.

Organometallic sources for tellurium and selenium are the alkyls diethyltelluride and diethylselenide, either directly as liquids or in diluted form in hydrogen gas. H$_2$S and H$_2$Se, diluted in hydrogen, are also used as dopant sources in both chloride and OMVPE processes. Relatively high doping levels, up to 7×10^{18} cm^{-3}, can be achieved with these sources.

By far the most commonly used n-type dopant for OMVPE-grown GaAs is silicon [99]. This dopant has a high solid solubility in GaAs, and free carrier concentrations approaching 10^{19} cm^{-3} have been achieved, using SiH$_4$ as the dopant source [100]. It is also completely free from memory effects, so that it is favored for use in the growth of multilayer structures.

Silicon is incorporated into GaAs by trapping, and has a low vapor pressure. It is an amphoteric impurity in GaAs; nevertheless, it is almost entirely incorporated as an *n*-type species on the Ga sublattice. The electron concentration of *n*-GaAs doped with SiH_4 increases with temperature, because of an increase in the cracking efficiency of this dopant. Recently, it has been shown that disilane, Si_2H_6, which is more readily pyrolysed at growth temperatures, does not exhibit this temperature variation [101].

Semi-insulating GaAs can be made by doping with oxygen, iron, or chromium. Of these, oxygen-doped GaAs is very sensitive to thermal processing and is of little technological importance. Iron doping, though relatively stable, results in *p*-type layers with a maximum resistivity in the 10^4- to $10^5 \Omega$-cm range. Chromium, on the other hand, can be used to produce epitaxial GaAs in the 10^8-Ω-cm range, and is also relatively stable.

Chromium doping can be accomplished by the addition of chromyl chloride (CRO_2Cl_2) during VPE by the halide process [102]. Vapors of this material are transported by means of a bubbler arrangement into the reactor to form chromium, which dopes the growing GaAs layer. The by-product of this reaction is water vapor, so it is possible that material doped in this way also has some incorporated oxygen. Nevertheless, its properties are relatively unchanged by thermal treatment, so that this oxygen is inactive (possibly in the form of Cr–O complexes), if at all present.

Chromium carbonyl, $Cr(CO)_6$, has been used in the growth of GaAs by OMVPE [103]. Here, this solid dopant source necessitates the use of a heated bubbler as well as heated lines for delivery to the reactor. The detailed nature of the incorporation reactions have not been studied at the present time.

5.5.7 Selective Epitaxy

Epitaxy in selected regions of a GaAs substrate can be performed by depositing through a mask etched in its surface. An alternative approach is to epitaxially grow GaAs in selected regions *without* the aid of a mask. Either approach, if successful, can give the designer increased versatility by allowing the use of nonplanar structures.

Growth by OMVPE, in the 600–700°C range, is relatively independent of crystal orientation (see Fig. 5.27). Moreover, growth is carried out under conditions of high supersaturation, with deposit of material on all hot surfaces. As a result, growth on a masked surface, followed by removal of the mask and its (polycrystalline) overgrowth, will result in selective epitaxial regions. The quality of this epitaxial material is generally quite poor, especially due to pile-up at the edges of the window in the mask.

Selectivity in the organometallic process can be achieved by operation at low pressures, and under conditions of low supersaturation [104]. Since the diffusivity is inversely proportional to pressure, the system now operates under kinetic control, with rates which are orientation-dependent, as seen in Fig. 5.27. Using this approach, epitaxial GaAs can be formed in windows cut in the surface of an SiO_2 covered substrate [105]. Here, although epitaxy was confined to the exposed surface, the quantity of material which grew in these windows was found to be almost equal to the amount that would have grown over the entire substrate, in the absence of SiO_2. The most probable reason for this is that gallium, formed on the SiO_2 surface, diffuses faster to the GaAs than it reacts with arsenic. Consequently, all growth is confined to the semiconductor surface, and the growth rate is a function of the ratio of masked to unmasked regions.

OMVPE growth of GaAs and AlGaAs can also be carried out so as to completely cover over a masked pattern on a GaAs substrate. This allows the formation of a coherent single-crystal layer which can be subsequently removed from its parent substrate by cleaving, and used in thin film form [106].

The chloride process exhibits distinctly different characteristics because of its reversible character [107]. It is highly sensitive to substrate orientation because of the presence of the volatile GaCl species, and it results in growth rates which are widely different for different crystal directions, as seen in Fig. 5.24.

Growth on a (100) surface, with a window whose sides are along the $\langle 011 \rangle$ directions, results in the formation of an epilayer which takes the form of a truncated pyramid [108]. The sides of this pyramid are slow-growing (111)As faces. Epitaxial growth thus occurs without contact with the mask edges, and is thus of excellent quality.

Selective epitaxial growth characteristics [109] are also observed in OMVPE growth using the alkyl diethylgallium chloride (DEGaCl). This alkyl breaks down by the β-elimination reaction at about 350°C, as follows:

$$(C_2H_5)_2GaCl \rightarrow GaCl + 2C_2H_4 + H_2 \qquad (5.59)$$

resulting in the formation of GaCl on the substrate. Thus, the growth proceeds in a fashion analogous to that obtained for the chloride processes. The use of diethylaluminum chloride (DEAlCl) results in the formation of AlCl, so that selective epitaxy of AlGaAs can also be achieved in this manner [92].

The halide process has also been used for selective epitaxy without masking the GaAs surface. Here V-grooves, etched in (100) substrates, were formed by delineating a pattern in the [110] direction, and etching to expose the low-growth (111)As faces. During epitaxy, the groove fills from the bottom until eventually the material reaches the (100) surface by the formation of successively shallow grooves, corresponding to planes of progressively slower growth [110].

5.5.8 Growth Imperfections

Presently available GaAs substrates are poorer in crystal quality than starting silicon, so that the quality of epitaxial layers can be expected to be correspondingly worse. Furthermore, GaAs is made up of two components with very different vapor pressures, so that defects are often nonstoichiometric in character. One such defect, the *hillock*, is often seen in the form of a well-defined fault surrounded by a streamline pattern of material. Hillocks have been observed in

GaAs grown by all the processes described earlier. They appear to originate at the start of layer growth, as evidenced by a uniformity of size and shape, much like stacking faults on silicon substrates. However, their shape is not as distinct, because of partial dissolution due to excess gallium. Typically, they are very nonuniform in composition, with regions having a gallium excess of as much as 35%.

Hillocks are usually formed when the rate of incident atoms from the gas phase is higher than the rate at which atoms are incorporated into the growing layer. They are usually controlled by lowering the growth rate—in particular by reducing the flux of the gallium reactant species.

Another common type of defect is the *growth pyramid*. This is due to twinning and is often initiated at the substrate surface, at the beginning of epitaxial growth. Pyramids also tend to have ill-defined sides, because of partial dissolution due to excess gallium. An *in situ* etch, prior to growth, greatly reduces their incidence.

Whisker formation is also encountered if an excessive amount of the gallium species is present. These whiskers are usually observed when growth is attempted at very low temperatures.

5.6 LIQUID-PHASE EPITAXY

Liquid-phase epitaxy (LPE) involves the growth of epitaxial layers on crystalline substrates by direct precipitation from the liquid phase [4]. Although many applications of this technique have been described in the literature, the widest use of LPE has been in the growth of compound semiconductors such as GaAs from solutions of gallium metal. The reason for its popularity is seen with reference to the phase diagram of Fig. 5.28. Here the conventional approach to growth of a crystal of composition B (or rather β) is direct freezing from a melt of this material, which is held at T_B. An alternative approach, often referred to as *solution growth*, is to begin with a melt of composition C_0 (i.e., a solution of B and A), and to cool it from an initial temperature of T_0, which is well below T_B. This considerable reduction in temperature is an important advantage in the growth of GaAs.

Growth rates by this technique are extremely slow compared to direct growth from a GaAs melt, so that the latter approach is usually preferred for bulk crystal growth. On the other hand, solution growth is ideally suited for thin epitaxial layers on single-crystal substrates, where a slow growth rate can be a distinct advantage. Epitaxial growth can be readily preceded by an *in situ* dissolution (or etch) step, so that the substrate–layer interface can be kept free of damage. This is an important requirement for the growth of high-quality layers. Additionally, the distribution coefficients of most deep-lying impurities are low compared to those for shallow impurities, as seen from Table 3.4. Consequently, epitaxial layers grown by LPE exhibit exceptionally long minority carrier lifetime. They are thus the preferred choice for optical devices at the present time.

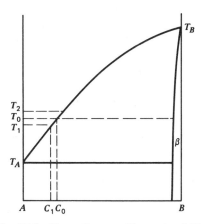

Fig. 5.28 Phase diagram illustrating LPE.

The disadvantages of LPE are primarily related to the difficulty of obtaining layers of uniform thickness, the difficulty of avoiding growth instabilities, and the necessity of wiping off the solvent after the layer is grown. Gallium inclusions, ridged and wavy surfaces, and incomplete surface coverage are typical problems encountered with this process. In addition, the technique is not suited for the growth of submicron layers with abrupt interfaces, of the type required for high-speed field effect transistors and integrated circuits.

The LPE growth of AlGaAs has the same problems as for GaAs, in addition to the added problem of composition nonuniformity. On the positive side, however, it is relatively more easy to obtain high photoluminescent efficiencies with LPE than with MBE or VPE, since growth takes place in a (primarily) liquid gallium medium which has excellent impurity-gettering properties.

LPE is not commonly used in silicon technology, since this semiconductor will tolerate high-temperature processing with relative ease. However, some device applications, which require the low-temperature growth of silicon layers with minimal autodoping, have brought about [111] an interest in this technique.

5.6.1 Choice of Solvent

It is necessary that the material to be grown be capable of dissolving in a solvent, and that this solution melt at a temperature that is well below the melting point of the semiconductor. A number of binary and pseudobinary* systems can be used for this purpose. The most suitable ones are those of the eutectic type which are free from compound formation.

Epitaxial layers, formed by freezing from the solution, will be saturated by the solvent. This is indicated as the β solution in Fig. 5.28. Although the com-

*Systems such as A–GaAs fall into this category.

positional range associated with β is very small, this requires that the solid solubility of the solvent in the grown crystal be low, or that the solvent be an electrically inactive species. Examples of systems with active solvents are Ga–Si, Al–Si, Ge–GaAs (all p-type), and Sn–GaAs (n-type). Inactive systems for silicon* include Sn–Si. Inactive systems for GaAs include Pb–GaAs and Ga–GaAs. Tin is the preferred inactive solvent for silicon, whereas gallium is almost universally used with GaAs.

The Ga–GaAs system can be considered as one in which gallium is the solvent and GaAs is the solute. An alternative viewpoint, that gallium is the solvent and arsenic is the solute, is more commonly used for describing the LPE of GaAs. For any given temperature, there is a specific amount of arsenic that can saturate gallium, to result in an equilibrium solution at that temperature. Typically, this is on the order of a few mole %. Figure 5.29 shows the amount of gallium arsenide that must be added to 100 g of gallium to provide the necessary arsenic for saturation at any given temperature [8]. Also shown in this figure are comparable data for the Sn–GaAs system.

*Systems such as In–Si, Au–Si, and Ag–Si are often considered inactive, even though the resulting silicon is doped to its solid-solubility limit with these deep impurities. This is because the concentration of these dopants at growth temperatures ($\simeq 700°$C) is extremely low.

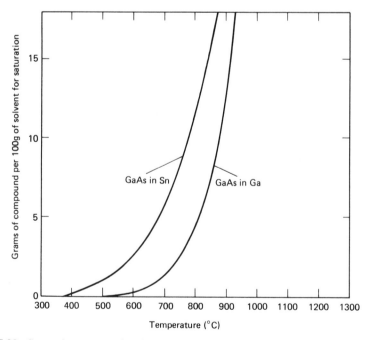

Fig. 5.29 Saturation curves for the Ga–GaAs and Sn–GaAs systems. Adapted from Pamplin [8].

The LPE growth of AlGaAs can be accomplished readily by the addition of aluminum to the melt. The composition of the resulting ternary alloy is related to that of the melt with which it is in equilibrium, and can be calculated on the basis of thermodynamic principles. Figure 5.30 shows the results of such a computation, as well as experimental data for a range of alloy compositions and growth temperatures [112].

5.6.2 Nucleation

In LPE, the substrate is held in contact with the solution, and conditions are established so that this solution becomes supersaturated by the solvent. Once a critical supersaturation is achieved, the solute will freely nucleate, resulting in nuclei formation in the solution and on the crucible walls, in addition to the substrate. Thus it is desirable to maintain a thermal gradient in the system, so that the solute can nucleate in the region where its solubility is the lowest (that is, the coldest region). Ideally, this is the immersed substrate, and growth proceeds by heterogeneous nucleation on it.

The most common technique for establishing conditions for epitaxy is to place the substrate in equilibrium with a saturated solution, and then to cool the system so as to achieve a supersaturated condition. The initial stages of nucleation are similar to those described for VPE. Using the same line of argument, embryos which achieve a critical radius will lower their energy by growing larger, until eventually the nuclei coalesce to form an atomic layer. The number

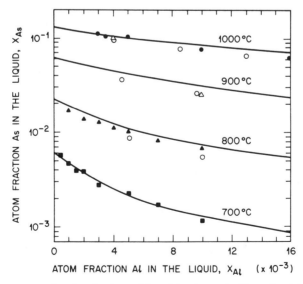

Fig. 5.30 Aluminum fraction in the solid and in the melt. From Panish and Hayashi [112].

of critical nuclei is given by Eq. (5.7) as

$$n_{crit} = n_0\, e^{-\Delta G_{crit}/kT} \tag{5.7}$$

where ΔG_{crit} is the free energy change required for heterogeneous nucleation. With VPE, the addition of atoms to the embryo is relatively rapid, since they come directly from the vapor phase. With LPE, on the other hand, the process is limited by diffusion through the solute. For GaAs, grown in an arsenic-saturated gallium solution, the diffusivity of arsenic through gallium is the important controlling factor [113]. Thus the rate of addition of atoms into the nucleus is reduced by $e^{-E_d/kT}$, where E_d is the activation energy for diffusion. As a consequence, the rate is given by

$$R_n = K e^{-(\Delta G_{crit} - E_d)/kT} \tag{5.60}$$

where K is a proportionality factor.

This behavior differs from that of VPE in two respects. First, since nucleation occurs close to the melting point, $\Delta G_v \rightarrow 0$ and the nucleation rate is much slower. Next, the limiting effects of diffusion through the solvent further reduce this nucleation rate. As a result, it is considerably easier to obtain control of nucleation in LPE than in VPE.

5.6.3 Growth

The growth of a layer by LPE is analogous to the process of doping from a melt, described in Section 3.4. Consider here the situation of growth of B (β) from a solution of A and B. The phase diagram of Fig. 5.28 illustrates this situation. A melt with an initial composition C_0 is used. Assume that a substrate is inserted into the melt at temperature T_0 and is brought in equilibrium with it. The temperature of the system is now lowered to T_1 so that a solid (β) freezes out of the solution.* At the same time, the composition of the solution shifts to C_1. With specific reference to GaAs, where Fig. 5.28 represents the Ga–GaAs phase diagram (A = Ga and B = GaAs), the process of cooling the arsenic-saturated solution of gallium causes the arsenic to precipitate in the form of GaAs. At the same time, the solution becomes more gallium-rich, shifting its composition from C_0 and C_1. Finally, since GaAs exists as a compound in a very narrow stoichiometric range, the β solution is essentially this semiconductor.

Growth proceeds by the transport of material from the bulk of the solution to the edge of the boundary layer through which it must diffuse before it can be adsorbed on the surface. Here it moves until it ultimately attaches to a kink site on the surface. Diffusion through the boundary layer is generally the limiting factor under most conditions of LPE growth.

---

*Thus the entire operation takes place in a relatively small region of the phase diagram.

Consider a GaAs system, in order to illustrate this process. Here a solution of gallium with a few atomic percent arsenic (the solute) is used, so that growth proceeds by the diffusion of arsenic in gallium. In the 700–950°C range, the diffusivity for this process [114] is given by

$$D = 8.6 \times 10^{-4} e^{-3240/kT} \text{ cm}^2 \text{ s}^{-1} \tag{5.61}$$

Let C_s be the concentration of the solute in the bulk solution, and C_e the equilibrium solute concentration at the surface of the substrate. Assuming a linear gradient through the boundary layer, of thickness δ, the flux of the solute is given by

$$j = \frac{D(C_s - C_e)}{\delta} \tag{5.62}$$

If ρ is the density of the grown layer, the growth rate is given by

$$R = \frac{DC_e\sigma}{\rho\delta} \tag{5.63}$$

where the supersaturation σ is given by $(C_s - C_e)/C_e$. The actual growth strategy will determine the appropriate values of C_e and σ, since these are a function of the temperature of the solution.

One basic approach for growing an LPE layer of GaAs is to begin with a solution of arsenic in gallium, which is saturated at a temperature T_0. The temperature is now lowered by a few degrees to T_1 so that the melt becomes supersaturated. Typically, $\Delta T \simeq 5$–20°C, although growth can be achieved with a much lower temperature differential. The sample is now inserted into the melt, where it is held at T_1. Initially, growth occurs because of the supersaturation in the melt. During growth, however, the solution becomes depleted of arsenic. Thus the supersaturation, as well as the growth rate, fall with increasing time.

This technique, known as the *step-cooling method*, is well suited for the growth of layers with controlled thickness, as well as for extremely thin layers. Growth rates are generally quite low ($\simeq 0.01 \ \mu\text{m}/$ min), and layers grown in this manner tend to be extremely smooth.

A second technique is known as *equilibrium cooling*. Here the solution is initially saturated with arsenic at T_0, and has a composition C_0 which is in equilibrium with it. A substrate, also at T_0, is inserted into the solution, and the temperature of the system is lowered slowly. This drives the solution into a supersaturated state, from which it proceeds toward equilibrium by the loss of arsenic—that is, by the growth of GaAs. Growth will thus continue until the substrate is removed from the solution.

A detailed analysis of these processes must take into consideration [113] the gradients in composition, temperature, and density of the gallium solvent which arise during growth. In addition, the outflow of latent heat from the surface can affect the thermal conditions in the system. These and other factors have been considered elsewhere [115, 116]. Here an approximate analysis is undertaken [117, 118], after the following assumptions have been made:

1. The solution and the substrate are isothermal.
2. The liquid and the substrate are in equilibrium at the surface, so that the solute (arsenic) concentration is given by the liquidus curve.
3. The bulk concentration and the diffusivity constant do not change during the growth.

These approximations are reasonable, because of (a) the large volume of the chamber and (b) the small temperature changes that are involved during LPE.

Figure 5.31 shows the concentration of the solute in the bulk (C_0) and at the liquid-substrate interface (C_1). The distribution coefficient of arsenic in gallium is less than unity. Consequently, the growing layer rejects this component, so that $C_1 > C_0$, as shown here.

Diffusion of arsenic takes place through the boundary layer and results in growth at the surface. The magnitude of $C(0, t)$ depends on the growth process used. The one-dimensional diffusion equation for this system is

$$\frac{\partial C}{\partial t} = D\frac{\partial^2 C}{\partial x^2} + v\frac{\partial C}{\partial x} \tag{5.64}$$

where v is the velocity of the moving interface. A reasonable approximation in a practical growth situation is to neglect the last term, so that

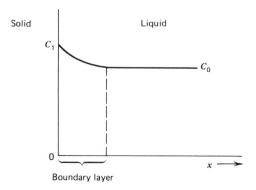

Fig. 5.31 Concentration of solute during LPE.

$$\frac{\partial C}{\partial t} \simeq D \frac{\partial^2 C}{\partial x^2} \qquad (5.65)$$

The amount of solute that is transported through the boundary layer, and deposits on the substrate, is given by M_t per unit area, where

$$M_t = \int_0^t D \left(\frac{\partial C}{\partial x} \right)_{x=0} dt \qquad (5.66)$$

Finally, the layer thickness d is given by

$$d = M_t / C_s \qquad (5.67)$$

where C_s is the solute concentration in the grown layer (=0.5 for GaAs). Application to specific growth processes now follow.

Step Cooling. Here the boundary conditions are

$$C(x, 0) = C_0 \qquad (5.68a)$$
$$C(0, t) = C_1 \qquad (5.68b)$$

where C_0 and C_1 are shown in Fig. 5.31. For this situation, the amount of supercooling is given by $T_0 - T_1$. The change in temperature is related to the change in composition by the slope of the liquidus curve, $m = dT/dC$, around $T = T_0$.

The diffusion equation can be solved [119] subject to the boundary conditions of Eqs. (5.68a) and (5.68b), so that

$$\frac{C - C_1}{C_0 - C_1} = \mathrm{erf} \left[\frac{x}{2(Dt)^{1/2}} \right] \qquad (5.69)$$

Substituting in Eq. (5.66), gives

$$M_t = 2(C_0 - C_1) \left(\frac{Dt}{\pi} \right)^{1/2} \qquad (5.70)$$

so that the layer thickness is

$$d = 2 \frac{T_0 - T_1}{C_s m} \left(\frac{D}{\pi} \right)^{1/2} t^{1/2} \qquad (5.71)$$

Thus the layer thickness is proportional to $t^{1/2}$. Moreover, the growth rate of the layer falls with increasing time, and is proportional to $t^{-1/2}$.

Equilibrium Cooling. Consider a cooling rate α, such that

$$C(0, t) = C_0 - \left(\frac{\alpha}{m} \right) t \tag{5.72a}$$

As before,

$$C(x, 0) = C_0 \tag{5.72b}$$

Solution of the diffusion equation [119], subject to these boundary conditions, gives the magnitude of solute concentration as

$$C(x, t) = C_0 - 4 \left(\frac{\alpha t}{m} \right) i^2 \operatorname{erfc} \left[\frac{x}{2(Dt)^{1/2}} \right] \tag{5.73}$$

where

$$i^2 \operatorname{erfc} x \equiv \int_x^\infty \int_y^\infty \operatorname{erfc} \xi \, d\xi \, dy \tag{5.74}$$

Combining with Eqs. (5.66) and (5.67) gives the layer thickness as

$$d = \frac{4}{3} \left(\frac{\alpha}{C_s m} \right) \left(\frac{d}{\pi} \right)^{1/2} t^{3/2} \tag{5.75}$$

so that the growth rate increases with time, and is proportional to $t^{1/2}$.

A number of different programming sequences have been employed, in an attempt to obtain improved layer morphology. Analysis of the growth rate for these sequences is conveniently done by computer simulation.

5.6.3.1 *Constitutional Supercooling*

The role of constitutional supercooling during crystal growth from a doped melt was described in Section 3.4.4. Here it was noted that, for impurities with $k < 1$, the freezing solid rejects the dopant, resulting in its pileup at the solid–liquid interface. Too great a pileup, as might result from rapid growth, can lead to the condition of constitutional supercooling, accompanied by spurious nucleation and polycrystalline growth. Growth could be stabilized, however, if the temperature gradient in the melt (i.e., the temperature increase with distance from the interface) was sufficiently large (as shown by *XB* in Fig. 3.12*b*) and positive.

A somewhat analogous situation occurs during LPE, where we are concerned with a pileup of arsenic (for the case of GaAs growth) at the solid–liquid interface. Here, too, stable steady-state growth requires a sufficient positive temperature gradient to be maintained. Unlike crystal growth from the melt, however, the temperature gradient in an LPE system is often negative; that is, the solution temperature can be lower than that of the substrate. As a result, the avoidance of constitutional supercooling in LPE is an extremely difficult task. In fact, at first glance it would appear to be impossible. This is not so, however, because of the transient nature of LPE growth.

Consider a GaAs substrate that is immersed in a solution with a finite solute concentration. Initially, the concentration gradient is zero. As the melt temperature is lowered, however, growth of the layer is accompanied by a depletion of the arsenic which gives rise to a concentration gradient. Solutions of this transient process show that the required temperature gradient must be continually increased during growth in order for stable conditions to be maintained [120]. As the melt cools toward room temperature, however, temperature gradients in the system tend to zero. This later stage of growth is generally accompanied by surface irregularities and gallium inclusions, which are characteristic of unstable conditions. In practice, therefore, it is customary to avoid this situation by growing the layer during a relatively small temperature change and to terminate growth by removal from the melt, accompanied by wiping.

5.6.4 *In Situ* Etching

The incorporation of an *in situ* etch step, just prior to epitaxy, is extremely important for the growth of high-quality layers. This can be readily incorporated into LPE systems, and will be described here as an extension of the equilibrium growth method. Referring again to Fig. 5.28, the solution is first saturated to T_0, and its temperature is *raised* by a small amount to T_2, so that it is now slightly *undersaturated*. About 10°C undersaturation is used in practice. The substrate, also at T_2, is inserted into the solution, and the temperature is lowered to T_1 at a controlled rate. Etching takes place as the temperature falls from T_2 to T_0, with growth occurring at temperatures from T_0 to T_1. The etch rate for this situation falls off as T_0 is approached, so that the amount that is removed cannot be precisely controlled. However, this is usually not an important process parameter for most devices.

An additional advantage of the *in situ* etch in LPE comes about because of the gettering action of the gallium solution, which tends to remove deep-lying impurities from the substrate just prior to layer growth. This is of prime importance in many electro-optical devices, where long minority carrier lifetimes are required.

5.6.5 Doping

Doping in LPE is commonly done by saturating the gallium solution with doped GaAs. Dopants which have been used successfully are those belonging to group II (cadmium and zinc) which are p-type, and to group VI (sulfur, selenium, and tellurium) which are n-type. These dopants generally have a distribution coefficient which is below unity (see Table 3.4), so that the layer becomes more heavily doped as growth proceeds.

Elements from group IV can also be employed for doping purposes. These include tin, germanium, and silicon. They can occupy gallium sites where they behave as donors, as well as arsenic sites where they are acceptor-like. However, the relative attachment to gallium and arsenic sites at the surface is strongly dependent on the nature of impurity–point-defect interactions. Unlike VPE however, the incorporation ratio of silicon into gallium and arsenic sites is close to unity. Growth at low temperatures (600–700°C) in a silicon-doped solution results in p-type material, whereas growth in the 900°C range produces n-type layers. The actual crossover point is a function of temperature, silicon concentration, and crystal orientation. Advantage has been taken of this ampho-teric doping behavior for the growth of successive n- and p-type layers from a single solution.

Amphoteric behavior has not been observed for other group IV impurities, such as tin or germanium. Tin-doped layers grown by LPE are invariably n-type, whereas germanium-doped layers are p-type. Interestingly, germanium-doped layers grown by VPE are always n-type. The reasons for these differences are not known; in fact this behavior only serves to emphasize the complex nature of impurity–defect interactions at the growing surface.

Dopants for AlGaAs are similar to those used for GaAs. As with GaAs, silicon can be used to grow both p- and n-type layers, depending on the growth temperature and on the composition. Here, too, n-type layers are grown in the 900°C range, with p-type at lower temperatures.

5.7 LPE SYSTEMS

The basic LPE process consists of placing the substrate in a solution, program-ming the system's temperature–time characteristics, removing the substrate, and wiping off the excess solvent. A great variety of systems have been used for carrying out these operations; these include schemes involving tilting furnaces, rotating substrates, and so on. In addition, some systems involve stirring of the solution while others do not.

Early systems utilized a furnace tube which was tilted so that the substrate could be inserted into the solution (and withdrawn from it) by the action of gravity. These were difficult to operate and control, and are rarely used today. Systems involving substrate rotation for this purpose are also occasionally encountered. They reduce some of the control problems of the tilting systems, but are relatively complex and prone to malfunction.

Both vertical dipping systems and horizontal sliding systems are in common use today. Most modern systems are of the horizontal slide type because of their greater flexibility and convenience. However, dipping systems are especially suited for the rapid growth of thick layers.

This section describes the important features of LPE systems. Next, both dipping and sliding systems are considered, together with their operational advantages and disadvantages.

5.7.1 Hydrodynamic Considerations

The main problems of LPE systems are (a) nonuniformity of layer thickness and (b) surface perturbations in the form of facets, ridges, and scallops. Some of these nonuniformities are inherent in the growth process. However, the main cause of growth irregularities is uncontrolled fluid flow in these systems [121]. Thus stirring, and systems which stimulate stirring during growth, are not generally favored because they produce unsteady streamlines across the surface. These encourage perturbations and instabilities, some of which can grow as the layer is formed.

Consider an LPE system, using a gallium–arsenic solution in a heated crucible. Here the growth of GaAs is to be accomplished by dipping a substrate vertically into this solution. This growth processes causes the formation of a vertical boundary layer which is depleted of arsenic. This layer is more dense* than the surrounding solution, so that it sinks to the bottom. This alters the solutal gradient across the face of the layer as the falling solution becomes increasingly depleted of arsenic, resulting in a wedge-shaped epitaxial growth, which is thinner at the bottom than at the top. In addition, circulating convection currents are created by this process, adding to those already present in the crucible due to the external heat source, and lead to further growth nonuniformities.

Convection-induced problems of this type can be avoided in systems where the substrate is horizontal. Hydrodynamic considerations are important here, also. Thus, the substrate can be placed *above* the melt as shown in Fig. 5.32 *a*, or *below* it as in Fig. 5.32 *b*. Consider first the case of Fig. 5.32 *a* for the gallium–arsenic system. Here, as growth proceeds, the boundary layer below the substrate becomes depleted of arsenic and thus more dense. This tends to sink to the bottom, setting up a fluid flow which augments the effects of natural convection. In the system of Fig. 5.32 *b*, on the other hand, the dense boundary layer is formed above the substrate and tends to remain there. Hence it acts to stabilize the growth process.

The LPE growth of AlGaAs is complicated by the fact that the segregation coefficients of aluminum and gallium are quite dissimilar, so that provision

---

*The density of arsenic is 5.727, whereas that of gallium is 6.095.

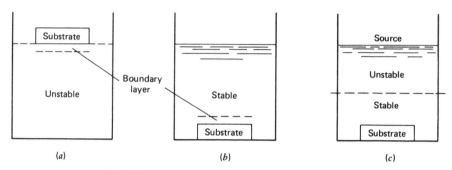

Fig. 5.32 Different horizontal growth situations.

must be made to supply extra aluminum during layer growth. Source wafers of AlGaAs, located as shown in Fig. 5.32c, are often used for this purpose. Here, following the above arguments, the region below the source is unstable, whereas that above the substrate is stable. One technique for stabilizing these systems is to apply a temperature gradient so that the source is at a higher temperature than the substrate. This procedure is referred to as *temperature-gradient zone melting* and has been the subject of much study [122].

5.7.2 Vertical Systems

These consist of a crucible, usually of high-purity graphite or alumina, mounted in a vertical furnace. The system is enclosed in a quartz chamber with provision for flushing with hydrogen gas. The substrate is held in a quartz or graphite holder attached to the end of a rod, which passes through the top cap. This rod is capable of both rotation and sliding while maintaining hermetic integrity. In operation, the substrate is lowered into the solution and its temperature adjusted in a controlled manner. Growth is terminated by raising the substrate out of the solution.

Convection currents play an important role in these systems, and lead to tapered growth across the substrate. The action of these currents greatly reduces the boundary layer thickness, with a subsequent increase in the growth rate. This feature makes them suited for the growth of thick layers, where a slight taper is acceptable. A variety of sophisticated features have been incorporated in these systems in order to extend the versatility of this technique. These include provisions for source wafers, multiple substrates, and substrate wiping. Elaborate heat shields are also used to minimize thermal gradients, and considerable care has been expended on process optimization.

Vertical dipping systems have also been used for the growth of silicon films which must be relatively thick, and have minimal autodoping [111]. Here tin is the usual solvent, although electronically active materials such as gallium and aluminum have also been used, to result in heavily doped p^+-layers. Both doping and saturation of the melt with silicon can be achieved by means of doped silicon wafers.

5.7.3 Horizontal Systems

By far the most popular LPE systems use horizontal substrates and are of the sliding boat type. These systems are relatively free of morphology problems, which are caused by uncontrolled convection currents. Their growth rate is much lower than that of dipping systems, so they are better suited for the fabrication of thin layers of the type required in microwave and electro-optical applications. A variety of such systems are in use; here we describe two schemes in order to outline their basic operational principles. Systems for GaAs are considered for specificity.

Figure 5.33 shows the simplest apparatus for this purpose [4]. Here a bin is machined in a high-purity graphite block, and serves to hold the solution. The temperature is monitored by a thermocouple (not shown) embedded in the block. A graphite slide is used as a substrate holder, in an arrangement by which it can be moved so as to be located under the well. The entire system is placed in a furnace, in a neutral carrier gas ambient. Hydrogen gas, passed through a palladium purifier, is in general use because of its availability in highly pure form.

In operation, the system is brought up to temperature with the substrate covered by part of the graphite block. At this point, a push rod is used to position the substrate under the bin, and the furnace temperature is changed in a programmed manner. A linear reduction of temperature with time is commonly used, with temperature gradients from 1°C/min to 1°C/hr. The slide is moved out from under the solution in order to terminate the growth. This also provides a wiping action to remove the solvent, so that often no further substrate cleanup is necessary. The system is now cooled, and the substrate removed. Expansion of the system to multiple slices is accomplished by providing a series of bins in a common graphite block, together with a slide having a corresponding series of indentations for holding the substrates.

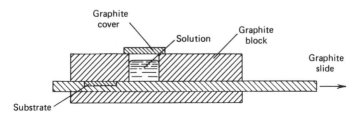

Fig. 5.33 Sliding boat apparatus for LPE.

There are many modifications of this basic system. Thus, Fig. 5.34 shows an arrangement by which the solution can be saturated with arsenic prior to epitaxial growth. This is done by first sliding a "source" wafer of GaAs under the well; partial dissolution of this wafer is a convenient way of incorporating arsenic into the solution. An important advantage of this approach is that the growth solution can be held in local thermodynamic equilibrium in the vicinity of the eventual growth interface. This results in a high degree of reproducibility in the fabrication of thin LPE layers. This approach can be readily extended to allow the successive growth of multiple layers by the use of multiple bins [123]. Systems have also been designed to grow LPE layers simultaneously on multiple substrates [124].

In summary, the sliding technique, using horizontal substrates, is the most versatile system for LPE. It is extremely flexible and permits the growth of multiple layers in a single operation. Presaturation of the melt just prior to layer growth can be readily accomplished without opening up the system, thus avoiding unnecessary contamination problems and giving a high degree of layer thickness control. Finally, it can be readily expanded to handle many substrates at a given time.

5.7.4 Growth Imperfections

Many different types of growth imperfections can occur during LPE. Some of these are due to the properties of the substrate material, and to imperfections in it. Yet others arise from hydrodynamic and thermal conditions during this process.

The nature of any substrate, no matter how carefully prepared, is that it is not flat. On a microscopic scale, a number of different faces are exposed to the solution during growth. In general, the free energy of these faces is a function of the orientation, with the {111} faces having a minimum free energy. During growth, the addition of extra atoms on these faces results in a large change in this energy. Generally, this causes clustering of atoms on these faces, and leads to growth irregularities such as faceting.

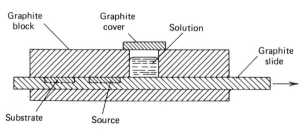

Fig. 5.34 Sliding boat apparatus with arrangement for presaturation with arsenic.

Imperfections in the substrate can act as sites for spurious nucleation during early stages of growth. Dopants such as silicon are particularly bad, since they tend to form inclusions which result in large spiral growth patterns. It goes without saying that substrate selection and surface preparation are particularly important for LPE. *In situ* etching is a necessity if uniformly flat surfaces are required.

It has been shown that improved morphology can be obtained when there are a large number of nucleation sites on the surface. Thus although it is possible to obtain growth with a supersaturation of as little as 0.25°C, the use of larger values (5–20°C) results in smoother epitaxial layers.

Many problems of layer growth come about from turbulences in the fluid flow, and from the unstable nature of the growth process. These can give rise to growing waves and ripples in the layer surface, when the system is subjected to arbitrarily small mechanical or thermal perturbations. These problems are closely allied to those caused by constitutional supercooling, and are especially severe under conditions of high growth rate. Additional imperfections are created due to the motion of the trailing edges of the liquid during the sliding operation.

In summary, therefore, the growth of thin layers by LPE, with a high degree of planarity, is an extremely difficult problem. Unfortunately, this is an important requirement for high-speed VLSI applications. Consequently, VPE and MBE are the most favored approaches at the present time.

5.8 HETEROEPITAXY

This section concerns itself with two examples of heteroepitaxy, both of which have potential advantages for integrated electronics. These are the growth of silicon on sapphire (SOS), and GaAs on silicon (GaAs:Si).

The use of silicon films on sapphire allows the formation of MOS transistors with greatly reduced source and drain capacitances. In addition, there is a significant reduction in area/device because of this simple dielectric isolation. A further advantage is that parasitic latch-up in CMOS circuits can be avoided by use of this material, so that devices can be spaced more closely than in conventional CMOS. Consequently, SOS technology has the potential for high-density, high-speed MOS circuits.

There are many benefits to be derived from the use of silicon as a substrate for the epitaxy of GaAs. First is the possibility of using large size wafers. At the present time, GaAs can only be grown and manipulated in slices of 100-mm diameter or less in order to prevent breakage in a manufacturing environment. A second advantage of using silicon as a substrate is that its thermal conductivity is about three times as large as that of GaAs. This allows a corresponding increase in the number of active devices on a chip. Finally, the cost of high-quality silicon substrate material is considerably lower than that of GaAs substrates. As a result, the successful implementation of high-quality GaAs:Si films will combine the advantages of these materials systems.

A brief treatment of each of these topics follows.

5.8.1 Silicon on Sapphire

The choice of sapphire as an insulating substrate is largely based on its mechanical and chemical stability at processing temperatures, and on its availability in high-grade commercial form. The lattice match between sapphire and silicon is not good on any plane. Nevertheless, long-range order appears to allow reasonable epitaxial growth to be carried out. Of greater importance, however, is the built-in stress in the grown films, resulting from differences in their TCEs. This produces a compressive stress in the films, so that they have a high defect density and extremely short minority carrier lifetimes (in the nanosecond range). This is of little consequence for MOS devices; however, the use of SOS films in bipolar transistors and microcircuits is precluded because of this fact. Table 5.3 at the end of this chapter provides a comparison of the properties of silicon and sapphire which are of significance for epitaxial growth.

Studies have been made of the orientation relationships of silicon and sapphire [125]. From these it is concluded that (001) silicon is best matched to $(1\bar{1}02)$ sapphire, whereas (111) silicon is most closely matched to $(11\bar{2}0)$ sapphire. The former combination is often used for (001)-based MOS microcircuits made on this material. Some work has also been done with MgO:Al_2O_3 (substrate) material, which has the advantage that it is cubic. However, it is not available in commercial quantities.

Silicon films have been grown on sapphire by both MBE and VPE. However, VPE processes are used almost exclusively because they are capable of producing SOS material in commercial quantities. Both chlorosilane and silane processes are used for this purpose. Of these, the silane process is preferred because it can be carried out at lower temperatures (900–1000°C), thereby minimizing aluminum doping from the substrate. In addition, this process avoids the resultant halogenic pitting of sapphire, which occurs at high temperatures and leads to poor crystal quality.

In situ cleaning of substrates, prior to epitaxy, is an important step in the growth process. Typically, this is done by firing the substrates in hydrogen at about 1000°C, for a few minutes. The silicon growth temperature is a compromise between crystal quality and doping. In general, films grown at higher temperature have better crystal morphology, but are more highly doped (*p*-type) with aluminum from the substrate. In addition, silicon growth on sapphire is very sensitive to the presence of even trace amounts of oxygen [126, 127].

Nucleation of silicon on sapphire is three-dimensional, and leads to island growth. These islands are of random character, with a predominance of ⟨100⟩ and ⟨110⟩ domains. As growth proceeds, they coalesce to form a continuous film, with a large number of misfit dislocations at these domain boundaries. Additionally, the wide mismatch in TCEs between silicon and sapphire results in a compressive stress in the silicon during the cool-down phase, and a further increase in the dislocation content. This latter effect can be reduced by lowering the growth temperature.

For VPE films, growth in the 950°C range yields material with the highest mobility. MBE films, grown at 700–750°C with an elemental silicon source, are slightly better in quality. However, the improvement is only slight, because of the large number of misfit dislocations in this material. An electron mobility of about 500 cm^2/V s is the highest that can be obtained in thin films (0.5–1 μm) of the type used for MOS-based circuits. This is only 33% of the bulk mobility which can be achieved in epitaxial silicon.

5.8.2 Gallium Arsenide on Silicon

The epitaxial growth of GaAs:Si is difficult because of the widely different properties of these materials. The first of these is that silicon is a nonpolar semiconductor whereas GaAs is polar. As a result, epitaxial growth takes the form of a series of islands with anti-phase domains. These coalesce to form a continuous film, with dislocation networks generated at the coalescence boundaries. Anti-phase domains of this type can be minimized by growth on misoriented substrates, and misorientations from 4° to 10° have been used for this purpose [128]. It is believed that this produces double atomic steps which inhibit their formation.

A second problem comes about because the surface of silicon is covered at all times with a layer of silicon dioxide. This is even true for freshly cleaved silicon, which develops a few monolayers of oxide upon exposure to air, and gradually thickens with time to an upper limit of 40 Å. Considerable care must be taken to remove this oxide, prior to GaAs growth. Substrates are first cleaned in hot trichloroethylene, acetone, and methanol to remove all traces of cutting oils and plasticizers. Next, they are treated in dilute HF (1 : 5 by volume in water) to remove the residual "dirty" oxide film, and transferred to deionized water *without exposure to air*. This ensures the growth of a "clean" oxide film, about 10 Å thick. These substrates are blown dry immediately prior to loading into the reactor. Next, a short heat treatment in flowing hydrogen gas, at around 700–900°C, is used to thermally remove the thin oxide film prior to epitaxial growth [129].

An alternative approach to cleaning the silicon substrate is to use a 5% solution of HF, followed by a rinse in de-ionized water [130, 131]. This removes surface oxides, and terminates the silicon with strong Si—H bonds. The surface hydrogen is subsequently desorbed in the growth system, upon heating above 400–500°C, at which point growth of the GaAs can proceed.

Both MBE and OMVPE can be used for the growth of GaAs:Si. With either technique, a multistep growth sequence has been found most successful. After the surface cleaning step, GaAs growth is carried out at around 425°C, in order to form a continuous amorphous layer, which is about 200 Å thick. The temperature at which this step is carried out is critical, to avoid island growth. Next, the amorphous layer is heat-treated at 700°C (in arsenic for MBE, and in AsH$_3$ for OMVPE) for about 10 min to promote its nucleation, and a thin layer of GaAs grown on it, at a low growth rate (typically 100 Å/min). The remainder of the layer is grown at a more conventional rate (≈500 Å/min) after a second, intermediate anneal step. Sometimes, this is followed by a post-growth anneal at a high temperature (800–900°C). Annealing at still higher temperatures is not practical because it results in loss of stoichiometry.

The lattice mismatch between GaAs and silicon is large (≈4.1%), so that the epitaxial layer is essentially relaxed at the growth temperature. During cooldown, the layer goes into biaxial tension because its TCE is larger than that of silicon (6×10^{-6}/K as compared to 4.2×10^{-6}/K). This results in a relatively large dislocation density, as evidenced by an x-ray full width at half-maximum (FWHM) of 150–200 arc sec for 3-μm-thick layers [(004) reflections]. By way of comparison, the theoretical value for bulk GaAs is 8.7 arc sec and a FWHM of 9–10 arc sec is routinely obtained in practice, when GaAs substrates are used.

Buffer layers of CaF$_2$ and Ca$_x$Ba$_{1-x}$F$_2$, which have extremely low plastic transition temperatures, have been used in attempts to overcome the large lattice and TCE mismatch in this system [132]. Here, too, HF surface treatments have been used prior to the growth of these alkaline earth fluorides on the silicon.

Many different types of electron devices have been fabricated in GaAs:Si, including field effect transistors (FETs) and modulation doped FETs [133]. Light-emitting diodes have also been made [134] on GaAs:Si, thus allowing the integration of optical devices with silicon microcircuits. Lasers have also been demonstrated [135] in films of GaAs:Si.

5.9 THE EVALUATION OF EPITAXIAL LAYERS

It is necessary to evaluate layer parameters such as mobility, carrier concentration, doping profile, and thickness in order to assess the quality of epitaxial silicon and GaAs. With epitaxial silicon, the quality is sufficiently high so that the carrier concentration can be determined by measuring the sheet resistance and the junction depth, and assuming an appropriate value of mobility for the resulting concentration. With GaAs, however, a wide range of compensation ratios is encountered. Thus a separate assessment is necessary for the mobility and the carrier concentration.

5.9.1 Sheet Resistance

Consider a layer of material which is grown on a substrate of opposite impurity type. This layer may be electrically characterized in terms of its sheet resistance, as described in Section 4.10.2. Consider a rectangle of the layer, of length l and width w. If its resistance is measured along its length, then

$$R = R_S \frac{l}{w} \tag{5.76}$$

where R_S is defined as the *sheet resistance* of the layer, in ohms. The sheet resistance of this layer can be measured by the four-point probe configuration shown in Fig. 5.35 a. Here four equally spaced collinear probes are placed on the layer. A current I is passed through the outer probes, and the voltage V developed across the inner probes is measured. The value of current is so chosen that V and I are linearly interrelated.

Since the epitaxial layer is of opposite impurity type to its substrate, it can be assumed that the current flow is restricted within it. Consider a layer of infinite dimensions compared to the probe spacing. For the configuration of Fig. 5.35 b comprising a positive source and a negative source of current, the potential at

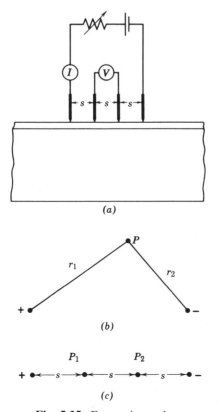

(a)

(b)

(c)

Fig. 5.35 Four-point probe.

any point P is given by

$$\psi_P = \frac{IR_S}{2\pi} \ln \frac{r_2}{r_1} + A \tag{5.77}$$

where r_1 and r_2 are the distances of the point P from the positive and negative source, respectively, and A is a constant of integration. For the four-point probe configuration of Fig. 5.35c, where the probes are collinear and equally spaced, we have

$$\psi_{P1} = \frac{IR_S}{2\pi} \ln 2 + A \tag{5.78}$$

$$\psi_{P2} = \frac{-IR_S}{2\pi} \ln 2 + A \tag{5.79}$$

and ψ_{P1} and ψ_{P2} are the potentials at points 1 and 2, respectively. Thus

$$\psi_{P1} - \psi_{P2} = V = \frac{IR_S}{\pi} \ln 2 \tag{5.80}$$

Rearranging Eq. (5.80) gives

$$R_S = \left(\frac{\pi}{\ln 2} \right) \frac{V}{I} \tag{5.81a}$$

$$= 4.5324 \frac{V}{I} \tag{5.81b}$$

Thus the sheet resistance may be directly obtained from the V/I ratio.

Formulae for layers of finite dimensions are available in the literature [136]. By way of example, for probes located centrally on a circular sample of finite diameter, the sheet resistance can be shown to be given by

$$R_S = C \frac{V}{I} \tag{5.82}$$

where C is a function of the probe spacing and the slice diameter, as shown in Fig. 5.36.

Four-point probe measurements are usually made on monitor slices that are placed in the reactor along with those on which the circuits are to be fabricated. This avoids the necessity for interpreting data on slices which already have diffused regions (e.g., buried layers) prior to epitaxial growth. These monitor slices are of the same starting material as that used for the microcircuit. In some situations, the epitaxial layer and the substrate are of the same conductivity type.

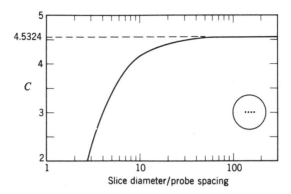

Fig. 5.36 Correction factor for circular slices.

Four-point probe measurements cannot be made on these layers, necessitating the use of monitor substrates of opposite conductivity type. With GaAs, it is customary to make measurements on layers grown on semi-insulating (SI) substrates.

The tacit assumption made in taking four-point probe measurements is that the probe–semiconductor contact is highly conductive, and linear in its V–I characteristic (i.e., ohmic). This assumption is only satisfactory for layers with a sheet resistivity under 20 Ω-cm, and the taking of four-point probe data is more difficult beyond this point.

5.9.2 Layer Thickness

Epilayer thickness can be measured by techniques of the type described in Chapter 4, which are used for the evaluation of diffused layers. These are destructive in character, in addition to being tedious. Interferometric techniques, based on measurement of the reflectance spectrum of an epilayer on a substrate, are an extremely convenient, nondestructive alternative [137]. This spectrum exhibits interference fringes due to small differences in refractive index between the epilayer and the substrate. Layer thickness is related to the distance between fringes and the refractive indices of the layer and the substrate. The success of this method is based on the assumption that the epilayer is uniform, and can be characterized by a constant refractive index. This is a reasonable approximation in practical situations.

Fourier transform infrared (FTIR) spectrometry is an elegant implementation of this principle, and allows rapid measurement of the infrared spectrum, from which the layer thickness can be deduced [138]. The use of a computer-driven stage permits areal layer thickness measurements across the wafer in a completely automated manner. This technique can also be used to determine the individual thicknesses of multi-layer films of the type used in quantum well devices [139]. Details of these methods, as well as other techniques such as microwave interferometry and surface-acoustic-wave probing, are beyond the scope of this book.

5.9.3 Mobility and Carrier Concentration

Silicon is generally uncompensated, so that a measurement of resistivity and junction depth (see Section 4.10) is usually sufficient for determining the mobility and carrier concentration. Curves for mobility versus carrier concentration and for resistivity versus carrier concentration in silicon have been provided in Figs. 1.3 and 1.4, respectively.

GaAs is usually compensated. Here the mobility and the carrier concentration in an epitaxial layer must be separately determined by means of the Hall effect. Mobility values are usually measured at room temperature (300 K) and at liquid nitrogen temperature (77 K). The value of the compensation ratio can be extracted by the techniques outlined in Section 1.1, which is applicable to layers which are not in communication with the substrate (e.g., *n*-layers on *p*-substrates, and vice versa). With GaAs, both *p*- and *n*-layers can be evaluated if they are grown on SI substrates.

Consider an infinitely long, uniformly doped layer through which a current flows. A magnetic field is applied across the layer, at right angles to the direction of current flow, as shown in Fig. 5.37. For this condition, there exists an electric field $\mathscr{E}_H$ which is mutually perpendicular to both the current and the magnetic field; the value of this electric field is given by

$$\mathscr{E}_H = \frac{R_H I B}{A} \tag{5.83}$$

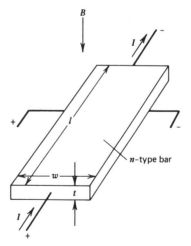

Fig. 5.37 Hall effect.

where R_H is the Hall coefficient, I the current, A the cross section, and B the magnetic field strength. If V_H is the voltage across the bar, and t is its thickness, then

$$R_H = \frac{t}{IB} V_H \qquad (5.84)$$

Finally, the Hall mobility μ_H is given by

$$\mu_H = \frac{R_H}{\rho} \qquad (5.85)$$

where ρ is the resistivity of the bar.

The Hall effect principle can be applied by the technique developed by van der Pauw [140], which can be used with arbitrarily shaped layers of uniform thickness and composition, with four point contacts at locations on the sample periphery. This method is sensitive to the size and placement of these contacts [141], unless the sample shape is modified by cutting isolating regions between the contacts to make a "clover-leaf" shape, as shown in Fig. 5.38. It is ideally suited for the evaluation of GaAs epitaxial layers, which can be isolated by patterning after they are grown on SI substrates, or on substrates with which they have no electrical communication.

The specific resistivity of samples of this type is determined by passing a known current through two adjacent terminals and measuring the voltage developed across the other two. Thus, if $R_{AB,CD}$ is the potential difference across CD resulting from unit current flow between the contacts A and B, then it can be shown that

$$\exp\left[-\frac{\pi R_{AB,CD} t}{\rho}\right] + \exp\left[-\frac{\pi R_{BC,DA} t}{\rho}\right] = 1 \qquad (5.86)$$

where ρ is the resistivity and t is the layer thickness.

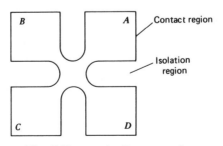

Fig. 5.38 van der Pauw sample.

The Hall mobility can be determined by passing a known current through a set of nonadjacent contacts, and measuring the voltage drop across the other pair (e.g., $R_{BD,AC}$), both with and without the application of a perpendicular magnetic field. Here

$$\mu_H = \frac{t}{B}\frac{\Delta R_{BD,AC}}{\rho} \tag{5.87}$$

where $\Delta R_{BD,AC}$ is the difference in these voltages.

The assumption of uniform carrier concentration is not always true, especially near the substrate–epitaxial-layer interface. This presents a serious problem in determining the mobility of very thin epitaxial layers. Techniques involve making a series of van der Pauw measurements as the layer is thinned,* and calculating the mobility and concentration gradient from the small differences in these measurements [142].

Highly conducting inclusions in a sample can often cause serious error in mobility measurements, and lead to artificially high values. Their absence can be ensured if data, taken at different values of magnetic field strength, result in the same values of mobility.

In summary, the van der Pauw technique is uniquely suited for small samples of arbitrary shape. Thus it is usually the preferred method for GaAs and other compound semiconductors, where samples of this type are often used for reasons of economy or availability.

Two-point, spreading resistance probes can also be used to measure [143] the sheet resistance of epitaxial layers. Although simpler in implementation, the measurement is based on the assumption that conduction paths due to charge injection can be ignored. Thus, it is sensitive to the nature of the metal–semiconductor contact, and also to the actual resistivity. For these reasons, the technique is primarily useful in comparing samples, rather than in making absolute measurements. It is an important technique for depth profiling on angle-lapped substrates, and is used for both silicon and GaAs.

5.9.4 Impurity

Direct techniques such as secondary ion mass spectrometry [144] can be used to measure the impurity profile in epitaxial layers. This technique is extremely powerful because it can give a quantitative measure of impurity concentration, over 5 to 6 orders of magnitude. Moreover, it has excellent sensitivity for most elements, especially when used with oxygen or cesium bombardment in order to enhance the ion yield. Its disadvantage is that it requires highly sophisticated equipment, so that it is usually performed by an outside vendor of this service. An alternative approach consists of forming a diode structure and measuring its

---

*This is done by chemical etching, or by depleting part of the sample by means of a Schottky diode structure.

capacitance versus reverse voltage characteristic [145]. This necessitates growth of the layer on a heavily doped substrate of the same type—that is, p on p^+ or n on n^+. Next, a shallow junction is formed on the surface of the layer to obtain this structure. This technique is most readily applicable to the evaluation of n-layers on n^+-substrates. Here a Schottky diode structure can be formed by vacuum evaporation of a suitable metal (see Chapter 8), thus avoiding a high-temperature diffusion step.

The principle behind this technique is described with reference to Fig. 5.39, which shows a diode of this type, having an arbitrary doping profile [139]. Here an increase of the reverse bias voltage from V to $V + \Delta V$ results in an increase in the electric field, $\Delta\mathscr{E}$, by $\Delta V/x$ and an increase in the depletion layer width from x to $x + \Delta x$. This uncovers an additional charge $qN(x)\Delta x$ such that

$$\Delta\mathscr{E} = \frac{\Delta V}{x} = \frac{q}{\epsilon\epsilon_0}N(x)\Delta x \tag{5.88}$$

At this bias, the small-signal capacitance C is given by

$$C = \frac{\epsilon\epsilon_0 A}{x} \tag{5.89}$$

Eliminating x and Δx from these equations gives

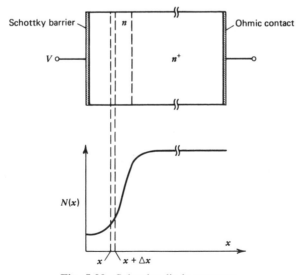

Fig. 5.39 Schottky diode structure.

$$N(x) = -\frac{C^3}{q\epsilon\epsilon_0 A^2}\Big/\left(\frac{\Delta C}{\Delta V}\right) \tag{5.90}$$

so that the measurement of the C–V characteristic of the diode allows calculation of the impurity profile in the epitaxial layer.

The Schottky diode technique can be made extremely convenient by using a mercury probe, of known diameter, to form the top contact. Additionally, a large-area mercury probe can be used to provide a return path for the diode on the top face, and eliminate the need for the bottom contact to the semiconductor. This allows measurement of n–p layer structures, as well as layers which are grown on SI GaAs substrates. The technique should not be considered to be nondestructive, since its use invariably leaves a fine residue of mercury oxides on the semiconductor surface.

Variations of this technique have been developed to provide greater accuracy and/or speed in taking this measurement [146]. Perhaps the most commonly used approach consists of a feedback technique for obtaining a direct indication of this impurity distribution in automated form [147].

The Schottky diode technique described above cannot be used to measure the carrier concentration in multilayer epitaxial structures which are often used with compound semiconductors. Here, an interesting extension is to use an electrolyte for the Schottky contact. The advantage of this approach is that, in addition to allowing capacitance measurements to be made, the semiconductor can also be etched by reversing the polarity—that is, by making it positive with respect to the electrolyte. This allows successive etch and measurement steps, so that a carrier concentration profile can be obtained through the depth of the multilayer structure [148, 149]. Commercial equipment for carrying out this process is available, and is extensively used in the evaluation of compound semiconductor epilayers.

TABLE 5.1   Dopants Used in the MBE of GaAs [25]

Dopant	Type	Controllable Doping Range (cm^3)	Effusion Cell Temperataure ($^\circ$C)	Substrate Temperature ($^\circ$C)
Si	n	10^{14} to 6×10^{18}	810–1140	490–700
S	n	10^{15} to 6×10^{18}	$\simeq 200$	<560
Se	n	10^{15} to 6×10^{18}	$\simeq 200$	<560
Be	p	10^{14} to 6×10^{19}	550–895	490–670

TABLE 5.2   Commonly Used Sources for the VPE of GaAs [83]

Compound	Acronym	Melting Point ($^{\circ}$C)	Boiling Point at 760 torr ($^{\circ}$C)	a	b	Vapor Pressure (torr)
Arsenic trichloride $AsCl_3$		-18	130.2	8.018	2072	40 at 50°C
Trimethylgallium $(CH_3)_3Ga$	TMGa	-15.8	55.7	8.5	1824	65.8 at 0°C
Triethylgallium $(C_2H_5)_3Ga$	TEGa	-82.3	143	9.172	2532	3.4 at 20°C
Diethylgalliumchloride $(C_2H_5)_2GaCl$	DEGaCl	-7		8.78	2815	0.5 at 60°C
Trimethylaluminum $(CH_3)_3Al$	TMAl	15.4	126	8.224	2134.8	8.7 at 20°C
Triethylaluminum $(C_2H_5)_3Al$	TEAl	-52.5	194	10.784	3625	0.22 at 44°C
Diethylaluminumchloride $(C_2H_5)_2AlCl$	DEAlCl	-90		8.943	3080	0.8 at 45°C
Trimethylindium $(CH_3)_3In$	TMIn	88.4	134	10.52	3014	1.7 at 20°C
Triethylindium $(C_2H_5)_3In$	TEIn	-32	184	8.69	2689	1.5 at 44°C
Dimethylzinc $(CH_3)_2Zn$	DMZn	-42	46	7.802	1560	124 at 0°C
Diethylzinc $(C_2H_5)_2Zn$	DEZn	-28	118	8.28	2190	6.4 at 20°C
Dimethylcadmium $(CH_3)_2Cd$	DMCd	-4.5	105.5	7.764	1850	28 at 20°C
Diethylselenide $(C_2H_5)_2Se$	DESe		108	7.905	1924	7.2 at 0°C
Diethyltelluride $(C_2H_5)_2Cd$	DETe		137–138	7.99	2093	7 at 20°C
Diisopropyltelluride $(C_3H_7)_2Te$	DIPTe			8.288	2309	2.56 at 20°C
Tertiarybutylarsenic $(CH_3)_3CAsH_2$	TBAs	-1	65	7.243	1509	32 at $-10$$^{\circ}$C

TABLE 5.3 Comparitive Properties of Silicon and Sapphire

Parameter	Sapphire	Silicon
Crystal structure	Rhombohedral	Zincblende
Lattice parameter, Å	$a = 4.76$	$a = 5.43$
	$c = 12.99$	
Permittivity	9.4	11.7
Linear coefficient of expansion ($^{\circ}C^{-1}$)	8.4×10^{-6}	3.1×10^{-6}
Thermal conductivity (cal/cm s deg)	0.06	0.30

REFERENCES

1. R. T. Tung, L. R. Dawson, and R. L. Gunshor, Eds., *Epitaxy of Semiconductor Layered Structures*, Materials Research Society Symposium Proceedings, Vol. 102, 1987.

2. A. Y. Cho and J. R. Arthur, Molecular Beam Epitaxy, *Prog. Solid State Chem.* **10**, 157 (1973).

3. A. Sherman, *Chemical Vapor Deposition for Microelectronics: Principles, Technology and Applications*, Noyes Publications, Park Ridge, NJ, 1987.

4. L. R. Dawson, Liquid Phase Epitaxy, in *Progress in Solid State Physics*, Vol. 7, H. Reiss and J. O. McCaldin, Eds., Pergamon, New York, 1972, p. 117.

5. L. M. Smith, D. J. Wolford, R. Venkatasubramanian, and S. K. Ghandhi, Photoexcited Carrier Lifetimes and Spatial Transport in Surface-Free GaAs Homostructures, *J. Vac. Sci. Technol.* **B8**(8), 787 (1990).

6. E. Kaldis, Principles of the Vapor Growth of Single Crystals, in *Crystal Growth*, C. H. L. Goodman, Ed., Vol. 1, Plenum, New York, 1974, p. 49.

7. R. E. Reed-Hill, *Physical Metallurgy Principles*, Van Nostrand, New York, 1973.

8. B. R. Pamplin, Ed., *Crystal Growth*, Pergamon Press, New York, 1980.

9. R. G. Sangster, Model Studies of Crystal Growth Phenomena in the III–V Semiconducting Compounds, in *Compound Semiconductors*, Vol. 1, R. K. Willardson and H. L. Goering, Eds., Reinhold, New York, 1962, p. 241.

10. L. J. Giling and H. H. C. de Moore, The Incorporation of Dopants During Growth of GaAs by CVD, in *Proceedings of 4th European Conference on CVD, Eindhoven, The Netherlands*, J. Bloem, G. Verspui, and L. R. Wolff, Eds., 1983, p. 184.

11. J. L. Weyher and J. van der Ven, Influence of Substrate Defects on the Structure of Epitaxial GaAs Grown by MOCVD, *J. Cryst. Growth* **88**, 221 (1988).

12. J. W. Matthews, Ed., *Epitaxial Growth*, Academic Press, New York, 1975.

13. J. W. Matthews and A. E. Blakeslee, Defects in Epitaxial Layers. I. Misfit Dislocations, *J. Cryst. Growth* **27**, 118 (1974).

14. H. J. Herzog, L. Csepregi, and H. Seidel, X-Ray Investigation of Boron- and Germanium-Doped Silicon Epitaxial Layers, *J. Electrochem. Soc.* **131**, 2969 (1984).

15. J. Bloem and A. H. Goemans, Slip in Silicon Epitaxy, *J. Appl. Phys.* **43**, 1281 (1972).

16. S. Timoshenko and J. N. Goodier, *Theory of Elasticity*, McGraw-Hill, New York, 1951.

17. L. D. Dyer, H. R. Huff, and W. W. Boyd, Plastic Deformation in Central Regions of Epitaxial Silicon Slices, *J. Appl. Phys.* **42**, 5680 (1971).

18. M. A. Herman and H. Sitter, *Molecular Beam Epitaxy: Fundamentals and Current Status*, Springer-Verlag, New York, 1988.

19. J. C. Bean, Recent Developments in Silicon Molecular Beam Epitaxy, *J. Vac. Sci. Technol.* **A1**, 540 (1983).

20. S. S. Iyer, Silicon Molecular Beam Epitaxy, in *Epitaxial Silicon Technology*, B. J. Baliga, Ed., Academic Press, New York, 1986, p. 91.

21. W. T. Tsang, Molecular Beam Epitaxy for III–V Compound Semiconductors, in *Semiconductors and Semimetals*, Vol. 22, Part A, R. K. Willardson and A. C. Beer, Eds., Academic Press, New York, 1985, p. 96.

22. J. H. Jeans, *An Introduction to the Kinetic Theory of Gases*, The University Press, Cambridge, 1967.

23. B. B. Dayton, Gas Flow Patterns at Entrance and Exit of Cylindrical Tubes, in *1956 National Symposium on Vacuum Technology Transactions*, E. S. Perry, T. H. Devant, Eds., Pergamon Press, Oxford, 1957, p. 5.

24. R. E. Honig and D. E. Kramer, Vapor Pressure Data for the Solid and Liquid Elements, *RCA Rev.* **30**, 2 (1969).

25. C. E. C. Wood, Dopant Incorporation, Characteristics, and Behavior, in *The Technology and Physics of Molecular Beam Epitaxy*, E. H. C. Parker, Ed., Plenum Press, New York, 1985, p. 61.

26. A. Y. Cho, Growth of III–V Semiconductors by Molecular Beam Epitaxy and Their Properties, *Thin Solid Films* **100**, 291 (1983).

27. R. Wunder, R. Stall, R. Malik, and S. Woelfer, Automated Growth of $Al_xGa_{1-x}As$ and $In_xGa_{1-x}As$ by Molecular Beam Epitaxy using an Ion Gauge Flux Monitor, *J. Vac. Sci. Technol.* **B3**(4), 964 (1985).

28. C. T. Foxon and B. A. Joyce, "Interaction Kinetics of As_4 and Ga On (100) GaAs Surface Using a Modulated Molecular Beam Technique, *Surf. Sci.* **50**, 434 (1975).

29. G. J. Davies and D. A. Andrews, Metal–Organic Molecular Beam Epitaxy (MOMBE), *Chemtronics* **3**, 3 (1988).

30. T. U. M. S. Murthy, N. Miyamoto, M. Shimbo, and J. Nishizawa, Gas- Phase Nucleation During the Thermal Decomposition of Silane in Hydrogen, *J. Cryst. Growth* **33**, 1 (1976).

31. G. B. Stringfellow, Non-hydride Group V Sources for OMVPE, *J. Electron. Mater.* **17**, 327 (1988).

32. S. Whitaker, *Introduction to Fluid Mechanics*, Prentice-Hall, Englewood Cliffs, NJ, 1968.

33. P. C. Rundle, The Epitaxial Growth of Silicon on a Horizontal Reactor, *Int. J. Electron.* **24**, 405 (1968).

34. S. K. Ghandhi and R. J. Field, A Re-examination of Boundary Layer Theory for a Horizontal CVD Reactor, *J. Cryst. Growth* **69**, 619 (1984).

35. F. C. Eversteyn, P. J. W. Severin, C. H. J. van den Brekel, and H. L. Peek, A Stagnant Layer Model for the Epitaxial Growth of Silicon from Silane in a Horizontal Reactor, *J. Electrochem. Soc.* **117**, 925 (1970).

36. H. Moffat and K. F. Jensen, Complex Flow Phenomena in MOCVD Reactors, *J. Cryst. Growth* **77**, 108 (1986).

37. P. B. Chinoy, P. D. Agnello, and S. K. Ghandhi, An Experimental and Theoretical Study of Growth in Horizontal Epitaxial Reactors, *J. Electron Mater.* **17**, 493 (1988).

38. C. Houtman, D. B. Graves, and K. F. Jensen, CVD in Stagnation Point Flow. An Evaluation of the Classical 1-D Treatment, *J. Electrochem. Soc.* **133**, 961 (1986).

39. K. Chen and A. R. Mortazavi, An Analytical Study of the Chemical Vapor Deposition (CVD) Processes in a Rotating Pedestal Reactor, *J. Cryst. Growth* **77**, 199 (1976).

40. P. B. Chinoy, D. A. Kaminski, and S. K. Ghandhi, A Novel Reactor for Large Area Epitaxial Solar Cell Materials, *Solar Cells* **30**, 323 (1991).

41. C. Van Opdorp and M. R. Leys, On the Factors Impairing the Compositional Transition Abruptness in Heterojunctions Grown by Vapour-Phase Epitaxy, *J. Cryst. Growth* **84**, 271 (1987).

42. G. S. Tompa, M. A. McKee, C. Beckham, P. A. Zawadzki, J. M. Colabella, D. Reinert, K. Capuder, R. A. Stall, and P. E. Norris, A Parametric Investigation of GaAs Epitaxial Growth Uniformity in a High Speed, Rotating Disk MOCVD Reactor, *J. Cryst. Growth* **93**, 220 (1988).

43. E. Krullman and W. L. Engel, Low Pressure Silicon Epitaxy, *IEEE Trans. Electron Dev.* **ED-29**, 491 (1982).

44. M. J-P. Duchemin, M. M. Bonnet, and M. F. Koelsch, Kinetics of Silicon Growth Under Low Hydrogen Pressure, *J. Electrochem. Soc.* **125**, 637 (1978).

45. C. L. Yaws, P. N. Shah, P. M. Patel, G. Hsu, and R. Lutwack, Physical and Thermodynamic Properties of Silicon Tetrachloride, *Solid State Technol.* (Feb. 1979) p. 65.

46. R. W. Borreson, C. L. Yaws, G. Hsu and R. Lutwack, Physical and Thermodynamic Properties of Silane, *Solid State Technol.* (Jan. 1978) p. 43.

47. V. S. Ban, Mass Spectrometric Studies of Chemical Reactions and Transport Phenomena in Si Epitaxy, in *Proceedings of the VI International Conference on Chemical Vapor Deposition*, L. F. Donaghey, P. Rai-Chaudary, and R. Tauber, Eds., The Electrochemical Society, Princeton, NJ, 1977, p. 66.

48. F. C. Eversteyn, Chemical Reaction Engineering in the Semiconductor Industry, *Philips Res. Rep.* **29**, 45 (1974).

49. W. B. deBoer and D. J. Meyer, Low-Temperature Chemical Vapor Deposition of Epitaxial Si and SiGe Layers at Atmospheric Pressure, *Appl. Phys. Lett.* **58**, 1286 (1991).

50. M. E. Jones and D. W. Shaw, Growth from the Vapor, in *Treatise on Solid State Chemistry*, Vol. 5, N. B. Hannay, Ed., Plenum, New York, 1976.

51. C. L. Yaws, J. R. Hopper and E. M. Swinderman, Physical and Thermodynamic Properties of Semiconductor Industry Chemical Materials: *P*-Type Gas Phase Dopants (Diborane and Pentaborane), *Solid State Technol.* (Nov. 1974) p. 31.

52. C. L. Yaws, H. S. N. Setty, J. R. Hopper, and E. M. Swinderman, Physical and Thermodynamic Properties of Semiconductor Industry Chemical Materials: N-Type Gas Phase Dopants (Arsine, Phosphine, and Stibine), *Solid State Technol.* (Jan. 1974) p. 47.

53. L. J. Stinson, J. A. Howard, and R. C. Neville, Sulfur Hexafluoride Etching Effects in Silicon, *J. Electrochem. Soc.* **123**, 551 (1976).

54. C. O. Thomas, D. Kahng, and R. C. Manz, Impurity Distribution in Epitaxial Silicon Films, *J. Electrochem. Soc.* **109**, 1055 (1962).

55. G. R. Srinivasan, Autodoping Effects in Silicon Epitaxy, *J. Electrochem. Soc.* **127**, 1334 (1980).

56. H-R. Chang, Autodoping in Silicon Epitaxy, *J. Electrochem. Soc.* **132**, 219 (1985).

57. M. W. M. Graef, B. J. H. Leunissen, and H. H. C. de Moor, Antimony, Arsenic, Phosphorus and Boron Autodoping in Silicon Epitaxy, *J. Electrochem. Soc.* **132**, 1943 (1985).

58. P. H. Langer and J. I. Goldstein, Boron Autodoping During Silane Epitaxy, *J. Electrochem. Soc.* **124**, 591 (1977).

59. P. H. Langer and J. I. Goldstein, Impurity Redistribution During Silicon Epitaxial Growth and Semiconductor Device Processing, *J. Electrochem. Soc.* **121**, 563 (1974).

60. R. Reif and R. W. Dutton, Computer Simulations in Silicon Epitaxy, *J. Electrochem. Soc.* **128**, 909 (1981).

61. G. R. Srinivasan, CVD Epitaxial Autodoping in Bipolar VLSI Technology, *J. Electrochem. Soc.* **132**, 3005 (1985).

62. G. R. Srinivasan, Kinetics of Lateral Autodoping in Silicon Epitaxy, *J. Electrochem. Soc.* **125**, 146 (1978).

63. J. O. Borland and T. Deacon, Advanced CMOS Epitaxial Processing for Latch-Up Hardening and Improved Epilayer Quality, *Solid State Technol.* (Aug. 1984) p. 123.

64. J. O. Borland and I. Beinglass, Selective Silicon Deposition for the Megabit Age, *Solid State Technol.* (Jan. 1990) p. 73.

65. W. A. P. Classen and J. Bloem, The Nucleation of CVD Silicon on SiO_2 and Si_3N_4 Substrates. II. The SiH_2Cl_2-H_2-N_2 System, *J. Electrochem. Soc.* **127**, 1837 (1980).

66. J. W. Osenbach, D. G. Schimmel, A. Feygenson, J. J. Bastek, J. C. C. Tsai, H. C. Praefcke, and E. W. Bonato, Selective Epitaxial Growth of Silicon, *J. Mater. Res.* **6**, 2318 (1991).

67. R. K. Smeltzer, Epitaxial Deposition of Silicon in Deep Grooves, *J. Electrochem. Soc.* **122**, 1666 (1975).

68. C. M. Drum and C. A. Clark, Geometrical Stability of Shallow Surface Depressions During Growth of (111) and (100) Epitaxial Silicon, *J. Electrochem. Soc.* **115**, 664 (1968).

69. S. P. Weeks, Pattern Shift Distortion During CVD Epitaxy on (111) and (100) Silicon, *Solid State Technol.* (Nov. 1981) p. 111.

70. C. M. Drum and C. A. Clark, Anisotropy of Macrostep Motion and Pattern Edge-Displacements During Growth of Epitaxial Silicon Near {100}, *J. Electrochem. Soc.* **117**, 1401 (1970).

71. R. H. Finch, H. J. Queisser, G. Thomas, and J. Washburn, Structure and Origin of Stacking Faults in Epitaxial Silicon, *J. Appl. Phys.* **34**, 406 (1963).

72. R. B. Marcus, M. Robinson, T. T. Sheng, S. E. Haszko, and S. P. Murarka, Electrical Activity of Epitaxial Stacking Faults, *J. Electrochem. Soc.* **124**, 425 (1977).

73. H. M. Liaw and J. W. Rose, Silicon Vapor Phase Epitaxy, in *Epitaxial Silicon Technology*, B. J. Baliga, Ed., Academic Press, New York, 1986, p. 1.

74. H. M. Manasevit and W. I. Simpson, The Use of Metal–Organics in the Preparation of Semiconductor Materials. I. Epitaxial Gallium-V Compounds, *J. Electrochem. Soc.* **116**, 1725 (1969).

75. J. Komeno, S. Ohkawa, A. Miura, K. Dazai, and O. Ryuzan, Variation of GaAs Epitaxial Growth Rate with Distance Along Substrate Within a Constant Temperature Zone, *J. Electrochem. Soc.* **124**, 1440 (1977).

76. J. R. Knight, D. Effer, and P. R. Evans, The Preparation of High Purity Gallium Arsenide by Vapor Phase Epitaxial Growth, *Solid State Electron.* **8**, 178 (1965).

77. M. J. Cardwell, Vapour Phase Epitaxy of High Purity III–V Compounds, *J. Cryst. Growth* **70**, 97 (1984).

78. D. J. Kirwan, Reaction Equilibria in the Growth of GaAs and GaP by the Chloride Transport Process, *J. Electrochem. Soc.* **117**, 1572 (1970).

79. D. W. Shaw, Influence of Substrate Temperature on GaAs Epitaxial Deposition Rates, *J. Electrochem. Soc.* **115**, 405 (1968).

80. J. L. LaPorte, M. Cadoret, and R. Cadoret, Investigation of the Parameters Which Control the Growth of {111} and {$\bar{1}\,\bar{1}\,\bar{1}$} Faces of GaAs by Chemical Vapour Deposit, *J. Cryst. Growth* **50**, 663 (1980).

81. J. J. Tietjen and J. A. Amick, The Preparation and Properties of Vapor-Deposited Epitaxial GaAsP Using Arsine and Phosphine, *J. Electrochem. Soc.* **113**, 724 (1966).

82. K. Grüter, M. Deschler, H. Jügensen, R. Reecard and P. Balk, Deposition of High Quality GaAs Films at Fast Rates in the LP–CVD System, *J. Cryst. Growth* **94**, 607 (1989).

83. G. B. Stringfellow, *Organometallic Vapor-Phase Epitaxy: Theory and Practice*, Academic Press, New York, 1989.

84. M. Yoshida, H. Watanabe, and F. Vesugi, Mass Spectrometric Study of Ga(CH$_3$)$_3$ and Ga(C$_2$H$_5$)$_3$ Decomposition Reaction in H$_2$ and N$_2$, *J. Electrochem. Soc.* **132**, 677 (1985).

85. D. H. Reep and S. K. Ghandhi, Deposition of GaAs Epitaxial Layers by Organometallic CVD, *J. Electrochem. Soc.* **130**, 675 (1983).

86. M. G. Jacko and S. J. W. Price, The Pyrolysis of Trimethylgallium, *Can. J. Chem.* **41**, 1560 (1963).

87. Tamaru, K., The Decomposition of Arsine, *J. Phys. Chem.* **59**, 777 (1955).

88. W. Richter, P. Kurpas, R. Lückerath, M. Motzkus, and M. Waschbüsch, Gas Phase Studies of MOVPE by Optical Methods, *J. Crystal Growth* **107**, 13 (1991).

89. T. F. Kuech and E. Veuhoff, Mechanism of Carbon Incorporation in MOCVD GaAs, *J. Cryst. Growth* **68**, 148 (1984).

90. K. Lindeke, W. Sack, and J. J. Nickl, Gallium Diethyl Chloride: A New Substance in the Preparation of Epitaxial Gallium Arsenide, *J. Electrochem. Soc.* **117**, 1316 (1970).

91. Y. Nakayama, S. Ohkawa, H. Hashimoto, and H. Ishikawa, Submicron GaAs Epitaxial Layer from Diethylgalliumchloride and Arsine, *J. Electrochem. Soc.* **123**, 1227 (1976).

92. T. F. Kuech, M. S. Goorsky, M. A. Tischler, A. Palevski, P. Solomon, R. Potemski, C. S. Tsai, J. A. Lebens, and K. J. Vahala, Selective Epitaxy of GaAs, Al$_x$Ga$_{1-x}$As and In$_x$Ga$_{1-x}$As, *J. Cryst. Growth* **107**, 116 (1991).

93. J. van de Ven, H. G. Schoot, and L. J. Giling, Influence of Growth Parameters on the Incorporation of Residual Impurities in GaAs Grown by Metalorganic Chemical Vapor Deposition, *J. Appl. Phys.* **60**, 1648 (1986).

94. S. K. Shastry, S. Zemon, D. G. Kenneson, and G. Lambert, Control of Residual Impurities in Very High Purity GaAs Grown by Organometallic Vapor Phase Epitaxy, *Appl. Phys. Lett.* **52**, 150 (1988).

95. J. V. DiLorenzo, Vapor Growth of Epitaxial GaAs: A Summary of Parameters Which Influence the Purity and Morphology of Epitaxial Layers, *J. Cryst. Growth* **17**, 189 (1972).

96. L. R. Weisberg and J. Blanc, Diffusion with Interstitial–Substitutional Equilibrium. Zinc in GaAs, *Phys. Rev.* **131**, 1548 (1963).

97. B. T. Cunningham, J. E. Baker, and G. E. Stillman, Carbon Tetrachloride Doped Al$_x$Ga$_{1-x}$As Grown by Metalorganic Chemical Vapor Deposition, *Appl. Phys. Lett.* **56**, 836 (1990).

98. T. F. Kuech, M. A. Tischler, P. -J. Wang, G. Scilla, R. Potemski, and F. Cardone, The Controlled Carbon Doping of GaAs by Metal–Organic Vapor Phase Epitaxy, *Appl. Phys. Lett.* **53**, 1317 (1988).

99. S. J. Bass, Silicon and Germanium Doping of Epitaxial Gallium Arsenide Grown by the Trimethylgallium-Arsenic Method, *J. Cryst. Growth* **47**, 613 (1979).

100. R. Venkatasubramanian, K. Patel and S. K. Ghandhi, Compensation Mechanisms in N$^+$ GaAs Doped with Silicon, *J. Cryst. Growth* **94**, 34 (1989).

101. E. Veuhoff, T. F. Kuech, and B. S. Meyerson, A Study of Silicon Incorporation in GaAs MOCVD Layers, *J. Electrochem. Soc.* **132**, 1958 (1985).

102. O. Mizuno, S. Kikuchi, and Y. Seki, Epitaxial Growth of Semi-insulating Gallium Arsenide, *Jpn. J. Appl. Phys.* **10**, 208 (1971).

103. S. J. Bass, Growth of Semi-Insulating Epitaxial Gallium Arsenide by Chromium Doping in the Metal-Alkyl and Hydride System, *J. Cryst. Growth* **44**, 29 (1978).

104. R. P. Gale, R. W. McClelland, J. C. C. Fan, and C. O. Bosler, Lateral Epitaxial Overgrowth of GaAs by Organometallic Chemical Vapor Deposition, *Appl. Phys. Lett.* **41**, 545 (1982).

105. J. P. Duchemin, M. Bonnet, F. Koelsche, and D. Huyghe, A New Method for the Growth of GaAs Epilayer at Low H_2 Pressure, *J. Cryst. Growth* **45**, 181 (1978).

106. R. P. Gale, R. W. McClelland, J. C. C. Fan, and C. O. Bosler, Lateral Epitaxial Overgrowth of GaAs and AlGaAs: Organometallic Chemical Vapor Deposition, in *Gallium Arsenide and Related Compounds*, 1982, G. E. Stillman, Ed., Conference Series No. 65, The Institute of Physics, London, 1983, p. 101.

107. L. K. Seidel-Salinas, S. H. Jones, and J. M. Duva, A Semi-empirical Model for the Complete Orientation Dependence of the Growth Rate for Vapor Phase Epitaxy: Chloride VPE of GaAs, *J. Cryst. Growth* **123**, 575 (1992).

108. D. W. Shaw, Selective Epitaxial Deposition of Gallium Arsenide in Holes, *J. Electrochem. Soc.* **113**, 904 (1966).

109. T. F. Kuech, M. Goorsky, A. Palevsky, P. Solomon, and T. A. Tischler, Selective Epitaxy of Compound Semiconductors in MOVPE Growth: Growth, Modelling, and Applications, *Proc. Mater. Res. Soc. Symp.* **198**, 23 (1990).

110. R. Sankaram, S. B. Hyder, and S. G. Bandy, Selective *In Situ* Vapor Etch and Growth of GaAs, *J. Electrochem. Soc.* **126** 1241 (1979).

111. B. J. Baliga, Silicon Liquid Phase Epitaxy, in *Epitaxial Silicon Technology*, B. J. Baliga, Ed., Academic Press, New York, 1986, p. 177.

112. M. B. Panish and I. Hayashi, Heterostructure Junction Lasers, in *Applied Solid State Science*, R. Wolfe, Ed., Vol. 4, Academic Press, New York, 1974, p. 235.

113. D. L. Rode, Isothermal Diffusion Theory of LPE: GaAs, GaP, Bubble Garnet, *J. Cryst. Growth* **20**, 13 (1973).

114. T. Brys'kiewicz, Investigation of the Mechanism and Kinetics of Growth of LPE GaAs, *J. Cryst. Growth* **43**, 101 (1978).

115. R. Ghez, An Exact Solution of Crystal Growth Rates Under Conditions of Constant Cooling Rate, *J. Cryst. Growth* **19**, 153 (1973).

116. R. Muralidharan and S. C. Jain, Improvements in the Theory of Growth of LPE Layers of GaAs and Interpretation of Recent Experiments, *J. Cryst. Growth* **50**, 707 (1980).

117. J. J. Hsieh, Thickness and Surface Morphology of GaAs LPE Layers, *J. Cryst. Growth* **27**, 49 (1974).

118. M. B. Small and J. F. Barnes, The Distribution of Solvent in an Unstirred Melt Under Conditions of Crystal Growth by Liquid Phase Epitaxy and Its Effect on the Rate of Growth, *J. Cryst. Growth* **5**, 9 (1969).

119. J. Crank, *Mathematics of Diffusion*, Oxford University Press, Oxford, 1955.

120. H. T. Minden, Constitutional Supercooling of GaAs Liquid Phase Epitaxy, *J. Cryst. Growth* **6**, 228 (1970).

121. M. B. Small and I. Crossley, The Physical Processes Occurring During Liquid Phase Epitaxial Growth, *J. Cryst. Growth* **27**, 35 (1974).

122. W. G. Pfann, *Zone Melting*, Wiley, New York, 1966.

123. H. C. Casey, Jr., M. B. Panish, W. O. Schlosser, and T. L. Paoli, GaAs-Al$_x$Ga$_{1-x}$As Heterostructure Laser with Separate Optical and Carrier Confinement, *J. Appl. Phys.* **45**, 322 (1974).

124. J. Heinen, Simultaneous Liquid Phase Epitaxial Growth of Multilayer Structures in a Multislice Boat, *J. Cryst. Growth* **58**, 596 (1982).

125. H. M. Manasevit, F. M. Erdmann, and A. C. Thorsen, The Preparation and Properties of (111) Si Films Grown on Sapphire by the SiH$_4$–H$_2$ Process, *J. Electrochem. Soc.* **123**, 52 (1976).

126. C. E. Weitzel and R. T. Smith, Silicon on Sapphire Crystalline Perfection and MOS Transistor Mobility, *J. Electrochem. Soc.* **125**, 792 (1978).

127. G. W. Cullen, M. T. Duffy, and R. K. Smeltzer, Recent Advances in Heteroepitaxial Silicon on Sapphire Technology, in *VLSI Science and Technology*, The Electrochemical Society, Pennington, NJ, 1984.

128. H. Kroemer, Polar-on-Nonpolar Epitaxy, *J. Cryst. Growth* **81**, 193 (1987).

129. J. W. Lee, J. P. Salerno, R. P. Gale, and J. C. C. Fan, Epitaxy of GaAs on Si: MBE and OMCVD, in *Heteroepitaxy on Silicon*, Vol. 67, J. C. C. Fan and J. M. Poate, Eds., Materials Research Society, Pittsburgh, PA, 1986, p. 33.

130. L. J. Huang and W. M. Lau, Effect of HF Cleaning and Subsequent Heating on the Electrical Properties of Silicon (100) Surfaces, *Appl. Phys. Lett.* **60**, 1108 (1992).

131. M. Grundner, D. Graf, P. O. Hahn and A. Schnegg, Wet Chemical Treatments on Si Surfaces: Chemical Composition and Morphology, *Solid State Technol.* (Feb. 1991) p. 69.

132. H. Zogg, S. Blunier, and J. Masek, Progress in Compound Semiconductor-on-Silicon Heteroepitaxy with Fluoride Buffer Layers, *J. Electrochem. Soc.* **136**, 775 (1989).

133. D. K. Arch, H. Morkoç, P. J. Vold, and M. Longerbone, High Performance Self-Aligned Gate (Al, Ga)As/GaAs MODFETs on MBE Layers Grown on (100) Silicon Substrates, *IEEE Electron. Dev. Lett.* **EDL-7**, 635 (1986).

134. H. K. Choi, J. P. Mattia, G. W. Turner, and B-Y. Tsaur, Monolithic Integration of GaAs/AlGaAs LED and Si Driver Circuit, *IEEE Electron. Dev. Lett.* **EDL-9**, 512 (1988).

135. S. K. Sakai, T. Soag, M. Takeyasu, and M. Umeno, Room Temperature Laser Operation of AlGaAs/GaAs Double Heterostructure Fabricated on Si-Substrates by Metalorganic Chemical Vapor Deposition, *Appl. Phys. Lett.* **48**, 413 (1986).

136. F. M. Smits, Measurements of Sheet Resistance with the Four-Point Probe, *Bell Syst. Tech. J.* **37**, 711 (1958).

137. W. G. Spitzer and M. Tannenbaum, Interference Method for Measuring the Thickness of Epitaxially Grown Films, *J. Appl. Phys.* **32**, 744 (1961).

138. Thickness of Epitaxial Layers of Silicon on Substrates of the Same Type by Infrared Reflectance, in *Annual Book of ASTM Standards*, Vol. 10.05, American Society of Testing and Materials, Philadelphia, PA, 1986, p. 212.

139. R. A. Moellering, L. B. Bauer, and C. L. Balestra, Epitaxial Layer Thickness Measurements Using Fourier Transform Infrared Spectroscopy (FTIR), *J. Electron. Mater.,* **19** 181 (1990).

140. L. J. van der Pauw, A Method for Measuring Specific Resistivity and Hall Effect of Discs of Arbitrary Shape, *Philips Res. Rep.* **13**, 1 (1958).

141. P. Chwang, B. J. Smith, and C. R. Crowell, Contact Size Effects on the van der Pauw Method of Resistivity and Hall Coefficient Measurement, *Solid State Electron.* **17**, 1217 (1974).

142. R. L. Petritz, Theory of an Experiment for Measuring the Mobility and Density of Carriers in the Space-Charge Region of a Semiconductor Surface, *Phys. Rev.* **110**, 1254 (1958).

143. R. G. Mazur, D. H. Dickey, A Spreading Resistance Technique for Resistivity Measurements on Silicon, *J. Electrochem. Soc.* **113**, 255 (1966).

144. J. I. Goldstein and H. Yakowitz, Eds., *Practical Scanning Electron Microscopy and Ion Microprobe Analyses,* Plenum, New York, 1975.

145. H. Hilibrand and R. D. Gold, Determination of the Impurity Distribution in Junction Diodes from Capacitance–Voltage Measurements, *RCA Rev.* **21**, 245 (1960).

146. R. R. Spiwak, Design and Construction of a Direct-Plotting Inverse-Doping Profiler for Semiconductor Evaluator, *IEEE Trans. Instrum. Meas.* **IM-18**, 197 (1969).

147. G. L. Miller, Feedback Method for Investigating Carrier Distributions in Semiconductors, *IEEE Trans. Electron Dev.* **ED-19**, 1103 (1972).

148. C. D. Sharpe, P. Lilley, C. R. Elliot and T. Ambridge, Electrochemical Carrier Concentration Profiling in Silicon, *Electron. Lett.* **15**, 622 (1979).

149. T. Ambridge, J. L. Stevenson, and R. M. Redstall, Applications of Electrochemical Methods for Semiconductor Characterization. 1. Highly Reproducible Carrier Concentration Profiling of VPE "Hi–Lo" *n*-GaAs, *J. Electrochem. Soc.* **127**, 222 (1980).

PROBLEMS

1. An effusion cell, containing gallium at 1000°C, is used to grow GaAs in an MBE system. Assuming a 2-cm-diameter cell and a cell substrate distance of 20 cm, calculate the arrival rate of Ga at the substrate, as well as the growth rate. Assume a partial pressure of 4×10^{-3} torr for the gallium.

2. Calculate the binding energy for the cube E in Fig. 5.2.

3. A bubbler, using hydrogen as a carrier gas, has an inlet tube which produces bubbles of 2-mm diameter. Assuming that the diffusivity of the reactor vapor in hydrogen is 0.05 cm^2/s, determine the minimum residence time for the bubbles, in order that they are saturated with the reactant vapor. Assume that saturation occurs when the vapor reaches the center of the bubble.

4. Given that $D/D_0 = (T/T_0)^{1.75}$, and $v/v_0 = T/T_0$, determine the dependence of the growth rate upon temperature for an endothermic process in the mass-transfer-limited region. Calculate the apparent activation energy at 700°C.

5. Calculate the width of the diffusion boundary layer at a point 1.0 cm down the susceptor, for an OMVPE reactor for GaAs growth, operating with 5-liter/min hydrogen carrier gas and a reaction tube diameter of 5.5 cm. What is the Reynolds number? Assume that the reactor is operated at 700°C and at atmospheric pressure, and that the diffusivity of TMGa in hydrogen is 0.31 cm^2 s^{-1}.

6. Repeat Problem 5, assuming operation at 0.1 atm. To a first-order approximation, the diffusivity varies as P^{-1}. Compare the flux for both cases. Assume identical partial pressures for the reactants.

7. Prove Eq. (5.40), starting with Eq. (5.38).

8. A phosphorus-doped layer, having a doping concentration of 10^{16} cm^{-3}, is grown on a boron-doped substrate, having a background concentration of 10^{15} cm^{-3}. Determine the junction lag resulting from this growth, assuming that $\phi = 2 \times 10^{-4}$ cm.

9. Repeat Problem 8, for an epilayer doping concentration of 10^{15} cm^{-3}.

10. A phosphorus-doped epitaxial layer of concentration N_E is grown on an arsenic-doped buried layer of concentration N_S. An abrupt doping profile results from this process. The slice is subjected to heat treatment at 1200°C for a period of 10 min. Sketch the doping profile resulting from this heat treatment for $N_S = 1000 N_E$. Assume diffusivities of 2.6×10^{-13} and 2.5×10^{-12} cm^2 s^{-1} for arsenic and phosphorus, respectively.

11. Repeat Problem 8, assuming a heat treatment period of 100 min. Compare the results for the two cases.

12. A susceptor is shaped so that the temperature gradient across a slice is parabolic with the center region cooler than the edge by T_0 degrees. Writing

$$T(r) = \frac{T_0}{R^2}(2Rr - r^2)$$

sketch the radial and tangential stress in this slice. Discuss your result.

CHAPTER 6

ION IMPLANTATION

Ion implantation, as applied to semiconductor technology, is a process by which energetic impurity atoms can be introduced into a single-crystal substrate in order to change its electronic properties. Implantation is ordinarily carried out with ion energies in the 50- to 500-keV range. Basic requirements for implantation systems are ion sources and means for their extraction, acceleration, and purification. This is followed by beam deflection and scanning prior to impingement on the substrate.

Ion implantation provides an alternative to diffusion as a means for junction fabrication in semiconductor technology. The technique, however, has many unique characteristics which have led to its rapid development from a research tool to an extremely flexible, competitive technology. Some of these unique characteristics are now considered.

1. Mass separation techniques can be used to obtain a monoenergetic, highly pure beam of impurity atoms, free from contamination. Thus a single machine can be used for a wide variety of impurities. Furthermore, the process of implantation is carried out under high-vacuum conditions—that is, in an inherently clean environment.

2. A wide range of doses, from 10^{11} to 10^{17} ions cm^{-2}, can be delivered to the target, and controlled to within $\pm 1\%$ over this range. In contrast, control of impurity concentration in diffusion systems is at best 5–10% at high concentrations, and becomes worse at low concentrations. Furthermore, dopant incorporation during diffusion is sensitive to variations in the electronic character of the surface, whereas this is not the case with ion implantation. Thus ion implantation provides inherently more uniform surface coverage than does

diffusion—particularly when low surface concentrations are required, and when the surface has a pattern of doped regions.

3. Ion implantation is usually carried out at room temperature. As a result, a wide variety of masks (such as silica, silicon nitride, aluminum, and photoresist) can be used for selective doping. This gives great freedom in the design of unique self-aligned mask techniques for device fabrication, which are not possible with diffusion technology.

4. Ion implantation provides independent control of dose and penetration depth. Many types of dopant profiles can be obtained by controlling the energy and dose of multiple implants of the same or different impurities. Both hyperabrupt and retrograde doping profiles can be made with relative ease in this manner.

5. Ion implantation is a nonequilibrium process, so that the resulting carrier concentration is not limited by thermodynamic considerations, but rather by the ability of the dopant to become electronically active in the host lattice. Thus it is possible to introduce dopants into a semiconductor at concentrations in excess of their equilibrium solid solubility.

6. Ion implantation can also be used for depositing a controlled amount of a charge species in a specific region of a semiconductor. Thus it has important applications in the threshold control of MOS devices.

There are some disadvantages to ion implantation as well. The equipment is highly sophisticated and expensive, so that the technology is at an economic disadvantage when compared to diffusion (in those areas where diffusion *can* be used). The competitive disadvantage is worse for GaAs (where energy requirements are in the 200- to 500-keV range) than for silicon (50- to 150-keV range). This, however, is offset by the fact that this technique lends itself to a high degree of automation, including in-line process monitoring as well as end-point determination. As a result, ion implantation is extensively used in both technologies today.

A second disadvantage is that ion implantation results in damage to the semiconductor. Annealing at elevated temperatures is necessary to heal some or all of this damage. This is not a problem with silicon since its vapor pressure is extremely low at annealing temperatures. Furthermore, silicon is often subjected to later high-temperature processes, where this damage can be completely annealed. With GaAs, however, it is necessary to use a cap, or to conduct the anneal in an arsenic overpressure, in order to avoid dissociation and loss of stoichiometry. Neither approach is completely satisfactory, and this problem has not been solved at the present time.

Yet another problem comes about because of the relationship of the penetration depth R_p and the Dt product associated with the anneal process. The movement of impurities during this process is dominated by point defect interactions, and is often three to four orders of magnitude larger than in undamaged, crystalline material. Thus, dopant motion during annealing can be significant.

Rapid thermal annealing techniques have already shown considerable promise for solving this problem. Although not fully understood, they have been incorporated into manufacturing processes for both these semiconductors.

In summary, the exploitation of ion implantation techniques for almost every aspect of silicon microcircuits fabrication has been phenomenal, because of the ease with which this material can be annealed. In contrast, its use is less common in GaAs circuits, because of some difficulties in this area.

6.1 PENETRATION RANGE

There are two basic stopping mechanisms by which energetic ions, upon entering a semiconductor, can be brought to rest [1]. The first of these is by energy transfer to the target nuclei. This causes deflection of the projectile ions, and also a dislodging of the target nuclei from their original sites. If E is the energy of the ion at any point x along its path, we can define a *nuclear stopping power* $S_n = (1/N)(dE/dx)_n$ to characterize this process. Nuclear stopping results in physical damage to the semiconductor, which takes the form of point as well as line defects. Often the semiconductor can become amorphous and/or semi-insulating as a result of this process.

A second stopping process is by the interaction of the ion with both bound and free electrons in the target. This gives rise to the transient generation of hole–electron pairs as energy is lost by the moving ion. We can define an *electronic stopping power* $S_e = (1/N)(dE/dx)_e$ to characterize this process.

The average rate of energy loss with distance is then given by

$$-\frac{dE}{dx} = N[S_n(E) + S_e(E)] \tag{6.1}$$

where N is the density of target atoms in the semiconductor. If the total distance traveled by the ion before coming to rest is R, then

$$R = \int_0^R dx = \frac{1}{N} \int_0^{E_0} \frac{dE}{S_n(E) + S_e(E)} \tag{6.2}$$

where E_0 is the initial ion energy. The quantity R is known as the *range*. A more significant parameter, of interest in semiconductor technology, is the projection of this range along the direction of the incident ion, as shown in Fig. 6.1a. Because of the statistical nature of this process, this *projected range* is characterized by its mean value R_p, as well as by a standard deviation ΔR_p along the direction of the incident ion. This latter term is referred to as the *straggle*.

In practice, the ion beam also has a spread at right angles to its incidence, as shown in Fig. 6.1b. This *transverse straggle* is denoted by ΔR_t, and is of

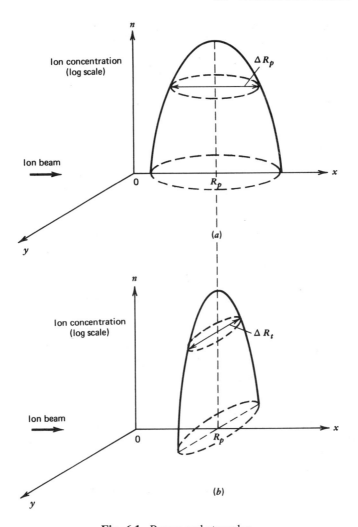

Fig. 6.1 Range and straggles.

importance in determining the doping distribution near the edge of a window which is cut in a mask.

Transverse straggle can be ignored if the width of this window is large compared to the implant depth. For this case, the statistical distribution for an amorphous target can be described by a one-dimensional gaussian distribution function of the form

$$\phi(x) = \exp\left[-\frac{1}{2}\left(\frac{x - \bar{x}}{\sigma}\right)^2\right] \qquad (6.3)$$

where $\bar{x}$ is the mean value and σ is the standard deviation. Integration of this function over the limits $\pm \infty$ results in the dose. Writing this dose as Q_0 ions cm^{-2} and noting that

$$\int_0^\infty e^{-z^2} dz = \frac{\sqrt{\pi}}{2}$$

(6.4)

the impurity distribution is given by

$$N(x) = \frac{Q_0}{(2\pi)^{1/2} \Delta R_p} \exp\left[-\frac{1}{2} \left(\frac{x - R_p}{\Delta R_p} \right)^2 \right]$$

(6.5)

This function has a maximum at R_p, and falls off rapidly on either side of this mean value.

In single-crystal semiconductors, if the incident beam is aligned to a major crystallographic axis, the ions can be steered for a considerable distance through the lattice with little energy loss, by a process known as *channeling*. Such channeled beams result in greatly increased penetration depth, as well as in reduced lattice disorder [2]. Unfortunately, the range is now critically dependent on the degree of alignment (or misalignment) of the beam and the crystallographic axis, and also on the ion dose.

Figure 6.2 shows an example of normalized doping profiles for phosphorus in silicon [3] for a beam aligned in the $\langle 110 \rangle$ direction, and for doses from 1.2×10^{13} to 7.25×10^{14} ions cm^{-2}. In all cases the position of the peaks is at 0.1 μm, and is only a function of the ion energy. The extent of the channeled region, on the other hand, is a sensitive function of the dose, and can result in a penetration depth that is almost a decade larger than would be obtained in the absence of channeling. It is seen here that the higher doses create more lattice disorder, and so the channeling effect is correspondingly reduced as the crystal quality is deteriorated.

Ion implantation for both silicon and GaAs microcircuit fabrication is usually carried out [4] so as to avoid any possibility of channeling. In both cases, the most common approach consists of tilting the slice by an angle of about 7–10° along the [110] direction in order to prevent the alignment of the beam with a major crystal axis. Additionally, the substrate is rotated by about 30–45° to further prevent any direct ion path through the crystal. Even so, it is extremely difficult to avoid channeling completely, especially with low-dose implants, since some of the ions can be deflected into channeling directions *after* entering the semiconductor. This residual channeling results in the formation of a tail as the dopant penetrates deeper into the semiconductor. Figure 6.3 shows typical doping profiles for boron and phosphorus implants into single-crystal silicon [3, 5] and illustrates this situation. Here, the doping profile described by Eq. (6.5)

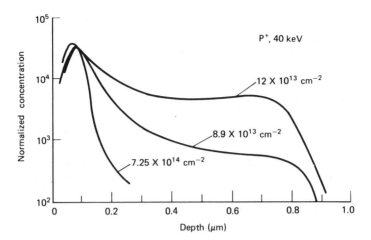

Fig. 6.2 Channeling of phosphorus in silicon, 50-keV ions. Adapted from Dearnaley et al. [3].

is closely followed for only one to two decades below the peak value. The tail is not observed in heavily damaged material, especially if it has been rendered amorphous, so that a gaussian profile is obtained in these situations [6].

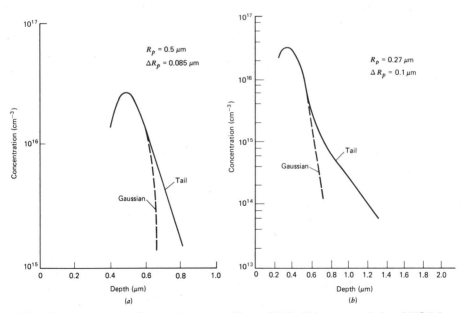

Fig. 6.3 Ion concentration. (*a*) Boron in silicon, 250-keV ions, annealed at 850°C for 30 min. Adapted from Moline [5]. (*b*) Phosphorus in silicon, 300-keV ions, annealed at 800°C for 30 min. Adapted from Dearnaley et al. [3].

Other approaches for suppressing channeling consist of pre-implanting the semiconductor with a neutral ion such as argon or neon, so as to make it amorphous before the active species is implanted. Implantation through a barrier layer, such as SiO_2, also serves to scatter the ion beam and reduce channeling problems.

The formation of a tail in the doping profile can also be caused by a rapid interstitial diffusion process undergone by the particles once they have lost their incident energy. Such particles continue diffusing, even at room temperature, until they encounter suitable trapping centers such as vacancies or a surface. If τ is an average trapping time and D is the diffusion coefficient associated with this process, it can be shown [7] that the impurity tail will be of the form given by

$$n = n_0 \, e^{-x/\sqrt{D\tau}} \tag{6.6}$$

In some situations, this tail region is enhanced when the semiconductor is annealed (by heat treatment) after the implantation. However, τ can be very short in heavily damaged material, thus suppressing its formation.

6.1.1 Nuclear Stopping

Some of the ions impinging on a semiconductor conductor surface are reflected by collision with the outermost layers of atoms. Some impart sufficient energy to target atoms which are then ejected from the substrate by a process known as *sputtering*. In this section, however, our interest focuses on the nuclear stopping process, by which the majority of the incident ions penetrate into the semiconductor and transfer their energy to the lattice atoms by elastic collisions. Detailed calculations for nuclear stopping were first advanced by Linhard, Scharff, and Schiøtt (LSS) and are available in the literature [8, 9]. Here the approach used in developing them is outlined in order to gain some insight into the implantation process.

Consider first the elastic collision of two hard spheres, each of radius R_0, as shown in Fig. 6.4. Let v_0 and E_0 be the velocity and kinetic energy of the moving sphere (projectile), of mass M_1. The mass of the stationary sphere (target) is M_2; after collision, its velocity and kinetic energy are v_2 and E_2, respectively. In like manner, let v_1 and E_1 be the velocity and kinetic energy of the projectile after impact. The distance between spheres is given by an impact parameter p. For this model, energy transfer only occurs if $p \leq 2R_0$. A head-on collision is represented by $p = 0$.

Upon collision, momentum is transferred along the line of centers of the spheres. In addition, the kinetic energy is conserved. Solving for these condi-

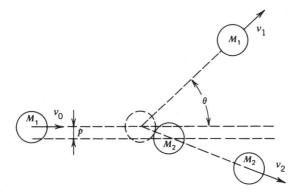

Fig. 6.4 Collision of hard spheres.

tions, the projectile deflects from its original trajectory by an angle θ, such that [10]

$$\cos \theta = \frac{1}{2}\left[\left(1+\frac{M_2}{M_1}\right)\left(\frac{E_2}{E_0}\right)^{1/2}+\left(1-\frac{M_2}{M_1}\right)\left(\frac{E_0}{E_2}\right)^{1/2}\right] \tag{6.7}$$

This equation shows that the energy transferred to the target particle is related to the scattering angle θ. In addition, the velocity v_1 of the projectile after impact is given by

$$v_1^2 = v_0^2\left[\frac{M_1 \cos \theta + (M_2^2 - M_1^2 \sin^2 \theta)^{1/2}}{M_1 + M_2}\right] \tag{6.8}$$

Maximum energy is lost by the projectile in a head-on collision with a target particle that is initially at rest. For this case, the final velocity of the target M_2 is

$$v_2 = \left(\frac{2M_1}{M_1 + M_2}\right) v_0 \tag{6.9}$$

so that the energy gained by it is T_m, where

$$T_m = \frac{1}{2}M_2 v_2^2 = \left[\frac{4M_1 M_2}{(M_1 + M_2)^2}\right] E_0 \tag{6.10}$$

This is also the energy lost by the projectile M_1.

The situation is more complex when the projectile and the target have an attractive (or repulsive) force between them. Associated with this force is a potential $V(r)$, which usually extends out to infinity. If this is the case, both particles will be continually moving (without physical encounter) as they approach each other, as shown in Fig. 6.5a, so that the impact parameter p also extends out to infinity. The energy transferred in this process is thus somewhat less than T_m, and depends on the impact parameter. The problem now reduces to a system of point masses moving under the influence of this potential, and has been classically treated [11] by two-particle elastic scattering theory. Here the approach consists of transforming from laboratory coordinates to a set of moving coordinates, whose center is located at the center of mass of the system. This renders the problem symmetric, as shown in Fig. 6.5b, with a deflection angle ϕ between the particles, where

$$\phi = \pi - 2p \int_{-\infty}^{R_M} \frac{dr}{r^2[1 - V(r)/E_r - p^2/r^2]^{1/2}} \tag{6.11}$$

where p is the impact parameter, R_M the minimum distance of separation between the particles, and $V(r)$ the interaction potential function. E_r is the energy of the ion in the center-of-mass system, and is given by

$$E_r = \frac{1}{2} \left(\frac{M_1 M_2}{M_1 + M_2} \right) v_0^2 \tag{6.12}$$

The integral of Eq. (6.11) can be evaluated for any given interaction potential

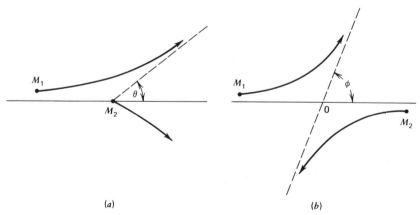

Fig. 6.5 Two-body scattering. (a) Laboratory coordinates. (b) Center of mass coordinates.

function. The magnitude of θ, in laboratory coordinates, can be shown to be given by

$$\tan \theta = \frac{\sin \phi}{\cos \phi + (M_1/M_2)} \tag{6.13}$$

and the energy transferred by this process is given by

$$T_n(E,p) = E_0 \frac{2M_1M_2}{(M_1 + M_2)^2}(1 - \cos \phi)$$

The choice of $V(r)$ is based on physical consideration of what happens if two atoms, with atomic numbers Z_1 and Z_2, approach each other at a distance r. The force between these atoms is coulombic in nature, and is given by

$$F(r) = \frac{q^2 Z_1 Z_2}{r^2} \tag{6.14}$$

where $q = 1.6 \times 10^{-19}$ coulombs. The potential is thus

$$V(r) = \frac{q^2 Z_1 Z_2}{r} \tag{6.15}$$

This situation holds only as long as the electrons of each atom are excluded. Their inclusion, however, results in a screening influence on the nuclear repulsion, which causes a modification of the potential function. One general form of such a potential function is

$$V(r) = \frac{q^2 Z_1 Z_2}{r} f\left(\frac{r}{a}\right) \tag{6.16}$$

where a is a screening parameter, and $f(r/a)$ is known as a *screening function*. One such function, shown in Fig. 6.6, is known as the *Thomas–Fermi screening function*, and has been found to be useful in calculations of nuclear stopping power. Often it is approximated by a/r, where

$$a = \frac{0.885 \, a_0}{(Z_1^{2/3} + Z_2^{2/3})^{1/2}} \tag{6.17}$$

and a_0 is the Bohr radius (0.53 Å). For this approximation, the potential is inversely proportional to the square of the distance.

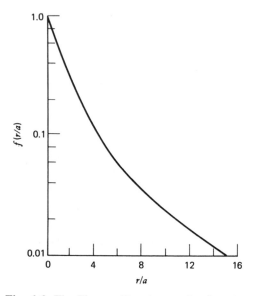

Fig. 6.6 The Thomas–Fermi screening function.

Using either the Thomas–Fermi screening function or its approximation, the deflection angle ϕ (and hence θ) can be computed for any impact parameter p. This, in turn, allows computation of $T_n(E,p)$, the energy lost by the incident particle in an elastic encounter with a single stationary particle.

In the practical situation of an amorphous target, of thickness Δx and having an atom density of N, energy is transferred to all particles. The total energy transferred is obtained by integrating over all possible values of impact parameter, so that

$$\Delta E = -N\Delta x \int_0^\infty T_n(E,p)2\pi p \, dp \qquad (6.18)$$

The nuclear stopping power is thus given as

$$S_n(E) = \int_0^\infty T_n(E,p)2\pi p \, dp \qquad (6.19)$$

Using this approach and the Thomas–Fermi screening function, the relationship for nuclear stopping takes the form shown in Fig. 6.7. Here, we note that as an ion enters a semiconductor (i.e., for large values of E), it first transfers energy to the lattice at a relatively slow rate. As the ion slows down, this rate

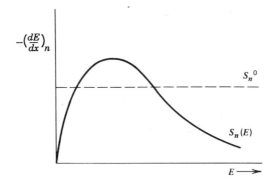

Fig. 6.7 Rate of energy loss as a function of ion energy.

of energy transfer increases, and then eventually decreases to zero as the ion finally comes to rest.

Use of the approximate form of the screening function, where the potential is inversely proportional to the square of the distance, leads to a constant rate of energy loss. This is shown by the dashed line Figure 6.7. The approximate expression for this rate of energy loss turns out to be

$$-\left(\frac{dE}{dx}\right)_n \simeq 2.8 \times 10^{-15} N \left(\frac{Z_1 Z_2}{Z^{1/3}}\right) \left(\frac{M_1}{M_1 + M_2}\right) \text{ eV cm}^{-1} \qquad (6.20)$$

where

$$Z^{1/3} = (Z_1^{2/3} + Z_2^{2/3})^{1/2} \qquad (6.21)$$

This corresponds to an energy loss rate of about 10–200 eV per angstrom for most ion–target combinations. The approximate stopping power S_n^0 is thus given by

$$S_n^0 \simeq 2.8 \times 10^{-15} \left(\frac{Z_1 Z_2}{Z^{1/3}}\right) \left(\frac{M_1}{M_1 + M_2}\right) \text{ eV cm}^2 \qquad (6.22)$$

A short listing of Z and M values is given in Table 6.1 at the end of this chapter.

6.1.2 Electronic Stopping

A comprehensive treatment of inelastic energy exchange processes, which accompany the excitation and ionization of electrons by collisions with incident ions, can only be provided by a quantum mechanical approach. However, semiclassical approaches can give reasonable estimates for the energy loss rate due to these processes. The fact that electron excitation and ionization do indeed occur is readily seen by considering a head-on elastic collision between an ion of mass M_1 and energy E_0, and an electron of mass m_0. Since $M_1 \gg m_0$, the maximum energy transfer due to this process is given by

$$\frac{4M_1 m_0}{(M_1 + m_0)^2} E_0 \simeq \frac{4m_0}{M_1} E_0 \qquad (6.23)$$

Excitation energies are a few electron volts, whereas $M_1/m_0 \simeq 1000\text{--}2000$. Consequently, both electron excitation and ionization are possible since the energy of incident ions during implantation is normally in the 50- to 500-keV range.

Approximate computations of the inelastic energy loss rate associated with this process have been made [10] by considering the projectile and target atoms as forming a quasi-molecule, with energy transferred due to electronic interaction as the particles undergo an encounter. As before, the total energy transferred to a target of thickness Δx can be obtained by integrating over all values of the impact parameter so that

$$\Delta E = -N\Delta x \int_0^\infty T_e(E,p) 2\pi p \, dp \qquad (6.24a)$$

and

$$S_e(E) = \int_0^\infty T_e(E,p) 2\pi p \, dp \qquad (6.24b)$$

The rate of energy loss can be shown to be given by

$$-\left(\frac{dE}{dx}\right)_e \simeq 2.34 \times 10^{-23} N(Z_1 + Z_2) v_0 \, \text{eV cm}^{-1} \qquad (6.25)$$

The projectile velocity is given by

$$v_0 = 1.412 \times 10^6 \times \left(\frac{E_0}{M_1} \right)^{1/2} \text{cm s}^{-1} \tag{6.26}$$

so that

$$S_e(E) \simeq 3.3 \times 10^{-17} \left(\frac{Z_1 + Z_2}{M_1^{1/2}} \right) E_0^{1/2} \text{ eV cm}^2 \tag{6.27}$$

Thus, the rate of energy loss due to electronic stopping is several (15–50) electron volts per angstrom, which is somewhat less than the energy loss rate for nuclear stopping.

An alternative approach is to consider the collision of an incident ion with a free electron gas. Using this approach, it can be shown that, if the ion velocity is less than the velocity of an electron having an energy equal to the Fermi energy, then the electronic stopping power is proportional to the ion velocity. This approach gives

$$S_e(v) = \left[\frac{2 q^2 a_0 Z_1^{7/6} Z_2 N}{\epsilon_0 Z} \right] \frac{v}{v_0} \tag{6.28}$$

where a_0 is the Bohr radius and Z is given by Eq. (6.21).

Using either approach, the electronic stopping power can be given in terms of ion energy as

$$S_e(E) = kE^{1/2} \tag{6.29}$$

The value of k depends on both the projectile and the target material. For an amorphous target, however, k is relatively independent of the projectile. For silicon,

$$k_{Si} \simeq 0.18 \times 10^{-15} (\text{eV})^{1/2} \text{ cm}^2 \tag{6.30a}$$

For gallium arsenide,

$$k_{GaAs} \simeq 0.3 \times 10^{-15} (\text{eV})^{1/2} \text{ cm}^2 \tag{6.30b}$$

Although both these approaches show a monotonic increase in $S_e(E)$ with energy, experiments have revealed a periodic dependence of this parameter on the atomic number of the ion. Other theories have attempted to treat this problem in more detail, and have resulted in equations with a better fit to this oscillatory behavior [12].

6.1.3 Range

Equations for computing the projected range, R_p, have been derived in the literature [13, 14], and evaluated for several different types of projectiles, substrates, and energies. A rough estimate of the range can be made by using the values of S_n^0 and $S_e(E)$ given in Eqs. (6.22) and (6.27), respectively. A sketch of these quantities, shown in Fig. 6.8, identifies a critical energy E_c at which nuclear and electronic stopping powers are equal. For silicon, E_c is about 17 keV for a light projectile such as boron, 140 keV for phosphorus, and as large as 800 keV for a heavy projectile such as arsenic. Comparable values for GaAs are about 15 keV for beryllium, 150 keV for sulfur, and 800 keV for selenium.

If the initial energy of the projectile is much lower than E_c, the dominant loss mechanism is nuclear stopping. Electronic stopping can thus be ignored in Eq. (6.2), resulting in a range

$$R \simeq \frac{E_0}{NS_n^0} \simeq K_1 E_0 \qquad (6.31a)$$

where K_1 is a constant for a particular projectile–target combination.

The range can also be roughly estimated for the situation where electronic stopping dominates—that is, if $E > E_c$. For this case, combining Eq. (6.2) with Eq. (6.27) gives

$$R \simeq K_2 E_0^{1/2} \qquad (6.31b)$$

where K_2 is a constant for a particular projectile–target combination.

Differential equations for computing the projected range R_p and straggle have been developed by LSS. Computer solutions of these equations, together with

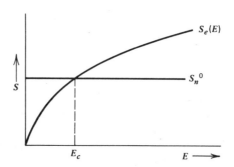

Fig. 6.8   Relative stopping power of an incident ion.

correction factors, are available in the literature for a variety of ion energies and ion–target combinations [13, 14]. To an approximation,

$$R_p = \frac{R}{1 + (M_2/3M_1)} \tag{6.32a}$$

and

$$\Delta R_p = \frac{2R_p\sqrt{M_1 M_2}}{3(M_1 + M_2)} \tag{6.32b}$$

We have shown that the ion profile of Eq. (6.5) is gaussian in character, and falls off symmetrically on either side of R_p. The extent of falloff, as a fraction of the magnitude at the peak value, is given in Table 6.2 at the end of this chapter. For some situations, such as boron in silicon, considerable skewing of this profile has been observed, with more impurity ions deposited in the region nearest the surface. Higher-order moments in range theory can be used to calculate more precise ion profiles for these situations. Figure 6.9 illustrates this behavior for a boron dose of 10^{15} cm^{-2} over a wide range of implantation energies [15]. For these experiments, both amorphous and polycrystalline silicon targets were used to avoid channeling effects.

Figures 6.10 and 6.11 give the projected range and straggle for dopants commonly used in silicon technology. Values for GaAs are given in Figs. 6.12 and 6.13.

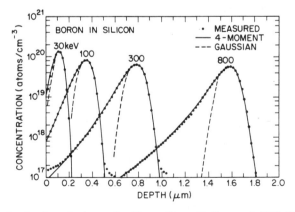

Fig. 6.9 Ion distributions for boron in silicon. From Hofker et al. [15]. With permission from the publisher, Gordon and Breach, Inc.

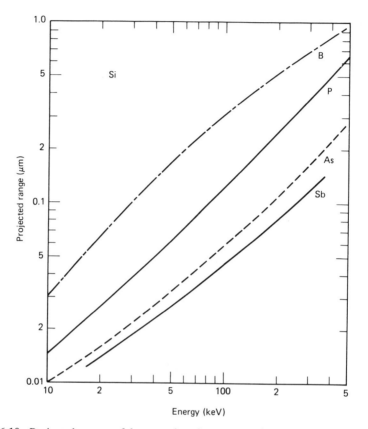

Fig. 6.10 Projected range of boron, phosphorus, arsenic, and antimony in silicon. Adapted from Gibbons et al. [13].

6.1.4 Transverse Effects

Transverse straggle is used to characterize the lateral spread of ions upon entering a target. This term is of importance in defining the ion penetration at the edge of a mask, and hence the curvature of a junction that may he formed by this process. It is also important when implantation is to be done through extremely narrow mask cuts, as in the case of microwave devices and in VLSI technology.

Consider an ion beam of infinitely small radius, which is incident on an amorphous target. If this beam enters the target in the x direction, then the spatial distribution function $f(x, y, z)$ is given by

$$f(x,y,z) = \frac{1}{(2\pi)^{3/2}\Delta R_p \Delta Y \Delta Z} \exp\left\{-\frac{1}{2}\left[\frac{y^2}{\Delta Y^2} + \frac{z^2}{\Delta Z^2} + \frac{(x - R_p)^2}{\Delta R_p^2}\right]\right\} \quad (6.33)$$

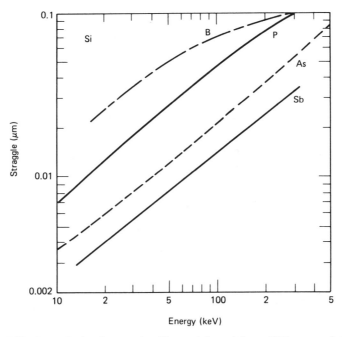

Fig. 6.11 Straggle for dopants in silicon. Adapted from Gibbons et al. [13].

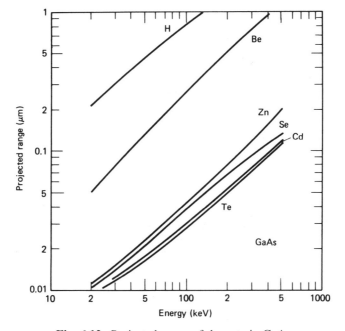

Fig. 6.12 Projected range of dopants in GaAs.

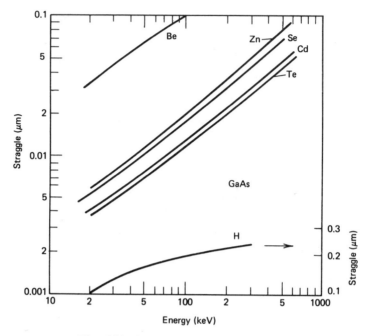

Fig. 6.13 Straggle for dopants in GaAs.

where ΔY and ΔZ are the standard deviations in the y and z directions. The transverse* straggle ΔR_t is then given by

$$\Delta R_t = \Delta Y = \Delta Z \tag{6.34}$$

because of symmetry. Figures 6.14 and 6.15 show the values of ΔR_t as a function of ion energy for implantation into silicon and GaAs, respectively, as calculated from the LSS theory.

Based on the above formulation, it can be shown that the spatial distribution of ions implanted through a narrow mask cut is given, for positive values of y [16], by

$$N(x, y, z) \simeq \frac{Q_0}{(2\pi)^{1/2}\Delta R_p} \left[\frac{1}{2}\text{erfc}\left(\frac{y - a}{\sqrt{2}\Delta R_t} \right) \right] \exp\left[-\frac{1}{2}\left(\frac{x - R_p}{\Delta R_p} \right)^2 \right] \tag{6.35}$$

where the width of the slit is $2a$, and the slit is parallel to the z direction, as shown in Fig. 6.16. Thus the ion concentration is 50% at the edge of the mask,

*Note that the radial straggle is $(\Delta Y^2 + \Delta Z^2)^{1/2} = \sqrt{2}\Delta Y$ because of symmetry.

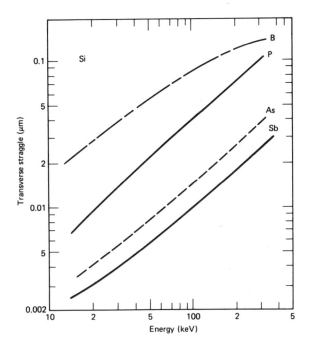

Fig. 6.14 Transverse straggle for dopants in silicon. Adapted from Furakawa et al. [16]. Reprinted with permission from the *Japanese Journal of Applied Physics.*

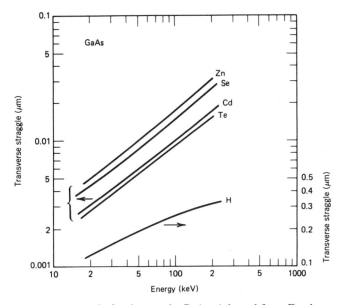

Fig. 6.15 Transverse straggle for dopants in GaAs. Adapted from Furakawa et al. [16]. Reprinted with permission from the *Japanese Journal of Applied Physics.*

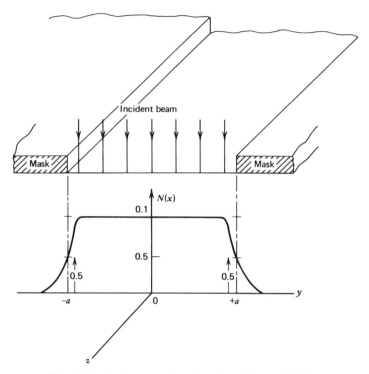

Fig. 6.16 Ion implantation through a slit of width $2a$.

and the falloff beyond this point follows a complementary error function profile. The extent of this falloff is shown in Table 6.3 at the end of this chapter.

Figure 6.17a shows equi-ion concentration contours for a variety of 70-keV implants into silicon through a 1-μm mask cut. Each contour represents the 0.1% concentration for this ion species. Figure 6.17b shows the 0.1% contours for a phosphorus implantation through the same mask cut, but over different energy ranges. These figures illustrate the magnitude of this effect, as well as its implications on device performance. As a rule of thumb, the transverse straggle is approximately 0.5 times the straggle for $M_2 = M_1$, but is equal to the straggle for $M_2 \simeq 10M_1$.

The theory outlined here must be applied with considerable caution in practical situations, for a number of reasons. First, the edge of the oxide is not vertical, but is generally tapered in character. Next, the 7–10° misalignment of the slice, together with its rotation by 30–45°, affects the impingement angle of the beam with respect to the substrate. Finally, the arguments presented here assume a gaussian ion beam spread, which is not true for all projectile–target combinations. Thus, large errors can occur with high-energy boron implants in silicon, which have been shown to exhibit considerable skew in their distribution characteristics.

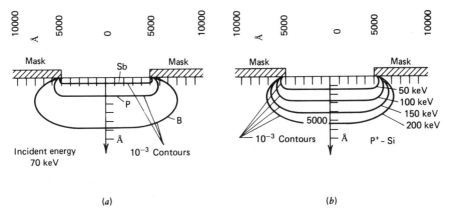

Fig. 6.17 Equi-ion concentration (0.1%) curves for implantation through a 1 μm mask cut. (*a*) 70-keV implants into silicon. (*b*) Variable-energy implants of phosphorus into silicon. Adapted from Furakawa et al. [16]. Reprinted with permission from the *Japanese Journal of Applied Physics.*

6.2 IMPLANTATION DAMAGE

During ion implantation, the energetic ion makes many collisions with the lattice atoms before finally coming to rest. Sufficient energy is transferred to the lattice so that many of its atoms are displaced, often with considerable energy. These in turn can displace other atoms, resulting in a cascade of collisions.

Consider the effect of a single energetic ion as it moves through the semiconductor, making successive collisions with target atoms. Let E_d be the energy required to dislodge one of these atoms from its site, and assume a hard-sphere model with elastic collisions. If a lattice atom receives less energy than E_d, it will not be displaced. Similarly, if the incident ion recoils from a collision with an energy less than E_d, it will not dislodge any additional atoms. Based on these assumptions, the incident ion must have an energy greater than $2E_d$ if there is to be an *increase* in the net number of displaced atoms. Assuming no additional energy transfer mechanisms, the number of displaced atoms per incident ion is roughly equal to $E_0/2E_d$, where E_0 is the initial ion energy [17]. The displacement energy for both silicon and GaAs is about 14-15 eV. Thus about $10^3 - 10^4$ lattice atoms are displaced as a result of each incident ion.

An understanding of lattice damage during ion implantation, and its removal by annealing, is of great importance to the practical exploitation of this technology. Experimental techniques for study of this damage involve direct observation by electron paramagnetic resonance, electron microscopy, backscattering, and channeling techniques [18]. Indirect techniques, based on the effect of damage on the semiconductor, are also used. Thus a crude but effective indicator of large amounts of damage is a visual change in surface reflectivity, accompanied by a milky appearance, with both silicon and GaAs. The sharpness of the optical attenuation characteristic with wavelength is yet another indicator. Mechanical properties, such as surface hardness, have also been used to study ion implantation damage.

Perhaps the most sensitive (and directly useful) technique involves studies of changes in the electronic properties resulting from ion implantation. For example, the minority carrier lifetime in silicon is reduced to below 1 ns for a low-dose (10^{12} cm^{-2}) boron implant at 100 keV. The mobility of majority carriers in both silicon and GaAs is reduced to below 1 cm^2/Vs after implant doses in the 10^{13}- to 10^{14}-cm^{-2} range. Theoretical computations of damage have also been made [19, 20], based on the assumption that the damage at any given depth is directly proportional to the amount of kinetic energy transferred to the lattice at that depth.

The nature of defects created in silicon by ion implantation are now summarized [21]. A large number of Frenkel (vacancy–interstitial) pairs are formed, but anneal by 100 K. Divacancies, created when an incident ion displaces adjacent atoms, are present in large number as well. These anneal by 550–600 K, and exist in material that has been implanted at room temperature. Vacancy–impurity pairs are also created: vacancy-group III defects anneal by 500K, whereas vacancy–group V defects anneal in the 400- to 500-K range.

Dislocations are also produced during ion implantation, and consist of clusters of point defects. They can also be formed by the strain produced by the implantation process, and disappear only upon annealing for several hours at 1000°C [22]. In addition, they can grow and multiply, by the Frank–Read mechanism described in Section 1.6.3, if they are pinned.

The nature of the damage path created by an incident ion will depend upon whether it is light or heavy relative to the lattice atoms. A light ion will generally transfer a small amount of energy during each encounter with the target, and will be deflected through a large scattering angle during this process. Each displaced target atom will have a small amount of energy imparted to it, and will probably not create many further displacements of its own. The damage produced by a single, light incident ion will thus take the form of a branching dislocation track, as shown in Fig. 6.18a. In addition, much of the energy from a light ion is transmitted to the lattice by electronic stopping, so that there is relatively little crystal damage. Finally, the range is comparatively large, so that the damage will be spread out over an extensive volume of the target.

The effect of a heavy ion is quite different. Here a large amount of energy is transferred with each collision, and the incident ion is deflected through a relatively small scattering angle. Each displaced atom is itself capable of producing a large number of displacements as it is moved away from the path of the incident ion. At the same time, the range is small, and most of the energy is transferred to the lattice by nuclear stopping, so that there is considerable lattice damage within a relatively small volume. The damage cluster now takes the form shown in Fig. 6.18b, and consists of a somewhat straight dislocation track, surrounded by a region in which a high density of interstitial atoms are

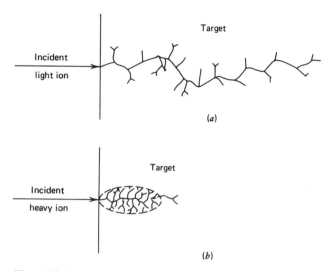

Fig. 6.18 Damage due to (*a*) light ions and (*b*) heavy ions.

clustered. In both situations, the volume of the target in which the incident ion energy is deposited is usually larger than the volume in which the actual lattice damage occurs.

The ion energy is transferred to the lattice in the form of a localized thermal spike. There is considerable interaction between the impurity ion and the target atoms during this process, often resulting in some self-annealing and in deviations from the gaussian profile. Such deviations are related to the nature of the charge associated with the ion species, and thus differ from one impurity to another.

The disordered region created by each incident ion has a central core in which the defect concentration is extremely large [22]. Otherwise, it consists primarily of vacancies, divacancies, vacancy–impurity pairs, and dislocations, with many of the impurity and target ions occupying substitutional as well as interstitial sites. With increasing dosage, the isolated regions of disorder begin to overlap. In the case of silicon, this eventually results in the formation of an amorphous layer, when all target atoms can be considered to be displaced from their lattice sites. From Fig. 6.18 and the discussion of the nature of damage, it follows that the lighter the ion, the larger the dose that is required to produce this amorphous layer. Typically, if the beam energy exceeds E_0 [the energy at which $S_n(E) = S_e(E)$], then the critical dose for creating an amorphous layer in silicon is given by

$$N_{\text{crit}} \simeq 5\Delta R_p \frac{NE_d}{E_0} \tag{6.36}$$

where N is the target atom density and E_0 is the beam energy [23].

The formation of an amorphous layer in silicon has been definitely established; for GaAs, however, there is evidence that the material becomes heavily polycrystalline under high-dose conditions, but does not transform to a truly amorphous state. It is possible that this comes about because of the tendency of this material to reorder during implantation, resulting in a crystalline material with a dense dislocation network. Often this highly disordered material is referred to as "amorphous" GaAs in the literature.

One advantage to achieving an amorphous state during silicon implantation is that the ion concentration profile now follows its predicted gaussian shape quite closely, because of the random nature of the host structure. In low-doping situations, the silicon is often implanted with an inert ion (argon, neon, or silicon, for example), to render it amorphous *before* the impurity ion is introduced. In some cases, the substrate is cooled to enhance the possibility of amorphization. Figure 6.19 shows data [24] for a low-dose 100-keV phosphorus implant into silicon which has been predamaged by 100-keV silicon implants of 10^{14} and 10^{15} ions cm^{-2}. The more heavily predamaged case corresponds to the situation in which the implant region was rendered amorphous prior to the phosphorus dose. Here the ion concentration profile is seen to closely follow the gaussian shape predicted by the LSS theory.

Damage effects in GaAs are considerably more complex. Here, because of their difference in atomic weights, gallium and arsenic recoil differently

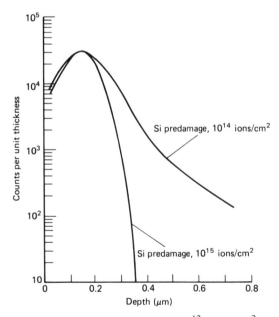

Fig. 6.19 Phosphorus implantation (100 keV, 7×10^{12} ions cm^{-2}) into silicon. Preimplanted. Adapted from Blood et al. [24].

[25], resulting in an excess of the heavier element (arsenic) near the surface. Annealing of the damage which results from this loss of stoichiometry is incomplete, and results in highly twinned material. Dopant activation effects are also more complex for this reason. In contrast to silicon, the best electrical activation is obtained when amorphization is avoided, and implantations are often made into heated substrates ($80–300°C$) for this reason.

In ion-implanted silicon, the mobility and minority carrier lifetime are extremely small. In addition, many of the incident ions do not come to rest in substitutional sites, so that the as-implanted carrier concentration is often several decades below the ion concentration. There exists, however, considerable long-range crystallographic order in this structure.

Ion implantation damage in GaAs makes this material semi-insulating (SI), with an active carrier concentration less than 10^{11} cm^{-3}. This characteristic can be used to advantage in order to isolate devices fabricated on the same substrate. In many situations, however, implantation damage in GaAs has been found to create unwanted layers of SI material within the substrate [26], resulting in the inadvertent formation of p–i–n or p–n^-–n structures instead of p–n junctions.

The energy of the incident ion falls off as it moves into the target, so that electronic stopping predominates in the initial penetration region. Consequently, the region of peak damage will be located at some distance from the target surface. In addition, this peak will occur further within the target if the incident ion beam has a higher initial energy, as will the range. In all cases, however, computations of the damage distribution show that it precedes the range distribution, and that the damage peak is (very approximately) at 75% of the projected range [27].

6.3 ANNEALING

We have noted that ion implantation damage results in a degradation of material parameters such as mobility and minority carrier lifetime. In addition, only a part of the as-implanted ions are located in substitutional sites, where they are electronically active. It is possible to heal some or all of this damage by annealing the semiconductor at an appropriate combination of time and temperature. Ideally, complete recovery of lifetime, mobility, and carrier activation would occur if the semiconductor could be returned to its original, single-crystalline state. This can only be achieved by melting and epitaxial regrowth.* Practical thermal anneal cycles fall far short of this stage, however, so that all annealing is essentially partial.

The actual details of the anneal treatment depend on the degree to which recovery of these properties are necessary for the satisfactory operation of the devices in a circuit. Thus they are a function of both device and circuit design.

---

*Solid-phase epitaxy is also possible, when the semiconductor is heated to somewhat below its melting point.

By way of example, the operation of silicon-based high-speed bipolar transistor logic circuits places great emphasis on attaining a high emitter carrier concentration, but actually requires a short minority carrier lifetime. On the other hand, the fabrication of a solid-state television camera target requires junctions with extremely low leakage—that is, a large space-charge generation lifetime. GaAs devices and microcircuits are most often based on majority carrier conduction, so that minority carrier lifetime is not important.* On the other hand, precise control of carrier activation and depth of the implanted region are primary requirements in devices such as field effect transistors.

Annealing requirements are also related to (a) the physical location of the damage in the semiconductor, and (b) its effect on device operation. Typically, the peak in the ion damage occurs closer to the surface of the semiconductor than does the peak in ion concentration. In diode structures of the type illustrated in Fig. 6.20, much of this damage is outside the depletion layer region, thus easing the requirement on maintaining long lifetime. In yet other situations, such as those where it is necessary to maintain an accurate, tail-free gaussian profile, the semiconductor must first be made amorphous by a preimplant. In order to be successful, both the preimplant type and the range must be chosen so that its region of damage coincides with the region where the impurity ion is to be deposited.

Thermal anneal cycles are usually 15–30 min in duration. The temperature is dictated by the application, and also by the ability of the semiconductor to tolerate such thermal treatment at this point in the fabrication process. Thus bare silicon can be readily heat-treated to 1200°C. On the other hand, a silicon microcircuit with aluminum metallization cannot be heat-treated above

---

*Important exceptions are electro-optical detectors and emitters.

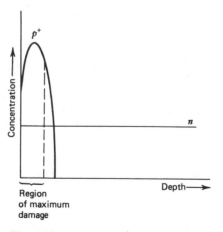

Fig. 6.20 Implanted p^+–n diode structure.

500°C if Al–Si alloy formation is to be avoided. GaAs will decompose if heated above 400°C for any length of time. Consequently, it must be annealed under an arsenic overpressure, or covered with a suitable capping layer during heat treatment.

In some situations, a suitable anneal-time/temperature cycle results in changes in the doping profile which can affect device performance. This is especially true in modern very-large-scale integration (VLSI) applications, where extremely shallow junctions are used. Rapid thermal annealing (RTA) processes, to be described in this chapter, are increasingly used in these situations.

During the anneal phase, displaced atoms tend to move so as to repair the damage created by implantation. At the same time, however, these atoms can coalesce, resulting in extended defects which require high temperatures to anneal. As a result, annealing behavior is strongly dependent on the nature of the implantation damage. In material which is heavily damaged, to the point of becoming amorphous, displaced atoms have an extremely short distance to move into lattice sites during anneal. Consequently, lattice repair is readily accomplished, and progresses by movement of the damage–crystalline interface towards the surface. Material with isolated defects, on the other hand, requires atom movement over long distances, and is thus considerably harder to anneal, requiring high temperatures and long anneal times. Here, too, the use of RTA techniques has proven to be highly successful.

6.3.1 Annealing Characteristics of Silicon

The annealing behavior of implanted silicon is reasonably well understood at the present time. Let us first consider the low-temperature annealing characteristics of a layer which has been made amorphous prior to the implant, or rendered amorphous by the implant itself. Individual disorder clusters begin to disappear at temperatures as low as 400°C, together with partial recovery (20–30%) of electronic properties. Actual epitaxial regrowth of this layer occurs within a narrow temperature range between 550°C and 600°C, with an activation energy of about 2.68 eV. Annealing occurs in a direction which is normal to the surface, and is ten times more rapid for (100) silicon than for (111) material. During this process, many of the impurity atoms move into lattice sites and become electronically active. This reordering is accompanied by the formation of dislocation loops, which are initiated at the end of the implantation range. However, considerable damage still remains in the lattice, because of the relatively slower rearrangement of silicon atoms on lattice sites.* Thus impurity activation occurs well before recovery of the lattice defect structure.

---

*Note that the movement of substitutional diffusers in damaged silicon is dominated by impurity–defect interactions. Thus it will be considerably more rapid than in undamaged material. By way of example, the activation energies of diffusion of boron and phosphorus in silicon are approximately 3.46 eV and 3.66 eV, respectively, whereas comparable values for ion implanted boron and phosphorus are approximately 0.62 eV and 0.34 eV, respectively.

Lattice damage serves as a sink for heavy metallic impurities such as copper, iron, and gold, all of which degrade the lifetime. Recovery of this parameter is thus difficult to achieve with annealing in the 600°C temperature range. Typically, the 600°C annealing of silicon for these implantation conditions results in an activation* of about 50–90% of the implanted ions, but very little recovery of the lifetime. The degradation of mobility is primarily caused by the unactivated dopant, so that its recovery goes hand in hand with that of the lifetime.

In many situations [metal-oxide-semiconductor (MOS) devices, for example], dopant activation and mobility are of primary interest. Often the silicon is preimplanted with a heavy, inert ion in order to make it amorphous before the active impurity implant. An alternative approach is to reduce the tendency to self-annealing by implantation at low substrate temperatures.

Dislocation loops, formed near the end of range during the annealing of amorphous layers, will grow with increasing temperature up to about 800°C. As a result, annealing of high-dose, heavy-ion implants at this temperature causes little improvement in lifetime, even though the dopant is 50–70% activated at this point. Recovery of this parameter only takes place at annealing temperatures beyond 950°C, with full recovery by 1000–1100°C.

Low-dose implantation leads to the formation of individual damage trees. Here, it is more difficult to obtain full dopant activation. Some evidence of annealing occurs by 350°C, however, as the individual dislocations begin to heal. Mobility values are extremely low at these temperatures. By about 500°C, however, the recovery of this parameter goes hand in hand with the dopant activation. Typically, 600°C annealing results in a 30–70% recovery of dopant activation. Recovery rates for light ions such as boron are considerably higher than for phosphorus or arsenic, because of the lesser extent of damage that is involved. The mobility typically recovers to 30–70% of the bulk value as well, with lifetime recovery to as much as 50% of its initial value.

With light-ion implants into preamorphized silicon, almost full recovery of all parameters is obtained by 800–950°C. A further increase in temperature, to 1000°C, results in 100% dopant activation, and full recovery of parameters for heavy ions. By this time, however, there is some change in the as-implanted profile because of diffusion effects.

In summary, therefore, it is relatively easy to remove damage caused by low-dose, light-ion implants in amorphized silicon by annealing at 800–950°C. However, the difficulty of annealing increases with both the dose as well as the ion mass. In all cases, dopant activation occurs at lower temperatures than does the recovery of either mobility or lifetime.

Annealing of implanted silicon is also carried out at 1100–1200°C, often as a process step in the formation of p-wells for complementary MOS (CMOS) technology. Here the final doping profile is essentially one-sided gaussian in shape, and is very similar to that obtained with conventional base diffusions. Full

---

*The lower value holds for high-dose, heavy-ion implants; conversely, the higher value is for low-dose, light ions.

recovery of carrier concentration, mobility, and lifetime also occurs, allowing the formation of low-leakage junctions.

Low-dose, low-energy ion implantation is an important step in MOS technology, where it is used to control the threshold voltage. Implantation is conducted into silicon substrates through the gate oxide, with ion doses in the 10^{11}- to 10^{12}-cm^{-2} range and with ion energies generally below 50 keV. This results in the formation of slow charge states at the Si–SiO$_2$ interface. For the same dose, the density of these charge states is linearly proportional to the incident ion energy, regardless of the ion type. Thus it is in all probability related to the density of displaced atoms $(\approx E_0/2E_d)$ created by this process. These can be annealed by heat treatment in the 200–300°C range, with complete recovery occurring within 30 min at 400°C. A small fraction of the impurity ion in the channel region is activated during this process.

The behavior of some commonly used dopants in silicon is now briefly described.

Phosphorus. This is a heavy ion, so that its projected range is almost linearly proportional to the implantation energy ($\approx 1.1\ \mu m/$ MeV). There is considerable evidence of tails in the implantation profile, particularly at low doses. It has been shown that these are formed during the actual implant process, probably by the steering of ions into channeling directions [6]. The activation of phosphorus in amorphous silicon (either preamorphized, or made amorphous by a sufficiently high dose) occurs rapidly at 525°C, to an 80% level. Doses which are not sufficient to make the material amorphous exhibit [28] monotonic increases in carrier activation between 500°C and 800°C, with 80% activation by 700°C.

Arsenic. The behavior of arsenic is very similar to that of phosphorus. Again, being a heavy ion, its projected range varies linearly with energy ($\approx 0.58\ \mu m/MeV$). Tails in the arsenic profile have also been observed, prior to annealing, and are caused by partial ion steering into channeling directions. Significant enhanced diffusion effects are also observed upon annealing, and there is some evidence of arsenic loss at the surface during this process. This loss can be reduced by using a deeper implant.

It is possible to implant arsenic into silicon in concentrations that exceed its maximum solid solubility, because of the nonequilibrium nature of this process [29]. Arsenic concentrations of this magnitude are metastable, and fall in a few minutes to their equilibrium value upon subsequent heat treatment. However, rapid thermal annealing has been shown to preserve this high conductivity state.

Boron. This is a light ion, with a projected range of approximately 3.1 $\mu m/MeV$ at energies of 10–100 keV. However, the range variation with energy becomes sublinear above 100 keV, due to the increasing contribution of electronic stopping. Boron implantation into an amorphous target exhibits considerable asymmetry in its ion profile at energies beyond 100 keV, as shown in Fig. 6.9. The complete description of its distribution function requires two additional parameters (skew and kurtosis) in addition to the usual range and standard deviation [15]. This "Pearson IV"-type distribution is further modified by the presence of an exponential tail to include enhanced diffusion effects during annealing. The extent of these effects depends on the quality of the starting material and the nature of the ion damage. As a result, the location of deep, boron-implanted junctions is difficult to control with any degree of accuracy.

Some annealing of boron implants occurs by 300–400°C. However, it is relatively difficult to obtain full activity with low-dose implants, unless annealing is carried out at 900–1000°C, or the substrate is made amorphous prior to the implantation.

Boron often exhibits an anomalous annealing characteristic [30], as shown in Fig. 6.21, with three clearly defined regions. In region I, point defect disorder is annealed out, with a rapid increase in the activation. An extended dislocation structure is evidenced in region II. Movement of boron atoms to relieve strain associated with these dislocations results in a rapid fall in activation, known as *reverse annealing*. Finally, the activation increases in region III, for annealing temperatures from 700°C to 900°C, as the boron precipitates dissolve into the silicon and become active.

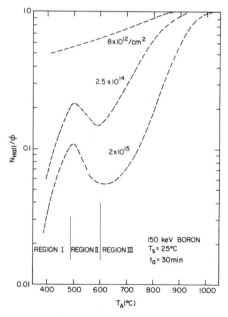

Fig. 6.21 The annealing of boron in silicon. Adapted from Seidel and MacRae [30].

6.3.2 Annealing Characteristics of Gallium Arsenide

GaAs devices and microcircuits are usually based on majority carrier transport, so that the minority carrier lifetime is generally not important. Consequently, emphasis in this area has focused on carrier activation and on the recovery of bulk mobility, both of which tend to go hand in hand. The primary application area is control of the channel doping in field effect transistors. Channels of this type are typically doped to 10^{17} cm^{-3}, with a requirement of uniformity across the wafer. A second important area is in the heavy doping of source and drain regions, in order to minimize the contact resistance of these devices. Here, dopant activation is a primary concern, rather than uniformity across the substrate.

Although the basic concepts of damage and its removal in silicon apply to GaAs as well, there are a number of important differences and complications which arise in work with this semiconductor. First, with the notable exceptions of beryllium and silicon, almost all dopants are heavy ions. They require considerable energy for implantation (in the 200- to 500-keV range), have short penetration depths, and create a large amount of lattice damage. Heavily damaged GaAs, even when implanted with an impurity ion, is semi-insulating because of its deep-level defect structure.

Ion implantation into GaAs results in a local loss of stoichiometry due to the different atomic weights of gallium and arsenic. A truly amorphous state does not appear to be created under these conditions; rather, the material becomes polycrystalline, with a very dense dislocation network. Reordering of the crystal structure requires displaced atoms to diffuse back to appropriate sites, often over considerable distance. This necessitates annealing at high temperatures ($\geq 900°$C), and there is no well-defined activation energy* associated with this process, since it requires the movement of two different host species, as well as a variety of defects, into their appropriate sites (e.g., gallium into gallium sites, arsenic into arsenic sites). In addition, there are a greater variety of defect interactions in GaAs than in silicon, and dopant activation is often less than 50%. Figure 6.22 shows results for a 40-keV tellurium implant [10] in GaAs with a 1×10^{15}-ion/cm^2 dose, and illustrates the gradual reduction in residual disorder that comes about with increasing annealing temperature. In contrast, silicon shows a rapid fall in disorder over a narrow ($50°$C) temperature range, indicating that annealing comes about by means of a single, dominant process.

Upon implantation, GaAs generally has no carrier activation. Annealing at temperatures up to $700°$C usually has no effect. Unlike silicon, activation of most dopants begins only *after* most of the implant damage has been removed, and then continues to increase with higher temperatures. Often significant activity is not achieved even at $900°$C, and there appear to be many stable defects beyond this temperature.

---

*Note that the activation energies of diffusion of gallium and arsenic in single-crystal GaAs are 3.2 and 5.6 eV, respectively. Activation energies in the presence of point defects are not known.

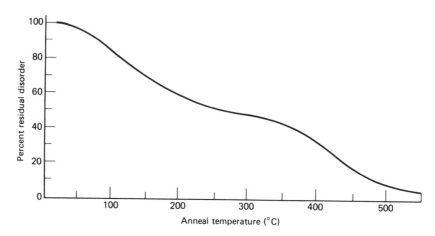

Fig. 6.22 Annealing behavior of lattice disorder in GaAs implanted at room temperature. Adapted from G. Carter and W. A. Grant, *Ion Implantation of Semiconductors* [10], John Wiley, 1976.

As a result of the above difficulties, it is common practice to implant GaAs at elevated temperatures (100–400°C) so that considerable self-annealing can occur during implantation. This prevents the material from reaching its "amorphous" state, from which it can only recover at temperatures near its melting point. The effectiveness of this approach is shown in Fig. 6.23, which illustrates the residual disorder for a heavier dose (40-keV tellurium implant at 5×10^{19} ions cm^{-2}) than for the situation in Fig. 6.22, as a function of the substrate temperature during implantation, and with *no* subsequent high-temperature annealing. It goes without saying that further reduction in disorder can be effected by a subsequent high-temperature anneal step. This approach is in striking contrast to that of silicon technology, where reduced temperatures are often used to make the substrate amorphous.

Substantial changes in doping profile have often been observed upon annealing. These are usually explained by assigning values for the diffusion constant which are three to four decades larger than those reported in the diffusion literature, and have been attributed to rapid interstitial-diffusion processes. In some cases, a rapid anomalous fall in carrier activation has been noted prior to an increase with annealing temperature. This phenomenon is not fully understood at the present time, and is probably due to competition for substitutional sites at different annealing temperatures, as has been observed with boron in silicon.

Changes in the doping profile from the expected gaussian shape have also been observed *before* annealing, when the implant is made into a substrate held at elevated temperatures. Thus fast diffusion has been noted for selenium implants [31] into substrates which were held at ≥150°C, even though 1000°C annealing *after* the implantation produced no additional diffusion.

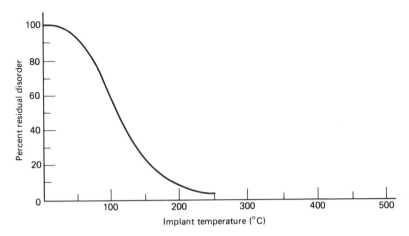

Fig. 6.23 Disorder in unannealed GaAs after implantation at elevated temperatures. Adapted from G. Carter and W. A. Grant, *Ion Implantation of Semiconductors* [10], John Wiley, 1976.

High-temperature anneals carry with them the problem of loss of stoichiometry, since implanted GaAs begins to dissociate by 350°C, with significant arsenic loss by 600°C. One approach to minimizing this problem is to use a capping layer during anneal. Often the implant is made through this cap, as a matter of convenience. This also serves to dope the cap with the impurity, and tends to reduce out-diffusion effects during the anneal step.

A variety of capping materials have been used, with moderate success. Here, an essential requirement is that the cap prevents the loss of both gallium and arsenic during the anneal step. Early work with silica and phosphosilicate glass has been superseded by silicon nitride, aluminum oxide, and aluminum oxynitride. Most success has been obtained with silicon nitride if it is especially prepared to be completely free of oxygen.

Capless annealing, in an arsenic overpressure, has also been used [32]. This approach has the important advantage of process simplification. However, control of the arsenic vapor pressure is important, since this controls the concentration of gallium and arsenic vacancies, and hence their occupation. As expected, dopants such as sulfur, which occupy arsenic sites, have higher activation with lower arsenic vapor pressure. Magnesium, on the other hand, occupies gallium sites and has higher activation at higher arsenic pressures. Silicon, although amphoteric, has a preferred location on gallium sites and is usually *n*-type upon annealing, with a considerable degree of compensation.

Finally, ion implantation in GaAs is sensitive to the quality of the layer in which the implant is made. The effect of substrate quality on the doping profile is shown in Fig. 6.24, where results are presented for two different Cr-doped SI substrates, as well as for implantation into an epitaxial high-resistivity buffer

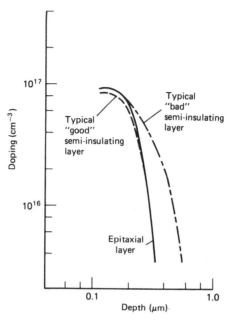

Fig. 6.24 Effect of selenium implantation into SI gallium arsenide. From Eisen [33]. With permission from the publisher, Gordon and Breach, Inc.

layer [33]. In all cases 1.8×10^{12}-cm^{-2}, 400-keV selenium ions were implanted at 350°C and annealed with a silicon nitride cap at 850°C for 30 min. Here implantation into a "good" substrate gave the same results as for an epitaxially grown buffer layer, in close agreement with LSS theory. Undoped material, made by *in situ* compounding techniques described in Chapter 3, also behaves as a "good" substrate with selenium implantation.

Results of implantation with different types of impurities now follow.

p-Type. Work with *p*-type impurities has included zinc, cadmium, beryllium, and magnesium, using room temperature implants [34]. In general, no carrier activation was found upon implantation at room-temperature, or with annealing up to 600°C. Above this point, however, there was a steady improvement in carrier activation; by 800°C almost 80–90% activation has been achieved for all dopants. Of these acceptors, beryllium has the lowest mass, and can be implanted at room temperature without "amorphizing" the GaAs. With other *p*-type impurities, implantation into heated substrates is highly desirable. Some workers have found that low-dose beryllium implants can be activated at temperatures as low as 500°C, with full activation by 800°C. This result is of particular significance since the decomposition of GaAs becomes increasingly rapid above this temperature.

The annealing behavior of beryllium at high temperatures has been found to be strongly dependent on the dose [35, 36]. This is seen from Fig. 6.25 for a 250-keV beryllium implant at 1×10^{15} ions cm^{-2}, followed by a 30-min anneal at a number of different temperatures. Here the concentration profile for this high-dose implant undergoes a rapid transition by 800°C, from a gaussian shape to that normally encountered in diffusion situations. In contrast, no such change was observed for a dose of 5×10^{13} ions cm^{-2}, even after annealing at 900°C for 30 min. Some similar results have been noted for zinc and cadmium. There is some evidence of a sharp reduction in carrier activation for zinc upon annealing at 700°C. However, 80–100% carrier activation is obtained by 900°C.

n-Type. Considerable work has been done with these dopants because of interest in high-speed GaAs devices, where the active region is *n*-type. Dopants under investigation have included sulfur, silicon, selenium, and tellurium. Implantation of these dopants has been found to be more difficult than for *p*-type, and requires critical control of implantation parameters, substrate temperatures, and capping layers. Their annealing behavior is more complex, because of their considerably slower diffusion rates. Of these impurities, both silicon and tin are amphoteric; however, their behavior after ion implantation is preferentially *n*-type.

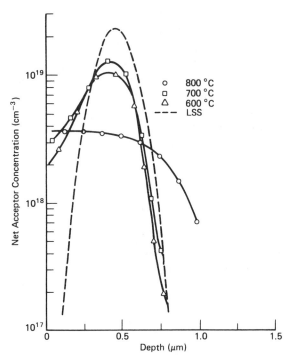

Fig. 6.25 Annealing of beryllium implants in gallium arsenide. From Levige et al. [35]. With permission from the *Journal of Applied Physics*.

Silicon is most commonly used because its light mass allows satisfactory implantations to be made at room temperature. This dopant is amphoteric in character, and incorporates into both Ga and As sites upon annealing. The compensation ratio thus increases with increasing anneal temperature, as does the $[V_{As}]/[V_{Ga}]$ ratio [37]. Typically, concentrations up to 2×10^{18} cm^{-3} can be achieved with 80% activation, beyond which point the doping level saturates. The use of an SiO$_2$ cap results in the preferential loss of gallium during annealing, since it is a rapid diffuser in this material. This, in turn, increases $[V_{Ga}]$, and hence improves silicon activation.

The activation of silicon in SI GaAs is also related to the background level of EL 2 [38], and has been found to increase by about 1% for each 10^{15} cm^{-3} of EL 2 in the substrate. It is believed that this is caused by an interchange reaction between As$_{Ga}$ and Si$_I$ during the annealing process. Detailed models of silicon implant activation are available in the literature [39].

Sulfur is an n-type dopant in GaAs. Results with ion implantation of this impurity have shown carrier activation as low as 20%, accompanied by enhanced diffusion, and considerable departure from the predicted gaussian profile. Studies of sulfur annealing have shown considerable out-diffusion into the cap layer, as well as some trapping at the substrate surface. It has also been noted that the doping efficiency is improved with deeper implants, which supports this argument.

Results for a 100-keV sulfur implant, with a dose of 9×10^{12} ions cm^{-2}, are shown in Fig. 6.26, together with the distribution calculated from LSS theory [33]. Also shown are comparable results for a 400-keV selenium implant, with a dose rate of 2×10^{12} ions cm^{-2}. Implants were followed by a 30-min anneal at 850°C with a silicon nitride cap. It is interesting to note that both implants resulted in the same peak carrier concentration. However, the selenium profile closely follows LSS theory, whereas that of sulfur is much deeper. Advantage can be taken of these differences in the fabrication of ohmic contact regions to shallow implanted layers, as will be described later.

Selenium can be activated to 90% and shows no significant diffusion during annealing. Thus, it is an extremely useful dopant for the implantation of n-type layers into SI substrates, which serve as a starting point for field effect transistors and integrated circuits, where control of both implant doping and depth must be precise. The behavior of tellurium [40] is relatively similar to that of selenium. However, it has a higher atomic weight, so that implantation to the same depth is accompanied by more damage than for selenium. Doping profiles are found to follow LSS theory at low ion doses, with increasing tail formation at higher levels.

Semi-insulating. Both p- and n-type GaAs can be made semi-insulating by proton bombardment. Here 10^{13} H$^+$ ions cm^{-2} in the 100-keV to 3-MeV range have been used to create a disordered layer whose thickness is about 10 μm/MeV of implant energy [41]. Layers as thick as 30 μm can be formed in this manner. Free carrier concentration in these layers is under 10^{11} cm^{-3},

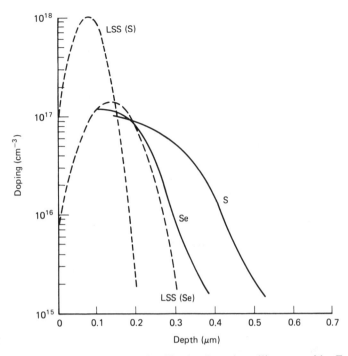

Fig. 6.26 Annealing of selenium and sulfur implants in gallium arsenide. From Eisen [33]. With permission from the publisher, Gordon and Breach, Inc.

so they can be considered to be semi-insulating for all practical purposes. Proton bombardment is useful for isolating devices on a common substrate, and is an important technique for microcircuit fabrication. Both single- and multiple-energy proton bombardments have been used in this application [42]. Deuteron bombardments have also been investigated for this purpose.

One disadvantage of device isolation by proton bombardment is that significant annealing can occur above 350°C. Oxygen implantation [43] has been used to avoid this problem, and introduces a number of damage-related trap levels, with the primary one being close to the midgap at $E_c - 0.75$ eV. Carrier removal is also due to the formation of complexes with the oxygen, its extent depending on the nature of the dopant species. In Si-doped GaAs, considerable carrier removal has been observed [44, 45], presumably by the formation of inactive Si–O complexes. No recovery of the dopant was seen with this dopant up to 800°C; however, implantation depths are much shallower than those obtained with proton bombardment, and are typically 0.7 μm for a 380-keV implant. Low-dose (10^{11}–10^{12} cm^{-2}) boron implants have also been used to produce localized region of SI GaAs for device isolation [46]. These regions are stable to 500°C.

6.3.3 Rapid Thermal Annealing

We have shown that it is often necessary to carry out the annealing step at temperatures around 1000°C in order to dissolve extended (end of range) defects in both silicon and GaAs. These have an activation energy of around 5 eV, and are associated with the diffusion of self-interstitials. There are two important problems associated with this procedure. First, the large thermal mass of slice carriers and insertion equipment necessitate process times of at least 15–30 min to obtain reproducible results. Significant changes in the doping profile can occur, making the precise placement of shallow junctions extremely difficult to control.

Next, loss of dopants such as arsenic can occur due to preferential evaporation effects. In GaAs, arsenic loss is more severe, with considerable deterioration of the material unless it is capped in an appropriate manner. These problems can be alleviated by the aid of RTA techniques. Experiments with RTA, using both coherent and incoherent sources, have shown little difference in annealing behavior [47]. As a result, switched, incoherent heat sources are commonly used for this purpose.

A schematic of a typical water-cooled RTA system, using tungsten halogen lamps, is shown in Fig. 6.27. Here, thermal mass effects are minimized by using point supports for the slice. At 1000°C, the thermal diffusion constant for silicon is about 0.1 cm^2/s, corresponding to a thermal diffusion length of 0.5 cm/s. Thus, heating of the slice is very uniform throughout its bulk [48], so that the incidence of slip is minimal. The absorption of radiant heat by the slice is related to the free carrier concentration, so that the heating rate for heavily doped material is more rapid than for slices with less doping. Temperature monitoring, coupled with feedback control, is therefore necessary in these systems. Optical pyrometers, as well as thermocouples (embedded in monitor slices) are used for this purpose.

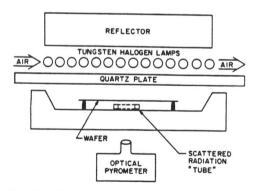

Fig. 6.27 Schematic of an RTA system. From Rimini [48]. With permission from Kluwer Publishers, Holland.

RTA has been extremely successful in silicon microcircuit technology, where the nature of extended defects, and their manner of removal, is well understood. Experiments have shown that these defects, especially those associated with the end of range, can be annealed in 10 s, at temperatures around 1000°C. This annealing schedule results in junction movements of under 100Å. Short annealing cycles of this type have also been found to be sufficient to activate boron implants in silicon, when the dose is not sufficient to make the material amorphous [49].

RTA systems usually do not have a provision for operation with an arsine atmosphere; this presents a problem for GaAs because of loss of arsenic during the anneal cycle. This is not serious if an uncapped, implanted layer is put "face to face" with a sacrificial slice of GaAs, provided that the anneal cycle is for 10 s or less at 900°C. Proximity annealing of this type is usually sufficient for p-type dopants, and for light n-type channel implants. For example, a 5-s anneal at this temperature can be used to anneal Be-implanted GaAs, with 100% activation [50].

Heavy implants, of the type needed as a base region for ohmic contacts, require annealing temperatures of 1000–1100°C, necessitating the use of a cap. Furnace anneals in an arsine ambient are more effective for this purpose, and moving furnaces are used to carry out this step in a "rapid" (about 5 min) manner.

An abrupt doping profile is especially important in modulation-doped GaAs FETs, where spillover of channel doping from AlGaAs into the GaAs can cause a fall in mobility. RTA methods have been very successful for avoiding this problem in these structures, using proximity annealing techniques. This approach has also been used for annealing of ion-implanted heterojunction bipolar transistors, where diffusional spread of thin regions cannot be tolerated [51].

6.4 ION IMPLANTATION SYSTEMS

The development of machines for ion implantation has been very rapid ever since their first application to semiconductor technology. Many recent improvements have been aimed at ease of operation, with sophisticated control electronics and automatic pump-down capability. Their performance has continually improved at the same time, with special emphasis on dose uniformity over large areas, accuracy of dose measurements, and multiple wafer-handling capability. In this section we outline some of the basic elements of these systems, without attempting to provide a complete catalog of all improvements and refinements.

Ion implants are usually specified in terms of two key parameters; the ion energy which determines the penetration depth, and the ion current which sets the dose (and thus the implantation time). The general rule is that higher beam energies go hand in hand with lower currents, and vice versa. However, there are exceptions to this rule.

The choice of beam parameters depends on the system application. In silicon technology, implants for use in predeposition, prior to drive-in, do not require beam energies above 25–50 keV; however, a high beam current (1–20 mA) reduces the implant time for this large-volume application. Implanters for deep doping applications, such as the formation of buried layers in bipolar technology and wells in CMOS circuits, require extremely high energies (3–5 MeV). Some special applications, such as the formation of silicon on insulating layers, to be described in Section 6.7.1, require both high currents (100 mA) and moderately high beam energies (250 keV).

GaAs systems generally require high beam energies, because the average atomic weight of this material (72.33), and hence its stopping power, is considerably larger than that of silicon (28.09). Current requirements are relatively modest, however. Typical implanters for GaAs operate at energies up to 500 keV, with beam currents of 0.1–0.5 mA.

The basic features of an ion implantation machine [52, 53] are a variety of ion sources, means for ion extraction, acceleration to high energies, and beam manipulation. Ion beams are generally produced in a discharge chamber, so that they are contaminated by atomic and molecular ion species which are sputtered from its walls and filaments. Purification of this beam, to select the desired implant species, is thus an essential requirement of these machines. This is carried out by means of a mass analyzer, which selects a single species of interest. This analysis can be done before or after the beam has been accelerated to the required energy. Analysis before acceleration results in the use of a smaller, low-cost machine and is often used for this reason. On the other hand, analysis after acceleration is preferred when it is required to implant molecular species, which might partially dissociate during the acceleration process. The presence of multiple species at this point leads to problems in dosimetry as well as in the implantation process itself.

A wide variety of beam manipulation techniques can be used. These include electrical, magnetic, and mechanical systems. Moreover, hybrid schemes are often used when many slices must be handled at any one time. Techniques for precise monitoring of the integrated ion current must also be provided. Perhaps the key advantage of ion implantation is in this area, where the dose repeatability can be held to better than 1%.

Finally, a wide variety of high-voltage and vacuum techniques are employed to result in a system that is convenient to use. These include (a) load-lock devices which allow substrate loading without breaking vacuum and (b) auxiliary ports which allow substrate loading in one chamber while implantation is being carried out in another.

Figure 6.28 shows a block diagram of one such system in which the analyzer is placed after the accelerator. Here, the ion source is at a high voltage and the rest of the system, beyond the accelerator, is at ground potential. All the ion sources, including their power supplies and control elements, are operated by remote means, as are the ion optical elements of the accelerator column. This allows the rest of the system be kept at ground potential, for ease of safe access.

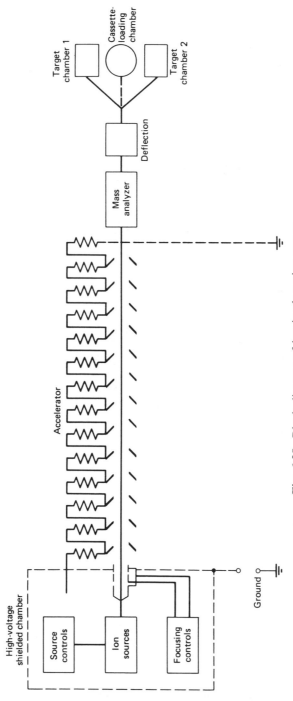

Fig. 6.28 Block diagram of ion implantation system.

Separate valves and pumps are provided to allow individual access to the target chambers, while keeping the rest of the system under high-vacuum conditions at all times. In some systems, a cassette loader, with load-lock capability, is used to facilitate slice handling. Particular care is taken to operate the target chambers with oil-free pumping systems (e.g., cryogenic, sorption, and turbomolecular), to avoid hydrocarbon cracking of residual oil vapors.

6.4.1 Ion Sources

Ion sources usually consist of (a) compounds of the desired species and (b) a means for their ionization prior to delivery to the accelerator column. The choice of materials for ion sources is very wide [54]; almost any ionizable compound can be used here. Gaseous materials are more convenient to use than solid ones since they avoid the necessity of using a vaporization chamber, and can be replenished without opening the system. Consequently, these are the preferred choice in implantation systems today.

Ionization of the source material is usually done by passing the vapor through a hot cathode electronic discharge. Cold cathode and r.f. discharges are also used in some machines. Electrons are accelerated towards an anode which is typically at 100 V. A magnetic field is provided so as to force the electrons to move in a spiral trajectory, thus increasing the ionizing efficiency of the source. Also provided is a means for extracting the positive ions from this discharge by means of an anode biased at 15–20 kV. As a consequence, the output of a source of this type consists of ions at an energy of 15–20 keV. The outlet of the ionizer is either circular or in the form of a rectangular slit, and defines the geometry of the ion beam. Figure 6.29 shows a sketch of a gas-fed ion source [55], and illustrates many of the features described here.

The effectiveness of an ion source is measured by the magnitude of ion current delivered to the accelerator, and ultimately to the target. If I is the ion

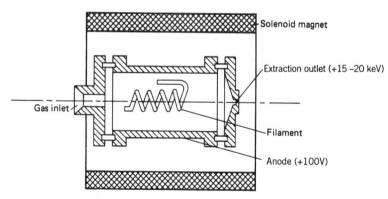

Fig. 6.29 A Nielsen-type gaseous source. Adapted from Nielsen [55].

current (in amperes) for a species of charge state r, then the rate at which ions arrive at the target is I/qr per second, where q is the electronic charge in coloumbs. Thus a singly ionized beam delivers $6.25 \times 10^{18} I$ ions/s to the target. Finally, if A is the target area in cm^2, and t is the implantation time in seconds, the ion dose is given by

$$Q_0 = 6.25 \times 10^{18} \frac{It}{A} \text{ ions cm}^{-2} \tag{6.37}$$

A large beam current is thus highly desirable in commercial systems where many wafers must be handled with a minimum of machine time. In practice, the beam current is actually a function of the ion source material, the ion extraction technique, and the machine design. By way of example, high-current arsenic or phosphorus ion beams can be more readily produced than beams of boron. Typical values for a general-purpose production machine are in the ratio of 8 : 8 : 5 for these ions, respectively.

6.4.1.1 Molecular Beams

Shallow implants of light ions are extremely difficult to make with any degree of precision, for two reasons. First, the beam energy at the output of the ion extractor is often too high (15–20 keV) for this purpose. While it is possible to use the accelerator in a reverse-biased mode to reduce the beam energy, lower energies are undesirable because they lead to ejection of target atoms by sputtering. Next, low-energy beams of light ions create almost no damage in the target, thus increasing the chance of channeling as well as the chance of departures from a gaussian impurity profile. Molecular beams can be used to advantage in this situation.

Consider, for example, the case of boron in silicon. Here, BF_3 is commonly used as a dopant source. Upon ionization, it breaks into the fragments shown in Fig. 6.30. Of these fragments, $^{11}B(^{19}F^+)_2$ is commonly selected for ion implantation purposes. Upon impact with the target surface, BF_2 further fragments into its boron and fluorine ions, with the energy divided according to their relative masses. Thus, a BF_2 molecule with a beam energy of E_0 fragments to one boron ion of energy $(11/49)E_0$ and two fluorine ions, each of energy $(19/49)E_0$. Thus, the energy of the boron ion is effectively 22% of that of the BF_2 molecule. Additionally, the relatively heavy, energetic fluorine ions increase the lattice damage, and thus reduce the possibility of channeling. Fluorine is inert in silicon [56], and is readily lost during subsequent thermal processing above 800°C. The critical dose required to amorphize silicon is under 2.5×10^{15} ions cm^{-2} for BF_2 ions in the 25- to 150-keV range, and is much lower than that required for boron [57].

The use of BF_2^+ can lead to problems in systems where the accelerator is situated after the analyzer, because it is possible to obtain some fragmentation

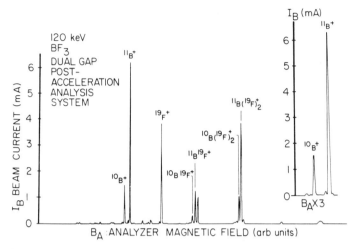

Fig. 6.30 Fragmentation spectrum of BF$_3$ at 120 keV. Adapted from Hanley [59].

at this point. Now, the beam consists of B$^+$ as well as BF$_2^+$, both of which* are accelerated to the same energy. However, the B$^+$ ions will penetrate considerably deeper than the BF$_2^+$ because of their much lower mass. This can result in departures from the gaussian implant profile, and also lead to errors in dose measurement.

Molecular ion sources are not used in GaAs technology, where the *avoidance* of amorphization is desirable for high activation. An exception is BeF$_3$, which is routinely used because it provides a larger beam current than is available from a sputtered Be source.

6.4.2 The Accelerator

Energy is imparted to the ion beam by passing it through a long column across which the accelerating potential is established by a series of biased, annular, ring electrodes. The output end of this tube is usually maintained at ground potential for safety reasons. Care must be taken in the design of the accelerator tube to minimize collisions with slits and apertures, in order to prevent the formation of secondary electrons. In systems where the accelerator follows the analyzer, these collisions can result in additional fragmentation products.

The beam energy determines the projected range of an ion. However, it is possible to increase this range by using multiply charged ions. This usually results in a lower beam current. For example, an arsenic source typically produces As$^+$ and As^{2+} ions in a 10:1 ratio. Thus, use of an As^{2+} beam results in a large penalty in beam current, although it doubles the effective energy. For this reason, it is customary to use singly charged ion beams whenever possible.

---

*Additional fragments are ignored in this discussion, for simplicity.

6.4.3 Mass Separation

We have noted that the raw output from an ion source consists of many species, in addition to contamination produced by sputtering from its walls and filaments. Moreover, the chemicals which are used to feed the ion source often contain impurities as contaminants, even if every effort is made to use chemically pure reagents. As a result, some form of mass separation is essential in a practical ion implanter. The use of mass separation techniques provides a unique distinction between ion implantation and diffusion, in that a variety of dopants can be handled in a single machine, with complete freedom from contamination with each other.

The most commonly employed technique utilizes a homogeneous-field magnetic analyzer. Its principle is based on the dynamics of charged particles, of mass m and velocity v, moving at right angles to a uniform magnetic field with a flux density B. These particles will experience a force F such that

$$F = q(v \times B) \tag{6.38}$$

This tends to move them in a circular path of radius r, thereby creating a centrifugal force mv^2/r. These forces must be equal and opposite. In addition, the velocity of the particles is related to their energy by

$$\tfrac{1}{2}mv^2 = qV \tag{6.39}$$

where V is the accelerating potential. Combining these relations, the radius of the ion path is given as

$$r = \frac{1}{B}\left(\frac{2mV}{q}\right)^{1/2} \tag{6.40}$$

Thus, for a given extraction voltage and magnetic flux density, the path radius is directly proportional to the square root of the mass. Trajectories for three different masses are shown in Fig. 6.31. From this figure it is seen that ions of any particular mass can be selected by the appropriate placement of an exit slit.

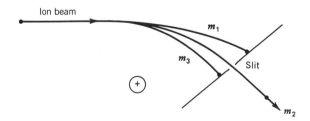

Fig. 6.31 Mass separation through a slit.

Velocity filters are sometimes used when extremely precise control of the range is desired. The most common is the Wien filter. This filter consists of a chamber through which the beam passes in a direction that is perpendicular to a crossed electric and magnetic field system. Such a beam will travel straight through if the electric and magnetic forces balance each other. If $\mathscr{E}$ and H are the electric and magnetic field intensities, then the condition for zero deflection is given by

$$q\mathscr{E} = Hqv \qquad (6.41)$$

where v is the ion velocity. Thus ions of velocity v pass through such a filter without disturbance, provided that

$$v = \frac{\mathscr{E}}{H} \qquad (6.42)$$

6.4.4 Beam Scanning

The primary requirements of beam scanners in commercial implantation systems are uniform coverage and the ability to handle a large number of slices in a single pump-down. Here, electrostatic scanning in both x and y directions is the simplest, and has the advantage of very high scanning speed. It is used in the majority of single-wafer, cassette-loading systems, where beam currents are below 1 mA and where beam voltages are 50 keV or higher. At higher beam currents (and lower beam energies), beam expansion due to space charge creates problems with these systems [58].

Scanning in high-current, low-energy systems, of the type used for pre-deposition, is usually accomplished in a hybrid deflection system where the beam is electrostatically (or electromagnetically) scanned in one direction while the slices are mechanically moved in the other.* One such system, shown in Fig. 6.32, consists of electrostatic deflection combined with a rotating carousel which can hold many wafers.

During implantation, practical considerations preclude the possibility of mounting wafers in good thermal contact with the slice carrier; often pressure fingers or slides are used to hold these slices. In addition, implantation is carried out in a vacuum environment so that the primary mechanism of heat loss from the slice is by radiation. Ion beam heating effects can thus be quite severe, and often limit the rate at which energy can be deposited in a slice. As a result, the temperature attained by a slice, during implantation, can be significant.

Figure 6.33 shows the equilibrium temperature that is reached by slices during implantation as a function of ion dose, on the assumption that heat loss is only by radiation [59]. This calculation shows that temperature excursions of many hundreds of degrees are possible in currently available systems. Fur-

---

*Neutral particles are not deflected by any of these techniques. A bend in the ion beam line, of 5–10°, is usually made so that they can be picked off by a collector plate.

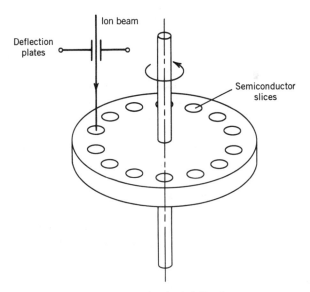

Fig. 6.32 Electromechanical deflection system.

thermore, the temperature across a slice is extremely uneven, since the slice is usually held at three or four points along its edge. This uneven heating creates a major problem when it is necessary to preimplant a semiconductor with a

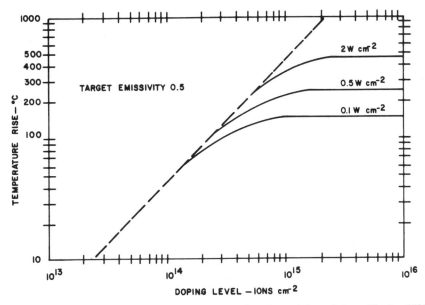

Fig. 6.33 Equilibrium target temperature versus ion dose. Adapted from Hanley [58].

heavy ion in order to make it amorphous, or when implantation is to be carried out in cooled substrates for the same purpose.

Thermal effects place a limit on the beam current density that can be used during implantation, and thus on the implantation time required to achieve a desired doping level, unless some means are provided for cooling the substrates. Both gas and liquid nitrogen cooling of the back surface have been attempted with some degree of success. Here, the slow scan speed of carousel-type scanners results in much higher instantaneous power densities than are obtained in fast, electrostatic deflection systems. However, carousel systems have provision to clamp slices to a cooled substrate, whereas substrates in cassette-loading systems are usually held in place by gravity. Consequently, cassette loading, with slices implanted one at a time, is the favored approach for wafer implantation in machines where the beam current is less than 1 mA. However, carousel scanning, of the type shown in Fig. 6.32, is used in the higher current "predeposition" machines, with provision for cooling the substrate.

6.4.5 Beam Current Measurements

Dosimetry is an important feature of ion implantation technology, and is done by measuring the ion integrated beam current to obtain the total dose. Accurate measurement is a difficult task, and is only achieved if care is taken to restrict the ions in the beam path to a single species. This can be a problem with molecular ions, which are subject to fragmentation when the accelerator follows the analyzer stage.

Accurate dose measurements necessitate the elimination of electrons, neutrals, and negative ions from the beam, as well as secondary particles which can be emitted as a result of the ion bombardment of the target. Improvements in machine design have resulted in minimizing the error sources generated from the beam, so that the major problems of measurement are those caused by the secondary particles. Faraday cage arrangements, with both electrostatic and electromagnetic suppression schemes, are used to minimize measurement errors from these secondaries. A schematic of such an arrangement is shown in Fig. 6.34. Often a small magnetic suppression flux ($\simeq 100$ gauss) is also incorporated by means of a permanent magnet. Techniques of this type allow dose measurements to be made over six decades of ion flux, with an accuracy of better than 0.5% on all ranges.

6.5 PROCESS CONSIDERATIONS

Ion implantation is an extremely flexible process. As such, it is used not only as an alternative to diffusion, but also in many ways that are quite unique to this approach. Thus a number of different process considerations must be understood if full advantage is to be taken of this technology. Some of these are now described.

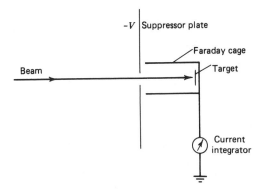

Fig. 6.34 Dosimeter arrangement.

6.5.1 Diffusion Effects During Annealing

Ion implantation into an amorphous target (or a crystalline target that is misoriented) results in a gaussian ion profile, at least for a few decades down from the peak ion concentration. This profile is modified by diffusion effects during annealing. In general, the diffusivity is dominated by point-defect interactions and can be enhanced by as much as three to four orders of magnitude during the early stages of annealing [60]. Let D and t be the diffusivity and the time associated with this process, respectively, and assume that the substrate is infinitely thick on either side of the implant. For this approximation, the ion concentration will still retain its gaussian shape, with the same mean value. However, the standard deviation will be altered from σ to $(\sigma^2 + 2Dt)^{1/2}$, so that the distribution is now of the form

$$\phi(x) = \exp\left[-\frac{1}{2}\left\{\frac{x - \bar{x}}{(\sigma^2 + 2Dt)^{1/2}}\right\}^2\right] \tag{6.43}$$

The ion concentration is thus given by

$$N(x) = \frac{Q_0}{(2\pi)^{1/2}(\Delta R_p^2 + 2Dt)^{1/2}} \exp\left\{-\frac{1}{2}\left[\frac{x - R_p}{(\Delta R_p^2 + 2Dt)^{1/2}}\right]^2\right\} \tag{6.44}$$

The presence of the surface, at a finite distance from R_p, can be taken into consideration by assuming perfect reflection at this boundary. A solution for this situation, assuming no out-diffusion effects, has been given [61] as

$$N(x,t) \simeq \frac{Q_0}{(2\pi)^{1/2}(\Delta R_p^2 + 2Dt)^{1/2}} \left(\exp\left\{ -\frac{1}{2}\frac{\Delta R_p}{R_p} \left[\frac{x - R_p}{(\Delta R_p^2 + 2Dt)^{1/2}} \right]^2 \right\} \right.$$
$$\left. + \exp\left\{ -\frac{1}{2}\frac{\Delta R_p}{R_p} \left[\frac{x + R_p}{(\Delta R_p^2 + 2Dt)^{1/2}} \right]^2 \right\} \right) \qquad (6.45)$$

A plot of this equation, for the specific set of values shown, results in the doping profiles of Fig. 6.35. It is interesting to note that, for sufficiently large Dt values, the bell shaped profile is replaced by the type normally expected from diffusion. However, appreciable movement of the peak ion concentration from R_p does not occur until Dt is at least equal to $2.5(\Delta R_p)^2$.

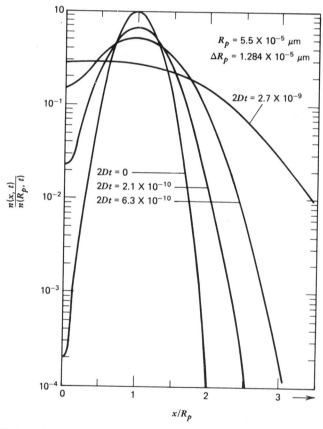

Fig. 6.35 Effect of diffusion on an implant profile, assuming no out-diffusion. From Seidel and MacRae [61]. With permission from the Metallurgical Society of the American Institute of Mechanical Engineers.

Equations (6.44) and (6.45) can also be used to evaluate the effect of successive processing steps, by replacing Dt with $\sum_{i=1}^{n} D_i t_i$, where D_i and t_i are the appropriate diffusion constant and time for the ith process step.

Finally, it should be emphasized that Eq. (6.45) applies only to situations where the diffusivity is constant during the annealing process. In general, however, this is not the case. A more complete solution, where the diffusivity is both concentration- and temperature-dependent, is available in the literature [62].

6.5.2 Multiple Implants

There are a number of situations where it is desirable to make multiple implants. One such case, described earlier, is preimplantation of silicon with an inert ion (such as silicon or neon) in order to make it amorphous, prior to deposition of the impurity. This allows close control of the doping profile for many decades below its peak concentration, and also permits nearly 100% carrier activation at annealing temperatures as low as 600°C. In situations where a deep amorphous region is required, it is necessary to make a series of such implants at varying energies and doses, to create a region of this type prior to implantation of the impurity ion.

Multiple H$^+$ implants, for the purpose of forming SI regions in GaAs, have been found to produce significantly more temperature-stable material than is obtained by a single implant [42]. This improvement comes about because the compensation of GaAs by ion bombardment most probably involves many defect levels which anneal at different temperatures. Consequently, annealing a singly implanted device results in many regions of different thermal behavior. On the other hand, uniformity of the defect level structure, as obtained by multiple implants, results in material whose thermal behavior can be better controlled.

Multiple implants are also used when it is desired [63] to obtain a deep, flat doping profile, as shown in Fig. 6.36. Here four boron implants into silicon were used to provide a composite doping profile, as shown. The actual carrier concentration, as well as that predicted by the LSS theory, are shown in this figure. Unique profiles, unavailable by conventional diffusion technology, can also be obtained by using various combinations of dose and energy for such implantations. One such is the profile for a high-efficiency impact ionization Avalanche Transit Time (IMPATT) diode using a double drift region [64]. Techniques for selecting appropriate ion implantation energies and doses to synthesize any desired impurity profile are available in the literature [65].

Multiple implants have been used [66] as a means for preserving stoichiometry during the implantation and annealing of GaAs. This approach, whereby equal amounts of gallium and an n-type dopant (or arsenic and a p-type dopant) are implanted prior to annealing, have resulted in obtaining higher carrier activation with a number of dopants in GaAs. This is especially true for group

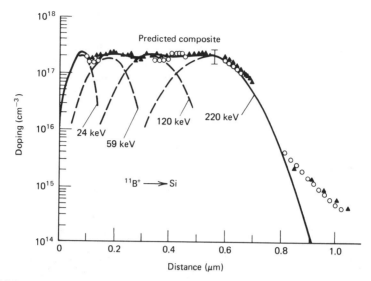

Fig. 6.36 Composite doping profile using multiple implants. From Lee and Mayer [63]. ©1974 by the Institute of Electrical and Electronic Engineers, Inc.

IV dopants which are amphoteric when implanted. Here, co-implantation with either gallium or arsenic results in *p*- or *n*-type GaAs, respectively, and greatly reduces the compensation ratio in both cases [67].

Co-implantation with an element such as fluorine has also been used [68] to improve the activation efficiency of implanted Be in GaAs. Improvement in the shape of the doping profile after RTA has also been observed in some situations.

6.5.3 Masking

A unique feature of ion implantation is that a wide variety of masking materials can be used, since it is a low-temperature process. The thickness required for masking is a function of the stopping parameters of the mask material, and can be readily estimated.

Consider the effect of a beam of incident ions on a mask material of infinite thickness, as shown in Fig. 6.37. For this material the dose which is deposited* in the semiconductor region beyond a depth *d* (shown shaded) is given by integration of Eq. (6.5) as

---

*It goes without saying that the actual profile of this species depends on its range statistics in the semiconductor, and is only approximated by the curve of Fig. 6.37.

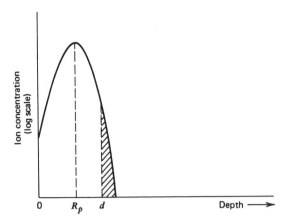

Fig. 6.37 Ion penetration beyond a depth d.

$$Q = \frac{Q_0}{(2\pi)^{1/2}\Delta R_p} \int_d^\infty \exp\left[-\frac{1}{2}\left(\frac{x - R_p}{\Delta R_p}\right)^2\right] dx \qquad (6.46)$$

where Q_0 is the incident ion dose, and R_p and ΔR_p refer to the mask material. Noting that

$$\int_d^\infty e^{-x^2} dx = \frac{\sqrt{\pi}}{2}\text{erfc}\, d \qquad (6.47)$$

the fraction of the dose that is deposited beyond a depth d is given by

$$\frac{Q}{Q_0} = \frac{1}{2}\text{erfc}\left(\frac{d - R_p}{\sqrt{2}\Delta R_p}\right) \qquad (6.48)$$

To an approximation, this is also the fraction of the incident dose that penetrates a mask of thickness d. From this equation it is seen that a mask of thickness $3.72\ \Delta R_p + R_p$ is required for a masking effectiveness of 99.99%, or $4.27\ \Delta R_p + R_p$ for 99.999% masking. These values should only be used as guidelines since the LSS parameters for most mask materials vary with the manner in which they are deposited or grown. Figures 6.38–6.40 show the minimum thicknesses, based on these considerations, for 99.99% effective masks made of photoresist, silicon dioxide, and silicon nitride, respectively. These materials are most commonly used in device processing, and can also provide the masking function for ion implantation.

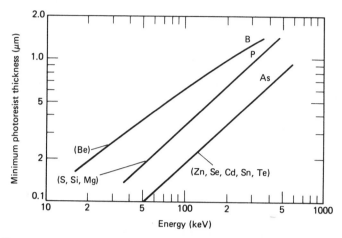

Fig. 6.38 Minimum photoresist thickness for a masking effectiveness of 99.99%.

Mask thicknesses given in these figures are for boron, phosphorus, and arsenic implants into silicon [52]. However, they can also be used as approximate guidelines for impurity masking in GaAs. Masking data for impurities in GaAs are shown in parentheses in these figures.

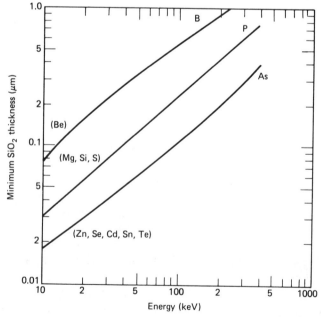

Fig. 6.39 Minimum SiO_2 thickness for a masking effectiveness of 99.99%.

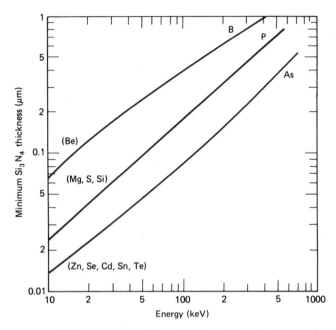

Fig. 6.40 Minimum Si₃N₄ thickness for a masking effectiveness of 99. 99%.

There are many practical considerations in the use of masking materials for ion implantation. For instance, the effect of the high-energy beam on an insulating surface is to cause considerable ionization. This prevents surface charging to high voltages and rupture of the insulator by breakdown, as long as there are conducting paths within 500 μm of this surface. This condition is readily met in microcircuits, so that no special precautions or grounding arrangements have to be provided for implantation of this type.

Ion bombardment of photoresist liberates hydrogen during the early stages of the implantation process. This outgassing is significant, and can affect the accuracy of the dose counting process, as well as the vacuum conditions. Its effect can be minimized by implanting wafers in two stages, the first at beam densities up to 10^{15} ions cm^{-2}, followed by the bulk of the implant.

The resulting carbonization of the photoresist can lead to difficulties in its subsequent removal by standard photoresist strippers. One solution here is to make the photoresist film much thicker than is normally dictated by masking requirements. This results in an undamaged layer in direct contact with the semiconductor. This region can be attacked by the stripper, so that the entire layer can be removed by undercutting and subsequent lift-off. An alternative approach is to use an oxygen plasma for photoresist removal.*

---

*This technique is described in Chapter 9.

In extreme cases, bombardment with high-dose, high-energy ion beams can lead to cracking of the photoresist. It goes without saying that this material cannot be used in these situations.

Ion implantation through an SiO_2 layer results in creating damage in this material. This has a number of consequences in practical situations. One is that photoresist adhesion is impaired, and there is often lift-off during subsequent masking operations, by capillary action of the etchants under the photoresist. Here, too, plasma etching of the oxide is sometimes used to avoid this problem. A second consequence is that the etch rate of the oxide is increased, often by a factor of 2–3. This enhancement is a function of the oxide growth technique, and also of the ion type and energy. It is caused by damage due to the implantation, and can be annealed [69] by heat treatment between 800°C and 1000°C.

The increased etch rate of implantation-damaged oxide can be used to form oxide cuts with sloping sides [70]. This improves the step coverage in the subsequent metallization, and is useful in MOS technology where large steps in oxide thickness are encountered, as is the case in going from the gate oxide (200–1000 Å) to the field oxide (usually >1 μm).

Finally, implantation through an SiO_2 layer results in the deposition of recoiled oxygen into the slice. This oxygen has a tendency to reduce the interstitial silicon which is present in the implant-damaged substrate. This, in turn, suppresses the diffusion tail associated with implanted boron [71], which moves by an interstitialcy mechanism, as described in Chapter 4.

Ion implantation will sometimes adversely affect the properties of silicon nitride. This material, which is often used as a cap during the annealing of GaAs, has been found to lose its adherence to the substrate with subsequent bubbling and lift-off. These problems have only been encountered in high-dose situations and for high-energy implants. In addition, the severity of the problem is related to the process by which the nitride layer is grown.

During ion implantation, the charge associated with the ion beam is removed by electrical conduction to the substrate holder. However, charge can build up if the surface of the wafer is covered, either totally or partially, with an insulating oxide or nitride layer. In high-current systems, voltage due to these charging effects can be sufficiently high so as to exceed the breakdown voltage of the insulator. In some systems, electron flooding is used in the vicinity of the substrate [72], in order to neutralize this charge buildup. Here, care must be taken with this approach to minimize the possibility of errors in dosimetry.

Deposited metal films can also be used as masks for ion implantation. However, some of the mask material can be driven into the substrate during implantation. One approach here, which prevents accidental doping from the residual metal, is to place a thin silicon dioxide or silicon nitride layer between the film and the underlying semiconductor.

Deposited metal films can also be used as a dopant source which can be driven into the semiconductor by means of an inert ion beam. This technique, known as *recoil bombardment*, can be used to form extremely shallow, high-concentration doping profiles in both silicon and GaAs.

It is also possible to implant a semiconductor with its capping layer already in place. In the case of GaAs, this avoids the problems of dopant loss during the deposition of the silicon nitride cap layer, which is usually performed at around 700–800°C. With silicon there are additional advantages in the form of surface passivation, so that most silicon implants are performed in this manner. Details of this important technique are discussed in Section 6.5.5.

6.5.4 Contacts to Implanted Layers

The conventional approach for making a contact to a semiconductor region is to deposit the contact metal on it by sputtering or vacuum evaporation, and to heat treat the combination to form a "microalloy" at this interface. This process tends, in practice, to be relatively uneven, with actual contact formation occurring by relatively deep penetration at a number of discrete alloying sites. With ion-implanted layers (or, for that matter, with any type of extremely thin layer), this results in punching through to the underlying semiconductor. This is not a problem if the implantation has been made in an SI layer as is often the case with GaAs; otherwise, the making of contacts to implanted layers presents a special situation.

One approach here consists of forming a deep-diffused region of the same impurity type, connected to the implanted layer, and making the metal contact to this region (see Fig. 6.41a). This technique is sometimes not possible, as in the case of a device where this might short out an underlying layer. Here ohmic contact to the shallow implanted region must be made by forming a Schottky barrier contact to the semiconductor, as shown in Fig. 6.41b. This requires the choice of a contact material of suitable work function, which will adhere to the semiconductor. Materials such as titanium and chromium are often used here. An alternative approach with silicon is to use a metal such as platinum or molybdenum, and form its silicides *in situ* by a suitable heat treatment.

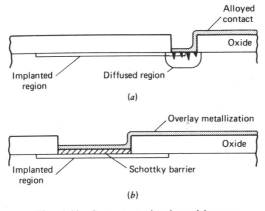

Fig. 6.41 Contacts to implanted layers.

An overlay metallization is next connected to this Schottky barrier region. The details of both alloy- and Schottky-type contacts are provided in Chapter 8.

6.5.5 Implantation Through an Oxide

One of the more significant advantages of silicon for microcircuit fabrication is the fact that its surface can be covered by a dense amorphous layer of grown SiO_2. The surface protection afforded by this layer can be maintained by making implantations through this oxide, so that the bare silicon surface is not exposed during processing. Although this results in the formation of charge states in this material, we have noted in Section 6.3.1 that they are readily annealed during subsequent processing.

Implants may be made through an oxide in a number of different ways. This is shown in Fig. 6.42 for the case of silicon covered with a 1200 Å oxide layer, for three different boron implant conditions [63]. Profile A represents a situation where the oxide layer allows only a small amount of the total ion dose to be deposited in the silicon. The actual fraction is not only small, but is also a function of the oxide quality as well as the implantation parameters. An oxide layer of this type serves essentially as a mask; in a practical application, a slightly thicker film would be used to almost completely prevent boron penetration into the silicon. Profile B deposits a large fraction of the dose in the silicon. This is useful in doping the semiconductor while still keeping its surface covered by a passivating film; in a practical situation, a somewhat higher ion energy would

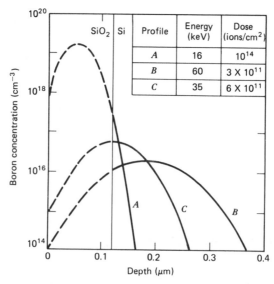

Fig. 6.42 Implantation through an oxide. From Lee and Mayer [63]. ©1974 by the Institute of Electrical and Electronic Engineers, Inc.

result in almost no dopant loss due to the oxide. In this case, the oxide layer serves as an implantation window.

Profile *C* illustrates the case where one-half of the ion distribution is deposited in the silicon, while the rest is retained in the oxide. Ideally, implantation of group III or V species (such as boron, phosphorus, or arsenic) in this manner results in the placement of a controlled quantity of sheet charge at the Si–SiO$_2$ interface. This charge, which is placed in the channel region, can be used to tailor the threshold voltage of MOS devices. Specific applications for implantation through an oxide now follow.

6.5.5.1 *Junctions*

In silicon microcircuit technology it is important that the region where the junction penetrates to the surface be covered by an oxide layer. This is achieved automatically during junction formation by diffusion techniques, since the lateral spread of the impurity beyond the edge of the window is approximately 75% of the junction depth. Implantation through an oxide window does indeed result in some junction penetration beyond the oxide edge, due to transverse straggle effects described in Section 6.1.4. However, this penetration is extremely small, and is often insufficient to protect the junction from contamination during further processing.

One approach to obtaining junction protection is to drive-in the implant with a high-temperature diffusion step. A second technique is to use a double photolithographic process, as shown in Fig. 6.43. Here a mask is used to delineate the implant, which is made through the oxide (see Fig. 6.43a). Next the oxide is cut to form the contact region for the implant, as shown in Fig. 6.43b. In this

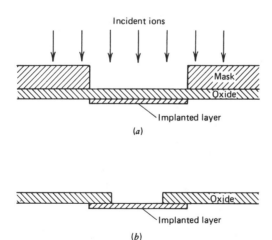

Fig. 6.43 Passivated junction formed by double photolithography.

manner, the junction edge is far beyond the contact window and is thus well protected from contamination. Contacts of the type illustrated in Fig. 6.41 are used to make connection to this region.

All oxide-masked junctions have curvature at the window edge. This increases the electric field over that obtained with a parallel-plane structure and causes premature breakdown [73]. This problem is especially severe with ion-implanted junctions which are shallow and have small radii of curvature. This radius of curvature is primarily due to transverse straggle, which is somewhat smaller than the projected range. The actual shape, however, is also determined by the profile of the oxide cut, which is never perfectly vertical. Thus its break-down voltage (BV) cannot be calculated with any precision. In general, how-ever, implanted junctions are found to behave as if they had a radius of curvature of 1–1.5 μm at the edge of the oxide cut. This serves as a good design rule for calculating their breakdown characteristics.

Alternative approaches have been used for BV enhancement. These include direct implant through an oxide window with intentionally tapered edges, and rely upon the variable penetration range provided by this oxide. Guard ring structures can also be used, where a deep diffusion region is provided all around the implant edge. Such a structure, shown in Fig. 6.44, has the additional advan-tage of producing depletion layer curvature in the direction opposite to that obtained by a single diffusion (or implant). As a consequence, breakdown volt-ages close to the bulk value can be achieved in this manner [74]. Structures of this type are widely used in avalanche photodiodes, which require controlled current multiplication characteristics.

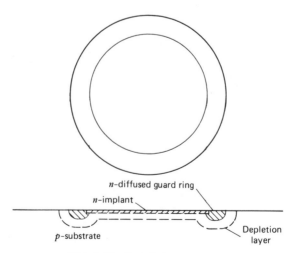

Fig. 6.44 Guard ring structure.

6.5.5.2 Resistors

An important use for ion implantation through an oxide is in the fabrication of precise resistors for microcircuit technology. Typically, diffused resistors are limited to low sheet resistances (125–180 Ω/square) and are, at best, only settable to within ±10% of their designed values. With implantation, however, almost any sheet resistance is possible, and settable to within ±1%. Practical considerations limit the value to about 4000 Ω/square, since the resistor becomes extremely sensitive to surface states beyond this value, because of its light doping.

Implanted resistors find wide use in analog circuits where a variety of resistor values, with close settability, provide great freedom to the circuit designer. A typical boron-implanted resistor, which is compatible with existing diffused resistors in the same silicon microcircuit, is shown in Fig. 6.45. Here contact is made to its end by p^+-regions of the type used to form the diffused base in n–p–n transistors made on the same chip. Again, direct alloying can be used in GaAs circuits, if the resistor implant is made into SI material.

Silicon microcircuits sometimes require extremely high-value resistors, which are used for capacitive discharge purposes. These can also be made by ion implantation. Here it is necessary to implant through a heavily doped shielding layer of opposite conductivity type, which serves as a cover and makes the resistor independent of the surface states in the oxide. Values as high as 100 kΩ/square can be obtained in this manner. These resistors are highly nonlinear with voltage, however, because of depletion layer pinching effects during operation.

Polycrystalline silicon films are often used for the fabrication of these resistors, in order to avoid such problems. Their resistivity is extremely difficult

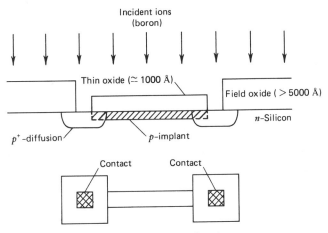

Fig. 6.45 Implanted resistor.

to control if they are doped by conventional diffusion technology, since their conductance depends on the relative doping of individual crystallites and their grain boundaries. Typically, variations by as much as four orders of magnitude are encountered in high-value (1 MΩ/square) diffused resistors of this type. With ion implantation, it is possible to dope the individual crystallites without scavanging effects created by their grain boundaries. As a result, a considerable improvement in their tolerance value, by as much as two orders of magnitude, can be obtained.

6.6 HIGH-ENERGY IMPLANTS

Ion implanters, capable of energies to as high as 3–5 MeV, have recently become available and have been used for a number of important, novel applications [75]. The majority of these depend on the ability to dope the semiconductor to many microns in depth, without the need for long diffusion times at high temperatures.

At 3 MeV, the projected ranges for boron, phosphorus, and arsenic ions in silicon follow a Pearson IV-type distribution and are 4 μm, 2.3 μm, and 1.5 μm, respectively. In each case, the damage density lags the dopant peak by a small amount. Consequently, it is possible to obtain a large volume of defect-free material *above* the implant layer by means of very little subsequent annealing.

In some cases, high-energy implants have been used to place a layer of damage in an optimum location for metallic impurity gettering [76]. Using this technique, the leakage current of *p–n* junctions formed above this layer can be greatly reduced. This is of significant benefit in the fabrication of dynamic random access memories (DRAMs) where this leakage current has an adverse effect on the refresh time.

High-energy implants can also be used to produce low-resistivity buried layers and grids, beneath the actual microcircuit. These have been used for interconnection purposes, and also for the reduction of alpha-particle-induced soft errors in silicon DRAMs [77] as well as in GaAs static random access memories [78].

An important application for high-energy implant technology is in the doping of wells for CMOS structures. The use of a well with retrograde doping allows an increase in the latch-up current threshold, without penalizing such device parameters as the channel mobility and the punch-through voltage. Here, a series of implants at different energies are used to develop the necessary doping profile, and also provide a gettering layer to reduce leakage current in the device structures [79].

The performance of many high-frequency GaAs devices which are used in monolithic microwave integrated circuits (MMICs) can be greatly improved by the use of a highly conducting buried layer. This can be achieved by the use of S^+ or Si^+ implants in the 5- to 6-MeV range. Doping concentrations of 2.5×10^{18} cm^{-3} have been achieved [80] by this approach, using a 10-s rapid thermal anneal cycle at 1000°C. In some applications, requirements are for n^+-layers 2.5 to 5 μm below the GaAs surface. These can be used to provide a gettering layer to reduce leakage current in the device structures and also serve as a high-conductivity return path.

An inherent problem with making high-energy implants is the difficulty of obtaining suitable masks. The upper limit of photoresist thickness that can be used commercially is 2–3 μm, which is suitable for masking 1.5-MeV implants of B^+ and P^+ as well as for masking ~3-MeV implants of As^+. Consequently, increasing use is being made of tungsten and other materials with high atomic number for this purpose.

6.7 HIGH CURRENT IMPLANTS

High-current implanters, operating in the 25- to 30-keV range, are routinely used for the predeposition step in diffusion technology. Recently, new developments in ion sources have resulted in high-current machines with energies in the 150- to 200-keV range. A major use for these machines is the formation of high-quality films of silicon, which are insulated from the substrate by the implantation of oxygen, thereby creating an intervening layer of silicon dioxide. This approach provides an alternative to the SOS technique for obtaining thin films of silicon on insulators, as described in Chapter 5. This "separation by implantation of oxygen" technique is commonly referred to as the *SIMOX* process.

6.7.1 The SIMOX Process

The SIMOX process uses [81] a high-energy O^+ beam, typically in the 150- to 200-keV range, to result in penetration under the silicon to a depth of 0.1–0.2 μm. Additionally, a heavy dose, $1-2 \times 10^{18}$ ions cm^{-2}, is used to produce an insulating layer of SiO_2 which is 0.1–0.5 μm thick.

The implantation step is carried out on slices which are heated to at least 600°C in order to keep the silicon surface crystalline. A subsequent high-temperature anneal, at 1250–1300°C, promotes solid-state crystallization of this layer from the top down, resulting in a single-crystal silicon film with a relatively sharp Si–SiO$_2$ interface, into which microcircuits are eventually fabricated. This step reduces the dislocation density in the layer by four to five orders of magnitude. An additional function of this high-temperature anneal is to coalesce the implanted oxygen–silicon matrix into a uniform layer of silicon dioxide, free from precipitates.

Silicon layers formed by the SIMOX process are excellent from the standpoint of material quality. Typically, because of the extended high-temperature anneal step, their minority carrier lifetime is about 100 μs, which is comparable to that of bulk silicon. In contrast, the lifetime of epitaxial silicon grown on sapphire by the SOS process is about three orders of magnitude lower. In addition, dislocation densities in these layers are around 10^4–10^5 cm^{-2}, so that they are suitable for the fabrication of both analog and digital integrated circuits. Recently, the use of multiple, low-dose implants in the 2–3 × 10^{17} atoms cm^{-2} range, each followed by an annealing step, has reduced [82] the dislocation content of these layers by an additional factor of 10.

The use of SIMOX material leads to a significant reduction of source/drain capacitances in MOS devices. Moreover, it reduces coupling between devices, and thus allows tighter packing without the problems of latch-up. As a result, it is widely proposed as the material of choice for advanced, high-speed CMOS microcircuits. The thickness of these layers, 0.1–0.2 μm, is suitable for this purpose. In some cases, conventional silicon epitaxy can be used to increase this thickness, if necessary.

Because of the long implantation times associated with the high dose, special attention must be paid to minimizing both particle and heavy metal contamination during the implantation process. Polysilicon shielding is often used to prevent inadvertent contamination from beam spillover effects during this process [81].

The SIMOX process is relatively expensive, because of the need for special implanters, high implant doses, and long anneal times. A number of approaches involving the use of short, multiple implant/anneal sequences are currently under investigation in order to bring down the cost of this process, by reducing either the total dose or the total anneal time [83].

Another important problem area is the possibility of plastic deformation of the slice after being subjected to a high-temperature anneal step. This represents a serious problem because of the volume expansion associated with the conversion of silicon to silicon dioxide. In some cases, it can result [84] in slip along with a concurrent deterioration in the substrate quality.

6.8   APPLICATION TO SILICON

The flexibility and the control available with ion implantation, combined with the ease of annealing, has made it rapidly accepted in silicon technology. This section outlines areas in which this approach has provided a measure of uniqueness, with no attempt at being all-inclusive.

6.8.1   Bipolar Devices

Ion implantation is routinely used as a means for predeposition, prior to base diffusion in bipolar transistors. This is especially true for high-voltage devices, where the base must be lightly doped. In addition, there are a number of unique areas for exploitation of this technology, such as the following:

1. Ion implantation permits selective placement of doped regions on the same slice. Thus it can be used to fabricate high-performance vertical *p–n–p*

transistors in the same microcircuit as n–p–n devices [85]. Conventional p–n–p devices are of the lateral type, and have extremely poor frequency performance as compared to their n–p–n counterparts.

2. Transistor fabrication by conventional means necessitates the formation of the emitter *after* the base. This results in the "emitter-push" effect described in Section 4.8.3. Ion implantation allows the formation of these regions in reverse order, so that this problem can be avoided.

3. Ion implantation can be used to make extremely rapid doping concentration changes in a semiconductor. This has been exploited in the design of hyperabrupt diode structures [86]. Here, the doping profile is shaped so that the edge of the depletion layer is in the high-concentration region at zero bias, but spreads into the lightly doped region upon the application of a reverse voltage. Structures of this type exhibit anomalously large variations in capacitance with reverse bias, and are used in electronic tuning applications.

4. Extremely small emitter regions can be formed by this technique due to the absence of lateral diffusion effects. Thus ion implantation lends itself to the fabrication of VLSI circuits. Here arsenic is the preferred impurity since its doping profile after heat treatment is relatively abrupt, and free from deep tailing effects.

5. Arsenic is the ideal dopant for buried layers in bipolar microcircuits, because its tetrahedral radius is the same as that of silicon. However, buried layer formation requires a long drive-in cycle, which brings with it the serious problem of dopant loss by surface evaporation. Often double deposition and masking are used to replenish this loss. This problem can be greatly reduced by using a deep arsenic implant for predeposition before this drive-in step.

6. IMPATT diodes, consisting of p^+–n–n^+ structures, are conventionally formed by making a p^+-diffusion into an n-epitaxial layer. Precise thickness control of this n-layer is essential for device operation in the 50- to 100-GHz range. Here the use of implantation for the p^+-region allows the width of the n-layer to be controlled by the epitaxial growth process, and not by a subsequent high-temperature p^+-diffusion step.

In summary, a number of unique characteristics of ion implantation have been exploited in these applications, and serve to indicate different ways in which this approach can be used with bipolar silicon technology. The list is by no means complete, but is indicative of the rapidly expanding use of this process.

6.8.2 MOS Devices

The greatest degree of success enjoyed by ion implantation today is in the area of MOS technology, because it presents a number of situations in which precise control of low-dose levels is required. In addition, the unique masking possibilities with implantation techniques allow considerable flexibility in the process steps, so that device parameters can be tailored to suit the needs of the circuit design.

The most straightforward use of ion implantation is in tailoring the resistivity of the starting silicon. This is done by implantation of an unmasked slice of higher resistivity than the desired value. This sets the resistivity of the surface region to a tight tolerance, even for wide variations in the starting resistivity.

An extremely important use of ion implantation is in the formation of the p-well in CMOS circuits. Here a precisely doped, deep p-region is required in n-type starting silicon. This region is used to form the n-channel transistor portion of the complementary pair. In this application, ion implantation is a direct replacement for the predeposition step in a conventional diffusion process. With predeposition, however, it is extremely difficult to obtain reproducible characteristics from run to run, because of the low carrier concentration requirements for this region. Ion implantation, on the other hand, accomplishes this with ease,* and has led to the successful implementation of CMOS technology in large scale integration. Furthermore, the use of retrograde doped wells, using high-energy implants, has greatly expanded the use of this technology.

The well that is formed by this technique must be quite deep in order to accommodate a complete MOS device. This requires a long drive-in step (typically 24 hr at 1200°C), so that special precautions must be taken to avoid loss of dopant during this process. These include (a) the use of high energies (150–200 keV) in order to deposit the boron ions deep into the semiconductor and (b) drive-in in a nonoxidizing ambient to minimize boron redistribution effects (see Section 4.8.1).

Ion implantation is ideally suited for controlling the threshold voltage of MOS devices. This is done by implantation through the gate oxide, and results in the deposition of an ionized charge species in the channel region. Use of a boron implant in an n-channel device (or a phosphorus implant in a p-channel structure) results in an increase in the magnitude of the threshold voltage. For the ideal case of a sheet charge at the Si–SiO$_2$ interface, which is approximated by profile C of Fig. 6.42, this increase is given by

$$\Delta V_T = \frac{Q}{C_{\text{ox}}} \tag{6.49}$$

where ΔV_T is the change in the threshold voltage, Q is the sheet charge at the Si-SiO$_2$ interface, and C_{ox} is the gate capacitance. This linear dependence of the change in threshold voltage with the dose allows reproducible adjustment of this important MOS device parameter.

The effect of a small change in the gate oxide thickness, due to lack of process control during oxidation, is to create a change in the gate capacitance as well as in the implanted charge. These terms have opposing effects on the

---

*A typical ion implant dose for this application is 1–5 $\times 10^{12}$ ions cm^{-2}.

threshold voltage and tend to compensate each other. However, it can be shown that, for each oxide thickness, there is an optimum implant energy which will minimize the effect of these variations [87].

MOS threshold parameter control can be used in a number of different situations. Thus the direct reduction of V_T with implant dose allows the fabrication of low-threshold devices, which can be operated at low power dissipation levels. Threshold adjustment of p- and n-channel devices on the same chip allows the fabrication of matched complementary devices [88], and is the cornerstone of practical CMOS technology. Finally, it is possible to use this technique to fabricate depletion-type loads on the same substrate as enhancement–type devices. This has resulted in significant improvements in the speed and voltage swing capability of MOS circuits.

Threshold control is usually accomplished at low dose levels. The surface state charge created at these levels can be readily annealed by heat treatment at 300°C. However, significant carrier activation to a reproducible fraction is only accomplished by heat treatment to 525°C or higher, so that it is customary to anneal at this temperature after the threshold-adjustment implant step.

Implantation through an oxide can also be used to increase the threshold voltage outside the channel region. This is desirable to avoid parasitic transistor action by metallization passing over the field oxide, or by surface inversion effects. Figure 6.46 shows how a channel stopper of this type may be formed by means of a single unmasked high-energy implant [89]. In this situation, a 200- to 300-keV boron implant is used for an n-channel MOS device. This implant has a large penetration depth under the thin gate oxide, so that it has no effect on the MOS threshold voltage. However, the thick field oxide results in shallow implant penetration in the p-type field regions, where it increases the threshold voltage required to cause parasitic action. Multiple implants can also be used for channel stopper regions. This represents a more conventional use of ion implantation techniques.

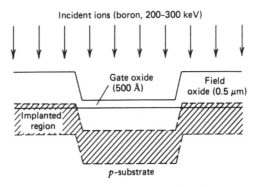

Fig. 6.46 Channel stopper implant.

In MOS devices, the overlap of the gate electrode over the drain region creates a parasite feedback capacitance and deteriorates the frequency response. Precision masking, combined with shallow source and drain diffusions, has been used to minimize this problem. Another approach, which avoids masking problems, is to use self-registered structures where a refractory gate acts as the mask for the source and drain diffusions (see Fig. 6.47a). This gate is conventionally made of polycrystalline silicon in order to withstand high diffusion temperatures, although a number of silicides and refractory metals can also be used (see Chapter 8).

Ion implantation offers the alternative approach of building the source and drain region up to the gate edge, after the diffusions have been conducted (see Fig. 6.47b). The primary advantage of implantation here is that the gate overlap in this structure is due to transverse straggle effects, which are much less than the overlap due to lateral diffusion of the drain region. This results in a significant improvement in the frequency response.

Charge-coupled devices are another area where ion implantation can be used to either simplify the fabrication process or provide improved device performance. In these devices, charge packets are transported along the $Si-SiO_2$ interface and serve to transfer data in this manner. The movement of these packets is conventionally achieved by manipulating electrode voltages, and by using different oxide thicknesses with a uniformly doped semiconductor. Ion implantation allows the possibility of introducing an asymmetry into semiconductor doping, by the placement of pockets of varying doping concentration. This allows much simplification in the oxide and metallization processes. In addition, the placement of buried channels by ion implantation permits a reduction of transmission losses in these devices. Details of a wide variety of such processes are available in the literature [90].

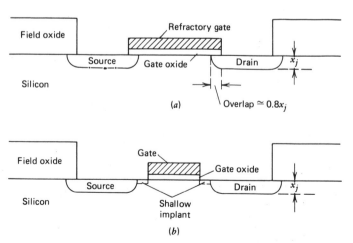

Fig. 6.47 High-speed MOS device using ion implantation.

6.9   APPLICATION TO GALLIUM ARSENIDE

It was initially predicted that ion implantation would be ideally suited for GaAs and other compound semiconductors, but would find little use in silicon technology. However, quite the opposite situation has developed, and success with ion implantation of GaAs has been relatively modest during this time. There are a number of reasons for this outcome. First, there is as yet no viable MOS technology for GaAs, so that the needs for well-controlled low doses have not been established. Next, the problems of forming heavily doped regions by ion implantation have not as yet been satisfactorily solved. As shown earlier, they necessitate high-dose implants on heated substrates, followed by high-temperature anneals using caps. Each of these technologies has many associated problems. Together they result in a relatively complex sequence of processing steps as compared to those required for processing silicon. Again, this problem area shows promise for being solved, as evidenced by the recent success of RTA annealing techniques [53]. Nevertheless, there are a number of areas in which ion implantation technology has already proved successful. These are now outlined briefly, with the emphasis on their unique features.

Proton bombardment has been used to produce isolation regions around a number of devices [91]. These include photodiodes and field effect transistors, as well as transferred electron devices and avalanche transit time oscillators. In many of these devices, the unique doping profiles provided by multiple implants have allowed improved performance, and have extended their operating frequencies as well as their power outputs. A unique application of proton bombardment has been in providing optical isolation in couplers and optical waveguides [92]. Oxygen and boron implantations, somewhat more recent techniques, have also been applied to a number of these devices.

Ion implantation has been used for the intentional introduction of damage into GaAs. Dislocations produced by this method are effective [93] in gettering impurities and result in improved photoluminescence for a distance of 25–100 μm. Locally induced damage, in the vicinity of the active region of stripe lasers, has resulted in a reduction in the lasing threshold current density by a factor of five. Ion implantation has also been used to make ohmic contacts to a number of microwave and optoelectronic devices, by providing localized regions of heavy doping. Often the uniquely different annealing characteristics of selenium and sulfur implants, as seen in Fig. 6.26, are exploited to form a deep n^+-contact region (sulfur implant) connected to the active n-region (selenium implant). This permits making an alloyed ohmic contact to the shallow implant region with relative ease (see Fig. 6.48). RTA techniques are extensively used in this application.

Counterdoping of heavily doped n-GaAs with acceptors can also be achieved by ion implantation. This technique has been used to increase the Schottky barrier height of some metal–GaAs combinations, and thus reduce the leakage current in these devices [94].

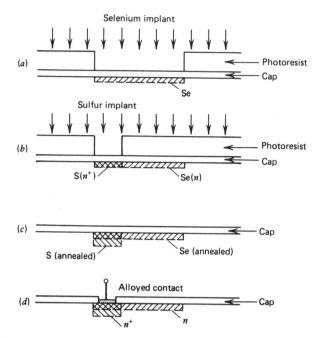

Fig. 6.48 Use of double implantation to form a region for making an alloyed contact to a shallow implant.

Hyperabrupt Schottky diodes, suitable for use as electronic tuning elements, have been fabricated [95] using silicon and sulfur implants into GaAs. The doping profile for these structures is of the same type as that for silicon devices.

6.9.1 Field Effect Transistor Devices

The full exploitation of any electronic technology depends on the development of high-performance three-terminal devices with significant power gain. These devices allow the isolation of input and output signals, and provide the circuit designer with flexibility in designing complex circuits and subsystems. With GaAs, majority carrier devices such as the field effect transistor (FET), based on n-type carrier conduction, can give a speed advantage of $3:1$ over silicon, assuming comparable technologies. This speed advantage has already been realized in discrete devices.

There are two types of FET that can be made: the Schottky gate device, of which the high-electron-mobility transistor (HEMT) is a subset, and the junction gate device. Consider first the Schottky gate FET. Here the conventional approach* begins with the epitaxial growth of a thin layer of suitable doped

---

*A number of process variations are employed in practice to ease the burdens of device fabrication, and are not considered here.

n-GaAs on an SI GaAs substrate. Typical parameters for this active region are a carrier concentration of $1-2 \times 10^{17}$ cm^{-3} and a thickness of 0.1-0.3 μm. Often an intervening buffer layer of high-impurity material is first grown to prevent out-diffusion effects from the substrate, and to produce a damage-free interface with the active layer. The Schottky gate FET is fabricated by making ohmic contacts for the source and drain, and using an evaporated metal contact for the gate, as shown in Fig. 6.49a.

Problems associated with this fabrication technique are largely due to a lack of sufficient control of the carrier concentration and the epitaxial layer thick-

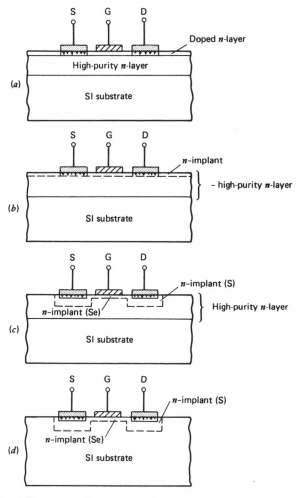

Fig. 6.49   Alternative schemes for junction gate FET devices.

ness. These can be greatly reduced by the use of ion implantation. In one approach a high-purity buffer layer is grown in the SI GaAs and the active layer is formed by selenium implantation in it (see Fig. 6.49b), followed by annealing under a Si_3N_4 cap [96]. Ion implantation allows the formation of this active layer with better than $\pm3\%$ uniformity over a 100-mm slice diameter.

Double implantation with sulfur and selenium can also be used. This results in the formation of deep source and drain regions, as well as the shallow channel region, after a single anneal step. Here sulfur is used for the source and drain regions, since it penetrates considerably more deeply than selenium during the annealing step (see Fig. 6.49c). An alternative approach, using silicon for the channel implant, has found wide acceptance because of the ease with which it can be annealed.

Ion implantation can also be used to eliminate the epitaxial growth steps completely, as seen in Fig. 6.49d. Here direct implantation is made into SI substrates to result in a relatively simple process [97]. This approach is commonly used for the manufacture of GaAs digital integrated circuits today. Undoped, LEC-grown GaAs is suitable for this purpose.

Refractory gates, which can also serve as masks for ion implantation and as caps during the anneal step, have allowed self-aligned gate (SAG) technology to be used for the fabrication of high-performance Schottky gate FET devices [98]. Here, SI GaAs is implanted with silicon to form the channel region. Next, a refractory gate is patterned and used as a mask for implantation of the source and drain regions. Silicon nitride is used for this purpose, since it can withstand the annealing temperatures which must be used, following the implantation step. Details of this and other SAG techniques will be provided in Chapter 11.

Enhancement mode FETs, of both Schottky and junction gate types, are of increasing importance in GaAs technology, because they operate in the normally "off" mode. Their design requires the use of a lightly doped channel with tight control over its doping concentration. This is commonly made by the use of a low-dose silicon implant into an SI GaAs substrate for the channel [99]. With junction gate FET devices, an additional p^+-implant must be made to form the gate region [100]. Both beryllium and magnesium have been used successfully in this application.

6.9.2 Integrated Circuits

Integrated circuits in GaAs are based on the use of FETs as active devices, with signal coupling and routing usually done by resistors and by Schottky diodes. The development of both analog and digital circuits has rapidly followed the development of the FET. Control of doping, provided by ion implantation, has allowed the fabrication of both enhancement mode circuits and the more usual depletion mode type [101].

Implantation into epitaxially grown n-layers has been used to fabricate these integrated circuits. Here sections of the layer are electrically isolated from each other by means of a mesa etch technique. These isolated regions are used for forming the various devices that are required in the circuit. Proton implantation has also been used as an alternative isolation technique [102].

Microcircuits can also be fabricated by direct implantation into SI GaAs substrates. This approach has the advantage that it results in a planar topology, since mesa etching is not necessary to obtain isolation between components. The resulting planar surface greatly simplifies the task of photolithography at the submicron level, and allows fabrication of ultrahigh-speed circuits [103, 104]. Advanced designs, using self-aligned refractory gate technology, have also been developed to exploit the high-speed capabilities of GaAs FETs in integrated circuits [105].

In all of these techniques, the main impetus for using ion implantation is its ability to provide precise control of the doping and depth of the active channel region. Channel parameters eventually set the threshold voltage of the FET, and thus the signal swing available in linear and digital integrated circuits. The main obstacle to ion implantation processes in this area is in the reproducible conversion of implanted impurities into active donors. As noted earlier, carrier activation in GaAs is far from complete, so that variations in the damage structure and in the starting material are reflected in run-to-run variations in the doping concentration [106]. Carrier activation is greatly improved by implantation into heated substrates. However, this can result in anomalous enhanced diffusion during the implantation, so that control of channel depth is not as tight as can normally be expected from this technique. Nevertheless, practical high-speed integrated circuits [107] as well as millimeter-wave integrated circuits (MMICs) [108] are already being successfully fabricated by ion implantation, and developments in the area are extremely rapid.

TABLE 6.1   Properties of Materials Used in Ion Implantation

Element/Compound	Atomic Number	Atomic Weight	Density (g/cm^3)
Al	13	26.98	2.70
As	33	74.92	5.72
B	5	10.81	2.34
Be	4	9.01	1.85
Cd	48	112.40	8.65
Ga	31	69.72	5.91
Ge	32	72.59	5.32
Mg	12	24.31	1.74
N	7	14.01	1.25[b]
O	8	16	1.43[b]
P	15	30.97	1.83
S	16	32.06	2.07
Sb	51	121.75	6.62
Se	34	78.96	4.79
Si	14	28.09	2.33
Sn	50	118.69	7.30
Te	52	127.6	6.24
Zn	30	65.37	7.13
GaAs	32[a]	72.32[a]	5.27
SiO$_2$	10[a]	20.03[a]	2.27
Si$_3$N$_4$	10[a]	18.04[a]	3.44

[a] Average.   [b] g/liter.

TABLE 6.2   Nature of the Concentration Profile Due to Ion Implantation

Distance from Mean Value	Magnitude of Ion Concentration (Normalized to Peak Value)
0	1
$\pm 1.18 \Delta R_p$	0.5
$\pm 2.14 \Delta R_p$	10^{-1}
$\pm 3.04 \Delta R_p$	10^{-2}
$\pm 3.72 \Delta R_p$	10^{-3}
$\pm 4.29 \Delta R_p$	10^{-4}
$\pm 4.80 \Delta R_p$	10^{-5}
$\pm 5.25 \Delta R_p$	10^{-6}
$\pm 5.67 \Delta R_p$	10^{-7}

TABLE 6.3 Nature of the Concentration Profile at a Window Edge

Distance from Edge	Magnitude of Ion Concentration (Normalized to Peak Value)
0	0.5
$1.28\Delta R_t$	10^{-1}
$2.33\Delta R_t$	10^{-2}
$3.09\Delta R_t$	10^{-3}
$3.72\Delta R_t$	10^{-4}
$4.30\Delta R_t$	10^{-5}
$4.78\Delta R_t$	10^{-6}
$5.22\Delta R_t$	10^{-7}

REFERENCES

1. J. F. Gibbons, Ion Implantation in Semiconductors—Part I. Range Distribution Theory and Experiments, *Proc. IEEE* **56**, 295 (1968).

2. J. F. Zeigler and R. F. Lever, Channeling of Ions Near the Silicon ⟨100⟩ Axis, *Appl. Phys. Lett.* **46**, 358 (1985).

3. G. Dearnaley, J. M. Freeman, G. A. Gard, and M. A. Wilkins, Implantation Profiles of ^{32}P Channeled into Silicon Crystals, *Can. J. Phys.* **46**, 487 (1968).

4. K. A. Pickar, Ion Implantation in Silicon, in *Applied Solid State Science: Advances in Materials and Device Research*, Vol. 5, R. Wolfe, Ed., Academic Press, New York, 1973, p. 152.

5. R. A. Moline, Ion Implanted Phosphorus in Silicon: Profiles Using C–V Analysis, *J. Appl. Phys.* **42**, 3553 (1971).

6. D. R. Zrudsky, Channeling Control in Ion Implantation, *Solid State Technol.* (July 1988) p. 69.

7. B. L. Crowder, Ed., *Ion Implantation in Semiconductors and Other Materials* (International Symposium on Ion Implantation, 1972), Plenum, New York, 1973.

8. J. Linhard, M. Scharff, and H. E. Schiøtt, Range Concepts and Heavy Ion Ranges, *Mat. Fys. Medd. Dan. Vidensk. Selskab* **33**, 1 (1963).

9. J. W. Mayer, L. Eriksson, and J. A. Davies, *Ion Implantation in Semiconductors*, Academic Press, New York, 1970.

10. G. Carter and W. A. Grant, *Ion Implantation of Semiconductors*, Wiley, New York, 1976.

11. H. Goldstein, Classical Mechanics, Addison-Wesley Publ. Co., Reading, MA (1956).

12. Y. Xia and C. Tan, Four-Parameter Formulae for the Electronic Stopping Cross-Section of Low Energy Ions in Solids, *Nucl. Instr. Methods* **B13**, 100 (1986).

13. J. F. Gibbons, W. S. Johnson, and S. W. Mylroie, *Projected Range Statistics: Semiconductors and Related Materials*, Dowden, Hutchinson, and Ross, Stroudsburg, 1975.

14. D. K. Brice, *Ion Implantation Range and Energy Deposition Distributions*, Plenum, New York, 1975.

15. W. K. Hofker, D. P. Oosthoek, N. J. Koeman, and H. A. M. De Grefte, Concentration Profiles of Boron Implantations in Amorphous and Polycrystalline Silicon, *Radiat. Eff.* **24**, 223 (1975).

16. S. Furukawa, H. Matsumura, and H. Ishiwara, Theoretical Considerations on Lateral Spread of Implanted Ions, *Jpn. J. Appl. Phys.* **11**, 134 (1972).

17. G. H. Kinchin and R. S. Pease, The Displacement of Atoms in Solids by Radiation, *Rep. Prog. Phys.* **18**, 2 (1955).

18. J. F. Zeigler, Determination of Lattice Disorder Profiles in Crystals by Nuclear Backscattering, *J. Appl. Phys.* **43**, 2973 (1972).

19. F. F. Morehead, Jr., and B. L. Crowder, in *Ion Implantation*, F. H. Eisen and L. T. Chadderton, Eds., Gordon and Breach, London, 1971, p. 25.

20. D. K. Brice, Ion Implantation Depth Distributions: Energy Deposition into Atomic Processes and Ion Locations, *Appl. Phys. Lett.* **16**, 103 (1970).

21. J. W. Corbett, Radiation Damage in Silicon and Germanium, in *Proceedings of the First International Conference on Ion Implantation*, L. Chadderton and F. Eisen, Eds., Gordon and Breach, New York (1971).

22. J. F. Gibbons, Ion Implantation in Semiconductors—Part II. Damage Production and Annealing, *Proc. IEEE* **60**(9), 1062 (1972).

23. S. Furukawa and H. Ishiwara, Mass Dependence of Critical Amorphizing Dose in Ion Implantation, *Jpn. J. Appl. Phys.* **11**, 1062 (1972).

24. P. Blood, G. Dearnaley, and M. A. Wilkins, The Depth Distribution of Phosphorus Ions Implanted into Silicon Crystals, in *Ion Implantation in Semiconductor and Other Materials*, International Symposium on Ion Implantation, 1972, B. L. Crowder, Ed., Plenum, New York, 1973, p. 75.

25. S. J. Pearton, W. S. Hobson, and C. R. Abernathy, Ion Implantation Processing of GaAs and Related Compounds, in *Ion Beam Processing of Advanced Electronic Materials*, N. W. Cheung, A. D. Marwick, and J. B. Roberts, Eds., *Proc. Mater. Res. Soc.* **147**, 261 (1989).

26. J. M. Woodcock and D. J. Clark, The Ion Implantation of Donors for n^+–p Junctions in GaAs, in *Gallium Arsenide and Related Compounds—1974*, Publication 24, Institute of Physics, London and Bristol, 1974, p. 331.

27. D. K. Brice, Spatial Distribution of Energy Deposited into Atomic Processes in Ion-Implanted Silicon, *Radiat. Eff.* **6**, 77 (1970).

28. B. C. Crowder and F. F. Morehead, Annealing Characteristics of *n*-Type Dopants in Ion Implanted Silicon, *Appl. Phys. Lett.* **14**, 313 (1969).

29. A. Lietolla, J. F. Gibbons, T. J. Magee, J. Peng, and J. D. Hong, Solid Solubility of As in Si as Determined by Ion Implantation and CW Laser Annealing, *Appl. Phys. Lett.* **35**, 532 (1979).

30. T. E. Seidel and A. U. MacRae, The Isothermal Annealing of Boron Implanted in Silicon, in *First International Conference on Ion Implantation*, F. Eisen and L. Chadderton, Eds., Gordon and Breach, New York, 1971.

31. A. Lidow, J. F. Gibbons, V. R. Deline, and C. A. Evans, Jr., Fast Diffusion of Elevated Temperature Ion-Implanted Se in GaAs, *Appl. Phys. Lett.* **32**, 149 (1978).

32. J. Kashahara, M. Arai, and T. Watanabe, Effect of Arsenic Partial Pressure on Capless Anneal of Ion-Implanted GaAs, *J. Electrochem. Soc.* **126**, 1997 (1979).

33. F. H. Eisen, Ion Implantation in III–V Compounds, *Radiat. Eff.* **47**, 99 (1980).

34. R. G. Hunsperger, R. G. Wilson, and D. M. Jamba, Mg and Be Ion Implanted GaAs, *J. Appl. Phys.* **43**, 1318 (1972).

35. W. W. Levige, N. J. Helix, K. V. Vaidyanathan, and B. G. Streetman, Electrical Profiling and Optical Activation Studies in Be-Implanted GaAs, *J. Appl. Phys.* **48**, 3342 (1977).

36. M. D. Deal and H. G. Robinson, Diffusion of Implanted Beryllium in Gallium Arsenide as a Function of Anneal Temperature and Dose, *Appl. Phys. Lett.* **55**, 996 (1989).

37. R. N. Thomas, H. M. Hobgood, G. W. E. Eldridge, D. L. Barrelt, T. T. Braggins, B. Ta, and S. K. Wang, in *Semiconductors and Semimetals*, Vol. 20, R. K. Willardson and A. C. Beer, Eds., Academic Press, New York, 1984, p. 1.

38. S. K. Brierley, T. E. Anderson, and A. K. Grabinski, Correlation Between Implant Activation and EL2 Concentration in Semi-insulating GaAs, in *Proceedings of 5th Conference on Semi-insulating III–V Materials*, Malmo, Sweden, 1988, G. Grossmann and L. Ledebo, Eds., Adam Hilger, New York, 1988, p. 21.

39. R. A. Morrow, Electrical Activation Curve of Silicon Implanted in GaAs, *Appl. Phys. Lett.* **55**, 2523 (1989).

40. F. H. Eisen, B. M. Welch, H. Muller, K. Gamo, T. Inada, and J. W. Mayer, Tellurium Implantation in GaAs, *Solid State Electron.* **20**, 219 (1977).

41. J. P. Donnelly, Ion Implantation in GaAs, *Inst. Phys. Conf. Ser.* **33b**, 166 (1977).

42. J. P. Donnelly and F. J. Leonberger, Multiple Energy Proton Bombardment on n^+-GaAs, *Solid State Electron.* **20**, 183 (1977).

43. P. N. Favennec, G. P. Pelous, M. Binet, and P. Baudet, Compensation of GaAs by Oxygen Implantation, in *Ion Implantation in Semiconductor and Other Materials*, International Symposium on Ion Implantation, 1972, B. L. Crowder, Ed., Plenum, New York, 1973, p. 621.

44. M. Berth, C. Venger and G. M. Martin, Selective Carrier Removal Using Oxygen Implantation in GaAs, *Electron. Lett.* **17**, 873 (1981).

45. T. E. Kazior, Isolation Implant Studies in GaAs, *J. Electrochem. Soc.* **137**, 2257 (1990).

46. F. Clauwert, P. Van Daele, R. Baets, and P. Lagasse, Characterization of Device Isolation in GaAs MESFET Circuits by Boron Implantation, *J. Electrochem. Soc.* **134**, 711 (1987).

47. T. O. Sedgwick, Short Time Annealing, *J. Electrochem. Soc.* **130**, 484 (1983).

48. E. Rimini, Ion Implantation, in *Microelectronic Materials and Processes*, R. A. Levy, Ed., Kluwer Academic Publishers, Boston, MA, 1989, p. 521.

49. T. E. Seidel, D. J. Lischner, C. S. Pai, R. V. Knoell, D. M. Maher, and D. A. Jacobson, A Review of Rapid Thermal Annealing (RTA) of B, BF_2 and As Ions Implanted into Silicon, *Nucl. Instr. Methods Phys. Res.* **B7/8**, 251 (1985).

50. Y. Lu, C. A. Paz de Araujo, and T. S. Kalkur, Activation of Be-Implanted GaAs by Using RTA with Proximity Contact, *J. Electrochem. Soc.* **137**, 1904 (1990).

51. P. M. Asbeck, D. G. Miller, E. J. Babcock, and C. G. Kirkpatrick, Application of Thermal Pulse Annealing to Ion Implanted GaAlAs/GaAs Heterojunction Bipolar Transistors, *IEEE Electron Dev. Lett.* **EDL-4**, 81 (1983).

52. G. Dearnaley, J. H. Freeman, R. S. Nelson, and J. Stephen, *Ion Implantation*, North-Holland, New York, 1973.

53. H. Ryssel and I. Ruge, *Ion Implantation*, John Wiley, New York, 1986.

54. A. Axmann, Ionizable Materials to Produce Ions for Implantation, *Solid State Technol.* (Nov. 1974) p. 36.

55. K. O. Nielsen, The Development of Magnetic Ion Sources for an Electromagnetic Isotope Separator, *Nucl. Instrum. Methods* **1**, 289 (1957).

56. G. S. Virdi, C. M. S. Rauthan, B. C. Pathak, W. S. Khokle, S. K. Gupta, and K. Lal, On the Role of Fluorine in BF_2^+ Implanted Silicon, *Solid State Electron.* **35**, 535 (1992).

57. J. Narayan and O. N. Holland, Rapid Thermal Annealing of Ion-Implanted Semiconductors, *J. Appl. Phys.* **56**, 2912 (1984).

58. G. Ryding, Evolution and Performance of the Nova NV-ID Predep Implanter, in *Ion Implantation: Equipment and Techniques*, Proceedings of the Fourth International Conference, Berchtesgarden, Germany, 1982, H. Ryssel and H. Glawischnig, Eds., Springer-Verlag, New York, 1982, p. 319.

59. P. R. Hanley, Physical Limitations of Ion Implant Equipment, in *Ion Implantation: Equipment and Techniques* (Proceedings of the Fourth International Conference, Berchtesgarden, Germany, 1982), H. Ryssel and H. Glawischnig, Eds., Springer-Verlag, New York, 1983, p. 2.

60. B. Baccus, Impact of Low-Temperature Transient Enhanced Diffusion of Dopants in Silicon, *Solid State Electron.* **35**, 1045 (1992).

61. T. E. Seidel and A. U. MacRae, Some Properties of Ion Implanted Boron in Silicon, *Trans. Met. Soc. AIME* **245**, 491 (1969).

62. D. S. Moroi and P. M. Hemenger, Exact Analytical Solution to Diffusion Equation for Ion-Implanted Dopant Profile Evolution During Annealing, *Appl. Phys. Lett.* **50**, 155 (1987).

63. D. H. Lee and J. W. Mayer, Ion Implanted Semiconductor Devices, *Proc. IEEE* **62**, 1241 (1974).

64. T. E. Seidel and D. L. Sharfetter, High Power Millimeter Wave IMPATT Oscillators Made by Ion Implantation, *Proc. IEEE* **58**, 1135 (1970).

65. A. J. Zaremba, G. T. Marcyk, and B. G. Streetman, Optimal Summation of Gaus-

sians for Ion Implantation Profile Control, *IEEE Trans. Electron Dev.* **ED-24**, 163 (1977).

66. R. Heckingbottom and T. Ambridge, Ion Implantation in Compound Semiconductors—An Approach based on Solid State Theory, *Radiat. Eff.* **17**, 31 (1973).

67. Y. K. Yeo, F. L. Pedrotti, and Y. S. Park, Modification of the Amphoteric Activity of Ge Implants in GaAs by Dual Implantation of Ge and As, *J. Appl. Phys.* **51**, 5765 (1980).

68. P. E. Hallali, H. Baratte, F. Cardone, M. Norcott, F. Legoues, and D. K. Sadana, Effect of F Co-implant during Annealing of Be-implanted GaAs, *Appl. Phys. Lett.* **57**, 569 (1990).

69. T. R. Cass and V. G. K. Reddi, Anomalous Residual Damage in Si After Annealing of Through-Oxide Arsenic Implantations, *Appl. Phys. Lett.* **23**, 268 (1973).

70. J. C. North, T. E. McGahan, D. W. Rice, and A. C. Adams, Tapered Windows in Phosphorus-Doped SiO_2 by Ion Implantation, *IEEE Trans. Electron Dev.* **ED-25**, 809 (1978).

71. T. O. Sedgwick, A. E. Michel, V. R. Deline and S. A. Cohen, Transient Boron Diffusion in Ion-Implanted Crystalline and Amorphous Silicon, *J. Appl. Phys.* **63**, 1452 (1988).

72. G. Ryding and M. Farley, A New Dose Control Technique for Ion Implantation, *Nucl. Instr. Methods* **189**, 295 (1981).

73. S. M. Sze and G. Gibbons, Effect of Junction Curvature on Breakdown Voltage in Semiconductors, *Solid State Electron.* **9**, 831 (1966).

74. S. K. Ghandhi, *Semiconductor Power Devices*, Wiley, New York 1977.

75. A. N. Saxena and D. Pramanik, Megaelectronvolt Implantations in Silicon Very Large Scale Integration, in *Proceedings of Symposium C on Deep Implants: Fundamentals and Applications*, E-MRS Spring Conference, Strassbourg, France, May 31–June 2, 1988.

76. H. Wong, N. W. Cheung, and P. K. Chu, Gettering of Gold and Copper with Implanted Carbon in Silicon, *Appl. Phys. Lett.* **52**, 889 (1988).

77. G. A. Sai-Halaz, M. R. Wordman, and R. H. Dennard, Alpha-Particle Induced Soft Error Rate in VLSI Circuits, *IEEE Trans. Electron Dev.* **ED-29**, 725 (1982).

78. Y. Umimoto, N. Masuda, J. Shigeta, and K. Mitsusada, Improvement of Alpha-Particle-Induced Soft-Error Immunity in a GaAs SRAM by a Buried *p*-Layer, *IEEE Trans. Electron Dev.* **ED-35**, 268 (1988).

79. K. Tsukamoto, T. Kuroi, S. Komori, and Y. Akasaka, High Energy Ion Implantation for VLSI: Well Engineering and Gettering, *Solid State Technol.* (June 1992) p. 49.

80. H. Kanber, J. C. Chen, and M. J. Barger, MeV Implantation of Gallium Arsenide, *Proc. Mater. Res. Soc.* **147**, 185 (1989).

81. M. A. Guerra, The Status of SIMOX Technology, *Solid State Technol.* (Nov. 1990) p. 75.

82. H. H. Hosack, T. W. Houston, and C. P. Pollack, SIMOX Silicon-on-Insulator: Materials and Devices, *Solid State Technol.* (Dec. 1990) p. 61.

83. T. F. Cheek and D. Chen, Silicon-on-Insulator and Buried Metals in Semiconductors, *Proc. Mater. Res. Soc.* **107**, 53 (1987).

84. A. K. White, K. T. Short, J. L. Batstone, D. C. Jacobson, J. M. Poate, and K. W. West, Mechanisms of Buried Oxide Formation by Ion Implantation, *Appl. Phys. Lett.* **50**, 19 (1987).

85. P. C. Davis, J. F. Graczyk, and W. A. Griffin, Design of an Integrated Circuit for the T1C Low-Power Line Repeater, *IEEE Trans. Commun.* **COM-27**, 367 (1979).

86. R. A. Moline and G. Foxhall, Ion-Implanted Hyperabrupt Junction Voltage Variable Capacitors, *IEEE Trans. Electron Dev.* **ED-12**, 267 (1972).

87. R. B. Palmer, C. C. Mai, and M. Hswe, The Effect of Oxide Thickness on Threshold Voltages of Boron Ion Implanted MOSFET, *J. Electrochem. Soc.* **120**, 999 (1973).

88. E. C. Douglas and A. G. F. Dingwall, Ion Implantation for Threshold Control in COS–MOS Circuits, *IEEE Trans. Electron Dev.* **ED-21**, 324 (1974).

89. H. J. Sansbury, Applications of Ion Implantation in Semiconductor Processing, *Solid State Technol.* (Nov. 1976) p. 31.

90. C. O. Séquin and M. F. Tompsett, *Charge Transfer Devices, Suppl. 8, Advances in Electronics and Electron Physics*, Academic Press, New York, 1975.

91. G. E. Stillman, C. M. Wolfe, J. A. Rossi, and J. P. Donnelly, Electro-absorption Avalanche Photodiodes, *Appl. Phys. Lett.* **25**, 671 (1974).

92. E. Garmire, H. Stoll, A. Yariv, and R. G. Hunsperger, Optical Waveguide in Proton-Implanted GaAs, *Appl. Phys. Lett.* **21**, 87 (1972).

93. W. Heinke and H. J. Queisser, Photoluminescence at Dislocations in GaAs, *Phys. Rev. Lett.* **33**, 18, 1082 (1974).

94. M. Eisenberg, A. C. Callegari, D. K. Sadana, H. J. Hovel, and T. N. Jackson, Enhanced Schottky Barriers by Recoil Implantation of Mg into n-GaAs, *Appl. Phys. Lett.* **54**, 1696 (1989).

95. N. Toyada, I, Niikura, Y. Shimura, T. Hozuki, H. Sugibuchi, M. Mihara, and T. Hara, Ion Implanted GaAs Varactor Diodes: Capacitance Uniformity, *Electron Lett.* **14**, 152 (1978).

96. R. L. van Tuyl, C. H. Liechti, R. E. Lee, and E. Gowen, GaAs MESFET with 4 GHz Clock Rates, *IEEE J. Solid State Circuits* **SC-12**, 485 (1977).

97. B. G. Bosch, Gigabit Electronics—A Review, *Proc. IEEE* **67**, 340 (1979).

98. K. Yamasaki, K. Arai, and K. Kurumada, GaAs LSI-Directed MESFETs with Self-Aligned Implantation for n-Layer Technology (SAINT), *IEEE Trans. Electron Dev.* **ED-29**, 1772 (1982).

99. Y. Kato, M. Dohsen, J. Kasahara, and K. Taira, Planar Normally-Off GaAs JFET for High Speed Logic Circuits, *Electron. Lett.* **17**, 951 (1981).

100. R. Zuleeg, J. K. Notthoff, and G. L. Troeger, IEEE Electron Dev. Lett., **EDL-5**, 21 (1984).

101. R. E. Lundgren, C. F. Krumm, and R. P. Pierson, Fast Enhancement-Mode GaAs MES-FET Logic, *IEEE Trans. Electron Dev.* **ED-26**, 1827 (1979).

102. D. D'Avanzo, Proton Isolation of GaAs Integrated Circuits, *IEEE Trans. Electron Dev.* **ED-29**, 1051 (1982).

103. R. C. Eden, B. M. Welch, and R. Zucca, Planar GaAs IC Technology: Applications for Digital LSI, *IEEE J. Solid State Circuits* **SC-13**, 419 (1978).

104. W. V. McLevige, C. T. M. Chang, and A. H. Taddiken, An ECL Compatible GaAs MESFET 1k-bit Static RAM, IEEE J. Solid State Circuits, **SC-22**, 262 (1987).

105. K. Yamasaki, K. Asai, T. Mizutani, and K. Kurumada, Self-Align Implantation for N^+ Layer Technology (SAINT) for High Speed GaAs ICs, *Electron. Lett.* **18**, 119 (1982).

106. C. A. Stolte, Ion Implantation and Materials, in *Semiconductors and Semimetals*, Vol. 20, R. K. Willardson and A. C. Beer, Eds., Academic Press, New York, Inc., 1984, p. 89.

107. R. Zuccca, B. M. Welch, R. C. Eden, and S. I. Long, GaAs Digital IC Technology/Statistical Analysis of Device Performance, *IEEE Trans. Electron Dev.* **ED-27**, 1109 (1980).

108. R. Williams, *Modern GaAs Processing Methods*, Artech House, Boston, MA, 1990.

PROBLEMS

1. Show that, to an approximation, the critical dose required to form an amorphous layer is given by N_{crit}, where

$$N_{crit} = \frac{2E_d}{S_n^0}\left(1 + \frac{M_2}{3M_1}\right)$$

where E_d is the displacement energy and S_n^0 is the stopping power.

2. Establish Eq. (6. 26).

3. A resistor is made by boron implantation into silicon at 200 keV. Assuming an ion dose of 5×10^{12} cm^{-2}, calculate: (a) the peak concentration; (b) the sheet resistance of the layer; (c) the specific resistivity. Assume $\mu = 250$ cm^2/Vs. In addition, provide an approximate algebraic expression for the resistivity.

4. A junction is formed by implanting boron at 100 keV through a window in an oxide. Assume a dose of 10^{15} ions/cm^2 and a background concentration of 5×10^{14} cm^{-3}. Calculate the location of the *p–n* junction so formed, and make a sketch of this junction. Also, sketch the *p–n* junction that would be formed by a diffusion to the same depth. Comment on the nature of your answer.

5. It is required to fabricate a p–n diode by ion implantation. A p-silicon substrate, impurity doped to 10^{16} cm^{-3}, is to be used. A window of $15\,\mu\text{m} \times 15\,\mu\text{m}$ is cut in the mask. What is the total number of phosphorus ions (at 100 keV) so that the maximum doping concentration does not exceed 10^{19} cm^{-3}. At what depth can you expect the junction to be formed? Draw the impurity core profile (both ideal and realistic).

6. Calculate the selenium ion implantation parameters that could be used to provide a channel that is suitable for a GaAs FET ($0.1\,\mu\text{m}$ thick with an average concentration of 10^{17} cm^{-3}). What is the average sheet resistance of this channel if $\mu_n = 4000$ cm^2/V s?

7. An ion implanter has a beam current of 1 mA. Implantation is carried out on 6-in.-diameter slices, using a cassette loader. Assume 100-keV ions and a total implantation time of 2 min. What is the dose? What is the slice equilibrium temperature for this dose? Assume a rectangular raster pattern.

8. A boron channel implant must be made through a 500-Å-thick silicon dioxide gate oxide. What is the masking power of a field oxide of 5000 Å for this implant condition? Assume that $\Delta R_p \simeq 0.3 R_p$.

CHAPTER 7

NATIVE FILMS

A wide variety of films are used in microcircuit technology today. Many of these are deposited on the semiconductor in which the microcircuit has been fashioned. Yet others are grown out of the semiconductor itself, and are referred to as *native* films. The principal advantage of native films is that they are relatively free from contamination, as is the semiconductor from which they are made. Inherently, their chemical composition is dictated by the starting semiconductor, and one must resort to deposition techniques if other film compositions are desired. Native grown films are widely used in semiconductor processing because of their ease of formation and their excellent interface with the underlying semiconductor. In silicon, the basis for all device passivation is native SiO_2, although other deposited insulating films are useful as secondary layers in microcircuit fabrication.

Native films of both silicon and GaAs form adherent, insulating deposits on the surface. Both nitride and oxide films can be grown, and used in a number of fabrication processes, depending upon their quality, stability, and electronic properties. Uses include diffusion masks for impurities, surfaces upon which electrical connection can be made between devices, gate oxides and passivation layers, and antireflective windows for photodevices.

Thermal, electrochemical, and plasma processes can be used to form these films. Their properties are quite different, so that each film type finds use in specific applications. With silicon, thermal oxidation results in the formation of relatively dense, trap-free films which can be used to protect the slice during its various high-temperature processing steps. Additionally, their high breakdown electric field strength, in excess of 10 MV/cm, makes them suitable for ultrathin gate insulators in MOS devices. Anodic oxides, on the other hand, are relatively porous. Their primary use has been in diagnostic situations where thin layers of silicon must be removed (by repeated anodization and oxide dissolution) without subjecting the slice to elevated temperatures. Highly dense, silicon nitride films can also be grown. These are impervious to the transport of light ions, and thus have potential advantages as gate insulators for metal-oxide-semiconductor (MOS) devices, where extremely thin layers are required.

Native oxides can also be grown on GaAs. With this semiconductor, however, thermal oxidation results in films which contain the oxides of arsenic and gallium as well as free arsenic, and are invariably nonstoichiometric. Thermal oxides are poor electrical insulators, and also poor in their ability to provide protection to the semiconductor surface. Consequently, they are rarely used in GaAs technology. On the other hand, oxides grown by plasma enhanced anodization, as well as electrochemically anodized films, have shown much promise for this area.

This chapter discusses the physics and technology of native films. Deposited films, including oxides and nitrides, will be discussed in Chapter 8. Tables of pertinent data are assembled at the end of this chapter, for easy reference.

7.1 THERMAL OXIDATION OF SILICON

The surface of silicon is covered at all times with a layer of SiO_2. This is even true for freshly cleaved silicon, which becomes covered with a few monolayers (≈ 15–20 Å) of oxide upon exposure to air, gradually thickening with time to an upper limit of about 40 Å. Films of this type are patchy in character and are of no technological value. Considerably thicker films are required in silicon micro-circuit fabrication. Here, an oxide film is grown on the slice by maintaining it in an elevated temperature in an oxidizing ambient, usually dry oxygen or water vapor.

An important property of thermally grown SiO_2 is its ability to reduce the surface state density of silicon by tying up some of its dangling bonds. Moreover, silicon oxides can be grown with good control over interface traps and fixed charge. Their use in controlling the leakage current of junction devices [1], and in the formation of a stable gate oxide for field effect devices, stems from these properties and makes them the cornerstone of modern silicon integrated circuit technology.

7.1.1 Intrinsic Silica Glass

Intrinsic silica glass consists of fused SiO_2, with a melting point of $1732°C$. It is thermodynamically unstable below $1710°C$ and tends to return to its crystalline form at temperatures below this value. However, the rate of this devitrification process can be neglected at $1000°C$ and lower.

The model for pure silica in the vitreous state consists of a random three-dimensional network of SiO_2, constructed from polyhedra of oxygen ions [2].

The centers of these polyhedra are occupied with Si^{2+} ions. The tetrahedral distance between the silicon and oxygen ions is 1.60 Å while the distance between oxygen ions is 2.27 Å. The O–Si–O bond angle is 109°.

Silica polyhedra are joined to one another by *bridging* oxygen ions, each of which is common to two such polyhedra. In crystalline silica all such oxygen ions play this role and all vertices of the polyhedra are tied to their nearest neighbors by these ions. In fused silica, however, some of the vertices have *nonbridging* oxygen ions which belong to only one polyhedron. The degree of cohesion between the polyhedra—and, hence, of the network as a whole—is thus a function of the ratio of bridging to nonbridging oxygen ions.

In pure silica, movement of the silicon atom is accomplished by the rupture of four Si–O bonds, whereas the movement of a bridging oxygen atom requires the rupture of only two Si–O bonds. It follows that oxygen is more free to move in this glass than is silicon. The movement of this oxygen from its polyhedral site can give rise to the formation of oxygen ion vacancies, which represent positively charged defects in the structure.

Silica films, as grown by oxidation of the silicon surface, are amorphous in nature, and consist of a random network of such polyhedra. Typically they have a density of 2.15–2.25 as compared to 2.65 for single-crystal quartz. Because of their open structure, it is possible for a number of impurities to diffuse through them quite readily.

Diffusion processes in intrinsic silica glass are similar to those described in Chapter 4, even though a random structure is involved. To a first order, the process can be described by a diffusivity D such that

$$D = D_0 \, e^{-E_A/kT} \tag{7.1}$$

where D_0 is the diffusion constant and E_A is the activation energy of the diffusion species in eV/molecule.* Figure 7.1 shows the diffusivity of hydrogen, oxygen, and water in silica glass as a function of temperature. These species are involved in the oxidation process. Also shown is the diffusivity of sodium, which is an important contaminant in thermal oxidation systems. Note that these data represent approximate values only, and are dependent on the nature of the glass network—that is, on the ratio of bridging to nonbridging oxygen ions in it.

There is much uncertainty about the nature of the species involved in the thermal oxidation of silicon. Some work seems to indicate that the species is a charged form of oxygen, either O_2^- or O_i^{2-}. In either event, it has been established that oxidation involves the transport of the oxidizing species *through* the silicon and to the Si–SiO$_2$ interface where oxidation occurs [3]. As a result, thermal oxidation of silicon leaves a clean interface, with ionic contaminants

---

*These values are often given in units of kcal/mole. Note that 1 kcal/mole = 0.0434 eV/molecule.

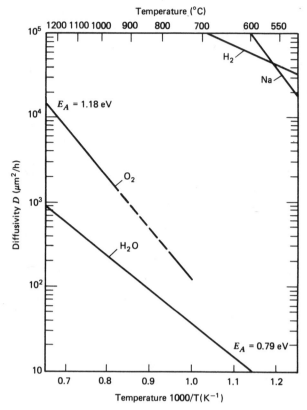

Fig. 7.1 Diffusivities of hydrogen, oxygen, sodium, and water vapor in silica glass.

transported to the exposed oxide surface. It is this characteristic which results in greatly improving the stability of planar junctions on which this oxide is grown.

7.1.2 Extrinsic Silica Glass

The properties of silica glass are greatly modified by the introduction of impurities [4]. These impurities are primarily of two types: substitutional and interstitial. In addition, water vapor can also be considered as an important impurity that is always present in silica glass.

7.1.2.1 Substitutional Impurities

A substitutional impurity is one which replaces the silicon in a silica polyhedron. The most common impurities of this type are B^{3+} and P^{5+}. They are called *network formers*, since it is possible to build vitreous structures using them instead of SiO_2 (i.e., glasses which are totally free of SiO_2). In microcircuit fabrication, however, our interest is centered on the effect of small amounts of these impurities in silica glass.

The valence of a substitutional cation is usually either 3 or 5. In the silica lattice, such cations result in charge defects. The presence of group V impurities gives rise to the formation of an excess of nonbridging ions, while group III impurities usually (but not always) reduce the nonbridging ion concentration.

7.1.2.2 Interstitial Impurities

These are usually the oxides of large alkali ions of low positive charge which enter into the network interstitially between the polyhedra. In so doing, they give up their oxygen to it, thus producing two nonbridging oxygen ions in place of the original bridging ion. This results in weakening the structure and rendering it more porous to other diffusing species. The reaction is shown schematically for Na_2O in the silica lattice as follows:

$$Na_2O + Si—O—Si = Si—O + O—Si + 2\ Na^+ \qquad (7.2)$$

Impurity oxides of this type are called *network modifiers*, since they are not capable of forming glasses by themselves. Ions such as Na^+, K^+, Pb^{2+}, and Ba^{2+} fall into this class. Aluminum sometimes plays a dual role of network modifier as well as network former.

Sodium is particularly important because it is widely present as part of the environment, especially around humans. It is also an important impurity in the firebrick that is used for holding furnace heating elements, and is even present in the fused quartz tubes that are used in diffusion and oxidation equipment. Since it is a charged, mobile species, its avoidance, removal, or immobilization are important in the fabrication of all types of stable devices, and especially for MOS-based microcircuits. Excessive amounts of sodium in an oxide can accelerate its ability to crystallize, and even result in cracking of the film.

7.1.2.3 Water Vapor

Water vapor is also widely present in the environment. It can reside in molecular form at interstitial sites in oxidized silicon slices to a depth of several hundred angstroms, if they are left unpackaged for as little as a week. Its presence in this form leads to problems of poor adhesion of photoresist in subsequent masking operations, and to instability in the magnitude of the reverse breakdown voltage of diodes which are covered with this oxide. Prolonged bake-out (48–72 hr) at low temperatures (200–250°C) is often done just before packaging to remove this water.*

Water vapor is incorporated during a "wet" oxidation process, and also as a contaminant in a "dry" process. On entering, it combines with bridging

---

*Sometimes, a brief surface etch is used for this purpose.

oxygen ions to form pairs of stable nonbridging *hydroxyl* groups. The reaction is described schematically by

$$H_2O + Si—O—Si \rightleftharpoons Si—OH + OH—Si \qquad (7.3)$$

The presence of these hydroxyl groups also tends to weaken the silica network and render it more porous to diffusing species. Thus their behavior is similar to that of interstitial impurities.

Figure 7.2 shows a schematic representation of fused silica glass together with the various types of defect structures that may be present in it [5].

7.1.3 Oxide Formation

Oxidation of the silicon slice is conveniently carried out by subjecting it to dry oxygen or water vapor, while it is maintained at an elevated temperature. In water vapor, or wet oxidation schemes, an inert gas (or oxygen) is bubbled through water which is usually held at 95°C, to avoid its excessive depletion. Direct oxidation of the silicon surface results in forming an oxide layer which is approximately 2.27 times the thickness of the consumed silicon. The silica layer so formed contains about 2.2×10^{22} molecules cm^{-3} of SiO$_2$.

The process of oxide formation require that the oxidizing species move through the growing oxide layer in order to reach the silicon surface. Thus, growth proceeds at an ever-decreasing rate as the thickness of the intervening oxide layer increases. This diffusion-controlled regime dominates the growth

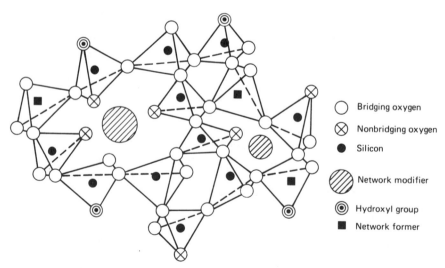

Bridging oxygen

Nonbridging oxygen

Silicon

Network modifier

Hydroxyl group

Network former

Fig. 7.2 Structure of silica glass. Adapted from Revesz [7].

process for layers greater than 40 Å thickness for dry oxidation, and 1000 Å for wet oxidation.

The chemistry of oxidation with dry oxygen is relatively straightforward. It is assumed that the species diffusing through the growing layer is molecular oxygen. The chemical reaction at the silicon surface is

$$Si + O_2 \rightarrow SiO_2 \tag{7.4}$$

Here *one* molecule of oxygen results in the formation of one molecule of SiO_2.

The overall process of oxidation of silicon with water vapor may also be considered as one in which the oxidizing species diffuses through the oxide and reacts with the silicon surface, so that

$$Si + 2H_2O \rightarrow SiO_2 + 2H_2 \tag{7.5}$$

Here, however, *two* molecules of water vapor are used to form one molecule of SiO_2. The hydrogen evolved by this reaction diffuses rapidly through the growing oxide and leaves the system at the gas–oxide interface. The detailed nature of this reaction, although somewhat more complex [6], is assumed to proceed in the following manner:

1. Water vapor reacts with the bridging oxygen ions in the silica structure to form nonbridging hydroxyl groups. This reaction, which results in weakening the silica structure, may be written as

$$H_2O + Si\!-\!O\!-\!Si \rightleftharpoons Si\!-\!OH + OH\!-\!Si \tag{7.6}$$

2. At the oxide–silicon interface, the hydroxyl groups react with the silicon lattice to form silica polyhedra and hydrogen. The reaction is

$$
\begin{matrix}
Si\!-\!OH & & Si\!-\!O\!-\!Si & \\
& +Si\!-\!Si \rightleftharpoons & & +H_2 \\
Si\!-\!OH & & Si\!-\!O\!-\!Si &
\end{matrix}
\tag{7.7}
$$

3. Hydrogen is a rapid diffuser in SiO_2, and leaves after being formed by this process. Some of this diffusing hydrogen reacts with bridging oxygen ions in the silica structure to form hydroxyl groups, as shown by

$$\tfrac{1}{2}H_2 + O\!-\!Si \rightleftharpoons Si\!-\!OH \tag{7.8}$$

thus further weakening the silica structure.

A somewhat more detailed picture of the role of hydrogen in silica is given elsewhere [7]. However, the relatively simple model, outlined above, is suitable for describing the basic mechanism of formation of these films.

7.1.4 Kinetics of Oxide Growth

The kinetics of oxide growth on silicon may be determined [8] with reference to the model of Fig. 7.3. Assume that a silicon slice is brought in contact with the oxidant, with concentration N_g in the gas phase, resulting in a surface concentration of N_0 molecules cm^{-3} for this species. In typical oxidation systems, the mass transfer coefficient is extremely high, so that the magnitude of N_0 is essentially the solid solubility of the species at the oxidation temperature. At 1000°C the solid solubility of these species is 5.2×10^{16} molecules cm^{-3} for dry oxygen and 3×10^{19} molecules cm^{-3} for water vapor at a pressure of 1 atm.

The oxidizing species is assumed to diffuse in molecular form through the silicon dioxide layer, resulting in a concentration N_1 at the surface of the silicon. Transport of the species may occur by both drift and diffusion. Writing D as the diffusivity and ignoring the effects of drift, the flux density of oxidizing species arriving at the gas–oxide interface is given by j, where

$$j = -D \frac{\partial N}{\partial x} \simeq \frac{D(N_0 - N_1)}{x} \qquad (7.9)$$

and x is the thickness of the oxide at a given point in time. This is sometimes referred to as *Henry's law*.

On arrival at the silicon surface the species enters into chemical reaction with it. If it is assumed that this reaction proceeds at a rate proportional to the concentration of the oxidizing species, then

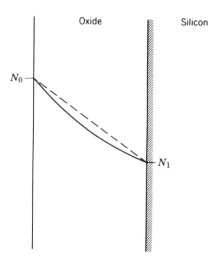

Fig. 7.3 Model for oxidation kinetics.

$$j = kN_1 \tag{7.10}$$

where k is the interfacial reaction rate constant. These fluxes must be equal under steady-state diffusion conditions. Combining Eqs. (7. 9) and (7. 10) gives

$$j \simeq \frac{DN_0}{x + D/k} \tag{7.11}$$

The reaction of the oxidizing species with the silicon results in the formation of SiO_2. Writing n as the number of molecules * of the oxidizing impurity that are incorporated into unit volume of the oxide, the rate of change of the oxide layer thickness is given by

$$\frac{dx}{dt} = \frac{j}{n} = \frac{DN_0/n}{x + D/k} \tag{7.12}$$

Solving this equation, subject to the boundary value that $x = 0$ at $t = 0$, gives

$$x^2 + \frac{2D}{k}x = \frac{2DN_0}{n}t \tag{7.13}$$

so that

$$x = \frac{D}{k}\left[\left(1 + \frac{2N_0 k^2 t}{Dn}\right)^{1/2} - 1\right] \tag{7.14}$$

Equation (7.14) reduces to

$$x = \frac{N_0 k}{n}t \tag{7.15}$$

for small values of t, and to

$$x = \left(\frac{2N_0 D}{n}\right)^{1/2} t^{1/2} \tag{7.16}$$

for large values. Thus in the initial stages of growth, in which the reaction

*$n = 2.2 \times 10^{22}$ cm^{-3} for dry oxidation, and 4.4×10^{22} cm^{-3} for wet oxidation.

is kinetically controlled, the oxide thickness varies linearly with time. In later stages the reaction is diffusion-limited and the oxide thickness is directly proportional to the square root of time.

Equation (7.13) is often written in the more compact form

$$x^2 + Ax = Bt \tag{7.17}$$

Using this form, Eqs. (7.15) and (7.16) can be written as

$$x = \frac{B}{A}t \tag{7.18}$$

for small values of t, and

$$x = B^{1/2}t^{1/2} \tag{7.19}$$

for large values. For this reason, the term B/A is referred to as the *linear rate constant* whereas the *parabolic rate constant* is given by B. Measured values for these rate constants as a function of temperature are given [9] in Figs. 7.4 and 7.5 for both dry and wet oxidation ($p_{H_2O} = 640$ torr). The data of Fig. 7.4 are shown for both (111) and (100) silicon. This orientation-dependence effect will be considered in Section 7.1.4.2. Mathematical relations, describing the behavior of these rate constants, are given in Table 7. 1 at the end of this chapter.

Insight into the physical processes involved during the thermal oxidation of silicon can be gleaned by study of these figures. Thus, Fig. 7.4 shows that the logarithm of the linear rate constant falls inversely with T at a slope of 2 eV/molecule for dry oxygen, and 2.05 eV/molecule for wet oxidation. This is in close agreement with the energy required to break Si–Si bonds, which is 1.83 eV/molecule. The logarithm of the parabolic rate constant (Fig. 7.5) also falls inversely with T, but at a slope of 1.23 eV/molecule for dry oxidation. The comparable activation energy for diffusion of oxygen in fused silica is about 1.18 eV/molecule, as shown in Fig. 7.1. The corresponding value for wet oxidation is 0.78 eV/molecule, which compares favorably with the activation energy of diffusion of H_2O in fused silica (0.79 eV/molecule).

To an approximation, the linear rate constant varies directly with the concentration of oxidizing species at the surface, and thus with its partial pressure. This has been experimentally verified for both wet and dry oxidation conditions. Finally, the parabolic rate constant for wet oxidation is found to be much larger than that for dry oxidation. This is primarily due to the significantly greater solid solubility of water over oxygen in silica glass (about 3 decades larger), which more than compensates for the slightly lower diffusivity as seen in Fig. 7.1.

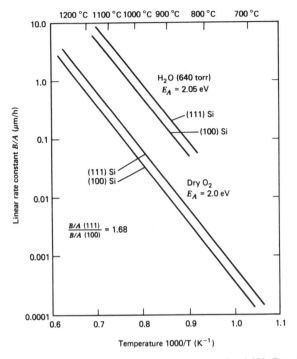

Fig. 7.4 Linear rate constant versus temperature. From Deal [8]. Reprinted with permission of the publisher, The Electrochemical Society, Inc.

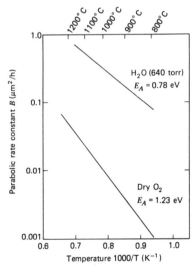

Fig. 7.5 Parabolic rate constant versus temperature. From Deal [8]. Reprinted with permission of the publisher, The Electrochemical Society, Inc.

7.1.4.1 Initial Growth Phase

The theory for the kinetics of oxide formation has been found to apply very well to growth in wet oxygen and steam. There is consistent evidence, however, to indicate the presence of an extremely rapid initial growth phase with dry oxygen, generally existing for the first 250 Å of growth. Considerable effort has been made to explain this rapid growth behavior, since oxides of this type and thickness are critically important for use as gate insulators in MOS devices. A number of theories have been proposed and are briefly summarized here.

One set of theories postulates the movement of charge species through the oxide. For example, it has been proposed [10] that the molecular oxygen, on entering the oxide, dissociates to form a negative-charged O_2^- or O_i^{2-}, and one or two holes respectively. The holes have considerably higher mobility than the oxygen ion, and run ahead of it; the result is the formation of a space-charge region. The resulting field enhances the diffusion of the oxygen in the layer, by providing an additional drift component. This enhancement occurs until the oxide thickness exceeds the thickness of the space charge region. This region of high space-charge density is near the gas–oxide interface, with the rest of the oxide layer being almost space-charge neutral. Its thickness region is thus on the order of the extrinsic Debye length, and is given by

$$L_D = \sqrt{\left(\frac{kT}{q}\right)\frac{\epsilon \epsilon_0}{2qN_0}} \tag{7.20}$$

where ϵ is the relative permittivity of the oxide (≈ 3.9). This Debye length is about 150 Å in dry oxygen, but only 5 Å for water vapor.* In practice, an accelerated oxidation rate is seen for a depth of about 230 ± 30 Å in dry oxygen processing, but is unnoticed for wet oxygen or steam processing.

Equation (7.17) can be rewritten as

$$x^2 + Ax = B(t + \tau) \tag{7.21}$$

in order to include this rapid oxidation phase. Here, τ is a time displacement which can be used to adjust the initial oxide layer (at $t = 0$).

A second set of theories, based on the structure of the oxide during the initial growth phase, have also been proposed [11]. Transmission electron microscope studies of silica structure have shown the presence of pores, about 10 Å in diameter, in very thin oxides grown in dry oxygen. These allow the oxidant to remain in direct contact with the silicon in the early phases of growth. Such pores have not been observed in wet oxides, presumably because their

---

*This is because the solid solubility of water in silica is three orders of magnitude larger than that of oxygen.

defect density is some four orders of magnitude larger than that obtained with dry oxygen (10^{20} cm^{-3} as compared to 10^{16} cm^{-3}, respectively). Yet another argument is that the basic assumption of Henry's law, as given by Eq. (7.9), only holds for infinitely thick layers, and that the solubility of oxygen in SiO$_2$ is greatly increased when the oxide thickness is less than the mean spacing between solute molecules.

A third set of theories postulate a two-stream oxidation process, with enhanced growth resulting from the presence of a thin region of material in which excess sites for oxidation are present [12]. Using this model, the experimental data were fitted to the theory by modifying Eq. (7.12) to the form

$$\frac{dx}{dt} = \frac{DN_0/n}{x + D/k} + Ce^{-\alpha/L} \tag{7.22}$$

where L is a characteristic length, of value 70 ± 10 Å, for growth in the 800–1000°C range. The pre-exponential term C has an Arrhenious type of temperature dependence, with an activation energy of 2.37 and 2.32 eV/molecule for growth on lightly doped (100) and (111) silicon respectively.

Dissolved oxygen, due to indiffusion in the very early phase of growth, is yet another possible candidate for the formation of a thin region of rapid oxide growth [13]. The diffusivity of oxygen in silicon is 2.8×10^5 Å^2/s at 1000°C, so that indiffusion to a 50-Å depth can occur within the first 10 ms of exposure of bare silicon to oxygen. However, its extent falls off with oxide growth because of a rapid reduction in this surface concentration as oxidation proceeds.

7.1.4.2 Doping Dependence Effects

Heavily doped silicon oxidizes at a faster rate than lightly doped material. However, detailed studies of boron- and phosphorus-doped material have shown considerable differences in oxide growth behavior. Thus, during oxidation, boron is preferentially incorporated into the silicon dioxide because of its relatively small segregation coefficient* ($\simeq 0.15$ to 0.3). This results in a weakening of the bond structure of the silica film, and an increase in the diffusivity of the oxidizing species through it. Consequently, there is an increase in the parabolic rate constant with boron doping concentration, but little change in the linear rate constant [14].

Phosphorus, on the other hand, has a large segregation coefficient ($\simeq 10$), is only slightly incorporated into the growing oxide, and piles up at the Si–SiO$_2$ interface. This causes an increase in the reaction rate, with a corresponding increase in the linear rate constant. On the other hand, the lack of phosphorus

---

*Defined as the ratio of the equilibrium concentration of the impurity in silicon to its equilibrium concentration in the oxide. See Section 4.8.1.

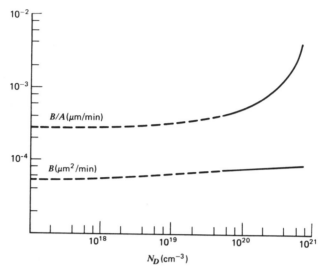

Fig. 7.6 Rate constant versus substrate phosphorus doping level for oxidation at 900°C. From Ho et al. [15]. Reprinted with permission of the publisher, The Electrochemical Society, Inc.

incorporation into the oxide results in the parabolic rate constant being relatively insensitive to doping. Experimental data, indicating the magnitude of these rate constants, are shown in Fig. 7.6 for a wet oxidation temperature of 900°C and illustrates these effects [15].

As a result of the above, we can expect that the rate of oxide growth on boron- and phosphorus-doped silicon will depend upon whether the oxidation process is operative in the linear or in the parabolic regime. With the increased packing density requirements of VLSI, there is an increased need for thin oxides grown at relatively low temperatures—that is, in the enhanced growth regime. This presents a special problem with devices such as n^+-p-n transistors where the oxide growth rate over the heavily phosphorus-doped emitter region can be as much as 2 to 5 times faster than the oxidation rate over neighboring regions where the doping is light. This can result in large steps in the oxide, with the attendant possibility of breaks in metal interconnections placed over them.

Oxide growth variations are generally not a problem in situations where the oxidation is diffusion-limited. This is partly because the resulting oxide is relatively thick over all regions of the semiconductor. In addition, the change of oxidation rate with doping concentration is only significant with boron-doped silicon, which cannot be doped as heavily as phosphorus, because of the large misfit factor of this impurity atom. As a result, oxidation rate enhancement with boron-doped silicon is generally no greater than 20% over the undoped silicon material.

7.1.4.3 Orientation-Dependence Effects

The parabolic growth rate has been found to be independent of crystal orientation during oxidation. This is reasonable, since this parameter is a measure of the diffusivity of the oxidizing species through an amorphous silica layer. The linear reaction rate, on the other hand, is related to the rate of incorporation of silicon atoms into the silica network. This, in turn, is a function of the atom concentration on the silicon surface, and is thus orientation-dependent. The areal density of silicon atoms on the (111) plane is larger than that on the (100) plane, so we can expect the linear rate constant on (111) silicon to be somewhat larger than that for (100) silicon. This is indeed the case, as is shown in Fig. 7.4.

The above argument is an oversimplification, however, since it does not take into consideration the three-dimensional nature of the silicon atomic structure, where atoms in a lower plane are partly shadowed by adjacent atoms in the upper plane, or the relative sizes of the silicon and oxygen atoms. These combined effects can be used to explain the somewhat larger change in the linear rate [(111):(100) = 1.68 : 1] than is given by atom density considerations (1.16 : 1) alone.

7.1.5 Oxidation Systems

Oxide growth is carried out in a quartz diffusion tube, in which the silicon slice is maintained at a temperature between 900°C and 1200°C. Often a high density ceramic liner is used to serve as a diffusion barrier to sodium, which is an impurity in the furnace heating elements. Some systems use diffusion tubes of pure silicon, which provides the highest degree of cleanliness and freedom from sodium contamination.

Wet oxidation processes are quite rapid, but result in relatively porous silica films. These oxides are used for most general-purpose applications such as surface coverage and diffusion masking. Experience has shown that the use of live steam for this purpose leads to poor grades of oxide because of the etching and pitting action of the excess water. Consequently, this step is usually accomplished by flowing a carrier gas through a water bubbler whose temperature is maintained below the boiling point, in order to prevent its undue depletion. A temperature of 95°C is commonly used and corresponds to a vapor pressure of about 640 torr (0.842 atm.). The carrier gas may be either oxygen or an inert species (nitrogen or argon), since the oxidation is almost entirely due to the water vapor.

Pyrogenic water systems are also used for wet oxidation, and are well suited to a manufacturing environment. Here, pure hydrogen and oxygen are directly fed to the diffusion tube, where they react to form water vapor. These systems allow a wide variation in the partial pressure of H_2O; in addition, they avoid the use of a bubbler and all the nuisance problems associated with its continual refilling and cleaning. Extensive safety features are built into these systems to avoid the possibility of explosions with this gas combination.

Oxides grown in dry oxygen are dense, and have a relatively low concentration of traps and interface states. Consequently, gate oxides for MOS-based circuits are made by this process, with elaborate precautions taken to ensure a clean, sodium-free system. Also, it is extremely important that the oxygen be truly dry, since as little as 25 ppm of water will significantly alter the growth rate, as well as the subsequent properties of the oxide. For this reason, a complete separate system is usually dedicated to growing gate oxides.

Figures 7.7a and 7.7b show the growth rates for oxide layers in 95°C H_2O, for (100) and (111) silicon respectively. Growth rates for dry oxygen are shown in Figs. 7.8a and 7.8b. It is observed that the reaction is diffusion-limited over most of the oxidation range, and that oxide growth with water vapor is considerably more rapid than with dry oxygen. This is primarily due to the considerably higher solid solubility of water vapor in silicon dioxide, as pointed out in Section 7.1.4.

Figure 7.9 shows the oxidation rate for (100) silicon in the enhanced oxidation regime, using dry oxygen [12]. This figure is relevant to the growth of gate oxides for MOS devices which are made on this orientation. Here, considerable attention must be paid in order to achieve uniform, high-quality oxides, which are free from pinholes. Techniques include careful pre-cleaning of the substrate, growth at low temperature, and the use of reduced pressures. Additional techniques, such as growth in the presence of halogens and rapid thermal processing, are also used for this purpose.

Extremely thick layers (>1 μm) are often required for field oxides in microcircuit fabrication [16]. Growth of these oxides at atmospheric pressure requires long growth times at elevated temperatures, and result in poor-quality, polycrystalline films which tend to crack during further processing. The use of high-pressure oxidation techniques can eliminate these problems, since they allow film growth to be carried out at lower temperatures where devitrification effects are not important.

Growth at low temperatures* is also desirable in VLSI schemes which require the local oxidation of exposed regions of an otherwise masked silicon slice. The movement of diffused impurities during low-temperature oxidation is greatly reduced, so that this technique is ideal for small geometries. Finally, high-pressure oxidation, at temperatures as low as 700°C, has been used to grow gate oxides whose dielectric breakdown properties are superior to those of oxides grown at atmospheric pressure [17].

Figure 7.10 shows growth rates which can be achieved during a 1-h high-pressure oxidation in water vapor, and indicates the extent of enhancement that can be obtained [18]. This is mainly due to an increase in the parabolic rate constant, which varies almost linearly with the partial pressure of H_2O. Thermal

---

*Fabrication aspects of this important technology are described in Chapter 11.

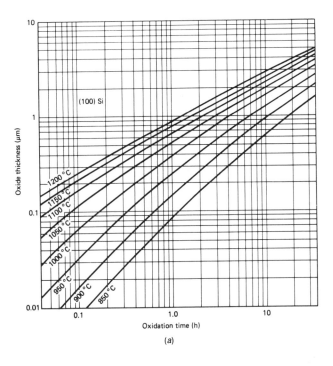

(a)

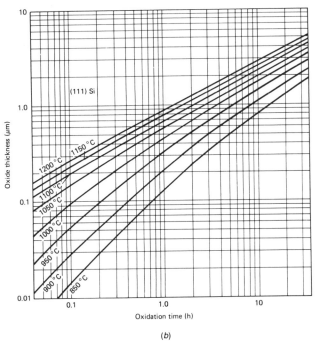

(b)

Fig. 7.7 Oxide growth rate for wet oxygen: (a) (100) silicon. (b) (111) silicon.

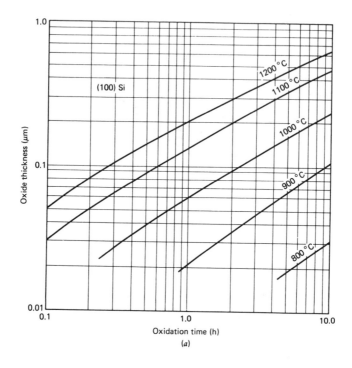

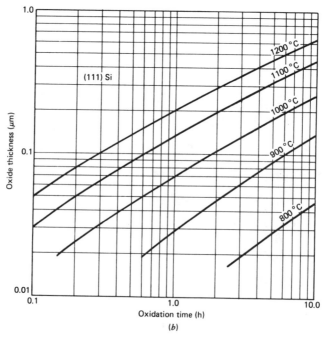

Fig. 7.8 Oxide growth rate for dry oxygen: (*a*) (100) silicon. (*b*) (111) silicon.

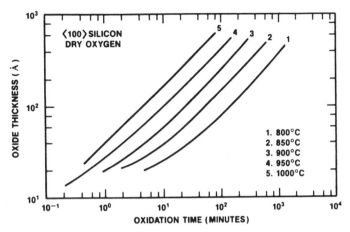

Fig. 7.9 Oxide growth rate for the early phase of dry oxidation. (100) silicon. Adapted from Massoud et al. [12].

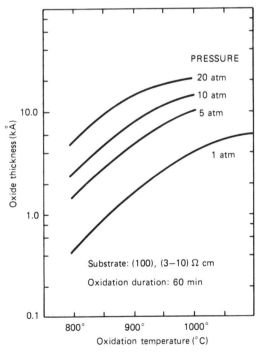

Fig. 7.10 Wet oxide growth at increased pressures. From Su [18]. Reprinted with permission from *Solid State Technology*.

oxidation of silicon has also been conducted with dry oxygen pressures as high as 140 atm. Here, growth rates at 800°C have been found comparable to those for wet oxidations at 1200°C.

Rapid thermal oxidation (RTO) followed by rapid thermal annealing (RTA) has been used for making thin, high-quality oxides for MOS gate applications [19]. Here, a two-step cycle is sometimes used, consisting of RTO in dry oxygen, followed by RTA in nitrogen for 30–60 s at 1050°C. The function of this second step is to densify the oxide, and results in an improvement of its breakdown electric field strength, to values above 10 MV/cm. Additionally, Si–SiO$_2$ interface states are reduced by about 25% as a result of this step. Figure 7.11 shows the oxide thickness as a function of oxidation time and temperature [20], by the two-step RTO/RTA process described here.

The density of dry oxides is typically 2.25 g cm^{-3}, whereas that for water vapor grown oxides is usually about 2.15. The dielectric strength of these wet oxidized films is correspondingly lower. In many oxidation schemes, it is common practice to use a combination of these processes, the oxidation procedure being both initiated and concluded in dry oxygen with an intervening wet oxidation step.

Dry oxidation is always used for MOS gate oxides. The breakdown voltage for such oxides, of thickness 500 Å, is typically 50 V provided that they are free from pinholes. The pinhole problem becomes increasingly severe when thinner oxides are used for these devices.

7.1.6 Halogenic Oxidation

The addition of a halogenic species during dry oxidation results in significant improvements in the electronic properties of the oxide and of the underlying silicon. Typically, a chlorine species is used for this purpose. Oxides with incorporated chlorine have been found to have increased dielectric field strength, often in excess of 10 MV/cm, and to result in MOS devices with improved threshold stability. In addition, an improvement in the lifetime of the silicon results from the use of these oxides as well as a reduction in the number of stacking faults on the silicon surface.*

A variety of chlorine-bearing species can be used for this purpose, the most common being chlorine gas and anhydrous HCl. Lately, there has been considerable interest in trichloroethylene (C$_2$HCl$_3$) and 1,1,1-trichloroethane (C$_2$H$_3$Cl$_3$), since they are less corrosive and can be readily handled in a bubbler arrangement. All of the above materials are available in semiconductor-grade purity.

Chlorine has been found to concentrate as a neutral species at or near the Si–SiO$_2$ interface during thermal oxidation, when one of these reactant species is involved. This has a number of consequences. First, it interacts with rapidly moving deep-lying impurities in the silicon, and removes them by conversion to their chlorides. This results in an improved lifetime in the material on which

---

*This topic will be discussed in the following section.

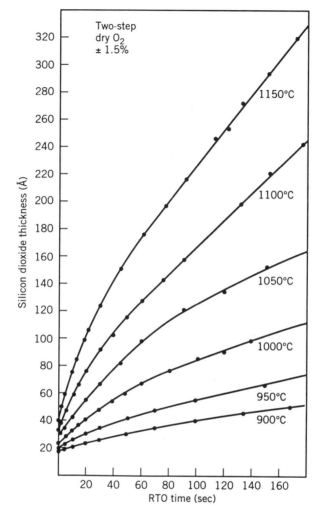

Fig. 7.11 Oxide growth rate with rapid thermal oxidation. From Nulman [20]. Reprinted with permission from *Solid State Technology.*

this halogenic oxide is grown. Next, at temperatures above 1060°C, it captures Na^+ ions to form a neutral species. Thus, it effectively removes this highly mobile ion and greatly improves the threshold stability of MOS devices. It has been shown that only a small percentage of the total chlorine at the interface is involved in this relatively complete capture process [21]. The introduction of a chlorine species during oxidation is also believed to create vacancies at the silicon surface, leading to a reduction of the stacking fault formation during oxidation.

Up to 2.3×10^{15} atoms cm^{-2} of chlorine can be incorporated into an oxide by any of the above means, with its location close to the Si–SiO_2 interface [22]. Beyond this level, however, a chlorine-rich phase is formed, and liquid agglomorates (probably chlorosilanes) are found to segregate at the interface, resulting in surface roughness and bubbles in the oxide. This renders the oxide useless for subsequent photolithography, so that the useful concentration of HCl is limited to about 2–3%.

The introduction of chlorine during dry oxidation results in an increase in the oxidation rate [23]. Comparable results have also been observed when TCE is used as the chlorine source [24]. Both the linear and parabolic rate constants increase with chlorine incorporation; however, the mechanisms involved in the growth rate are not fully understood.

Thermodynamic studies of the decomposition products of HCl and TCE have shown almost equal partial pressures for chlorine [24], although the HCl and H_2O partial pressures are a factor of 10 lower in the case of TCE. Thus the active gaseous species which results in an enhanced rate is chlorine, with H_2O and HCl playing a second-order role.

Oxidation in H_2O/HCl mixtures proceeds at the same rate as oxidation in pure H_2O alone, and apparently no chlorine is incorporated in the oxide. This is because there is no reaction between the H_2O and HCl. Here, too, the results indicate that HCl does not enter into the oxidation process [25], and the growth rate is consequently unchanged.

Experiments with HCl oxidation have shown [26] that oxides which are grown rapidly at high temperatures have a higher defect density than those grown more slowly at reduced temperatures ($\approx 900°C$). On the other hand, it is necessary to use elevated temperatures ($\geq 1060°C$) in order to obtain good passivation properties with chlorine-bearing species. As a result, a number of two-step schemes have been developed [27, 28] for oxides grown with HCl and with TCE. In these schemes, an initial oxide is grown (with HCl or TCE) at 900°C. This is followed by a high-temperature step at around 1100°C, and consists of an anneal in nitrogen gas and/or an oxidation in the chlorine species. Thin oxides with breakdown strengths in excess of 10 MV/cm have been obtained, using these approaches. Additionally, the fixed charge density in these oxides is lower than that obtained by growth involving a single chlorine-free low-temperature step, by a factor of two.

7.1.7 Mechanical Stress

The oxidation of silicon involves a large increase in its volume, much of which is taken up by expansion in the direction at normal to the surface. However, considerable biaxial compressive stress has been noted [29], when oxide growth is carried out at or below the temperature ($\approx 950°C$) at which viscous flow occurs. Cool-down to room temperature results in a further increase in this stress because of the differences in the TCEs of silicon and SiO_2 (2.6×10^{-6}/K and 5×10^{-7}/K, respectively). The resulting interfacial stress is compressive, and about $2-4 \times 10^9$ dyn cm^{-1}. This is sufficient to produce dislocations, as well as to cause enhanced impurity diffusion, in the underlying silicon. Moreover, this stress tends to concentrate at the edges of windows cut in the oxide, resulting in segregation of (a) fast-moving impurities such as gold (which is often used for lifetime control) or (b) transition elements which may be present as contaminants.

Compressive stress due to oxidation causes slices to bow, so that the oxidized surface becomes convex. Optical measurements of its curvature can be used [30] to quantify this stress (see Section 7.6).

7.1.8 Oxidation-Induced Stacking Faults

Stacking faults can be generated during the thermal oxidation of silicon by wet or dry processes. It is believed that they are caused by incomplete oxidation at the Si–SiO$_2$ interface, which results in the formation of excess interstitial silicon in this region. This interstitial silicon causes fault formation by nucleation at strain centers in the bulk or at the surface [31]. These centers are primarily associated with oxygen precipitates in the silicon. Moreover, mechanical damage resulting from this strain can also act as nucleation centers for these oxidation-induced stacking faults (OISFs).

Both wet and dry oxidation result in the formation of OISFs, which can be as much as $40-50$ μm long at their point of penetration to the silicon surface. Their growth is thermally activated [32], with an activation energy of 2.55 eV for dry oxidation and 2.37 eV for wet oxidation. These values are close to the energy of the Si–O bond (2.2 eV), after allowances are made for the interfacial stress.

OISFs can be shrunk, and even totally annihilated, by sufficiently long heat treatments at high temperatures in a nonoxidizing atmosphere such as argon. The activation energy of shrinkage is [33] about 5.2 eV. This is very nearly equal to the activation energy of self-diffusion in silicon (5.13 eV), so that shrinkage is probably caused by the extra plane of atoms diffusing out towards vacancies in the silicon.

Onset of the point in time at which shrinkage occurs is enhanced by the use of a chlorine-bearing species during oxidation [34, 35]. This is seen in Fig. 7.12 for dry oxidation at 1150°C in the presence of trichloroethane (TCA).

Support for the above arguments is provided in studies of impurity diffusion during oxidation. Thus, boron diffuses by an interstitialcy mechanism, so that its diffusion rate is enhanced in situations which encourage the formation of interstitials. In addition, both enhanced diffusion and stacking fault formation are more strongly affected by steam oxidation than by dry oxidation. Finally, both are dependent on crystal orientation.

Both arsenic and phosphorus diffusion occur partly by the interstitialcy mechanism; some enhanced diffusion is observed with these dopants, but to a smaller extent than for boron. Antimony, on the other hand, diffuses in sil-

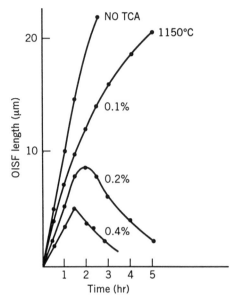

Fig. 7.12 OISF length as a function of time and TCA concentration. From Claeys et al [35]. Reprinted with permission of the publisher, The Electrochemical Society, Inc.

icon by a purely substitutional mechanism, so that enhanced diffusion behavior is not observed.

The shrinkage of OISFs can also be brought about by heavy impurity diffusion into the semiconductor. This is because the vacancy concentration in a semiconductor is increased by heavy doping, partly due to the increased stress in the material, and partly (in the case of donor impurities) due to the interaction of the extrinsic donor with the charged vacancy during diffusion (see Section 4.2.3.1).

The beneficial effect of a chlorine species during oxidation can also be explained in terms of the mechanisms outlined here. Thus, chlorine appears to promote vacancy formation at the silicon surface, and thus provides a sink by which interstitial silicon atoms can be removed. As a result, OISF formation is greatly reduced by oxidation in a chlorine species. Additional support for this argument comes from the observation that both boron and phosphorus diffusion in silicon are retarded if the surface oxide has chlorine incorporated in it.

7.1.9 Properties of Thermal Oxides of Silicon

The thermally grown native oxide of silicon has been shown to be suited for a large number of processing situations. Properties of this oxide which are relevant to these situations are now discussed, together with their effect on the underlying silicon.

7.1.9.1 *Diffusion and Masking Properties*

The diffusion of impurities in SiO_2 is concentration-dependent, since their presence leads to the creation of charge defects and alters the properties of the oxide layer. Consequently, their diffusivities are a function of their impurity concentration and also of the defect structure of the film. Some of these are listed [36] in Table 7.2 at the end of this chapter, and must be taken as very approximate. Network modifiers such as gallium and aluminum move by diffusion through the oxide layer. The diffusion constant of gallium is extremely large, so that it cannot be masked by the use of silica layers. Aluminum not only moves rapidly through the oxide, but vigorously attacks it, converting to Al_2O_3 in the process. Silicon nitride can be used as a mask for these dopants (see Section 8.3.3).

The masking properties of impurities such as B, P, and As are more correctly explained in terms of their ability to behave as network formers in the silica structure. Thus B_2O_3, P_2O_5, and As_2O_3 form mixed glasses with the silica layer with which they come into contact. The boundary between the mixed glass phase and the silica is quite sharp; the masking properties of the layer are excellent until this boundary extends down to the silicon–oxide interface. As a result, the phase diagrams of these binary systems are useful in evaluating the masking properties of SiO_2 against these impurities.

Masking data have been obtained empirically for the more commonly used diffusion systems by a number of workers. Representative results for boron and phosphorus are presented in curves shown in Figs. 7.13 and 7.14 and are typical of commercial practice in silicon device and microcircuit fabrication. An interesting point, shown by these data, is that the masking properties of

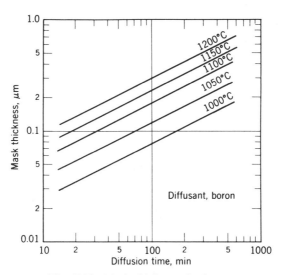

Fig. 7.13 Mask thickness for boron.

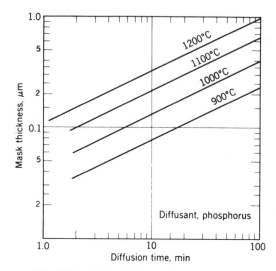

Fig. 7.14 Mask thickness for phosphorus.

silica films is about 10 times greater for boron diffusion than for phosphorus diffusion. This is primarily due to the lower temperatures associated with the respective glass compositions, as is seen by a comparison of Figs. 2.17 and 2.18. Typically, a 0.5- to 0.6-μm oxide thickness is adequate for most conventional diffusion steps, and is used as the starting point in the microcircuit fabrication process.

7.1.9.2 Charge States

The electronic properties of the oxide, and of the oxide–silicon interface, have a profound effect on the properties of devices in the underlying semiconductor. In some cases, these effects must be accounted for in device design; in others (such as MOS transistors) they form the very basis of device operation and long-term reliability [37].

Charge states in the silicon dioxide are intimately associated with the nature of the oxide growth process, and with interaction between the oxide and the silicon surface. This silicon surface presents a major discontinuity in an otherwise periodic crystal lattice; its electronic properties are largely determined by the defect nature of this discontinuity. Even if atomically clean surfaces were possible, their behavior would be governed by the large number of dangling bonds at the silicon–oxide interface. From purely quantum-mechanical considerations, it can be shown that the discontinuity in the periodic potential at a clean semiconductor surface gives rise to a number of allowed states within the forbidden gap. These states, the so-called *Tamm* or *Shockley states*, are associated with unsaturated covalent bonds at the surface discontinuity. In freshly cleaved silicon, their density is roughly equal to the density of dangling bonds on the silicon surface ($\approx 10^{15}$ cm^{-2}).

In microcircuits, in which an oxide layer is grown out of this silicon surface, a number of surface atoms are bound to the oxygen in the form of silica polyhedra, with considerable bond stretching due to the differences in lattice parameter. The net result is a partial coherence of the silicon lattice, accompanied by a reduction in the density of surface states to about 10^{11}–10^{12} cm^{-2}. The time constant associated with these *interface traps* is on the order of 1 μs or less. Consequently, they represent fast states. They reside within the first 25 Å of the silicon surface, and play an important role in altering the electronic properties of the semiconductor surface regions.

Interface traps have energy levels which can be represented by a broad, U-shaped spectrum of surface states, located throughout the energy gap, as shown in Fig. 7.15 for a thermally oxidized (111) silicon surface [37]. The density of these traps (N_{it}) is a function of both the process conditions and crystal orientation. Thus, N_{it} is larger for steam oxidation, it is reduced by oxide growth at elevated temperature, and it is lower for (100) silicon than for (111) silicon.

Being deep, these traps are responsible for both generation and recombination effects at the surface. Thus, their presence results in increased leakage currents in the region where the junction penetrates to the surface, shorter minority carrier lifetimes, and early falloff in transistor current gain at low levels.

The density of interface traps can be reduced by a wet hydrogen heat treatment after metallization of the surface. Typically, this process is carried out at 350–500°C for 15 min, and reduces the trap density by forming Si–H bonds at this interface. It is believed that atomic hydrogen, caused by the reaction of aluminum with moisture, plays an important role in this process. Support for

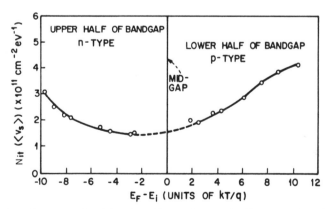

Fig. 7.15   Interface trap densities for silicon. From Nicollian and Goetzberger [37]. ©1967. AT&T. All rights reserved. Reprinted with permission.

this argument lies in the fact that wet oxidation with a nonreactive metallization, such as gold, is ineffective in reducing N_{it}. Values of N_{it} in the low 10^{10} cm^{-2} eV^{-1} region can be achieved in this manner.

Slow states are also present in the oxide. With chemically etched silicon, the surface of the adsorbed layer may have a positive or negative charge state. In thermally oxidized silicon, the first 100 Å of surface layer represents a silicon-rich transition between the silicon and the oxide. Consequently, there is a net fixed positive charge associated with this layer. The density of fixed *slow surface states*, N_f, is on the order of 10^{11} cm^{-2} in thermally grown oxides on (111) silicon, and about 10^{12} cm^{-2} for chemically treated silicon surfaces.* Values for (100) silicon are typically lower than those for (111) silicon by a factor of three to four.

Slow surface states primarily behave as traps with trapping times on the order of a few seconds to many months. Consequently, they do not enter directly into electronic processes with the semiconductor, but do so indirectly by establishing the surface potential, by pinning the Fermi level at the surface trap level. This results in an electric field normal to the semiconductor surface, a lowering of the surface mobility in this high field region, and a change in the surface conductance [38]. With thermally oxidized silicon, the positive charge in the oxide results in an *n*-shift in the semiconductor; that is, an *n*-type semiconductor behaves more *n*-type whereas a *p*-type material behaves less *p*-type, and may even invert to *n*-type. Thus, the role of slow surface states is relatively slight on heavily doped silicon, but becomes increasing significant with lighter doped material. It is especially important in high-voltage transistors and MOS-based microcircuits.

The density of slow surface states can be reduced by processing the samples in such a manner as to reduce the excess silicon in this interface layer. High-temperature oxidation, in dry O_2, is effective if carried out at 1100–1200°C. Alternatively, oxidation at relatively low temperatures (around 700°C or higher), followed by a post-anneal treatment in nitrogen gas, can be used for this purpose [39]. Annealing temperatures from 950°C to 1200°C are effective, with corresponding anneal times from 120 min to under 5 min, respectively, and can reduce N_f by a factor of 5 in both (100) and (111) silicon. The optimum time for this process is quite critical at 1200°C, but not at 1100°C and lower. Heat treatment at 1100°C, for about 20 min, is generally used for this reason, and results in values of N_f which would be obtained by direct oxidation at 1200°C.

Figure 7.16 shows the density of both slow surface states and fast interface traps, as a function of crystal orientation [40]. The close correspondence of these data indicates their common origin—that is, the formation of Si–O bonds during the oxidation process. Variations in the bond density and bond angles for the different orientations have been proposed to explain the relatively low

---

*All values are referenced to the silicon surface.

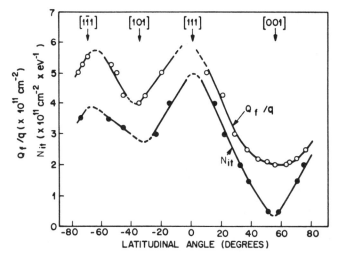

Fig. 7.16 Fixed charge and interface trap density as a function of orientation. From Arnold et al. [40]. With permission from *Applied Physics Letters*.

values for (100) Si, and high values for (111) Si. However, this explanation is, at best, a tentative one.

Slow surface states are also caused by alkali ions throughout the bulk of the oxide layer. Their presence in significant amounts affects the reliability of device operation. They enter the silica in ionized form (Na^+, K^+, Li^+) and are highly mobile at temperatures as low as 125°C. Consider what happens if these ions are present in the gate oxide of an MOS transistor. The net result of applying a field across the oxide directed towards the silicon surface is an accumulation of a positive space-charge layer close to the silicon–oxide interface and a corresponding *n*-type shift in the surface potential. This will make the threshold voltage less positive in an *n*-channel device; that is, the magnitude of this voltage decreases. On the other hand, a reverse field moves this charge away from the silicon surface, and makes the magnitude of the threshold voltage larger in this device. Similar arguments can be applied to explain the threshold instability in *p*-channel devices. In both cases, this effect impairs the long-term stability of the MOS device.

There are a number of correctives for this problem, which are often used singly or together in microcircuit fabrication. First, gate oxide diffusions must be carried out with dry oxygen, and in ultraclean, dedicated systems having ceramic liners to prevent in-diffusion of sodium from the furnace walls. The use of silicon diffusion tubes has been found to be excellent for this situation; however, their excessive cost has restricted their use to only a few commercial installations at the present time. An alternative approach consists of using a double-walled quartz oxidation tube, with a chlorine-bearing species flowing between the walls for scavanging purposes.

Yet another approach is to use a P_2O_5-doped glass for the outer surface of this oxide. This can be readily accomplished by introducing P_2O_5 during the oxidation process by procedures commonly used for making phosphorus diffusions. The effectiveness of this approach depends on the gettering action of sodium by the phosphosilicate glass [41]. In addition, the introduction of a pentavalent network former such as P_2O_5 provides an excess of oxygen to the silica structure and makes the glass more dense, and thus more resistant to the transport of ionic impurities through it. In line with the above arguments, it is reasonable to expect that a surface layer of B_2O_3 is not useful in this application because it creates an oxygen deficiency in the oxide. This has indeed been found to be the case in practice.

Excessive amounts of P_2O_5 cannot be used for this purpose since they cause large, fixed threshold shifts due to polarization effects. In addition, they tend to make the gate oxide hygroscopic and hence even more unstable. One approach, often used in the formation of MOS oxides, is to grow this oxide thermally, incorporate a heavy surface layer of P_2O_5, remove this layer in a selective oxide etching solution, and replace it with a lightly doped layer of $P_2O_5 \cdot SiO_2$ glass.

Finally, the use of a chlorine-bearing species during gate oxidation greatly reduces the mobile alkali ion concentration in the oxide. As pointed out in Section 7.1.6, this results in a pile-up of chlorine at the $Si–SiO_2$ interface, which captures these ions to form neutral species.

7.1.9.3 Recombination Velocity and Lifetime

The presence of deep-lying interface traps leads to the enhancement of recombination processes at the surface. To characterize this effect, it is useful to think of the surface as a sink for minority carriers that approach it at a velocity S, the *surface recombination velocity*. Then S is defined as the ratio of the number of minority carriers recombining per second per unit surface area to the excess carrier concentration in the bulk just beneath the surface.

The surface recombination velocity is related to the density of interface traps, and also to their charge state. Thus, the effectiveness of these traps in reducing lifetime is a function of the density of slow surface states, which establishes the surface potential.

The interaction of both N_{it} and N_f in determining S can be obtained by assuming a simple model with interface traps of equal capture cross section to electrons and holes, located at a discrete energy level which is ΔE_t eV from the intrinsic level. It is also assumed that the effect of the slow surface states is to create a surface band bending, given by ΔE_s from the intrinsic level. For this model [42] it can be shown that

$$\frac{S}{S_{max}} \simeq \frac{1 + \cosh \Delta E_t}{\cosh \Delta E_s + \cosh \Delta E_t} \tag{7.23}$$

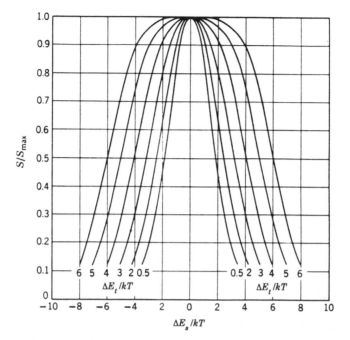

Fig. 7.17 Surface recombination velocity versus surface band bending. From Many et al. [42].

From this equation, which is illustrated in Fig. 7.17, it is seen that S is maximized when there is no surface band bending. Thus, the interface traps are most effective in reducing the minority carrier lifetime when there are no slow surface states.

This somewhat surprising result can be explained by noting that surface band bending is accompanied by an electric field which is normal to the semiconductor. This limits the access of one type of carrier to the surface, where it can recombine with the other. In many practical situations this band bending can be established by grading the doping concentration via diffusion or ion implantation.

Values of S range from as high as 10^5 cm/s on sandblasted surface to 1–100 cm/s on properly etched or otherwise-treated silicon surfaces. The value of S for thermally oxidized surfaces is typically below 10 cm/s. Note that since S is the velocity with which minority carriers (of effective mass m^*) approach the surface (and are recombined), its upper value is given [43] by the unilateral mean velocity, $\sqrt{kT/2\pi m^*}$.

The relationship between S and the lifetime is relatively straightforward. Thus, consider the motion of carriers in an infinitely wide channel of depth $2d$. For low values of S, the movement of carriers towards the surface by drift will be the limiting factor. Thus, the surface lifetime is given by

$$\tau_s = \frac{d}{S} \tag{7.24}$$

since the average position of these carriers is half the channel depth.

For high values of S, however, diffusion will limit the movement of carriers to the surface. From transistor theory, the transit time for these carriers is given by

$$\tau_s = \frac{d^2}{2D} \tag{7.25}$$

where D is their diffusivity. Finally, the effective lifetime is given by

$$\frac{1}{\tau_{eff}} = \frac{1}{\tau_s} + \frac{1}{\tau_b} \tag{7.26}$$

where τ_b is the bulk lifetime.

7.1.9.4 Hot Electron Effects

In thermally oxidized silicon, the barrier height between the silicon and the silicon dioxide at both the valence and the conduction band edges is about 3.2 eV. This is sufficiently low so that it is possible for electrons (or holes), generated in the silicon by device action, to occasionally have enough energy to surmount it, and get trapped into the silica layer. This phenomenon of "hot carrier" trapping is encountered in any high-field semiconductor region which is in contact with the silicon dioxide. One situation where it presents a serious problem is during avalanche breakdown of p–n junctions, since the energy of avalanching carriers has a mean value of about 3 eV.

Junctions made by microcircuit technology have curved sides. During avalanche breakdown, the region of high field is very near the surface. In addition, there is a strong fringing field which extends out into the oxide and aids in the trapping of electrons in it. Once trapped, they create a space charge which affects the behavior of the junction. Typically, the value of the breakdown voltage changes by an amount which depends on the extent of avalanching, and consequent carrier trapping. Furthermore, the device recovers over a period of days, at a rate which depends critically on the quality of the surface oxide, and especially on its water content. This problem, known as *junction walkout*, cannot be eliminated, but it can be reduced by careful processing [44]. In particular, a long-low temperature bake-out (48–72 hr at 250°C), which reduces the water content in the oxide, has been found beneficial.

Hot electron effects also result in breaking some of the satisfied bonds (Si–O, Si–H, or Si–OH) which are present on the silicon surface, so that the interface trap density increases. This gives rise to the well-known phenomenon of deterioration in the low-level current gain of bipolar transistors whose emitter–base junctions have been subjected to avalanche breakdown [45].

Hot electron effects can also occur, but to a lesser extent, in high-field regions which are not avalanching. These include the channel region of MOS transistors, as well as the depletion layer at the drain regions. Both of these effects are of increasing importance in VLSI devices, which require the use of ultrathin gate oxides and short channel lengths [46]. They result in a slow, long-term change in the threshold voltage, as well as a fall in the transconductance. This long-term stability problem is as yet unsolved, although it can be minimized by careful attention to gate oxide formation. The use of halogenic oxidation has been shown to reduce these effects. More recently, orders of magnitude improvements have been achieved by the use of ultrathin gate films of silicon oxynitride. This topic will be covered in the following section.

Hot electron effects can also be used to advantage in the design of programmable read-only memory (PROM) elements. One such structure consists of an MOS transistor with a floating gate which is buried in the oxide to prevent leakage [47]. Information is entered by causing a momentary avalanche breakdown between the source (or drain), and the substrate. This injects hot electrons into the oxide which causes the floating gate to become charged, and thus triggers device action. With proper processing, the leakage rate of such a gate can be very long, typically many years, so that data are essentially stored permanently in this manner [48].

7.2 THERMAL NITRIDATION OF SILICON

Native films of silicon nitride, grown by the direct nitridation of silicon, are candidates for meeting the need for thin gate insulators in VLSI applications. These films are somewhat more dense than SiO_2; moreover, they are excellent barriers to the diffusion of light alkali ions such as Na^+ and K^+ [49].

The direct nitridation of silicon in nitrogen gas is difficult because of the high strength of the triple N–N bond (9. 8 eV/molecule), and requires extended growth temperatures in the 1200–1300°C range. However, ammonia gas can be used [50] for growth at temperatures as low as 950°C. The activation energy of this process ranges from 0.23 to 0.62 eV/molecule, depending on the degree of stoichiometry of the resulting film, and is associated with the energy for breaking of the N–H bond ($\approx$3.5 eV/molecule). Amorphous as well as polycrystalline films are obtained by this process, with films grown at lower temperatures being amorphous.

Figure 7.18 shows growth data for silicon nitride films grown in a cold wall reactor by ammonia nitridation [51]. Here, the strong self-limiting character is an advantage, since it allows control of the film thickness in a manufacturing environment.

Thermal nitridation of silicon usually results in films whose nitrogen content varies (falls) with depth; that is, they are nonuniform as well as nonstoichio-

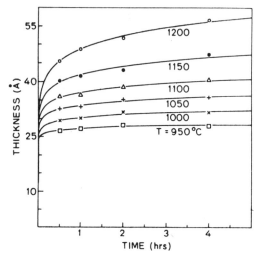

Fig. 7.18 Nitride growth rate in NH_3. From Moslehi and Saraswat [51]. ©1985. The Institute of Electrical and Electronic Engineers.

metric. In addition, growth must be carried out in systems which are extremely leak-tight, to prevent the incorporation of oxygen.

7.2.1 Oxynitride Films

Films of silicon oxynitride (SiO_xN_y) have the potential of combining the advantages of both oxides and nitrides, and many different processes have been used for their growth. All result in films with a variable N/O ratio, with a pile-up of nitrogen ($\approx 3\%$) at the silicon interface. This nitrogen reduces the concentration of strained Si–O bonds, and suppresses the generation of interface states during electrical stressing of the insulator. Hot electron effects are greatly reduced as a result, often by as much as three orders of magnitude. For this reason, oxynitrides have great potential in MOS devices for ultra large-scale integration.

Techniques for fabricating native oxynitride films include the oxidation of silicon nitride, the nitridation of SiO_2 with NH_3, and the direct growth of SiO_xN_y using nitrous oxide as the oxidant species. Conventional thermal, as well as rapid thermal, techniques have been used with all approaches, in addition to atmospheric- and low-pressure processes. It serves no useful purpose to catalog all of these variations; however, they share many common features.

The reaction of NH_3 with thermal SiO_2 results in the formation of an oxynitride with an N/O ratio of about 0.3 after a 2-hr nitridation time [52]. Moreover, this ratio varies with layer thickness. Studies of the kinetics of this process have shown a fast surface reaction of SiO_2 with NH_3 as well as rapid diffusion of NH_3 to the interface, so that nearly stoichiometric silicon nitride is formed in these regions. The bulk of these layers, however, is only partially converted to nitride during this process, presumably because the surface nitrogen saturates after the rapid initial exchange reaction phase.

The nitridation of 60-Å-thick thermal oxides has been carried out by heat treatments in NH_3. Here, a significant reduction was observed in the density of pinhole defects [52], which are normally a very serious problem with ultrathin gate oxides. Rapid thermal processing has also been used for the nitridation of thermally grown SiO_2 at temperatures of 950–1150°C, using N_2O/O_2 mixtures [53]. No movement of the Si–SiO_2 interface was detected, as a result of the short nitridation times.

The most direct approach for the formation of an oxynitride involves [54, 55] the oxidation of silicon with N_2O. This method has the advantage of being a single-step process; in addition, no hydrogen is involved in the reaction, so that the resulting film has no OH groups. Films grown in this manner have a variable N/O ratio, with a lower N_2 content at the surface, and a peak in N_2 concentration ($\approx$3%) at the interface. Oxidation rates are lower than for dry oxygen [56], probably because of the reduced diffusivity in the interfacial nitrogen-rich layer. At a growth temperature of 1150°C, film thicknesses ranged from 63 Å to 243 Å, for growth periods of 30–300 s.

7.3 THERMAL OXIDATION OF GALLIUM ARSENIDE

The thermal oxidation of GaAs can be carried out by heat treatment in oxygen; however, the oxidation process is very slow at temperatures below 450°C. Between this temperature and 530°C, it is possible to grow amorphous oxide films with good uniformity [57]. A linear growth rate characteristic was observed [58] for thermally oxidized films over this temperature range, with thicknesses up to 3.4 μm. At higher oxidation temperatures, the films tend to crystallize and become porous.

Oxidation proceeds by the breaking of Ga–As bonds, reaction with chemisorbed oxygen at the surface, and the subsequent nucleation of gallium and arsenic oxides. These nuclei form oxide islands which grow outwards until the surface is covered. Further growth of the oxide proceeds in all probability by (a) the in-diffusion of oxygen and (b) the diffusion of both gallium and arsenic *outwards* through the oxide layer to the surface, where they are converted to the appropriate oxides [59]. This is quite different from what occurs during the oxidation of silicon, where oxygen diffuses *inwards* through the growing oxide layer. One important consequence of this difference is that thermal oxidation of silicon results in a fresh interface region, with surface contaminants on the original silicon ending up on the oxide surface. With GaAs, however, it is highly probable that contaminants on the original semiconductor surface will remain at the interface, even after the oxidation process.

The quality of the thermal oxide of GaAs is dominated by the widely different properties of the elements comprising this semiconductor, and of their oxides. Thus, the oxidation potential of gallium is +0.65 V, whereas that of arsenic is −0.25 V. As a result, gallium will oxidize more readily than arsenic. Next, the free energy of formation of Ga_2O_3 is −214 kcal/mole, whereas that of As_2O_3 is −151 kcal/mole, so that the oxide of gallium is thermodynamically more stable than that of arsenic. For a typical oxidation temperature of 450°C, the vapor pressure of gallium is 10^{-11} torr, whereas that of arsenic is 10 torr; thus, evaporation of arsenic from the surface of the oxide is rapid, whereas that of gallium is negligible. Finally, the vapor pressure of As_2O_3 at 450°C is almost 760 torr, whereas that of Ga_2O_3 is negligible at this temperature.

As a consequence of the above factors, the arsenic content of a thermally grown oxide of GaAs rapidly falls with distance from the oxide–semiconductor interface, and most of the oxide layer consists of Ga_2O_3, with as little as 5–10% of As_2O_3, in addition to elemental arsenic [60]. Moreover, any subsequent heat treatment results in further loss of As_2O_3 from the oxide, and in conversion of the surface oxide to β-Ga_2O_3 or to $GaAsO_4$. These films are brittle and crystalline, and extremely porous to the diffusion of impurities such as zinc. They are unsatisfactory as cover layers for GaAs, and cannot be used for masking purposes.

Superior-quality thermal oxides can be grown under an overpressure of As_2O_3, to prevent loss of this component [61]. This is done by using a sealed tube with two zones. The GaAs to be oxidized is placed at one end, at 500°C, whereas an excess of As_2O_3 is placed at the other, at 450°C. This arrangement establishes a vapor pressure of As_2O_3 over the GaAs, which is oxidized at a rate of about 0.1 μm/hr. Films grown by this technique have high breakdown field strengths (5–7 MV/cm) and consist of both As_2O_3 and Ga_2O_3. Their composition tends to be graded, however, becoming slightly depleted of As_2O_3 as the surface is approached. Both crystalline and amorphous arsenic have been observed at the oxide–GaAs interface in these films [62].

7.3.1 Interface Trap Density

Compared to silicon, the density of interface traps in GaAs is extremely high, and results in a surface recombination velocity of about 10^6 cm/s. The origin of these traps is not clear at the present time, but it is possible that they are associated with a loss of stoichiometry at the interface [63]. Some workers have proposed that vacancy defects (both site and antisite) are the origin of these traps [64].

A high density of interface traps near the midgap has been observed for both p- and n-type GaAs. This results in pinning the fermi level near the center of the gap and has prevented the development of an MIS technology for this semiconductor. Various chemical treatments have been explored in an attempt to remove this pinning effect [65, 66]. Alternative techniques include the use of cladding layers of AlGaAs [67] and of n^+-GaAs [68], both of which lower the value of S to about 10^3 cm/s by serving as minority carrier mirrors. Yet another approach consists of the growth [69] of a pseudomorphic layer of ZnSe as an extension of the GaAs lattice, in order to reduce the value of S to around 10^3 cm/s. None of these approaches, however, are sufficient to allow the use of GaAs in enhancement mode MIS devices at the present time.

Midgap pinning of the fermi level is an advantage in the fabrication of GaAs Schottky diodes, since it reduces the curvature of the depletion layer at the edge of the metal. As a result, premature breakdown is avoided, so that their breakdown voltage (BV) is comparable to the value for bulk material. In contrast, the BV of silicon-based Schottky diodes is about 33% of the value for bulk material.

7.4 ANODIC OXIDATION

A unique characteristic of anodic oxidation is that it is carried out at or around room temperature. Thus, impurity profiles in the semiconductor are not altered during this process. As a result, this technique is a useful means for the controlled removal of layers of silicon and GaAs at room temperature, and is often used as a diagnostic tool.

Anodic oxidation can also be used to grow reasonably high-quality oxides on GaAs, since problems associated with the widely different characteristics of Ga_2O_3 and As_2O_3 are minimized at low temperature. Both the dielectric breakdown strength and the electrical resistivity of these oxides is superior to that obtained by thermal oxidation. As a result, they are useful in technological processes for GaAs, such as masking and coating. In addition, they have been used as gate oxides in MOS transistors, and in GaAs integrated circuits based on these devices.

Anodic oxidation, or *anodization*, is carried out by placing the semiconductor in an electrolytic cell as shown in Fig. 7.19, where it is connected to the positive terminal of a power supply so that it serves as the anode. A noble metal such

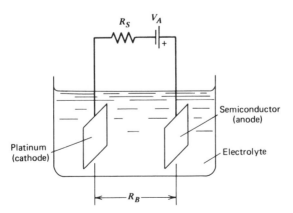

Fig. 7.19 Circuit diagram of an anodization cell.

as platinum is connected to the negative terminal of the supply, and serves as a cathode. The source impedance of the supply can be kept low, so that anodization occurs at constant voltage; alternatively, a high source impedance allows anodization to proceed under constant current conditions.

A large variety of electrolyte formulations can be used; the primary oxidizing component of all of these is water, which dissociates into H^+ and $(OH)^-$. The dissociation re   )n can be written as

$$H_2O \rightleftharpoons H^+ + (OH)^- \tag{7.27}$$

so that

$$k = [H^+][OH^-] \tag{7.28}$$

The equilibrium constant for deionized water at 24°C is 10^{-14}. For this case, since $[H^+] = [OH^-]$, the concentration of $[H^+]$ is 10^{-7}, resulting in a pH of 7.* Often, a conductivity/pH modifier is added to this system to vary the resistance of the electrolyte, as well as the dissolution rate of the oxide in it. Typical modifiers are H_3PO_4 (acidic), NH_4OH (basic), and $(NH_4)_2HPO_4$ (almost neutral). Usually, very small amounts of these are required for pH control.

It is possible to carry out an anodization process in a nonaqueous solution. Here, the pH is set before addition of the nonaqueous component. Again, the requirement for this electrolyte is that one of its dissociation products be the $(OH)^-$ ion. Both ethylene and propylene glycol are often used as the nonaqueous medium. Additional components of the electrolyte serve to control its viscosity and dielectric constant and improve the reproducibility of the process, as well as its sensitivity to ambient conditions and to contaminants.

Oxides formed by anodization are generally porous in character, and have water incorporated in them. However, they can be used to advantage in a number of situations. By way of example, anodic oxidation, followed by chemical dissolution, provides an excellent method for removing a controlled amount of semiconductor by a low-temperature, damage-free process. The technique is reproducible and nearly independent of doping concentration or crystallographic orientation. In addition, it can be carried out many times without significant loss of surface planarity, so that it is an ideal means for profiling the impurity concentration in semiconductor layers [70, 71].

Anodic oxides of silicon are generally poorer than their thermally grown counterparts, so that their use is restricted to singular applications of the type just described. A notable exception is the anodization of polycrystalline silicon [72], which results in oxides with higher breakdown electric field strengths than are obtained by thermal processes. With GaAs, the thermal oxide is extremely

---

*By definition, pH $= -\log_{10} [H^+]$.

poor and anodic oxides are the only native-grown choice for device and processing applications.

Reactions leading to the anodic oxidation of silicon are as follows:

1. Water in the electrolyte medium dissociates into H^+ and $(OH)^-$:

$$2H_2O \rightleftharpoons 2H^+ + 2(OH)^- \qquad (7.29)$$

2. The difference in electrochemical potentials between the silicon and the electrolyte results in charge transfer from the silicon until equilibrium is established. This leaves the surface layer partially depleted of electrons. During anodization, holes are supplied from the bulk of the semiconductor to the semiconductor–electrolyte interface, thus promoting the silicon surface atoms to a higher oxidation state:

$$Si + 2h^+ \rightarrow Si^{2+} \qquad (7.30)$$

3. The Si^{2+} combines with $(OH)^-$ to form the hydroxide.

$$Si^{2+} + 2(OH)^- \rightarrow Si(OH)_2 \qquad (7.31)$$

4. $Si(OH)_2$ subsequently forms SiO_2, thereby liberating hydrogen in the process.

$$Si(OH)_2 \rightarrow SiO_2 + H_2 \qquad (7.32)$$

The overall reaction is given by

$$Si + 2h^+ + 2H_2O \rightarrow SiO_2 + 2H^+ + H_2 \qquad (7.33)$$

The anodization of p-type material is relatively straightforward since there is no problem with delivering holes to the semiconductor surface by means of a battery. The situation is somewhat more complex for an n-type semiconductor, however. Here, the initial charge transfer from the semiconductor into the electrolyte (in order to establish thermal equilibrium) creates (a) a depletion layer in the semiconductor and (b) a barrier to the flow of holes. In effect, then, the electrolyte–semiconductor system behaves much like a Schottky diode with the electrolyte serving the role of the "metal." Here, oxidation will not proceed beyond a few monolayers, unless provision is made to supply holes to the semiconductor surface. One approach is to illuminate the sample to provide these holes by photogeneration. Alternatively, the anodization cell can be operated at a voltage which exceeds the junction breakdown voltage, so that avalanche-generated holes can allow the oxidation to proceed. The build-up of this growing oxide reduces the available voltage across the depletion layer; anodization eventually ceases when this voltage falls below the breakdown voltage of the electrolyte–semiconductor system.

The H$^+$ drifts to the cathode where it is evolved as molecular hydrogen by the addition of electrons.

$$2H^+ + 2e^- \rightarrow H_2 \tag{7.34}$$

The growth of anodic oxides of GaAs proceeds along somewhat similar lines. Here, the ionic states for gallium and arsenic are Ga^{3+} and As^{3+} and are formed by the anodic reaction

$$GaAs + 6\,h^+ \rightarrow Ga^{3+} + As^{3+} \tag{7.35}$$

The separate reactions for the gallium components are

$$3H_2O \rightleftharpoons 3H^+ + 3(OH)^- \tag{7.36a}$$
$$Ga^{3+} + 3(OH)^- \rightarrow Ga(OH)_3 \tag{7.36b}$$
$$2Ga(OH)_3 \rightarrow Ga_2O_3 + 3H_2O \tag{7.36c}$$

The reactions for arsenic are

$$2H_2O \rightleftharpoons 2H^+ + 2(OH)^- \tag{7.37a}$$
$$As^{3+} + 2(OH)^- \rightarrow AsO_2^- + 2H^+ \tag{7.37b}$$
$$2AsO_2^- + 2H^+ \rightarrow As_2O_3 + H_2O \tag{7.37c}$$

so that the overall reaction is

$$2GaAs + 12\,h^+ + 10H_2O \rightarrow Ga_2O_3 + As_2O_3 + 4H_2O + 12H^+ \tag{7.38}$$

Once again, water plays the primary role in forming the oxidation products, and continuous oxidation is sustained by the supply of holes to the anode surface, via the external battery. Finally the delivery of electrons to the cathode (from the battery) is accompanied by the evolution of molecular hydrogen at this electrode.

The description of the anodization process provided here is at best elementary. The effect of early stages of anodic growth, as well as the role of double layers at the electrolyte–oxide interface, is considered elsewhere in the literature [73] and will not be treated here.

7.4.1 Oxide Growth

Consider the anodization cell shown in Fig. 7.19, under conditions of no illumination. In this system, initial oxide growth is by the formation of islands on the semiconductor surface. Once these nucleate to form a continuous film, further oxidation proceeds by the transport of the $(OH)^-$ ion through the oxide to the semiconductor, or by the transport of semiconductor material through the oxide to the electrolyte–oxide interface. The anodization of GaAs most probably proceeds by the transport of Ga^{3+} and As^{3+} through the oxide, rather than by the motion of $(OH)^-$ through the oxide.

The consumption of semiconductor material (with its conversion to the oxide) is governed by Faraday's law of electrolysis. This law states that the number of grams (W) of material consumed per coulomb of charge is given by

$$W = \frac{W_e}{qN} \tag{7.39}$$

where W_e is the electrochemical equivalent weight (in grams) of the material involved, q is the electronic charge, and N is Avogadro's number ($qN = 96,483$). The electrochemical weight of an element or compound is its atomic or molecular weight, respectively, divided by the change in valence involved in the reaction. GaAs undergoes a valence change of $+6$ during anodization. From Faraday's law, the amount of GaAs consumed per coulomb of charge is thus 2.5×10^{-4} g. A typical experimental value for this quantity is about 2.4×10^{-4} g/coulomb. The ratio of experimental to theoretical values, quoted as a percentage, is referred to as the *current efficiency*. For GaAs, this is [74] about 95%, whereas the current efficiency for silicon is only 1–3%.

The formation rate of an oxide, α, is defined as the increase in film thickness per unit charge, passing through unit area. This term is closely related to the current efficiency defined above. A second term, of importance during anodization, is the oxide dissolution rate. This is the change in film thickness as a function of time, if left in the electrolyte *without* any applied voltage. The dissolution rate for oxides of GaAs depends on the electrolyte, and ranges from 2 Å/s to as low as 4.2×10^{-2} Å/s at room temperature.

Assume unit area for the electrodes, and that R_S and R_B are the resistance of the circuit and the bath, respectively. The initial current flow is given by

$$I_0 = \frac{V_A - V_r}{R_S + R_B} \tag{7.40}$$

where V_A is the applied voltage, and V_r is the rest potential of the anodization cell (i.e. the reverse electromotive force generated when the cell is used as a battery), for the case of a *p*-type semiconductor. With an *n*-type semiconductor, however, V_r is the sum of the rest potential and the breakdown voltage associated with its depletion layer. If f_d is the dissolution rate of the film, the equivalent dissolution current is given by

$$I_d = \frac{f_d}{\alpha}$$

(7.41)

During anodization, current flow through the oxide is primarily ionic in nature, and can be described by an equation of the form $J \alpha\, e^{KE_{ox}}$. For a high-field situation, as in the case for the anodization of GaAs (where $E_{ox} \simeq 5 \times 10^6$ V/cm), we can assume that there is a constant electric field across the oxide during film growth, as an approximation. For this case, the final thickness of oxide which is formed is given by

$$x_\infty = \frac{V_A - V_r}{E_{ox}} \left(1 - \frac{I_d}{I_0} \right)$$

(7.42)

The growth rate of the film can be determined by writing Kirchhoff's law for the anodization circuit of Fig. 7.19. Thus,

$$V_A - V_r = i(R_S + R_B) + \alpha E_{ox} \int_0^t i\, dt$$

(7.43)

Solution of this equation, along with insertion of the final value of oxide thickness given in Eq. (7.42), gives

$$x = x_\infty \left[1 - \exp\left(\frac{-t}{\tau_f} \right) \right]$$

(7.44)

where the time constant for growth, τ_f, is given by

$$\tau_f = \frac{R_S + R_B}{\alpha E_{ox}}$$

(7.45)

Thus, the oxide thickness builds up exponentially until a final value, dictated by the applied voltage, is reached. Typically, the final oxide thickness is 20 Å/V for gallium arsenide and about 3 Å/V for silicon.

Anodization can also be carried out from a constant current source. Here, Faraday's law requires that the oxide thickness vary linearly with time. In addition, since the field across the oxide is constant, the voltage drop across it will also increase linearly with time. In many practical situations, it is common practice to exploit this behavior to monitor the film thickness. This is done by using a constant current source, monitoring the linear increase of voltage with time, and then stopping the process at some preset voltage (i.e., at some preset oxide thickness).

One application of anodization is in control of the pinch-off voltage of GaAs junction gate field effect transistors. This voltage is determined by the product of the thickness and the doping concentration of the epitaxial layer, so that this parameter must be controlled across the wafer, for a practical device fabrication process. This can be done by anodization of the slice in the dark [75], or under uniform controlled lighting [76] if a lower value of $(t_{epi} N_d)$ is required. In the latter situation, this product is set by this illumination.

In either case, anodization of the n-GaAs film [on the semi-insulating (SI) substrate] proceeds by avalanche breakdown of the substrate. As the layer is consumed, the depletion layer moves closer to the substrate–epitaxial layer interface, and eventually sweeps rapidly into the SI substrate. As a result, insufficient voltage is left to sustain breakdown so that anodization ceases. Thus, semiconductor removal is locally achieved until the pinch-off voltage is constant across the slice.

Quantities of interest to the anodization of silicon and GaAs are listed in Table 7.3 at the end of this chapter.

7.4.2 Anodic Oxidation Systems

Anodization systems for GaAs are of three forms: aqueous, nonaqueous, and mixed. The aqueous systems are primarily based on water, or on water/H_2O_2 mixtures [77]. Water has an extremely high resistivity ($\approx 16 \times 10^6 \, \Omega$ cm) in pure form, so it is necessary to use small amounts of additives for the purpose of pH/conductivity control [78]. These additives may be acidic, such as H_3PO_4, which gives good oxides in the 2.5–3.5 pH range. They may also be basic such as NH_4OH; here, excessive etching is usually observed, and it is difficult (if not impossible) to obtain reproducible results. Neutral additives can also be used; thus, the use of $(NH_4)_2HPO_4$ shifts the pH very slightly. For example, a 0.1 N solution of this reagent will have a pH of 7.8 as compared to a pH of 7 for pure water.

All of the above electrolytes operate successfully over a very narrow pH range, which must be tightly controlled to avoid excessive dissolution of the oxide while it is being formed. Thus, the water/$(NH_4)_2HPO_4$ solution produces better-quality oxides, with less dissolution in the electrolyte, if a small amount of H_3PO_4 is added to lower its pH from 7.8 to 7.15. Aqueous electrolytes have also been found to be very sensitive to reagent contamination, with consequent nonreproducibility of oxide formation.

Nonaqueous solutions have also been used for anodization of GaAs. Here, the pH value loses its meaning in the classical sense; however, it is often used to set up the process.

Considerable work has been done on one such system using a mixed solution of water and propylene glycol containing tartaric acid [74]. This system has been found to be relatively insensitive to the presence of contaminants in it, and to consistently give oxides with a resistivity that is 10^3–10^5 higher than that obtained with aqueous solutions. In this system, water is the oxidant and tartaric acid is used for pH/conductivity control. Typically, a 3% aqueous solution of tartaric acid is used, with ammonia to adjust its pH to a suitable value. This solution, which is aqueous, can be used by itself, if its pH is in the 5–7 range. However, it is very sensitive to contamination and gives results which are difficult to reproduce. The same solution, however, if added to glycol in a 1 : 2 parts by volume ratio (aqueous/glycol), is useful for starting pH values from 2 to 9, and results in highly reproducible anodizations. This mixture, known as AGW, has been used over an aqueous/glycol range from 1:1 to 1:4 with little variation in anodization quality.

The oxide dissolution rate for AGW solutions is very low, typically 4–6 $\times 10^{-2}$ Å/s, as compared to dissolution rates of 1–3 Å/s for aqueous oxides. Furthermore, AGW oxides have been found to have a higher As_2O_3 content than those made with aqueous electrolytes. Electron microprobe studies of AGW oxides have shown that it approaches stoichiometry more closely than those formed by most of the other methods.

Systems for the anodization of silicon have been primarily developed for use in the controlled removal of layers for diagnostic purposes. A nonaqueous solution of *N*-methylacetamide with 0.4 N KNO_3 has been found useful for this application [79]. With this system the oxide thickness is in the 3- to 5-Å/V range, with measured values of ionic current efficiency from 1.6% to 3%. It should be noted that this electrolyte does have a small amount of water incorporated in it because of the hygroscopic nature of NMA; sometimes, additional water (up to 9%) is added to modify its anodization characteristics.

A solution of ethylene glycol in 0.04 N KNO_3, with 0.5% water, has also been used [70]. The addition of about 1–2 g of $Al(NO_3)_3 \cdot 9H_2O$ per liter of this solution has been found to greatly improve the uniformity of this electrolyte with successive anodizations. Oxide growth rates for this system are typically 2.2 Å/V.

7.4.3 Properties of Anodic Oxides

Anodic oxidation of silicon results in a relative porous oxide, which is not suited for use as a passivation coating. Its porosity also makes it more readily etched than thermal oxides, typically as much as five times faster. Densification of the oxide, by holding it at 450–500°C for a period of time, results in some improvement in these properties, and also serves to remove trapped water from the oxide.

Anodic silicon dioxide has an interface state charge density of about a decade larger than that of conventional thermal oxide. As a result, it cannot be used as a gate oxide in MOS applications. Because of all these disadvantages, the anodization of silicon is generally restricted to its use in diagnostic situations, requiring the repeated removal of thin layers.

Anodic oxides of GaAs have considerably better electrical properties than the thermal oxides. Here, anodization has the advantage over thermal oxidation, since it is conducted at room temperature. Even so, some departure for stoichiometry is present, because of the significant vapor pressure difference between As_2O_3 and Ga_2O_3.

To date, the closest approach to stoichiometry has been achieved by the AGW formulation, which results in an oxide having a high specific resistivity (10^{14} Ω cm) and also a high breakdown field strength ($\approx$5–6 MV/cm). The trap density at the oxide–GaAs interface is so large that this oxide cannot be used as a gate in an MOS device. In addition, this density is critically dependent on the preanodization treatment given to the GaAs [80]. However, a thermal anneal at 300°C in hydrogen for 30–60 min has been found to greatly reduce this trap density.

Anodic oxides of GaAs have also been used as anti-reflection coatings in solar cells, since their refractive index provides a reasonable match between air and GaAs. In addition, they have been used as masks for low-temperature diffusions in GaAs.

7.5 PLASMA PROCESSES

Plasma oxidation and anodization of both silicon and GaAs can be accomplished by using activated oxygen, created by an electrical discharge, as the oxidizing species. This is usually carried out in a low-pressure (0.05 – 0.5 torr) system, where a plasma can be sustained by d.c. or r.f. excitation [81]. Both inductive and capacitive coupling have been used in r.f. plasma oxidation systems. In addition, a solenoidal magnetic field is sometimes used to confine the plasma to the region around the semiconductor [82]. Figure 7.20 shows a schematic diagram of an anodization system for silicon.

In plasma oxidation systems, the sample is usually held at ground potential. Anodization systems, on the other hand, are usually arranged for a d.c. bias to be maintained between the sample and an electrode, with the sample positive with respect to the cathode. Platinum electrodes are commonly used as cathodes; however, silicon can be used for the growth of SiO_2 films.

As many as 34 different elementary reactions have been known to occur in an oxygen plasma [83]. However, many of these result in an insignificant amount of active species. Additionally, many of these active species have extremely short lifetimes so that they recombine before arriving at the substrate. In practice, only a few percent of the active oxygen molecules are useful, with the primary species being O^- and O_2^+, which are present in nearly equal concentrations. Of these, O^- is relevant in plasma anodization. It is possible that this species enters directly into the oxidation process. An alternative possibility, however, is that oxygen molecules interact with electrons at the semiconductor surface to result in the active oxygen species.

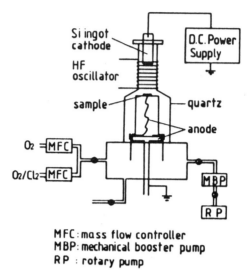

Si ingot cathode

HF oscillator

sample

quartz

anode

D.C. Power Supply

O₂ =MFC

O₂/Cl₂=MFC

MBP

RP

MFC: mass flow controller
MBP: mechanical booster pump
RP : rotary pump

Fig. 7.20 Typical plasma anodization apparatus for silicon. From Haneji et al. [89].
©1985. The Institute of Electrical and Electronic Engineers.

The potential advantage of plasma methods lies in the fact that the electron temperature of the ionized gas is about 10,000 K. As a result, oxidation can be carried out at low thermal temperatures, thus minimizing impurity redistribution during this process [84]. With GaAs, the use of plasmas avoids problems associated with massive loss of As_2O_3 during the high-temperature thermal oxidation of this semiconductor.

The primary disadvantage of plasma processes is that damage is created as a result of the high electric fields which are present during the oxidation process. This usually results in material with a low breakdown electric field strength and a high density of interface traps. In some cases, post-annealing can be used to improve film quality.

Plasma processes usually result in sputtering of the walls of the chamber in which the active species are generated. Both silicon and oxygen are contaminants in systems where quartz is used for this purpose. This is of minor consequence in silicon oxidation, but represents a problem with GaAs as well as with silicon nitridation, where both of these species are active.

During plasma oxidation of silicon, we can expect a greatly increased linear rate constant because of the use of excited species. Experiments along these lines have shown that the parabolic rate constant is also increased [85], indicating that the species diffusing through the oxide is probably not molecular oxygen. For these experiments, the growth rate at 500°C was comparable to that for thermal oxidation at 1100°C, but the stress in the films was lower by a factor of two. It follows that the built-in stress at the growth temperature (500°C) is small compared to the compressive stress which is produced by cooling these films to room temperature.

A disadvantage of direct plasma oxidation of silicon is the relatively low growth rate associated with this process, typically 1000 Å/hr. Plasma anodization, on the other hand, with a current density of 25 mA cm^{-2}, has been shown to increase [86] this growth rate by a factor of 10. This can be further enhanced if 1–3% chlorine is used during the anodization process.

Both plasma oxidation and anodization of silicon have usually resulted in films of poorer quality, and lower breakdown strength, than thermally oxidized films. In general, this can be traced to the difficulty of maintaining a vacuum environment which is sufficiently clean for this purpose. Recent results, however, have shown that these problems can be surmounted, and breakdown field strengths as high as 7–8 MV/cm have been achieved in anodized SiO$_2$ films, grown in both oxygen and oxygen–chlorine mixtures.

Plasma anodization of GaAs is attractive since it results in cleaner oxides than those obtained by liquid anodization, as well as in oxides which are more stoichiometric than those formed by thermal oxidation. Typically, oxide growth rates are about 40 Å/V of applied bias, which is twice that obtained by wet anodization techniques [87]. This enhanced value is probably related to the self-bias associated with the plasma potential, and is a function of both system and plasma parameters.

Experimental results with plasma anodization of GaAs show that the growth rate [88] is independent of oxide thickness, for a constant bias current and at temperatures below 300°C. This suggests that ion transport in the oxide is by ionic drift at high electric fields (about 2.7 MV/cm).

The impetus for plasma anodization of GaAs has been directed towards the development of a viable MOS technology for this material. High-speed MOS devices and integrated circuits, using this type of gate oxide, have been successfully operated [89]. These circuits are based on depletion-mode devices, since inversion of the substrate has not been achieved at the present time.

Plasma processes have also been used for the nitridation of silicon, in order to increase the growth rate which is normally obtained by thermal nitridation. The use of active nitrogen species [90], generated by a microwave plasma at 2.45 GHz, resulted in growth rate enhancement by a factor of two. In this work, SF$_6$ was used to promote the dissociation of nitrogen, and possibly serve as a catalyst for the nitridation process.

Plasma anodic nitridation of silicon has also been studied [91], for this reason. Here, a growth rate of 0.75 Å/V was obtained at 900°C, with system pressures of 0.5–2 torr, using nitrogen–hydrogen gas mixtures. A relatively low dielectric breakdown field strength (3.2 MV/cm) was obtained with as-grown films, after annealing in nitrogen gas to reduce plasma damage effects. Annealing in oxygen, however, improved this property by a factor of two.

The oxynitridation of GaAs can be carried out in an N–O mixture at temperatures as low as 100°C [92]. Here, an increase in the N/O ratio from 0 to 3 caused a reduction in the interface trap density by a factor of 10. Additionally, the static dielectric constant fell from 11 (which is close to that for Ga_2O_3) to 5, which approximates that of pure GaN [93].

7.6 EVALUATION OF NATIVE FILMS

The most commonly used methods for evaluating native films are visual in nature. Thus a check is made to ensure that the film is smooth and free from pinholes, crystallites, and other surface blemishes. Often, the slice is immersed for a few minutes in a 10% copper sulfate solution which penetrates the pinholes. This results in displacement plating of copper around them, and aids in their visual observation.

Layer thickness can be measured by the angle-lap and interferometric technique described in Section 4.10.1. This technique has largely been replaced by ellipsometric methods which are both rapid and nondestructive. Here, a monochromatic beam of plane-polarized light is reflected off the film and also off the film–semiconductor interface. These two components will usually experience different amounts of phase shift upon reflection, and also have different reflection coefficients.

The basic ellipsometer system is shown in Fig. 7.21. Details of system operation and interpretation of the results have been discussed elsewhere [94]. Here, it suffices to point out that the data can be processed to obtain both the film thickness and its refractive index. In addition, the technique can be extended to determine the thicknesses and refractive indices of multiple layers (such as SiO_2 followed by Si_3N_4) on a semiconductor substrate. A major advantage of this technique is that areal information can be obtained in a nondestructive manner, by scanning the beam over the surface.

Film thickness may also be determined to a fair degree of accuracy by visual means. If a thin film with a reflecting back surface is viewed by monochromatic light in an almost perpendicular direction, it can be shown that intensity enhancement occurs at wavelengths λ_k, where

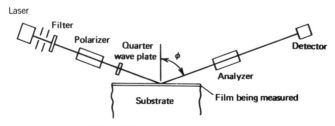

Fig. 7.21 Ellipsometer optics.

$$\lambda_k = \frac{2nd}{k} \qquad (7.46)$$

Here, d is the film thickness (in Å), n is the index of refraction of the film (1.46 for SiO_2), and $k = 1, 2, \ldots$.

If the film is viewed in white light, it will exhibit a brightly colored appearance at one of the wavelengths given by Eq. (7.46). Once the film color is known, its actual thickness can be determined from Table 7.4 at the end of this chapter, which has been derived for SiO_2 layers [95]. The uncertainty on the value of k is generally resolved on the basis of known growth conditions, which allow a rough estimate of film thickness.

Some care is required to make visual measurements of this kind. Thus, the chart shown here is for SiO_2 films which are observed normal to their surface. If the viewing angle is θ, the true thickness of the film is given by $d_0/\cos \theta$, where d_0 is the value read off the color change. In addition, it must be emphasized that this chart is for fluorescent light, and that color quality will be different under other conditions of illumination.

Finally, the color chart can also be used for materials of different refractive index. Thus, the thickness d_f for a film other than SiO_2 is given by

$$d_f = d_0 \, n_0 / n_f \qquad (7.47)$$

where d_0 is the thickness as read from the chart for SiO_2 films (of refractive index n_0), and n_f is the refractive index of the new film. For anodic oxides of GaAs, $n_f = 1.75 - 1.8$, whereas $n_f = 1.98 - 2.05$ for Si_3N_4 films. The color chart of Table 7.4 can thus be applied to these films as well, provided that all listed film thicknesses are multiplied by $1.46/n_f$.

Stress caused by the growth of a native film on a substrate can be measured in a number of different ways. Perhaps the most accurate (and sophisticated) is the use of double-crystal x-ray diffraction by which the change in lattice parameter and the line broadening can be used to determine the strain [96]. The advantage of this approach is that it can be used for small samples. Detailed discussions of this technique are beyond the scope of this book.

Stress causes a circular wafer to become distorted to a paraboloidal shape. If R is the radius of curvature which results from the growth of a film of thickness d_f, on a substrate of thickness D, it can be shown [97] that the stress in the film is given by

$$\sigma_f = \frac{E D^2}{6(1 - \nu) R d_f} \qquad (7.48)$$

where E and ν are the Young's modulus and Poisson's ratio for the substrate, respectively. Values of E and ν are given in Table 7.5 for silicon and GaAs.

Another measure of film quality is its dielectric breakdown strength. In the absence of pinholes in the film, this is usually linearly related to the film density. High-quality native films of silicon dioxide have a breakdown electrical field strength of about 10 MV/cm. A comparable value for GaAs films is 6 MV/cm [98].

With oxide films, the charge states in the oxide and at the oxide–semiconductor interface can be evaluated by evaporating a metal dot (aluminum or gold) on the surface to form a MOS capacitor on which electrical measurements can be conducted. No attempt will be made to treat this important topic here, since it has been covered thoroughly in many papers as well as in texts [37, 99] on device physics.

TABLE 7.1 Mathematical Relationships for the Rate Constants [9]

Parabolic:	$B = C_1 e^{-E_1/kT}$
Linear:	$B/A = C_2 e^{-E_2/kT}$
(111) Silicon	
Dry O_2:	$C_1 = 7.72 \times 10^2 \ \mu m^2/hr$
	$C_2 = 6.23 \times 10^6 \ \mu m/hr$
	$E_1 = 1.23 \ eV$
	$E_2 = 2.0 \ eV$
H_2O (640 torr):	$C_1 = 3.86 \times 10^2 \ \mu m^2/hr$
	$C_2 = 1.63 \times 10^8 \ \mu m/hr$
	$E_1 = 0.78 \ eV$
	$E_2 = 2.05 \ eV$
(100) Silicon	
Dry O_2:	$C_2 = 3.71 \times 10^6 \ \mu m/hr$
H_2O (640 torr):	$C_2 = 0.97 \times 10^8 \ \mu m/hr$
All other parameters:	Same as for (111) silicon

TABLE 7.2 Diffusivities of Elements in SiO$_2$ [36]

Element	D (cm^2 s-1) at 1100°C	D (cm^2 s^{-1}) at 1200°C
Boron	3×10^{-17} to 2×10^{-14}	2×10^{-16} to 5×10^{-14}
Gallium	5.3×10^{-11}	5×10^{-8}
Phosphorus	2.9×10^{-16} to 2×10^{-13}	2×10^{-15} to 7.6×10^{-13}
Antimony	9.9×10^{-17}	1.5×10^{-14}
Arsenic	1.2×10^{-16} to 3.5×10^{-15}	2×10^{-15} to 2.4×10^{-14}

TABLE 7.3 Parameters of Interest for Anodization of Si and GaAs

Parameter	Si	GaAs
Atomic or molecular weight	28.09	144.64
Valence change	2$^+$	6$^+$
Current efficiency	1–3%	90–95%
Electric field during anodization (Mv/cm)	10	5
Density, semiconductor (g/cm^3)	2.33	5.32
Density, oxide (g/cm^3)	2.2	4.2
Dielectric constant, oxide	3.2	5.4
Refractive index, oxide	1.46	1.80

TABLE 7.4   Color Chart for Thermally Grown SiO₂ Films Observed Perpendiculary Under Daylight Fluorescent Lighting [95]

Film Thickness (μm)	Color and Comments	Film Thickness (μm)	Color and Comments
0.05	Tan	0.63	Violet-red
0.07	Brown	0.68	"Bluish" (not blue but borderline between violet and blue-green). It appears more like a mixture between violet-red and blue-green and looks grayish.
0.10	Dark violet to red-violet		
0.12	Royal blue		
0.15	Light blue to metallic blue		
0.17	Metallic to very light yellow green		
0.20	Light gold or yellow, slightly metallic	0.72	Blue-green to green (quite broad)
0.22	Gold with slight yellow-orange	0.77	"Yellowish"
0.25	Orange to melon	0.80	Orange (rather broad for orange)
0.27	Red-violet	0.82	Salmon
0.30	Blue to violet-blue	0.85	Dull, light red-violet
0.31	Blue	0.86	Violet
0.32	Blue to blue-green	0.87	Blue-violet
0.34	Light green	0.89	Blue
0.35	Green to yellow-green	0.92	Blue-green
0.36	Yellow-green	0.95	Dull yellow-green
0.37	Green-yellow	0.97	Yellow to "yellowish"
0.39	Yellow	0.99	Orange
0.41	Light orange	1.00	Carnation pink
0.42	Carnation pink	1.02	Violet-red
0.44	Violet-red	1.05	Red-violet
0.46	Red-violet	1.06	Violet
0.47	Violet	1.07	Blue-violet
0.48	Blue-violet	1.10	Green
0.49	Blue	1.11	Yellow-green
0.50	Blue-green	1.12	Green
0.52	Green (broad)	1.18	Violet
0.54	Yellow-green	1.19	Red-violet
0.56	Green-yellow	1.21	Violet-red
0.57	Yellow to "yellowish" (not yellow) but is in the position where yellow is to be expected. At times it appears to be light creamy gray or metallic.	1.24	Carnation pink to salmon
		1.25	Orange
		1.28	"Yellowish"
		1.32	Sky blue to green-blue
		1.40	Orange
0.58	Light orange or yellow to pink borderline	1.45	Violet
		1.46	Blue-violet
		1.50	Blue
0.60	Carnation pink	1.54	Dull yellow-green

TABLE 7.5 Elastic Constants of Silicon and
Gallium Arsenide at Room Temperature

	$\nu(001)$	$E(001)$ $(\mathrm{dyn/cm}^2)$
Si	0.265	12.94×10^{11}
GaAs	0.312	8.49×10^{11}

REFERENCES

1. C. J. Frosh and L. Derick, Surface Protection and Selective Masking During Diffusion in Silicon, *J. Electrochem. Soc.* **104**, 547 (1957).

2. J. M. Stevels and A. Kats, The Systematics of Imperfections in Silicon–Oxygen Networks, *Philips Res. Rep.* **11**, 103 (1956).

3. W. A. Pliskin and R. P. Gnall, Evidence for Oxidation Growth at the Oxide–Silicon Interface From Controlled Etch Studies, *J. Electrochem. Soc.* **111**, 872 (1964).

4. J. M. Stevels, New Light on the Structure of Glass, *Phillips Tech. Rev.* **22**, 300 (1960/61).

5. A. G. Revesz, The Defect Structure of Grown Silicon Dioxide Films, *IEEE Trans. Electron Dev.* **ED-12**, 97 (1965).

6. J. R. Ligenza, Oxidation of Silicon by High Pressure Steam, *J. Electrochem. Soc.* **109**, 73 (1962).

7. A. G. Revesz, The Role of Hydrogen in SiO_2 Films on Silicon, *J. Electrochem. Soc.* **126**, 122 (1979).

8. B. E. Deal, The Oxidation of Silicon in Dry Oxygen, Wet Oxygen, and Steam, *J. Electrochem. Soc.* **110**, 527 (1963).

9. B. E. Deal, Thermal Oxidation Kinetics of Silicon in Pyrogenic H_2O and 5% HCl/H_2O Mixtures, *J. Electrochem. Soc.* **125**, 576 (1978).

10. B. E. Deal and A. S. Grove, General Relationship for the Thermal Oxidation of Silicon, *J. Appl. Phys.* **36**, 3770 (1965).

11. J. M. Gibson and D. W. Dong, Direct Evidence of 1 nm Pores in Dry Thermal SiO_2 from High Resolution Transmission Electron Microscopy, *J. Electrochem. Soc.* **127**, 2722 (1980).

12. H. Z. Massoud, J. D. Plummer, and E. A. Irene, Thermal Oxidation of Silicon in Dry Oxygen: Growth Rate Enhancement in the Thin Regime II—Physical Mechanisms, *J. Electrochem. Soc.* **132**, 2693 (1985).

13. V. Murali and S. P. Murarka, Kinetics of Ultrathin SiO_2 Growth, *J. Appl. Phys.* **60**, 2106 (1986).

14. E. A. Irene and D. W. Dong, Silicon Oxidation Studies: The Oxidation of Heavily B- and P-Doped Single Crystal Silicon, *J. Electrochem. Soc.* **125**, 1146 (1978).

15. C. P. Ho, J. D. Plummer, J. D. Meindl, and B. E. Deal, Thermal Oxidation of Heavily Phosphorus-Doped Silicon, *J. Electrochem. Soc.* **125**, 665 (1978).

16. N. Tsubouchi, M. Miyoshi, H. Abe, and T. Enomoto, The Application of High Pressure Oxidation Process to the Fabrication of MOS LSI, *IEEE Trans. Electron. Dev.* **ED-26**, 618 (1979).

17. M. Hirayama, H. Miyoshi, N. Tsubouchi, and H. Abe, High Pressure Oxidation for Thin Gate Insulator Process, *IEEE Trans. Electron. Dev.* **ED-29**, 503 (1982).

18. S. C. Su, Low Temperature Silicon Processing Techniques for VLSIC Fabrication, *Solid State Technol.* (March 1981) p. 72.

19. J. Nulman, J. P. Krusius, and A. Gat, Rapid-Thermal Processing of Thin Gate Dielectrics. Oxidation of Silicon, *IEEE Electron Dev. Lett.* **EDL-6**, 205 (1985).

20. J. Nulman, Rapid Thermal Processing of High Quality Silicon Dioxide Films, *Solid State Technol.* (June 1986) p. 189.

21. A. Rohatgi, S. R. Butler, and F. J. Feigl, Mobile Sodium Ion Passivation in HCl Oxides, *J. Electrochem. Soc.* **126**, 149 (1979).

22. H. L. Tsai, S. R. Butler, D. B. Williams, H. W. Kraner, and K. W. Jones, Cl Incorporation at the Si/SiO_2 Interface During the Oxidation of Si in HCl/O_2 Ambients, *J. Electrochem. Soc.* **131**, 411 (1984).

23. B. E. Deal, D. W. Hess, J. D. Plummer, and C. P. Ho, Kinetics of Thermal Oxidation of Silicon in O_2/H_2O and O_2/Cl_2 Mixtures, *J. Electrochem. Soc.* **125**, 339 (1978).

24. M. B. Das, J. Stack, R. E. Tressler, and W. H. Grubs, A Comparison of HCl- and Trichloroethylene-Grown Oxides on Silicon, *J. Electrochem. Soc.* **131**, 389 (1984).

25. B. R. Singh and P. Balk, Thermal Oxidation of Silicon in O_2-Trichloroethylene, *J. Electrochem. Soc.* **126** 1288 (1979).

26. C. Osburn, Dielectric Breakdown Properties of SiO_2 Films Grown in Halogen and Hydrogen-Containing Environments, *J. Electrochem. Soc.* **121**, 809 (1974).

27. J. Steinberg, Dual HCl Thin Gate Oxidation Process, *J. Electrochem. Soc.* **129**, 1778 (1982).

28. B. Y. Liu and Y. C. Cheng, Growth and Characterization of Thin Gate Oxides by Dual TCE Process, *J. Electrochem. Soc.* **131**, 683 (1984).

29. E. P. EerNisse, Viscous Flow of Thermal SiO_2, *Appl. Phys. Lett.* **30**, 290 (1977).

30. L. I. Maissel and R. Glang, Eds., *Handbook of Thin Film Technology*, McGraw-Hill, New York, 1970.

31. S. M. Hu, Formation of Stacking Faults and Enhanced Diffusion in the Oxidation of Silicon, *J. Appl. Phys.* **45**, 1567 (1974).

32. R. B. Fair, Oxidation, Impurity Diffusion and Defect Growth in Silicon—An Overview, *J. Electrochem. Soc.* **128**, 1360 (1981).

33. T. Hattori and T. Suzuki, Elimination of Stacking-Fault Formation in Silicon by Preoxidation Annealing in $N_2/HCl/O_2$ Mixtures, *Appl. Phys. Lett.* **33**, 347 (1978).

34. T. Hattori, Elimination of Stacking Faults in Silicon by Trichloroethylene Oxidation, *J. Electrochem. Soc.* **123**, 945 (1976).

35. C. L. Claeys, E. E. Laes, G. J. Declerck, and R. L. Van Overstraeten, Elimination of Stacking Faults for Charge-Coupled Device Processing, in *Semiconductor Silicon*, H. R. Huff and E. Sirtl, Eds., The Electrochemical Society, Princeton, NJ, 1977, p. 773.

36. M. Ghezzo and D. M. Brown, Diffusivity Summary of B, Ga, P. As, and Sb in SiO_2, *J. Electrochem. Soc.* **120**, 146 (1973).

37. E. H. Nicollian and A. Goetzberger, The Si-SiO_2 Interface: Electrical Properties as determined by the Metal–Insulator–Silicon Conductance Technique, *Bell Syst. Tech. J.* **46**, 1055 (1967).

38. C. E. Young, Extended Curves of Space Charge, Electric Field, and Free Carrier Concentration at the Surface of a Semiconductor, *J. Appl. Phys.* **32**, 329 (1961).

39. B. E. Deal, The Current Understanding of Charges in the Thermally Oxidized Silicon Structure, *J. Electrochem. Soc.* **121**, 198C (1974).

40. E. Arnold, J. Ladell, and G. Abowitz, Crystallographic Symmetry of Surface State Density in Thermally Oxidized Silicon, *Appl. Phys. Lett.* **13**, 413 (1968).

41. E. Yon, W. H. Ko, and A. B. Kuper, Sodium Distribution in Thermal Oxide on Silicon by Radiochemical and MOS Analysis, *IEEE Trans. Electron Dev.* **ED-13**, 276 (1966).

42. A. Many, Y. Goldstein, and N. B. Grover, *Semiconductor Surfaces*, North-Holland, Amsterdam, 1965.

43. E. Spenke, *Electronic Semiconductors*, McGraw-Hill, New York, 1958.

44. S. K. Ghandhi, *Semiconductor Power Devices*, John Wiley and Sons, New York, 1977.

45. J. F. Verwey, On the Emitter Degradation by Avalanche Breakdown in Planar Transistors, *Solid State Electron.* **14**, 775 (1971).

46. S. A. Abbas and R. C. Dockerty, Hot Carrier Instability in IGFET's, *Appl. Phys. Lett.* **27**, 147 (1975).

47. D. Frohman-Bentchkowsky, FAMOS—A New Semiconductor Charge Storage Devices, *Solid State Electron.* **17**, 517 (1974).

48. V. N. Kynett, M. L. Fandrich, J. Anderson, P. Dix, O. Jungroth, J. A. Kreifels, R. A. Lodenquai, B. Vajdic, S. Wells, M. D. Winston, and L. Yang, A 90-ns One-Million Erase/Program Cycle 1-Mbit Flash Memory, *J. Solid-State Circuits* **24**, 1259 (1989).

49. T. Ito, T. Nakamura, and H. Ishikawa, Advantages of Thermal Nitride and Nitroxide Gate Films in VLSI Process, *IEEE Trans. Electron Dev.* **ED-29**, 498 (1982).

50. S. P. Murarka, C. C. Chang, and A. C. Adams, Thermal Nitridation of Silicon in Ammonia Gas: Composition and Oxidation Resistance of the Resulting Films, *J. Electrochem. Soc.* **126**, 996 (1979).

51. M. M. Moslehi and K. Saraswat, Thermal Nitridation of Si and SiO_2 for VLSI, *IEEE Trans. Electron Dev.* **ED-32**, 106 (1985).

52. Y. Hayafugi and K. Kajiwara, Nitridation of Silicon and Oxidized Silicon, *J. Electrochem. Soc.* **129**, 2102 (1982).

53. Y. Okada, P. J. Tobin, R. I. Hegde, J. Liao, and P. Rushbrook, Oxynitride Gate Dielectrics Prepared by Rapid Thermal Processing, Using Mixtures of Nitrous Oxide and Oxygen, *Appl. Phys. Lett.* **61**, 3163 (1992).

54. T. Hori, H. Iwasaki, Y. Naito, and H. Esaki, Electrical and Physical Characteristics of Thin Nitrided Oxides Prepared by Rapid Thermal Nitridation, *IEEE Trans. Electron Dev.* **ED-34**, 2238 (1987).

55. W. Ting, H. Hwang, J. Lee, and D. L. Kwong, Composition and Growth Kinetics of Ultrathin SiO_2 Films Formed by Oxidizing Si Substrates in N_2O, *Appl. Phys. Lett.* **57**, 2808 (1990).

56. G. Weidner and K. Krüger, Nitrogen Incorporation in SiO_2 by Rapid Thermal Processing of Silicon and SiO_2 in N_2O, *Appl. Phys. Lett.* **62**(3), 294 (1993).

57. S. P. Murarka, Thermal Oxidation of GaAs, *Appl. Phys. Lett.* **26**, 180 (1975).

58. B. Schwartz, GaAs Surface Chemistry—A Review, *Crit. Rev. Solid State Sci.* **5**, 609 (1975).

59. I. Shiota, N. Miyamoto, and J. Nishizawa, Auger Analysis of Thermally Oxidized GaAs Surfaces, *J. Electrochem. Soc.* **124**, 1406 (1977).

60. C. W. Wilmsen, Chemical Composition and Formation of Thermal and Anodic Oxide III–V Compound Semiconductor Interfaces, *J. Vac. Sci. Technol.* **19**, 279 (1981).

61. H. Takagi, G. Kano, and I. Termoto, Thermal Oxidation of GaAs in Arsenic Trioxide Vapor, *J. Electrochem. Soc.* **125**, 579 (1978).

62. G. P. Schwartz, J. E. Griffiths, D. DiStefano, G. J. Gualtieri, and B. Schwartz, Arsenic Incorporation in Native Oxides of GaAs Grown Thermally Under Arsenic Trioxide Vapor, *Appl. Phys. Lett.* **34**, 742 (1979).

63. T. Sawada and H. Hasegawa, Interface State Band Between GaAs and Its Anodic Naive Oxide, *Thin Solid Films* **56**, 183 (1979).

64. W. E. Spicer, I. Lindau, P. Skeath, and C. Y. Su, Unified Defect Model and Beyond, *J. Vac. Sci. Technol.* **17**, 1019 (1980).

65. C. J. Sandroff, R. N. Nottenburg, J.-C. Bischoff, and R. Bhat, Dramatic Enhancement in the Gain of a GaAs/AlGaAs Heterostructure Bipolar Transistor by Chemical Surface Passivation, *Appl. Phys. Lett.* **51**, 33 (1987).

66. J.-L. Lee, L. Wei, S. Tanigawa, H. Oigawa, and Y. Nannicki, Evidence for the Passivation Effect of $(NH_4)_2S_x$-Treated GaAs Observed by Slow Positrons, *Appl. Phys. Lett.* **58**, 1167 (1991).

67. R. J. Nelson and R. G. Sobers, Minority Carrier Lifetime and Internal Quantum Efficiency of Surface-Free GaAs, *J. Appl. Phys.* **49**, 6103 (1978).

68. L. M. Smith, D. J. Wolford, J. Martinsen, R. Venkatasubramanian, and S. K. Ghandhi, Photoexcited Carrier Lifetimes and Spatial Transport in Surface-Free GaAs Homostructures, *J. Vac. Sci. Technol.* **B8**, 787 (1990).

69. S. K. Ghandhi, S. Tyagi, and R. Venkatasubramanian, Improved Photoluminescence of GaAs in ZnSe/GaAs Heterojunctions Grown by Organometallic Epitaxy, *Appl. Phys. Lett.* **53**, 1308 (1988).

70. H. D. Barber, H. B. Lo, and J. E. Jones, Repeated Removal of Thin Layers of Silicon by Anodic Oxidation, *J. Electrochem. Soc.* **123**, 1404 (1976).

71. H. Muller, F. H. Eisen, and J. W. Mayer, Anodic Oxidation of GaAs as a Technique to Evaluate Carrier Concentration Profiles, *J. Electrochem. Soc.* **122**, 651 (1975).

72. G. Mende and J. Wende, Breakdown Field Strengths of Anodically and Thermally Grown Metal/Oxide/Semiconductor Structures of Polycrystalline Silicon, *Thin Solid Films* **142**, 21 (1986).

73. S. R. Morrison, *The Chemical Physics of Surfaces*, Plenum Press, New York, 1977.

74. H. Hasegawa and H. L. Hartnagel, Anodic Oxidation of GaAs in Mixed Solutions of Glycol and Water, *J. Electrochem. Soc.* **123**, 713 (1976).

75. D. L. Rode, B. Schwartz, and J. V. DiLorenzo, Electrolytic Etching and Electron Mobility of GaAs for FET's, *Solid State Electron.* **17**, 1119 (1974).

76. A. Shimano, H. Takagi, and G. Kano, Light Controlled Anodic Oxidation of n-GaAs and its Application to Preparation of Specific Active Layers of MESFET's, *IEEE Trans. Electron Dev.* **ED-26**, 1690 (1979).

77. R. A. Logan, B. Schwartz, and W. J. Sundburg, The Anodic Oxidation of GaAs in Aqueous H_2O_2 Solution, *J. Electrochem. Soc.* **120**, 1385 (1973).

78. B. Schwartz, F. Ermanis, and M. H. Brastad, The Anodization of GaAs and GaP in Aqueous Solution, *J. Electrochem. Soc.* **123**, 1089 (1976).

79. E. F. Duffek, C. Mylorie, and E. A. Benjamin, Electrode-Reactions and Mechanism of Silicon Anodization in N-Methylacetamide, *J. Electrochem. Soc.* **111**, 1042 (1964).

80. P. A. Breeze and H. L. Hartnagel, An Assessment of the Quality of Anodic Native Oxides of GaAs for MOS Devices, *Thin Solid Films* **56**, 51 (1979).

81. R. P. H. Chang and A. K. Sinha, Plasma Oxidation of GaAs, *Appl. Phys. Lett.* **29**, 56 (1976).

82. N. Yokoyama, T. Mimura, K. Odani, and M. Fukuta, Low-Temperature Plasma Oxidation of GaAs, *Appl. Phys. Lett.* **32**(1), 58 (1978).

83. J. R. Hollahan and A. T. Bells, Eds., *Techniques and Applications of Plasma Chemistry*, John Wiley and Sons, New York (1974).

84. R. P. H. Chang, A. J. Polak, D. C. Allara, and C. C. Chang, Physical Properties of Plasma-Grown GaAs Oxides, *J. Vac. Sci. Technol.* **116**, 888 (1979).

85. A. K. Ray and A. Reisman, The Formation of SiO_2 in an RF Generated Oxygen Plasma: The Pressure Range above 10 mTorr, *J. Electrochem. Soc.* **128**, 2454 (1981).

86. N. Haneji, F. Arai, K. Asada, and T. Sugano, Anodic Oxidation of Si in Oxygen/Chlorine Plasma, *IEEE Trans. Electron Dev.* **ED-32**, 100 (1985).

87. K. Yamasaki and T. Sugano, Anodic Oxidation of GaAs Using Oxygen Plasma, *Jpn. J. Appl. Phys.* **17**, 321 (1978).

88. A. Matsuzawa, T. Itoh, Y. Ishikawa, and Y. Yanagide, Direct Anodization of GaAs and Si at Extremely Low Substrate Temperature by Low Pressure Oxygen Plasma, *J. Vac. Sci. Technol.* **17**, 793 (1980).

89. N. Yokoyama, T. Nimura, and M. Fukuta, Planar GaAs MOSFET Integrated Logic, *IEEE Trans. Electron Dev.* **ED-22**, 1124 (1980).

90. R. V. Giridhar and K. Rose, SF_6 Enhanced Nitridation of Silicon in Active Nitrogen, *Appl. Phys. Lett.* **45**, 578 (1992).

91. M. Hirayama, T. Matsukawa, H. Arima, Y. Ohno, N. Tsubouchi and N. Nakata, Plasma Anodic Nitridation of Silicon in N_2-H_2 System, *J. Electrochem. Soc.* **131**, 663 (1984).

92. F. I. Hsieh, J. M. Borrego, and S. K. Ghandhi, The Effect of Processing Conditions on the GaAs/Plasma-Grown Insulator Interface, *J. Appl. Phys.* **59**, 1 (1986).

93. T. Haria, T. Usuba, H. Adachi, and Y. Shibata, Reactive Sputtering of Gallium Nitride Thin Films for GaAs MIS Structures, *Appl. Phys. Lett.* **32**, 252 (1978).

94. W. R. Runyan, *Semiconductor Measurements and Instrumentation*, McGraw-Hill, New York, 1975.

95. W. A. Pliskin and E. E. Conrad, Nondestructure Determination of Thickness and Refractive Index of Transparent Films, *IBM J. Res. Dev.* **8**, 43 (1964).

96. W. H. Zachariasen, *Theory of X-Ray Diffraction in Crystals*, John Wiley and Sons, New York, 1950.

97. S. P. Murarka and M. C. Peckerar, *Electronic Materials Science and Technology*, Academic Press, New York, 1989.

98. H. Hasegawa, K. E. Forward, and H. L. Hartnagel, New Anodic Native Oxide of GaAs with Improved Dielectric and Interface Properties, *Appl. Phys. Lett.* **26**, 567 (1975).

99. S. M. Sze, *Physics of Semiconductor Devices*, John Wiley and Sons, New York, 1981.

PROBLEMS

1. A slice of silicon has a 5000-Å thick oxide on its surface. A window is cut in this oxide, and a diffusion is made through it by the two-step process. The resulting oxide over this window is 1500 Å thick. Sketch the cross section of the silicon, indicating clearly the location of the oxide relative to the substrate. Assume (111) silicon, and a 1100°C process in wet oxygen.

2. Calculate the oxide thickness which would be obtained by the following sequence: 20 min in dry O_2, followed by 20 min in wet O_2, both at 1100°C. Use the graphs of Figs. 7.7 and 7.8, and assume (100) silicon.

3. A slice of (100) silicon has small phosphorus-doped islands, each with a surface concentration of 6×10^{20} cm^3. The slice is given a 60-min wet oxidation at 900°C. Calculate the thickness of the oxide over doped and undoped regions, respectively.

4. Using Fig. 7.10, determine a mathematical expression for the pressure dependence of the oxide growth rate at 800°C. What would you expect at higher temperatures?

5. An *n–p–n* transistor is made on (100) silicon by sequential boron and phosphorus diffusions. The boron drive-in step is 30 min at 1100°C in steam. The resulting oxide has a window cut in it for the emitter diffusion, which is carried out at 1000°C for 15 min in dry oxygen. Will the oxide resulting from the base diffusion be sufficiently thick to serve as a mask for the emitter diffusion?

6. It is required to produce a 4400-Å thick layer of silicon dioxide by anodization. Assuming a constant current source of 20 mA/cm² and a current efficiency of 2%, determine the anodization time.

7. Repeat Problem 6 for GaAs.

8. Prove that the oxidation of the silicon surface results in an oxide layer which is about 2.27 times the thickness of the consumed silicon.

9. Silicon dioxide is grown on (001) Si at 1000°C. Determine the oxide thickness which represents the transition between a linear and a parabolic growth rate. Solve for both dry and wet oxidation.

CHAPTER 8

DEPOSITED FILMS

A large number of different kinds of deposited films are used in microcircuit technology today; moreover, the functions they perform are continually increasing, as are their variety. In recent years, this list has greatly expanded because of the new requirements imposed by different VLSI schemes. Both insulating and conducting films are used. The division is not rigid here, since some films are deposited in one of these forms, and later converted to the other by processing. Even the term "deposited" is not a rigid one, since some films are deposited in elemental form and later converted to their final form by chemical reaction.

Deposited films are conveniently categorized by the primary functions they are required to perform. Some of these functions are electronic in nature, yet others are process-oriented. Thus, a number of different types of films are used to provide an insulating, protective cover on which interconnections can be made. Some of these can be patterned for use as diffusion or ion implantation masks, while others can be used to provide the doping function itself. Yet other films make possible the use of greatly simplified processing schemes such as self-aligned gate technologies for MOS microcircuits.

Many films and film combinations can be used to form ohmic contacts, as well as interconnections between devices in a microcircuit. Some of these can also be used to perform device functions as in Schottky diodes, and in Schottky-gate field effect transistors.

The films to be described in this chapter can be deposited by an equally wide variety of techniques; indeed, most films can be formed by more than one method. For each film, the emphasis will be on those techniques which are favored in semiconductor usage.

Physical deposition techniques of practical interest include vacuum evaporation, as well as d.c. and r.f. sputtering. The recent development of magnetron sputtering represents improvements over conventional sputtering methods, and provides increased growth rates. In addition, reactive sputtering is also used in some applications where the desired film is not elemental in nature, but is a compound. All of these sputtering techniques share the advantage that substrates can be cleaned (by back-sputtering) prior to film deposition. Ion beam deposition is yet another technique and has the following advantages: (a) relative freedom from contamination and (b) independent adjustment of system parameters such as substrate temperature, beam energy, and current density. Equipments for physical deposition of films form a highly specialized branch of manufacturing technology and have been considered in detail elsewhere [1].

The primary emphasis on chemical techniques has been on chemical vapor deposition (CVD) processes, which were considered in detail in Chapter 5. Many of the systems and techniques developed for vapor phase epitaxy can be directly used (or modified for use) to deposit a wide variety of insulating and conducting films. These systems are often operated at low pressure (≤ 0.1 torr); this results in increasing the diffusivity of the reactant species, and leads to greater uniformity and throughput than does operation at atmospheric pressure. Consequently, recent emphasis on production techniques is in this area. Finally, the use of plasma enhancement allows reactions to be carried out by energetic species with an electron temperature of around 10,000 K. As a result, growth can be accomplished at greatly reduced temperatures by these techniques.

In this chapter, the salient features of various deposition techniques will be described in order to highlight their advantages and disadvantages. Next, we shall consider the characteristics of importance in the selection of these deposition methods. Finally, we shall consider the many films which are used in microcircuit fabrication, in terms of the functions which they provide in an integrated circuit.

Tables of relevance to this topic are assembled at the end of this chapter for ready reference.

8.1 FILM DEPOSITION METHODS

The principal techniques for depositing films are vacuum evaporation, sputtering, and chemical vapor deposition. Wet chemical plating is also used in some applications. Each of these techniques is a highly specialized field in itself, and has been the subject of much research and development. Here, we shall confine ourselves to their basic operating principles, in order to assess their applicability in microcircuit fabrication.

8.1.1 Vacuum Evaporation

Perhaps the simplest technique for depositing a material in thin film form is by vacuum evaporation. In this process, the material to be evaporated is heated in an evacuated chamber so that it attains a gaseous state. Vapors of this material traverse the space from the source to the substrate, on which they eventually land. In principle, therefore, it is identical to molecular beam epitaxy (MBE). The key differences, however, are that vacuum evaporators operate at higher base pressures, and handle multiple slices and with much higher deposition rates than their MBE counterparts. These differences will determine the suitability of these systems for any given application.

Metal deposition rates of 0.5 μm/min are typically desired in vacuum evaporation. This necessitates that the equilibrium pressure of the evaporant [see Eq. (5.22)] be in the 10-mtorr range, and restricts the use to materials which can be evaporated from heated crucibles. Materials such as aluminum and gold are nearly ideal for this technique. On the other hand, materials such as tungsten and titanium have extremely low vapor pressures, and cannot be delivered in this manner. Extensive data on the vapor pressure versus temperature characteristics of a number of materials are available in the literature [2]. Figure 8.1 shows vapor pressure data for some materials of interest for microcircuit fabrication.

Electron-beam (e-beam) evaporation, whereby an intense beam of energy is applied locally to a target, can be used to develop a sufficiently large flux of evaporant from refractory materials. Here, the source is held in a water-cooled hearth, and evaporation occurs at a highly localized point on this source, while its bulk remains solid. Thus, there is no contamination from the crucible, as long as the e-beam is well designed, and only strikes the target during this process.

The use of an electron gun results in a point source, so that the flux pattern is Lambertian in nature, resulting in a nonuniform film thickness across the substrate. In multiple-slice systems, this nonuniformity can be minimized by rotating the slices during evaporation, and sophisticated planetary motion systems are used for this purpose. In addition, the point of impact of the e-beam is moved around electronically during the evaporation, so as to simulate a large area source.

Electron-beam sources can be used with refractory as well as nonrefractory materials. Often, multiple hearths are incorporated into these systems, to permit the deposition of multilayer films in a sequential manner, without the need to break vacuum and with very little system contamination.

The path taken by the evaporant to the substrate should be one that is collision-free, to prevent agglomeration of material in the gas phase. Thus, its length should be less than the mean free path. From Eq. (5.19b), it can be shown that this mean free path is about 10^3 cm, for a system pressure of 10^{-5} torr. However, the typical vacuum evaporator must be capable of being pumped down to 10^{-7} torr, to minimize the residual contamination level in the system. Many pumping techniques can be used for this purpose. Their technology [3] is an art in itself and is beyond the scope of this book.

Vacuum conditions imposed by the mean free path requirement place additional restraints on the choice of materials which can be deposited by this method. For example, the vapor pressure of materials such as cadmium and zinc are so high that they evaporate too rapidly, and tend to contaminate the entire system during this process.

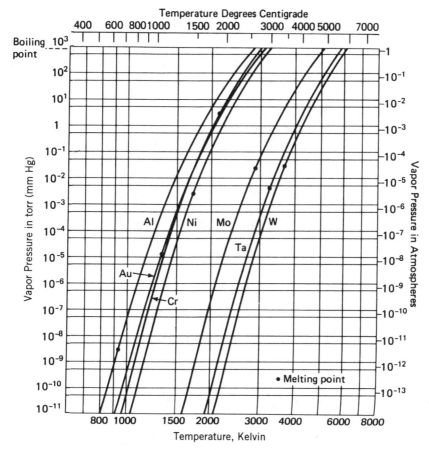

Fig. 8.1 Vapor pressure curves for selected materials. Adapted from [2].

During vacuum evaporation, substrates are essentially unheated, except for radiation from the evaporant source. In some systems, internal infrared heaters are used when it is desired to conduct the evaporation at elevated temperature.

The energy associated with thermal evaporation is in the 0.1- to 0.15-eV range. Thus, this process results in no damage to the substrate surface. This is a significant advantage in the deposition of films over gate oxides. However, this advantage is not present with e-beam evaporation, where relatively high voltages (≥10 kV) are used for beam formation and direction. This produces x-rays, which cause damage in the substrate on which the film is deposited. With gate oxides, there is a significant increase in the fixed oxide charge as well as in the density of interface traps when e-beam evaporation is used.

Vacuum evaporation of alloys with accurately controlled composition is extremely difficult because of the vapor pressure dissimilarity of their separate components. In general, therefore, the composition of the alloy film will be quite different from that of the source, and this difference will vary with time as the source gets preferentially depleted. Multiple crucible sources, one for each element, have been used but these require extremely precise temperature control for satisfactory operation. Although used in MBE systems, they are not practical in commercial evaporators. The control of evaporant flux from an electron gun is sufficiently poor so that the alloy composition control is not satisfactory by this means, either.

Finally, *in situ* thermal substrate cleaning, although used in MBE, is not practical in vacuum evaporators because of their multiple-slice capability. Thus, adhesion of the film to the substrate is sometimes imperfect. This is not a problem with aluminum films on SiO_2, since bonding is accomplished by local reaction of these materials. With noble metals such as gold, however, it is customary to first evaporate a thin layer of a reactive material, which serves as a "glue" between the evaporant and the substrate. Chromium and titanium are often used for this purpose.

8.1.2 Sputter Deposition

Sputter deposition, or *sputtering*, is considerably more versatile than vacuum evaporation, and is extensively used in the deposition of thin films in microcircuit technology. The reasons for its popularity are (a) the ability to deposit a wide variety of metals and insulators, as well as their mixtures, (b) the replication of target composition in the deposited film, and (c) the capability for *in situ* cleaning of the substrate prior to film deposition.

Sputter deposition is carried out in a self-sustained glow discharge which is created by the breakdown of a heavy inert gas such as argon. The physics of this process is illustrated in Fig. 8.2a. Here, a d.c. electric field is impressed across two water-cooled electrodes which are located in this gas. At sufficiently low electric field intensities, a very small current will flow, primarily by the transport of electrons between these electrodes. These electrons may be produced by photoemission or by cosmic ray stimuli, and are always present to some degree in any gaseous medium.

Transport of electrons between the electrodes will result in some collisions with the gas molecules, so that ionization will occur in those encounters where a sufficiently large amount of energy is transferred to the gas. The products of these few ionizing collisions (namely, Ar^+ and electrons) are themselves accelerated (in opposite directions) because of the electric field, so that they can also enter into collisions with neutral argon molecules, thus resulting in an avalanche multiplication effect. The actual magnitude of this effect* will depend on the electric field intensity since this determines the acceleration of the electrons and Ar^+, and on the gas pressure which establishes the mean free path for collisions. It will depend also upon the distance between electrodes,

---

*About 5 argon ions cm^{-1} for an electron energy of 1 keV, at a pressure of 1 torr.

since the number of collisions per unit length traveled is reciprocally related to the mean free path.

Figure 8.2a shows this situation, where multiple electrons reach the anode and the Ar$^+$ ions reach the cathode. At sufficiently high applied voltages, some of these Ar$^+$ ions can eject secondary electrons from the cathode, thus adding to the supply of electrons which contribute to the avalanche multiplication process. This can eventually lead to gaseous breakdown [4] with the current being limited by the circuit impedance. For this condition, Ar$^+$ ions bombard the cathode, resulting in sputtering* of its surface material by momentum transfer. In addition, secondary electrons** emitted at the cathode participate in sustaining the discharge by ionizing collisions with argon molecules.

Figure 8.2b shows the nature of the discharge. The most important region here is the Crooke's dark space across which nearly all of the applied voltage is dropped. Both ions and electrons created at breakdown are primarily accelerated across this region. Its thickness d is related to the chamber pressure p; for argon the pd product is about 0.4 torr-cm.

---

*This sputtering process is of technological importance for the etching of materials, and will be taken up in detail in Chapter 9.
**About one electron per ten incident argon ions.

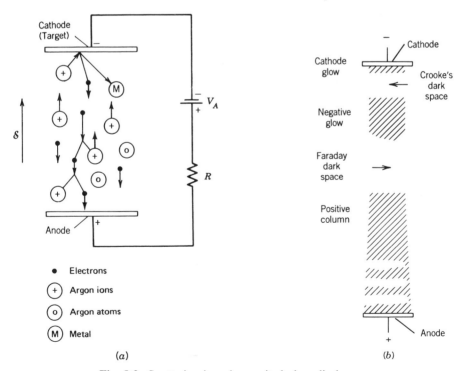

Fig. 8.2 Sputtering in a d.c.-excited glow discharge.

The positive column and the Faraday dark space are not necessary for the operation of the discharge; for all practical purposes, they serve merely as electrical connections between the anode and the negative glow region. Consequently, it is possible to shorten the anode–cathode spacing by moving the anode into this negative glow region. This results in a higher breakdown voltage, since it eliminates part of the region where collisions can take place. The principles outlined here for the d.c. glow discharge can also be extended to discharges with r.f. excitation.

It should be emphasized that nearly all of the voltage drop is in the vicinity of the cathode, resulting in its bombardment by high-velocity Ar^+ ions. On the other hand, the anode, on which deposition of the material which constitutes the cathode (target) occurs, is in a relatively field-free region. Atoms ejected from the cathode are mostly neutral, although both positive- and negative-charged species are also produced. They have an energy spread of 0–20 eV, and diffuse to the anode or the chamber walls, where they arrive in a relatively nondirected manner. Sputter deposition of the material which constitutes the cathode (target) occurs on substrates placed on the anode; good adhesion is achieved because of ion mixing, resulting from this energy spread.

With d.c sputter deposition, it is necessary that positive ions impinging on the target surface have an opportunity to recombine with electrons to prevent charge accumulation on the surface. While discharging arrangements can be provided to minimize this problem, it can be avoided if the discharge is maintained by an r.f. field. This approach is most commonly used, since it is equally effective with both conducting and insulating substrates.

Radio frequency sputtering is usually carried out with a 1- to 3-kV peak-to-peak r.f. potential. Excitation is commonly provided at 13.56 MHz, which is internationally assigned for equipment usage. A schematic arrangement of this type of system is shown in Fig. 8.3. Here, advantage is taken of the diode-like character of the discharge to produce a self-bias, so that the electrodes take on the role of anode and cathode (target) as shown. Sputter deposition is carried out by placing substrates on the anode, as before.

A wide variety of conducting and insulating materials, as well as their mix-

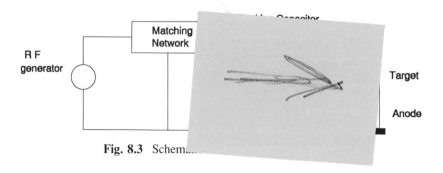

Fig. 8.3 Schema...

tures, can be deposited by sputtering. In most cases, targets for this purpose are generally made of mixtures of the components to be deposited, which are hot-pressed into the form of a disc. This disc is bonded to a copper cathode which is water-cooled; elaborate shielding is provided to prevent sputter deposition of copper from this cathode.

Upon system evacuation, a large amount of out-gassing occurs from the target because of its porous character. Further out-gassing occurs upon bombardment during operation. This results in considerable variability from run to run, unless the sputtering system is maintained under vacuum at all times, and substrates introduced via a load-lock arrangement.

A key advantage of sputtering over vacuum evaporation is that the alloy content of a multicomponent target is replicated in the source film, even if it consists of materials with widely different sputter yield. By way of example, consider a two-component target, A–B, in which the sputtering yield of A is much larger than that of B. On initial use, the film will have an excessively large concentration of A, and the surface layers of this target will become preferentially depleted of this material. Consequently, the concentration of A in the film will fall as sputter deposition proceeds. This process will continue until steady state is reached, at which point the composition of the film and the bulk of the target will become identical. Typically, the thickness of the altered layer is a few tens of angstroms for metal alloys, and can be as thick as 1000 Å for targets of oxide mixtures [5]. Thus, the steady-state condition for the target is reached shortly after it is put into operation for the first time [6].

In a d.c. sputtering system, it is possible to reverse system polarity, and thereby etch the substrate prior to film deposition. This process, known as *back-sputtering*, is extremely useful for removing thin surface layers such as residual oxides, which can affect the electrical and mechanical properties of the film–substrate combination, or prevent adhesion of the film to the substrate.

Sputtering systems are normally operated at pressures around 10^{-2} torr, which are considerably higher than those encountered during vacuum evaporation. Thus, they are relatively "dirty" in comparison, unless considerable care is taken to provide a clean pumping arrangement. Often, this includes the incorporation of cryogenic traps and the use of turbomolecular pumps, which are oil-free. In addition, high gas flows are commonly used to continually sweep impurities out of the sputtering chamber. Base pressures of these systems are comparable to those used with vacuum evaporation—that is, in the 10^{-7} torr range.

The biggest disadvantage of sputtering is that it results in damage to the substrate, which is not completely free from impact by high-energy species. Sputtering is thus to be avoided in situations such as the deposition of materials on a gate oxide. In some cases, heating of the substrate (in addition to that which is intrinsic to the sputter deposition process) is used to anneal this damage as it is created. This is usually provided by resistive or infrared elements.

8.1.2.1 Magnetron Sources

In sputtering systems of the type described here, only a small fraction of electrons emitted from the target collide with argon ions, whereas the rest are collected at the anode, where they cause unwanted heating. The use of a magnetic field, to confine these electrons near the target surface, greatly increases the possibility of ionizing collisions with the argon gas molecules. This, in turn, increases the Ar^+ concentration, and hence the film deposition rate. This process is known as *magnetron sputtering*.

Detailed descriptions of a wide variety of both cylindrical and planar magnetron sputtering targets are provided in the literature [1]. Their principle is illustrated in Fig. 8.4 for two particular configurations. With both circular and rectangular sources, a magnetic field strength of 100–200 gauss is used, resulting in field lines which confine electrons to the region embraced by them. The discharge region is thus in the shape of a toroid or an elongated ring, with extensive sputtering of the target in these regions. Substrates are usually placed on a rotating anode in order to obtain uniform film deposition.

8.1.3 Chemical Vapor Deposition

This topic has been covered in Chapter 5 in the context of epitaxial growth. As with vapor-phase epitaxy (VPE), chemical vapor deposition (CVD) involves transport of reactant species into the reactor, reaction at the substrate, and removal of reaction products. The process can be carried out in hot or cold wall reactors, at atmospheric pressure (APCVD) as well as at low pressure (LPCVD).

APCVD is usually conducted in cold wall reactors, since many of the reactions involved in the deposition of thin films are pyrolytic in character. Resistive heating of the susceptor is often used, since growth temperatures are usually

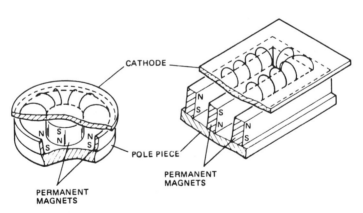

Fig. 8.4 Circular and rectangular magnetron sputtering targets. The curved lines represent magnetic field lines. From Vossen and Kern [1]. Reprinted with permission from Academic Press.

lower than those encountered with VPE. Systems of the type shown in Fig. 5.13*a* are used for this application. Continuous deposition systems, using muffle furnaces, are also available for high-volume applications.

In VPE, the emphasis is on the growth of single-crystal films with a high degree of crystal quality. For applications considered in this chapter, however, uniformity and high throughput are the overriding factors. Films are generally polycrystalline or amorphous so that high growth rates can be used; although a clean reactor environment is always desirable, the levels of purity and freedom from contamination which are essential for VPE are seldom required (or met) in CVD systems. LPCVD is a very attractive approach for the growth of these films.

Following along the lines of Chapter 5, the growth rate of a film is given by

$$\frac{dx}{dt} = \frac{N_g}{n}\left(\frac{hk}{h+k}\right) \tag{5.30}$$

where N_g is the reactant concentration in the gas phase, n is the number of atoms (or molecules) of the grown material in unit volume, k is the reaction rate constant, and h is the mass transfer coefficient which is directly proportional to the diffusivity, D. Here,

$$k = k_0\, e^{-E_a/RT} \tag{8.1a}$$

$$h = \frac{D}{\delta_D} \tag{8.1b}$$

where E_a is the activation energy of the reaction at the substrate, and δ_D is the thickness of the diffusive boundary layer.

The diffusivity, and hence h, varies inversely with system pressure. Operation at 0.5 torr, for example, results in an increase in h by a factor of 1520 from its value at atmospheric pressure. Now,

$$\frac{dx}{dt} \simeq \frac{N_g\, k}{n} \tag{8.2a}$$

$$\simeq \frac{N_g\, k_0}{n} e^{-E_a/RT} \tag{8.2b}$$

since $h \gg k$.

There are two important consequences of this result. First, the growth rate across the slice is only a function of the temperature, and is thus independent of flow conditions in the reactor. Thus, films can be deposited on a very large number of slices, with a high degree of uniformity, provided that tight control is placed on temperature. This favors growth in a hot wall reactor, with multiple temperature-controlled zones. Slices can be closely spaced, to obtain a high throughput in this system. A schematic of a LPCVD reaction chamber is shown in Fig. 8.5. Front-end loading is illustrated here, since it is commonly used for operational convenience.

The second consequence is that operation at low pressure results in no penalty in the growth rate. In fact, this can actually *increase*, provided that N_g is kept unaltered. In practice, low-pressure operation is accomplished by reducing, or even eliminating, the inert carrier gas which is used in systems which operate at atmospheric pressure. Typical LPCVD systems operate in the 0.25- to 1-torr range; this sets an upper limit to the reactant partial pressure, and hence to the growth rate.

For practical reasons, LPCVD systems are of the hot wall type, where uniform temperature control is relatively easy to accomplish over a large operating length. Unfortunately, many of the reactions involving the deposition of thin films are of the pyrolytic nature. As a result, dust formation due to deposits on the hot wall is a continual problem. This is kept within manageable bounds by reducing the partial pressure of reactants, so as to reduce the possibility of gas phase reactions. LPCVD systems are generally operated at lower reactant pressures (i.e., at lower growth rates) than APCVD systems for this reason. In addition, reaction chamber clean-up is a frequent step in the routine maintenance of these systems.

A consequence of maintaining a large throughput is that reactant depletion often occurs as gases are transported down the reaction chamber. The resulting falloff in growth rate is usually compensated by adjusting the reactor temperature so that it gradually increases from inlet to outlet by a few degrees (typically 25–50°C).

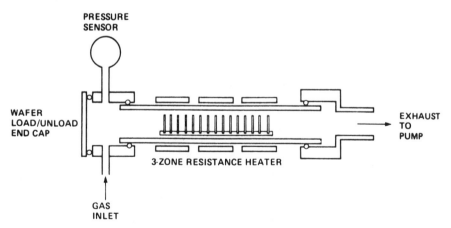

Fig. 8.5 LPCVD reaction chamber schematic. From Vossen and Kern [1]. Reprinted with permission from Academic Press.

8.1.3.1 Plasma-Enhanced CVD

Plasma enhancement represents a natural extension of LPCVD, since system pressures for both techniques are comparable. The principles underlying the use and advantage of plasma processes for the growth of native films (described in Section 7.5) hold for deposited films as well. Thus, the high electron temperature allows growth at greatly reduced temperatures compared to those which are required for thermally activated processes.

Growth at reduced temperature greatly expands the number of situations where deposited films can be used. By way of example, Si_3N_4, grown at 800–900°C by conventional CVD, cannot be deposited on top of an aluminum metallization, which melts at 660°C. Plasma-grown nitride, at 350°C, is quite suitable for this application.

Plasma-enhanced chemical vapor deposition (PECVD) systems have been built in a number of different configurations. Of these, the planar, radial flow reactor shown in Fig. 8.6 is the most commonly used [7]. Here, the plasma is struck between two electrodes which are held a few centimeters apart. The rotating substrate electrode is grounded, with provision for reactant feed at its center. Internal parts of the reactor are usually made of anodized aluminum, so that they have a protective coating of alumina. Substrate rotation is provided for improving the uniformity of the grown layers.

An inherent characteristic of deposition by this technique is that it is accompanied by some sputtering. This makes it especially suited for

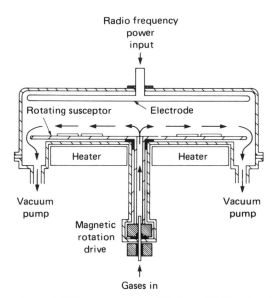

Fig. 8.6 Plasma-enhanced CVD system. From Mattson [7]. Reprinted with permission from *Solid State Technology*.

filling high-aspect-ratio geometries, such as deep trenches, by control of the deposition/removal ratio. In addition, the impact of energetic species results in excellent adhesion of the film to the substrate, by ion mixing. By the same token, however, it creates damage in the underlying semiconductor. This is especially apparent when PECVD processes are carried out at the gate oxide level.

Relatively large quantities of hydrogen are present as dissolved species in all films grown by PECVD. The actual amount, typically about 5–10 at.% in SiO_2 and 15–30 at.% in Si_3N_4 films, is a function of reactant concentrations as well as r.f. power and frequency. Typically, the etch rate of PECVD films is higher than for those grown by LPCVD, because of the incorporation of hydrogen in them.

8.2   FILM CHARACTERISTICS

Deposited films must possess a number of specific characteristics in order to be useful in microcircuit fabrication. Thus, they should be capable of being laid down with a smooth morphology, free from voids and pinholes, and adhere well to the surface on which they are placed. Since this surface often has a stepped character because of previous patterning, the film should be capable of covering these steps in a conformal manner.

These properties should be sustained during the many subsequent heat treatments to which the slice is subjected. Thus, stress in the film should be controlled, in order to prevent cracking or loss of adhesion during these treatments.

Finally, the electrical properties of the film should be optimized for their function. Thus, insulating films should have high resistivity and high breakdown electric field strength; conducting films should have low resistivity, and be capable of operating at high current densities without failure.

This section will consider some of these characteristics, and the manner in which they relate to deposition methods.

8.2.1   Growth Habit

The growth of a thin film, by condensation from the vapor phase, proceeds along the lines outlined in Chapter 5. As mentioned, stable nuclei are formed, and move on the substrate as they coalesce into clusters. Here, however, the substrate is non-crystalline, so that there are no preferred kink sites; rather, randomly oriented clusters grow until they meet to form a continuous film which spreads across the surface. The size of the individual cluster, prior to meeting a neighboring cluster, is related to the surface mobility, and also to the density of nuclei. A high surface mobility and/or a low nucleation rate promotes the formation of relatively large clusters, so that resulting films are polycrystalline in nature, with short-range order. On the other hand, extremely low surface mobility, combined with a high nucleation rate, can lead to amorphous film growth. Such a film can be looked upon as a single crystal in which every atom is displaced from its lattice site; that is, it has no short-range order. For example, the distance between silicon atoms in amorphous SiO_2 is about 3.6 Å, and there is no order between atoms at distances above 10 Å.

At the present time, there is no definitive way of predicting whether film growth will be amorphous or crystalline. Most deposited films tend to be polycrystalline in nature, and crystallite size tends to be larger with films which are grown on heated substrates, because of their high surface mobility. On the other hand, oxide and nitride films tend to be amorphous under normal growth conditions, but become polycrystalline if grown at elevated temperatures.

Films which are grown at low temperatures tend to be less dense, and have poorer structural properties, than their high-temperature counterparts. A short heat treatment at the higher temperature is often sufficient to "densify" these films in order to improve their properties. Rapid thermal anneal methods are used increasingly for this purpose.

The adhesion of a film to a surface is determined by the initial stages of growth. Films will generally adhere strongly to clean surfaces with which they enter into chemical bonding, and to surfaces with which they alloy. Oxygen-active metals such as chromium and titanium are especially useful as "glues" for this reason. Additionally, excellent adhesion is obtained between materials which intermix during deposition (e.g., copper on gold). Sputtering is an important preparative technique because argon bombardment results in ion mixing between the deposited film and its substrate during the early stages of growth. Sputtered films generally adhere quite strongly for this reason. The adhesion of a film is also a function of its interfacial stress with the substrate, which is strongly process-dependent.

8.2.2 Step Coverage

In microcircuit fabrication, conformal coverage of a stepped surface is a highly desirable film property. Figure 8.7a shows an example of this type of step coverage; also shown in Fig. 8.7b is an undesirable situation which can occur in practice. In extreme cases, cracks in the film can occur in regions where it is extremely thin, resulting in circuit failure.

The growth of a film which covers a step in a conformal manner requires that two conditions be met. First, the film material or the reactant species which produce it must impinge on the substrate from many directions. A large-area source will tend to promote this situation. Next, the deposited species must have a high surface mobility, so that it can freely move over the surface, regardless of its topography, before becoming immobilized.

Problems of step coverage are especially severe with vacuum evaporation, where the evaporant approximates a point source, travels in straight lines without collisions, and impinges on an unheated substrate. Movement of the e-beam over the source area during evaporation, together with (a) the use of a planetary drive for the substrates and (b) the heating of substrates during vacuum evaporation, reduces this problem to some extent. Increasingly, computer simulation [8] is used to model and optimize this process.

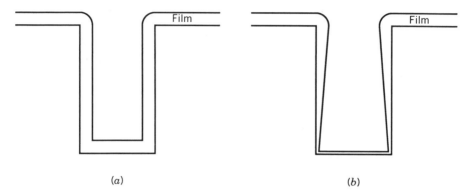

(a)   (b)

Fig. 8.7 Examples of step coverage: (*a*) conformal and (*b*) nonconformal.

With both sputtering and CVD, material is delivered from a large-area source, and arrives at the substrate in an undirected manner, after many gas-phase collisions. Moreover, substrate temperatures are relatively high in both cases. Thus, step coverage is usually excellent.

Many collisions also occur during LPCVD and PECVD processes, so that material impinges on the substrate in an undirected manner. Moreover, their relatively low growth rate increases the time for surface diffusion of species before they are covered by the next layer of material. As a result, step coverage is usually excellent with LPCVD. It is sometimes a problem with PECVD, because of the lower temperature associated with this method.

A unique feature of CVD is the wide variety of reactions that can be used to form the same film material. In some cases, the reaction pathway for growing a film involves intermediate compounds whose properties may dominate the flow process. Thus, both silane and tetraethylorthosilane (TEOS) can be oxidized to form SiO_2. However, because of the different species involved in the reaction pathways, their step coverages are very different, even when growth is carried out at the same temperature.

An alternative approach to obtaining conformal step coverage is to reduce the abruptness of the step, prior to film deposition. This can be accomplished by elevating the substrate film beyond the temperature at which plastic flow occurs, and holding it at this point for a specific time period (usually 30 min). As a rule of thumb, this temperature is about two-thirds the melting point, with both temperatures given in Kelvin.

8.2.3 Mechanical Stress

All deposition techniques result in the formation of films in which there is stress. Some of this stress is intrinsic, and is associated with condensation from the vapor phase. Volume changes during this process, together with the incorporation of gases from the system environment, can greatly affect its nature and magnitude. In vacuum evaporation, the use of a clean system with a low base pressure is highly desirable to minimize this problem.

Vacuum-evaporated films are usually in tensile stress. This stress is lessened if the films are deposited on heated substrates, because some annealing occurs during the growth process. Its magnitude is related to film parameters such as the crystal structure and yield strength, as well as to process parameters such as the growth rate. No firm theoretical understanding of this subject is available at the present time.

Sputtered films usually contain relatively large amounts of argon gas, whereas dissolved hydrogen is usually present in PECVD films. As a consequence, the intrinsic stress in these films can be tensile or compressive, and is a strong function of process conditions.

Also present in deposited films is thermal stress, which occurs when films are grown at elevated temperatures. This stress is directly related to differences in thermal coefficient of expansion (TCE) of the film and the substrate, and to the temperature excursion, ΔT. If it is assumed that all the stress is taken up by the thin film (i.e., the substrate is thick), then the thermal stress in the film is

$$\sigma_{th} \simeq E_f(\alpha_f - \alpha_s)\Delta T \qquad (8.3)$$

where E_f is the Young's modulus of the film, and α_f and α_s are the TCEs of the film and substrate, respectively.

Stress in a film can deform the substrate on which it is deposited. If the film is in biaxial tension, the substrate will curve upwards; that is, its surface will become concave. By the same token, compressive stress will result in a convex substrate. The amount of this deformation can be used to determine the magnitude of this stress. This can be done by using a rectangular specimen [9]; alternatively, nondestructive measurements can be made on complete slices for this purpose.

A circular disc, stressed in this manner, will take up a paraboloidal shape. If the Young's modulus and the Poisson's ratio of the film are comparable to that of the substrate, it can be shown that the film stress is given [10] by

$$S = \frac{E_s}{6(1 - \nu_s)} \frac{t^2}{RD} \qquad (8.4)$$

where E_s and ν_s are the Young's modulus and Poisson's ratio, respectively, for the substrate t is the film thickness, D is the substrate thickness, and R is the radius of curvature. This formula becomes increasingly inaccurate with thick substrates, and sometimes necessitates the use of more sophisticated approaches [11].

8.2.4   Electromigration

The term *electromigration* refers to the transport of mass in metals when stressed to high current densities [12]. It occurs during passage of d.c. current through thin metal conductors in integrated circuits, and it results in (a) the piling up of metal in some regions and (b) void formation in others.

Failures due to electromigration take many forms. The most dramatic is the formation of voids, which reduce the effective cross section of the conductor. This, in turn, increases the current density, resulting in a cumulative effect until the conductor becomes an open circuit [13]. Additional effects are the accumulation of matter in the form of hillocks and whiskers which can cause shorts between adjacent lines.

Electromigration sets a fundamental limit to the dimensions of current-carrying conductors, and many studies have been made to understand this phenomenon. Although no clear picture has emerged, it has been shown that this effect is a strong function of the choice of material, and of such material properties as the grain size. It is also related to the dimensions of the conductor and the manner in which it is terminated. Finally, numerous additives have been studied in order to increase the mean time to failure (MTF) of films which operate at high current densities.

Electromigration is only one of several transport phenomena that occur in in the presence of current flow [14]. However, it is the most significant failure mode in interconnection technology for microcircuits where current densities can often exceed 10^6 A cm^{-2} without catastrophic effects such as melting.

Figure 8.8 shows a metal conductor in which the electric field is directed from left to right, as is the current flow. Here, there are two forces on the (+ve) metal ions in the conductor. The first, a field-ion force (F_1), is directed towards the negative terminal, and is proportional to the electric field strength and the valence of the metal ion. A second force (F_2) comes about because of the exchange of momentum between the electrons which comprise the current flow and the metal ions. These electrons move from right to left, towards the positive terminal, and cause metal ion motion in this direction by an "electron wind" effect. In general, this second force dominates, so that the net result is a movement of metal ions towards the positive terminal (i.e., with the elec-

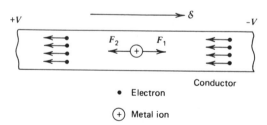

Fig. 8.8   Forces on a metal ion during current transport.

tron flow). This cannot cause void formation if all the metal is moved simultaneously. In practice, however, local variations in mobility along the length of the conductor will cause the metal to be moved at different rates, with the eventual formation of voids.

Ignoring the field-ion effect, the rate of metal transport (R) is directly proportional to the electron momentum and to the electron flux. It is also directly proportional to the target cross section (ϕ) of the metal. Metal motion is also governed by its activation energy of self-diffusion, E_d. The electron momentum and the electron flux are, in turn, each proportional to the current density. It follows that

$$R \propto J^2 \phi e^{-E_d/kT} \tag{8.5}$$

where J is the current density. The MTF is commonly used as a measure of R, with which it is inversely related. Thus, a curve of $\log \{(MTF) J^2\}$ versus $1/T$ results in a straight line from which the value of E_d can be extracted.

Metal interconnections are polycrystalline in character, so that diffusion is primarily via grain boundaries. As a consequence, the value of E_d is well below that which would be obtained for single-crystal material [15], and is highly dependent on the grain size. Values of E_d are typically $0.4-0.5$ eV for deposited aluminum metal films which are commonly used as metallization in integrated circuits. The value for single crystal aluminum is 1.4 eV.

8.3 FILMS FOR PROTECTION AND MASKING

Insulating films are usually deposited on top of a semiconductor in order to provide protection to its surface, or to serve as masks through which selective diffusions and implants can be made. They also serve as the base for electrical connections between semiconductor devices in a microcircuit. Often, they provide the role of an insulator between two levels of metallization. In all these situations, it is highly desirable that they be free from pinholes and cracks, both when grown and when subjected to heat treatments during subsequent processing. Thus, their built-in stress, as well as the stress during thermal cycling, must be sufficiently low to maintain their integrity. These requirements become increasingly important in VLSI technology, as wafer size increases and devices become more densely packed.

Films that are useful in masking applications must be capable of preventing the transport of dopant species through them, in addition to maintaining their integrity at diffusion temperatures. Furthermore, they must be capable of being etched into fine-line patterns by photolithographic techniques. Often, they are left in place after providing the masking function. Thus, they should be either highly insulating if used as cover layers, or highly conducting if used in the subsequent metallization scheme.

Deposited films can be used for protection of microcircuits during manufacture, and also for improving their reliability in use. In addition, they can be used to block the movement of light alkali ions such as sodium, or else getter them so as to render them immobile. These films are usually placed over the metallization, to prevent its damage during handling. At the present time, their quality is sufficiently high so that microcircuits with protective films can be used without further packaging in many commercial applications.

8.3.1 Silicon Dioxide

Silica films can be grown by the pyrolytic oxidation of a variety of alkoxysilanes [16]; of these the most commonly used compound is tetraethylorthosilane (TEOS) which is a liquid at room temperature, and can be transported to the reaction chamber by means of a bubbler arrangement. The oxidation reaction is commonly carried by APCVD and LPCVD at 650–750°C, and proceeds as follows:

$$Si(C_2H_5O)_4 + 12O_2 = SiO_2 + 8CO_2 + 10H_2O \tag{8.6}$$

An alternative source, tetraethylcyclotetrasiloxane (TMCTS), can be used at about 100°C lower growth temperatures than TEOS. Both are popular because they have excellent step coverage, due to the formation of high-mobility reaction intermediates, known as *oligomers*, during the process of oxidation.

TEOS can be used for the growth of SiO_2 films at temperatures as low as 400°C, when oxygen containing 4% ozone is used [17]. High growth rates (≈ 1000 Å/min) are obtained by this process, with step coverage that is comparable to that which is obtained with high temperature growth in the absence of ozone.

Temperatures as low as 300°C can be used in PECVD systems. These films generally have a large amount of dissolved hydrogen incorporated in them. Their built-in stress can be made either tensile or compressive, depending on reactor and growth parameters. A compressive stress, in the range of 10^8–10^9 dyn cm^{-2}, is typical for these films.

Growth of silica films can also be carried out by the oxidation of silane [18]. Both N_2O and O_2 can be used as oxidants in APCVD and LPCVD systems operated at 450°C. These reactions proceed as follows:

$$SiH_4 + O_2 \rightarrow SiO_2 + 2H_2 \tag{8.7a}$$
$$SiH_4 + 2N_2O \rightarrow SiO_2 + 2N_2 + 2H_2 \tag{8.7b}$$

resulting in high-quality silica films with a built-in tensile stress of about 3×10^9 dyn cm^{-2} for a 450°C growth temperature. Growth proceeds by the strong adsorption of oxygen on the silicon surface, and its subsequent reaction with silane to form SiO_2. This can lead to a retardation in the growth rate when oxygen is present in high concentration [19].

Pure SiH_4 is a highly pyrophoric, unstable gas which burns on exposure to air. Consequently, it is commonly supplied in a low dilution (typically 2–10% by volume) in argon or nitrogen, in order to make it more safe to handle. LPCVD systems often require silane to be used in high concentration, because of the small amount (or complete absence) of diluent carrier gas. Provision must be made to store this gas in a bunker at some distance from the building in which it is used, with coaxial gas feed lines for additional safety.

The popularity of the silane process is due to its low deposition temperature and high growth rate, typically 500–1000 Å/min for APCVD. Consequently, these films can be used in many applications where rapid, low-temperature growth is essential. These include:

1. Thick field oxides for MOS microcircuits as well as for high-voltage devices. Films are usually deposited over a base layer of thermally grown oxide to avoid a high interface trap density.
2. Films where the previously grown layer has been removed—for example, deep diffusions of the type used for buried layers and isolation walls. These often utilize fresh masking and redoping, part-way through the diffusion process.
3. Insulating layers over a metallization film, to form a base for the next layer of metal.
4. Cover layers to protect the microcircuit from physical abuse during mounting and packaging.
5. Diffusion masks for GaAs. As shown in Chapter 7, the native oxide rapidly deteriorates at high temperatures ($>600°C$) and cannot be used in this application.
6. Cap layers for regions of silicon and GaAs substrates which must not be exposed during processing (the back side of a slice, for example).

LPCVD systems are also used for the growth of SiO_2, and result in films with a reduced pinhole density. However, the growth rate is significantly slower (100–150 Å/min) than for APCVD grown films. Silicon dioxide can also be grown by PECVD, by the reaction of SiH_4/O_2, SiH_4/CO_2 and SiH_4/N_2O mixtures [7]. The SiH_4/N_2O system can be operated at low temperatures (300–350°C) with growth rates of about 600 Å/min. About 3% nitrogen is incorporated in the films during this process.

PECVD results in a built-in compressive stress in the deposited silica films. This greatly reduces the tendency to cracking during subsequent thermal cycling, allowing the growth of thick films. Films grown by this technique are almost completely free of pinholes, so that they are suited for cap layers in VLSI applications.

Notwithstanding their high quality, deposited oxide films have a higher contamination level (and an associated trap density) than do native oxides. Often, they are grown over an initial thin (200–400 Å) native oxide to avoid direct contact with the silicon surface. This is especially true when coverage is required over lightly doped silicon, or over regions where a junction is exposed at the Si–SiO$_2$ interface. However, recent work with CVD-grown oxides has shown that it is possible to deposit extremely high-quality SiO$_2$ if sufficient care is taken with both substrate preparation and system cleanliness. Thus, they are sometimes used in situations which normally call for native oxide films.

8.3.2 Phosphosilicate and Borosilicate Glass

The addition of P$_2$O$_5$ to silica, to form phosphosilicate glass (PSG), results in many improvements in its physical properties. Films of this type can be grown by the simultaneous pyrolysis of trimethylphosphate (TMPO) and TEOS at 750°C in oxygen. The oxidation reaction for TMPO is

$$2(CH_3O)_3P + 10O_2 \rightarrow P_2O_5 + 6CO_2 + 9H_2O \qquad (8.9)$$

resulting in the formation of P$_2$O$_5$ which is incorporated as a network former in the resulting glass. The phase diagram for the P$_2$O$_5$–SiO$_2$ system is shown in Fig. 2.18.

PSG can also be grown by the simultaneous oxidation of silane and phosphine [19, 20], in the 350–450°C range. Phosphine gas is used for this application in a 5–10% dilution in argon or nitrogen. The phosphine oxidation reaction is as follows:

$$2PH_3 + 4O_2 \rightarrow P_2O_5 + 3H_2O \qquad (8.10)$$

so that water is produced as a byproduct of this process. Here too, growth of the film proceeds by the strong adsorption of oxygen on its surface, so that retardation effects become dominant at high oxygen concentrations. Film properties depend on the specifics of the CVD process which is used for growth [21].

Almost any amount of P$_2$O$_5$ can be incorporated by this technique. However, the films become increasingly hygroscopic, so that their P$_2$O$_5$ content is limited to about 4 wt.% for films which are left permanently in place on the finished product, although as much as 20 wt.% can be tolerated in films which are used temporarily during device processing.

The incorporation of P$_2$O$_5$ into silica films causes a reduction in the built-in tensile stress from 3×10^9 dyn cm^{-2} for the undoped film, to zero for films with 20 wt.% P$_2$O$_5$. This improves their integrity during thermal cycling. A further advantage comes about because the TCE of silica rapidly increases with the incorporation of P$_2$O$_5$. Consequently, PSG films can be tailored to provide a more suitable thermal match to the underlying semiconductor. By way of example, silica films, deposited on silicon, will crack upon thermal cycling to 1200°C if they are thicker than 1.5 μm. Silica films, deposited on GaAs, will break this semiconductor upon thermal cycling to 800°C, if they are thicker than 2000 Å. In both situations, PSG films can be used to avoid this problem.

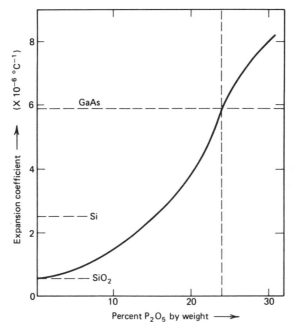

Fig. 8.9 Thermal expansion coefficient of PSG films. From Baliga and Ghandhi [22]. ©1974 The Institute of Electrical and Electronic Engineers.

The TCE of PSG is shown in Fig. 8.9 as a function of P_2O_5 content, over the composition range of practical interest [22]. Note that the expansion coefficient* of GaAs ($5.9 \times 10^{-6}/°C$) can be matched at a P_2O_5 concentration of about 24 wt.%. Films with 15 wt.% P_2O_5 have been used in layers as thick as 6000 Å, with no tendency towards cracking when cycled to temperatures as high as 1000°C.

PSG is more dense and void-free than silica, so that it can be used as a mask to such dopants as zinc and tin which are used for diffusion in GaAs [22, 23], whereas silica films are relatively transparent to these dopants. For the same reasons, PSG films make better cap layers for GaAs than do silica films in this application as well.

PSG layers have been found to be effective in immobilizing sodium ions in MOS technology (see Section 7.1.9.2) and are commonly used in this important application area. They are also used to enhance the stability of bipolar devices and microcircuits [24]. Often these layers are automatically present, as occurs with n–p–n transistors, where the last process is a masked phosphorus diffusion. In those cases, however, when the last diffusion is not phosphorus (p–n–p tran-

---

*Values for undoped silica ($6 \times 10^{-7}/°C$) and for silicon ($2.6 \times 10^{-6}/°C$) are also shown in this figure.

sistors, for example), it is customary to strip the surface oxide which consists of B_2O_3–SiO_2, and grow a PSG film prior to packaging.

PSG layers are often used as coatings on finished microcircuits, to protect the aluminum metallization from scratches during the final bonding operation and to provide a permanent protection against alkali ion migration. These passivation layers must contain no more than 4 wt.% of P_2O_5 to prevent corrosion reactions with the aluminum metallization in the presence of moisture [25].

PSG layers can also be used to improve the step coverage of metal films which are deposited by vacuum evaporation on patterned, oxide-coated substrates, as part of the fabrication process. One approach to improving this step coverage is to round off the sharp steps in the patterned oxide by heat treatment. The temperature at which this *reflow process* is carried out is related to its softening point, which occurs at about two-thirds of the melting point (both temperatures in Kelvin). The incorporation of P_2O_5 into the oxide can thus reduce the temperature at which this process can be carried out [26]. However, this approach has limited effectiveness, because of the need to hold the P_2O_5 content to a maximum of 4 wt.%.

B_2O_3 can be used instead of P_2O_5 to avoid this problem by the formation of borosilicate glass (BSG). Moreover, B_2O_3 is even more effective in reflow processes, on a wt.% basis, than P_2O_5. Its disadvantage lies in the fact that it is not a good barrier to alkali ions, because of the trivalent character of the network former.

8.3.2.1   Borophosphosilicate Glass

The incorporation of both B_2O_3 and P_2O_5 into SiO_2, to form borophosphosilicate glass (BPSG), allows a reduction in the reflow temperature with a much lower P_2O_5 concentration than is required for PSG alone. Moreover, it greatly improves the ability of the film to block the transport of alkali ions [27]. Reflow data for BPSG is shown in Fig. 8.10, where ϕ is the angle of a step (originally 90°) after a 30-min anneal in a steam ambient. This figure indicates the improvement which results from adding increasing amounts of B_2O_3 to PSG with a 5 wt.% P. Similar results were obtained for annealing in dry nitrogen; however, a 70°C higher anneal temperatures is required when reflow is carried out in this ambient [28]. Typically, reflow temperatures for BPSG films are around 100°C lower than for PSG films, for the same reflow angle.

BSPG films have been grown by all CVD methods, using the co-oxidation of SiH_4, B_2H_6, and PH_3 with O_2 and N_2O, in a nitrogen carrier gas. Typical growth temperatures for these films are 350–400°C for films grown by PECVD, and 400–450°C for films grown in the absence of plasma enhancement. Details of the deposition kinetics for these different situations are provided elsewhere [29].

Finally, BPSG films have been grown by LPCVD, using the oxidation of TEOS, TMB, and trimethylphosphite (TMP), in the 525–680°C temperature

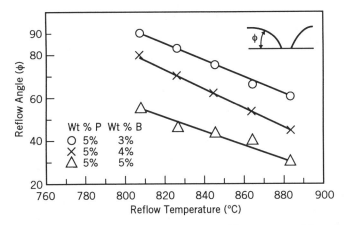

Fig. 8.10 Reflow angle versus reflow temperature, with a steam ambient. From Tong et al. [28]. Reprinted with permission from *Solid State Technology*.

range. As-grown films were highly comformable, and had excellent reflow characteristics at temperatures as low as 850°C.

8.3.3 Silicon Nitride

Silicon nitride films are extensively used in both silicon and GaAs device technology. Thermally grown films require a higher deposition temperature than do films of SiO_2 or doped SiO_2, so that their use is dictated in situations where they provide improved properties over silica. Often, they are used in conjunction with SiO_2 films to obtain a combination of characteristics which neither can provide alone. PECVD films, on the other hand, are grown at temperatures which are comparable to those for SiO_2.

Silicon nitride, in its stoichiometric form, has a composition given by Si_3N_4. Considerable departure from stoichiometry is often encountered in deposited films, with Si/N atom ratios from 0.7 to 1.1 being commonly encountered. They are sometimes referred to as SiN films for this reason. In general, their properties (density, resistance to ion migration, etc.) improve as they become more stoichiometric, and also when they have less built-in stress.

Unlike SiO_2, Si_3N_4 is an excellent barrier to alkali ion migration, and is used extensively as a cover layer in MOS technology for this reason. It is superior to PSG since it is not hygroscopic. Its use has allowed unencapsulated circuits to be practical in many consumer applications. Although impervious to alkali ion migration, it is poorer than SiO_2 in its ability to block electron movement through it. Consequently, it is often used in conjunction with SiO_2 films to provide a barrier to both electron and ion transport.

Films of Si_3N_4 are also excellent diffusion masks for gallium, and are used for making junctions with this dopant in silicon power device applications. This ability to restrict the diffusion of gallium results in their use as a capping material for GaAs during the high-temperature (900°C) anneal process which must be carried out after ion implantation. This has allowed the development [31, 32] of a highly successful self-aligned technology for both GaAs devices and microcircuits.

Si_3N_4 is about 100 times more resistant to thermal oxidation than silicon, so that it can be used as a mask in a number of silicon-based VLSI schemes which require selective oxidation of the semiconductor.* Finally, it is used in radiation-hardened devices because of its superior radiation resistance over SiO_2.

Amorphous films of Si_3N_4 can be deposited by r.f. sputtering using a silicon target with a nitrogen discharge [33]. Direct sputtering from an Si_3N_4 target with an argon/nitrogen background can also be accomplished. However, the most commonly used process is CVD involving the reaction of SiH_4 and NH_3, with N_2 as the diluent [34]. This reaction is usually carried out at 700–900°C, and proceeds along the following lines:

$$3SiH_4 + 4NH_3 \rightarrow Si_3N_4 + 12H_2 \tag{8.11}$$

Both hot- and cold-wall systems can be used, with deposition at atmospheric as well as at reduced pressures. In all cases, film composition and properties are controlled by the ratio NH_3/SiH_4 in the gas stream. Typical mole fraction ratios of 150 or higher are used for hot-wall systems, with growth rates in the 100- to 200-Å/min range.

The present trend in Si_3N_4 film deposition for large commercial applications is to use hot-wall, LPCVD systems [35], which allow close stacking of slices while still providing uniform coverage. Because of the stability of the NH_3 molecule, gas-phase reactions rarely occur. As a result, dusting in these systems is less of a problem than for the LPCVD of SiO_2 films. Operation at low pressure is achieved by greatly reducing or eliminating the use of carrier gases, so that these systems can be run with reactant partial pressures (and growth rates) which are comparable to those achieved with atmospheric systems.

PECVD systems can also be used [36, 37] for the growth of Si_3N_4. Here, the deposition reaction can be conducted at relatively low thermal temperatures (275–325°C), because of the high electron temperature ($\approx 10,000$ K) associated with the plasma. This allows films to be grown directly on finished microcircuits. Growth rates for these systems are in the 200 Å/min range.

Plasma-enhanced Si_3N_4 films have large quantities of hydrogen incorporated in them, with the H/Si atom ratio often exceeding unity. This greatly affects properties such as their etch rate. Growth at elevated temperatures (300–400°C) results in a reduction in this ratio to about 0.5, with a resulting improvement in film properties. Another approach is to densify the grown film by heat treatment

*This important technology, known as the LOCOS process, is described in Chapter 11.

at 700°C. Unfortunately, this step obviates the initial advantage of low-temperature growth for this approach, unless RTA is used for this purpose.

In Table 8.1 at the end of this chapter, the relative properties of Si_3N_4 films grown at 900°C by LPCVD are compared to those grown at 300°C by PECVD [38]. Here, we note many similarities in their intensive parameters. However, their built-in stress and their wet chemical etch properties are strikingly different. LPCVD films of Si_3N_4 have a large amount of built-in tensile stress, and are usually grown in thicknesses below 1000 Å to avoid breakage or peeling, or damage to the underlying semiconductor. This property has been used to advantage in the fabrication of large area silicon devices [39]. Here, a thick layer of Si_3N_4 is deposited on the back surface of the slice and results in the formation of a dense dislocation network. Gettering of impurities by these dislocations (i.e., damage gettering) occurs during subsequent high-temperature processing of microcircuits which are located on the opposite face of the slice.

The stress in PECVD films is generally lower and can be either tensile or compressive, depending on both (a) the plasma power that is used and (b) the post-deposition annealing treatment. Relatively thick (0.5–1 μm) films can be grown with a compressive stress, with excellent adhesion to the semiconductor surface.

As seen from Table 8.1, LPCVD films etch much slower in HF than do PECVD films. Boiling phosphoric acid is often used to selectively etch these films in the presence of silica layers. Details of this etch process are given in Chapter 9.

8.3.3.1 Silicon Oxynitride

Films of silicon oxynitride can also be grown, and have physical properties that are intermediate between SiO_2 and Si_3N_4 [40]. Growth is carried out at 800–900°C in a Si_3N_4 deposition system, with the introduction of NO (or N_2O) in addition to NH_3 gas. Plasma-enhanced growth [41] can also be used at 200–250°C. The growth reaction results in the partial oxidation of the SiH_4, and produces films which are essentially glassy mixtures of silicon, oxygen, and nitrogen. Typically, the introduction of about 10% of NO results in films with $SiO_2 : Si_3N_4$ in a 1 : 1 mixture. The etch rate (in buffered HF) of these films is about 35 times that for silicon nitride.

There are some indications that films of this type may be more suited as passivating layers for GaAs than either SiO_2 or Si_3N_4. In addition, it should be noted that the initial 20- to 30-Å layer of silicon nitride on either a silicon or GaAs surface has most probably a graded silicon oxynitride composition, because of the native oxide that is always present on these semiconductors.

8.3.4 Self-Aligned Masks

Refractory materials, capable of withstanding diffusion and post-implant annealing temperatures, are often used for masking purposes in microcircuit fabrication. In some cases they provide unique advantages, as in MOS technology. By way of example, a major factor that limits the high-frequency gain of MOS transistors has been the parasitic capacitance caused by overlapping of the gate and drain electrodes. This overlap is necessary to accommodate the tolerances in mask alignment during device fabrication; some overlap is essential for MOS device operation.

Consider the MOS transistor, shown in schematic form in Fig. 8.11a. Here, the overlap L is governed by the gate length, the distance between the oxide cuts for the source and drain, and the lateral diffusion of these regions. All of these must be carefully controlled if this distance is to be kept to a minimum. This necessitates that the source and drain diffusions be made quite deep ($\approx 2\,\mu$m) in this structure, to ensure that there will be a finite overlap due to lateral diffusion effects. This increases the parasitic drain capacitance with a further deterioration in device performance.

These problems can be alleviated if the gate material can be used as a mask during the source and drain implantations and/or diffusions. The use of a refractory material of this type ensures that there *always* will be some overlap between the gate and these regions, regardless of the amount of misalignment or lateral diffusion. Extremely shallow depths ($\approx 0.2\,\mu$m) can be used for the source and drain regions, with a concurrent reduction in the overlap. In addition, these junctions have lower parasitic capacitances and result in smaller area devices, which produce additional speed advantages.

Another advantage of using the gate material as a mask is that the overlap is not dependent on its accurate placement with respect to the field oxide. This is shown in Fig. 8.11b, where a grossly displaced mask still yields a satisfactory device. Masks of this type are referred to as *self-aligned* for this reason.

Refractory gate technology is ideally suited for VLSI structures, and can be readily implemented by a number of different materials. Polycrystalline silicon films are extensively used for this purpose, because they can be formed with a high degree of purity; in addition, all the processes developed for silicon technology can be used with them. Doping of these films allows control of their fermi level, and hence control of the MOS threshold voltage. In addition, both *p*- and *n*-type doping can be carried out simultaneously in the same microcircuit. A number of silicide films, with sheet resistances between that of polysilicon and metals, are also used for this purpose. A unique characteristic of both polysilicon and silicide films is that a native silica layer can be subsequently *grown* on their surface, allowing considerably simplification in forming an upper layer of metallization.

Refractory gates, which can also serve as masks for ion implantation into GaAs and as caps during the anneal step, have allowed self-aligned gate (SAG) technology to be used for the fabrication of high-performance Schottky gate FET devices. Here, SI GaAs is implanted with silicon to form the channel region. Next, a refractory gate is patterned and used as a mask for implantation of the source and drain regions. Typically, tungsten and titanium and their silicides are used for this purpose, since they can withstand the annealing tem-

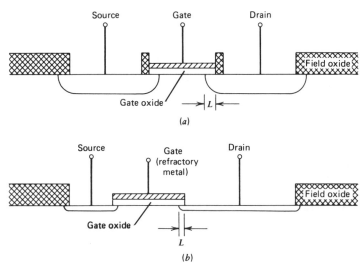

Fig. 8.11 Comparison of masking techniques. (*a*) Conventional and (*b*) self-Aligned.

peratures which must be used after the implantation step. Details of this and other SAG techniques will be provided in Chapter 11.

This section will describe and compare these approaches and will outline the considerations involved in the selection of this masking technology.

8.3.4.1 *Polysilicon*

Polycrystalline silicon (polysilicon) films can be formed by sputtering or by evaporation. However, CVD methods have the advantage of uniform deposition over oxide steps, and thus are most commonly used. These films can be grown in an atmospheric pressure, cold-wall reactor, or by LPCVD in a hot-wall system by the decomposition of SiH_4, over a wide range of temperatures. Both growth rate and crystallite size increase with increasing growth temperature. Films grown at 575°C are amorphous in character; however, reasonable growth rates are obtained at 650–700°C and result in mirror finish, fine grained polycrystalline material.

LPCVD is the most common approach for the growth of these films [42]. Growth is carried out in the 625–650°C range at system pressures from 0.25 to 1 torr, to allow uniform growth (±2% across a slice) with multiple slice capability (as many as 200 in a single load). Silane gas, in a dilution of 5–20%, is used for reasons of safety; however, 100% SiH_4 is used in some systems.

Growth rates in these systems is kinetically controlled [43], with an activation energy of 40 kcal/mole. The yield of this reaction approaches the maximum thermodynamic limit for growth at 700°C and higher, but falls off rapidly at lower temperatures. This is probably due to the competitive adsorption of reaction products (SiH_2 and H_2) which are formed as intermediates during growth (see Section 5.4.1). Precise temperature control, which is readily obtained in a hot-wall system, is thus extremely desirable. Often, the temperature is adjusted to gradually increase from the front end towards the exit, in order to compensate for reactant depletion effects down the reaction tube [44].

Films grown in undoped form have a resistivity in excess of 500 Ω-cm, and thus are essentially semi-insulating ($\approx 5 \times 10^6 \, \Omega$/square for a 1-$\mu$m-thick film). These can be used directly to provide resistive and field shaping layers as well as surface passivation in high-voltage devices. However, they must be doped for use as gates in MOS transistors. Doping can be carried out in the gas phase, by means of B_2H_6, AsH_3, or PH_3 which can be introduced in dilute form (200–300 ppm) in a hydrogen carrier gas. Use of these dopants alters both the resistivity and the growth rate. Results obtained with them, for a typical set of experimental conditions [45], are now summarized:

1. Introduction of B_2H_6 causes a monotonic increase in the growth rate, which doubles as the B/Si atom ratio increases from zero to 2.5×10^{-3}. At the same time, the resistivity falls from 500 Ω-cm to about 0.005 Ω-cm. Both the growth rate and the resistivity will level off for doping ratios in excess of this value. Films remain mirror smooth until the B/Si atom ratio exceeds 3.5×10^{-3}, at which point the signs of surface roughness become apparent.

2. The introduction of AsH_3 during polysilicon growth results in a precipitous fall in growth rate (by a factor of 70), along with a drop in resistivity from 500 to 0.01 Ω-cm, as the As/Si atom ratio is increased from zero to 2.5×10^{-5}. Both the growth rate and the resistivity will level off beyond this point.

3. Results with PH_3 doping are qualitatively similar to those for AsH_3, with a rapid fall in growth rate by a factor of 25. This is accompanied by a fall in resistivity from 500 to 0.02 Ω-cm. Both resistivity and growth rate will level off beyond this point, and will be unchanged with further doping. The deposition uniformity (which is typically ±2% in undoped films) deteriorates [46], with the growth rate in the perimeter regions being twice as large as that in the center. It has been proposed that the PH_3 suppresses the heterogeneous decomposition of the SiH_4, so that growth proceeds in the mass transport regime. Similar arguments can be suggested for AsH_3 doping as well.

The use of polysilicon in diffusion processes results in its becoming doped by the impurity which it is masking. From the above results, however, it is seen that the polycrystalline nature of these films makes it almost impossible to control their doping concentration to any extent. This is not a disadvantage, since these films are subsequently used for the gate material, where the actual sheet resistance is unimportant, as long as it is very low. Consequently, films of this type are intentionally used in heavily doped p^+- or n^+-form. Heavy boron incorporation can be satisfactorily obtained by gaseous doping. Sufficient n-type doping cannot be achieved by this method, as outlined earlier; moreover, it results in greatly reduced growth rates.

A practical alternative for n^+-doping is to grow an undoped film, use it as a diffusion mask (during which process it becomes doped), and subsequently dope it further by a predeposition process of the type used in diffusion technology [47]. Typically, $POCl_3$ is used for this purpose, with the polysilicon maintained at 900–1000°C for 30–60 min. This results in sheet resistances of about 10 Ω/square (for a 0.5-μm-thick film), whereas the sheet resistance is about 75 Ω/square if this second doping step is omitted. It should be noted that this is still an order of magnitude higher than what would be obtained for a $POCl_3$ predeposition into single-crystal silicon for the same time and temperature. The reasons for this difference are primarily due to the nature of conduction processes in polycrystalline materials. Here, conduction between grains is dominated by the trap density at each grain boundary, which results in the creation of an energy barrier [48]. This barrier can be quite narrow under conditions of heavy doping, so that tunneling through it is possible. Still, it serves to limit the effective conductivity of the material to a value below that obtained with single crystal material.

The conductivity of polysilicon films can be further increased by performing the predeposition at a higher temperature ($\simeq$1150°C), or by using growth techniques which result in larger grain sizes. Unfortunately, wet chemical etching of these films becomes difficult to control, and results in ragged edges due to the uneven etching of the grains and the grain boundaries. Increasingly, dry etching processes, to be described in Chapter 9, are used for this reason. Finally, films can also be doped by ion implantation; after annealing at 600–700°C, they have a sheet resistance of about 10^2 Ω/square.

A unique advantage of polycrystalline silicon is that a native oxide can be grown on its surface [49], and used as an insulating layer for crossovers. The quality of thermally grown native oxides is dependent on film texture and is somewhat poor, with a breakdown electric field strength of 6 MV/cm, as compared to 10 MV/cm for high-quality native oxides on single-crystal silicon. However, anodization of the polysilicon results in the formation of native oxides with breakdown characteristics which are comparable to those of native oxides on single-crystal material [50].

Polysilicon films can be partially oxidized to render them semi-insulating by the addition of controlled amounts of N_2O during deposition from SiH_4 at 650–700°C [51], with nitrogen as a carrier gas. These films can be used for the passivation of high-voltage silicon devices [52], where they have significant advantage over grown silica films. Partially nitrided polysilicon films can also be grown by the addition of NH_3. These have been used as cover layers because of their ability to inhibit the transport of both sodium ions and water.

8.3.4.2 Silicides

Although polysilicon is well suited for self-aligned masking technology, it suffers from the disadvantage that, even when heavily doped, it has a significantly high resistivity of about 500 $\mu\Omega$-cm. With the trend to finer lines, this results in an increasingly large resistive contribution to the RC delay associated with active devices. Consequently, alternative approaches are required for VLSI technology.

The transition metal silicides are immediately attractive, since many of these have resistivities that are ten or more times lower than that of heavily doped polysilicon. As a result, it is possible to define much thinner lines in these materials without paying a speed penalty in the operating device. In practice, a *polycide* structure, consisting of a silicide film on top of doped polysilicon, is commonly used. This takes advantage of the high conductivity of the silicide and the high purity of the underlying polysilicon. Finally, the resistivity of many silicides is sufficiently low so that it is possible to consider their use for interconnections as well, resulting in an overall simplification of the technology.

A number of silicides are possible for each combination or silicon and transition metal. For example, molybdenum and silicon can react to form Mo_3Si, Mo_5Si_3, and $MoSi_2$. From energetic considerations, however, the formation of monosilicides and disilicides has been shown to be more favorable than the more complex compounds [53]. The optimum choice, however, must include other considerations as well. Thus, a useful silicide should be stable on silica and in oxidizing ambients. It should not be so highly stressed as to crack, or cause cracking of the underlying material, during thermal processing. It should be capable of being etched into fine line patterns by readily available techniques. It should be capable of oxidation, so as to provide an insulating layer over which the next level of metallization can be placed. Finally, it should be capable of being joined to the contact metal (aluminum or gold) which connects to the package terminations.

The resistivity of a silicide film is related to the manner in which it is prepared, and to its subsequent heat treatment. In general, films which are free from contamination, and of stoichiometric composition, have the lowest resistivity. Annealing [54] at high temperatures for 30–40 min reduces the resistivity, as seen in Fig. 8.12 for $MoSi_2$, $TaSi_2$, and $TiSi_2$ films. Table 8.2 , at the end of this chapter, lists a number of silicides which are highly conducting, and whose lowest binary eutectic temperature is sufficiently high [55] for device processing.

The simplest method for silicide formation is to deposit a metal on bare silicon and to heat treat this combination. These silicides usually exhibit the lowest electrical resistivities. There are, however, significant problems with this approach. Thus, silicon is consumed in the process, and the resulting surfaces tend to be rough. Moreover, the process is extremely sensitive to preparation of the silicon surface, prior to metal deposition. This technique is restricted to contact formation for these reasons, and will be described in Section 8.6.

The preferred approach to silicide preparation involves its direct deposition as a compound, or co-deposition of the metal and silicon, which are converted to the silicide by subsequent heat treatment. Of the many deposition techniques described earlier, sputtering has been found to produce films which combine

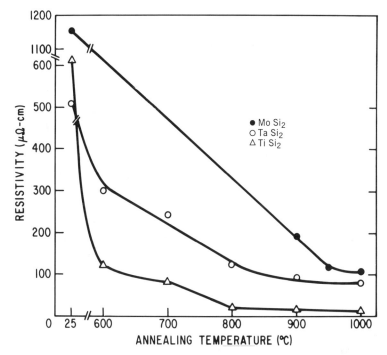

Fig. 8.12 Resistivity versus annealing temperature for MoSi$_2$, TaSi$_2$ and TiSi$_2$. Adapted from [54].

most of the characteristics which are desired in silicides. This process can be carried out by using a single hot-pressed silicide target, or by co-sputtering with separate targets for the metal and the silicon, followed by sintering. The latter approach is commonly used, because it allows experimental adjustment of the metal–silicon composition to give the lowest specific resistivity for the silicide. Thermal conversion of a co-sputtered layer to the silicide can be achieved by heat treatment at 900–1000°C for 10–30 min. Alternatively, RTA methods have been successfully applied at 1200°C for 90–120 s, for this purpose. The general characteristics of silicides, grown by these techniques, are listed [56] in Table 8.3 at the end of this chapter.

Silicide formation can also be accomplished by the *sequential* deposition of the separate components, followed by their thermal conversion. This approach allows patterning of these films (which can often be readily wet etched) prior to conversion to the considerably more inert silicide form. However, etching of silicides is usually done by reactive ion milling, or by plasma methods, so that this is not a significant advantage. On the other hand, films made by sequential deposition are extremely hard to convert to the silicide, and require heat treatments for long times and at elevated temperatures.

CVD techniques have also been explored for the deposition of silicides [57], because of their ability to provide excellent step coverage. These techniques are based on the reaction of a silicon bearing species (SiH_4 or SiH_2Cl_2) with a metal-bearing species (e.g., $TiCl_4$, $MoCl_5$, WF_6, etc). A problem that is common to all of these reactions [58] is the formation of mixed phase silicides. Thus, the reaction of WF_6 and SiH_4 can be used in the 350–450°C range for the LPCVD of tungsten silicide, and results in the formation of W_5Si_3 and WSi_2, in addition to tungsten and silicon. Multiphase film growth has also been observed [59] for the CVD of tungsten silicide films by the dichlorosilane reduction of WF_6.

The control of stress in a silicide film is important, if it is to maintain its integrity during subsequent thermal processing. Typically, formation from its components results in a significant volume decrease, which should lead to a large tensile stress in the film. At growth temperatures, however, much of this stress is relaxed. Thus, the residual stress in the film is primarily created during the cool-down phase, and is a function of the TCEs of the silicide and the silicon. As seen in Table 8.2, the TCE of most silicides is considerably larger than that of silicon, so that these films are usually in tension upon cooling. As expected, the tensile stress is lower when silicides are formed at reduced temperatures. Typical values of stress in these films are around 1×10^{10} dyn cm^{-2}.

Stress in a silicide film is also a function of the ambient in which it is annealed. Both oxygen and nitrogen ambients, for example, react with the silicide and cause volume expansion, resulting in compressive stress. Finally, it has been demonstrated that films which are slightly silicon-rich tend to have lower stress than those which are silicon-deficient. This is true for monosilicides as well as for disilicides.

An important requirement of silicide films is a convenient means for covering them with an insulating film over which the second metal layer can be run. This can be done by CVD techniques described earlier, or by direct oxidation of the silicide. Silicides such as $TiSi_2$ oxidize congruently, so that a stable film can be formed by the oxidation of its components. Generally, however, the silicides have varying oxidation rates for their separate components, so that this process results in their break-up or in the formation of cover films which are nonstoichiometric and often leaky.

The oxidation properties of silicide films which are formed on a silicon base layer are uniquely different from those described above [60]. With rare exceptions, oxidation results in the production of SiO_2 rather than the metal oxide. Moreover, the morphology and the silicide film thickness are unchanged, although the underlying silicon base layer is consumed during this process.

Two different models have been proposed to explain this behavior [61]. The first postulates that oxidation takes place by the diffusion of silicon (from the base) *through* the silicide, followed by conversion of this silicon to SiO_2, during which process the silicide layer remains essentially unchanged. A second theory is that the silicide itself is converted to SiO_2 at the silicide–SiO_2 interface. This liberates the metal which is driven through the silicide to the Si–silicide interface, where it reacts with silicon to replenish the silicide. Combinations of these models have also been proposed. In either case the driving force for this diffusional process appears to be the free energy of formation of SiO_2 (-192 kcal/mole). The diffusion process is rapid, since both silicon and transition metals are fast diffusers in their silicides.

The actual species whose movement dominates the oxidation process depends on the silicide phase. Thus, silicon is the dominant diffuser in the oxidation of CoSi, whereas metal diffusion dominates with $CoSi_2$, as well as for $CrSi_2$, NiSi, $NiSi_2$, PdSi, and PtSi [62]. Characteristic of silicides in which metal diffusion dominates the oxidation process is a lower oxidation temperature and smoother final surface morphology than is obtained with silicides in which silicon movement is dominant. Thus, these silicides are the preferred choice for VLSI, where pattern retention and linewidth control are of prime importance.

A typical process sequence for a WSi_2 gate is as follows: A 2500-Å mixture of tungsten and silicon, in a $1:2$ atom ratio, is co-deposited on a 1500-Å layer of n^+-polysilicon. The silicide is formed by heat treatment at 900–1100°C in an inert gas ambient for 10–30 min, resulting in the formation of a layer with a sheet resistance of 15 Ω/square. Oxidation is next performed in wet H_2O (95°C) and proceeds at approximately the same rate as for conventional heavily doped silicon. This results in a silicide film, with approximately 1000 Å of SiO_2 over its surface. The sheet resistance of this silicide film is unchanged from that of the original silicide layer, as long as the underlying polysilicon is not fully depleted by this process.

Silicide films are invariably deposited on doped polysilicon gates. As a consequence, dopant redistribution during silicide formation and subsequent heat treatment is an important area where much work needs to be done before this technology becomes fully exploited [63]. The role of dopants in promoting (or retarding) the homogenization of the silicide, as well as the use of inert ion bombardment [64] for this purpose, is an additional area for extended research.

8.3.4.3 Refractory Metals

Table 8.2 shows the melting points and resistivities for refractory metals which are potentially useful in self-aligned masking technology. The choice is not wide, since the specific resistivity of some of these materials (Hf, V, Ti, Zr) is almost as large as for their silicides. In addition, it is somewhat difficult to achieve the quoted resistivities in thin film form if the metal is highly reactive (such as chromium, for example), because of oxygen contamination during the deposition step. Most work at the present time has been centered around the use of molybdenum and tungsten since their resistivities are considerably lower than that of the silicides. Additionally, they can be used as interconnections in microcircuit fabrication.

Both molybdenum and tungsten can be deposited by e-beam evaporation or by sputtering methods. They do not enter into any significant reaction with SiO_2, so that they adhere somewhat poorly to layers of this material. Molybdenum is superior to tungsten in this respect [65, 66]; however, its adhesion can be significantly increased if it is deposited on heated substrates (200–300°C) or if a barrier "glue" layer is placed between it and the SiO_2. Films of molybdenum and tungsten, grown by CVD, have excellent step coverage, and good adhesion to SiO_2. In some cases, a "glue" layer of TiN, also deposited by CVD, is used to make them even more adherent.

The growth of molybdenum can be accomplished by the hydrogen reduction of molybdenum hexafluoride (MoF_6) [67] or molybdenum pentachloride ($MoCl_5$) [68], at around 600°C. Step coverage is generally better with the $MoCl_5$ process; additionally, its use results in HCl as a reaction product, and thus avoids the problems of fluoride pitting of the quartz reaction chamber when MoF_6 is used. Molybdenum can also be formed by the decomposition of molybdenum carbonyl [$Mo(Co)_6$]. This reaction usually results in films with as much as 30% carbon incorporated in them. These are generally not useful in device technology.

The CVD of tungsten can be accomplished by the hydrogen or silane reduction of WF_6, at a temperature around 400–800°C. Both hot- and cold-wall systems are used for this purpose. The reactions proceed along the following lines:

$$WF_6 + 3H_2 \rightarrow W + 6HF \tag{8.12a}$$

$$2WF_6 + 3SiH_4 \rightarrow 2W + 3SiF_4 + 6H_2 \tag{8.12b}$$

PECVD can also be used for this purpose. However, the resistivity of films grown by this technique is somewhat higher than that of CVD-grown tungsten.

Considerable interest in tungsten has resulted from the fact that it can be deposited selectively on materials such as silicon, silicides, and metals. This allows its use in filling via holes in microcircuits, thus improving their planarity. Selective deposition of tungsten on exposed silicon surfaces can be achieved at 300–400°C [69], by the following reaction:

$$2WF_6 + 3Si \rightarrow 2W + 3SiF_4 \tag{8.13}$$

In addition to filling via holes, this reaction can be used to selectively deposit tungsten into holes cut in a masked silicon surface, and provides a simple technique for the formation of W–Si ohmic contacts [70]. Tungsten can also be selectively deposited over polysilicon gates, thus reducing their resistivity. Typically, gate resistances as low as 1 Ω/square can be obtained in this manner. However, it is necessary to use spacers to prevent encroachment of the film under the oxide, since this can result in leaky junctions.

The reactions of Eq. (8.13) are self-limiting, since the exposed silicon surface gets covered up during this process. Consequently, a two-step technique must be used, with selective deposition on bare silicon to a thickness of about 100 Å using an argon carrier gas, followed by growth [71] along the lines of Eq. (8.12a) using H_2 as the carrier gas. Low-pressure CVD is used at this point to promote selective growth of tungsten on the initial deposit. As can be expected, the process is very sensitive to surface conditions, necessitating elaborate clean-up procedures prior to deposition.

8.3.4.4 Barrier Layers

A problem, common to all silicides, is that the diffusivity of silicon through them is quite rapid. Thus, heat treatment of a metal–silicide-silicon system results in interdiffusion of the metal and silicon through the silicide [72], and eventual loss of its integrity. Measurements of aluminum/silicide/silicon systems, using equally thick silicide layers of Pd_2Si, $PtSi$, $NiSi$, and $TiSi_2$, showed that system integrity was destroyed by 200°C, 400°C and 550°C, respectively. As a result, it is usually necessary to place a barrier film between the silicide and the metallization layer [73]. In order to be effective, this barrier should be adherent to both the silicide and the metal and should prevent their diffusive mixing. In addition, it should have low electrical resistivity, and must be compatible with the fabrication process.

Metal alloys can be used for this function. Of these, an alloy of titanium and tungsten, with 10 wt.% Ti (30 at.% Ti), is commonly used and has a resistivity of 50–80 $\mu\Omega$-cm. Here, the barrier function is primarily due to the tungsten with the titanium providing the "glue" function. This alloy is usually prepared by single-target Ti–W alloy sputtering, which is compatible with the formation of the silicide which is used for the contact. Typical barrier thicknesses range from 450 Å to 1800 Å, and can withstand [74] thermal processing for 30 min at temperatures from 400° to 500°C, respectively.

Metal nitrides, carbides, and borides have also been studied for use as barrier materials; most current work is with titanium nitride, which has been found to be significantly superior to Ti–W alloys.

Titanium nitride can exist as Ti_2N and TiN. Moreover, nitrogen can form a solid solution with these compounds [75], so that wide departures from their stoichiometric compositions can result from variations in its preparative process. Stoichiometric TiN has excellent barrier properties and low resistivity (50–100 Ω-cm). It is usually prepared by the reactive evaporation or sputtering of titanium in an N_2–Ar mixture [76], or by LPCVD [77], using organometallic source chemicals. The preferred technique is sputtering, since it allows the growth of the ohmic contact material (the silicide) in the same system, without breaking vacuum. Layers of this type, 700 Å in thickness, have been used in Al–TiN–Si structures and can withstand anneals of 20-min duration, at temperatures as high as 550°C. Temperatures as high as 600°C can be used in ohmic contact systems of the type described in Section 8.6.3, where a metal silicide is used as the contact layer.

Tungsten nitride films can be deposited by reactive sputtering of W in N_2–Ar mixtures [78]. These films are of $W_{1-x}N_x$ with $0 < x < 0.25$, and are excellent diffusion barriers between GaAs and metals such as gold, silver, or aluminum. They can be subjected to 550° heat treatments without degradation, and have been used in both contact and gate technology in silicon and GaAs microcircuits.

8.4 FILMS FOR DOPING

Deposited films, incorporating dopant impurities, can be used in microcircuit technology. These generally consists of small quantities of the dopant oxide in an inert binder such as silica, and are formed by CVD techniques. The simplest approach here involves the simultaneous oxidation of silane and a dopant hydride or alkyl, in a process that is similar to that used for the deposition of silica films.

The use of doped oxides has many advantages over conventional predeposition methods. First, the surface concentration obtained by making diffusions from a doped oxide is a function of the concentration of the doping source, and also of the relative diffusion rates of the dopant in the oxide and in the semiconductor [79, 80]. Thus, use of these sources allows high diffusion temperatures and relatively low surface concentrations at the same time. As shown in Section 4.3.1.3, this is only possible by means of a two-step process when conventional sources are used.

A second advantage is that ultra-shallow diffusions can be made using these sources; furthermore, they can be carried out with closely spaced slices in an inert gas ambient. Thus, the diffusion equipment involves no complex gas handling or effluent disposal features, and is extremely simple.

Finally, the doped oxide serves as a capping layer during the diffusion. This is especially significant for GaAs diffusion technology, where the widely disparate partial pressures of gallium and arsenic lead to decomposition of this semiconductor at diffusion temperatures. The conventional approach necessitates the use of sealed ampoules in which a carefully controlled partial pressure of arsenic is established to prevent this decomposition; close control of the dopant partial pressure is necessary as well, so that these techniques are quite impractical outside of the laboratory. On the other hand, the use of a doped oxide source allows precise control of diffusions in GaAs, with retention of the mirror-like surface quality that is essential for subsequent photolithographic processes.

Doped oxide sources usually incorporate only a few percent of the dopant oxide in the silica. As a result, they can be easily removed by conventional etchants, or patterned by fine-line photolithographic techniques. They can be used in situations where simultaneous p- and n-type diffusions are required in a microcircuit, as in the fabrication of complementary devices on the same chip. Moreover, the patterned placement of doped and undoped oxide regions on a microcircuit allows the fabrication of relatively simple self-aligned structures in MOS and bipolar technology.

Doped oxide sources can also be formed by spin-on techniques. These have been described in Chapter 4, and will not be considered here.

8.4.1 Dopant Sources for Silicon

Borosilicate glass is used as a p-type dopant source for silicon. Films of this material can be grown by the techniques outlined in Section 8.3.2. The reactions associated with this process are

$$2(CH_3O)_3B + 9O_2 \rightarrow B_2O_3 + 6CO_2 + 9H_2O \tag{8.14}$$

and

$$Si(C_2H_5O)_4 + 12O_2 = SiO_2 + 8CO_2 + 10H_2O, \tag{8.15}$$

An alternative system, which can be used in the 300–450°C range, involves [81] the oxidation of B_2H_6 and SiH_4. These separate reactions proceed as follows:

$$B_2H_6 + 3O_2 \rightarrow B_2O_3 + H_2O \tag{8.16}$$

$$SiH_4 + O_2 \rightarrow SiO_2 + 2H_2 \tag{8.17}$$

Subsequent drive-in at 1200°C results in diffusions with surface concentrations from 10^{17} to 10^{20} cm^{-3}.

Similar approaches have been used for the formation of n-type doped oxide sources for silicon. Here, TEOS and TMP mixtures have been used around 750°C, as well as SiH_4 and PH_3 mixtures at 300–350°C. The surface concentration varies from 2×10^{17} to 2×10^{20} cm^{-3}, after diffusion from these sources at 1200°C.

It has been observed that the surface concentration achieved by all of these doped oxide sources is a function of diffusion temperature, but is independent of the diffusion time. This implies that these oxides, usually 0.1–0.2 μm thick, serve essentially as infinite sources for the dopant. The resulting diffusion profiles, which are of the complementary error function type, confirm this observation.

Doped polysilicon can also be used as a source for the formation of extremely shallow diffusions with high surface concentration [82]. These have been used to obtain improved emitter characteristics in bipolar transistors. Their process technology is described in Chapter 11.

8.4.2 Dopant Sources for Gallium Arsenide

Doped oxide sources are especially useful in GaAs technology because they can serve as caps during the subsequent diffusion step. Here, p-type diffusions into GaAs can be made by using a mixture of ZnO and SiO_2 as the dopant source. This source is conveniently prepared [83] by CVD, involving the simultaneous pyrolysis of diethylzinc (DEZn) and silane in oxygen, at temperatures of 250–500°C. The oxidation reaction of DEZn proceeds along the following lines:

$$(C_2H_5)_2Zn + 7O_2 \rightarrow ZnO + 4CO_2 + 5H_2O \tag{8.18}$$

Growth of a $ZnO:SiO_2$ dopant source is carried out at around 350°C in a resistance-heated, cold-wall reactor. Diffusion from this source results in excellent surface characteristics, and has been used at temperatures of up to 700°C, for the formation of shallow (0.03–1.5 μm) junctions with high surface concentration ($\approx 10^{20}$ cm^{-3}) and excellent control of junction depth ($\pm$.01 μm) [84].

In all of these open-tube diffusions, it is important to prevent the out-diffusion of zinc from the top surface of the dopant source. This is done by using a cap layer* of PSG, 0.2 μm thick, since this material is impervious to the diffusion of zinc through it [23]. Use of this cap makes the diffusion relatively independent of thickness and composition of the dopant source, and also of the gas ambient in which the diffusions are conducted.

A mixture of SnO_2 and SiO_2 can be used as a dopant source for n-type diffusions into GaAs. Doped oxides of this type can be deposited by the simultaneous oxidation of tetramethyltin (TMSn) and silane in the 350–450°C temperature range [85], using a resistance-heated, cold-wall reactor. Here, the oxidation of TMSn proceeds along the following lines:

$$(CH_3)_4Sn + 4O_2 \rightarrow SnO_2 + CO_2 + 6H_2O \qquad (8.19)$$

$SnO_2:SiO_2$ films of this type can be used for open tube n-type diffusions into GaAs at temperatures of up to 900°C, provided that they are capped with a layer of PSG as described earlier. Here, too, a mirror-like surface is preserved because of the low concentration of dopant in the source.

8.5 FILMS FOR INTERCONNECTIONS

The most important application area for conductive films is to provide interconnections between contacts which are made to devices in a microcircuit. Both single-metal and multimetal films are used for this purpose. In addition, the complexity of VLSI circuits has brought about the need for *multi-level* metallization schemes. These provide additional surface area for making interconnections between devices, and have unique problems of their own. Here, too, both single- and multi-metal systems are in common use today.

Films used for interconnections must adhere firmly to the semiconductor contact and also to the insulating layer that is placed over the semiconductor surface. They should be readily deposited by a relatively low-temperature process, since metallization is one of the last steps in microcircuit fabrication. They should be easily patterned with high resolution, without etching the insulating layer on which they are placed. They should be relatively soft and ductile, so that they can withstand cyclic temperature variations during processing and in service. They must be highly conductive, and capable of handling high current

---

*This PSG layer is also grown on the back surface of the GaAs slice to prevent its decomposition during the diffusion process.

densities while still maintaining their electrical integrity. Finally, they should be easily connected to external terminations.

Most of these requirements are met by gold and aluminum films. Wires of these metals, ranging from 0.5 to 2 mils in diameter, are also used for connecting these films to the terminal posts. Gold wires are usually drawn from 99.999% pure gold and are work-hardened to provide enough stiffness for handling purposes. With aluminum, however, it is necessary to use a 99% Al–1% Si alloy to provide this stiffness.

Gold welds readily to both aluminum and gold bond pads by the simultaneous application of heat and pressure. This process, known as *thermocompression bonding*, is commonly carried out with the aid of a nail-head bonder. Figure 8.13 shows the various steps in its operation [86]. A gold wire with a sphere at one end is fed through a capillary from a spool. The sphere is aligned over the bonding pad and then lowered. During this process the semiconductor die, mounted on its header, is maintained at 280–300°C in a nitrogen gas ambient. The gold sphere is pressed against this pad for a few seconds until a weld is formed; this also results in its plastic flow into the form of a nail head. On raising the capillary, gold wire is drawn out from the spool, resulting in a lead. The capillary is now aligned over the terminal post, and the procedure is

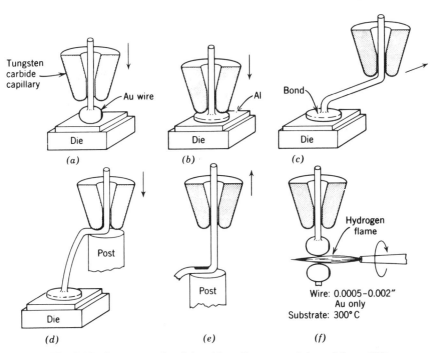

Fig. 8.13 Sequence of nail head bonding steps. Adapted from [86].

repeated. Finally, a hydrogen torch is used to break the wire, resulting in the formation of spheres at its ends. Thus the wire in the spool is ready for the next bonding operation.

A disadvantage of this type of bonding is the presence of the cantilever "tail" formed with each bond. This tail can be broken off within the package, and is thus an incipient cause of circuit failure. A stitch bonder is more commonly used since it avoids this problem. Steps in the bonding operation with this tool are shown in Fig. 8.14, and cutters are used to avoid the formation of the dangling lead.

Ultrasonic methods are also used for bonding aluminum leads to microcircuit pads. The bonding operation is accomplished as shown in Fig. 8.15. Here the wire is placed in contact with the aluminum pad and pressure is brought to bear on this combination by an ultrasonically driven tool. Welding occurs as the tough surface layers of Al_2O_3 are broken by the ultrasonic vibrations.

Ultrasonic bonding is done at, or slightly above, room temperature. It is commonly used with VLSI circuits which have a large number of bonding pads, since this avoids subjecting the chip to elevated temperatures during this process. Its success is a sensitive function of the pressure as well as of the ultrasonic power level. In contrast, the thermocompression bonding of gold leads is a considerably easier technique, and requires less operator skill.

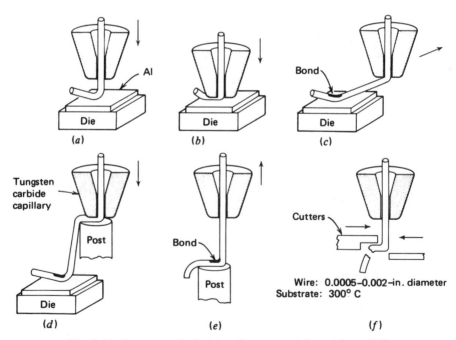

Fig. 8.14 Sequence of stitch bonding steps. Adapted from [86].

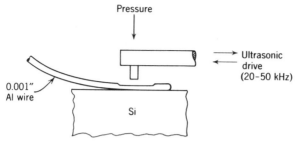

Fig. 8.15 Ultrasonic bonding.

A number of techniques have been developed to avoid the expensive process of individually bonding leads to a microcircuit. These include gang bonding, as well as solder bump and multiple weld methods, and involve a high degree of automation. These techniques are of a specialized nature, and will not be considered here.

8.5.1 Single-Metal Interconnections

Figure 8.16a shows an example of a single-level interconnection scheme, in which a single metal is used. Here, aluminum is a viable choice for the metal, since it bonds well to SiO_2 as well as to Si_3N_4 insulating layers, by a relatively short heat treatment. During this process, the aluminum reacts with the surface layer on which it is placed, at a number of localized points. Thus, it becomes firmly attached to this surface, while still providing for stress relief during subsequent thermal cycling.

The bonding of aluminum to SiO_2 comes about by chemical reaction, with the formation of Al_2O_3. The reason for the bonding of aluminum to Si_3N_4 is less clear, however. In all probability, actual bonding occurs via a thin interfacial layer of SiO_2 on the Si_3N_4 film.

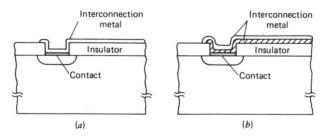

Fig. 8.16 Single-level interconnections. (a) Single metal and (b) multimetal.

There are many additional advantages to the use of aluminum [87]. It can be readily deposited by well-developed vacuum evaporation techniques, and both resistance-heated sources and e-beam evaporation methods are commonly used. Sputtering techniques are also used for this purpose. These are especially useful when more than one level of interconnect metal is required, since they allow the use of back-sputtering for cleaning purposes, prior to deposition of the next interconnection level.

Aluminum films are readily etched into fine-line patterns by chemicals which do not attack the insulating layers on which they are deposited. A variety of of such etches are described in Chapter 9. Finally, aluminum welds readily to gold leads by thermocompression bonding, and to aluminum leads by ultrasonic welding. Both of these techniques result in strong, reliable, low-resistance connections to the package terminals.

In silicon-based microcircuits, the biggest advantage of aluminum interconnections is that they can also be used to form the ohmic contacts. This allows the deposition of both the contact and interconnect metal in a single step. In addition, a single heat treatment serves to form this contact, and also to bond the metal to the silica surface layer. Thus, the overall process is an extremely simple one, and results in a reliable, low-cost technology which is widely used today.

Evaporated aluminum is generally free from contamination, because of the high rate at which it can be deposited, relative to the impingement rate of impurities in the vacuum system (as governed by the background pressure). As a result, its electrical resistivity in thin film form (3.0 Ω-cm) is not very different from its bulk resistivity (2.7 $\mu\Omega$-cm). A typical aluminum interconnection film, 1 μm thick, has a sheet resistance of about 0.03 Ω/square.

Step coverage can be a problem, especially in MOS circuits where oxide thicknesses can vary from 200 Å for the gate to over 15,000 Å for the field oxide. One method of reducing this problem is to reflow the oxide in order to avoid sharp corners at steps. In addition, evaporation of aluminum on heated substrates ($\approx$300°C) tends to promote increased coverage; this method can be used when the step profile is not too severe. Increasingly, sputtering techniques have been used because they provide excellent step coverage.

The use of aluminum interconnect metal has some disadvantages, however. Although it has a natural protective coating of native oxide, it is electronegative so that it is prone to corrosion in both acidic and basic environments [88]. In addition, it is soft and relatively easily scratched during wafer and chip handling. Both of these problems can be greatly reduced by coating the complete microcircuit slice with a film of either PSG or Si_3N_4. This additional step, which involves an extra masking operation to cut holes for the bonding pads, greatly improves the processing yield and also the long-term reliability of the packaged microcircuit.

Another problem with aluminum is that of compound formation where it interfaces with gold, during storage at temperatures in excess of 300°C. This phenomenon has been described in Section 2.2.6, together with methods by which its effects may be minimized.

Finally, an important problem comes about because of the requirements for VLSI technology. Here, the emphasis on high-speed, densely packed structures has resulted in small dimensions, and an increase in the operating current densities of these films. Thus, electromigration, which results in metal transport, is an important failure mode in them.

The rate of metal transport in a thin film, due to electromigration, has been given previously:

$$R \propto J^2 \phi e^{-E_d/kT} \tag{8.5}$$

where E_d is the activation energy of self-diffusion. For single-crystal aluminum [15], $E_d = 1.4$ eV, whereas E_d varies from 0.4 to 0.5 in polycrystalline aluminum depending on the grain size. The importance of a large value of E_d is seen from experiments with both single crystal and polycrystalline aluminum films. Single-crystal films, grown epitaxially on MgO substrates, have been operated at 175°C with a current density of 5×10^6 A cm^{-2}, for a period exceeding 26,500 hr. Under these same conditions, evaporated polycrystalline aluminum films have failed in less than 30 hr!

A number of approaches can be used to reduce electromigration effects in aluminum films. Primarily, they include control of deposition parameters and seeding with impurities such as copper, since these techniques critically affect the grain size. Thus, aluminum doped with 4 wt.% copper results in films with an improvement in the MTF by a factor of 10 [89]. In practice, however, problems with dry etching of these films, as well as corrosion effects [90], place an upper limit of 2 wt.% Cu content.

Cyclic stresses during processing as well as during circuit operation have been shown to induce voids and wedge-shaped cracks in aluminum metal by the movement (and eventual coalescence) of vacancies. Some success at reducing this problem has been achieved by sandwiching these lines between layers of conductive materials, in order to control this stress.

In the final analysis, the best solution is to avoid the problem entirely by restricting the current density to about 10^5 A cm^{-2}. In VLSI technology, where space for the metallization is restricted, this places a premium on the use of devices which can operate at low current levels.

8.5.2 Multimetal Interconnections

The problems of electromigration can be greatly reduced by the use of a metal having a larger activation energy of diffusion than aluminum. One such metal, which meets many of the requirements of an interconnect layer, is gold. This is a ductile, weldable metal which can be deposited by vacuum evaporation with a thin film resistivity that is almost identical to its bulk value (2.44 $\mu\Omega$-cm). It can be readily patterned into fine lines by a number of different etchants, some of which are listed in Table 9.5. In addition, a series of cyanide-based etches are available, as well as a variety of pre-packaged proprietary formulations.

Evaporated gold films have an activation energy of self-diffusion of about 0.85–1.0 eV, so that they can be operated at considerably higher current densities than aluminum films without suffering from electromigration effects. Typically, for the same current density and operating temperature (10^6 A cm^{-2} at 170°C), the MTF for gold-based interconnection systems is about 25–40 times higher than that for aluminum films [91]. At higher operating temperatures, the advantages of gold over aluminum are even more significant.

Gold is highly electropositive, and is not subject to the corrosion problems of aluminum. Unfortunately, its inert nature prevents adhesion to the insulating layer by chemical bonding. As a consequence, it can only be used in a multimetal system (as illustrated in Fig. 8.16b for a single-level interconnection scheme), where an additional "glue" layer is used for adhesion of the gold to the insulator. Transition metals such as chromium, tantalum, and titanium have all been used for this purpose. The use of these materials to promote adhesion to the gold film results in the possibility of reactions with it to form intermetallic compounds during subsequent processing at 400–500°C, and necessitates the use of a barrier layer as well. One popular approach is to combine these functions in a film of Ti–W (10 wt.% Ti), which can be easily deposited by r.f. sputtering. This alloy has excellent adhesion to both the gold and the insulating layer, and also serves as an interdiffusion barrier between the gold and the semiconductor.

Gold films can also be used over aluminum metallization, and can be plated up to form thick beam leads by which the chip can be directly bonded to the circuit board. Compound formation can be avoided [92], if interfacial barriers of platinum and titanium are used. The resulting system consists of Au–Pt–Ti–Al, where the platinum prevents Au–Ti reactions, and the titanium prevents Pt–Al reactions.

8.5.3 Multilevel Interconnections

Multilevel metallization schemes are important in VLSI technology, because they provide additional surface area on which interconnections can be made. Moreover, they give flexibility to the circuit layout, since crossovers can now be made in the vertical direction as well as in the plane of the chip. Two-level schemes will be described here, for simplicity. However, there is increasing need for three (and possibly four) layers of metallization, especially in high-density circuits which require customized wiring. The principles outlined here can be applied to higher levels of metallization as well.

In a two-level metallization scheme, the first level (M-1) is commonly used for connections to devices. This is usually aluminum, which is ideal for ohmic contacts and which bonds readily to silicon, polysilicon, silicides, and surface silica layers.

The second-level metal (M-2) can also be aluminum, as shown in Fig. 8.17a, following the placement of an insulating layer (L-2), whose growth temperature is sufficiently low so as not to damage M-1, or its contact to the circuit. CVD grown films, as well as organic films, can be used for this insulating layer.

Two problems arise if M-2 is made of aluminum. First, connections to M-1 consists of Al–Al. These are difficult to make because of the protective native oxide (sapphire) on the surface of M-1. This problem can be solved by using back-sputtering to remove the oxide from the metal in the via holes, prior to sputter deposition of M-2.

Next, connections of M-2 to the circuit are deep, so that the problems of step coverage become quite serious. This step coverage can be improved if the profile of the contact hole has been shaped by reflow of the insulators. Here, CVD grown aluminum is desirable because of its inherent ability to conform to surface features.

Organometallic alkyls such as trimethylaluminum [$(CH_3)_3Al$] have not been satisfactory for this application, because of carbon incorporation which results in high-resistivity films. However, low resistivity films have been obtained

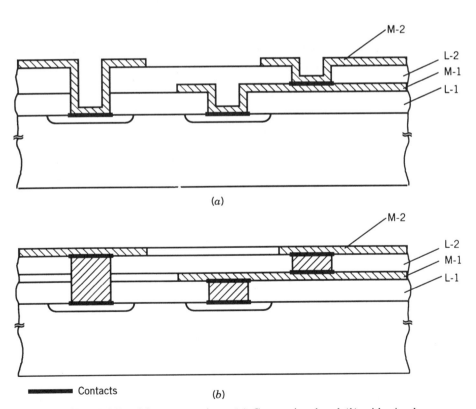

Fig. 8.17 Multilevel interconnections. (a) Conventional and (b) with via plugs.

[93] by the use of tri-isobutylaluminum [$(C_4H_9)_3Al$] and trimethylamineallane [$(CH_3)_3NAlH_3$]. These films can be deposited by CVD at temperatures in the 250–450°C range and have conductivities of about 90% of the bulk value, in addition to excellent step coverage. An inherent problem with this approach is that thin film processes for the CVD of alloy films of Al–Cu and Al–Cu–Si have not been developed at the present time.

An alternative approach is the use of sputter-deposited Al–Ge for M-2. The eutectic temperature of this alloy is about 424°C, which is considerably lower than the melting point of aluminum (660°C). Consequently, it can be reflowed at a lower temperature.

Perhaps the most promising approach to this problem is to use tungsten plugs to fill via holes, resulting in the interconnection scheme of Fig. 8.17*b*. This tungsten can be deposited selectively by the process described in Section 8.3.4.3, or deposited as a blanket layer, followed by an etch to remove the excess metal. Both approaches have the advantage of resulting in a planar structure, as illustrated in this figure. Moreover, the effect of thermal treatments during processing is minimized, since the TCE of tungsten is closely matched to that of silicon.

A number of process options are possible with this approach. For example, molybdenum with a Ti–W overlayer, or $MoSi_2$/Ti–W, have been used for M-1 in order to promote adhesion of the tungsten plug. An alternative approach is the use of a TiN barrier layer, grown by the reaction of $TiCl_4$ and NH_3 gas at 450–600°C. An additional 50–100°C reduction in growth temperature can be obtained by CVD growth in the presence of UV radiation [94].

Copper is also being investigated as an alternative to Al-based alloys for higher levels of metallization, because of its low electrical resistivity and its excellent resistance to electromigration. Both blanket and selective copper deposition are under study [95, 96]. Many problems concerning both its low-temperature growth and its stability during processing remain to be solved. For example, copper reacts with oxides, silicon, and silicides, and is a rapid diffuser in these materials. Moreover, it is a deep-level impurity in silicon, where its presence degrades device leakage current and minority carrier lifetime. However, progress in the development of suitable barrier layers should make the use of this metal viable in future VLSI applications.

8.6 FILMS FOR OHMIC CONTACTS

An ohmic contact to a semiconductor region must have good mechanical properties. It must adhere firmly, both during formation and subsequent processing, and also in service. It must not cause excessive stress in the underlying semiconductor, since this can result in a change in its electronic character. It must have low electric resistance, and be "ohmic"; that is, its voltage–current characteristic should approximate a straight line going through the origin and extending over the entire range of voltages and currents to which the contact is subjected.

It must serve purely as a means for getting current into and out of the semiconductor, but play no part in the active processes occurring within the device; that is, it must not be an injecting contact. Finally, it must be compatible with the metal system for the interconnection technology which is used to make contact to it.

There are many ways for forming an ohmic contact that are suitable for use in a laboratory environment. Thus, almost any contact to a highly damaged semiconductor region will be noninjecting because of the extremely high recombination rate associated with this region. Contacts which are made remote from the active region (e.g., contacts to a Hall bar) are also "ohmic" in the sense that the region of device operation is many diffusion lengths away from the contact, so that injection effects can be neglected.

The need for small dimensions and smooth surfaces precludes the use of techniques of this type for VLSI applications. Consequently, other approaches must be used to form contacts which meet the necessary mechanical requirements and which also offer a minimum barrier to the flow of electrons and holes into the device. One such approach consists of a two-step process, which can often be performed in a single operation. First, a heavily doped p^+- or n^+-region is formed on (or in) the semiconductor. Next, a metallic contact is made to this region. This two-step process results in great latitude in the choice of dopants and contact metals, and is commonly used for both silicon and GaAs devices.

The behavior of an ohmic contact of this type can be studied by considering each step separately. **Step 1** results in the following possibilities:

n^+-*Contact on* p^+-*Semiconductor, or* p^+-*Contact on* n^+-*Semiconductor.* A contact of this type forms a symmetric tunneling junction whose built-in contact potential is sufficient to cause breakdown at 0 V. It results in a low-resistance "ohmic" contact with the voltage–current characteristics of Fig. 8.18*a*. Moreover, this contact is made between highly doped, low-lifetime regions, so that minority carrier injection into the semiconductor is negligible.

p^+-*Contact on* p-*Semiconductor, or* n^+-*Contact on* n-*Semiconductor.* This also results in "ohmic" behavior. In effect, the "high–low" junction formed by this combination has an extremely large leakage current which completely masks the usual diode-like characteristics [97]. Its voltage–current characteristic is of the form of Fig. 8.18*b*, and approximates a straight line, representative of ohmic behavior.

p^+-*Contact on an* n-*Semiconductor, or* n^+-*Contact on a* p-*Semiconductor.* These situations result in the formation of a well-defined p–n junction diode whose characteristics are quite nonohmic. However, ohmic contact to a lightly doped region can be achieved by the series connection of the two schemes outlined earlier. For example, a p^+-contact to an n-region can be made by first forming an n^+-region on it, followed by a p^+-region, resulting in a sandwich of p^+–n^+ (tunneling junction) and n^+–n (high–low junction) regions, which are

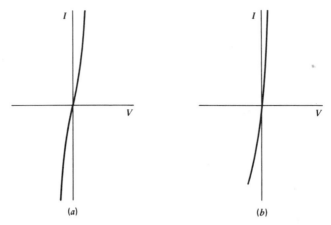

Fig. 8.18 Ohmic contact characteristics. (*a*) High-low junction and (*b*) tunneling junction. The voltage axis is greatly expanded.

both ohmic in character. In like manner, an n^+–p^+ junction, followed by a p^+–p junction, results in an n^+-contact to a lightly doped p-region.

Step 2, which consists of the placement of a metal in intimate contact with a semiconductor, creates a Schottky barrier. The properties of this barrier are a function of the choice of metal and the semiconductor doping, and can be exploited in a variety of diode structures (see Section 8.7). Here, however, it suffices to point out that the voltage–current characteristic of a Schottky barrier made with *any* metal will approximate ohmic behavior if the metal layer is in contact with a region of very high doping, ($>10^{19}$ cm^{-3}), regardless of the barrier height [98]. This is because field-emission dominated conduction, i.e., tunneling through the barrier, prevails so that it is almost transparent to the carrier flow. Thus, a tunneling junction is formed in this manner. The use of this principle allows great latitude in the choice of contact metal, which can be now selected on metallurgical considerations alone.

In summary, the overall strategy for making an ohmic contact to any semiconductor is to first terminate it with a heavily doped region of appropriate impurity type. This is followed by the contact metal, which may be single-layered or multilayered. In some instances, both of these steps are incorporated in a single process.

8.6.1 Single-Layer Contacts

In silicon technology, the use of aluminum metal, which belongs to group III of the periodic table, allows the combination of these steps in a single process. The use of aluminum for the interconnection also results in a simple, low-cost process that is both highly reproducible and highly reliable.

In order to form an ohmic contact, aluminum is deposited on the silicon and subjected to a heat treatment at a temperature that is somewhat *below* the eutectic temperature of the combination. Consider the metallurgical processes that occur for this condition. Figure 8.19a illustrates this situation. On raising its temperature, some of the aluminum combines with the silicon to form a melt (see Fig. 8.19b). On cooling, a solid solution of aluminum in silicon freezes out epitaxially as an extension of the single-crystal substrate. This is followed by a polycrystalline Al–Si alloy and, finally, by aluminum metal, as shown in Fig. 8.19c. Thus a highly aluminum-doped, p^+-region is formed on the silicon; connection is made to it by a polycrystalline Al–Si alloy, and eventually by the aluminum metal.

It would at first appear necessary to perform this alloying operation *above* the eutectic temperature. This is not the case because, on a microscopic scale, physical contact between the deposited aluminum film and the silicon only occurs at a discrete number of points. Here, melting occurs below the eutectic temperature because there is considerable localized pressure at these points.* With a *microalloy contact* of this type, bonding occurs at a large number of isolated points on the film surface. Stresses created during circuit operation are relieved by plastic flow of the metal, thus greatly reducing the possibility of failure due to thermal cycling. It can be shown that such a contact, if made over 10% of the surface, has as low an electrical resistance as one made over the full area of the contact region [99].

In this contact technology, the aluminum and p^+-silicon form a tunneling Schottky contact, without the necessity of further steps or the deposition of additional layers. One particularly important advantage of this system lies in the fact that trace oxides (15–30 Å), which are normally present on all silicon surfaces, are readily dissolved by reaction with the aluminum during contact formation. Thus, elaborate cleaning procedures, involving back-sputtering or ion milling, are not necessary prior to aluminum deposition.

---

*Typically, this heat treatment is carried out at 400–500°C, whereas the eutectic temperature of the Al–Si system is 577°C.

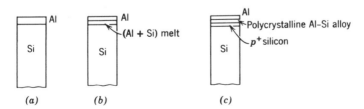

Fig. 8.19 Steps in the formation of an aluminum–silicon alloyed contact.

Figure 8.20*a* illustrates the types of ohmic contact which are made to an n^+–p–n silicon bipolar transistor using this contact technology. Here, the emitter contact consists of metal–p^+–n^+ combination. The base contact is a metal–p^+–p structure. The collector contact consists of metal-p^+-n^+ in series with an n^+–n junction. The n^+-region is formed by diffusion into the collector; however, it is made at the same time as the emitter diffusion, so that it requires no additional process steps. In contrast, connection to a p^+–n–p silicon transistor, as shown in Fig. 8.20*b*, requires an additional process step for the n^+-diffusion which is necessary for ohmic contact to the n-base.

Single-layer systems can also be used in simple contacting schemes for GaAs [100]. Here, a successful contact material for n-type GaAs is an Au–Ge alloy, of eutectic composition (12 wt.% Ge). Deposition of this material, followed by a 1-min heat treatment at 450–500°C, serves to form an n^+–n contact, with the gold acting as a Schottky barrier connection to it. Gold–metal interconnections can be made readily to this contact.

The exact mechanism of contact formation with this system is not known. In fact, its success is somewhat surprising since germanium can be incorporated on both gallium and arsenic sites. One argument is that the presence of the gold serves as a getter for gallium during the heat treatment [101]. This results in the increased formation of $[V_{Ga}]$ in the semiconductor, accompanied by its preferential occupation by the germanium, so that the material becomes heavily n-type. Sometimes, a little nickel ($\approx 1\%$) is added to this material to enhance the diffusion of germanium into the lattice.

An alternative argument is that heat treatment of the Au–GaAs system lowers the barrier height [102], and the in-diffusion of germanium seems to further enhance this effect. In addition, the germanium does substitute for both gallium and arsenic atoms, but appears heavily n-type because of the strain in the lattice which is created by the in-diffusion of both these elements (Au and Ge).

This single-layer system is also used for ohmic contacts to AlGaAs. Here, adhesion is sometimes a problem because of the presence of trace amounts of Al_2O_3 at the interface. RTA has been found to be especially effective in forming a mechanically strong bond in this situation. Often, a cap layer of heavily doped GaAs is grown over the AlGaAs to facilitate this contact formation process.

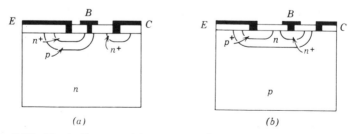

Fig. 8.20 Ohmic Contacts. (*a*) n–p–n transistors and (*b*) p–n–p transistors.

Gold, doped with 5–10 wt.% Sn or with 1–2 wt.% Se, has also been used successfully in the formation of ohmic contacts to *n*-GaAs. Contacts to *p*-GaAs are often made by means of an Au–Zn alloy (2 wt.% Zn) or an Au–Be alloy (1 wt.% Be). In all cases, the gold forms a Schottky barrier having suitable metallurgical properties for use as a contact metal.

Contact formation with these Au-based alloys is usually carried out by vacuum evaporation. In all cases, selective evaporation of the alloy constituents occurs because of their different vapor pressures. Consequently, the entire charge of metal must be evaporated, to ensure sufficient transport of the dopant species in each case. This is a special problem with Au–Zn because of the high volatility of zinc.

8.6.2 Kirkendall Effects

The single-layer contacts described above are based on microalloying of the contact metal to the semiconductor, by means of heat treatment. This results in the dissolution of the semiconductor, which tends to proceed more slowly along the ⟨111⟩ directions, resulting in the formation of voids. These eventually become back-filled with the contact metal or its alloy, to form conducting spikes as shown in Fig. 8.21. These are of little consequence in deep junction structures. With shallower structures, however, they often result in leaky or shorted junctions. They represent a special problem in VLSI technology where the requirement for densely packed structures mandates the use of shallow junction devices.

The tendency to spike formation comes about because of the differences in the solid solubilities of the contact metal in the semiconductor, and vice versa. Upon heat treatment, this gives rise to the differential transport of material [103], known as the *Kirkendall effect*. This can result in the formation of voids in the region in which there is a net loss of material.

Kirkendall effects present a serious problem in the formation of ohmic contacts to silicon, when aluminum is used for this purpose. Here, connection is made by heat treatment at 400–500°C as described previously, resulting in alloying at discrete points. From Fig. 8.22 it is seen that the solid solubility of silicon in aluminum is quite significant at these alloying temperatures [87].

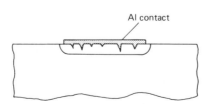

Fig. 8.21 Alloy spiking effects.

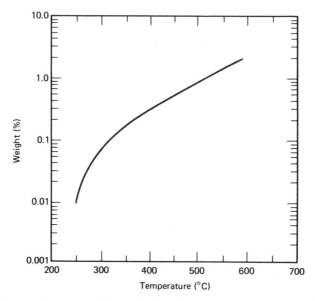

Fig. 8.22 Solid solubility of silicon in aluminum. From Schnable and Keen [87]. ©1971 The Institute of Electrical and Electronic Engineers.

On the other hand, the solid solubility of aluminum in silicon is much lower, $\simeq 0.0011$ wt.% at 600°C (see Fig. 2.6). This difference in solid solubility results in the net dissolution of silicon into the aluminum, with the resultant spike formation. These spikes are pyramidal in shape on (001) Si, and appear as rectangular etch pits on the surface. The incidence of spikes is usually highest at the edges of windows, where stress concentration can lead to enhanced dissolution of silicon in aluminum.

A second mechanism, which greatly enhances the Kirkendall effect, is the high diffusivity of silicon in aluminum films, as shown in Fig. 8.23 [104]. This mechanism allows the silicon to be transported a considerable distance into the aluminum during heat treatment. Furthermore, void formation will be especially severe if the contact is made in a region where there is a large aluminum sink, such as a bonding pad.

One approach to reducing the penetration depth of these spikes is to carry out the microalloying operation at temperatures below 400°C. Unfortunately, this often results in high-resistivity ohmic contacts. A second approach is to use RTA techniques for contact formation.

It is possible to minimize Kirkendall effects by the use of aluminum metal which is already presaturated with silicon. Typically, about 3–4 by wt.% of silicon is necessary for this purpose; however, dissolved silicon comes out of solution during contact formation, and takes the form of precipitates. These are formed at grain boundaries in the film, and at steps in the dielectric on

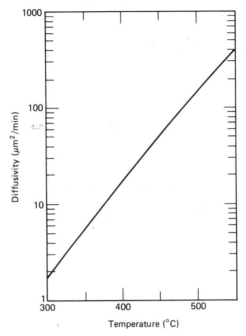

Fig. 8.23 Diffusivity of silicon in deposited aluminum films. Adapted from [104].

which they are placed. In order to minimize precipitate formation, 1% Si is commonly used for this VLSI application. Combinations of these approaches are used in VLSI technology. Other approaches, using multilayer contact schemes, are described in the next section.

Contacts between aluminum and polysilicon represent a special problem. Here, dissolution of silicon in aluminum is enhanced in the vicinity of grain boundaries, so that it eventually precipitates in these regions. It is important to note that this problem is not alleviated by the use of aluminum metallization which is saturated with silicon [105], since it is a grain-boundary effect. Moreover, it is not present with phosphorus-doped polysilicon [106], where the grain boundaries are heavily saturated with phosphorus.

Kirkendall effects are also present in the Au–Ge–GaAs system when it is heated above 250°C. Here, the solubility of gallium in gold is considerably higher than that of gold in GaAs, resulting in the formation of gallium vacancies. This process eventually leads to void formation and to the "balling up" of contact metal, resulting in rough, poorly defined contact regions. The addition of nickel to the Au–Ge alloy reduces this tendency. Finally, multilayer contacting schemes are used to minimize these problems in VLSI technology.

Both electromigration and Kirkendall effects can cooperate in a number of

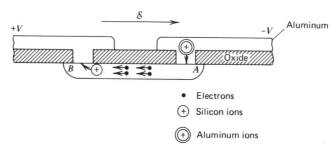

Fig. 8.24 Cooperation of electromigration and Kirkendall effects.

situations. One such situation is shown in Fig. 8.24, which depicts a diffused resistor with aluminum connections. The bias conditions for this resistor are also shown, with the path of electron flow from interface A to interface B. Here, both of these effects aid in transporting silicon into the aluminum at interface B. At A, however, electromigration effects tend to transport aluminum into the silicon. However, very little actual material transfer is involved, because of the low solid solubility associated with this process. As a result, the interface at B will steadily degrade during device operation, as the voids and alloy spikes continue to grow. On the other hand, the interface at A will be relatively unaffected.

8.6.3 Multilayer Contacts

Contact systems of this type are more complex than those described earlier. Consequently, they are used when single-layer contacts have poor electrical performance or are unreliable. By way of example, the aluminum contact to n-silicon, described in Section 8.6.1, resulted in nonohmic behavior and necessitated the formation of an additional n^+-layer between the aluminum and the contact. Thus, it was, in effect, a multilayer contact.

Figure 8.25 shows the resistivity of Al–n^+-Si contacts made in this way. Here, it is seen that the specific resistivity of this system is initially high, but falls rapidly with heat treatment between 350° and 450°C. This fall is, unfortunately, accompanied by spike formation during alloying. Spike formation can be greatly reduced by the use of an intermediate polysilicon layer [107, 108]. A small improvement in contact resistance is achieved; here, too, rapid improvement only occurs when the alloying spikes penetrate through the polysilicon.

Figure 8.25 also shows the effect of using an n^+-polysilicon intermediate layer, which serves as a diffusion source during subsequent heat treatment at 200°C, and lowers the contact resistance. Here, the specific resistivity is significantly lower, even when there is no spike penetration. This system is particularly convenient if aluminum is used as the top contact metal. In this case, heat

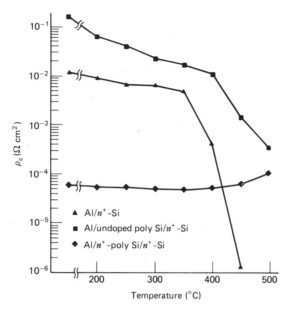

Fig. 8.25 Ohmic contacts. (*a*) single layer and (*b*) self-aligned multilayer. From Finetti et al. [107].

treatment of the combination results in some epitaxial regrowth of the silicon, as well as alloying of the aluminum contact material.*

The direct deposition of aluminum metal in a contact window, illustrated in Fig. 8.16*a*, is only possible when relatively deep diffusions are used. This is because the correspondingly large lateral diffusion results in a junction which is protected by the oxide layer. With shallow junctions, of the type required for VLSI, this penetration is greatly reduced. This necessitates that the contact be made through a hole cut in the oxide, as shown in Fig. 8.26*a*, if a leakage-free junction is to be preserved. This hole is smaller than the oxide cut, and results in a penalty in the contact resistance. In high-density schemes, this penalty can be as large as a factor of nine to ten.

A technique for avoiding this problem is to deposit a transition metal in the hole, after previously removing its oxide cover [109]. Next, the system is heat-treated to convert this metal to its silicide. Consumption of the silicon during the silicide formation process is an advantage, since it provides a clean interface with the silicon.

Historically, PtSi was the first silicide to be explored for use as a contact layer, and it has found its way into many operating systems. Here, a layer of

---

*Note that epitaxial silicon cannot be directly grown for this purpose, because of the high growth temperature (≃1200°C). In contrast, the growth temperature for polysilicon can be as low as 600°C, or 400°C if PECVD is used.

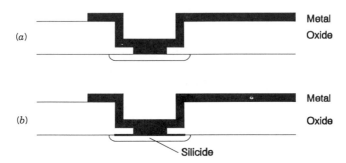

Fig. 8.26 Contact resistivity of a variety of aluminum–silicon systems.

platinum, about 500 Å thick, is sputter-deposited on the silicon surface which is masked except for openings where the contacts must be made. Next, the combination is heat-treated at $500-600°C$ for about 10 min, in a nonoxidizing atmosphere, to convert the platinum to PtSi. This conversion results in planar penetration into the silicon and avoids alloy spiking [110]. In addition, it consumes silicon and produces a clean interface with the semiconductor. Typically, 500 Å of platinum consumes 665 Å of silicon to form 990 Å of PtSi. Eventually, aluminum or Ti–W/Al is placed over the PtSi film.

The formation of a PtSi film necessitates the use of back-sputtering to clean the contact regions prior to platinum deposition, and also to etch away the unwanted platinum from the rest of the surface. The presence of even trace amounts of oxygen or water during the silicide formation process must be avoided here, so that it is sometimes carried out in a vacuum. Many of these processing difficulties can be avoided by the use of palladium [111], which can be more easily deposited than platinum; in addition, it can be readily etched by wet chemical techniques.

Of the many silicides listed in Table 8.2, both WSi_2 and $TiSi_2$ are preferred choices because they share common components with Ti–W and TiN barrier layers. They are usually deposited by co-sputtering, in the same system which is used for the silicide contact, and in a single continuous process. A shallow, ohmic contact using Mo/Ti–W is an example [112] of this approach, and has been used in VLSI technology based on the use of molybdenum gate transistors.

A unique example of process simplification by this approach [113] consists of sequentially sputtering 750 Å of titanium and 1000 Å of tungsten, followed by annealing at 650°C in pure nitrogen, for a period of 20 min. This heat treatment results in both $TiSi_2$ and TiN formation, while leaving the tungsten unchanged. A final metallization provides an $Al/W/TiN_x/TiSi_y/Si$ contact system with a minimum of processing.

Silicide films can be used in both silicon and GaAs systems where the upper metal is gold. Gold does not adhere to these silicides, so the intermediate layer is essential in this case. Sputtered TiN films have been used as barriers to gold in compound semiconductor systems, and can withstand heat treatments up to 430°C [114]. Barrier layers of titanium metal, followed by platinum to prevent the formation of intermetallic compounds, have also been used in silicon technology. The beam lead system, for example, uses Au–Pt–Ti–PtSi–silicon and can withstand thermal processing to 500°C [115].

Multilayer schemes are also used to form low-resistivity contacts to n-GaAs, in order to avoid balling up problems which arise from the use of Au–Ge eutectic. Here, the most common approaches are based on the use of a nickel layer, which greatly reduces the tendency of Au–Ge eutectic to ball up during contact formation [116]. Heat treatment results in thorough intermixing, so that very little difference in contact resistance is observed if this nickel layer is deposited before or after the Au–Ge. Separate layers of Au and Ge are often used for convenience and flexibility, although no improvement is seen in the ohmic contact by their use. Here too, the actual sequence of Au, Ge, and Ni layers is not important. For example, a sandwich of Ni/Ge/Au/Ni/Au of thicknesses 100, 500, 1200, 200, 1000 Å respectively, has been used with n-GaAs, where the top gold layer was used as the metallization [117]. A second arrangement of Ni/Ge/Au/Ni/Ti/Au in a 100, 500, 1200, 200, 1000 and 1000 Å, respectively, was used to provide improved adhesion of the gold metallization layer. RTA was used to alloy the contact in both cases. In other work [118], the titanium layer has been replaced by Ti–W alloy. This layer also serves to provide an adherent film over the microcircuit surface, so that the gold contact can be extended to the interconnections and the bonding pads. Both tungsten and $W_{1-x}N_x$ have been used as diffusion barriers between gold contacts and GaAs, and have allowed system integrity up to 550°C [119]. Finally, intermediate layers of WSi_2 have also been used to obtain contacts which are stable when operated at 300°C [120]. Here, the silicide layer was formed by sequential deposition of multiple layers of tungsten and silicon. RTA was used at 640°C to homogenize these layers, after the GaAs was covered with a protective coating of Si_3N_4.

It is also possible to make an ohmic contact directly to the semiconductor by means of a Schottky barrier, and avoid the alloying step entirely. This allows contact to extremely shallow junctions; however, it greatly restricts the choice of both the contact metal and the doping concentration (and type) of the underlying semiconductor. For example, both PtSi and Pd_2Si have large barrier heights to n-silicon, and thus make excellent ohmic contacts to only p-type material. Chromium and titanium have a low barrier height ($\simeq 0.5$ eV) to both n- and p-silicon, and can be used to make ohmic contacts directly to moderately doped silicon, in addition to adhering to the surface insulator. These reactive metals are generally deposited by e-beam evaporation techniques and are subsequently followed by evaporated layers of gold to which they make strongly adherent contact. Systems of this type, Au–Ti–silicon or Au–Cr–silicon, are considerably simpler than the Au–Pt–Ti–PtSi–silicon system described previously, and are in common use at the present time [121].

It is difficult to make simple Schottky ohmic contacts to GaAs, unless this material is heavily doped. An alternative approach is to use an intermediate, epitaxially grown n^+-Ge layer, so as to form an n^+-Ge/n^+-GaAs heterojunction [122]. The Schottky metal contact can be readily made to the germanium, and is ohmic because of its lower energy gap. Contacts of this type require no heat treatment; in fact, the ideal contact metals are molybdenum, titanium, or tungsten, as well as their silicides and nitrides, which are relatively inert to the GaAs. Alloy spiking effects are avoided in this manner, since no heat treatment is used other than during the epitaxial growth of n^+-Ge on the GaAs.

It is more difficult to make ohmic contacts to AlGaAs than to GaAs, because of its wider energy gap. In addition, the presence of trace amounts of Al_2O_3 on the surface prevent the formation of an adherent bond, unless RTA is used for this purpose. Very often, the AlGaAs layer is capped with a thin layer of n^+-GaAs in order to avoid both of these problems. This GaAs is removed from all regions other than those where the contact is made, by selective etching techniques which will be described in Chapter 9.

8.6.4   Die Bonds

Contact schemes are also required for forming a mechanically strong bond between the chip and the header.* Here, problems of spike formation are not important, since the active region of the device or microcircuit is 100–200 μm away from the contact region. Consequently, single-layer contacts are invariably used here. Dies are usually bonded by means of a preform which is placed between a plated header and the chip, as shown in Fig. 8.27. The combination is raised to the bonding temperature and pressure is applied to the die in conjunction with a scrubbing motion. The process is carried out in an inert or a slightly reducing (N_2/H_2) gas ambient.

Gold is often used to make a strong bond to the chip in silicon technology; vacuum evaporation of this metal on the back side of the slice avoids both the cost and the handling complexities of a separate preform. In some cases, as in discrete devices, this bond must make an ohmic contact as well. Typically, this is obtained by using gold that is doped with about 1% gallium or arsenic; upon alloying, this results in a Schottky connection to p^+- or n^+-doped silicon respectively. Often, a predoped Au–Ge alloy preform, of eutectic composition (12 wt.% Ge), is used. This provides a better wetting action with the silicon, and results in a Au–Ge–Si bond at a slightly reduced bonding temperature.

Bonds of this type are commonly made at about 390–420°C, which is well above the eutectic temperature for these systems. As a result, they form strong, large-area bonds, with considerable damage to the silicon surface. As mentioned earlier, this is of little consequence since the damage is far from the active regions.

---

*Note the microcircuits are one-sided, and do not require an "ohmic" contact to the header; epoxy bonding is used in many applications.

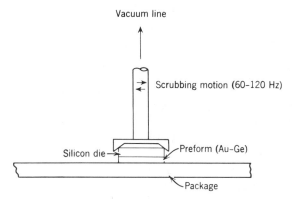

Fig. 8.27 Die bonding arrangement.

Soft solder preforms, consisting of lead–tin alloys, are used for bonding large-area power devices where thermal cycling effects are important [123]. These solders provide stress relief during device operation, and prevent fatigue by plastic deformation of the interface layer between the silicon chip and the header. Typical soft solders are 95% Pb and 5 wt.% Sn, and have melting points of 310–314°C. Alloys of Pb–In–Ag are also used because of their improved thermal cycling characteristics. One example is an alloy of 92.5% Pb, 5% In, and 2.5 wt.% Ag.

Nickel-plated steel headers are commonly used for power devices. The silicon is also nickel-plated, the deposit being commonly formed by electroless plating [124, 125]. This technique, sometimes known as *autocatalytic plating*, depends on the action of a reducing agent in the bath to convert the metallic ions to the metal. The most commonly used reducing agent for electroless nickel baths is NaH_2PO_2, which results in the deposition of nickel films with as much as 3–15% phosphorus. A short heat treatment after deposition greatly improves the adherence of these films, and forms an n^+-contact to the silicon, which becomes phosphorus-doped during this process. Electroless nickel films can also be formed, using dimethylamine borane as the chemical reducing agent. These films have about 0.3–10% dissolved boron, and can be used for contacts to p-silicon.

A single-layer bonding scheme, consisting of Au–Ge, Au–Se, Au–Zn, or Au–Be, can be used to form an ohmic contact bond to the reverse side of GaAs chips. Bonding is usually done by heating the combination to 450°C in a flowing N_2–H_2 gas mixture (10–20% H_2 by volume) for 1–2 min. An alternative approach is to evaporate indium metal on this back surface. Bonding is accomplished by heat treatment at 350°C for 5 min in a flowing N_2/H_2 gas mixture. This results in the formation of a thin regrowth layer of GaInAs; this material has a lower bandgap than GaAs, and thus provides a heterojunction contact to both p- and n-type materials.

In GaAs millimeter-wave integrated circuits, the die bond serves as a ground plane, and all connections are made individually to it. These grounding paths must be extremely short, and of low electrical and thermal resistance. This is accomplished by etching via holes in the GaAs, after the slice has been thinned [126]. The slice is then metallized over its entire back side, and makes separate contacts to the circuit through the via holes. This also serves as a plated heat sink, and reduces the thermal resistance of the active devices to the package.

Back-side metallization is usually of gold, which is sputtered because of its excellent step coverage. A barrier layer of chromium, titanium, Ti–Pt, or Ti–W is used to promote adhesion to the GaAs. Finally, the sputtered gold layer is built up by electroplating. The basic steps in the formation of this contact are shown in Fig. 8.28.

8.7 FILMS FOR SCHOTTKY DIODES

The Schottky diode is firmly established in both silicon and GaAs technology, since it performs functions which cannot be achieved by conventional junction devices. Thus, it has a higher conductance than is possible with p–n structures. In addition, it is a majority carrier device so that it has extremely fast recovery. This combination of characteristics allows its use in the circuit arrangement [127] of Fig. 8.29. Here, the turn-on voltage of the diode is lower than that of the collector–base junction, so that its conduction prevents the transistor from going into saturation. This allows fast switching of the transistor, but still preserves the clamping action that is desirable in saturated logic circuits. As a result, this transistor–diode configuration has a twofold advantage [128] over a conventional gold-doped silicon transistor. First, it is much faster; second, it avoids the need for gold doping and thus eliminates one of the most poorly controlled steps in silicon microcircuit technology.

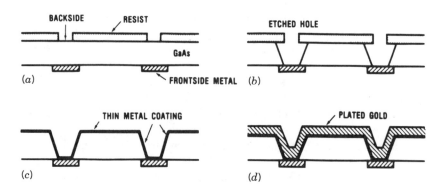

Fig. 8.28 Back side bonding arrangement for gallium arsenide devices. From Williams [126]. With permission from Artech House, Inc.

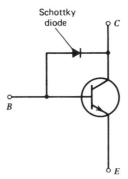

Fig. 8.29 Schottky clamp circuit.

An extremely important advantage of Schottky diode technology with GaAs is that it avoids the high-temperature steps that are associated with junction formation by diffusion. Furthermore, since this structure requires only the deposition of one or more contact materials followed by photolithographic delineation, it is possible to precisely control its physical dimensions. Thus, it can be used as the gate in short-channel, high-speed field-effect transistors. These transistors, with Schottky gates, are the almost universal choice for VLSI applications in GaAs.

The factors which determine diode characteristics can be evaluated by noting that the fermi levels of a metal and a semiconductor will line up if they are placed in intimate contact under conditions of thermal equilibrium. This gives rise to band bending in the semiconductor, and a resultant barrier ϕ_{Bn} or ϕ_{Bp}, as shown in Fig. 8.30 for both p- and n-type material. The height of this barrier is related to the work function of the metal, the resistivity and electron affinity of the semiconductor, the Schottky barrier reduction due to image force lowering of the barrier height, and the nature of the semiconductor surface states [129].

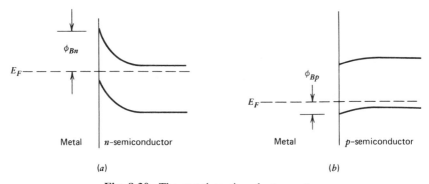

Fig. 8.30 The metal-semiconductor system.

Experimental values of ϕ_{Bn} are relatively insensitive to the doping level, provided that it is below 10^{17} cm^{-3}. These are given in Table 8.4 at the end of this chapter, for a variety of metals and silicides on n-silicon. To an approximation,

$$\phi_{Bn} + \phi_{Bp} \simeq E_g \tag{8.19}$$

where E_g is the energy gap in eV.

From Table 8.4, it is seen that ϕ_{Bn} is considerably less than the contact potential of a p–n junction ($\simeq 1$ eV), so that the "threshold voltage" for significant current conduction (1 A cm^{-2}) will be considerably lower for these diodes than for junction silicon diodes. Typical values range from 0.2 V for Al/n-Si to 0.35 V for PtSi/n-Si devices, as compared to 0.7 V for junction diodes.

The room-temperature value of reverse leakage current of a Schottky diode increases by one decade for each 0.06 eV reduction in the barrier height. It follows, therefore, that this leakage current will be many decades higher than that of a p–n junction structure. Consequently, there are only a few materials which result in acceptable reverse current values for silicon devices, from the point of view of circuit design.

From Eq. (8.19) and Table 8.4, it follows that barrier heights to p-type silicon are considerably lower than those to n-type. Thus, good rectifiers are usually made to n-type silicon, whereas ohmic contacts are more easily made to p-type silicon. Note that Mo and Ti, as well as MoSi$_2$ and TiSi$_2$, have relatively low values of ϕ_{Bn}. These materials find use in Schottky ohmic contacts to silicon that is not heavily doped, as described earlier.

It is worth remembering at this point that a Schottky diode made on a highly doped semiconductor ($>10^{19}$ cm^{-3}) will have an extremely narrow depletion width. Here, conduction is primarily by field-assisted tunneling *through* the barrier, rather than *over* it. As a result, these barriers are relatively transparent to carrier flow regardless of their height, and are used for ohmic contacts, as described in Section 8.6.

GaAs has a relatively large energy gap (1.43 eV) compared to that of silicon (1.11 eV), so that barrier heights to this semiconductor are comparably larger. In consequence, the combinations which result in diodes with acceptable leakage current are much greater. An extensive catalog of such combinations is available in the literature [101], but the practical choice is actually quite small because of lack of good adhesion or because of metallurgical incompatibility with the GaAs. A few of the more common systems are listed in Table 8.5. Of these, materials such as aluminum, chromium, titanium, tungsten, Ti–W, and WSi$_2$ are most suited for GaAs Schottky diodes.

A silicon Schottky diode will have a high electric field at its corners, due to depletion-layer curvature (see Fig. 8.31a). This causes premature avalanche breakdown when reverse-biased, with a "soft" characteristic which varies from device to device, depending on the exact contour of the metal edge. The breakdown voltage of these devices is usually about one-third of what would be obtained for a parallel-plane diode made on the same bulk semiconductor.

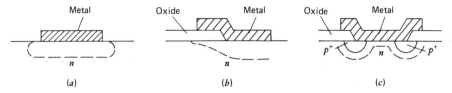

Fig. 8.31 Depletion layer curvature for a Schottky diode.

Two approaches are used for improving this reverse characteristic, and near-ideal behavior can be obtained by these means. The first consists of using a metal field plate which extends over the surface insulator. This reduces the curvature in the depletion-layer edge as shown in Fig. 8.31*b*, and thus delays the onset of premature breakdown. In practice, slight softening of the junction characteristic near breakdown is observed with this structure.

A second approach is to use a deep-diffused guard ring. This curves the depletion layer in the opposite direction (see Fig. 8.31*c*) to what is normally obtained, and reduces the electric field at the edge to a value below that for the parallel-plane region. In consequence, bulk breakdown prevails for this structure, with a near-ideal reverse characteristic. The diffused guard ring approach is conventionally integrated into the transistor-clamp circuit of Fig. 8.29 by means of the configuration of Fig. 8.32. This configuration is commonly used in silicon-based Schottky T^2L logic circuits.

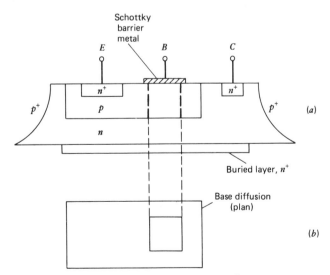

Fig. 8.32 Schottky diode realization in T^2L logic gates.

The breakdown voltage of a GaAs Schottky diode often approaches the bulk breakdown of a *p–n* junction. The reason for this behavior is that the fermi level in GaAs is pinned at the surface, to a value close to its intrinsic level. Depletion-layer curvature is controlled by this pinning effect, and not by the contour of the metal edge.

At first glance, it would appear that the Schottky diode is an extremely simple structure to make, since it can be formed by physical deposition methods, with no high-temperature steps. However, these devices need great care in their fabrication, with particular emphasis placed on cleaning the interface prior to deposition of the contact material. In addition, many combinations are precluded for reasons of metallurgical incompatibility, which is evidenced by changes during subsequent heat treatment or in circuit operation. Thus the fabrication of a reliable Schottky diode can often place severe restrictions on the subsequent processing steps that can be employed.

Aluminum has been considered for silicon diodes, since it can also be used for the ohmic contacts and the metallization. An important problem here is that these diodes are extremely sensitive to subsequent heat treatment [130]. This results in microalloying, thus partially converting the Schottky diode to a $p^+–n$ junction structure. As a result, the forward voltage increases if the device is subjected to heat treatment in excess of 350°C.

Both PtSi and Pd$_2$Si can be used for Schottky diodes on silicon, and are formed by the same method as that described in Section 8.4.3.2. An important advantage of forming these silicides by reaction of the metal with the silicon is that their interface is clean and stable, in addition to being free from alloy spiking. The use of these materials on *n*-silicon results in diodes with threshold voltages of 0.35 V (for PtSi) and 0.25 V (for Pd$_2$Si), respectively.

A cover metal of aluminum can be used here, and forms an adherent contact to the silicide, in addition to serving as the interconnect metal. However, rapid diffusion of aluminum through this silicide limits thermal processing to 350°C. A Ti–W barrier layer is commonly used to prevent this interdiffusion, and allows heat treatment at 500°C without degradation [131]. Chromium has also been used [132] as a barrier layer in this application.

Schottky diodes, using PtSi and Pd$_2$Si, can also be used with a cover metal of gold. A Ti–W layer is used to provide an adherent contact between the gold and the silicide, and also to allow the gold interconnection to bond to the surface insulating layer.

Reliable Schottky diodes, which can tolerate subsequent heat treatment, are considerably more difficult to make with GaAs, since metallurgical compatibility with both gallium and arsenic is a requirement that must be met simultaneously. For example, gold and silver make excellent Schottky barriers to GaAs. However, heat treatment to 260°C results in dissolution of the gallium into the contact metal [118]. This causes morphologically irregular contacts, considerable alloy spiking, and changes in the barrier height. Often, some of the gallium reaches the metal–air surface to form a thin layer of Ga$_2$O$_3$ which makes bonding difficult.

Metals such as aluminum, molybdenum, titanium, and tungsten have relatively inert interfaces with GaAs, and can be used for Schottky diodes. GaAs devices with aluminum metal can be heat treated to 450°C for a few minutes without any change in their diode characteristics. Molybdenum and platinum, on the other hand, forms stable contacts with GaAs to 600°C. They adhere poorly to GaAs, necessitating the use of an interlayer of palladium, titanium, or TiN. Using this approach [133, 134], excellent Schottky diodes can be fabricated which can withstand subsequent heat treatment up to 500–550°C for 30 min.

The use of tungsten improves the situation [135], since the W–Ga eutectic temperature is around 860°C; multilayer Schottky diodes of Au–W–Ti–GaAs are stable with heat treatment up to 650°C. A layer of Ti–W can also be used as the Schottky metal, with an upper layer of gold. This fabrication technique is extremely simple; however, separate layers of these metals have been found to be slightly more stable with subsequent thermal processing.

Both WSi_x and $W_{1-x}N_x$ have been the subject of much study [136–138] since they have stable interfaces with GaAs up to 850°C. Thus, they can be used as post-implantation masks with GaAs. They have been used as gates [139] in the fabrication of HEMT-based integrated circuits made by self-aligned techniques. Interlayers of titanium are sometimes used with these refractory gate materials.

Cleaning procedures, prior to Schottky diode formation, are extremely important since device performance is highly sensitive to the nature of the metal–semiconductor interface. Back-sputtering is generally used for silicon devices. However, its use with GaAs is undesirable since it results in selective loss of surface arsenic, and hence in arsenic vacancy formation. Moreover, such damaged surfaces cannot be annealed by subsequent heat treatment; rather, the vacancy distribution becomes spatially altered during annealing. As a result, wet chemical etching procedures are commonly used for this purpose. Details of some of the more successful of these are given in Chapter 9.

Up to the present time, most research has been aimed at developing GaAs Schottky diodes with the ability to withstand high temperatures without metallurgical degradation. Recent work with deep level transient spectroscopy [140] has shown that a number of the materials used with these devices introduce multiple deep levels into the semiconductor, with a consequent degradation of its electrical properties. Thus, an "ideal" Schottky diode technology for GaAs has not been realized at the present time.

TABLE 8.1   Properties of Silicon Nitride Films[a]

Property	Growth Process	
	LPCVD	PECVD
Si/N ratio	0.75	0.8–1
Density (g cm^{-3})	2.8–3.1	2.5–2.8
Refractive index	2.0–2.1	2.0–2.1
Dielectric constant	6–7	6–9
Dielectric strength (MV/cm)	10	6
Thermal expansion coefficient (C$^{°-1}$)	4×10^{-6}	4–7×10^{-6}
Step coverage	Fair	Conformal
Stress at 23°C on silicon (dyn cm^{-2})	1.2–1.8×10^{10} (tensile)	1–8×10^{9} (tensile or comprehensive)
Wet chemical etch rate:		
49% HF at 23°C	80 Å/min	1500–3000 Å/min
85% H$_3$PO$_4$ at 155°C	15 Å/min	100–200 Å/min
85% H$_3$PO$_4$ at 180°C	120 Å/min	600–1000 Å/min

[a] Adapted from Kern and Rosler [38].

TABLE 8.2   Some Properties of Silicides of Interest[a]

Silicide	Lowest Binary Eutectic Temperature (°C)	Specific Resistivity ($\mu\Omega$-cm)	TCE $\times 10^{-6}$ (deg^{-1})
CoSi$_2$	900	10–18	10.14
HfSi$_2$	1300	45–50	
MoSi$_2$	1410	100	8.25
NbSi$_2$	1295	50	8.4
NiSi$_2$	966	50–60	12.06
Pd$_2$Si	720	30–35	
PtSi	830	28–35	
TaSi$_2$	1385	35–45	8.8
TiSi$_2$	1330	13–25	12.5
VSi$_2$	1385	50–55	11.2
WSi$_2$	1440	70	6.25
ZrSi$_2$	1355	35–40	8.3

[a] Adapted from Murarka [55].

TABLE 8.3   Properties of Silicides Formed by Sintering Co-Sputtered Metal and Silicon[a]

Considerations	
Silicides can be formed on	Any substrate
Control of metal-to-silicon ratio	Yes
Purity of the silicide	Good
Applicability to all metals	Yes
Depositional environmental sensitivity	Yes
Sintering environmental sensitivity	Not as sensitive
After sintering, surface is	Smooth
Possibility of selective etching of metal after sintering	No
Ability to deposit sandwich of metal and silicon	Yes
Back-sputtering to clean surface prior to deposition	Yes
Silicide on back surface	No

[a]Adapted from Murarka [56].

TABLE 8.4   Barrier Heights of Some Metals and Silicides on n-Si[a]

Metal	ϕ_{Bn} (eV)
Al	0.7–0.75
Au	0.78–0.80
CoSi	0.68
CoSi$_2$	0.65
CrSi$_2$	0.57
Mo	0.57
MoSi$_2$	0.55
Ni	0.58
NiSi	0.66–0.7
Pd	0.75
Pd$_2$Si	0.74
Pt	0.83
PtSi	0.85–0.87
TaSi$_2$	0.59
Ti	0.50
TiSi$_2$	0.60
RhSi	0.68–0.70
W	0.65
WSi$_2$	0.65
ZrSi$_2$	0.55

[a]Adapted from Murarka [55].

TABLE 8.5   **Barrier Heights of Some Metals and Silicides on n-GaAs**[a]

Metal/Silicide	ϕ_{Bn} (eV)
Ag	0.88 – 093
Al	0.8 – 0.85
Au	0.89 – 0.95
Cr	0.67 – 077
Cs	0.7
Cu	0.87 – 0.96
Ni	0.77
Pd	0.85 – 0.91
Pt	0.86 – 0.94
Ru	1.09
Ti	0.83
W	0.64 – 0.75

[a] Adapted from Gupta and Khokle [101].

REFERENCES

1. J. L. Vossen and W. Kern, Eds. , *Thin Film Processes*, Academic Press, New York, 1978.

2. R. E. Honig and D. E. Kramer, Vapor Pressure Data for the Solid and Liquid Elements, *RCA Rev.* **30**, 285 (1969).

3. J. F. O'Hanlon, *A Users Guide to Vacuum Technology*, John Wiley, 1980.

4. J. D. Cobine, *Gaseous Conductors*, Dover, New York, 1958.

5. H. M. Naguib and R. Kelly, On the Increase in the Electrical Conductivity of MoO_3 and V_2O_5 Following Ion Bombardment, *J. Phys. Chem. Solids* **33**, 1751 (1972).

6. H. W. Pickering, Ion Sputtering of Alloys, *J. Vac. Sci. Technol.* **13**, 618 (1976).

7. B. Mattson, CVD Films for Interlayer Dielectrics, *Solid State Technol*, (Jan. 1980) p. 60.

8. A. R. Neureuther, C. H. Ting, and C-Y. Liu, Application of Line-Edge Profile Simulation to Thin-Film Deposition Processes, *IEEE Trans. Electron Dev.* **ED-27**, 1449 (1980).

9. L. I. Maissel and R. Glang, *Handbook of Thin Film Technology*, McGraw-Hill, New York, 1970.

10. S. P. Timoshenko and J. M. Gere, *Theory of Elastic Stability*, McGraw-Hill, New York, 1961.

11. A. G. van Nie, A Method for the Determination of the Stress in, and Young's Modulus of, Silicon Nitride Passivation Layers, *Solid State Technol.*, (Jan. 1980) p. 81.

12. H. B. Huntington and A. R. Grone, Current Induced Marker Motion in Gold Wires, *J. Phys. Chem. Solids* **20** (1/2), 76 (1961).

13. M. J. Altardo, R. Rutledge, and R. C. Jack, Statistical Metallurgical Model for Electromigration Failure in Aluminum Thin Film Conductors, *J. Appl. Phys.* **42**, 4343 (1971).

14. F. d'Heurle, Electromigration and Failure in Electronics: An Introduction, *Proc. IEEE* **59**, 1409 (1971).

15. H. -U. Schreiber, Activation Energies for the Different Electromigration Mechanisms in Aluminum, *Solid State Electron* **24**, 583 (1981).

16. B. Gelernt, Selecting an Organosilicon Source for LPCVD, *Semicond. Int.* (March 1990) p. 92.

17. K. Fujino, Y. Nishimoto, N. Tokumasu, and K. Maeda, Silicon Dioxide Deposition by Atmospheric Pressure and Low Temperature CVD Using TEOS and Ozone, *J. Electrochem. Soc.* **137**, 2883 (1990).

18. N. Goldsmith and W. Kern, The Deposition of Vitreous Silicon Dioxide from Silane, *RCA Rev.* **28**, 153 (1967).

19. B. J. Baliga and S. K. Ghandhi, Growth of Silica and Phosphosilicate Films, *J. Appl. Phys.* **44**, 990 (1973).

20. W. Kern, G. L. Schnable, and A. W. Fisher, CVD Glass Films for Passivation of Silicon Devices: Preparation, Composition, and Stress Properties, *RCA Rev.* **37**, 3 (1976).

21. Y. Shioya and M. Maeda, Comparison of Phosphosilicate Glass Films Deposited by Three Different Chemical Vapor Deposition Methods, *J. Electrochem. Soc.* **133**, 1943 (1986).

22. B. J. Baliga and S. K. Ghandhi, Lateral Diffusion of Zinc and Tin in Gallium Arsenide, *IEEE Trans. Electron Dev.* **ED-21**, 410 (1974).

23. B. J. Baliga and S. K. Ghandhi, PSG Masks for Diffusion in Gallium Arsenide, *IEEE Trans. Electron Dev.* **ED-19**, 761 (1972).

24. M. Yamin, Observations on Phosphorus Stabilized SiO_2 Films, *IEEE Trans. Electron Dev.* **ED-13**, 256 (1966).

25. R. B. Comizzoli, Aluminum Corrosion in the Presence of Phosphosilicate Glass and Moisture, *RCA Rev.* **37**, 483 (1976).

26. R. M. Levin and K. Evans-Lutterodt, The Step Coverage of Undoped and Phosphorus-Doped SiO_2 Glass Films, *J. Vac. Sci. Technol.* **B1**, 54 (1983).

27. W. Kern and G. L. Schnable, Chemically Vapor-Deposited Borophosphosilicate Glasses for Silicon Device Applications, *RCA Rev.* **43**, 423, (1982).

28. J. E. Tong, K. Schertenleib, and R. A. Carpio, Process and Film Characterization of PECVD Borophosphosilicate Films for VLSI Applications, *Solid State Technol.* (Jan. 1984) p. 161.

29. D. S. Williams and E. A. Dein, LPCVD of Borophosphosilicate Glass from Organic Reactants, *J. Electrochem. Soc.* **134**, 657 (1987).

30. S. M. Fisher, H. Chino, K. Maeda, and Y. Nishimoto, Characterizing B-, P-, and Ge-Doped Silicon Oxide Films for Interlevel Dielectrics, *Solid State Technol.* (Sept. 1993) p. 55.

31. K. Yamasaki, K. Asai, and K. Kuramada, GaAs LSI-Directed MESFETs with Self-Aligned Implantation for n^+ Layer Technology (SAINT), *IEEE Trans. Electron. Dev.* **ED-29**, 1772 (1982).

32. T. Enoki, K. Yamasaki, K. Osafune, and K. Ohwada, 0.3 μm Advanced SAINT FETs Having Asymmetric n^+-Layers for Ultra-High-Frequency GaAs MMICs, *IEEE Trans. Electron. Dev.* **ED-35**, 18 (1988).

33. K. Matsuzaki, H. Hirabayashi, and M. Saga, Characterization of Reactively Sputtered Silicon Nitride, *J. Electrochem. Soc.* **139**, 3259 (1992).

34. R. Ginsburgh, D. L. Heald, and R. C. Neville, Silicon Nitride Chemical Vapor Deposition in a Hot Wall Diffusion System, *J. Electrochem. Soc.* **125**, 1557 (1978).

35. W. Kern and G. L. Schnable, Low-Pressure Chemical Vapor Deposition for Very Large-Scale Integration Processing—A Review, *IEEE Trans. Electron Dev.* **ED-26**, 647 (1979).

36. R. S. Rosler, W. C. Bensing, and J. Baldo, A Production Reactor for Low Temperature Plasma-Enhanced Silicon-Nitride Deposition, *Solid State Technol.* (June 1976) p. 45.

37. A. K. Sinha, H. J. Levinstein, T. E. Smith, G. Quintana, and S. E. Haszko, Reactive Plasma Deposited Si–N Films for MOS-LSI Passivation, *J. Electrochem. Soc.* **125**, 60 (1978).

38. W. Kern and R. S. Rosler, Advances in Deposition Processes for Passivation Films, *J. Vac. Sci. Technol.* **14**, 1082 (1977).

39. P. M. Petroff, G. A. Rozgonyi, and T. T. Shen, Elimination of Process-Induced Stacking Faults by Preoxidation Gettering of Si Wafers, *J. Electrochem. Soc.* **123**, 565 (1976).

40. D. M. Brown, P. V. Gray, F. K. Heumann, H. R. Philipp, and E. A. Taft, Properties of $Si_xO_yN_z$ Films on Si, *J. Electrochem. Soc.* **115**, 211 (1968).

41. W. R. Knolle and J. W. Osenbach, Plasma-Deposited Silicon Oxynitride from Silane, Nitrogen, and Carbon Dioxide or Carbon Monoxide or Nitric Oxide, *J. Electrochem. Soc.* **139**, 3310 (1992).

42. R. S. Rosler, Low Pressure CVD Production Processes for Poly, Nitride and Oxide, *Solid State Technol.* (April 1977) p. 63.

43. R. J. Gieske, J. J. McMullen, and L. F. Donaghey, Low Pressure Chemical Vapor Deposition of Silicon, in *Chemical Vapor Deposition*, L. F. Donaghey, P. Rai-Choudhury and R. N. Tauber, Eds. , The Electrochemical Society, Princeton, NJ, 1977, p. 183.

44. D. Foster, A. Learn, and T. Kamins, Silicon Films Deposited in a Vertical-Flow Reactor, *Solid State Technol.*, (May 1986) p. 227.

45. F. C. Eversteyn and B. H. Put, Influence of AsH_3, PH_3 and B_2H_6 on the Growth Rate and Resistivity of Polycrystalline Films Deposited from a SiH_4–H_2 Mixture, *J. Electrochem. Soc.* **120**, 106 (1973).

46. B. S. Myerson and W. Olbricht, Phosphorus-Doped Polycrystalline Silicon Via LPCVD: 1. Process Characterization, *J. Electrochem. Soc.* **131**, 2361 (1984).

47. H. Yamamoto, T. Wada, O. Kudoh, and M. Sakamoto, Polysilicon Interconnection Technology for IC Device, *J. Electrochem. Soc.* **126**, 1415 (1979).

48. J. Y. W. Seto, The Electrical Properties of Polycrystalline Silicon Films, *J. Appl. Phys.* **46**, 5247 (1975).

49. K. Ohyu, Y. Wade, S. Iijima, and N. Natsuaki, Highly Reliable Thin Silicon Dioxide Layers Grown on Heavily Doped Poly-Si by Rapid Thermal Oxidation, *J. Electrochem. Soc.* **137**, 7, 2261 (1990).

50. G. Mende and J. Wende, Breakdown Field Strengths of Anodically and Thermally Grown Metal/Oxide/Semiconductor Structures on Polycrystalline Silicon, *Thin Solid Films* **142**, 21 (1986).

51. H. R. Maxwell, Jr. and W. R. Knolle, Densification of SIPOS, *J. Electrochem. Soc.* **128**, 576 (1981).

52. T. Matsushita, T. Aoki, T. Ohtsu, H. Yamoto, H. Hayashi, and M. Okayama, Highly Reliable High-Voltage Transistors by Use of the SIPOS Process, *IEEE Trans. Electron Dev.* **ED-23**, 826 (1976).

53. K. N. Tu and J. W. Mayer, Silicide Formation in *Thin Films–Interdiffusion and Reactions*, J. M. Poate, K. N. Tu, and J. W. Mayer, Eds. , John Wiley and Sons, New York, 1978, p. 359.

54. T. P. Chow and A. J. Steckl, Refractory Metal Silicides: Thin Film Properties and Processing Technology, *IEEE Trans. Electron Dev.* **ED-30**, 1480 (1983).

55. S. P. Murarka, *Silicides for VLSI Applications*, Academic Press, New York, 1983.

56. S. P. Murarka, Refractory Silicides for Integrated Circuits, *J. Vac. Sci. Technol.* **17**, 775 (1980).

57. C. Bernard, R. Madar and Y. Pauleau, Chemical Vapor Deposition of Refractory Metal Silicides for VLSI Metallization, *Solid State Technol.* (Feb. 1989) p. 79.

58. K. C. Saraswat, D. L. Brors, J. A. Fair, K. A. Mounig, and R. Beyers, Properties of Low Pressure CVD Tungsten Silicide for MOS VLSI Interconnections, *IEEE Trans. Electron Dev.* **ED-30**, 1497 (1983).

59. T. Hara, T. Miyamoto, H. Hagiwara, E. I. Bromley, and W. R. Harshbarger, Composition of Tungsten Silicide Films Deposited by Dichlorosilane Reduction of Tungsten Hexafluoride, *J. Electrochem. Soc.* **137**, 2955 (1990).

60. S. Zermsky, W. Hammer, F. d'Heurle, and J. Baglin, Oxidation Mechanisms in WSi_2 Thin Films. *Appl. Phys. Lett.* **33**, 76 (1978).

61. M. Bartur and M. A. Nicolet, Thermal Oxidation of Transition Metal Silicides on Si: A Summary, *J. Electrochem. Soc.* **131**, 371 (1984).

62. R. M. Pretorius, M. A. E. Wandt, J. E. McLeod, A. P. Botha, and C. M. Comrie, Determination of the Diffusing Species and Diffusion Mechanism During CoSi,

NiSi and PtSi Formation by Using Radioactive Silicon as a Tracer, *J. Electrochem. Soc.* **136**, 839 (1989).

63. O. Thomas, P. Gas, F. M. d'Heurle, F. K. LeGoues, A. Michel, and G. Scilla, Diffusion of Boron, Phosphorus and Arsenic Implanted in Thin Films of Cobalt Disilicide, *J. Vac. Sci. Technol.* **A6**(3), 1736 (1988).

64. G. A. Mattiussi, Titanium Silicide Formation by Ion Beam Mixing and Rapid Thermal Anneal, *J. Vac. Sci. Technol.* **B4**, 1352 (1986).

65. D. M. Brown, W. E. Engeler, M. Garfinkel, and P. V. Gray, Self-Registered Molybdenum-Gate MOSFET, *J. Electrochem. Soc.* **115**, 874 (1968).

66. F. Yanagawa, K. Kiuchi, T. Hosoya, T. Tsuchiya, T. Amazawa, and T. Mano, A 1 μm Mo-Gate 65 kbit MOS Ram, in *Transactions of the IEEE Electron Devices Conference*, Washington, D.C., 1979, p. 362.

67. J. G. Donaldson and H. Kenworthy, *Vapor Deposition of Molybdenum–Tungsten Alloys*, Bureau of Mines, No. 6853 (1966).

68. T. Sugano, H-K. Chou, M. Yoshida, and T. Nishi, Chemical Deposition of Mo on Si, *Jpn. J. Appl. Phys.* **7**, 1028 (1968).

69. J. M. Shaw and J. A. Amick, Vapor-Deposited Tungsten as a Metallization and Interconnection Material for Silicon Devices, *RCA Rev.* **30**, 306 (1970).

70. E. K. Broadbent and W. T. Stacy, Selective Tungsten Processing by Low Pressure CVD, *Solid State Technol.*, (Dec. 1985) p. 51.

71. Y. Pauleau and P. Lami, Kinetics and Mechanism of Selective Tungsten Deposition by LPCVD, *J. Electrochem. Soc.* **132**, 2779 (1985).

72. C. Y. Ting and M. Wittmer, The Use of Titanium-Based Contact Barrier Layers in Silicon Technology, *Thin Solid Films* **96**, 327 (1982).

73. M. Wittmer and H. Melchior, Applications of TiN Thin Films in Silicon Devices, *Thin Solid Films* **93**, 397 (1982).

74. S. A. Eshraghi, G. E. Georgiou, R. Liu, C. Beairsto, and K. P. Cheung, Electrical Degradation of Al/TiW/CoSi$_2$ Shallow Junctions, *J. Vac. Sci. Technol.* **B9**(1), 69 (1991).

75. D. Pramanik, V. Jain, Barrier Metals for ULSI: Processing and Reliability, *Solid State Technol.* (May 1991) p. 97.

76. A. Sherman, Growth and Properties of LPCVD Titanium Nitride as a Diffusion Barrier for Silicon Device Technology, *J. Electrochem. Soc.* **137**, 1982 (1990).

77. G. S. Sandhu, S. G. Meikle, and T. T. Doan, Metalorganic Chemical Vapor Deposition of TiN Films for Advanced Metallization, *Appl. Phys. Lett.* **62**(3), 240 (1993).

78. E. Kolawa, F. C. T. So, J. L. Tandon, and M-A. Nicolet, Reactively Sputtered W–N Films as Diffusion Barriers in GaAs Metallizations, *J. Electrochem. Soc.* **134**, 1759 (1987).

79. M. L. Barry and P. Olofsen, Doped Oxides as Diffusion Sources–I. Boron into Silicon, *J. Electrochem. Soc.* **116**, 854 (1969).

80. M. L. Barry, Doped Oxides as Diffusion Sources–II. Phosphorus into Silicon, *J. Electrochem. Soc.* **117**, 1405 (1970).

81. A. W. Fisher, J. A. Amick, H. Hymann and J. H. Scott, Jr. , Diffusion Characteristics and Applications of Doped Silicon Dioxide Layers Deposited from Silane, *RCA Rev.* **29**, 533 (1968).

82. G. L. Patton, J. C. Bravman, and J. D. Plummer, Physics, Technology and Modelling of Polysilicon Emitter Contacts for VLSI Bipolar Transistors, *IEEE Trans. Electron Dev.* **ED-33**, 1754 (1986).

83. J. R. Shealy, B. J. Baliga, and S. K. Ghandhi, Preparation and Properties of Zinc Oxide Films Grown by the Oxidation of Diethylzinc, *J. Electrochem. Soc.* **128**, 558 (1981).

84. S. K. Ghandhi and R. J. Field, Precisely Controlled p^+-Diffusion in GaAs, *Appl. Phys. Lett.* **38**, 267 (1981).

85. B. J. Baliga and S. K. Ghandhi, Planar Diffusion in Gallium Arsenide from Tin-Doped Oxides, *J. Electrochem. Soc.* **126**, 135 (1979).

86. Integrated Circuit Engineering Corp., Integrated Circuits Course, *Electron Eng.* 63–102 (1966).

87. G. L. Schnable and R. S. Keen, Aluminum Metallization—Advantages and Limitations for Integrated Circuit Applications, *Proc. IEEE* **57**, 1570 (1971).

88. A. J. Learn, Evolution and Current Status of Aluminum Metallization, *J. Electrochem. Soc.* **123**, 894 (1976).

89. I. Ames, F. M. d'Heurle, and R. Horstmann, Reduction of Electromigration in Aluminum Films by Copper Doping, *IBM J. Res. Dev.* **14**, 461 (1970).

90. S. Mayumi, I. Murozono, H. Nanatsue, and S. Udea, Corrosion-Induced Contact Failures in Double Level Al–Si–Cu Metallization, *J. Electrochem. Soc.* **137** 1861 (1990).

91. L. E. Terry and R. W. Wilson, Metallization Systems for Silicon Integrated Circuits, *Proc. IEEE* **57**, 1580 (1969).

92. S. P. Murarka, H. J. Levenstein, I. Bleck, T. T. Sheng, and M. H. Reed, Investigation of the Ti-Pt Diffusion Barrier for Gold Beam Leads on Aluminum, *J. Electrochem. Soc.* **125**, 156 (1978).

93. H. O. Pierson, Aluminum Coatings by the Decomposition of Alkyls, *Thin Solid Films* **45**, 257 (1972).

94. S. Motojima and H. Mizutani, Preparation of TiN Films by Photochemical Vapor Deposition, *Appl. Phys. Lett.* **54**, 1104 (1989).

95. D.-H. Kim, R. H. Wentorf, and W. N. Gill, Low Pressure Chemically Vapor Deposited Copper Films for Advanced Device Metallization, *J. Electrochem. Soc.* **140**, 3273 (1993).

96. N. Away and Y. Arita, Double-Level Copper Interconnections Using Selective Copper CVD, *J. Electron Mater.* **21**, 959 (1992).

97. R. W. Lade and A. G. Jordan, A Study of Ohmicity and Exclusion in High–Low Semiconductor Devices, *IEEE Trans. Electron Dev.* **ED-10**, 268 (1963).

98. C. Y. Chang, Y. K. Fang, and S. M. Sze, Specific Contact Resistance of Metal Semiconductor Barriers, *Solid State Electron* **14**, 541 (1971).

99. R. Holm, *Electric Contacts: Theory and Application*, Springer-Verlag, New York, 1967.

100. B. L. Sharma, Ohmic Contacts to III–V Compound Semiconductors, *Semiconductors and Semimetals*, Vol. 15, R. K. Willardson and A. C. Beer, Eds., Academic Press, New York, 1981.

101. R. P. Gupta and W. S. Khokle, Gallium-Vacancy-Dependent Diffusion Model of Ohmic Contacts to GaAs, *Solid-State Electron.* **28**, 823 (1985).

102. C. R. M. Grovenor, Au/Ge Based Ohmic Contacts to GaAs, *Solid State Electron.* **24**, 792 (1981).

103. R. E. Reid-Hill, *Physical Metallurgy Principles*, 2nd Edition, Van Nostrand, New York, 1981.

104. J. O. McCaldin and H. Sankur, Diffusivity and Solubility of Si in the Al Metallization of Integrated Circuits, *Appl. Phys. Lett.* **19**, 524 (1971).

105. K. Nakamura, M-A. Nicolet, J. W. Mayer, R. J. Blattner, and C. A. Evans, Jr., Interaction of Al Layers with Polycrystalline Silicon, *J. Appl. Phys.* **46**, 4678 (1975).

106. A. J. Learn and R. S. Nowicki, Methods for Minimizing Silicon Regrowth in Aluminum Films, *Appl. Phys. Lett.* **35**, 611 (1979).

107. M. Finetti, P. Ostoja, S. Solmi, and G. Soncini, Aluminum–Silicon Ohmic Contact on Shallow n^+/p Junctions, *Solid State Electron* **23**, 255 (1980).

108. J. M. Andrews, The Role of the Metal–Semiconductor Interface in Silicon Integrated Circuit Technology, *J. Vac. Sci. Technol.* **11**, 972 (1974).

109. Y. Pauleau, Interconnect Materials for VLSI Circuits: Part II: Metal to Silicon Contacts, *Solid State Technol.*, (April 1987) p. 155.

110. J. Middelhoek and A. Kooy, Polycrystalline Silicon as a Diffusion Source and Interconnect Layer in I²L Realizations, *IEEE J. Solid-State Circuits* **SC-12** (2), 135 (1977).

111. C. J. Kircher, Metallurgical Properties and Electrical Characteristics of Palladium Silicide Contacts, *Solid State Electron.* **14**, 507 (1971).

112. M. J. Kim, D. M. Brown, S. S. Cohen, P. Piacente, and B. Gorowitz, Mo/TiW Contact for VLSI Applications, *IEEE Trans. Electron Dev.* **ED-32**, 1321 (1985).

113. S. W. Sun, J. J. Lee, B. Boeck, and R. L. Hance, Al/W/TiN$_x$/TiSi$_y$/Si Barrier Technology for 1. 0 μm Contacts, *IEEE Electron. Dev. Lett.* **9**, 71 (1988).

114. J-P Noel, D. C. Houghton, G. Este, F. R. Shepherd, and H. Plattner, Characteristics of d. c. Magnetron, Reactively Sputtered TiN$_x$ Films for Diffusion Barriers in III–V Semiconductor Metallization, *J. Vac. Sci. Technol.* **A(2)**, 284 (1984).

115. M. P. Lepselter, Beam-Lead Technology, *Bell Syst. Tech. J.* **45**, 233 (1966).

116. W. T. Anderson, Jr. , A. Christou, and J. E. Davey, Development of Ohmic Contacts for GaAs Devices Using Epitaxial Ge Films, *IEEE J. Solid State Circuits* **SC-13**, 430 (1978).

117. S. S. Gill, J. R. Dawsey and A. G. Culls, Contact Resistivity of IR Lamp Alloyed Au–Ge Metallization on GaAs, *Electron. Lett.* **20**, 944 (1984).

118. D. C. Miller, The Alloying of Gold and Gold Alloy Ohmic Contact Metallizations with Gallium Arsenide, *J. Electrochem. Soc.* **137**, 467 (1980).

119. F. C. T. So, E. Kolawa, J. L. Tandon, and M-A. Nicolet, Solid-Phase Ohmic Con-

tact to *p*-GaAs with W and W-N Diffusion Barriers, *J. Electochem. Soc.* **134**, 1755 (1987).

120. K. Fricke, H. L. Hartnagel, R. Schütz, G. Schweeger, and J. Würfl, A New GaAs Technology for Stable FETs at 300°C, *IEEE Electron Dev. Lett.* **10**, 577 (1989).

121. P. H. Holloway, Gold/Chromium Metallizations for Electronic Devices, *Solid State Technol.* (Feb. 1980) p. 109.

122. W. J. Delvin, C. E. C. Wood, R. Stall, and L. F. Eastman, A Molybdenum Source, Gate and Drain Metallization System for GaAs MESFET Layers Grown by Molecular Beam Epitaxy, *Solid State Electron.* **23**, 823 (1980).

123. S. K. Ghandhi, *Semiconductor Power Devices*, John Wiley and Sons, New York, 1977.

124. F. A. Lowenheim, Ed. , *Modern Electroplating*, John Wiley and Sons, New York, 1974.

125. A. Brenner and G. Riddell, Deposition of Nickel and Cobalt by Chemical Reduction, *J. Res. NBC* **39**, 385 (1947).

126. R. Williams, *Modern GaAs Processing Methods*, Artech House, Boston, MA, 1990.

127. R. H. Baker, Maximum Efficiency Switching Circuits, *MIT Lincoln Lab Rep.* **TR-110**, (1956).

128. A. Tarui, Y. Hayashi, H. Teshima, and T. Sekigawa, Transistor Schottky-Barrier-Diode Integrated Logic Circuit, *J. Solid State Circuits* **SC-4**, 3 (1969).

129. S. M. Sze, *Physics of Semiconductor Devices*, 2nd edition, John Wiley and Sons, New York, 1981.

130. K. Chino, Behavior of Al–Si Schottky Barrier Diodes Under Heat Treatment, *Solid State Electron.* **16**, 119 (1973).

131. P. C. Parekh, R. C. Sirrine, and P. Lemieux, Behavior of Various Silicon Schottky Barrier Diodes Under Heat Treatment, *Solid State Electron.* **19**, 493 (1976).

132. J. O. Olowolafe, M.-A. Nicolet, and J. W. Mayer, Chromium Thin Film as a Barrier to the Interaction of Pd_2Si with Al, *Solid State Electron.* **20**, 413 (1977).

133. C. Y. Nee, C-Y Chang, T. F. Cheng, and T. S. Huang, An Improved Mo/*n*-GaAs Contact by Interposition of a Thin Pd Layer, *IEEE Electron. Dev. Lett.* **EDL-9**, 315 (1988).

134. A. K. Sinha, T. E. Smith, M. H. Read, and J. M. Poate, *n*-GaAs Schottky Diodes Metallized with Ti and Pt/Ti, *Solid State Electron.* **19**, 489 (1976).

135. Y. T. Kim, C. W. Lee, C. W. Han, J. S. Hong, and S. K. Min, Characteristics of Plasma Deposited Schottky Contacts to GaAs, *Appl. Phys. Lett.* **61**, 1205 (1992).

136. S. Takatani, N. Matsuoka, J. Shigeta, N. Hashimoto, and N. Nakashima, Thermal Stability of WSi_x/GaAs Interface, *J. Appl. Phys.* **61**, 220 (1987).

137. A. E. Geissberger, R. A. Sadler, F. A. Leyenaar and M. L. Balzan, Investigation of Reactively Sputtered Tungsten Nitride as High Temperature Stable Schottky Contacts to GaAs, *J. Vac. Sci. Technol.* A(4), 3091 (1986).

138. N. Braslau, Contact and Metallization Problems in GaAs Integrated Circuits, *J. Vac. Sci. Technol.* A(4), 3085 (1986)]

139. N. C. Cirillo, Jr. , H. K. Chung, P. T. Vold, M. K. Hibbs-Brenner, and A. M. Fraasch, Refractory Metal Silicides for Self-Aligned Modulation Doped n^+-(Al, Ga)As/GaAs Field-Effect Transistor Integrated Circuits, *J. Vac. Sci. Technol.* **B3**, 1680 (1985).

140. D. L. Partin, A. G. Milnes, and L. F. Vassamillet, Hole Diffusion Lengths in VPE GaAs and GaAs$_{0.6}$P$_{0.4}$ Treated with Transition Metals, *J. Electrochem. Soc.* **126**, 1584 (1979).

CHAPTER 9

ETCHING AND CLEANING

Etching and cleaning processes are involved at many points in the microcircuit fabrication process. Thus, saw-cut slices of suitably oriented semiconductor material are first mechanically lapped to remove gross damage, and then chemically etched and polished to obtain an optically flat, damage-free surface. Often, this involves the removal of many microns of surface material. Next, slices are chemically cleaned and scrubbed to remove contaminants produced by handling and storing, before being covered with an initial protective layer of thermally grown SiO_2 (for silicon-based microcircuits) or deposited Si_3N_4 (for GaAs circuits). Etching processes, in conjunction with patterning, are used to cut openings in this protective film through which implants or diffusion are made to form the semiconductor regions. This process is repeated until all the components are formed.

Further etching and patterning processes are used to delineate one or more layers of metallization. This is often followed by protective coatings of deposited phosphosilicate glass (PSG) or Si_3N_4. Again, both etching and patterning processes are used to cut holes for access to the bonding pads, to which leads can be attached.

The term *etching* is used to describe all techniques by which material can be uniformly removed from a wafer as in surface polishing, or locally removed as in the delineation of a pattern for a microcircuit. It also includes chemical machining of a semiconductor as part of the fabrication process, as well as the delineation of surface features such as defects.

In addition to semiconductors, wet chemical etching is used for the dissolution of a wide variety of grown and deposited films of the type described in Chapters 7 and 8. This process involves simple dissolution of the material, as well as its conversion to a soluble species which can be dissolved in the etching medium.

Dry processes are also used for material removal, and can be physical or chemical in character. Physical etching processes involve the removal of material by momentum transfer from a rapidly moving inert projectile (usually argon). *Ion milling* is an important mechanical etching process of this type, and has been [1] appropriately referred to as 'sandblasting on an atomic scale." Another process that is often used is *sputtering*, which is similar in its essential features to ion milling. Both of these techniques require the formation of a gas discharge to produce high-velocity argon ions. However, chemically reactive energetic species are not involved in these processes, which are essentially physical in nature.

Dry chemical etching processes are sometimes carried out in the gas phase, and are restricted to special situations, such as the *in situ* etching of silicon prior to epitaxy, as described in Chapter 5. They can be made considerably more versatile if they can be carried out at low temperature, by means of energetic species of the gaseous reactants. These species can be produced by passing the reactant through a gas discharge, and delivering it downstream to the material to be etched.

Finally, the advantages of both dry physical and chemical processes can be combined by immersing the material to be etched directly in the gas discharge, so that it is subjected to both reactive chemical species *and* ion bombardment. The dominant effect of ion bombardment is to weaken the chemical bonding of the surface material, so that it can be chemically attacked by the energetic species. Here, the atom–substrate reactions of the energetic species are enhanced by the ion bombardment, so that the etch rate is usually greater than the sum of the two separate processes. This process is known as *reactive ion etching*.

This chapter will describe the many etching processes that are used in the fabrication of microcircuits. Both wet and dry processes will be considered, the latter being especially important for VLSI fabrication technology. Pattern transfer processes, which are often used in conjunction with etching, will be described in the following chapter.

The cleaning of semiconductor surfaces will also be treated here. This is necessary to remove a variety of organic and inorganic contaminants, often of unknown origin, which are present on the as-purchased semiconductor surface. These include (a) films of native oxide, airborne contaminants such as common salt and bacteria and (b) chemical residues such as oil and plasticizer films from the packages in which wafers are shipped. Cleaning between processing steps is also important in order to avoid the introduction of contaminants.

Many of these cleaning processes are closely related to etching, in that they involve some removal of the semiconductor surface; however, this is not their primary intent. They play an important role in microcircuit fabrication technology.

9.1 WET CHEMICAL ETCHING

Wet chemical etching of any material can be considered as a sequence of three steps: transport of the reactant to the surface, reaction at the surface, and movement of reaction products into the volume of the etchant solution. Each of these can serve as the rate limiter and thus dominate the overall process.

An etch process which is limited by the rate of surface reaction will tend to enhance surface roughness and promote faceting, since the surface activity is a strong function of localized defects and crystallographic orientation. On the other hand, etching can be limited by the rate of diffusion of the etchant through a stagnant layer which covers the surface. This results in an etch which is polishing in character; now, a rough surface with many protuberances and facets will tend to become smooth in this process.

Removal of the reaction products can also be a rate-limiting factor. In practice, this determines the "equilibrium" amount of reaction product that is left on the substrate. It also serves as an additional layer to the stagnant film through which reactant must diffuse before etching occurs; thus, it can slow down the etch rate.

In all cases, the etch rate can be altered by rapid stirring to aid in removal of reaction products and gas bubbles, or by increasing the temperature of the etch solution. Typically, a doubling of etch rate occurs with each $10°C$ rise in temperature. As a result, the development of etch processes for use in a manufacturing environment requires that attention be paid to these factors, in addition to the choice of chemicals and their concentrations.

The temperature-dependent characteristics of etching can be used to advantage in situations where it is essential to keep residual surface contamination to a minimum. Here, the strategy is to use a high dilution with deionized water, and maintain a reasonable etch rate by operating at an elevated temperature. In other situations, where precise etch control is desired, the use of a cooled etch solution is indicated.

A number of chemical reagents, and their mixtures, are used for etching purposes. Many of these are available in "transistor-grade" purity and are preferred in order to minimize contamination of the semiconductor during processing. Water is an intrinsic component of all of these reagents. Moreover, deionized water is invariably used as a diluent. The compositions of commonly used aqueous reagents are given in Table 9.1. This table, as well as others on this topic, are collected at the end of this chapter for ready reference.

9.1.1 Crystalline Materials

Single-crystal silicon and GaAs fall into this category; both exhibit long-range order, and their etching can be either isotropic or anisotropic in character. Etching by wet chemicals usually proceeds by their oxidation, followed by the chemical dissolution of the oxide (or oxides). Both of these processes are carried out simultaneously by a mixture of the reagents in the same etching solution. The oxidation chemistry is identical to that of anodic oxidation described in Section 7.4. Here, however, no clearly defined anode or cathode is established by a battery. Rather, points on the surface of the semiconductor behave randomly as localized anodes and cathodes. The oxidation reaction proceeds from the action of these localized electrolytic cells and gives rise to the flow of relatively large corrosion currents, often in excess of 100 A cm^{-2}.

Over a period of time, each localized area (which is large compared to atomic dimensions) adopts the role of both anode and cathode. Uniform etching occurs if the proportion of time allocated to each role is roughly equal. Conversely, selective etching occurs if these times are very different. Such factors as the doping concentration and/or the defect nature of the semiconductor surface, the etchant temperature, impurities in the etchant, and adsorption processes at the semiconductor–etchant interface play an important role in determining the degree of selectivity of the etchant as well as its etch rate. A brief review [1] of the oxidation process now follows.

Consider a localized anodic site. Here, the semiconductor is promoted from its initial oxidation state to some higher oxidation state, as given by

$$M^0 + xh^+ \rightleftharpoons M^{x+} \tag{9.1}$$

where M^0 represents the semiconductor in its neutral charge state. This oxidation reaction requires holes for its execution.

Reduction occurs simultaneously at a localized cathodic site, and is accompanied by the liberation of holes. Writing R^0 as the oxidizing species, we obtain

$$R^0 \rightleftharpoons R^{x-} + xh^+ \tag{9.2}$$

The entire reduction–oxidation reaction, which is charge neutral, is given by

$$M^0 + R^0 \rightleftharpoons M^{x+} + R^{x-} \tag{9.3}$$

In semiconductor etching, the primary oxidizing species is $(OH)^-$. Often this is formed by the dissociation of water, which is present in the etchant, as given by

$$H_2O \rightleftharpoons (OH)^- + H^+ \tag{9.4}$$

The formation of an oxide presents a barrier to further oxidation of the semiconductor, so that it is necessary to add additional chemicals for its dissolution into compounds or complexes, which are soluble in water. Stirring removes these from the semiconductor surface so that further oxidation can proceed. The choice of complexing agents for this purpose is quite wide; both acids and bases can be used, in addition to salts involving (CN_x) and (NH_x) groupings. In practical systems, the choice is limited by the availability of high-purity reagents and by the desire to avoid metallic ion contamination. Thus, hydrofluoric acid is the invariable choice for silicon etching systems. GaAs systems often use sulfuric, phosphoric, and citric acid, or ammonium hydroxide.

9.1.2   Application to Silicon

The most commonly used etchants for silicon are mixtures of HNO_3 and HF in water or acetic acid. Here, the anode reaction is given [2] by

$$Si + 2h^+ \rightarrow Si^{2+} \qquad (9.5)$$

Holes which are required for this reaction are produced by the reduction of NO_2 at a localized cathode. This reaction is autocatalytic [3], in that the reaction products promote the reaction itself. It proceeds in the presence of trace impurities of HNO_2 in the HNO_3, as follows:

$$HNO_2 + HNO_3 \rightarrow N_2O_4 + H_2O \qquad (9.6a)$$
$$N_2O_4 \rightleftharpoons 2NO_2 \qquad (9.6b)$$
$$2NO_2 \rightleftharpoons 2NO_2^- + 2h^+ \qquad (9.6c)$$
$$2NO_2^- + 2H^+ \rightleftharpoons 2HNO_2 \qquad (9.6d)$$

The HNO_2 generated in Eq. (9.6d) reenters into reaction with HNO_3 in Eq. (9.6a), and the process is thus autocatalyzed.* The first of these reactions is the rate-limiting one; in some cases NO_2^- ions are deliberately added (in the form of NH_4NO_2) to induce the reaction. Since the HNO_2 is regenerated in this reaction, the oxidizing power is a function of the amount of undissociated HNO_3.

Water can be used as a diluent for this etchant. However, acetic acid is preferred because of its lower dielectric constant (6.15 as compared to 81 for water). Use of this diluent results in less dissociation of the HNO_3, and hence in a higher concentration of the undissociated species. This preserves the oxidizing power of the HNO_3 for a wider range of dilution than if water were used. Thus, the etching properties of the solution tend to remain relatively constant during its operating life, which is an advantage in commercial microcircuit fabrication.

Reaction (9.6d) requires H^+ for it to take place. This is provided by the dissociation of water, as given by Eq. (9.4). Here, $(OH)^-$ is also formed by this reaction and combines with Si^{2+}, so that

$$Si^{2+} + 2(OH)^- \rightarrow Si(OH)_2, \qquad (9.7)$$

---

*Note that reaction (9.6c) proceeds at the silicon surface, and not in the volume of the etchant.

which subsequently liberates hydrogen to form SiO_2

$$Si(OH)_2 \rightarrow SiO_2 + H_2 \qquad (9.8)$$

Finally, the role of HF is to dissolve this SiO_2, the reaction being given by

$$SiO_2 + 6HF \rightarrow H_2SiF_6 + 2H_2O \qquad (9.9)$$

Stirring serves to remove the soluble complex, H_2SiF_6, from the vicinity of the silicon slice. This reaction is referred to as a "complexing reaction" for this reason.

The overall etching process is preceded by an induction period during which the autocatalysis of HNO_3 is initiated. This is followed by the cathodic reduction of HNO_3, resulting in a supply of holes which enter into the oxidation reaction to produce SiO_2, which reacts with HF to form the soluble complex, H_2SiF_6. All of these steps occur within a single etch mixture, resulting in the overall reaction

$$Si + HNO_3 + 6HF = H_2SiF_6 + HNO_2 + H_2O + H_2 \qquad (9.10)$$

Extensive studies [4–7] of the HF–HNO_3 system have been made using both water and acetic acid as the diluents. Rapid stirring was used to prevent the formation of localized hot spots and to present the silicon surface with fresh etchant at all times. The uncertain induction period (for certain compositions) was avoided by the addition of NO_2^- ions in the form of NH_4NO_2. Figure 9.1 shows their results as a family of isoetch curves for the various constituents by weight. It should be emphasized here that normally available concentrated acids are 49.2 wt. % HF and 69.5 wt. % HNO_3, respectively, with water as the second component. In addition, the etch rates given in this figure represent silicon removal on both sides of the wafer, and should be halved to give the thickness removed from each surface.

Either water or acetic acid may be used as the diluent for the system. Qualitatively, both show similar behavior. Common to both systems are the following characteristics:

1. At low HNO_3 and high HF concentrations, corresponding to the region near the upper vertex of Fig. 9.1, the etching contours run parallel to the lines of constant HNO_3. Thus the etch rate is controlled by the HNO_3 concentration in this region. This is due to the fact that there is an excess of HF to dissolve the SiO_2 formed during the reaction. Etching with these formulations is sometimes difficult to initiate because of an uncertain induction period. In addition, they result in relatively unstable silicon surfaces, which proceed to slowly grow a layer of SiO_2 over a period of time. Finally, the etch is limited by the rate of the oxidation–reduction reaction, so that it tends to be somewhat orientation-

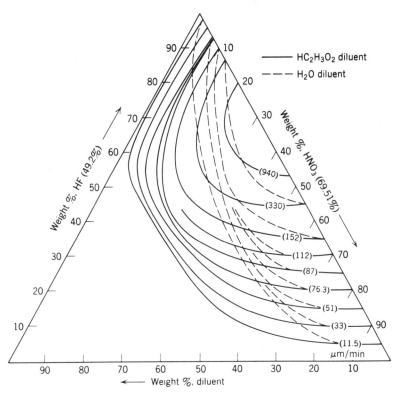

Fig. 9.1 Isoetch curves for silicon (HF : HNO₃ : diluent system). From Robbins and Schwartz [5]. Reprinted with permission of the publisher, The Electrochemical Society, Inc.

dependent. However, this effect is slight, since etching is relatively rapid and is accompanied by the liberation of heat.

2. In the region of the lower right vertex (low HF and high HNO₃ concentrations), the etch-rate contours are parallel to the lines of constant HF. Here there is an excess of HNO₃, so that the etch rate is governed by the ability of the HF to dissolve the SiO₂ as it is formed and by the removal of soluble complexes by diffusion into the volume of the etchant. These etches are "self-passivating" in that a freshly etched surface is already covered with a relatively thick layer of SiO₂ (30–50 Å). They are used extensively in the fabrication of nonplanar microwave and power devices, where they are known as *mesa* etches because of the physical shape of the resulting structure.

3. Etch formulations in the HF:HNO₃ = 1 : 1 range are initially insensitive to the addition of diluent. Upon dilution, their etch rate falls off very sharply, until the system becomes critical with respect to the diluent. They are generally hard to control, and are not used in practice.

As seen from this figure, an almost infinite choice of formulations can be used for silicon etching. A number of these are available in premixed form, and have been well characterized. Some are listed in Table 9.2 at the end of this chapter. Also included in this table are a few etchant formulations which are based on such oxidizing reagents as Br_2 [8], I_2 [9], and $KMnO_4$ [10].

9.1.3 Application to Gallium Arsenide

A wide variety of etches have been investigated for GaAs; however, very few of them are truly isotropic. This is because, unlike silicon, the surface activity of the (111) Ga and (111) As faces is quite different. The As face, terminated on arsenic, has two unsatisfied bonds per atom. Consequently, although some reconstruction occurs in the surface layer, it is still considerably more reactive than the Ga face, and thus etches at a faster rate. As a result, most etches give a polished surface on the As face. The Ga face etches much slower, because it has no unsatisfied bonds, and tends to show up surface features and crystallographic defects. Often, the appearance of an etched (111) Ga face appears cloudy or frosted for this reason.

The details of the oxidation reactions have been given in Chapter 7, but will be summarized here [2]. Immersion into the electrolyte system results in electronic charge transfer from the semiconductor to the electrolyte, as the separate fermi levels line up, so that

$$GaAs + 6h^+ = Ga^{3+} + As^{3+} \tag{9.11}$$

Reaction with $(OH)^-$ ions in the electrolyte occurs in a sequence of steps, outlined in Eqs. (7.36) and (7.37), to produce Ga_2O_3 and As_2O_3. Dissolution of these oxides occurs in acids or bases which are part of the etchant formulation, to form soluble salts of complexes.

One of the earliest (and still very popular) etching systems for GaAs is based on the use of small concentrations of bromine in methanol [11, 12]. Here, the $(OH)^-$ ion is provided by the methanol, while bromine serves the dual role of being a strong oxidizing agent and also dissolving the oxidation products to form soluble bromides.

The Br_2–CH_3OH system can be used over a wide range of concentrations. At low concentrations, its removal rate is linearly proportional to the Br_2 content. Typically, a removal rate of about 0.075 μm/min can be obtained with 0.05 vol.% Br_2 in CH_3OH. Higher Br_2 concentrations, up to 10 vol.%, can also be used to obtain a high degree of surface polish, and are used in the removal of saw-cut damage.

The etch rate for the Br_2–CH_3OH system is different for the various planes, with the (111) As plane etching at the fastest rate, and the (111) Ga at the slowest [13]. Typically, etch rates are (111) As : (100) : (111) Ga = 6 : 5 : 1. More uniform etch rates are observed with the faster etching formulations—that is, those with a high Br$_2$ content [14].

This etch can be used to polish all principal crystal faces except the (111)Ga. Here, its relatively slow etch rate results in preferential etching and feature delineation. The addition of Syton* allows polishing of this face as well. A 1-ml Br$_2$, 20-ml CH$_3$OH, 300-ml Syton formulation has been used for this purpose, with an etch rate of about 0.2 μm/min [15].

Many etch systems for GaAs are based on the use of H$_2$O$_2$, which is strongly oxidizing, in combination with an acid or base to dissolve the oxidation products. Sulfuric, phosphoric, nitric, hydrochloric, and citric acids, as well as ammonia and sodium hydroxides, have all been used for this purpose. The properties of some of these etching systems are now described.

The H$_2$SO$_4$–H$_2$O$_2$–H$_2$O System. This system is widely used in many formulations, which are referred to as *Caro's etch*. Its properties are illustrated by the 0°C isoetch curves of Fig. 9.2 [16]. Etch rates for this system double every 10°C, so that room temperature values are about five times as large as those in this figure. H$_2$SO$_4$ is highly viscous, so that diffusion-limited etching

*Colloidal silica suspension (Remet Chemical Corp., New York).

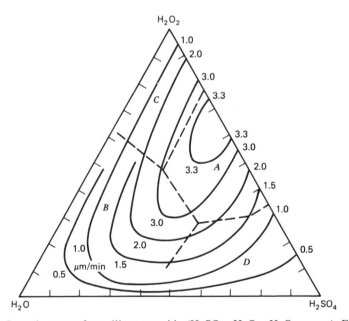

Fig. 9.2 Isoetch curves for gallium arsenide (H$_2$SO$_4$: H$_2$O$_2$: H$_2$O system). From Iida and Ito [16]. Reprinted with permission of the publisher, The Electrochemical Society, Inc.

occurs with formulations in which high concentrations of this acid are used. Formulation which fall into region D in this figure result in surfaces with mirror finish. Formulations in region A, involving large amounts of both these chemicals, etch very rapidly and result in a cloudy appearance. Region B results in very slow etch rates, and can be used to delineate surface defects such as etch pits.

Typically, an $8 : 1 : 1$ volume ratio of $H_2SO_4 : H_2O_2 : H_2O$ results in an etch rate of 0.8 μm/min for the (111) Ga face and 1.5 μm/min for all other faces. Etches with a low $H_2SO_4 : H_2O_2$ ratio tends to be somewhat more anisotropic. Thus a $1 : 8 : 1$ volume ratio of $H_2SO_4 : H_2O_2 : H_2O$ results in etch rates of $3 : 8 : 8 : 12$ μm/min for the (111) Ga, (100), (110), and (111) As faces, respectively.

The H_3PO_4–H_2O_2–H_2O System. Phosphoric acid is also quite viscous, and thus etches based on it tend to be polishing. Figure 9.3 shows isoetch curves for this system [17], and indicates essentially four regions of interest: A, B, C, and D. Region B formulations, with a high concentration of H_3PO_4, result in a polishing etch.

Etch formulations in regions A, B, and C have approximately equal etch rates on all principal planes except the (111) Ga, which etches at approximately one-half the rate. Thus, an etch consisting of $3 : 1 : 50$ by volume of

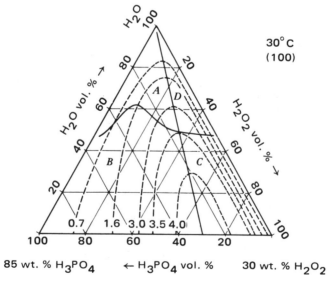

Fig. 9.3 Isoetch curves for gallium arsenide ($H_3PO_4 : H_2O_2 : H_2O$ system). From Mori and Watanabe [17]. Reprinted with permission of the publisher, The Electrochemical Society, Inc.

$H_3PO_4 : H_2O_2 : H_2O$ results in an etch rate of 0.4 μm/min for the(111) Ga plane and 0.8 μm/min for the other principal planes. Etches in region C are somewhat more anisotropic.

The $C_3H_4(OH)(COOH)_3H_2O–H_2O_2–H_2O$ System. Etches based on citric acid have also been studied. These etches, consisting of a mixture of a 50 wt.% citric acid aqueous solution with concentrated H_2O_2 in a volume ratio $k : 1$, have been used over a wide compositional range, with a correspondingly large range of etch rates. Figure 9.4 shows [18] the etch rate as a function of k, over a range from 1 Å/s to 100 Å/s. It is interesting to note that etches with $k < 2$ are relatively sensitive to stirring. Here, the rate-limiting step is the availability of citric acid for removing the oxidation products from the surface of the GaAs. Etch rates for the principal crystallographic planes are essentially identical, except for the (111) Ga plane, which etches at a rate of about 60% of the others.

The $NH_4OH–H_2O_2–H_2O$ System. This system can be used to etch the (111) Ga plane nonpreferentially, in addition to the other principal planes [15]. Here, a formulation consisting of 1 ml NH_4OH and 700 ml H_2O_2 results in an etch rate of about 0.3 μm/min for the (100), (111) As, and (111) Ga faces, so that the etch can be considered to be truly isotropic. Other formulations can be prepared, however, which exhibit strong anisotropy. An advantage of this system, of importance to wafer manufacturing, is its considerably superior aging

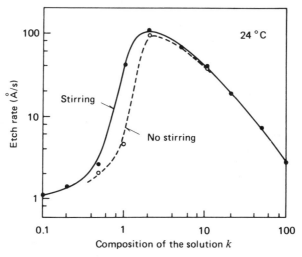

Fig. 9.4 Etch rate versus composition for gallium arsenide (citric acid: $H_2O_2 : H_2O$ system). From Otsubo et al. [18]. Reprinted with permission of the publisher, The Electrochemical Society, Inc.

qualities over the Br_2-CH_3OH or the $Br_2-CH_3OH-Syton$ systems. Yet another advantage is that its constituents are nitrogen, hydrogen, and oxygen, so that its use does not result in impurity contamination.

The $NaOH-H_2O_2-H_2O$ System. Very little work has been done with this system, but indications are that it has etch characteristics similar to those of ammonia-based formulations [19]. Typically, a 1 M NaOH, 0.76 M H_2O_2 formulation has been shown to result in an etch rate for the (100) plane of about 0.2 μm/min at 30°C [16].

Etches based on HF present an unusual situation since this acid, of itself, can chemically dissolve GaAs. This results in relatively rounded edges to the etch profile [20]. Mesa regions, which are delineated in GaAs to form isolated islands, can be made with rounded edges in this manner. This greatly improves the step coverage of deposited films which are used for ohmic contact and interconnection purposes. A typical formulation, consisting of HF : H_2O_2 : H_2O in a 1 : 2 : 25 formulation by volume, is used for this purpose and has an etch rate of about 3 μm/min.

Etches using $HCl-H_2O_2-H_2O$ have been investigated, but little is known of their detailed behavior. A variety of etchants, usually modifications of the systems described here, have also been developed [21, 22]. These involve the substitution of CH_3OH instead of H_2O, HNO_3 instead of H_2O_2, and such oxidizing agents as $HClO_4$, NaOCl [23], and $KMnO_4$. Their use has generally been confined to special-purpose applications. For example, NaOCl : H_2O solutions are widely used for polishing GaAs slices after they have been cut from the boule.

9.1.4 Anisotropic Effects

The etching of a semiconductor proceeds by the successive dissolution of layers of this material. As a consequence, it is reasonable to expect that this process will be slowest on the {111} planes, since they have the lowest bond density.* This has indeed been found to be true for etches which are reaction-rate-limited, provided that they are slow so that they do not generate much heat. Etches which are diffusion-limited, as well as those which are fast and result in a localized rise in temperature, tend to etch uniformly in all directions.

A number of etch formulations for silicon are available, which show extremely high anisotropy [24], as much as 600 : 1; this is well in excess of what can be expected from considerations of bond density alone. Common to these etches is the absence of a strong complexing agent (such as HF) for removing the SiO_2 which is formed as part of the etch process.

A working theory for this high degree of anisotropy is that {111} silicon oxidizes much faster than other planes. This effect is much stronger at reduced temperatures, such as those encountered in wet chemical etching (0–100°C),

---

*Bond densities for {100} : {110} : {111} planes are in the ratio 1 : 0.707 : 0.557. The etch rates for silicon usually fall in this sequence.

as compared to those used for thermal oxidation ($\approx 900-1100°C$). As a result, the {111} surfaces are rapidly covered with an oxide film, which blocks further dissolution. An interesting point, in support of this theory, is that the etch rate of {111} Si and the dissolution rate of SiO_2 grown on this face are roughly comparable, ≈ 85 Å/min at 85°C.

An additional factor arises with GaAs, since it is a polar semiconductor and has different surface activity in the (111) Ga and (111) As directions. Specifically, the (111) As face, terminated on arsenic atoms, is highly reactive and will usually etch faster than any other plane. As a consequence, the sequence of etch rates (in descending order) for silicon orientations is invariably {100}, {110}, and {111}; for GaAs it is usually (111) As, (100), (110), and (111) Ga.

Work with anisotropic etches has focused on their use for the chemical machining of semiconductor materials. Additionally, these etches are used for cutting apart semiconductor chips for beam lead devices [25], and for texturing the surface of GaAs solar cells [26]. The etching of both V-shaped [27] and vertical grooves [28] has been accomplished by these methods, and devices as well as microcircuits have been based on this technology. Anisotropic etching has been used in purely nonelectronic applications as well, such as the development of precision nozzles for ink-jet printers [29, 30].

The effect of anisotropic etching of {100} silicon is shown in Fig. 9.5, where the orientations of a window cut in a mask are as indicated. Here, it is assumed that the etch rate for all {111} planes is identical, and negligibly small. This is usually the case, and etch ratios for {100} : {111} planes are typically greater than 100. For this situation, the etch profile is trapezoidal after some fixed period of time. Eventually, however, the etchant will delineate a V-groove as shown

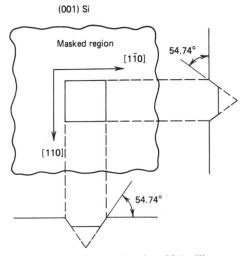

Fig. 9.5 Etch profiles for (001) silicon.

by the dashed line in this figure. In practice, the etch rate on the {111} planes is very small, but finite, so that some degree of undercutting will also occur.

The situation for GaAs is considerably more complex, since the (111) Ga planes etch the slowest, and the (111) As the fastest.* As a result, the etch profile takes on a trapezoidal shape for one direction and is dove-tailed in the other, as shown in Fig. 9.6a. Furthermore, many of the higher-order planes present

*The $(1\bar{1}\bar{1})$, $(\bar{1}1\bar{1})$ and $(\bar{1}\bar{1}1)$ planes all have gallium atoms on their surface and are thus Ga faces. The $(\bar{1}11)$, $(1\bar{1}1)$, and $(11\bar{1})$ planes, on the other hand, have arsenic atoms and are thus (111) As faces.

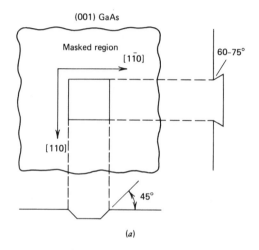

(a)

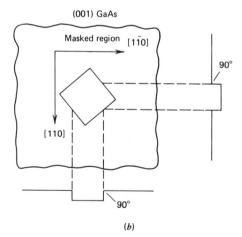

(b)

Fig. 9.6 Etch profiles for (001) gallium arsenide. From Tarui et al. [13]. Reprinted with permission of the publisher, The Electrochemical Society, Inc.

either Ga or As faces. Consequently, the etch angles are not precisely 54.74°, as are obtained for silicon. It has been shown, however, that nearly rectangular etch profiles can be obtained if the window is cut at 45° to the (110) and ($1\overline{1}0$) directions [16], as in Fig. 9.6b.

A variety of more complex profiles can be obtained depending on the orientation of the edge which defines the etch step, as well as on the etch formulation. Figure 9.7 shows the cross-sectional profiles obtained for a series of Caro's etches with $H_2SO_4 : H_2O_2 : H_2O = 1 : 8 : x$, for x values from 1 to 1000 [31]. All of these cross sections are normalized to the same depth for ease of comparison. It should be noted here that these profiles are unchanged with etch time, although the amount of undercutting increases.

A graphical procedure [32], based on experimental determination of the minimum etch rate for any given formulation along various crystal directions, has been developed to predict the nature of these etch profiles. Figure 9.8 shows a minimum etch rate diagram of this type, for the formulations used in Fig. 9.7. Some anisotropic etches are now described, together with their etching characteristics.

9.1.4.1 Silicon

A commonly used anisotropic etch for silicon consists of a mixture of 44 wt.% KOH in water [24]. Typically, etching is carried out at 80°C, and results in an etch ratio of 300 : 600 : 1 for (100) : (110) : (111), with an etch rate of 1.20 μm/min on the (110) plane. Dilution reduces this ratio, but results in a more controllable etch rate. The use of isopropanol alcohol in addition to water results in reduced etching of oxide masks. However, it changes the etch behavior

ACID	VOLUME RATIOS *	ETCH RATE (100) (μm min^{-1})	CROSS-SECTIONAL PROFILES	
			(011) SECTION	($01\overline{1}$) SECTION
H_2SO_4	1:8:1	14.6		
H_2SO_4	1:8:40	1.2		
H_2SO_4	1:8:80	0.54		
H_2SO_4	1:8:160	0.26		
H_2SO_4	1:8:1000	0.038		

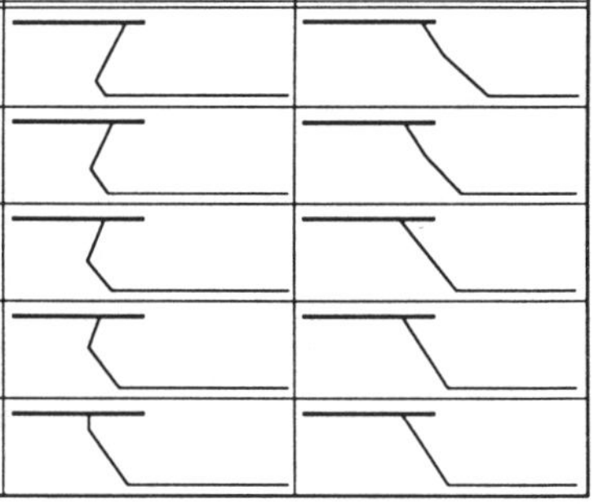

*H_2SO_4:H_2O_2:H_2O

Fig 9.7 Etch profiles in the H_2SO_4–H_2O_2–H_2O system. Adapted from Shaw [31].

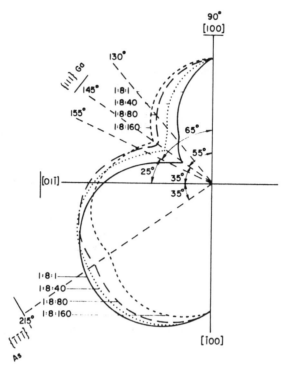

Fig. 9.8 Etch rate as a function of GaAs orientation. From Shaw [31]. Reprinted with permission of the publisher, The Electrochemical Society, Inc.

so that {100} planes etch more rapidly than the {110}. In addition, a condenser arrangement is needed to avoid evaporation of the alcohol during the long etching process.

Etches using hydrazine and water have also been used [33] for the anisotropic etching of V-grooves for microcircuits. Here, an etching solution consisting of 100 g N_2H_4 and 50 ml H_2O is used at 100°C to result in an etch rate of 3 μm/min on the {100} planes, with an etch rate of about 0.3 μm/min on the {111} planes.

A series of etchants, consisting of mixtures of ethylenediamine, pyrocatecol, and water (EPW), have been developed for use with silicon [34]. A typical formulation of this type, consisting of 17 ml ethylenediamine, 3 g pyrocatecol, and 8 ml water, results in etch rates of {100} : {110} : {111} = 50 : 30 : 1 μm/hr at 110°C. Lower etch rates can be obtained by reducing the water content of this etch. Catalysts such as diazine or pyrazine are sometimes added to EPW solutions, and have been shown to double its etch rate. The use of hydrazine instead of ethylenediamine as the oxidant has also been investigated.

EPW etches are normally used near their boiling point, so that etching necessitates a condenser arrangement. Operation at lower temperatures results in lower etch rates, and is sometimes employed for fine-line delineation of polysilicon. However, one problem here is that of residue formation, and modified formulations have been developed to address this issue.

Almost all machining applications for silicon etches are based on the fact that the sides of a window, lined up in the $\langle 110 \rangle$ directions on (100) Si, will be vertical and present $\{111\}$ faces. This allows cutting of deep vertical grooves, with aspect ratios as large as 400 : 1. A number of mechanical, electrical, and optical devices have been fabricated [29, 30] to date, based on this unique property.

9.1.4.2 Gallium Arsenide

Almost all of the polishing etches described for GaAs show some degree of anisotropy, because of its polar nature. At the same time, they are usually not as anisotropic as for silicon, since the oxides of Ga and As are readily removed by acids as well as bases. By way of example, Br–methanol formulations, which are normally considered to be polishing, show some degree of anisotropic behavior [14]. A 1 wt.% solution of bromine in methanol results in etch ratios of 6 : 5 : 4.6 : 1 for (110) : (111) As: (100) : (111) Ga, with an etch rate of about 1 μm/min in the (110) plane.

Anisotropic etching has also been investigated for the NH_4OH–H_2O_2–H_2O system. Here, an etch consisting of $NH_4OH : H_2O_2 : H_2O = 3 : 1 : 140$ results in etch rates [35] of $0.037 : 0.12 : 0.2 \mu$m/min for the (111) Ga, (100), and (111) As planes, respectively, with sharply delineated faces. Other etches, with a 1 : 1 : 8 formulation, have more strongly anisotropic character, and can be used to etch grooves in GaAs with vertical walls and no undercutting [36].

A disadvantage of ammonia-based etchants is that they destroy photoresist, so that a SiO_2 mask must be used with them. This is not a problem with etches containing phosphoric acid [17], which also show preferential etching behavior. Thus, the H_3PO_4–H_2O_2–H_2O system can be used in region C of Fig. 9.3 as an anisotropic etch. Here, a 1 : 9 : 1 formulation by volume results in etch rates of 5 : 4.2 : 3 : 1.5 μm/min for the (110) :(111)As: (100) :(111)Ga planes respectively.

9.1.5 Selective Etches

In both silicon and GaAs technology, there is a need for etches which are selective on the basis of (a) impurity type and/or concentration and (b) the material composition. These etches are used in a variety of micromachining applications, such as the formation of thin membranes. Additionally, they are an intrinsic part of processes for fabricating device structures such as high-electron-mobility transistors (HEMTs) and oxide isolated microcircuits.

An etch [38], consisting of KOH, water and isopropyl alcohol in a 7 : 19 : 4 ratio by weight, has been used at 80°C for the selective etching of silicon. Its etch rate on {100} silicon is 0.94 μm/min, for either p- or n-material doped in the 10^{14} to 10^{18}-cm^{-3} range. This rate falls off exponentially for doping levels above 10^{18} cm^{-3}, to 0.02 μm/min for doping concentrations of 10^{20} cm^{-3} or higher. It is often used in structures where a thin p^+ (or n^+) layer is intentionally inserted to serve as an etch stop. It should be pointed out that, in addition to being selective to doping, it is highly anisotropic, with negligible etching on {111} surfaces.

The system HF–HNO$_3$–CH$_3$COOH, in a 1 : 3 : 8 ratio, is also highly selective [39] with an etch rate 0.7 to 3 μm/min for resistivities below 0.015 Ω-cm, but no appreciable etching for n^+ or p^+ silicon with resistivities above 0.068 Ω-cm. In this system, this high selectivity falls off as etching progresses, due to an increase in the HNO$_2$ content. This can be remedied by the addition of an oxidant such as H$_2$O$_2$, which converts the HNO$_2$ to HNO$_3$ and allows the selectivity to be maintained over the entire etching process.

Formulations, based on the use of EPW, exhibit [40] an abrupt change in the etching characteristics of p-silicon, with no etching at doping concentrations above 7×10^{19} cm^{-3}, and an etch rate of 0.83 μm/min for lower p-type doping concentrations, or for n-type silicon.

The primary need for selective etches in GaAs technology is in AlGaAs–GaAs structures, where it is often necessary to etch off one of these layers while leaving the other intact. Redox (i.e., reduction–oxidation) reactions, of the type described by Eq. (9.3), can be used for etches of this type with the aid of suitable pH modifiers. Of special importance are the systems KI–I$_2$ and K$_3$Fe(CN)$_6$–K$_4$Fe(CN)$_6$, which are stable in both acid and alkaline solutions [41]. Organic systems, such as quinone–hydroquinone (C$_6$H$_4$O$_2$–C$_4$H$_6$O$_2$), also fall in this class. For the potassium ferricyanide–potassium ferrocyanide system, for example, selective etching of AlGaAs is obtained for pH values of 5–9. At lower pH values, etching is nonselective.

An extremely popular etch for the removal of GaAs, with a selectivity of 33 over AlGaAs, consists of H$_2$O$_2$ to which NH$_4$OH is added to adjust the pH value between 8.2 and 8.4 [42]. Yet another formulation [43], consisting of a 10 : 1 ratio of citric acid (50 wt.%) and H$_2$O$_2$, has a selectivity of 95 and results in a smoother surface than the NH$_4$OH–H$_2$O$_2$ formulation.

9.1.6 Crystallographic Etches

Crystallographic etches can be used to delineate the regions where dislocations intersect with the semiconductor surface. These dislocations, together with their associated strain fields, are present in starting material, and are also created during strain-inducing processes such as dopant incorporation and oxide growth. They result in highly localized shifts in the surface potential, which will etch selectively if the etch is slow and reaction rate limited. These etches are thus generally slow, and are often composed of the same constituents as polishing etches. Frequently, one or more heavy metal ions are added; they tend to plate out during etching, and give further contrast to the etched regions.

The most commonly used crystallographic etches for silicon, together with their properties, are compared in Table 9.3 at the end of this chapter. Of these, Dash etch [44] has historical interest because it was used to definitely establish a correlation between dislocations within a material and the pattern of etch pits created by them at the surface. This etch requires several hours of immersion to be fully effective.

Sirtl [45], Secco [46], and Wright [47] etches all utilize chromium salt oxidizers, which also provide the heavy metal ion. All delineate etch pits in a few minutes and are used extensively. Sirtl etch tends to be anisotropic, and has found most use on (111) silicon, where triangular pits are delineated by its action. Secco etch, on the other hand, is isotropic, and results in circular or elliptical etch pits. It is equally effective on (100) as well as on (111) silicon. Wright etch uses copper nitrate as a plating agent so as to more sharply delineate etch pits. It is anisotropic in its etch characteristics and can be used on (100) and (111) silicon, where the geometric shape of these pits gives information about the crystallographic orientation of the defects. Other etch formulations have been developed for special applications such as junction delineation [9]. A few of these are also listed in Table 9.3.

A number of crystallographic etches have been developed for use with GaAs. Here, because of its polar nature, it is relatively easy to distinguish between the slow-etching (111) Ga face and the fast-etching (111) As. Etch pits on these faces tend to be triangular in shape. On the other hand, etch pits on the (001) face are rectangular (instead of square) because of the difference in etch rates for the Ga and As faces which become exposed during etching [48].

Delineation of etch pits in the $\langle 100 \rangle$ and $\langle 110 \rangle$ directions is particularly useful for GaAs, since these faces are exposed in the formation of lasers and other electro-optical structures. An etch consisting of CrO_3 as the oxidizer, together with $AgNO_3$ and HF, has been found useful for this application [49]. Successful use of this etchant requires $65°C$ operation, and rapid stirring to remove surface residues.

All of the crystallographic etches developed for silicon can also be used with GaAs. For example, Sirtl etch is used for delineating features in the $\langle 100 \rangle$ and $\langle 110 \rangle$ directions. An unusual feature of this etch is that mounds and hillocks are formed at defect sites, rather than the usual etch pits. A dislocation etch that has found wide use in the delineation of interfaces in GaAs and other compound semiconductors is known as the *A–B* etch [49, 50]. This is a two-part solution with CrO_3 and $AgNO_3$, and is used on cleaved faces to indicate compositional changes resulting from multilayer epitaxy.

Etchant solutions, of the type developed for anisotropic etching, can also be used for etch pit delineation. Thus, dilute Br–methanol formulations can be used for delineating etch pits in GaAs. Etches based on the use of H_3PO_4 and H_2O_2 have also been used here. The properties of a few crystallographic etches for GaAs are compared in Table 9.4 at the end of this chapter.

9.1.7 Noncrystalline Films

The patterned removal of a variety of thin film materials is necessary during microcircuit fabrication. These films, formed by the techniques described in Chapters 7 and 8, are either vitreous, amorphous, or polycrystalline in nature. They lack long range order, so that etching by wet chemicals is usually isotropic in character; that is, the etchant spreads out under the mask layer by an amount roughly equal to the etched thickness. In many instances, stresses at the resist–film interface, combined with capillary action of the wet chemical, can cause excessive undercutting at this point. Often, lifting or tearing of the photoresist mask can occur during this step.

Excessive undercutting can also occur because of stresses at the interface between the layer being etched and the semiconductor. One such example is the SiO_2–Si interface. Here, the process of opening a window in the oxide can lead to undercutting of the SiO_2 near this interface. In extreme situations this may result in uncovering the oxide protection over a junction edge, as shown in Fig. 9.9b. This problem is sometimes encountered in the fabrication of shallow-diffused high-speed transistors in which the emitter–diffusion window is used also for the emitter metallization (the so-called *washout* emitter).

Etching is commonly done with the same chemicals which dissolve these materials in bulk form, and usually involves their conversion into soluble salts or complexes. Many etching recipes are available for each material; their characteristics will depend upon such film parameters as its microstructure, its porosity, how it is formed, and the nature of the previous processes to which it has been subjected. Thus:

1. Film materials will etch more rapidly than their bulk counterparts, so that reagents must be used in dilute form to reduce the etch rate to manageable proportions.

2. Films which have been irradiated will generally etch rapidly. This includes films which have been ion-implanted, those which have been grown by e-beam evaporation, and even those which have been subjected to an e-beam evaporation environment at some previous step. The exception here is

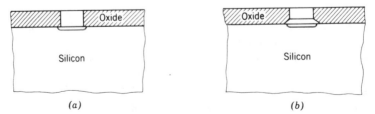

(a)   (b)

Fig. 9.9 Undercutting effects: (*a*) ideal and (*b*) actual.

certain resists which toughen by polymerization under these conditions. These materials, referred to as *negative* resists, are described in the following chapter.

3. Films which have a high built-in stress will etch rapidly. Often, the stress in a film can be controlled by the rate at which it is deposited, the deposition technique, and the substrate temperature.

4. Films with poor microstructure will etch rapidly. This includes films which are porous, or loosely structured. Such films can often be densified by a short heat treatment at a temperature in excess of their growth temperature. These densified films will etch at a slower rate than as-grown films.

5. Films of compounds will etch faster if their preparative technique results in departures from stoichiometry. Si_3N_4 is a good example of a material which falls in this category.

6. Mixed phase films will etch faster than films consisting of a single phase. This is because the preferential etching of one component often causes a rapid increase in the film porosity with a consequent increase in the wetting surface. PSG and BPSG are film materials which fall in this category.

The etch behavior of materials which are important to microcircuit fabrication will now be described. A brief formulary of useful etchants is provided in Table 9.5 at the end of this chapter; an encyclopedic listing of this type is provided in Ref. 51.

9.1.7.1 Silica Films

Silica films are readily etched by HF, as shown by the complexing reaction of Eq. (9.9). In practice, this reaction is performed in a dilute solution of HF, buffered with NH_4F to avoid depletion of the fluoride ion. Addition of this NH_4F results in etching characteristics that are consistent from run to run. Furthermore, it also lessens attack of the photoresist during etching. This etch formulation is commonly referred to as buffered HF, or BHF.

Often, BHF is used in diluted form in order to slow down its etch rate. The nature of the reaction kinetics is quite complex, and is related to the concentration of HF and HF_2^- ions which are present in this etchant.* The relative concentration of these species is related to the pH of the solution; typical BHF solutions have a pH of around 3, and are almost entirely made up of the HF_2^- ion.

Both thermally grown films and deposited SiO_2 films can be etched in BHF. However, etching of deposited films (grown by the SiH_4 process at 350–450°C) proceeds much more rapidly than for thermal oxides, which are normally grown at 900–1100°C. Typically, the etch rate in BHF is about 3000 Å/s for silane oxides grown at 450°C, but only 900–1000 Å/min for dry thermal oxides.**

---

*The reaction rate of HF_2^- with SiO_2 is about 4–5 times that of HF.
**Wet oxides etch slightly faster than dry oxides because they are somewhat more loosely structured.

Densification of these films, by heat treatment at 1000–1200°C for about 15 min, results in a fall in the etch rate to a value that is approximately the same as that for the thermally grown oxide [52].

Both HF and BHF have poor wetting characteristics on silicon surfaces. This presents a severe problem in the wet cleaning of silicon. It also leads to problems of penetration uniformity, when deep holes are required to be etched in the oxide. The use of surfactants represents an important approach to reducing these problems [53].

9.1.7.2 Silicate Glasses

Both PSG and BPSG have a lower as-deposited stress than does SiO_2; they etch readily in both HF and BHF, the etch rate increasing with the amount of P_2O_5 incorporated in them. This is because the HF attacks the SiO_2, rendering the glass porous, with the rapid dissolution of the P_2O_5 in the water. Typically [53], the etch rate of PSG with 8 mole % P_2O_5 is 5500 Å/min in BHF, whereas that for undoped SiO_2 is about 2500 Å/min. The same films, upon densification at 1100°C, have a greatly reduced etch rate of 3000 Å/min for the PSG and 800 Å/min for the SiO_2.

In addition to BHF, an etch formulation based on the combination of HF and HNO_3, known as P-etch, has also found much use [54] in the etching of both SiO_2 and PSG films. The etch rate of P-etch increases logarithmically with the phosphorus content of the film. Typically, the etch rate for a PSG film ($\approx$16 mole % P_2O_5) is about 500 Å/s for this formulation, whereas undoped thermal oxide etches at about 1.8 Å/s. Thus, P-etch is especially useful in removing layers of PSG which are formed on the surface of silica films during phosphorus diffusion, and also in some diagnostic situations. Some care is required in its use, however, because of the possibility of dissolution of exposed silicon with which it may come in contact.

The dissolution of borosilicate glass (BSG) in dilute BHF has been investigated [53]. Typically, the etch rate is seen to increase monotonically with the B_2O_3 content. In some instances, however, anomalous characteristics have been observed during the etching of BSG films.

A thin layer of BSG is also formed on the SiO_2 surface during the boron doping of masked silicon wafers. Selective etches, known as R-etch and S-etch, can be used to remove this film while leaving the underlying SiO_2 unattacked [55]. Both etches are modifications of the P-etch formulation, and they are about 5–6 times more rapid in their ability to remove the BSG layer.

9.1.7.3 Mixed Oxides

These include $ZnO–SiO_2$ and $SnO_2–SiO_2$, which are used as *p*- and *n*-type dopant sources for GaAs, respectively. They can be readily etched in HF or in BHF, since SiO_2 is their primary constituent. Again, etching proceeds by the dissolution of the SiO_2 in HF, rendering the structure porous and thus more susceptible to dissolution. Dilute BHF has been found useful for this purpose.

The etch characteristics of both $ZnO-SiO_2$ and SnO_2-SiO_2 have not been studied. However, they are comparable to undoped SiO_2 grown by the SiH_4 process.

9.1.7.4 Silicon Nitride

Si_3N_4 is widely used as a protective coating for silicon microcircuits, as a cap during the annealing of ion-implanted GaAs, and as a surface stabilization coating for GaAs microcircuits. Its protective characteristics are generally superior to those of SiO_2 and PSG. However, its patterned removal is more difficult and is highly dependent on the growth technique.

Both HF and BHF can be used [56] to etch these films [56]. However, even at elevated temperatures, the etch rates are extremely slow, so that photoresist films are destroyed during this process. Typically, the etch rate of CVD films (SiH_4-NH_3 process) is a function of the growth temperature [57]. The etch rate in concentrated HF is about 1000 Å/min for films grown at 800°C. The etch rate in BHF is considerably lower, about 5 Å/min for films grown at 1100°C.

The problem of photoresist destruction can be avoided by depositing a molybdenum film between the nitride and the photoresist. This film can be readily etched with excellent edge definition by standard photolithographic techniques. Once patterned, it can be used as a mask for etching the underlying Si_3N_4 film, after the resist has been stripped.

Si_3N_4 is often used as a cover layer over a film of SiO_2. In these situations, neither HF nor BHF can be used, since this would result in deep undercutting of the SiO_2 layer, which is etched rapidly by these solutions. Instead, etching is carried out in boiling* H_3PO_4, and a reflux boiler apparatus is used to avoid changes in the etchant composition during this process. Typically, the etch rate for CVD grown Si_3N_4 films is about 100 Å/min, while that for thermally grown SiO_2 layers is about 10–20 Å/min [58]. For these same conditions, the etch rate of silicon is under 3 Å/min, so that this technique can be used in the presence of exposed silicon surfaces. A curve of comparative etch rates of these films is shown in Fig. 9.10.

There are many situations in which consecutive layers of Si_3N_4 and SiO_2 must be etched at the same rate (the gate oxide of a metal–nitride–oxide field effect transistor, for example). Here, mixtures of HF and glycerol are used at 80–90°C, and experimentally adjusted to provide equal etch rates. Typically, a 1 to 3 mole/liter concentration of HF etches SiO_2 and CVD Si_3N_4 films at equal rates of about 100 Å/min at 80°C [59].

The etch rate of Si_3N_4 films is extremely sensitive to the incorporation of even trace amounts of oxygen in them. In general, the etch rate in HF and BHF increases with oxygen incorporation, whereas the etch rate in H_3PO_4 decreases. Typically, the etch rate in concentrated HF varies [57] from 350 Å/min for $Si_xO_yN_z$ films grown at 1000°C with 7% SiO_2 incorporated in them, to as high

---

*The etching temperature is related to the concentration of the H_3PO_4.

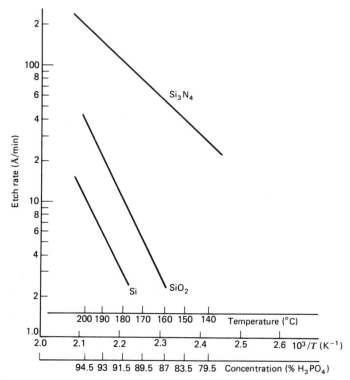

Fig. 9.10 Comparative etch rates of silicon, silicon dioxide, and silicon nitride in phosphoric acid. From van Gelder and Hauser [58]. Reprinted with permission of the publisher, The Electrochemical Society, Inc.

as 5000 Å/min for films with 50% SiO_2 incorporation. Also, it should be noted that Si_3N_4 films, grown at low temperatures by PECVD, contain a large amount of incorporated hydrogen and have considerably faster etch rates than those grown by conventional techniques.

9.1.7.5 Polysilicon and Semi-insulating Polysilicon

Chemical formulations used for etching silicon can be used for both undoped and doped polysilicon. In general, their etch rate is considerably faster, so that their use results in films with poor edge definition. However, these etches can be modified to be suitable for polysilicon. This usually consists of greatly reducing the amount of HF, using large ratios of HNO_3/HF and large amounts of diluent. Typical etch rates for these formulations are about 1500–7500 Å/min for undoped films. Etch rates for doped films are strongly dependent on the crystallite size and the doping concentration.

Formulations, based on the use of EPW, can also be used for etching polysilicon [34]. Originally developed as anisotropic silicon etches, their slow etch rate makes them attractive for this application.

Semi-insulating polysilicon (SIPOS) is formed by oxygen doping of silicon during growth, so that it can be readily etched in dilute BHF, which dissolves the SiO_2 content. The etch rate for SIPOS films is highly dependent on the amount of oxygen incorporated in them. Generally, about 20 at.% O_2 is used, and dilute BHF can be used to provide a reasonable etch rate of around 2000 Å/min.

9.1.7.6 Barrier Layers and Silicides

Barrier layers such as TiN and Ti–W alloys are extensively used in both silicon and GaAs technology; TiN can be readily etched [60] by wet chemical techniques, using the same etch system as that for silicon. Typically, $HF : HNO_3 : CH_3COOH$ in a 1 : 20 : 20 formulation by volume will etch TiN at 500–600 Å/min with excellent edge definition. The etch rate of SiO_2 with this formulation is about 40 Å/min, so that this is a suitable etch for use in the presence of SiO_2 layers.

Ti–W alloys can be etched at room temperature in 40% KOH solution, H_2O_2, and H_2O in a 2 : 1 : 4 ratio by volume. An alternative etch, which is alkali-free, consists [61] of a saturated solution of ethylenediaminetetraacetate (EDTA) in water and H_2O_2 in a 2 : 1 ratio by volume, which is used at 65°C.

Silicides are difficult to etch in most chemicals except mixtures of HF and HNO_3. This presents a problem, since these solutions attack SiO_2 as well. Plasma-assisted etching techniques are commonly used for this reason.

9.1.7.7 Metals

A large variety of metal films are employed in microcircuit fabrication today. A brief discussion of chemical etches for some of these metals now follows. Suitable etchant formulations are listed in Table 9.5.

Aluminum. This is certainly the most important metal, and is used for ohmic contacts, Schottky barriers, and interconnections. It can be etched readily by a wide variety of acidic formulations. The most commonly used etchant systems are based on mixtures of H_3PO_4 and HNO_3. Formulations based on HCl and H_2O are used for etching aluminum in the presence of GaAs, which would otherwise be etched by the HNO_3.

Gold and Silver. Gold is used in multilayer interconnection systems, in beam lead microcircuits, and as a Schottky barrier and ohmic contact metal for gallium arsenide microcircuits. It can be etched in aqua regia ($HCl:HNO_3$ in a 3 : 1 ratio by volume), but this approach is not practical since it destroys most resist and metal masks. A common etchant for gold consists of a dilution of KI and iodine in water. Photoresist can be used to pattern gold films using this etchant; however, it is opaque, so that visual observation of its progress is only possible by rinsing off the wafer prior to inspection. A number proprietary etch formulations are also available which do not have this problem. Most of these are cyanide-based, and are highly toxic.

Silver has the same barrier height to n-GaAs as gold, and is sometimes used for Schottky devices. Other applications include its use in multilayer metallization systems, where it is used together with titanium. Unlike gold, it can be readily etched in both acidic and basic etchants.

Chromium. Chromium is used in metal systems and ohmic contacts, usually in combination with gold. It is also used in hard surface photomasks. This metal is characterized by a surface that is passivated with a thin oxide film, which must be partially destroyed before etching can take place. This "depassivation" can be brought about by momentarily touching the film with a wire of aluminum or zinc. This results in the brief evolution of hydrogen gas, followed by ready etching of the film in many acids. Etchants which do not require this depassivation treatment are also available.

Molybdenum and Tungsten. Both these metals are used in refractory gate MOS devices (silicon), in Schottky diodes for GaAs, and in multimetal systems. A variety of successful etchants are available for each metal.

Platinum and Palladium. These metals are used for the formation of Schottky barriers and ohmic contacts to GaAs. Their deposition, followed by thermal conversion to the silicide, is used in forming Schottky barriers and ohmic contacts to shallow junction silicon devices. In this application, the unreacted metal is subsequently removed after the heat treatment. Dilute aqua regia can be used for platinum, since the platinum silicide is unaffected by it. Palladium, on the other hand, can be readily etched in a number of different solutions.

9.1.8 Problem Areas

Wet chemical processes are highly effective for the removal of material, and are used extensively in semiconductor technology for this purpose. In the area of pattern delineation, however, they have a number of inherent problems which become increasingly important with the need for ultrahigh-resolution patterns. First, capillary action tends to cause penetration of the liquid etchant under the pattern-defining film (usually a photoresist) and produces ragged edges, or even breakage of fine-line features. Second, their essentially isotropic character results in etching below the pattern-defining film, so that the resulting etch width is a function of the etch time. This is illustrated in Fig. 9.11 for both a lightly etched and an overetched situation. Finally, wet etching always produces a large amount of chemical waste which presents a disposal problem which will become increasingly difficult with time.

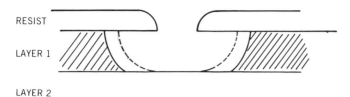

RESIST

LAYER 1

LAYER 2

Fig. 9.11 Schematic diagram of isotropic etching profiles.

Dry etching techniques have the promise of eliminating all these problems. They involve physical bombardment by atomic species, chemical reaction by energetic gaseous radical species, and combinations of these approaches. Physical bombardment, being highly directed, leads to anisotropic removal of material. Energetic species allow etching reactions to be carried out at low temperature, and are selective with respect to material removal. Both approaches allow penetration into small areas because of the atomic dimensions of the species which are involved. Moreover, it is possible to incorporate their advantages by combining them.

Finally, dry etching processes eliminate, to a large extent, the problems of chemical waste disposal. Thus, their cost and convenience advantages have been so great that they have won ready acceptance in a short period of time. These techniques will be covered in the following sections.

9.2 DRY PHYSICAL ETCHING

One technique for removing material from a substrate is by physically bombarding it with projectiles such as atoms or ions. In this approach, a gas discharge is used to impart energy to a chemically inert projectile so that it is moving at high velocity when it impinges on the substrate. Upon so doing, momentum is transferred by elastic collision to the substrate atoms; dislodging of these substrate atoms will occur when the projectile energy exceeds the binding energy. Figure 9.12 shows the threshold energy for a number of different elements as a function of their heat of sublimation [62], and illustrates this point.

This process is known as *sputtering* or *ion etching*, and a variety of plasma-assisted schemes can be used to bring it about. Here, the function of the plasma discharge is to impart energy to an inert projectile, and sputtering is accomplished by momentum transfer to the substrate atoms. The rate of sputtering, or ion etching, will therefore be related to the projectile momentum, flux density, and angle of incidence [63]. In addition, it will be related to the sputtering yield of the target. This sputtering yield is a characteristic parameter for a given material and is defined as the number of atoms ejected from the substrate per incident projectile. The sputtering yield for materials of interest is shown in Table 9.6

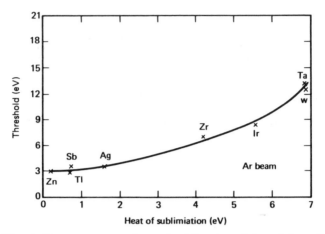

Fig. 9.12 Threshold energy versus heat of sublimation. Adapted from Spencer and Schmidt [69].

at the end of this chapter, for bombardment with argon ions in the 0.6-keV range.

The collision process which occurs is essentially an elastic one. Thus, the maximum amount of energy T_m that can be transferred from a projectile of mass M_1 and energy E_0 to a substrate of mass M_2 was shown in Chapter 6 to be

$$T_m = \left[\frac{4M_1M_2}{(M_1 + M_2)^2} \right] E_0 \tag{6.10}$$

Consequently, it is advantageous to use a heavy, inert gas such as argon for the projectile.

The sputtering yield of most materials is relatively independent of ion energy. This is because, at high energies, the projectile has an increasing penetration depth and hence increasing energy losses below the surface. Consequently, not all ejected atoms are able to reach the surface in order to escape. With sufficiently high energy, projectiles can penetrate quite deeply into the substrate, causing considerable damage.* For these reasons, projectile energies are usually kept below 2 keV in order to minimize this problem.

Another parameter of interest is the angle of incidence of the projectile. Typically, for polycrystalline targets, the yield increases as this angle is increased with a maximum at around 60° (where 90° represents grazing incidence). This is because the penetration depth increases with smaller angles, so that the dis-

---

*This process is known as *nuclear stopping*, and has been described in Chapter 6.

lodging process is less efficient. For ion etching, however, a normal incidence of the beam to the substrate is desirable for anisotropic etching, with vertical sidewalls. In some cases, the substrate is held at an angle to the beam to produce tapered walls, or to selectively remove material from sidewalls.

As seen from the above, this etching technique is strikingly similar to ion implantation, with the main difference being that low-energy ions are used as the projectile in this case, in order to avoid implantation damage. For this reason, the sputtering yield is around 1–3 for most projectile–target combinations.

Physical etching techniques share a number of common characteristics. Thus, they generally lead to highly anisotropic etching since material removal is primarily by momentum transfer. The etch rates for most materials are comparable; as a result, this technique is ideal for etching multicomponent films such as silicides and nitrides. It is particularly suited for systems with multiple layers, where all layers have to be etched with a minimum of undercutting.

These etching techniques cause varying amounts of damage to the underlying material. In some instances, as with MOS devices, this can cause a large shift in the threshold voltage. Schottky-type structures, made by depositing a metal film on an ion-cleaned substrate, have deteriorated characteristics because of this damage, even when low ion energies are used. Finally, photoresist becomes extremely difficult to remove by chemical means after ion etching.

9.2.1 Ion Beam Etching

Argon is most commonly used in this technique, and an ion source [63] is employed to produce the projectiles. In one such system [64], a hot filament provides the source of electrons which travel to an anode. A solenoidal magnetic field increases the electron path, and thus their ability to ionize argon gas which is then accelerated by means of an electric field (see Fig. 9.13). This allows independent control of the ion energy and the ion density. Next, the accelerated Ar^+ ions are made to leave the ion source through aligned, multiapertured grids. A hot filament, outside the ion source, injects electrons into this beam so as to neutralize the ion beam space charge. This results in the delivery of a highly directed, collimated beam of argon ions (and some neutral argon atoms) into a working chamber in which the substrates are located on a rotating platform.

Ion beam etching (IBE) depends upon momentum transfer for the removal of material from the substrate. Both conductors and insulators can be milled by this technique, because of the essentially charge-neutral character of the beam. Both the ion energy and ion density can be independently controlled by separate adjustment of the filament current and the accelerating voltage. In addition, the substrate temperature can be independently controlled, so that heated or chilled substrates are possible. The angle of incidence of the beam can be independently adjusted as well.

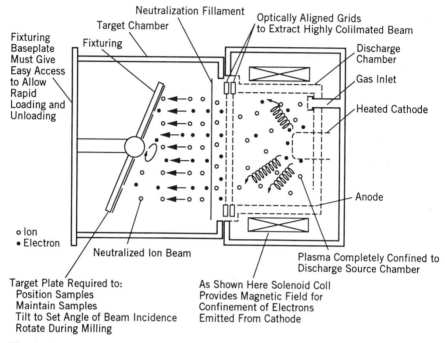

Neutralization Fillament

Target Chamber

Optically Aligned Grids
to Extract Highly Colilmated Beam

Fixturing
Baseplate
Must Give
Easy Access
to Allow
Rapid
Loading and
Unloading

Fixturing

Discharge
Chamber

Gas Inlet

Heated Cathode

Anode

o Ion
• Electron

Neutralized Ion Beam

Plasma Completely Confined to
Discharge Source Chamber

Target Plate Required to:
Position Samples
Maintain Samples
Tilt to Set Angle of Beam Incidence
Rotate During Milling

As Shown Here Solenoid Coil
Provides Magnetic Field for
Confinement of Electrons
Emitted From Cathode

Fig. 9.13 Ion milling equipment. From Thompson [64]. Reprinted with permission from *Solid State Technology*.

Finally the working chamber can be operated at a considerably lower pressure than the ion source (1–2 decades) so that the argon beam travels in a collision-free path, and thus retains its highly directed character after leaving the discharge chamber. However, sputtering of the walls of the plasma chamber produces a number of impurities, which contaminate the ion beam. Some of these can become lodged into the substrate during the milling process.

The milling rate for a number of materials of interest to microcircuit fabrication technology is shown in Table 9.7 at the end of this chapter. This table highlights one of the major advantages of ion milling, namely, the relative lack of selectivity in this process. This allows the etching of multilayer films with almost no undercutting of any of them.

The high degree of anisotropic etching that can be achieved by IBE makes it extremely attractive for use in VLSI applications. Consequently, a number of studies have been made to determine the role of substrate orientation and of the mask thickness in determining the etch profiles [65]. These have also included consideration of such important problems as the redeposition of sputtered material [40] onto the sidewall of a milled step. A treatment of these topics is beyond the scope of this work.

9.2.2 Sputter Etching

Sputter etching is carried out in a self-sustained glow discharge which is created by the breakdown of a heavy inert gas such as argon. The physics of this process has been detailed in Section 8.1.2 and is briefly summarized here for the d.c. excited glow discharge shown in Fig. 9.14a. Here, an electric field is impressed across electrodes which are located in this gas. Transport of electrons between these electrodes will result in some collisions with the gas molecules, so that ionization will occur in those encounters where a sufficiently large amount of energy is transferred to the gas. The products of these few ionizing collisions, Ar^+ and electrons, are of themselves accelerated (in opposite directions) because of the electric field, so that they can also enter into collisions with neutral argon molecules, thus resulting in an avalanche multiplication effect. These multiplied electrons reach the anode and the Ar^+ ions bombard the cathode, and hence the substrates which are placed on it, resulting in sputtering by momentum transfer. In addition, secondary electrons* emitted at the cathode participate in sustaining the discharge by ionizing collisions with argon molecules.

*About one electron per ten incident argon ions.

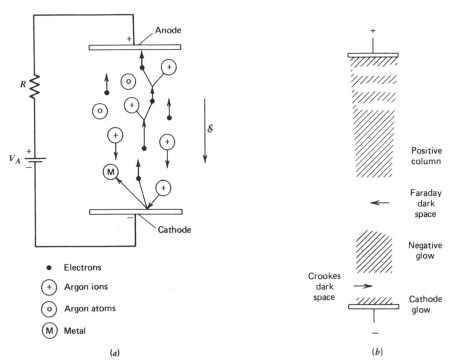

Fig. 9.14 Sputtering in a direct-current excited glow discharge.

Figure 9.14*b* shows the nature of the discharge. The most important region here is the cathode sheath (known as the *Crooke's dark space*), across which nearly all of the applied voltage is dropped, with the rest of the plasma providing electrical contact to it.

Current flow in a plasma is by both electrons and ions of nearly equal concentration. However, the electron velocity is about 1800 times that of the ions. As a result, electrons will rapidly pile up at the electrodes upon initiating the discharge, until a steady-state situation is reached, and a retarding field (for electrons) is established at both electrodes. This results in the potential diagram shown in Fig. 9.15*a*. Here, it is interesting to note that the plasma potential (V_p) is not intermediate between that of the anode–cathode potential, but is higher. It can be shown [66] that

$$V_p = \frac{kT_e}{q} \ln \left(\frac{v_e}{v_i} \right) \qquad (9.12)$$

where v_e and v_i are the electron and ion velocities, respectively, and T_e is the electron temperature. Typically, $V_p \simeq 15$ V. It follows that the majority of the voltage drop is across the cathode plasma sheath. Substrates placed on this cathode will thus be impacted by a high-velocity beam of Ar^+ ions, resulting in sputtering by momentum transfer. In practical situations, the sputtering of the cathode (on which the substrates are placed) is minimized by covering all exposed parts with a relatively pure, low-sputtering-yield material such as quartz.

The directionality of the ion beam arriving at the substrate is a function of the number of collisions encountered in the cathode plasma sheath. Thus, it is highly advantageous to operate at low pressures to keep these collisions to a minimum. Sputtering is usually carried out at pressures around 50 mtorr, corresponding to a mean free path of about 0.1 cm, so that some directionality is lost in these systems.

With d.c. sputter etching, it is necessary that positive ions impinging on the substrate surface have an opportunity to recombine with electrons to prevent charge accumulation on the surface. While discharging arrangements can be provided to minimize this problem, it can be completely avoided if the plasma is maintained by an r.f. field. This approach is most commonly used, since it is equally effective with both conducting and insulating substrates.

Radio-frequency sputtering is usually carried out with a 1- to 3-kV peak-to-peak r.f. potential, and at a pressure of 2–5 mtorr. Excitation is commonly provided at 13.56 MHz, which is internationally assigned for equipment usage. The symmetric nature of the applied voltage results in equal voltage drops across each plasma sheath, with both sputter deposition and sputter removal occurring with each successive half cycle. However, the system can be made asymmetric by varying the relative size of the electrodes. Thus, if A_C and A_A are the areas of the cathode and anode, respectively, and V_C and V_A are the respective voltage

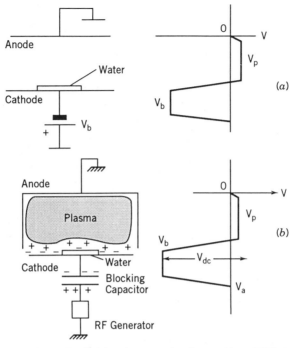

Fig. 9.15 Voltage distribution in a reactive ion etching (RIE) system.

drops at these electrodes, it can be shown (by assuming space-charge-limited current flow, and treating the sheath regions as capacitors) that

$$\frac{V_C}{V_A} = \left(\frac{A_A}{A_C} \right)^n \tag{9.13}$$

where $n = 4$. In practice, however, typical measured values for n are between 1 and 2. In any event, the basic strategy for r.f. sputtering is to make the cathode, on which substrates are placed, much smaller in area than the anode, so as to maximize the voltage drop across the cathode sheath. Usually, the chamber walls serve as the anode, and back-sputtering from them is minimized by making their area large compared to the cathode. In addition, they are often made of heavily anodized aluminum to further reduce these effects. Figure 9.15*b* shows a schematic of such a sputtering system. Here, a blocking capacitor is used to prevent shorting out the self-bias which results from the nonlinear characteristic of the plasma. The time-averaged voltage distribution in this system is also shown. The r.f. signal, superimposed on this potential, is omitted for the purposes of clarity.

In addition to allowing sputter etching of insulators, r.f. discharges have been shown to be more efficient in promoting and sustaining the plasma discharge. Thus, they can be operated at lower pressures than d.c. systems, by a factor of 20. For a system pressure of 4 mtorr (by way of example), this results in a mean free path of about 1.25 cm. Thus, collisions in the cathode sheath region are greatly reduced, resulting in a highly directed ion beam. This, in turn, leads to increased anisotropy in the etching characteristics of these systems.

Further increases in anisotropy can be obtained by operation at still lower pressures; however, this is accompanied by a fall in the plasma density and hence in the etch rate. Thus, there is considerable effort in the development of new plasma sources which can operate at low pressures, and still maintain high plasma densities. The electron cyclotron resonance (ECR) source is an example of one solution to this problem, in which the plasma generation and plasma interaction volumes are separated. In this manner, a high-intensity plasma can be produced with very little ion bombardment. A detailed description of this source, and of other alternative sources [67], is beyond the scope of this book.

9.3 DRY CHEMICAL ETCHING

Dry chemical etching is extremely attractive for a number of reasons. First, a dry process avoids problems of undercutting of the patterning film by capillary action of liquid etchants.* Next, the amount of reagent gases, and hence the effluent, is quite small. Although it may include some dangerous and toxic species, they are relatively easy to handle because of the small quantities involved. Finally, dry etching systems can be automated readily, with cassette loading, automatic pumpdown, and load-lock features.

Gaseous chemical species can be used for this purpose. For example, anhydrous HCl and SF_6 have been used for *in-situ* etching of silicon prior to epitaxy, whereas HCl and $AsCl_3$ vapor have been used for GaAs (see Chapter 5). In both cases, etching is carried out at high temperatures, comparable to those used for epitaxial growth.

Gaseous etching can also be accomplished at room temperature [68], by the use of interhalogenic compounds such as ClF_3, BrF_3, BrF_5, and IF_5. These etch at rates from 0.5 to 5 μm/min, are highly selective to SiO_2, and are isotropic in their characteristics.

An F_2/H_2 gas mixture, in the presence of ultraviolet excitation, has been used to remove SiO_2 from the surface of oxidized silicon wafers [69]. This technique can be used at atmospheric pressure and room temperature, and is highly selective with respect to the underlying silicon.

Yet another example of the use of gaseous species is the stripping of photoresist by means of UV and/or ozone gas. Removal rates as high as 1400 Å/min can be obtained at 300°C, by the combined use of ozone and UV excitation.

---

*An additional advantage is that the etching process is not unintentionally prolonged, wherever droplets of wet chemicals cling to the wafer upon removal from the etching solution.

A much wider range of materials, of interest to both silicon and GaAs micro-circuit technology, can be etched if energetic reactant species are used for this purpose. These can be produced by the use of plasmas, and have been known and widely used by the chemical industry for many years. They were first applied to microcircuit fabrication technology for the removal of polymerized photoresists. Here, their cost and convenience advantages have been so great that they have won ready acceptance in a short period of time.

Plasma chemistry deals with the conduct of reactions in a partially ionized gas composed of ions, electrons, and free radicals. These reactive species are produced by feeding the etchant gas into a glow discharge at pressures from 0.001–10 torr. In this discharge, free electrons gain energy from discharge, which they lose by collision with the gas molecules. Energy transfer from the electrons to the gas molecules can occur by elastic as well as inelastic collisions. A very small amount of energy is transferred by elastic collisions, because of the relatively low ratio of their masses ($\approx 10^{-5}$). On the other hand, inelastic collisions involve much larger energy transfer, accompanied by excitation of the etchant gas molecules.

A broad range of chemical and physical reactions can occur under these conditions [70]. Some of the more important electron–gas and gas–gas reactions are:

$$\text{Excitation:} \quad A_2 + e^- \rightarrow A_2^* + e^- \quad \text{(9.14a)}$$

$$\text{Dissociation:} \quad A_2 + e^- \rightarrow 2A + e^- \quad \text{(9.14b)}$$

$$\text{Electron Attachment:} \quad A_2 + e^- \rightarrow A_2^- \quad \text{(9.14c)}$$

$$\text{Dissociative Attachment:} \quad A_2 + e^- \rightarrow A + A^- \quad \text{(9.14d)}$$

$$\text{Ionization:} \quad A_2 + e^- \rightarrow A_2^+ + 2e^- \quad \text{(9.14e)}$$

$$\text{Photoemission:} \quad A_2^* \rightarrow A_2 + h\nu \quad \text{(9.14f)}$$

$$\text{Abstraction:} \quad A + B_2 \rightarrow AB + B \quad \text{(9.14g)}$$

where A_2^* is the excited molecule A_2.

Many species generated during this process are metastable and have short lifetime. Usually, free radicals are a major reactant species which participate in dry chemical processes. The most commonly used chemical species, in descending order of reactivity, are $O \simeq F > Cl > Br$.

If the lifetime of a radical species is sufficiently large, it can diffuse to the substrate where it adsorbs, and react with its surface, to produce reaction products. The etching process will proceed if these reaction products can be pumped away by the system. On the other hand, the process will be arrested if these products are not desorbed readily.

The essential process of dry chemical etching thus consists of the generation of active species, their transport to the substrate where the etching reaction occurs, and eventual removal of the reaction products from the surface. Each of these steps can be the rate limiter in the overall etching reaction. In addition, some reactants can polymerize on surfaces to form films whose removal is difficult, and is thus the rate limiter for the process.

Figure 9.16a shows the schematic of a "downstream" etching system, in which production of active species by r.f. excitation is separated from their reaction at the substrate. A more common approach is to integrate these functions as in Fig. 9.16b. This system consists an r.f.-excited quartz reaction chamber which can be pumped down to the requisite pressure. A gas, or gas mixture, is fed into this chamber through a controlled leak; the reaction products are exhausted through the pump.

Although plasma etching in systems of the type shown in Fig. 9.16b is primarily due to reaction with the energetic chemical species, some electron and ion bombardment also occurs [71], since the substrates are immersed in a region in which there is an electric field. This can result in (a) radiation damage due to bombardment by ionized species, and (b) uneven etching across the wafer. Here, the etching characteristics can be greatly improved by enclosing the wafers in a conductive mesh tunnel, to shield them from these species. In addition, it eliminates the effects of fringing fields at the wafer edge which would otherwise produce nonuniform etching due to localized heating, as well as localized bombardment by high-velocity particles [72]. This, in turn, greatly reduces the plasma damage in the device structure.

A commercial tunnel reactor, involving multiple gas inlets, is shown in Fig. 9.16c. Characteristic of this type of reactor is the fact that a high degree of selectivity can be obtained, depending on the choice of chemical species. In addition, the etching is isotropic because of the random direction of arrival of energetic species to the substrate. Finally, the process results in negligible damage to the substrate. As a result, tunnel reactors are widely used in industry today.

9.3.1   Photoresist Removal

Dry chemical etching techniques, using energetic species, achieved their first major acceptance in microcircuit fabrication when they were used for the removal of exposed photoresist. Conventionally, two wet chemical approaches have been used for this purpose. In the first, this tough, polymeric material is "burned" by treatment in hot oxidizing agents, such as H_2SO_4–H_2O_2 mixtures. Alternatively, the wafer is soaked in one of many hot chlorinated hydrocarbon mixtures (trichloroethylene, chlorobenzene, etc.) which induce swelling of the polymer and loss of adhesion to the substrate. In either case, mechanical scrubbing is often employed in order to remove remnants of these materials.

Both these approaches require the handling and eventual disposal of large amounts of corrosive chemicals. In addition, the combination of rough treatments described here can result in damage to underlying films, and especially to fine-line metallization films which are soft (gold and aluminum) and easily attacked (aluminum) by acids and bases. The requirements of VLSI technology

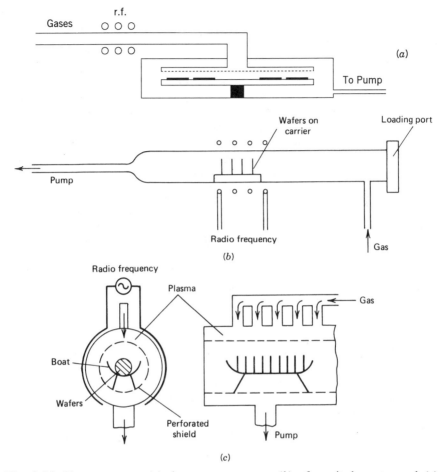

Fig. 9.16 Plasma reactor: (*a*) downstream reactor, (*b*) r.f. excited reactor and (*c*) modern tunnel unit. Adapted from Kern and Vossen, [51].

have made these problems extremely severe, so that plasma removal techniques are almost universally employed here.

Plasma removal of photoresist, or *plasma ashing* as it is often called, consists of placing the resist-covered substrates in a tunnel reactor, in the presence of an oxygen plasma. A large number of energetic species are generated in this plasma [73]. Of these, the most predominant are formed by

$$e + O_2 \rightarrow 2O + e^- \tag{9.15a}$$
$$\rightarrow O + O + e^- \tag{9.15b}$$

Temperatures as low as 40–50°C are sufficient to cause oxidation, or "burning" of the resist by these free radicals. The reaction products consist mostly of water, carbon monoxide, and carbon dioxide, which are rapidly desorbed.

The rate of removal of photoresist depends on the gas pressure, the gas flow rate (which establishes its residence time in the reactor), the r.f. power, the wafer temperature, and the number of substrates in any given load. Typically, 100–300 watts of r.f. power* for 5–10 min can be used to process a batch of wafers coated with 1-μm photoresist in a single operation.

A large number of closely spaced substrates are usually cleaned in a single operation, so that this process is diffusion-controlled. As a result, the removal rate is a function of the spacing between the wafers and also of their diameter [74]. In addition, the etch rate is faster at the perimeter, and slower at the center of the wafer because of this diffusional process. These problems are of little consequence, however, since almost all of the materials upon which the photoresist is placed are unaffected (or very slightly affected) by the oxygen plasma. These include silicon, GaAs, aluminum, gold, SiO_2, and Si_3N_4. As a result, plasma ashing can be carried out until all the resist is removed without fear of etching the underlying substrate film. In addition, the use of modern resists leaves almost no residue, so that the wafers are ready for the next process step after a minimal amount of clean-up and rinsing.

9.3.2 Pattern Delineation

A number of additional requirements are placed on dry etching systems when they are required to transfer photographically produced fine-line patterns to an underlying film. Here, if a photoresist pattern is used, it is necessary that it be unattacked by the plasma, or attacked at a much lower rate than the underlying film. If this is not possible, it is necessary to use an additional layer of a suitable material (such as SiO_2 or a refractory metal) in which the pattern has been previously delineated by the photoresist. Next, the etching of the film must not be accompanied by inadvertent etching of subsequent layers, so that selectivity is an important consideration. Finally, undercutting must be avoided. This is especially important when it is required to etch holes through a layer of varying thickness, as illustrated in Fig. 9.17. Here, a high degree of anisotropy is necessary to avoid overetching and accidental contact with closely spaced neighboring regions.

We have noted that dry physical processes, such as ion beam etching and sputtering, exhibit a high degree of anisotropy, but are nonselective in character. On the other hand, the use of active species results in highly selective etches, which are isotropic. In the following section, we shall consider combinations of these approaches, in order to take advantage of their unique features.

---

*This is the r.f. power output of the generator. The actual power coupled to the reactants or to the substrates is rarely known, but is considerably less.

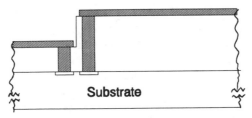

Fig. 9.17 Anisotropic etching profiles.

9.4 REACTIVE ION ETCHING

Reactive ion etching (RIE) can be accomplished by replacing the neutral gas in a r.f. sputtering system by one or more chemical species. Plasma interactions with these chemicals result in the production of both neutral and ionized energetic species which can etch the substrate.

A common system configuration for RIE is that of a planar reactor, as shown in Fig. 9.18. Here, the substrates are at normal to the gas flow, and are immersed in the plasma, allowing the use of energetic species which have short lifetimes. In addition, they are normal to the r.f. field, so that the movement of ionized species is both highly directional and rapid. Consequently, a high degree of anisotropic etching can occur. This anisotropy can be enhanced by using chemicals whose reaction products have a large ionized component. Additionally, reactors can be operated with a large voltage drop across the cathode sheath to increase their impingement velocity and aid in this process. Finally, operation at reduced pressure minimizes collisions in the sheath, thus preserving the directionality of the ionized species and enhancing further the anisotropic nature of the etch process.

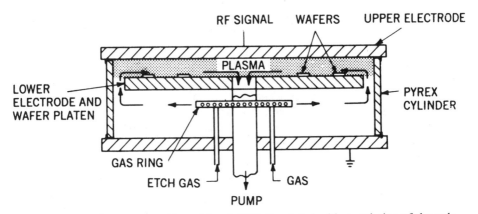

Fig. 9.18 A planar reactor. From Mogab [75]. Reprinted with permission of the publisher, The Electrochemical Society, Inc.

In RIE, the sputter-removal nature of physical processes such as ion bombardment results in relatively poor selectivity, since the removal rate is now related to the sputtering yield. On the other hand, energetic neutral chemical species also participate in the etch process. Their etching properties tend to be isotropic in nature, and highly selective with respect to the material being etched. Thus, successful etching is obtained when there is a suitable balance between physical and chemical processes. This balance depends on the choice of chemical species, and such system parameters as the r.f. voltage and the substrate self-bias.

RIE results in etching products which desorb and are removed by the pumping system. An important benefit of ion bombardment during this process is that it enhances the desorption of these products, and thus greatly improves the efficiency of their removal, with a concurrent increase in the etch rate. The directed ion beam also breaks up thin films of native oxides and polymeric products, some of which may be present on the substrate surface, or are formed during etching. This helps sustain the etching reaction, which would otherwise be arrested by their presence.

The formation of "passivating" films of this type during etching can be used to advantage with this technique. These films are removed from the surface by bombardment with ions which are normal to the substrate. However, the sidewalls of an etched region are not subjected to significant ion bombardment if the beam is highly directed, and thus remain relatively unetched. This prevents undercutting of the film during RIE, resulting in steep, vertical walls.

9.4.1 Loading Effects

Most of the process parameters which control the etch rate can be optimized, once and for all, by experimental runs of the system. Unfortunately, however, the etch rate is *also* related to the number of wafers in a load, or more precisely, to the area of the surface being etched. Specifically, it increases as this area is reduced, so that an adjustment of the etching time must be made to compensate for each load.

The magnitude of this loading effect can be determined if some simplifying assumptions are made concerning the reactant species and the nature of the reaction [75]. Thus, assume that a single reactant species is produced by the plasma, with a generation rate G per unit volume and unit time. Let τ be the recombination rate associated with this process.

Assume further that the reaction can be characterized by a linear reaction rate constant k at any given wafer temperature, and that a single reaction product is formed. Let N be the concentration of the reactant species, and let A be the surface area of the material that remains to be etched. If j is the flux density of this species, then the flux is given by

$$jA = kNA \qquad (9.16)$$

Let V be the volume of plasma from which this species is delivered to the wafer. From considerations of continuity, it follows that

$$\frac{\partial N}{\partial t} = G - \frac{kNA}{V} - \frac{N}{\tau} \tag{9.17}$$

In steady state,

$$\frac{\partial N}{\partial t} = 0 = G - \frac{kN_0 A}{V} - \frac{N_0}{\tau} \tag{9.18}$$

where N_0 is the steady-state concentration of the active species. The etch rate R is given by j/n, where n is the number of atoms per unit volume of the layer being etched. Combining with Eqs. (9.16) and (9.17) gives

$$R = \frac{G}{n} \left\{ \frac{\tau k}{1 + (\tau k A/V)} \right\} \tag{9.19}$$

From this equation it is seen that if $\tau k A/V \ll 1$, the etch rate is independent of the area being etched. However, the etch rate rises with decreasing A, when $\tau k A/V > 1$. Experimental results, for the etching of silicon and silicon nitride in a CF_4 plasma [76], are shown in Fig. 9.19 and illustrate this point.

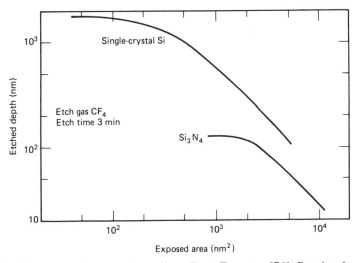

Fig. 9.19 Illustration of the loading effect. From Enomoto [76]. Reprinted with permission from *Solid State Technology*.

This changing etch rate creates problems in the patterning of fine-line structures for VLSI. Consider, for example, the etching of an aluminum metal film using a photoresist pattern. Here, operation of the reactor in the region where $\tau k A/V > 1$ results in an etch rate which increases as the aluminum film clears (i.e. as A falls). Consequently, any undercutting that occurs during this phase will proceed at an accelerated rate, and lead to loss of anisotropy.

One way of avoiding the problem of overetching due to loading effects is to load the reactor very lightly ($\tau k A/V \ll 1$) so that these effects are unimportant. This can be done by increasing the interelectrode spacing, so as to increase the plasma volume, and by reducing the surface area. A disadvantage of this procedure is the lowered throughput. Nevertheless, single-wafer processing is increasingly used to avoid this problem.

9.4.2   End-Point Detection

The poor etching selectivity of RIE systems, combined with loading effects, makes it extremely important to terminate the etch procedure at the precise point when the entire film has been removed. End-point detection schemes, which allow this to be done, are thus an integral part of the RIE process. In principle, this requires monitoring any parameter which undergoes a rapid change at the end of a particular etch step. One such parameter is the optical reflectivity of the surface, as the film is removed. A more sensitive approach is to measure the film thickness during the etch process, by study of its optical interference pattern.

Both of these techniques can be used for probing small areas of the film. There are also a number of alternative techniques which provide information averaged over the slice, or over the batch of slices, which are being etched [77, 78]. One such approach involves residual gas analysis of the reactor environment or effluent [79]. Although relatively complex, this is a valuable tool for elucidating the nature of species involved during the plasma process.

Yet another approach is to measure [80, 81] the change in optical emission during plasma etching. An appropriate wavelength for this purpose is usually selected by making spectrometer scans during the process, and selecting the wavelength associated with a maximum change in intensity when the process is terminated. Table 9.8, at the end of this chapter, lists the important characteristic emission signatures for a few species which are produced during the plasma etching of materials of interest, and which have been used for monitoring the etch process [81].

Other techniques for end-point detection include monitoring parameters such as the change in r.f. coupling, the change in system pressure, or the number density of the plasma by means of Langmuir probing [82]. All tend to be machine specific, and are relatively simple to implement in a production environment.

9.4.3 Etch Chemistries

Many of the considerations for dry chemical etching hold for RIE as well. Thus, chemicals should be volatile, for ready transport into the RIE chamber; moreover, they should not be excessively corrosive or unstable. Next, products of the etching reaction should have sufficiently high vapor pressures at the substrate operating temperature (typically 75–100°C) so that they can rapidly desorb and be pumped away. Gaseous halogen compounds are often used for these reasons, the fluoride and chloride species being preferred because they are less corrosive and easier to handle than the bromides and iodides. Elemental halogens are also used in some applications. Figure 9.20 shows the vapor pressure of the halides of a number of materials of interest to microcircuit fabrication.

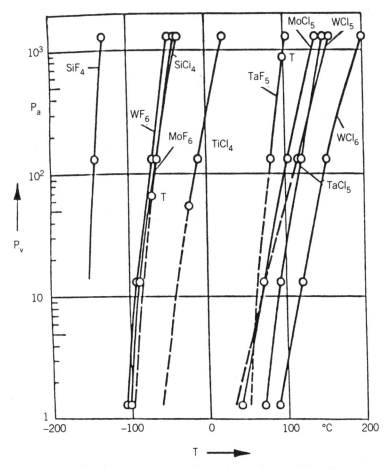

Fig. 9.20 Vapor pressures of some transistion metal halides.

Two classes of halogen compounds must be considered; those which include carbon (such as the fluorocarbons and chlorocarbons) and those which do not. An important characteristic of the former is the tendency to form polymeric products in the plasma environment. These polymers land on the substrate and prevent desorption of reaction products, so that the etch reaction could be arrested. Ion bombardment serves to remove these species, thus allowing the reaction to continue. However, if the ion beam is well directed, it will be ineffective in removing polymers from the sidewall. Highly anisotropic etching, with minimum undercutting, can be achieved in this manner.

The extent of sidewall "passivation" obtained in this manner is related to the carbon:halogen ratio. With the fluorocarbons, gases such as CF_4, C_2F_6, and C_3F_8 are used to adjust this ratio [83]. A large self-bias and a reduced system pressure both result in improvements in anisotropy, by increasing the intensity of directed ion bombardment.

A number of different etching systems are now considered in order to illustrate the conditions for their selection.

9.4.3.1 Silicon and Silicate Glasses

Both silicon and silicate glasses are extensively used in microcircuit fabrication technology. These films include polysilicon, SiO_2, PSG, BPSG, doped oxides, and Si_3N_4. Fluorocarbons, such as CF_4, C_2F_6, and CHF_3, have been used as etchant gases for these materials because of the high volability of the reaction product, SiF_4. Of these, CF_4 is often used because of its stability. The reaction of CF_4 which is most probable in a plasma environment is

$$e^- + CF_4 \rightleftharpoons CF_3 + F + e^- \tag{9.20}$$

The fluorine radical is the primary active species which etches the silicon by converting it to SiF_4 which is a stable, volatile reaction product. In this reaction, the steady-state concentration of F is relatively low because of strong homogeneous recombination with SF_3 in the gas phase. As a result, the etch rate is extremely low, as seen in Fig. 9.21.

Addition of small quantities of oxygen (1–20 vol.%) results in a rapid increase in the amount of atomic fluorine [84], by reaction with the CF_3 in the plasma. Both CO and COF_2 are also produced during this process. This is shown in the experimental data of Fig. 9.21, as evidenced by the emission intensity of F at a wavelength of 704 nm, and CO at 482 nm [85]. Also shown in this figure is the etch rate for silicon which tends to follow the same trend as the fluorine emission intensity.

The etch characteristics of SiO_2 in a CF_4–O_2 plasma are generally similar to those shown for silicon in Fig. 9.21. Here, however, the peak in the etch rate occurs at about 20% O_2 (as compared to 10% O_2 for silicon). It has been proposed that F is the active ingredient in the etch process. However, additional factors such as ion bombardment of the surface, or etching by CF_3, are probably

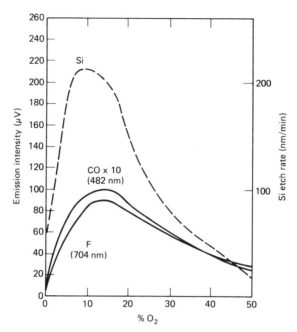

Fig. 9.21 Emission intensity and silicon etch rate as a function of oxygen concentration. Adapted from Harshbarger and Porter [85].

operative in this situation as well. The difference in the position of the peak is probably due to the stronger adsorption of oxygen on the silicon surface, which competes with fluorine for adsorption sites. This difference can be used to provide selectivity during the etching of contiguous layers of these materials.

Silicate glass films, such as SiO_2, PSG, and BPSG, can be readily etched in a CF_4-O_2 plasma [86]. The etch rate in this plasma is relatively independent of the temperature at which the glass is grown (450–650°C) or the amount of oxygen in the CF_4 plasma (0–66%). However, it tends to increase linearly with amount of dopant incorporated in the glass. Typical etch rates in a CF_4-O_2 plasma for heavily doped oxides (10^{21} dopant atoms cm^{-3}) are 2–10 times the etch rate for CVD-grown layers of undoped silica.

The etching of Si_3N_4 films occurs by the direct reaction of F with them [87]. Thus, the overall reaction can be given as

$$12\,F + Si_3N_4 \rightarrow 3SiF_4 + 2N_2 \tag{9.21}$$

The etching characteristics in a CF_4-O_2 plasma are thus very similar to those of SiO_2, with etch rates that are slightly higher. Typical etch rates [72] for the $CF_4-4\% \, O_2$ plasma are $Si : Si_3N_4 : SiO_2 : AZ \, 1350$ photoresist $= 17 : 3 : 2.5 : 1$.

The use of a CF$_4$ plasma for etching windows in a protective cover layer of PSG or Si$_3$N$_4$ (through which contacts are made to bonding pads in a microcircuit) is relatively straightforward. This is because aluminum is not etched by fluorine plasmas since they result in the formation of a tough, nonvolatile aluminum fluoride film which prevents further attack of the metal.* Thus the aluminum bonding pad effectively stops the etching process.

The etching of windows in an oxide or nitride, prior to diffusion, is considerably more difficult, since the etch rate of silicon is about 3–10 times faster than that of the SiO$_2$. Consequently, care must be taken to precisely terminate this etch step by end point detection methods. An alternative approach lies in altering the selectivity of the etch process by controlling the oxygen content of the etchant gas and thus varying the relative amounts of CF$_3$ and F that are produced. Finally, it is possible to use a CF$_4$–H$_2$ gas mixture, or CHF$_3$ gas for this purpose. The introduction of hydrogen into the system results in the subsidiary reaction

$$2F + H_2 \rightarrow 2HF \tag{9.22}$$

thus suppressing the fluorine concentration in the system, and hence the etch rate of silicon. On the other hand, the strongly reducing character of CF$_3$ is important in etching SiO$_2$ by the intermediate step of reducing it to SiO. Using this approach, etch ratios as large as 35 have been achieved for SiO$_2$: Si. An additional advantage here is that this plasma has no oxygen so that photoresists are relatively unattacked by it. Figure 9.22 shows etch rates for one type of resist, SiO$_2$, and silicon as a function of the concentration of hydrogen in the plasma [88].

Inorganic halides, such as SF$_6$, SiF$_6$, NF$_3$, and XeF$_3$, have also been used in RIE systems, for silicon and silicate glasses, and each has unique properties. Thus, the use of SF$_6$ results in passivation by sulfur polymers, and leads to characteristics similar to those of CF$_4$. Both NF$_3$ and XeF$_3$ contain only inert gases in addition to fluorine, so that no polymers are formed with them. A unique feature of XeF$_3$ is that it will [89, 90] attack silicon spontaneously. As a result, its etch rate for silicon is much more rapid in RIE than that of other fluorides. At the same time, however, it is more isotropic in its etch characteristics.

NF$_3$ is also free of surface passivating effects [91–93]; moreover, it does not spontaneously react with silicon, so that its etch characteristics are highly anisotropic. Extremely high etch rates (≈ 1.5 μm/min) can be obtained with this gas, together with a selectivity to SiO$_2$ of 15 : 1. The absence of oxygen allows the ready use of photoresist masking. During plasma etching, this species dissociates to NF$_2$ and NF, in addition to F.

Chlorine chemistry is also used in RIE systems for etching silicon and silicate glasses. Its success is based in part on the relatively lower volatility of SiCl$_4$ as

---

*Aluminum slice carriers are often used in plasma etching systems for this reason.

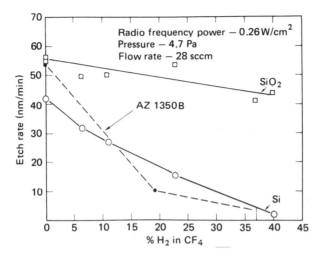

Fig. 9.22 Etch rates for resist, silicon, and silicon dioxide in a CF_4–H_2 plasma. From Ephrath [88]. Reprinted with permission of the publisher, The Electrochemical Society, Inc.

compared to SiF_4. Here, ion bombardment plays a more significant role in the etch process so that a higher degree of anisotropy can be achieved. This is also evidenced in the extremely small loading effect observed with these plasmas as compared to fluorine plasmas [94]. Typical chlorine-bearing species in common use are Cl_2 and CCl_4 [95].

Gas mixtures can also be used to take advantage of the separate properties of the individual species. For example, the inclusion of an inert gas [96–99] such as argon results in an increase in the ion-directed component in an RIE system. Chlorofluorocarbons such as $CClF_3$ and $C_2Cl_2F_4$ have been used to combine the unique features of both chlorine and fluorine chemistries. However, mixtures of chloride and fluoride species are preferred, because they provide control of the Cl/F ratio during RIE.

Polysilicon and SIPOS films are very rapidly etched in fluorine plasmas, with etch rates that are 3–10 times that for single-crystal silicon. Etching of these films is commonly carried out in a CF_4–O_2 mixture that has been diluted with nitrogen for this reason. The etch characteristic is isotropic, because of the high reactivity of silicon with F. Anisotropy can be achieved by the addition of chlorine; SF_6/Cl_2 or CF_4/Cl_2 gas mixtures have been used, as well as chlorofluorocarbons such as $C_2Cl_2F_4$.

9.4.3.2 Refractory Metals

Refractory metal films of molybdenum, tantalum, and tungsten cannot be etched in chlorine-based plasmas, because of the low vapor pressure of the resulting halides (see Fig. 9.20). However, fluorine-based plasmas can be used for this purpose, and CF_4 as well as CF_4-O_2 are commonly used. Etch rates for these metals are comparable, and are about $100-150$ Å/min for 100 W of input r.f. power [100]. The vapor pressure of $TiCl_4$ is sufficiently high so that chlorine-based plasmas can be used for the RIE of titanium, in addition to fluorine chemistries.

Compounds such as CCl_2F_2 have also been used in situations where a high degree of anisotropy is required. With these, the formation of chloride reaction products tends to preferentially arrest the etching products on the side walls, although they can be removed from the surface by ion bombardment.

9.4.3.3 Barrier Layers and Silicides

Barrier layers such as Ti–W and TiN can be readily etched in fluorine-based plasma, because of the reasonably high vapor pressure of the fluorides of both titanium and tungsten. Gases such as SF_6, CF_4, and CHF_3 are often used for this purpose. Chlorine-based plasmas can also be used for TiN. Their use with Ti–W results in a high degree of anisotropy, because of the relatively low vapor pressure of the chlorides of tungsten. Gases such as Cl_2, CCl_4, and BCl_3 have been used in this situation, as well as fluoride–chloride gas mixtures.

The RIE of silicides necessitates the breaking of Si–Si bonds, metal–Si bonds, and metal–metal bonds. Moreover, in order to obtain smooth morphology, the reaction products should desorb at relatively similar rates.

From Fig 9.20 it is seen that chlorine-based chemistries cannot be used with $MoSi_2$, $TaSi_2$, and WSi_2. Fluorine-based chemistries, using SF_6 and NF_3, have been used for this reason, in addition to CF_4-O_2. Note, however, that there is still a wide disparity in the vapor pressures of the reaction products. Thus, at a given temperature, the vapor pressure of SiF_4 is about two decades larger than that of MoF_6. As a result, desorption of the Mo species is the rate limiter during the RIE of $MoSi_2$ in these plasmas [101, 102].

The desorption rate of the metal silicide can be enhanced by a number of techniques. These include system operation at increased temperature and/or reduced pressure, and an increase in the ion bombardment component of RIE by increasing the self-bias. Yet another approach is the use of an RIE system which operates at low frequency (380 kHz) to enhance the ion bombardment effects [103].

In microcircuit technology, a silicide is often placed on polysilicon which has been deposited on a thin SiO_2 gate oxide. A successful RIE process for these "polycides" requires anisotropic etching of both the silicide and the polysilicon, with little or no undercutting. Additionally, the process should be highly selective to SiO_2, to prevent etching through the thin gate oxide. This is indeed a challenging task which can be handled with some degree of success.

One approach for the RIE of these structures is to use a nonstoichiometric silicide, in order to enhance its etch rate so that it is comparable to that of the polysilicon. In one set of experiments, an $SF_6 + 35\%$ O_2 mixture has been used to reduce the etch rate of the SiO_2, and thus obtain a selectivity ratio of as high as 10 : 1 [103].

The use of $TiSi_2$ films in polycide structures is advantageous, since $TiCl_4$ has a reasonably large vapor pressure (see Fig. 9.20). Here, a chlorine-based system has been used with a selectivity ratio to SiO_2 of 15 : 1 for polycide etching [104].

Mixed chemistries, using Cl- and F-based compounds, can also be used for etching polycide films [105]. With these mixtures, the chlorine compound retards the removal of the silicide relative to the polysilicon, and enhances the selectively to SiO_2 at the same time. Careful adjustment of the gas ratio is necessary for the success of this approach. Proper adjustment of the self-bias is yet another means for obtaining control of these processes.

9.4.3.4 Aluminum Metallization

The RIE of aluminum is a particularly difficult task. Fluorine chemistry cannot be used, since the residue, AlF_3, has a very low vapor pressure. This necessitates the use of chlorine and/or bromine compounds [106].

A second problem comes about because of the tough, patchy, native oxide which is always present on the surface of aluminum. Typically 10–30 Å thick, its initial removal during the etch process results in an induction period whose time depends on slice history prior to loading into the RIE system. Both BCl_3 and CCl_4 are successful in scavanging this native oxide; in addition, they can etch the aluminum, to produce its volatile chloride. However, their etching characteristics are quite dissimilar [107] and are related to the nature of boron and carbon residues which form on the surface, and alter the etch process. The addition of a small amount of oxygen to the etch gas is found to improve its controllability, probably by preventing the build-up of these residues. Chlorine gas is often used to avoid this problem. Some aluminum etching recipes use $CCl_4/BBr_3/Cl_2/O_2$ mixtures as a compromise for this reason.

The primary reaction product during the etching of aluminum by these gases is $AlCl_3$, which is hygroscopic. Moisture absorption, upon opening the RIE system to air, results in a high degree of variability from run to run. As a result, a load lock system is, for all practical purposes, essential. Both start- and end-point detection are often used, to correct for the variable induction period during aluminum etching.

The problems of etching aluminum metallization are greatly complicated when small amounts of copper and silicon are added to the film to improve its electromigration properties and reduce hillock formation, respectively. Here, the silicon does not represent a problem, because its reaction product ($SiCl_4$) is volatile and can be rapidly pumped away. The chlorides (and bromides) of copper, on the other hand, are relatively involatile. These react with the aluminum film (which has no protective Al_2O_3 layer at this point) and initiate corrosion. Further corrosion occurs when they hydrolyze upon eventual exposure of the substrates to air.

Post-etching corrosion can be minimized in a number of ways. Reactor operation under conditions which result in a large negative substrate bias (i.e., conditions which lead to high ion energy) can reduce the build-up of these products during etching. A short post-etch rinse in water, to remove these chlorides, is also useful. A post-etch rinse in concentrated HNO_3 can also be used for passivation purposes. Finally, some systems provide a second chamber in which a short RIE step can be conducted in an SF_6/O_2 or CF_4/O_2 plasma for the purpose of passivating the aluminum with an AlF_3 coating [108, 109].

Photoresist problems are especially severe with aluminum etching, since $AlCl_3$ is particularly harsh on these materials. Overbaking of the resist, for toughening purposes, has been found to help in some situations. Alternatively, both PSG and Si_3N_4 films can be used for masking purposes with aluminum.

9.4.3.5 Gallium Arsenide and Aluminum Gallium Arsenide

As with aluminum, fluorine chemistry cannot be used for GaAs and AlGaAs, because of the involatile nature of the fluoride residues [110]. Chlorine and bromine chemistries are thus necessary for these materials as well. The vapor pressures for $AlCl_3$, $GaCl_3$, and $AsCl_3$ are 0.122, 14.5, and 199 torr, respectively, at 75°C [111]. As a result, it is advantageous to etch at temperatures above 120°C in order to assist the desorption of these species.

An additional problem arises because of the presence of thin native oxides on these semiconductors. The removal of these oxides is facilitated by the use of BCl_3 and $SiCl_4$ as etchant gases, which tend to scavenge them as well as residual moisture which may be present in the system. Mixtures with argon gas are sometimes used to increase the directed ion component for this purpose [112].

Selective etching of GaAs over AlGaAs is technologically important for the fabrication of heterojunction structures. Here, advantage is taken of the extreme difficulty of removing fluorine compounds of aluminum to achieve a high degree of selectivity. One etch, with a mixture of CCl_2F_2 and He, was used [113] to obtain selectivity ratios from 2 to 200 depending on system pressure, with the higher ratio at a system pressure of 3.75 mtorr. This is to be expected, since low pressures result in more ready removal of the fluorine compounds.

Improved control of selectivity can be obtained by using mixtures of fluorine- and chlorine-bearing species. For example, a mixture of SF_6 and $SiCl_4$ can be used [114] to give a wide range of selectivity, from 1 to 500, as seen in Fig. 9.23.

9.5 CHEMICALLY ASSISTED ION BEAM TECHNIQUES

Reactive chemical species can also be combined with ion beam etching, to greatly widen its field of application. Here, the species can be introduced directly into the plasma source chamber, to result in reactive ion beam etching

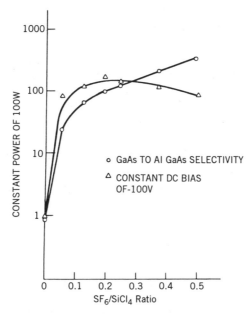

Fig. 9.23 Etch rates of GaAs and AlGaAs at a pressure of 50 mtorr. From Salimian et al. [114]. With permission from *Applied Physics Letters*.

(RIBE). Alternatively, chemical species can be introduced into the chamber where the etching takes place, to result in chemically assisted ion beam etching (CAIBE). This latter approach is generally used, since the problems of building reliable, trouble-free RIBE systems have not been solved at the present time.

One advantage of these approaches is that the pressure at which etching takes place is about 0.1 mtorr or less, and almost all bombardment is by ionized species. Thus, a very high degree of anisotropy can be achieved in their etching characteristics. Control of etch profiles can be obtained by adjustment of the impingement angle of the beam in these systems.

The nonselective character of IBE carries over in large measure, so that these systems are ideally suited for etching multilayer films which have highly different properties, such as aluminum/silicide/polysilicon [115]. Moreover, the degree of non-selectivity can be controlled by varying the amount of chemical species in the ion beam. For example, the use of CH_3F in RIBE can etch SiO_2 over silicon with a selectivity ratio of 12 to infinity, depending on beam energy and ion current density.

Etching of AlCuSi metallization can also be accomplished by RIBE, using CCl_4. Here, control of the beam energy is used to remove $CuCl_3$ residues, thus avoiding the need for post-etch treatments which are necessary with RIE techniques.

A major advantage of these techniques over IBE is their ability to cut deep trenches. With IBE, the aspect ratio is limited because of redeposition from the sidewalls on to the base of the trench. With both RIBE and CAIBE, this does not occur since etching products are gaseous in nature.

CAIBE, using chlorine which is injected into the etching chamber, has been used to define facets in GaAs lasers. Here, the use of this gas greatly increases the etch rate over IBE. This, in turn, allows etching to be conducted at low ion energies ($\leq$500 eV), to result in surfaces which are damage-free and which are capable of use as mirrors in this application [116].

In summary, both CAIBE and RIBE are extremely promising alternatives to RIE, especially in the area of pattern transfer of multilayer films with highly different properties. However, major improvements in design are necessary before these systems can be fully accepted in the microcircuit manufacturing environment.

9.6   ETCHING-INDUCED DAMAGE

Although wet chemical etching is sometimes used to delineate damage in a semiconductor (see Section 9.1.6), it does not, of itself, create any. This also true for dry chemical etching, especially when the energetic species are created upstream from the substrate, and care is taken to avoid exposing it to ultraviolet radiation. As indicated earlier, some damage can be created if the substrate is immersed in the plasma; however, this can be minimized by the use of mesh tunnels of the type shown in Fig. 9.16 c. For example, photoresist removal in a barrel asher creates a small amount of damage to the underlying silicon, with a resultant increase in the sheet resistance of its surface region.

In contrast, etch processes which involve the directed flow of species can cause damage in both silicon and GaAs. Much of this damage is created by the bombardment of the semiconductor surface, which leads to the formation of vacancy complexes, dislocations, and implanted impurities. In silicon, oxygen induced stacking faults can also be created; often, the top 50-Å layer is rendered amorphous. This can result in low minority carrier lifetime, and high reverse currents in the resulting devices. Much of this damage can be annealed in subsequent thermal process. On the other hand, gate oxide breakdown, which is caused by plasma nonuniformity during the etch process, is irreparable.

In GaAs, the different dislodging rates of gallium and arsenic lead to local departures from stoichiometry [117], and to the formation of arsenic vacancies. These effects cannot be subsequently annealed.

With silicon, RIE results in increasing the density of interface traps, N_{it}. This increase is roughly proportional to the ion beam energy, and hence to the d.c. bias voltage which typically varies from 50 to 500 V.

At this point, it is worth emphasizing that RIE is a relatively dirty process by semiconductor standards, and that all ion beam etching techniques are even dirtier! Contamination can arise from the sputtering of materials off system filaments, walls, targets and beam focusing plates. In addition, polymeric compounds often result from the breakdown of fluorocarbons which are used as etchants. As a result, N_{it} is also a function of the specific etchant gases which are used.

Minority carrier lifetime in silicon is adversely affected by RIE, the extent of deterioration increasing with the bias voltage. Even low-bias voltages ($\approx 50-100$ V) can result in a fall in this parameter by as much as a factor of 10. Again, the deterioration is due to both contamination and damage, and is a function of etchant gases as well as system design.

Ion bombardment of silicon by a low-energy Ar^+ beam has been shown [118] to produce positive charges in the surface region. This is probably due to the radiation induced damage and to the direct implantation of Ar ions [119].

Ion-induced damage can also be created in GaAs. This is evidenced by carrier removal, an increase in the barrier height of Schottky diodes, an increase in their ideality factor, and the formation of trap levels. Low levels of damage can be annealed by thermal treatments. However, damage created by ion energies above 200 eV, which induces changes in stoichiometry, cannot be removed in subsequent processing [120].

In both silicon and GaAs, the depth of the damaged layer is considerably larger than the projected ion range. This is probably due to enhanced diffusion effects created by the ion bombardment.

It follows from the above that ion-induced damage in RIE can be minimized by the use of low ion energies. This can be accomplished by using low r.f. power and low-d.c.-bias voltage, or by going to higher frequencies. Additionally, careful attention should be paid to the design of the electrodes and walls of the chamber, to minimize contamination from them. Aluminum, which is heavily anodized, is often used for the construction of these parts in order to keep contamination to a minimum.

Damage in GaAs, created by IBE, can be greatly reduced by etching with a low ion energy, and by flooding the substrate with a flux of chlorine, to result in an order-of-magnitude increase in the etch rate [121]. As a result, low beam energies (<200 eV) can be used, thus minimizing damage created in the material. This CAIBE process has been used for the formation of reflecting facets in GaAs lasers, as well as for the etching of deep via holes in high-frequency GaAs circuits, and represents an important advancement of the IBE process.

9.7 CLEANING

Semiconductor wafers are subjected to physical handling during the processes of cutting, lapping, polishing, and packaging. This leads to large amounts of molecular contamination, much of which is of unknown origin. Included in the list of such contaminants are (a) airborne bacteria, (b) grease and wax from cutting oils and from physical handling, (c) abrasive particulates (usually silicon carbide, alumina, and diamond dust) from the grinding and sawing operations, and (d) a variety of plasticizers which come from the containers and wrapping in which the wafers are handled and shipped. Removal of these contaminants is the first step in wafer cleaning, and is usually done by rinsing in hot organic solvents such as trichloroethane or xylene, accompanied by mechanical scrubbing, ultrasonic agitation, or compressed gas jets.

Ionic contaminants are also present on these wafers, and result from improper cleaning or from etch steps during processes, such as the opening of a window or the delineation of a metal interconnection pattern. These ions (usually light ions such as sodium and potassium) are due to trace impurities in the etchants and adhere to the semiconductor surface by physical adsorption or by chemisorption.

Light-metal-ion contamination is particularly undesirable in gate oxides of MOS-based microcircuits, where they can cause instabilities in the threshold voltage. However, they can often be tolerated to a greater extent in GaAs as well as in bipolar silicon microcircuits. Desorption of a large fraction of these ions can be accomplished by rinsing in hot deionized water for a few minutes [122]. However, their complete elimination generally requires chemical reaction with the semiconductor surface, accompanied by flushing to remove these soluble reaction products. HCl treatments are especially effective for this purpose.

Heavy metal impurities can also be present as contaminants on the semiconductor surface, and usually come about by electrodeposition out of the etchant solutions during device fabrication. Copper, gold, iron, nickel, and silver are common impurities found in most "electronic"-grade chemicals which are used in microcircuit fabrication technology. Although present in a few parts per million, this represents a significant amount of chemical contamination, especially in circuits where control of minority carrier lifetime is important. Etching processes will generally leave only small amounts (less than 10^{13} cm^{-2}) of these impurities on the surface, since they involve removal of semiconductor material, with continual flushing away of the surface reaction products. Moreover, oxidizing agents (such as H_2O_2 and HNO_3) which are a constituent of these etches tend to form a thin protective coating on the semiconductor, so that heavy metal contamination is not severe when these chemicals are used.

Heavy metal contamination primarily comes about during processes which dissolve the semiconductor protective oxides, but leave the surface unattacked. An example of one such process is the opening of a window in an oxide or nitride layer, prior to diffusion. Here, uninhibited electrochemical displacement plating can occur [122], because of the large difference between the oxidation potentials of the semiconductor and the metal ions in solution, resulting in contamination levels of as high as 10^{16} cm^{-2}.

The tolerance to ionic and atomic contamination, and hence the appropriate surface cleaning treatment, is a function of the device operation. Thus cleaning solution involving the use of sodium-based materials (NaOH, NaOCl, etc.) which may be quite acceptable for devices in GaAs cannot be tolerated for MOS-based silicon circuits, unless a second cleaning solution is used to remove the resultant sodium contamination. It follows that surface cleaning is best carried out by the use of chemicals which only contain volatile species* in their composition. Chemicals such as HCl, HF, HNO_3, NH_4OH, H_2O_2, and H_2O are favored for this purpose.

All semiconductor wafers must receive an initial cleaning in hot organic solvents. In addition, further cleaning of wafers must be done after each processing step in the fabrication sequence, and especially before each high-temperature operation. As far as possible, the same cleaning process should be used at each point, and made as routine as possible, in order to avoid operator error.

One highly successful approach [123, 124] to cleaning silicon, known as "RCA clean," is to use two solutions in sequence. The first of these consists of $1:1:5$ to $1:2:7$ volumes of $NH_4OH:H_2O_2:H_2O$. Here, the H_2O_2 functions to oxidize all remaining organic contaminants on the surface, which are present because of incomplete removal of photoresist,** and also because of airborne materials and physical handling. The NH_4OH is effective in removing heavy metals such as cadmium, cobalt, copper, iron, mercury, nickel, and silver by forming amino complexes with them. Next, a solution consisting of $HCl:H_2O_2:H_2O$ in a $1:1:6$ to $1:2:8$ volume ratio is used to remove aluminum, magnesium, and the light alkali ions, and to prevent displacement replating from the solution.

Each of these steps is carried out for $10-20$ min at $75-85°C$, under conditions of rapid agitation. Nitrogen gas bubbling through the etchants is often used for this purpose. Finally, wafers are blown dry and stored in a clean environment until further processing. Results with this cleaning technique make it quite suitable for bipolar as well as MOS microcircuits, so that it is in wide use in industry at the present time.

A number of variations of the RCA clean process have been studied, with a view towards meeting the increasingly stringent requirements of VLSI fabrication. Thus surface roughness of the resulting film has been reduced by the use of a lower concentration of NH_4OH in the first etch step. In some studies, the second etch step has been replaced by a dip in dilute ($\approx 5\%$) HF in order to terminate the silicon with hydrogen [125]. This approach is widely used as a clean-up step prior to growth in MBE. Finally, dilute $HF-H_2O_2$ mixtures have been found to reduce metallic contamination on the silicon even further, and have been proposed as a replacement for the dilute HF process [126, 127].

Special-purpose techniques are also used in individual situations. Many of these are unique to each manufacturer, and are often closely guarded trade secrets. A few of the more well-known cleaning steps will be briefly described.

Heavily doped silicon will occasionally become stained by the formation of a sub-oxide on its surface. A 10- to 15-sec dip in a solution consisting of 2 volumes HF to 1 volume $KMnO_4$ (6% solution) will remove this stain [10]. A small amount of silicon is removed during this process.

---

*This does not preclude the presence of ionic or atomic contaminants in them.
**Neither wet chemistry nor plasma stripping does a complete job in this area. Residues from the plasma process also present a problem here.

An excellent clean-up step for silicon consists of $HNO_3 : HF : HClO_4 : CH_3COOH$ in a $10 : 1.4 : 1 : 1$ volume ratio. This is particularly effective in removing heavy metal impurities, and has been used prior to the oxidation of silicon in high-power semiconductor circuits requiring long lifetime [128].

Oxidized silicon wafers that are stored for any length of time will often become contaminated within the top 50 Å of the oxide layer. This contamination is best eliminated by actual removal of this upper layer of oxide in a short etch consisting of $HF : H_2O$ in a $1 : 50$ volume ratio. The etch rate of this formulation is about 70 Å/min, so that an immersion time of 45–60 s is satisfactory [129]. An alternative approach consists of a 250°C heat treatment for 24–64 hr, just prior to packaging, or to the deposition of the final cover layer of PSG or Si_3N_4. Junction walkout problems, and other instabilities in diode reverse characteristics, can be greatly reduced by this treatment.

In principle, the cleaning of GaAs is very similar to that of silicon, and proceeds along the same lines. The starting point here is the use of strong solvents for removing organic residues which are present on the as-purchased wafer. The successive use of hot trichloroethane, acetone, and methanol, each for 5–10 min, is very effective for this purpose. Wafers cleaned in this manner are blown dry and can be stored under methanol for a short time until ready for further cleaning or processing.

Most of the cleaning formulations for GaAs are of an etching nature; that is, they remove some of the GaAs during the cleaning process. Typically, a quick clean-up etch is used, and involves one of the many formulations described in Section 9.1.3. Again, heavy metal impurities have a tendency to plate out of the reagents during this process. This displacement plating can be minimized if the etch is rapid and is accompanied by quick flushing away of the reaction products. In addition, the etch should involve large amounts of deionized water so as to minimize the concentration of these metals. One formulation, consisting of $HCl : H_2O_2 : H_2O$ in a $1 : 10 : 80$ ratio by volume, used at 70°C, has been found [129] highly effective for this application [130]. Its relatively high etch rate, 4.0 μm/min, requires only a short immersion time for clean-up purposes.

Potassium cyanide has also been used to remove residual surface metals such as copper, as a final step in the cleaning process. Its use has been restricted to the laboratory environment because of its extremely poisonous character.

9.7.1 "Dry" Cleaning Processes

A major disadvantage of all wet cleaning systems is the use of large amounts of hazardous/toxic liquids. Increasingly, interest has focused on the use of "dry" cleaning processes, with their inherent advantage of greatly reduced chemical usage. Vapor-phase cleaning and SiO_2 removal using anhydrous HF, for example, has allowed a reduction in chemical usage by a factor of 200. An additional advantage is that they can be readily integrated into automated manufacturing tools.

An important disadvantage of dry cleaning is its inability to remove particulates which are present on the slice, as well as those produced during the cleaning process. Here, a proposed solution is to use a combination of dry cleaning processes, followed by a short rinse in deionized water.

A number of different chemicals can be used in dry cleaning processes. Anhydrous HF, as well as HF/H_2O vapor, has been used for oxide removal. Anhydrous NH_4OH, NF_3, and HCl gases, followed by a rinse in de-ionized water, have also been used for removing heavy metal and alkali ions. Oxygen, irradiated with ultraviolet light to produce ozone and atomic oxygen, is useful [131] for removing organic photoresist residues, and is often used for cleaning the surface of GaAs. Ozone has also been used in conjunction with HF/H_2O vapor for light etching the surface of silicon slices [132]. Finally, plasma discharges have also been used for cleaning purposes. Here, the potential hazard of causing damage to the gate oxide is a serious one [133].

9.7.2 Schottky Diodes and Gates

Cleaning procedures of the above type are satisfactory prior to epitaxy, diffusion, alloying, and post-implant annealing. In all of these processes, the primary requirement is to avoid surface contamination with heavy metal impurities, which degrades lifetime during the subsequent high-temperature step. The fabrication of Schottky diodes and gates, however, poses an additional problem since these devices are highly sensitive to the nature of the metal–semiconductor interface [134]. Special procedures are required for cleaning the semiconductor surface prior to metal deposition.

Silicon slices are normally covered with a 10- to 30-Å layer of SiO_2. However, when subjected to a dilute ($\approx 5\%$) HF rinse, they become essentially passivated by the formation of Si–H bonds [135]. Exposure to air results in replacement of this hydrogen by oxygen after a relatively long induction period. Oxides of this type can be readily desorbed by short heat treatments, prior to the vacuum evaporation of Schottky metal films. The main requirement, therefore, is to transfer HF-passivated slices into the vacuum evaporator in a reasonably short period of time (≤ 15 min).

Sputter deposition provides a simple approach to Schottky diode fabrication, because of the relative ease of cleaning the semiconductor by back-sputtering prior to deposition of the Schottky metal film. Moreover, a wide variety of metals, some of which were listed in Table 8.3, can be deposited by this means, so that this process is commonly used. A disadvantage of this process is that it results in physical damage to the surface.

Silicides of platinum and palladium are favored for the Schottky barrier, because of their stability under subsequent processing conditions. These are formed by the deposition of the elemental metal on a sputter-cleaned silicon surface, followed by thermal conversion to the silicide. This ensures a clean interface with the silicon, and anneals the damage caused by back-sputtering.

Vacuum-cleaved GaAs will rapidly develop a surface oxide of about 8 Å upon exposure to air. This oxide increases in thickness to about 15–20 Å within an hour, and eventually to about 30 Å over a period of days. Cleaning in organic solvents such as trichloroethane and acetone, followed by a rinse in water or methanol, leaves this oxide layer thickness essentially unchanged [136]. A surface oxide of this type will generally consist of a nonstoichiometric mixture of As_2O_3 and Ga_2O_3, with an excess of As_2O_3. This is believed to produce traps in the energy gap of the Ga_2O_3; consequently, Schottky diodes made by the evaporation of a metal on surfaces of this type will generally have poor ideality factors ($n > 1$) and soft reverse V–I characteristics.

The problems with cleaning GaAs surfaces prior to Schottky gate deposition are somewhat more severe, since sputter etching causes loss of stoichiometry due to unequal displacement of gallium and arsenic atoms. Congruent sublimation of the surface layer, at temperatures around 650°C, is commonly used for cleaning GaAs prior to growth by MBE. However, wet chemical techniques are more commonly used to remove the residual oxides prior to deposition of the Schottky barrier film.

A variety of surface treatments can be used for this purpose. The effect of some of these will now be considered. GaAs etchants usually result in the formation of a thick surface oxide ($\simeq 40$–50 Å) which is As_2O_3-rich and which is unsuitable for Schottky diode formation. Consequently, the etch step must be followed by an additional process to reduce its thickness. Either HCl or NH_4OH, in a 50% concentration by volume, can be used for this purpose. A 30-s immersion in either of these, followed by a methanol rinse, results in an 8- to 12-Å residual oxide which is slightly Ga-deficient. Ga/As ratios of 0.78–0.82 have been observed with the HCl treatment [137]; the oxide with the NH_4OH treatment is somewhat more stoichiometric, with Ga/As ratios of about 0.84–0.94. Additionally, the HCl-treated surface is covered with a significant amount of excess amorphous arsenic. On the other hand, the arsenic concentration is much lower with NH_4OH-treated surfaces. Presumably, this is due to the fact that this treatment oxidizes the surface arsenic, which is then rapidly dissolved. An alternative theory for this improved surface character is based on the fact that the solubilities of Ga, As, and their oxides is a function of the pH of the etchant solution [138].

Diodes made on these surfaces have ideality factors that are close to unity [139]. Etches based on the use of H_2SO_4 result in highly As-rich surfaces, and are unsuitable for Schottky diode applications [140].

The Br–methanol etch is somewhat unusual, in that it results in a surface oxide that is deficient in arsenic. This is because the Br preferentially brominates the Ga, inducing its migration to the surface where the $GaBr_3$ converts to the oxide. This process leaves behind a conducting film of elemental arsenic at the oxide–GaAs interface, so that further treatment is necessary to remove this layer. Both the HCl and NH_4OH treatments can be used for this purpose. These act by dissolving the surface oxide, oxidizing the underlying arsenic, and then removing it. In both cases, the residual oxide is As_2O_3-rich.

In summary, therefore, a successful etch procedure preceding Schottky gate formation is to use a rapid etch (such as $H_2SO_4 : H_2O_2 : H_2O$ in a $5 : 1 : 1$ ratio by volume), followed by a 1 min treatment in HCl or NH_4OH in a 10–20% concentration by volume in order to thin the surface oxide to about 10 Å. Often, the wafer is immersed under the solution at this point, and subsequent gate formation carried out by rapid transfer to the deposition apparatus, after it has been blown dry. A short heat treatment, under vacuum conditions, serves to desorb these residual oxides prior to metal deposition.

The final oxide thinning step results in an almost bare GaAs surface, so that heavy metal plating is a strong possibility during this process. Often, however, this does not present a problem since the Schottky metal is formed by a low-temperature process, and the device is not subjected to further heat treatments. In some situations, however, it is important to minimize this type of contamination. This can be done by using a final rinse in KCN. Considerable care should be used, however, because of the highly toxic nature of this chemical. An alternative approach is to pass the final cleaning reagents through a column containing crushed, "sacrificial" GaAs where these impurities can plate out, prior to using it for the GaAs wafer [136]. A two- to three-decade reduction in metal contamination can be achieved by either of these techniques.

TABLE 9.1 Compositions of Commonly Used Concentrated Aqueous Reagents

Reagent	Weight %
HCl	37
HF	49
H_2SO_4	98
H_3PO_4	85
HNO_3	70
$HClO_4$	70
CH_3COOH	99
H_2O_2	30
NH_4OH	29
	(as NH_3)

TABLE 9.2 Some Polishing Etches for Silicon

Formulation	Remarks
3 ml HF 5 ml HNO_3 3 ml CH_3COOH	CP-4A, 80 μm/min
2 ml HF 15 ml HF 5 ml CH_3COOH	Planar etch, 5 μm/min
10 ml H_2O_2 3.7 g NH_4F	0.7 μm/min, almost neutral etchant
2 ml HF 1 ml $KMnO_4$ (6%)	0.3–0.4 μm/min
50 ml HF 100 ml HNO_3 110 ml CH_3COOH 3 g I_2	

TABLE 9.3 Some Crystallographic Etches for Silicon

Formulation	Remarks
1 ml HF 3 ml HNO$_3$ 10 ml CH$_3$COOH	Dash etch, 8 hr
1 ml HF 1 ml CrO$_3$ (5 M in H$_2$O)	Sirtl etch, for (111) silicon, 5 min
2 ml HF 1 ml K$_2$Cr$_2$O$_7$ (0.15 M in H$_2$O)	Secco etch, for (100) and (111) silicon, 5 min
60 ml HF 30 ml HNO$_3$ 60 ml CH$_3$COOH (glacial) 60 ml H$_2$O 30 ml solution of 1 g CrO$_3$ in 2 ml H$_2$O 2 g (CuNO$_3$)$_2 \cdot$3H$_2$O	Wright etch, for (100) and (111) silicon, 5 min. long shelf life
2 ml HF 1 ml HNO$_3$ 2 ml AgNO$_3$ (0.65 M in H$_2$O)	Silver etch, for faults in epitaxial layers
200 ml HF 1 ml HNO$_3$	For *p–n* junction delineation, 1 min

TABLE 9.4   Some Crystallographic Etches for Gallium Arsenide

Formulation	Remarks
1 ml Br_2 100 ml CH_3OH	Distinguishes between (111) Ga and (111) As planes
1 ml HF 2 ml H_2O 8 mg $AgNO_3$ 1 g CrO_3	Etch pits on (100) and (110) planes

A: 40 ml HF
   40 ml H_2O
   0.3 g $AgNO_3$
B: 40 ml H_2O
   40 g CrO_3
	A–B dislocation etch, separate parts store indefinitely, used for delineation of epitaxial layers. Mixed in a 1 : 1 ratio before use.

1 g $K_3Fe\,(CN)_6$ in 50 ml H_2O 12 ml NH_4OH in 36 ml H_2O	Used for the delineation of epitaxial layers. Mixed in a 1 : 1 ratio before use.

TABLE 9.5   Etchants for Noncrystalline Films[a]

Material	Etchant	Remark
SiO_2	28 ml HF 170 ml H_2O 113 g NH_4F	BHF, 1000–2500 Å/min at 25°C
	15 ml HF 10 ml HNO_3 300 ml H_2O	P-etch, 128 Å/min at 25°C
	1 ml BHF 7 ml H_2O	800 Å/min
BSG	1 ml HF 100 ml HNO_3 100 ml H_2O	R-etch, 300 Å/min for 9 mole % B_2O_3, 50 Å/min for SiO_2
	4.4 ml HF 100 ml HNO_3 100 ml H_2O	S-etch, 750 Å/min for 9 mole % B_2O_3, 135 Å/min for SiO_2
PSG	28 ml HF 170 ml H_2O 113 g NH_4F	BHF, 5500 Å/min for 8 mole % P_2O_5
	15 ml HF 10 ml HNO_3 300 ml H_2O	P-etch, 34,000 Å/min for 16 mole % P_2O_5, 110 Å/min for SiO_2
	1 ml BHF 7 ml H_2O	800 Å/min
Si_3N_4	HF	140 Å/min, CVD at 1100°C 750 Å/min, CVD at 900°C 1000 Å/min, CVD at 800°C
	28 ml HF 170 ml H_2O 113 g NH_4F	BHF, 5–10 Å/min
	H_3PO_4	100 Å/min at 180°

[a]Listed in the order in which they are described in Section 9.1.7 [51].

TABLE 9.5 *(Continued)*

Material	Etchant	Remark
Polysilicon	6 ml HF 100 mol HNO_3 40 ml H_2O	8000 Å/min, smooth edges
	1 ml HF 26 ml HNO_3 33 ml CH_3COOH	1500 Å/min
SIPOS	1 ml HF 6 ml H_2O 10 ml NH_4F (40%)	2000 Å/min for 20% O_2 film
Al	1 ml HCl 2 ml H_2O	80°C, fine line, can be used with gallium arsenide
	4 ml H_3PO_4 1 ml HNO_3 4 ml CH_3COOH 1 ml H_2O	350 Å/min, fine line, will attack gallium arsenide
	16–19 ml H_3PO_4 1 ml HNO_3 0–4 ml H_2O	1500–2500 Å/min, will attack gallium arsenide
	0.1 M $K_2Br_4O_7$ 0.51 M KOH 0.6 M $K_3Fe(CN)_6$	1 μm/min, pH 13.6, no gas evolved during etching
Au	3 ml HCl 1 ml HNO_3	Aqua regia, 25–50 μm/min
	4 g KI 1 g I_2 40 ml H_2O	0.5–1 μm/min, can be used with resist
Ag	1 ml NH_4OH 1 ml H_2O_2 4 ml CH_3OH	3600 Å/min, can be used with resists, must be rinsed rapidly after etching

TABLE 9.5   (*Continued*)

Material	Etchant	Remark
Cr	1 ml HCl 1 ml glycerine	800 Å/min, needs depassivation
	1 ml HCl 9 ml saturated $CeSO_4$ solution	800 Å/min, needs depassivation
	1 ml, 1 g NaOH in 2 ml H_2O 3 ml, 1 g $K_3Fe(CN)_6$ in 3 ml H_2O	250–1000 Å/min, no depassivation resist mask can be used
Mo	5 ml H_3PO_4 2 ml HNO_3 4 ml CH_3COOH 150 ml H_2O	0.5 μm/min, resist mask can be used
	5 ml H_3PO_4 3 ml HNO_3 2 ml H_2O	Polishing etch
	11 g $K_3Fe(CN)_6$ 10 g KOH 150 ml H_2O	1 μm/min
W	34 g KH_2PO_4 13.4 g KOH 33 g $K_3Fe(CN)_6$ H_2O to make 1 liter	1600 Å/min, high resolution, resist mask can be used
Pt	3 ml HCl 1 ml HNO_3	Aqua regia, 20 μm/min, precede by a 30-s immersion in HF
	7 ml HCl 1 ml HNO_3 8 ml H_2O	400–500 Å/min, 85°
Pd	1 ml HCl 10 ml HNO_3 10 ml CH_3COOH	1000 Å/min
	4 g KI 1 g I_2 40 ml H_2O	1 μm/min, opaque, must be rinsed before visual inspection

TABLE 9.6 Sputtering Yields for Materials Bombarded by Argon at 0.6 keV[a]

Target	Sputtering Yield
Al	1.2
Au	2.8
Mo	0.9
Ni	1.5
Pd	2.4
Pt	1.6
Si	0.5
Ta	0.6
Ti	0.6
W	0.6
GaAs	1.2 molecules/ion
SiO_2	0.1 molecules/ion
Si_3N_4	0.05 molecules/ion

[a]See Ref. 62.

TABLE 9.7   Ion Milling Rates for Argon[a]

Material	Milling Rate (Å/min)	Energy (eV)
Al	300–700	500
	450–750	1000
Au	1050–1500	500
	1600–2150	1000
Mo	230	500
	400	1000
Ta	130–330	500
Ti	200	500
	200	1000
W	180	500
SiO_2	280–420	500
	380–670	1000
Si	215–500	500
	360–750	1000
GaAs	650	500
	2600	1000
AZ 1350 (Shipley photoresist)	200–420	500
	600	1000
KTFR (Kodak photoresist)	390	1000
PMMA (positive electron resist)	840	1000
COP (negative electron resist)	860	5000

[a]See Ref. 62.

TABLE 9.8   Species and Their Emission Wavelengths

Film	Species	Wavelengths (nm)
Resist	CO*	297.7, 483.5, 519.8
	OH*	308.9
	H*	656.3
Polysilicon	F*	704
	SiF*	777
	N*	674
Aluminum	AlCl*	261.4
	Al	396

* = excited species.

REFERENCES

1. H. C. Gatos and M. C. Lavine, Chemical Behavior of Semiconductors: Etching Characteristics, in *Progress in Semiconductors*, A. F. Gibson and R. E. Burgess, Eds., Vol. 9, Temple Press, London, 1965.

2. H. Gerischer and W. Mindt, The Mechanisms of the Decomposition of Semiconductors by Electrochemical Oxidation and Reduction, *Electrochem. Acta* **13**, 1329 (1968).

3. D. R. Turner, On the Mechanism of Chemically Etching Ge and Si, *J. Electrochem. Soc.* **107**, 810 (1960).

4. H. Robbins and B. Schwartz, Chemical Etching of Silicon, I. The System HF, HNO_3, and H_2O, *J. Electrochem. Soc.* **106**, 505 (1959).

5. H. Robbins and B. Schwartz, Chemical Etching of Silicon, II. The System HF, HNO_3, and H_2O, *J. Electrochem. Soc.* **107**, 108 (1960).

6. B. Schwartz and H. Robbins, Chemical Etching of Silicon, III. A Temperature Study in the Acid System, *J. Electrochem. Soc.* **108**, 365 (1961).

7. B. Schwartz and H. Robbins, Chemical Etching of Silicon, IV. Etching Technology, *J. Electrochem. Soc.* **123**, 1903 (1976).

8. P. J. Holmes, Ed. , *The Electrochemistry of Semiconductors*, Academic Press, London, 1962.

9. W. R. Runyan, *Semiconductor Measurements and Instrumentation*, McGraw-Hill, New York, 1975.

10. D. G. Schimmel and N. J. Elkind, An Examination of the Chemical Staining of Silicon, *J. Electrochem. Soc.* **125**, 152 (1978).

11. C. S. Fuller and H. W. Allison, A Polishing Etchant for III–V Semiconductors, *J. Electrochem. Soc.* **109**, 880 (1962).

12. M. W. Sullivan and G. A. Kolb, The Chemical Polishing of Gallium Arsenide in Bromine-Methanol, *J. Electrochem. Soc.* **110**, 585 (1963).

13. Y. Tarui, Y. Komiya, and Y. Harada, Preferential Etching and Etched Profile of GaAs, *J. Electrochem. Soc.* **118**, 118 (1971).

14. P. D. Green, Selective Etching of Semi-insulating Gallium Arsenide, *Solid State Electron.* **19**, 815 (1976).

15. J. C. Dyment and G. A. Rozgonyi, Evaluation of a New Polish for Gallium Arsenide Using a Peroxide–Alkaline Solution, *J. Electrochem. Soc.* **118**, 1346 (1971).

16. S. Iida and K. Ito, Selective Etching of Gallium Arsenide Crystals in H_2SO_4–H_2O_2–H_2O System, *J. Electrochem. Soc.* **118**, 768 (1971).

17. Y. Mori and N. Watanabe, A New Etching System, H_3PO_4–H_2O_2–H_2O for GaAs and Its Kinetics, *J. Electrochem. Soc.* **125**, 1510 (1978).

18. M. Otsubo, T. Oda, H. Kumabe, and H. Miki, Preferential Etching of GaAs Through Photoresist Masks, *J. Electrochem. Soc.* **123**, 676 (1976).

19. D. W. Shaw, Enhanced GaAs Etch Rates Near the Edges of a Protective Mask, *J. Electrochem. Soc.* **113**, 958 (1966).

20. S. Adache and K. Oe, Chemical Etching Characteristics of (001) GaAs, *J. Electrochem. Soc.* **130**, 2427 (1983).

21. D. J. Stirland and B. W. Straughan, A Review of Etching and Defect Characterization of Gallium Arsenide Substrate Material, *Thin Solid Films* **31**, 139 (1976).

22. W. Kern, Chemical Etching of Silicon, Germanium, and Gallium Arsenide, *RCA Rev.* **39**, 278 (1978).

23. A. Khoukh, S. K. Krawczyk, R. Olier, A. Chabli, and E. Molva, Chemomechanical Polishing and Etching of GaAs:In and GaAs in Aqueous Solutions of NaOCl, *J. Electrochem. Soc.* **134**, 1859 (1987)

24. D. L. Kendall, Vertical Etching of Silicon at Very High Aspect Ratios, *Am. Rev. Mater. Sci.* **9**, 373 (1979)

25. M. P. Lepselter, Beam Lead Technology, *Bell Syst. Tech. J.* **45**, 233 (1966).

26. R. J. Roedel and P. M. Holm, The Design of Anisotropically Etched III–V Solar Cells, *Solar Cells* **11**, 221 (1984).

27. T. J. Rodgers, W. R. Hiltpold, B. Frederick, J. J. Barnes, F. B. Jenne, and J. D. Trotter, VMOS Memory Technology, *IEEE J. Solid State Circuits* **SC-12**, 515 (1977).

28. B. W. Wessels and B. J. Baliga, Vertical Channel Field Controlled Thyristors with High Gain and Fast Switching Speeds, *IEEE Trans. Electron Dev.* **ED-25**, 1261 (1978).

29. E. Bassous, Fabrication of Novel Three-Dimensional Microstructures by the Anisotropic Etching of (100) and (110) Silicon, *IEEE Trans. Electron Dev.* **ED-25**, 1178 (1978).

30. E. Bassous and E. F. Baran, The Fabrication of High Precision Nozzles by the Anisotropic Etching of (100) Silicon, *J. Electrochem. Soc.* **125**, 1321 (1978).

31. D. W. Shaw, Localized GaAs Etching with Acidic Hydrogen Peroxide Solutions, *J. Electrochem. Soc.* **128**, 874 (1981).

32. D. W. Shaw, Morphology Analysis in Localized Crystal Growth and Dissolution, *J. Cryst. Growth* **47**, 509 (1979).

33. M. J. Declercq, L. Gerzberg, and J. D. Meindl, Optimization of the Hydrazine–Water Solution for Anisotropic Etching of Silicon in Integrated Circuit Technology, *J. Electrochem. Soc.* **122**, 545 (1975).

34. R. M. Finne and D. L. Klein, A Water-Amine-Complexing Agent System for Etching Silicon, *J. Electrochem. Soc.* **114**, 965 (1967).

35. J. J. Gannon and C. J. Nuese, A Chemical Etchant for the Selective Removal of GaAs through SiO_2 Masks, *J. Electrochem. Soc.* **121**, 1215 (1974).

36. S. H. Jones and D. K. Walker, Highly Anisotropic Wet Chemical Etching of GaAs Using $NH_4OH:H_2O_2$:H_2O, *J. Electrochem. Soc.* **137**, 1653 (1990).

38. J. B. Price, Anisotropic Etching of Silicon with $KOH–H_2O–$ Isopropyl Alcohol, in *Semiconductor Silicon*, H. R. Huff and R. R. Burgess, Eds. , The Electrochemical Society, Princeton, NJ, 1973, p. 339.

39. H. Muraoka, T. Ohkashi, Y. Sumitomo, Controlled Preferential Etching Technology, in *Semiconductor Silicon, 1973,* H. R. Huff and R. R. Burgess, Eds. , The Electrochemical Society, Princeton, NJ, 1973, p. 327.

40. A. Bogh, Ethylene Diamine–Pyrocatechol–Water Mixtures Shows Etching Anomaly in Boron-Doped Silicon, *J. Electrochem. Soc.* **118**, 401 (1971).

41. R. P. Tijburg and T. van Dongen, Selective Etching of III–V Compounds with Redox Systems, *J. Electrochem. Soc.* **123**, 687 (1976).

42. K. Kenefick, Selective Etching Characteristics of Peroxide/Ammonium Hydroxide Solutions for GaAs/$Al_{0.16}Ga_{0.84}As$, *J. Electrochem. Soc.* **129**, 2380 (1982).

43. C. Juang, K. J. Kuhn, and R. B. Darling, Selective Etching of GaAs and $Al_{0.3}Ga_{0.7}As$ with Citric Acid/Hydrogen Peroxide Solutions, *J. Vac. Sci. Technol.* **B8**, 1122 (1990).

44. W. C. Dash, Copper Precipitation on Dislocations in Silicon, *J. Appl. Phys.* **27**, 1193 (1956).

45. E. Sirtl and A. Adler, Chromsäure-Flussäure als spezifisches System zur Ätzgruben entwicklung auf Silizium, *Z. Metallkunde* **52**, 529 (1961).

46. F. Secco d'Aragona, Dislocation Etch for (100) Planes in Silicon, *J. Electrochem. Soc.* **119**, 948 (1972).

47. M. W. Jenkins, A New Preferential Etch for Defects in Silicon Crystals, *J. Electrochem. Soc.* **124**, 757 (1977).

48. W. R. Wagner, L. I. Greene, and L. I. Koszi, Defect-Revealing Etches on GaAs: A Comparison of the AHA with the A/B and KOH Etches, *J. Electrochem. Soc.* **128**, 1091 (1981).

49. M. S. Abrahams and C. J. Buicchi, Etching of Dislocations on the Low Index Planes of GaAs, *J. Appl. Phys.* **36**, 2855 (1965).

50. G. H. Olsen and M. Ettenberg, Universal Stain/Etchant for Interfaces in III–V Compounds, *J. Appl. Phys.* **45**, 5112 (1974).

51. W. Kern and J. L. Vossen, Eds. , *Thin Film Processes*, Academic Press, New York, 1978.

52. A. S. Tenney and M. Ghezzo, Etch Rates of Doped Oxides in Solutions of Buffered HF, *J. Electrochem. Soc.* **120**, 1091 (1973).

53. H. Kikuyama, N. Miki, K. Saka, J. Takano, I. Kawanabe, M. Miyashita, and T. Ohmi, Surface Active Buffered Hydrogen Fluoride Having Excellent Wettability for VLSI Processing, *IEEE Trans. Semi. Manufacturing*, **3**, 99 (1990).

54. W. A. Pliskin and R. P. Gnall, Evidence for Oxidation Growth at the Oxide-Silicon Interface from Controlled Etch Studies, *J. Electrochem. Soc.* **113**, 263 (1966).

55. L. P. Plauger, Etching Studies of Diffusion Source Boron Glass, *J. Electrochem. Soc.* **120**, 1428 (1973).

56. C. A. Deckert, Etching of CVD Si_3N_4 in Acidic Fluoride Media, *J. Electrochem. Soc.* **125**, 320 (1978).

57. D. M. Brown, P. V. Gray, F. K. Heumann, H. R. Philipp, and E. A. Taft, Properties of $Si_xO_yN_z$ Films on Si, *J. Electrochem. Soc.* **115**, 311 (1968).

58. W. van Gelder and V. E. Hauser, The Etching of Silicon Nitride in Phosphoric Acid with Silicon Dioxide as a Mask, *J. Electrochem. Soc.* **114**, 869 (1967).

59. C. A. Deckert, Pattern Etching of CVD Si_3N_4/SiO_2 Composites in HF/Glycerol Mixtures, *J. Electrochem. Soc.* **127**, 2433 (1980).

60. M. Wittmer and M. Melchior, Applications of TiN Thin Films in Silicon Devices, *Thin Solid Films* **93**, 397 (1982).

61. Quick Reference Manual for Semiconductor Engineers, Vol. 1, Bell Laboratories, Reading, PA, 1981.

62. E. G. Spencer and P. H. Schmidt, Ion-Beam Techniques for Device Fabrication, *J. Vac. Sci. Technol.* **8**, S52 (1971).

63. P. D. Reader and H. R. Kaufman, Optimization of an Electron-Bombardment Ion Source for Ion Machining Applications, *J. Vac. Sci. Technol.* **12**, 1344 (1975).

64. G. R. Thompson, Ion Beam Coating—A New Deposition Method, *Solid State Technol.*, (Dec. 1978) p. 73.

65. L. Maeder and J. Hoepfner, Ion Beam Etching of Silicon Dioxide on Silicon, *J. Electrochem. Soc.* **123**, 1893 (1976).

66. B. Chapman, *Glow Discharge Processes*, John Wiley and Sons, New York, 1980.

67. P. Singer, Trends in Plasma Sources: The Search Continues, *Semicond. Int.*, (July 1992) p. 52.

68. D. E. Ibbotson, J. A. Mucha, D. L. Flamm, and J. M. Cook, Plasmaless Dry Etching of Silicon with Fluoirine Containing Compounds, *J. Appl. Phys.* **56**, 2939 (1984).

69. T. Aoyama, T. Yamazaki, and T. Ito, Removing Native Oxide from Si(001) Surfaces Using Photoexcited Fluorine Gas, *Appl. Phys. Lett.* **59**, 2576 (1991).

70. H. W. Sawin, A Review of Plasma Processing Fundamentals, *Solid State Technol.* (April 1985) p. 212.

71. J. L. Vossen, Glow Discharge Phenomena in Plasma Etching and Plasma Deposition, *J. Electrochem. Soc.* **126**, 319 (1979).

72. R. J. Poulsen, Plasma Etching in Integrated Circuit Manufacture—A Review, *J. Vac. Sci. Technol.* **14**, 266 (1977).

73. J. R. Hollahan and A. T. Bell, Eds. , *Techniques and Applications of Plasma Chemistry*, John Wiley and Sons, New York, 1974.

74. J. F. Battey, The Effects of Geometry on Diffusion-Controlled Chemical Reaction Rates in a Plasma, *J. Electrochem. Soc.* **124**, 437 (1977).

75. C. J. Mogab, The Loading Effect in Plasma Etching, *J. Electrochem. Soc.* **124**, 1263 (1977).

76. T. Enomoto, Loading Effect and Temperature Dependence of Etch Rate of Silicon Materials in a CF_4 Plasma, *Solid State Technol.* (April 1980) p. 117.

77. L. I. Maissel and R. Glang, Eds., *Handbook of Thin Film Technology*, McGraw-Hill, New York, 1970.

78. D. M. Manos and D. L. Flamm, Eds., *Plasma Etching: An Introduction*, Academic Press, 1989.

79. W. R. Harshbarger, R. A. Porter, and P. Norton, Optical Detector to Monitor Plasma Etching, *J. Electron. Mater.* **7**, 429 (1978).

80. K. Hirobe and T. Tsuchimoto, End Point Detection in Plasma Etching by Optical Emission Spectroscopy, *J. Electrochem. Soc.* **127**, 234 (1980).

81. P. J. Marcoux and P. D. Foo, Methods of End Point Detection for Plasma Etching, *Solid State Technol.* (April 1981) p. 115.

82. J. D. Swift and M. J. R. Schwar, *Electrical Probes for Plasma Diagnostics*, Illiffe, London, 1970.

83. R. A. H. Heinecke, Control of Relative Etch Rates of SiO_2 and Si in Plasma Etching, *Solid State Electron.* **18**, 1146 (1975).

84. C. J. Mogab, A. C. Adams, and D. L. Flamm, Plasma Etching of Si and SiO_2—The Effect of Oxygen Conditions to CF_4 Plasmas, *J. Appl. Phys.* **49**, 3796 (1978).

85. W. R. Harshbarger and R. A. Porter, Spectroscopic Analysis of RF Plasmas, *Solid State Technol.* (April 1978) p. 99.

86. K. Jinno, H. Knoshita, and Y. Matsumoto, Etching Characteristics of Silicate Glass Films in CF_4 Plasma, *J. Electrochem. Soc.* **124**, 1258 (1977).

87. H. Abe, Y. Sonobe, and T. Enomoto, Etching Characteristics of Silicon and Its Compounds in a Gas Plasma, *Jpn. J. Appl. Phys.* **12**, 154 (1973).

88. L. M. Ephrath, Selective Etching of Silicon Dioxide Using Reactive Ion Etching with CF_4-H_2, *J. Electrochem. Soc.* **126**, 1419 (1979).

89. D. L. Flamm, D. E. Ibbotson, J. A. Mucha, and V. M. Donnelly, XeF_2 and F-Atom Reactions with Si: Their Significance for Plasma Etching, *Solid State Technol.* (April 1983) p. 117.

90. H. F. Winters and J. W. Coburn, The Etching of Silicon with XeF_3 Vapor, *Appl. Phys. Lett.* **34**, 70 (1979).

91. J. A. Barkanic, D. M. Reynolds, R. J. Jaccodine, H. G. Stenger, J. Parks, and H. Vedage, Plasma Etching Using NF_3: A Review, *Solid State Technol.* (April 1989) p. 109.

92. N. J. Ianno, K. E. Greenberg, and J. T. Verdeyen, Comparison of the Etching and Plasma Characteristics of Discharges in CF_4 and NF_3, *J. Electrochem. Soc.* **128**, 2174 (1981).

93. K. E. Greenberg and J. E. Verdeyen, Kinetic Processes of NF_3 Etchant Gas Discharges, *J. Appl. Phys.* **57**(5), 1596 (1985).

94. A. C. Adams and C. D. Capio, Edge Profiles in Plasma Etching of Polycrystalline Silicon, *J. Electrochem. Soc.* **128**, 366 (1981).

95. J. A. Mucha, The Gases of Plasma Etching: Silicon-Based Technology, *Solid State Technol.*, (March 1985) p. 123.

96. V. M. Donnelly and D. Flamm, Anisotropic Etching of SiO_2 in Low Frequency CF_4-O_2 and NF_3-Ar Plasmas, *J. Appl. Phys.* **55**, 242 (1984).

97. L. Ta, An Evaluation of Etching in NF_3 and NF_3/Argon Plasma, *Semicond. Int.* **6**, 25 (1985).

98. R. Sellamuthu, J. A. Barkanic, and R. Jaccodine, A Study of Anisotropic Trench Etching of Si with NF_3-Halocarbon, *J. Vac. Sci. Technol.* **B5**(1), 342 (1987).

99. D. H. Bower, Planar Plasma Etching of Polysilicon Using CCl_4 and NF_3, *J. Electrochem. Soc.* **129**, 795 (1982).

100. K. Maeda and K. Fujino, The Patterning of Metal Films by Gas Plasma Technique, *Denki Kagaku* **43**, 22 (1975).

101. T. P. Chow and A. J. Steckl, Plasma Etching of Sputtered Mo and MoSi$_2$ Thin Films in NF$_3$ Gas Mixtures, *J. Appl. Phys.* **53**, 5531 (1982).

102. C. S. Korman, T. P. Chow, and D. H. Bower, Etching Characteristics of Polysilicon, SiO$_2$, and MoSi$_2$ in NF$_3$ and SF$_6$ Plasmas, *Solid State Technol.* (Jan. 1983) p. 115.

103. M. E. Coe and S. H. Rogers, Low Frequency Planar Plasma Etching of Polycide Structures in an SF$_6$ Glow Discharge, *Solid State Technol.* (Aug. 1982) p. 79.

104. K. L. Wang, T. C. Holloway, R. F. Pinizotto, Z. P. Sobczak, W. R. Hunter, and A. F. Tasch, Jr. , Composite TiSi$_2$/n^+ Poly-Si Low-Resistivity Gate Electrode and Interconnect for VLSI and Device Technology, *IEEE Trans. Electron Dev.* **ED-29**, 547 (1982).

105. H. J. Mattausch, B. Hasler, and W. Beinvogl, Reactive Ion Etching of Ta-Silicide/Polysilicon Double Layers for the Fabrication of Integrated Circuits, *J. Vac. Sci. Technol.* **B1**, 15 (1983).

106. K. Tokunaga and D. W. Hess, Aluminum Etching in Carbon Tetrachloride Plasmas, *J. Electrochem. Soc.* **127**, 928 (1980).

107. K. Tonuga, F. C. Redeker, D. A. Danner, and D. W. Hess, Comparison of Aluminum Etch Rates in Carbon Tetrachloride and Boron Trichloride Plasmas, *J. Electrochem. Soc.* **128**, 851 (1981).

108. L. Columba and F. Illuzzi, Plasma Etching of Aluminum Alloys for Submicron Technologies, *Solid State Technol.*, (Feb. 1990) p. 95.

109. A. A. Chambers, The Application of Reactive Ion Etching to the Definition of Patterns in the Al–Si–Cu Alloy Conductor Layers and Thick Silicon Oxide Films, *Solid State Technol.* (Aug. 1982) p. 92.

110. J. W. Coburn and H. F. Winters, Plasma Etching—A Discussion of Mechanisms, *J. Vac. Sci. Technol.* **16**, 391 (1979).

111. J. C. Bailor, H. J. Emelus, R. Nyholm, and A. P. Trotman-Dickinson, *Comprehensive Inorganic Chemistry*, Vol. 1, Pergamon Press, New York, 1973.

112. E. L. Hu and R. E. Howard, Reactive-Ion Etching of GaAs and InP Using CCl$_2$F$_2$/Ar/O$_2$, *Appl. Phys. Lett.* **37**, 1022 (1980).

113. K. Hikosaka, T. Nimura, and K. Joshin, Selective Dry Etching of AlGaAs–GaAs Heterojunctions, *Jpn. J. Appl. Phys.* **20**, L847 (1981).

114. S. Salimian, C. B. Cooper III, R. Norton, and J. Bacon, Reactive Ion Etch Process with Highly Controllable GaAs to AlGaAs Selectivity Using SF$_6$ and SiCl$_4$, *Appl. Phys. Lett.* **51**, 1083 (1987).

115. M. Zhang, J. Z. Li, I. Adesida, and E. D. Wolf, Reactive Ion Etching for Submicron Structures of Refractory Metal Silicides and Polycides, *J. Vac. Sci. Technol.* **B1** 1037 (1983).

116. P. Tikanyi, D. K. Wagner, A. J. Roza, J. J. Vollmer, C. M. Harding, R. J. Davis, and E. D. Wolf, High Power AlGaAs/GaAs Single Quantum Well Lasers with Chemically Assisted Ion Beam Etched Mirrors, *Appl. Phys. Lett.* **50**, 1640 (1987).

117 S. W. Pang, Drying Etching Induced Damage in Si and GaAs, *Solid State Technol.* (April 1984) p. 249.

118. R. Singh, S. J. Fonash, S. Ashok, P. J. Caplan, J. Shappiro, M. Hage-Ali, and J.

Ponpon, Electrical Structure and Bonding Changes Induced in Silicon by H, Ar and Kr Ion Beam Etching, *J. Vac. Sci. Technol.* **A1**, 334 (1983).

119. S. W. Pang, G. A. Lincoln, R. W. McClelland, P. D. DeGraff, and M. W. Geiss, Effects of Dry Etching of GaAs, *J. Vac. Sci. Technol.* **B1**, 1334 (1983).

120 S. K. Ghandhi, P. Kwan, K. N. Bhat, and J. M. Borrego, Ion Beam Damage Effects During the Low Energy Cleaning of GaAs, *IEEE Electron Dev. Lett.* **EDL-3**, 48 (1982).

121. G. A. Lincoln, M. W. Geiss, S. Pang, and N. N. Efremow, Large Area Ion Beam Assisted Etching of GaAs with High Etch Rates and Controlled Anisotropy, *J. Vac. Sci. Technol.* **B1**, 1043 (1983).

122. W. Kern, Radiochemical Study of Semiconductor Surface Contamination, II. Deposition of Trace Impurities on Silicon and Silica, *RCA Rev.* **31**, 234 (1970).

123. W. Kern and D. A. Poutinen, Cleaning Solutions Based on Hydrogen Peroxide for Use in Silicon Semiconductor Technology, *RCA Rev.* **31**, 187 (1970).

124. W. Kern, The Evolution of Silicon Wafer Cleaning Technology, *J. Electrochem. Soc.* **137**, 1887 (1990).

125. L. J. Huang and W. M. Lau, Effects of HF Cleaning and Subsequent Heating on the Electrical Properties of Silicon (100) Surfaces, *Appl. Phys. Lett.* **60**, 1108 (1992).

126. T. Ohmi, T. Imaoka, I. Sugiyama, and T. Kezuka, Metallic Impurities Segregation at the Interface Between Si and Liquid During Wet Cleaning, *J. Electrochem. Soc.* **139**, 3317 (1992).

127. T. Ohmi and T. Shibata, Scientific ULSI Manufacturing in the 21st Century, *Interface* (Winter 1992) p. 32.

128. R. E. Blaha and W. R. Fahrner, Passivation of High Breakdown Voltage $p-n-p$ Structures by Thermal Oxidation, *J. Electrochem. Soc.* **123**, 515 (1976).

129. M. Polinsky and S. Graf, MOS-Bipolar Monolithic Integrated Circuit Technology, *IEEE Trans. Electron. Dev.* **ED-20**, 239 (1973).

130. D. L. Partin, A. G. Milnes, and L. F. Vassamillet, Effect of Surface Preparation and Heat Treatment on Hole Diffusion Lengths in VPE GaAs and GaAs$_{0.6}$P$_{0.4}$, *J. Electrochem. Soc.* **126**, 1581 (1979).

131. S. J. Pearton, F. Ren, C. R. Abernathy, W. S. Hobson, and H. S. Luftman, Use of Ultraviolet/Ozone Cleaning to Remove C and O from GaAs Prior to Metalorganic Molecular Beam Epitaxy and Metalorganic Chemical Vapor Deposition, *Appl. Phys. Lett.* **58**, 1416 (1991).

132. M. Wong and R. A. Bowling, Silicon Etch Using Vapor Phase HF/H$_2$O and O$_3$, *J. Electrochem. Soc* **140**, 567 (1993).

133. J. Ruzyllo, A. M. Hoff, D. C. Frystak, and S. D. Hossain, Electrical Evaluation of Wet and Dry Cleaning Procedures for Silicon Device Fabrication, *J. Electrochem. Soc.* **136**, 1474 (1989).

134. G. Goldfinger, Ed, *Clean Surfaces: Their Preparation and Characterization for Interfacial Studies*, Dekker, New York, 1970.

135. M. Grundner, D. Graf, P. O. Hahn, and A. Schnegg, Wet Chemical Treatments of Si Surfaces: Chemical Composition and Morphology, *Solid State Technol.*, (Feb. 1991) p. 69.

136. W. Kern, Radiochemical Study of Semiconductor Surface Contamination, III. Deposition of Trace Impurities on Germanium and Gallium Arsenide, *RCA Rev.* **32**, 64 (1971).

137. C. C. Chang, P. H. Citrin, and B. Schwartz, Chemical Preparation of GaAs Surfaces and Their Characterization by Auger Electron and X-ray Photoemission Spectroscopies, *J. Vac. Sci. Technol.* **14**, 943 (1977).

138. H. J. Yoon, M. H. Choi, and I. S. Park, The Study of Native Oxide on Chemically Etched GaAs (100) Surfaces, *J. Electrochem. Soc.* **139**, 3229 (1992).

139. C. M. Garner, C. Y. Su, A. Saperstein, K. G. Jew, C. S. Lee, G. L. Pearson, and W. E. Spicer, Effect of GaAs or $Ga_xAl_{1-x}As$ Oxide Composition on Schottky Barrier Behavior, *J. Appl. Phys.* **50**, 3376 (1979).

140. Y. Aydinli and R. J. Mattauch, The Effects of Surface Treatments on the Pt/n-GaAs Schottky Interface, *Solid State Electron.* **25**, 551 (1982).

PROBLEMS

1. Silicon is to be etched in $HF : PHNO_3 : CH_3COOH = 1 : x : 1$. Describe the characteristic of etches with $x = 0.1, 1.0$, and 10. Assume that all compositions are in wt.% of the concentrated chemicals.

2. A 10-cm-diameter silicon wafer has microcircuits on it of size 4 mm × 4 mm. It is required to chemically cut this wafer into chips, by anisotropic etching. Outline a scheme for doing this, and calculate the fraction of the surface area that is lost in the process. Assume 300-μm-thick wafers, of $\langle 100 \rangle$ orientation.

   Suggest a scheme with less waste, assuming the same wafer thickness.

3. A slice of (111) silicon is anisotropically etched through a triangular window, whose sides are in the $\langle 110 \rangle$ directions. Sketch this window, the shape of the hole, and the planes delineated by the etch.

4. Repeat Problem 3 for (111)As GaAs, and identify the surface atoms on the etch planes.

5. What are the directions of the window edges in Fig. 9.6b. Show that the planes delineated by this etch are the {100}.

6. Sketch the (233) plane of GaAs, and identify the atoms in it.

7. Calculate the plasma potential for a system where the electron temperature is 2 eV, the electron velocity is 9.5×10^7 cm/s, and the ion velocity is 5×10^4 cm/s.

8. Verify Eq. (9.13). Note that $j \propto V^{3/2}/D^2$ under conditions of space-charge-limited current flow, where V is the voltage and D is the width of the sheath.

CHAPTER 10

LITHOGRAPHIC PROCESSES

Microcircuit fabrication requires the precise positioning of a number of appropriately doped regions in a slice of semiconductor, together with a series of interconnection patterns. These regions includes a variety of implants and diffusions, cuts for gates and metallizations, and windows in protective cover layers through which connections can be made to the bonding pads. A sequence of steps is required, together with a specific layout pattern, for each of these regions. This is followed by one or more levels of interconnection patterns between them.

Lithographic processes are used to perform these operations, and are carried out at various points in the circuit fabrication process. Typically, five to twenty complete lithographic operations are required on each wafer. By way of example, a conventional silicon-based microcircuit, using bipolar transistors, requires seven separate lithographic processes to define openings for the buried layer, the isolation wall, the base, the emitter/collector, the ohmic contact, the metallization, and the cover layer. As many as twenty complete sequences, with three to four levels of metallization, are often necessary for more complex microcircuit configurations such as those involving bipolar complementary integrated circuit technology (BICMOS). Thus lithographic processes play a central role in microcircuit technology.

There are many different pathways from circuit design to pattern placement on the wafer. Some of these are illustrated in Fig. 10.1 [1]. For example, it is seen that the placement of a pattern on a wafer can be done by direct electron beam (E-B) writing in a single step (path E in this figure), and some research in this area is along these lines. Typically, however, the process is broken into

662

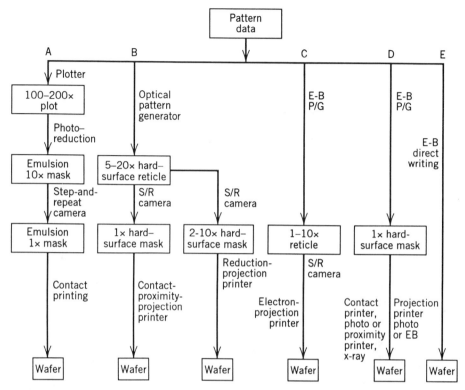

Fig. 10.1 Pathways from circuit design to pattern transfer. From Wolf and Tauber [1]. Reprinted from *Silicon Processing for the VLSI Era*, Vol. 1, Lattice Press, CA.

two separate operations: the generation of a reticle mask, and the transfer of its pattern to a large number of wafers.

The design of the reticle mask usually involves circuit and device designers, who together produce the layout of the circuit and of its individual masks. This layout is usually in the form of a set of drawings for circuits of low complexity, and computer tapes for the more complex VLSI circuits. Next, these are converted into masks by photo-reduction of the large drawings, or by the use of computer-controlled pattern generators using optical or electron beams. The sophistication involved here increases rapidly with circuit complexity. As a result, reticle masks for circuits of low complexity are sometimes made with in-house capability, whereas those for VLSI circuits are usually made by specialty houses. This is warranted on economic grounds, since one master mask set serves for the entire production run of a particular circuit for which it is designed.

The placement of mask patterns on the wafer is an in-house activity, since this must be done on each wafer, and at many stages in the circuit fabrication process. Optical printing methods are commonly used, but it does not appear that these will meet the minimum feature size requirements for microcircuits over the long term. Electron, x-ray, and ion techniques are under active consideration for this reason.

The technology driver for silicon-based microcircuits is the dynamic random access memory (DRAM). Because of the ever-increasing need for memory in modern computers, the thrust of lithography is driven by these microcircuits. Table 10.1, at the end of this chapter, lists parameters for these circuits, and includes those which are available in the present time frame as well as those projected for the future [2]. It is generally accepted that optical pattern transfer will meet the needs of DRAMs to the 64-Mbit size, but alternative technologies will have to be used beyond this point.

Mask making will probably continue to be done by e-beam lithography, since this technique can already achieve feature sizes of 0.15 μm and less. On the other hand, the options for pattern transfer technology are many. These include advanced optical methods, as well as x-ray and ion beam printing. Here, many problems in the areas of sources, masks, and resists will have to be solved before these become practical at the commercial level.

This chapter describes both pattern generation and pattern transfer processes, together with the new directions for these technologies. Central to these technologies is the use of thin films of photoreactive material, usually polymers, whose properties are altered by optical radiation over a specified wavelength range. The properties of these resist materials will also be considered in this chapter.

10.1 PHOTOREACTIVE MATERIALS

The earliest photoreactive materials, or *photoresists*, all had the same basic characteristic; upon exposure to light, they became hardened and could not be dissolved and washed away in solvents. Resists of this type are known as *negative* resists because the image formed in them is the inverse of the object, as seen in Fig. 10.2 *a*.

The *positive* resist is a more recent development. Here, the image is of the same type as the object, as shown in Fig. 10.2 *b*. In use, the material is initially insoluble, but degrades upon exposure to radiation. Both positive and negative resists are used in microcircuit technology today [3].

There are many requirements that a resist must satisfy before it can be useful for a specific technology [4]. Above all, it must be capable of resolving the required minimum feature size. Fortunately, this requirement is met, and exceeded, by all modern resists today. Other factors, over which there are some questions, are as follows:

Sensitivity. Resist materials are sensitive to radiation, which alters their chemical properties sufficiently so that a pattern can be delineated in them. A high sensitivity is desirable since this reduces the time for exposure, and

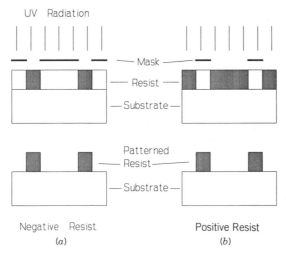

Fig. 10.2 Schematic of pattern transfer: (*a*) Negative resist and (*b*) positive resist.

thus the operating cost. Resists which are used for optical printing respond to radiation of 330–430 nm, and are designed for maximum absorption (and hence sensitivity) over this UV range. Recently, interest has focused on the deep UV region (150–300 nm) because of the higher resolution that can be obtained at these shorter wavelengths. Here, the region of practical interest is ≥ 200-nm, since a vacuum environment is necessary for operation in the 150- to 200-nm range. This range is further restricted by the availability of high-intensity deep UV sources. The most practical source today is an Hg–Xe lamp, which has a peak output at 250–290 nm. Excimer lasers are used as sources for the deep UV range.

The use of e-beam techniques for reticle mask generation has resulted in the need for resists which are sensitive to electron irradiation in the 10- to 30-kV range. Finally, resists are also required for use in printing systems which operate in the soft x-ray region (5–50 Å).

Adhesion. A resist must adhere sufficient well to the underlying film so that it does not lift off during subsequent processing. This is usually not a problem, provided that appropriate procedures are used for surface preparation. This requirement has been somewhat relaxed with VLSI technology because of the increasing use of dry processes such as plasma etching and ion milling.

Etch Resistance. This requirement has become harder to meet since many resist materials degrade in plasma or ion-beam environments. In fact, a tenuous balance between sensitivity to radiation during imaging, and insensitivity to radiation during subsequent processing, must be struck by the resist designer in order to produce a useful product.

Other features of a resist that are highly desirable are stability in storage, uniformity from batch to batch, and the absence of large particulates. It is fair to say that the optimum combination of all of these features is not available in any resist at the present time.

Most early negative resists consisted of polymeric organic materials, such as poly(vinylcinnamate), with long-chain molecules having a molecular weight range of 10^4–10^6. Irradiation of these molecules by an energetic beam results in bonding or cross-linking of adjacent polymer chains, and thus in an increase in the average molecular weight. These materials are sometimes referred to as *cross-linking* resists.

In more recently developed negative photoresists, cross-linking is done by means of photogenerated reactive species. These resists are usually chemically inert synthetic poly-isoprene rubbers, typically cyclized *cis*-poly-isoprene, in combination with a bisazide cross-linking component which is photoactive. They are commonly applied in layers of 0.3- to 1-μm thickness, and are sensitive to UV light in the 330- to 430-nm range. A common rule of thumb here is that the resist thickness should be one-third the minimum feature size. They are readily soluble in aromatic solvents such as benzene, toluene, and xylene, *prior* to cross-linking by exposure to UV light. These chemicals (as well as proprietary formulations) are used as developers in the photoengraving process.

Negative photoresists are well established in conventional microcircuit fabrication. However, their use is characterized by a solvent-induced swelling of unexposed regions that takes place during development. This results in ragged edges or a loss of pattern fidelity that makes them unsuited for resolving elements under 2 μm in size. Moreover, they can only be removed with great difficulty once they have been exposed to UV light. Consequently, optical printing techniques have turned almost exclusively to positive photoresists for VLSI applications.

Positive photoresist materials are designed so that the effect of radiation is to make them increasingly capable of dissolution by chemical solvents. Since the exposed photo-resist is not removed subsequently, it does not suffer from solvent-induced swelling problems, and can readily define elements which are much smaller than the limits set by the lithography system. They can also be used in much thicker coatings than negative photoresists (by a factor of three or more). As a result, they are more resistant to chemical etches, and are more reliable in problem areas such as steps, which lead to resist thinning. They are also relatively free from pinhole formation, since they can be applied in thick layers.

Positive resists can be made up of polymeric materials with weak links, which degrade upon irradiation by the process of *scission*. However, the most commonly used photoresists do not operate on this principle. Rather, they consist of three basic components: a low-molecular-weight, alkali-soluble resin such as phenol formaldehyde novolac, a photoactive dissolution inhibiter which usually consists of orthoquinone diazide, and a solvent such as xylene or cellosolve acetate. They are usually applied in 1- to 3-μm thick layers, and are sensitive to the same UV range as negative resists. Exposure to light degrades the photoactive component by converting it to carboxylic acid, so that the resin becomes readily soluble* in alkalis such as NaOH or KOH, both of which can be used as developers. The recent emphasis on MOS microcircuits has resulted in the use of nonionic developers such as tetramethylammonium hydroxide [$(CH_3)_4NOH$] as well as a number of proprietary formulations [5], which are free from ionic contaminants. The development process is one of the direct dissolution, without any swelling of unexposed material because it is not attacked by the developer. This accounts in large measure for their resolution capability.

Positive resists are more difficult to use than the negative type, and tight control of manufacturing processes must be maintained to obtain consistent adhesion to the mask film. Often the wafer is given a bake in an oxidizing environment such as air, and is spin-coated with a coupling agent to promote this adhesion. In addition, a pre-exposure bake is carried out in order to drive off excess solvents from the resist, and to promote its adhesion to the wafer. This *softbake* is a critical step since the photoactive component is subject to thermal decomposition. Thus excessive baking temperatures result in a loss of sensitivity during subsequent exposure. However, some form of softbake step is necessary in order to densify the resist and make it more resistant to subsequent chemical etching processes.

Unexposed resist can be removed by chemicals such as hot acetone or methylethylketone. Proprietary chemicals are available for removing these photoresists as well. In some situations, an alternative approach consists of flooding the slice with UV irradiation to degrade the remaining resist, prior to its removal.

Deep UV-sensitive photoresists are increasingly used in VLSI applications in order to exploit the development of optical systems in this wavelength range. Most of these are positive resists involving photoactive dissolution inhibitors. Thermally degrading resists, which operate by the process of scission, are increasingly finding use in this application, even though originally developed for use with e-beam systems. However, materials such as polymethylmethacrylate (PMMA) and its derivatives have a maximum absorption at about 215 nm so that they are not well matched to the range of available deep UV light sources. Typically a 1-min exposure is necessary for a 1-μm thick film, using an Hg–Xe arc lamp. Materials such as polymethyl isopropenyl ketone can extend this absorption range out to 290 nm, resulting in improved sensitivity by a factor of 5. This can be further improved by the addition of sensitizing agents, such as *p*-terbutyl benzoic acid.

Very few negative photoresists are available for the deep UV range, although a number are under active development. One proprietary formulation, whose principle ingredient is cyclized rubber, is commercially available for use in this region. This resist is about ten times more sensitive than the best positive resists,

---

*The dissolution rate of exposed positive resists is about three decades faster than that of unexposed materials.

and does not appear to have the severe swelling problems associated with other cross-linking materials [6].

The technology of electron resists has been the subject of intense research in the last 10 years. Here, the goal is to develop materials which can operate with a minimum incident charge density per unit area. A high sensitivity (10^{-6} coulombs cm^{-2} or less) to electrons in the 10- to 30-kV range is required in order to obtain reasonable exposure times. Moreover, since most of these resists will be used in conjunction with plasma etching or ion milling of the underlying films, the ability to withstand these processes is extremely important.

A large number of electron resists have been investigated. However, most of the work has been done with positive resists such as PMMA and its derivatives [7]. Electron irradiation results in chain scission, with a fall in the average molecular weight of the polymer. As a consequence, it can now be developed in those chemicals which are nonsolvent to the PMMA, prior to radiation. PMMA has a low sensitivity (3×10^{-5} C/cm^2 for a 20-keV beam). It also has a poor resistance to ion milling, and a tendency to flow at elevated temperatures. Techniques for improving its sensitivity include the introduction of chemical configurations which tend to weaken the chain stability of the polymer by copolymerization. One such cross-linked PMMA has been reported to have a sensitivity of 0.5×10^{-6} C/cm^2.

Electron resist systems, based on the poly(olefin sulfones) have also been developed. One such system, (PBS) poly(butenesulfone), is commercially available and has a sensitivity of 2×10^{-6} C/cm^2. Its unique characteristic is that it can be developed by gaseous means. This allows an all-dry process, which greatly reduces the adhesion requirements of the resist and the underlying layer.

Some work has been done with negative electron resists in recent years. These are generally prone to solvent-induced swelling effects and cannot resolve submicron features. On the other hand, they tend to be quite sensitive, and research is aimed at minimizing this problem. One such negative electron resist, based on a co-polymer of glycidylmethacrylate and ethylacrylate (COP), has a sensitivity of 6×10^{-7} C/cm^2 at 20 keV, but is limited to resolving a minimum element size of 1.5 μm.

Resists developed for e-beam lithography have all been found to be relatively insensitive in their x-ray absorption capability. Attempts to increase this sensitivity have involved the inclusion of heavy metal ions such as cesium and thallium [8], in order to increase the absorption capability of the resist at the wavelength of interest. Typical resists that have been used to date include both PMMA (a positive resist) and COP (a negative resist). COP is about 360 times more sensitive than PMMA, but has much poorer resolution and chemical resistance properties. Typical exposure times are quite large, as much as one to five minutes for the more sensitive resists. Research in this area is very active, so that considerable improvements in sensitivity should be forthcoming. A highly abbreviated list of e-beam and x-ray resists, which are available [9] at the present time, is given in Table 10.2 at the end of this chapter.

Resist improvements have also focused on the development of ultrathin layers which can be used with e-beam and x-ray lithography, and are capable of development by dry processes [10]. One approach is to silylate the resist after exposure, by the vapor diffusion of a silicon-containing species, to a depth of about 1000 Å. During O_2–plasma etching, the silylated regions are converted to rapidly form an SiO_2-rich layer, which stops further ion beam erosion effects. A variety of organosilanes have been used for this purpose.

Yet other approaches are based on inorganic resist systems. One such system [11] uses a 2000-Å thick film of $GeSe_2$ on which is placed a 100-Å film of $AgSe_2$. Here the upper layer is photoactive, and extremely thin so as to provide high resolution. The $GeSe_2$ layer is normally quite soluble in a number of alkaline solutions. Irradiation of the top layer produces silver which rapidly migrates into the $GeSe_2$ and renders it almost insoluble in these solutions. This photoinduced process is highly anisotropic in character, with migration almost entirely in the direction of the beam, and results in a high-contrast image. The sensitivity of these resists is relatively low, and needs to be improved by one or two orders of magnitude in order to be commercially viable.

10.1.1 Image Reversal

In microcircuit technology, a mask can often consist of small opaque regions on a predominantly clear background. Masks of this type are especially prone to light-scattering problems from dust and defects, which are more likely to be present on regions that are clear. In these situations, a dark-field mask would be advantageous. Some positive photoresists have been especially developed to meet this need. With these resists it is possible to reverse the image during the pattern transfer process, so that a negative image is produced with a conventional positive resist. After the initial mask and exposure steps, reversal is accomplished by means of a chemical treatment and a bake step, followed by flooding the resist with UV. Figure 10.3 illustrates this process, where this bake step results in the release of CO_2 from the resist.

In addition to providing the convenience of using a dark-field mask, this process allows a mask to serve dual functions, and has been used in the precise control of wall angles during pattern transfer [12].

10.2 PATTERN GENERATION AND MASK-MAKING

The complexity of a microcircuit is limited by three factors. The first is the ingenuity of the circuit designer in reducing the number of devices required to perform any given electronic function. Although there have been many dramatic advances along these lines, it appears likely that further progress will be slight, and only in specialized areas of circuit development. The second is the maximum size of the chip that can be made with a reasonable processing yield. Materials and process technologies, outlined in previous chapters, have a strong bearing on this size. The third limit is the size of the *minimum feature* which

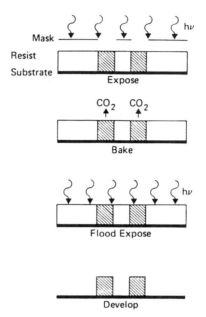

Fig. 10.3 The image reversal process. Adapted from Willson [4].

can be placed on the chip. This is determined by lithographic techniques, which are used in conjunction with pattern transfer processes to delineate the various regions in a microcircuit. The size of each region is set by device design and by the minimum feature size.

The most critical part of the lithographic process is conversion of the layout pattern into a *reticle* or *master mask*. This mask can be used directly to transfer the pattern to the substrate. Alternatively, *working masks* are made from this master by contact printing, and are subsequently used for defining the pattern on each wafer to be processed. It goes without saying that the quality of the reticle mask is important in determining the process yield for the microcircuit over its entire production life. Consequently, extreme care is taken to ensure that it will be free from defects.

As observed earlier, e-beam writing techniques are extensively used for pattern generation, since they already meet the requirements for both present and future generations of microcircuits. Optical techniques are also used for this purpose, especially when circuits of moderate complexity are required, and larger feature sizes can be tolerated. This approach is considerably less expensive than e-beam writing, and it is often used in-house in a number of companies, because it is highly flexible.

Finally, the reticle mask making step can be avoided by combining pattern generation and transfer processes into a single step, as shown by path E in Fig. 10.1, and writing the pattern directly on the wafer. This approach can meet the minimum feature requirement of 0.15 μm readily; its disadvantages lie in its high cost and low throughput.

10.2.1 Pattern Generation

Historically, the first microcircuits were made by manually drafting a series of large drawings, each 250 times the final pattern size ($\times 250$). One such drawing was required for each of the regions (the source/drain diffusions, for example) in the microcircuit. This drawing was made on a plastic laminate, consisting of a dimensionally stable mylar sheet bonded to a thin veneer of red plastic. The red plastic was cut and peeled off to form the desired pattern of clear and red regions.

Next, the pattern was illuminated and reduced in size to form a glass *reticle mask* which was usually $\times 10$ or $\times 5$. This mask consists of a polished glass plate,* coated with a high-resolution gelatin emulsion, in which a high-contrast pattern can be replicated. Hard-surface reticle masks, consisting of a thin (1000–2000 Å) film of chromium** and covered by a photoresist, are increasingly used for this purpose. After exposure, this resist is used to delineate the pattern in the chromium film, which provides better edge resolution since it is extremely thin. In addition, it is more hard-wearing than the soft gelatin-based emulsion, so that it is less prone to wear and tear during use.

Path A of Fig. 10.1 shows a flow diagram for this approach. Also shown (path B) is an optical-mask-making process which is in present use, in which a computer tape is used to drive an optical pen which directly writes the pattern on the reticle mask. This approach has two advantages over the last. First, mask errors can be readily corrected by altering the computer tape instead of the drawing. Second, it avoids the need for making unmanageably large drawings, as the chip size increases. Almost all VLSI masks are made from computer-generated patterns for these reasons. Reticle masks of $\times 5$ to $\times 20$ are commonly used in this approach.

The fundamental limit to optical techniques (i.e., the diffraction limit) is set by the wavelength of light. This limit can be slightly extended by the use of deep UV (200 nm). Electron beams, on the other hand, do not suffer from a practical diffraction limitation [13]. Thus, they are ideally suited for mask-making. During e-beam writing, however, the incident beam suffers from both forward and back scatter as it penetrates through the resist and underlying layers of material. The back-scatter component returns through the resist, and thus results in additional exposure. Back-scatter effects of this type are referred to as *proximity effects*, since they can cause problems with features in close proximity. For example, they can result in contact between two lines which run close to each other.

---

*Although optically flat over small areas, these plates may have as much as 5- to 10-μm warpage over their entire working surfaces.

**Borosilicate glass films can be used if subsequent printing is to be done with deep UV.

Proximity effects can be controlled by reducing the beam energy, or by controlling its dwell time near the boundary of an element during the writing process. They are especially severe when thin photoresist films are used. Positive resists, on the other hand, are relatively thick, so that this problem is of less consequence. At the present time, the use of e-beams for mask-making allows the fabrication of reticle masks with a 0.15-μm minimum feature size. Thus, this approach represents a workable technology for the submicron range when coupled with advanced printing methods such as those employing e-beam and x-ray techniques.

The flow chart for e-beam pattern generation is shown as paths C and D in Fig. 10.1. Here, too, the layout information is provided in the form of computer tapes which are used to control an e-beam which can be driven to directly produce a hard surface reticle mask at a $\times 1$ to $\times 10$ magnification. The larger magnification has the advantage that feature sizes in the reticle can be made larger than those in the actual circuit. In addition, the ability to precisely align one pattern over the next is easier, since errors are demagnified in the final reduction on the chip.

In opposition to these advantages is the fact that, since the reticle area increases with the square of the magnification, the probability of defects on this mask are multiplied by the same factor. Moreover, the area to be inspected is larger with these masks. Finally, reticle masks commonly use redundant fields, so that defects can be readily sensed and corrected. This is more easily accomplished in a $\times 1$ mask with low magnification than in a $\times 5$ or $\times 10$ mask.

The basic components of an e-beam pattern generator are shown in Fig. 10.4. The system resembles, in many ways, a scanning electron microscope with the addition of beam blanking and computer-controlled deflection [14]. Important additional features are the use of laser-driven interferometers and fiducial mark detectors [15]. This combination, together with fiducial marks which are printed during the first pattern writing, allows the system to be precisely positioned for each successive pattern transfer operation. Typically, the beam has a deflection field of about 2 mm $\times$ 2 mm. Consequently, a complete VLSI mask usually requires "stitching" together a number of such fields to form the entire pattern [16, 17].

Two types of scan systems are in use today—the raster scan and the vector scan [18]. In the raster scan system, rectangular strips of the circuit are scanned by a series of lines in order to form the complete chip pattern. In a vector scan system, on the other hand, the e-beam is controlled to scan a feature, move directly to the next feature, and so on. While this often requires wider scan deflection and considerably more complex data-handling and beam-blanking techniques, it results in a much faster system since the beam does not spend time scanning featureless regions. This is an important advantage, since one of the main limitations of these systems is their low throughput.

The writing time of mask-making equipment of this type is set primarily by limitations in the intensity of the electron beam, the sensitivity of the e-beam resist, and the speed of the associated electronics. Factors such as system

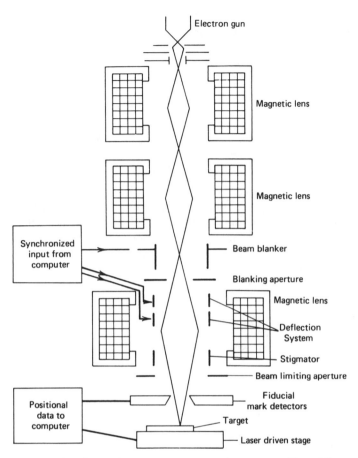

Fig. 10.4 Schematic of an e-beam pattern generation system. From Thornton [14]. Reprinted with permission from *Advances in Electronics and Electron Physics*.

pump-down must also be considered, but are relatively unimportant due to the availability of load-lock techniques and high-speed vacuum systems. Typically, a writing time of about 4 h for a 20-cm × 20-cm mask size is considered acceptable for this task. This is due to the fact that a single master is used to define one particular set of regions (source/drain diffusions, for example) for an entire microcircuit production run. On the other hand, multiple working masks are required for the many printers that are used to perform the pattern transfer operations at the wafer level.

10.2.2 Mask-Making

Reticle masks, made in the manner described in the previous section, can be used (see Fig. 10.1) in subsequent steps in the photographic process, up to the eventual pattern transfer step on the wafer surface. Starting with a ×10 or ×5 reticle mask, the conventional approach is to use a *step-and-repeat* (S/R) camera to form the master mask (paths A and B in this figure). This camera is essentially an inverted microscope, and projects a ×1 image on an emulsion-coated glass plate, or on an emulsion/hard-surfaced glass plate if this type of master is required. This plate is mounted on a mechanical stage which is programmed to move after each pattern is exposed, in an S/R manner. In this way, multiple images can be formed on a photographic plate of almost any size that is required to accommodate the wafer, while still permitting the optics to cover a relatively small field. The S/R process can be speeded up by the use of a multibarrel camera which projects four images of the reticle at one time. Multibarrel systems are in common use today.

Contact printing of the master, or of an intermediate submaster, is used to make multiple working masks which are used to define the pattern on each microcircuit wafer. Pattern transfer can be done by contact printing, or by printing with a small gap between the mask and the wafer.

Pattern transfer can also be accomplished by projection printing of the master mask, or by directly imaging the reticle mask on to the wafer in an S/R sequence (paths D and C in Fig. 10.1, respectively). Sometimes, only a part of the reticle mask is imaged on the wafer in each step, and a series of such parts or fields are "stitched" together to form each complete chip pattern. The main disadvantage to the S/R approach was the increased exposure time for processing each wafer; however, improvements in optical systems have eliminated this problem. For this reason, *stepper* systems are now invariably used for large-area VLSI circuits.

10.3 PATTERN TRANSFER

As seen in Fig. 10.1, the pattern transfer process consists of taking a suitably patterned mask, and imaging it on the surface of a wafer which is precoated with a resist. This is followed by the engraving process, where the exposed resist pattern is used for opening windows in an underlying layer to define semiconductor regions, or to remove metal from a coated wafer in order to delineate the interconnection pattern. These processes must be carried out on each wafer, and for each of the five to twenty masking operations that are necessary for any particular microcircuit. Thus the amount of time that this step takes is an important consideration. At the present, it is generally accepted that the time required to insert, align, and expose the wafer should be 1 min. or less in a practical situation. This requires resists of high sensitivity, as well as automatic (and accurate) means for alignment of the image on the wafer. This section describes the basic processes for pattern transfer. Optical printing processes are described in some detail since they are the most commonly used.

Four types of printers are in general use today: contact, proximity gap, projection and step-and-repeat (stepper). Of these, the contact printer gives the highest resolution because the pattern and the wafer are as close to each other as possible. Features under 0.15 μm can be replicated in this manner.

Contact printing is extensively used in GaAs microcircuits, where small dimensions are critical. A disadvantage of this technique is that mask damage occurs during this process. Physical contact between the mask and the wafer causes damage to the soft gelatin emulsion. In addition, this damage results in defects which are then transferred to *all* successive slices using this mask. As a result, mask life is very short. Depending upon the density and resolution requirements, these 'working masks' must be replaced after every 5–25 operations.

In proximity gap printing, a small gap, 2.5–25 μm wide, is always maintained between the mask and the slice. With a collimated light source, it can be shown that the minimum resolution increases with the square root of the gap, so that there is some loss of sharpness in this approach. The width of this gap is dictated by mask and wafer flatness, and by the precision of the mechanical system. Typically, a gap of 10–15 μm limits the minimum feature size to 2–3 μm, so that this method cannot be used in advanced VLSI applications.

In projection printing, the mask pattern, consisting of multiple chips, is imaged on the plane of the wafer. This technique allows greater flexibility in wafer handling. Moreover, no damage occurs to the reticle mask during this process.

In early projection printers, the complete wafer pattern was directly imaged on the plane of the slice. This became impractical with the trend to larger wafers. In modern systems, a small portion of the mask is imaged, with high resolution, over a small region of the wafer. Next, both the mask and the wafer are scanned so as to obtain full coverage over the wafer while still maintaining the resolution of the well corrected zone. In a typical projection aligner [19], this zone is annular in nature, and about 1 mm wide.

A major disadvantage of projection printers is the need to image the entire wafer in a single operation. However, very high resolution can be achieved [20] if the image field is limited to a small area, about 100-300 mm^2. Advantage is taken of this fact in the stepper system in which the reticle mask pattern consists of a single chip, from ×1 to ×10 in size, which is imaged sequentially over the entire wafer. Often the image field of each chip is divided into contiguous areas, which are processed sequentially, and "stitched" on the wafer to form the complete pattern.

The throughput of a stepper is comparable to that of a projection printer, even though this machine requires multiple exposures over a single wafer. This is because the light collection optics can be made more efficient when the image field is small. In addition, all of the light from the optical source is imaged on one field at a time, thus allowing a greatly reduced exposure time for each field.

In order to achieve a high throughput, it is essential to use rapid, accurate alignment techniques, which are effective with both metal and dielectric films. These involve the use of sophisticated alignment markers and laser-illuminated spatial filters [21, 22], which can achieve position errors of under ±0.02 μm for each field.

An important problem, common to all pattern transfer systems, is the lack of wafer planarity, combined with its process-induced curvature and changes in surface topology, which vary as the wafer progresses from the as-purchased state to the finished product. These in-plane distortions are a function of the specific wafer process, and can be as large as 1–2 μm. Thus, the success of printing schemes of this type depends on the use of processes which tend to maintain wafer planarity throughout its fabrication steps. A number of such *planarization* techniques are described in Chapter 11.

10.3.1 Optical Printing

This process can be described by considering first the problem of cutting windows in a film of the type that is used for masking purposes [23]. Typical films are silica, polysilicon, silicon nitride, silicides, and metals. Steps in this operation are detailed in the flow chart of Fig. 10.5, which illustrates the pattern transfer process for a single window. The contact printing process is shown, by way of example. Details of these steps are shown in Fig. 10.6, and now follow.

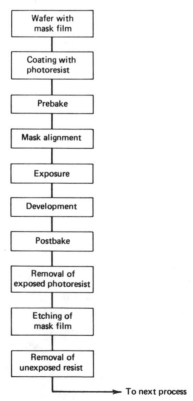

Fig. 10.5 Flow chart for the pattern transfer process.

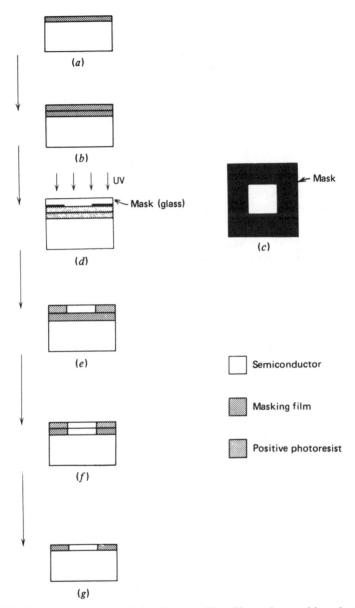

Fig. 10.6 Steps in opening a window in a masking film, using positive photoresist.

Coating with Photoresist. This consists (see Fig. 10.6b) of laying a film of a photoresist material on the surface of the wafer which is covered by the masking film. Ideally, such a film should be uniform, highly adherent, and completely free from dust or pinholes. A positive photoresist is used in this example, for illustrative purposes [5]. This photoresist becomes degraded upon exposure to ultraviolet light.

The coating process consists of spinning the wafer at high speed after a small quantity of prefiltered photoresist has been placed on it. The film thickness is inversely proportional to the square root of the spin rate; typically, spinning speeds range from 1000 to 6000 rpm and result in films that are about 0.5–3 μm thick. Consistent results are obtained only if the viscosity of the photoresist is maintained constant on a run-to-run basis.

Extreme care must be taken to use clean, dry slices to obtain good adhesion of the photoresist. Freshly prepared wafers may be coated directly; however, slices that have been stored must be subjected to cleaning and drying procedures before coating. Adhesion to some surfaces, such as PSG with a high phosphorus content, often presents a serious problem. A dip in a coupling agent, just prior to photoresist application, is routinely used to avoid such problems. Hexamethyldisilizane, $(CH_3)_3SiNHSi(CH_3)_3$, commonly abbreviated to HMDS, is often used for this purpose [24]. Other adhesion promoters which can be used include trichlorophenylsilane, trichlorobenzene, and xylene. The coupling agent can also be applied by vapor-plating, in a batch process.

Although ultraclean conditions should be maintained during the entire operation, the coating step is the most critical one from the point of view of dust contamination. This is because the spinning action creates an air suction along the axis of the slice and promotes the delivery of any airborne particles of dust to its surface. Moreover, the photoresist is sticky at this point.

Softbake. After coating, the slice is baked to drive out all traces of solvent from the photoresist. Typically, the film thickness shrinks to about 85% of its spun-on value during this *softbake* step. The softbake temperature specified by most resist manufacturers is 90–100°C, it must be carefully controlled since it affects the subsequent exposure time, as well as the time of development.

About 20–30 min are required for this process, if it is carried out in a convection oven. This is because, during drying, a skin forms over the surface of the resist and impedes the rapid loss of volatile components from it. Increasingly, softbake is carried out by placing the wafer on a hot plate. Now, heat transfer through the wafer results in loss of resist solvent from the volume of the resist without the formation of a barrier skin on its surface. This reduces the softbake time to under 1 min, resulting in a high throughput, and also minimizes particulate contamination during the baking process.

The softbake process also reduces the solubility of the unexposed film in the developer. For a typical unexposed positive photoresist, the dissolution rate in the developer is about 50 Å/s if no softbake is used, and about 17 Å/s with softbake.*

---

*For the same photoresist, the dissolution rate in the developer is 1000–2000 Å/s in the exposed regions.

Mask Alignment. There are many variations here. In manual systems, a mask consisting of a gelatin photographic emulsion on a glass plate is placed over the slice, brought into contact with it, and then backed off slightly to produce an air gap. Next it is manipulated into its desired position by micrometer adjustment. This alignment process is performed with the aid of a microscope. Finally, physical contact is reestablished between the mask and the slice, and the exposure is made. This is known as *contact printing* (see Fig. 10.6*c* and 10.6*d*).

Most of the variations in this approach have come about because of the needs for improved resolution, increased throughput, or economy. For example, a hard surface mask is sometimes used instead of the gelatin photographic plate. Masks of this type were originally introduced because they were three to four decades more abrasion resistant than gelatin masks, and were dimensionally stable during wet chemical processing. In VLSI applications, their primary advantage lies in their superior edge resolution characteristics and absence of shadowing effects, because of their extreme thinness. (Typically, a gelatin mask must be 4 μm thick in order to have the same opacity as a 1000- to 2000-Å chromium mask.)

Both soda-lime and borosilicate glass are used for masks in the near UV (330–450 nm) and UV (260–330 nm) ranges, but borosilicate glass is preferred because of its low thermal coefficient of expansion. Plates of synthetic quartz are used for masks in deep UV systems (200–260 nm) since they are transparent at these wavelengths. A thin film of borosilicate glass or chromium can be used to define the masking pattern on these plates, since these materials are opaque to deep UV light.

Printing of VLSI circuits requires rapid, accurate mask alignment in addition to high resolution. A rule of thumb in the industry is that alignment accuracy should be about one-fourth the minimum feature size. This can only be met by automatic mask alignment systems, which can be built into projection and stepper systems in which the mask does not come into contact with the wafer. Techniques for automatic mask alignment are similar to those used in e-beam systems, and have been described earlier.

Exposure. Photoresists are exposed by means of collimated UV light. Some filtering of the source is necessary to prevent undue heating of the masks during contact printing. On the other hand, a monochromatic light source is required for projection and stepper systems, where the mask is imaged on the wafer. The wavelength of this source usually corresponds to one of the strong emission lines from an Hg–Xe vapor lamp or from an excimer laser, and ranges from near UV to deep UV.

During exposure with such a light source, the optical wave propagates through the resist and is reflected backwards, resulting in a standing wave intensity pattern. Such a pattern is shown in Fig. 10.7*a* for a 6300 Å layer of positive resist, with a 404.7-nm light source. It is especially strong when the resist is placed on a highly reflecting surface, such as metal or bare silicon. Intensity variations of this type result in periodic changes in the development rate profile of the resist [25], as shown in Fig. 10.7b, causing striations of its side-wall pattern after exposure and development. In addition, this causes variations in the width of lines over profiled surfaces [26].

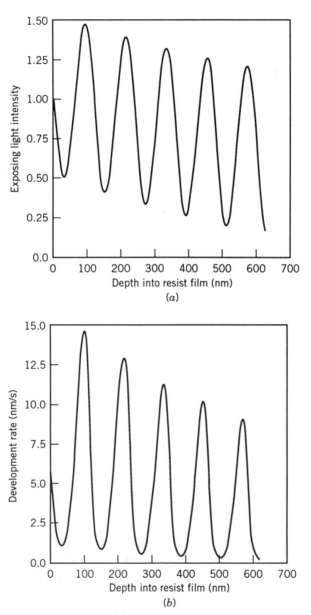

Fig. 10.7 (a) Intensity of light in a resist and (b) resulting striations in a developed line.

The extent of this striation effect on the line width increases rapidly with the reflectivity of the underlying layer. Thus, the problems of resist patterning on aluminum metal are especially severe. They can be greatly reduced by the use of an antireflective coating (ARC) which may be placed between either the resist-layer interface or the air–resist interface. Both bottom and top ARCs are used for this purpose.

Postbake. The function of this step, after exposure and before development, is to further densify the resist so as to reduce the dissolution rate of its undissolved regions. In addition, it improves resist adhesion, and prevents its undercutting during development. It also toughens the resist so that it can better withstand the harsh environment associated with reactive ion etching (RIE).

An important benefit of this postbake (sometimes called *hardbake*) step is that it reduces striations caused by standing wave effects, by redistributing the photoactive component which had been destroyed by optical radiation [27]. Typically, this step is carried out at 100–120°C, which is well below the glass transition point of the resist.

Development. The slice is now rinsed in an appropriate developer, specified by the photoresist manufacturer. This is usually a proprietary formulation which does not contain light alkali ions, but contains $(CH_3)_4NOH$ as an active constituent [5]. The development step results in dissolution of the exposed positive photoresist but does not affect the unexposed regions (see Fig. 10.6e).

It is often found preferable to provide an additional postbake step, especially if RIE is used for subsequent etching purposes. Care must be taken to avoid plastic flow at this point, since this hardbake can be carried out at temperatures up to 170°C. In some cases [28], a high-temperature hard bake can be used to promote flow of the resist, as part of the fabrication process.

Etching. The slice is now etched in order to remove those parts of the underlying film that are not covered by the photoresist. This results in the formation of windows in the mask film, as shown in Fig. 10.6f. To avoid unnecessary undercutting, this process is monitored and arrested as soon as full etching is accomplished. Increasingly, RIE or ion milling is used to remove this unwanted material.

Stripping. The final step consists of removal of the exposed photoresist. Positive photoresists are usually removed by means of a chemical solvent such as acetone or methylethylketone. These are highly inflammable; moreover, their effective use requires application at about 80°C, which is close to their flash point. Increasingly, 1-methyl-2-pyrollidinone is used, in addition to proprietary stripping solutions, for safety reasons.

Negative resists are considerably harder to remove. Here one approach is to immerse the slice in a 1:1 mixture of concentrated H_2SO_4 and H_2O_2 at 150°C.

This mixture is often referred to as "piranha" etch because of its virulent qualities. Problems with the rapid depletion of H_2O_2 have led to its replacement by peroxydisulfuric acid ($H_2S_2O_8$), commonly called PDSA, in order to increase its useful life. Another removal technique for negative photoresists consists of using hot chlorinated hydrocarbons to swell the polymer, together with acids to loosen its adhesion to the substrate. Solvent mixtures of trichlorethylene, methylene chloride, and dichlorobenzene, combined with formic acid or phenol, are used in the form of proprietary mixtures for this purpose.

Photoresist removal by plasma oxidation, commonly called *plasma ashing* and described in Chapter 9, has many advantages over wet chemical methods. Thus undercutting by capillary action of the liquids is avoided, so that photoresist adhesion is not critical. Waste disposal problems are eliminated, since only a small amount of plasma reaction products (water, carbon dioxide, and carbon monoxide) are produced. These factors add up to a major cost advantage for this approach, so that it has received wide acceptance by industry.

Plasma methods present a problem in the area of silicon MOS circuits. Here radiation damage to the gate oxide can cause significant threshold shifts, especially for short-channel, low-threshold devices. Recently, the use of UV, as well as the use of UV combined with ozone, has been explored for resist removal when ion damage to the underlying layer must be avoided.

The pattern transfer process for contacts and metallization is only slightly different from the above procedure. Here aluminum and gold/Ti-W are the most commonly used materials, as outlined in Chapter 8. The insulating layer covering silicon microcircuits usually consists of silicon dioxide, whereas silicon nitride is commonly used with GaAs circuits. Contacts and interconnections are made as shown in Fig. 10.8.

Windows are cut in the insulating layer over appropriate regions, by use of a mask with a contact pattern. This is shown in Fig. 10.8a for an n^+–p diode. The metallization film (or films) are deposited over this insulating layer, and make contact to the semiconductor, as in Fig. 10.8b. The wafer is now covered with photoresist and exposed through a mask carrying the interconnection pattern (Fig. 10.8c). Positive photoresists are almost universally used in this step. After development, the metallization that is not protected by the photoresist is removed by suitable etching techniques, as described in Chapter 9. The resulting metal connects to the contact regions, and runs over the insulating layer, as shown in Fig. 10.8d.

Interconnection lines may be widened once contact is made to the device. In addition, it is possible to widen these lines further to form large bonding pads to which leads may be attached. Both situations are shown in the plan view of Fig. 10.8e. Next, the resist is removed, and the slice is heated in an inert gas ambient to microalloy the contact to the semiconductor, and also to bond the metallization to the insulating surface.

The removal of resist over metallization is extremely difficult when negative resists are used, since swelling of the resist leads to damage of the fine

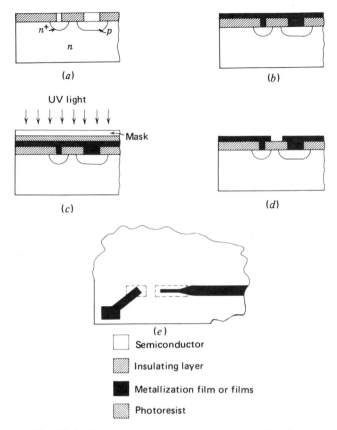

Fig. 10.8 Steps in making contacts and metallization.

lines. With positive resists, swelling does not occur, so that resist removal is a relatively easy matter.

10.3.1.1 Lift-Off Techniques

The *lift-off technique* is an important variation of the above process [29]. Here the delineation of, say, a metal pattern on an insulating substrate proceeds along the following lines. First a positive photoresist film is placed on the substrate, and patterned so that it covers all those regions where *no* metal film is desired, as in Fig. 10.9a. Thus, the mask is the inverse of what is required for conventional pattern transfer with a positive resist. Next, the metal film is deposited over the substrate–resist combination, as shown in Fig. 10.9b, and contacts the substrate *only* in those regions where it is required. Finally, the photoresist is removed by a solvent which does not attack the metal film. In so doing, it "lifts off" the material which is on its surface, leaving behind the patterned

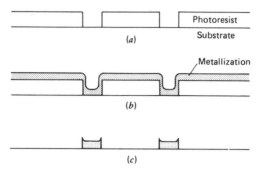

Fig. 10.9 The lift-off process.

metal film, as in Fig. 10.9c. The success of this approach requires the use of a relatively thick photoresist film so that the deposited metal film is very thin at the sides of the step, or even discontinuous.* This allows rapid dissolution of the photoresist mask, and makes it easier to lift off the raised portions of the film without breakage.

Two advantages are realized by this process. First, it can be used with films such as platinum, gold, silicides, and refractory metals, which are very difficult to etch by conventional means. Next, the use of a thick photoresist avoids problems due to pinhole formation. Unfortunately, however, lift-off is always accompanied by slight tearing at the film edges, as shown in Fig. 10.9c. Thus, it is extremely difficult to carry out this process with fine-line structures. One solution to this problem is to spin on a thick layer of a single photoresist, and toughen its surface by a pretreatment [29]. Positive photoresists such as AZ1350-J can be surface-toughened by immersion in an aromatic solvent such as chlorobenzene for a fixed period of time after the softbake period, and before or after pattern exposure. This process results in greatly reducing the dissolution rate of exposed resist in the surface layer. Control of the soak time and prebake temperature permits photoresist cuts with the overhang pattern shown in Fig. 10.10. Here the photoresist** was modified to a depth of 2000 Å by the use of chlorobenzene prior to development. Vacuum evaporation on this patterned resist results in a discontinuous film and eliminates problems of tearing. At the same time, the thick photoresist (1–3 μm) permits the use of thick metal films of sufficiently large cross section, so as to handle the current requirements of high-speed circuits. Thus the approach represents a practical alternative to the anisotropic RIE methods described in Chapter 9. Furthermore, it avoids the problems of damage and oxide charge associated with plasma processes, and is extensively used in VLSI applications for this reason.

---

*Techniques such as vacuum evaporation from a point source, which provides poor step coverage, are ideal for this purpose.
**Shipley Company, Newton, MA.

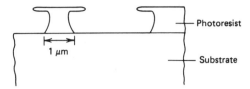

Fig. 10.10 Profile of photoresist using a chlorobenzene modification step prior to development. Adapted from Hatzakis et al [29].

Multilayer resists can also be used with the lift-off process. Now, the separate layers can be optimized individually; in addition, it is possible to form the image in the thin top layer, with an improvement in resolution. Details of this approach are provided in Section 10.4.2.

10.4 ADVANCED TECHNIQUES

We have noted that e-beam lithography can readily achieve the critical feature size which will be required in the foreseeable future. Thus, it is the basic method for reticle mask-making. Improvements in this area are now directed towards increasing the throughput, and improving the ease of operation of these systems.

Pattern transfer techniques, on the other hand, present many areas for improvement. Advances in pattern transfer by conventional optical methods involve the use of shorter wavelengths, the use of multilayer resist technology, and the use of improved techniques for sharpening image contrast.

Totally new systems for pattern transfer are also being investigated. Here, it is important to note that multiple stations are used at each manufacturing location, and handle a high volume of mask sets for each application. Thus, overriding factors in the development of these systems are high throughput and reduced cost. Their minimum feature size is, however, dictated by the reticle mask which is made by e-beam lithography.

In this section, we briefly described these new approaches, with a view towards assessing their usefulness in pattern transfer systems.

10.4.1 Short Wavelengths

The limitations of optical lithography are a consequence of basic physics. In any single wavelength optical system, the minimum feature size is given by

$$\sigma = \frac{k\lambda}{NA} \tag{10.1}$$

where λ is the wavelength, NA is the numerical aperture, and k is a proportionality factor. For a diffraction-limited optical system, $k = 0.5$. In practice, values of k are somewhat larger because of manufacturing uncertainties. The numerical aperture is given by sin α, where 2α is the acceptance angle of the lens at its point of focus. Thus, NA is a measure of the light-gathering power of the lens. From this equation, it follows that the minimum feature size can be reduced by operation at shorter wavelengths, and/or with larger values of NA.

The depth of focus of a lens is an important parameter, because the surface of a microcircuit has topological features and is not flat. Its magnitude is given by

$$d = \frac{\lambda}{(NA)^2} \tag{10.2}$$

Increasing the NA results in reducing the depth of focus; moreover, the field of view of the lens is reduced. Thus, the number of steps required to cover a microcircuit is increased, resulting in an increase in the time required to process an entire slice through the mask aligner. It follows that the use of shorter wavelengths is eventually essential in order to reduce the feature size.

Modern optical pattern transfer equipment uses high-pressure Hg–Xe vapor lamps, which exhibit a spectrum of sharply defined lines as shown in Fig. 10.11 [30]. Here, the lines at 546 nm, 436 nm, 405 nm, and 365 nm (denoted as the E, G, H, and I lines, respectively) are especially intense, and can be used for stepper applications.

The thrust of microcircuit technology has steadily shifted towards shorter

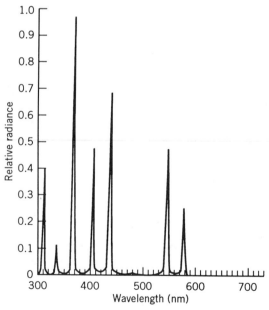

Fig. 10.11 Emission spectrum of an Hg-Xe vapor lamp.

wavelengths, from soft UV to UV, in order to reduce the minimum feature size. A number of mask aligners utilizing the I-line are now available, with a minimum feature size of 0.25 μm. Standard optical glasses can be used in these systems, although their properties rapidly degrade below 350 nm. Alternative glasses have also been investigated for I-line applications.

The optical absorption of conventional photoresists is much greater for the I-line than for the G-line, leading to problems with linewidth control. However, new resists have been developed with spectral absorption characteristics which are matched to this wavelength.

Even shorter wavelengths, in the deep UV range, can be obtained by the use of excimer laser light sources [31]. Light output at 351-nm, 308-nm, 248-nm, and 193-nm wavelength can be obtained using XeF, XeCl, KrF, and ArF source gases, respectively. All of these lasers are operated in the pulsed mode, with typical peak power outputs of about 10^7 times that from the individual lines from high pressure Hg–Xe vapor sources. Thus, single-shot exposures, which simplify mounting requirements, are a practical possibility with these sources. Additionally, materials capable of ablation can be used as positive photoresists, because of the high energy density that is involved. Carbon films have been found to be suitable for this application.

Excimer laser mask aligners are still under development, and 0.13-μm feature sizes have been demonstrated. However, many problems have to be solved before these systems become practical. For example, there is a need for deep UV optics which are free from aberration and which can withstand high peak powers without catastrophic failure. Quartz is the material of choice for this application and is also used for mask plates in this wavelength range. There is also a critical need for new resists which can take advantage of the short exposure times associated with these systems. Both organic and inorganic materials are being actively considered for this purpose. Finally, excimer lasers are notoriously unreliable and require maintenance after a few hours of operation. Reliable lasers are necessary for use in a commercial situation.

10.4.2 Multilayer Resists

Modern VLSI technology requires the definition of fine lines with high aspect ratios. This is especially true for metallization, which is 0.8–1 μm thick. This, in turn, necessitates the use of thick resist layers to serve as masks for defining these films.

During pattern transfer, the light intensity falls off with depth as the beam penetrates the photoresist, so that its top regions are excessively overexposed. With positive resists, this results in undercutting of the mask, and in nonvertical resist walls, as shown in Fig. 10.12. This is especially true for deep UV resists, which have a large absorbance at their operating wavelengths.

The use of multilayers of resist can avoid this problem. Consider, for example, the situation where two layers are used. Here, there are two distinctly different possibilities: Case 1, where both layers are exposed to optical radia-

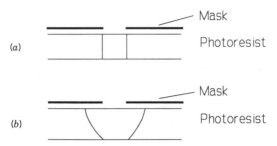

Fig. 10.12 Resist profiles after development: (*a*) ideal and (*b*) actual.

tion and then developed; and Case 2, where the top layer is used to form the photographic image, which is then used as a protective mask during removal of the second layer.

Figure 10.13*a* shows Case 1, where the bottom resist layer (R_1) is a material with high radiation sensitivity, and the top (thinner) layer (R_2) has low radiation sensitivity. Upon exposure and subsequent development, R_2 will etch slowly. R_1, on the other hand, will etch rapidly, thus undercutting R_2 as in Fig. 10.13*b*. This result is essentially the same as that obtained in the lift-off process of Section 10.3.1.1, where the top layer was formed by a chlorobenzene soak. However, the two-layer technique provides more control. Still more control can be obtained, at the expense of added complexity, if the developer for R_1 will not dissolve R_2, and vice versa.

Case 2 is quite different. Here, R_1 is sensitive to light of wavelength λ_1, whereas R_2 is sensitive to λ_2 but opaque to λ_1. In this approach, exposure at λ_2, followed by development, is used to define the pattern in R_1, which comprises a thin layer. Next, the wafer is exposed to light at λ_1. This is a flood exposure, without a photomask; here, the pattern in R_2 serves this purpose. This is followed by development of R_1.

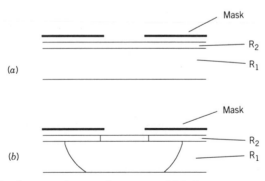

Fig. 10.13 Profiles for a two-layer resist process: (*a*) before exposure (*b*) after exposure and development.

The approach outlined here [32, 33], referred to as *portable conformable masking* (PCM), is superior to that of Case 1 since it allows a high-resolution pattern to be defined in a thin (2000–4000 Å) layer, although the overall thickness of the resist is 2–4 μm.

One problem with double-layer resists comes about because of the mixing of resists at the R_1–R_2 interface. This results in a transition layer, which may be difficult to remove. A solution to this problem is to use a layer of spin-on SiO_2 between the photoresist layers, resulting in a trilevel mask. RIE can be used readily with a system of this type, with an oxygen plasma for resist removal, and a CF_4–O_2 plasma for removal of the SiO_2 layer. Recently, new resist combinations have been developed [34] which do not have this intermixing problem.

Lift-off techniques have been successfully applied to PMMA resists, using a double–layer technique [35]. The lower layer consists of a 1- to 2-μm thick film of PMMA dissolved in chlorobenzene. After baking, a 4000 Å film of MMA dissolved in ethyl cellosolve acetate is spun on the wafer. This upper layer defines the pattern, while the lower layer is undercut during development so that easy lift-off is achieved.

Contrast enhancement lithography (CEL) is an alternative approach [36, 37] to obtaining improved pattern transfer. Here, a conventional resist R_1 is covered by a layer of a polymer containing a photobleachable dye, R_2. Upon exposure to light of wavelength λ_1, this layer bleaches in those regions which are exposed through the mask, and thus sharpens the intensity profile of light passing through the photomask. This process is illustrated in Fig. 10.14. After

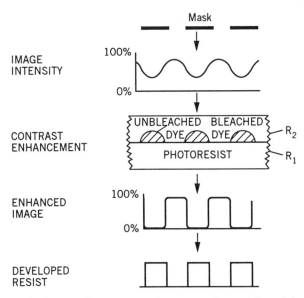

Fig. 10.14 Steps in the use of a contrast enhancement layer. Adapted from Halle [32].

exposure, the CEL layer is dissolved away, and the photoresist developed. Water-soluble CEL materials are available which greatly simplify this process.

Photolithography is perhaps the most expensive step in microcircuit manufacturing. As a result, the processes described here must be evaluated critically, and the simplest method selected to do the job at hand. Advances in single-layer resists have made these the choice for many situations. Among the multilayer resists, the CEL approach is the most promising, because it results in greatly improved performance with small additional complexity. Moreover, it can be designed for use at all wavelengths [38], including those which are provided by excimer laser sources.

10.4.3 Phase-Shifting Masks

Diffraction effects represent a fundamental limit on the resolution of lines in optical lithography [39]. This is illustrated in Fig. 10.15*a*, which shows two adjacent slits in an optical mask. The light intensity from each of these slits is diffracted into the obscured regions which surround them. As these slits are made closer, the diffracted regions overlap until the individual intensities cannot be resolved.

Figure 10.15*b* shows the same mask pattern. Here, however, a transparent layer is placed across one of the openings between slits. Its thickness is such as to provide a 180° phase shift to light passing through it. The diffracted signal from *this* slit will now be out of phase with the signal through the second slit. As a result, destructive interference occurs at the wafer, and both slits can be resolved.

The use of phase-shifting masks of this type has the potential of extending the resolution capabilities of modern optical systems by a factor of 2–4 [40]. A

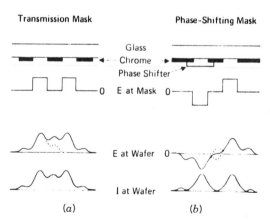

Fig. 10.15 Intensity patterns: (*a*) without a phase shifting mask and (*b*) with a phase shifting mask. From Levenson et al. [40].

central problem with this technique is that it is limited to highly repetitive patterns. More recently, however, a number of modifications have been proposed which allow the method to be applied to arbitrary mask patterns [41].

The CEL process can also be used to enhance the contrast in the optical image. Although it is an alternative approach to phase-shifting masks, it can be combined with them.

10.4.4 Electron-Beam Techniques

The concept of writing a pattern directly on a chip by e-beam lithography is attractive, since it represents the ultimate in resolution capability. However, this approach is severely limited by the time taken to write a single substrate because of the sequential nature of this process. One area where this approach has been implemented is in the direct writing of gates in GaAs devices and integrated circuits [42]. Here, wafer sizes are relatively small (50- to 100-mm diameter), and each circuit usually has a small device count. In each device the gate length is the most critical parameter, which must be kept as small as possible. Using direct e-beam writing, a value of 0.15–0.25 μm has been achieved for the gate length in this manner.

The speed of direct e-beam writing can be greatly improved if advantage is taken of the fact that the minimum feature size is critical in defining the edge of a pattern, but not for its inner parts [43]. Thus, most of the pattern can be written with a beam having a large spot size (of rectangular shape), with a large stepping pitch. At the same time, a minimum spot size (0.1 μm) can be used to define the pattern edge. Using this approach, GaAs monolithic microwave integrated circuits have been processed at the rate of 6–8 slices/hr.

The sequential nature of e-beam writing makes it extremely difficult to use for pattern transfer, even if significant improvements are made in resist sensitivity, high-intensity electron guns, and electronics. An alternative approach, e-beam projection printing, is presently under consideration [13, 44]. A schematic of a demagnifying projection system for this purpose is shown in Fig. 10.16. Here a self-supporting foil mask is used with a flood electron gun source which provides a collimated beam of electrons. The reticle mask, usually ×10, is imaged directly on a photoresist-coated wafer which is located on a laser-controlled table for positioning purposes.

Systems of this type are considerably simpler than e-beam systems which are used for pattern generation, since the beam is not required to maintain its precise convergence under scan conditions. This greatly simplifies the electro-optical system that can be used, in addition to the digital electronics required for data handling and beam deflection. Most of the present research with these systems involves step-and-repeat techniques for printing one chip at a time. Even with this limitation, electron projection printing is considerably faster than direct writing, where the pattern is written serially. This can be a major advantage in a development situation, where it can provide quick turnaround for prototype circuits.

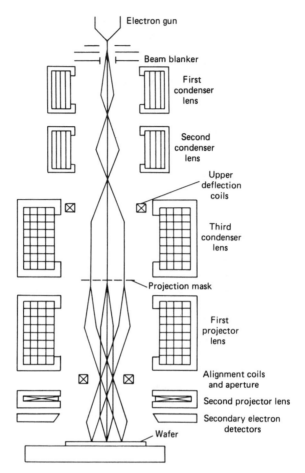

Electron gun

Beam blanker

First
condenser
lens

Second
condenser
lens

Upper
deflection
coils

Third
condenser
lens

Projection mask

First
projector
lens

Alignment coils
and aperture

Second projector lens

Secondary electron
detectors

Wafer

Fig. 10.16 Schematic of a demagnifying electron projection printing system. From Thornton [13]. Reprinted with permission from *Advances in Electronics and Electron Physics*.

Problems of mask fabrication for this system are severe, since the pattern consists of a large number of isolated structures. However, they must be physically connected for a mask to exist. One approach is to use a fine grid or mesh on which the pattern is placed. The resulting fine bars which connect isolated structures produce a small distortion at the edge of each structure, which can be kept within acceptable limits if a ×10 mask is used.

10.4.5 Ion-Beam Techniques

The use of ions instead of electrons in the lithographic process has two potential advantages [45]. First, ion beams do not suffer from as much scattering as do electron beams, when they penetrate a resist. In addition, secondary electrons produced by ion bombardment have very low energy so that scattering caused by them is limited. The ultimate limit to the minimum feature size is set by these scattering processes. Next, the ion sensitivity of resists is a function of the ion energy and mass, and is many orders of magnitude greater than that for electrons. Thus ion-beam techniques are potentially more rapid than e-beam methods.

Ion projection printing systems, utilizing an r.f. ion source and focusing optics, have been developed with ion sources using H^+, He^+, and Ar^+ in the 100-keV range, to image a field 5 mm × 5 mm in a single exposure [46]. Important problems with this system are its large size, so that it is prone to vibration. Moreover, considerable work needs to be done to improve the resolution of the ion-optical system in order for it to meet the 0.15-μm minimum feature requirement.

Direct writing by focused ion beams [47] is also under consideration. Here the critical elements are finely focused ion sources of the desired impurity. Although much progress has been made in their development, the writing speed of these systems is four to six orders of magnitude slower than for ion projection printers. This has confined their use to special applications such as (a) mask repair and (b) the making of small cuts and vias [48].

10.4.6 X-Ray Printing

The attractiveness of this approach stems from the extremely short wavelengths that are possible with x-ray sources [49]. This eliminates the diffraction limitation of optical and UV lithographic techniques, and has the potential of providing a means for fabricating submicron structures at a cost that is competitive with optical projection. There are, however, major differences in the design of these systems. Thus optical systems can use collimated light and focusing optics, whereas x-ray systems are generally restricted to a point source. Here, proximity gap printing must be used, with ×1 masks. In an x-ray system this results in a slight magnification of the pattern. This is seen in the proximity arrangement of Fig. 10.17, where the blur (σ), is related to the size of the x-ray source (S) and the gap (g), as well as to the distance between the source and the mask (D), by

$$\sigma = \frac{gS}{D} \qquad (10.3)$$

Usually, σ is small enough to be neglected; however, this equation emphasizes the need for keeping the magnitude of the gap (and more important, its spatial variation) constant to prevent distortion effects. This necessitates a somewhat larger gap than is used in optical proximity gap printers. Typically, systems of this type can be operated with a gap of about 25–50 μm and still give satisfactory performance at the submicron level.

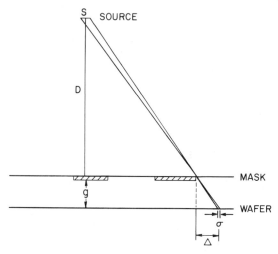

Fig. 10.17 X-ray proximity arrangement.

The ideal source for x-ray lithography is a point source. This can be produced by a high-intensity electron beam impinging on a target, which is water-cooled and rapidly rotated to prevent overheating. The choice of x-ray wavelengths ranges from 4.7 Å and upward. An aluminum target and a 10- to 20-keV beam can be used to produce x-rays with a wavelength of 8.34 Å. These are filtered to remove stray radiation, and are fed out of the vacuum through a beryllium window [50], as shown in Fig. 10.18. All processing of slices in these systems can be done outside the vacuum, which is a distinct advantage over e-beam lithography. The beam is usually confined to a helium gas environment to prevent unnecessary absorption; however, x-ray lithography is relatively insensitive to airborne particles, so that cleanliness requirements can be readily met. An additional advantage is that x-rays are not easily scattered in a resist during pattern transfer, because of their extremely short wavelengths. This results in nearly vertical resist profiles, which are highly advantageous for VLSI applications.

A severe problem with x-ray sources of this type is that no photoresists exist with efficient absorption in the 4- to 8-Å wavelength range. Alternative x-ray sources are being developed [51, 52], based on laser-induced plasmas which produce short pulses of "soft" x-rays with high irradiance in the 12-to 14-Å range, which are more closely matched to the absorption of available resists. These can also be used for "one-shot" exposures, thus avoiding vibration problems during pattern transfer.

Electron storage ring technology is also used for the generation of x-rays with an average irradiance of 10–50 times that of an x-ray tube, resulting in a reduction in the exposure time [53]. In this approach, x-rays are generated in

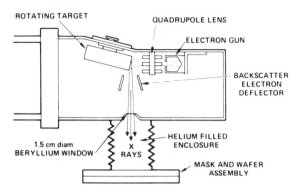

Fig. 10.18 Schematic of an x-ray printing system. From Stover et al. [50]. Reprinted from *Solid State Technology.*

a synchotron ring structure and can be brought out at multiple ports from the ring. A number of "low cost" rings of this type are under development at the present time [54, 55].

Masks for x-ray lithography must be extremely thin so as to be transparent at these wavelengths. Membranes of both inorganic (Si, SiC, Si_3N_4, Al_2O_3, and BN, for example) and organic materials have been investigated. Both Si and SiC films, 1–2 μm thick, are preferred for this purpose since they exhibit no radiation-induced discoloration. These materials are transparent to x-rays and also optically transparent, so that alignment of the mask to the wafer can be done by conventional means.

A gold or tungsten film, in which the pattern is defined, is used to absorb the soft x-rays. Typically, this is vacuum-evaporated on the membrane, to a thickness of about 0.5–1 μm. The mask pattern is made in this film by conventional e-beam lithography, and the excess gold is removed by ion milling.

A major problem with x-ray printing is the lack of suitable resists, in order to keep the exposure time to an acceptable value. The design of these resists is critical to the success of this technology. Here, it has generally been found that resists which are useful for e-beam lithography can also be used, and that resists which have high sensitivity for e-beam applications can also have a high sensitivity for x-rays.

Resists for x-ray lithography have been described in Section 10.1. Most work has been done with the methacrylates, and in particular with polymethylmethacrylate (PMMA) and its derivatives, which are based on scission. Resists, based on the dissolution inhibiting principle, can also be used with x-ray lithography. These have a chemistry very similar to that of positive resists which are used for conventional photolithography. A disadvantage of these resists is their relatively low x-ray absorption coefficient.

High-power x-ray sources are also crucial to the commercial realization of this technique, and are under continual development. Nevertheless, a number of practical systems have already been developed [56], and microcircuits have been fabricated using this approach.

At the present time, the biggest challenges to x-ray lithography come from both conventional and unconventional optical processes. These techniques are capable of 0.25-μm features at the research level, and 0.5-μm features in production [2]. These feature sizes can be obtained with I-line optics, and are suitable for the manufacture of 64-Mbit DRAMs. The case for x-ray lithography is thus one which must be made when feature sizes of 0.15 μm are necessary. These sizes will be necessary for memory fabrication in the 256 Mbit to 1 Gbit ranges, and can only be realized by the use of proximity printing using soft x-rays.

10.5   PROBLEM AREAS

A variety of problems can be encountered at all points in the lithographic process. One is that of unwanted particulate matter. This ranges from glass particles left on the plates by the manufacturer, to airborne particles which are always present to some extent in a clean room. Additional problems come about because of operator-induced errors during processing. In general, these result in random defects in the microcircuit.

Consider a microcircuit chip of area A. To an approximation, the number of such chips on a slice of diameter d is given by N, where

$$N \simeq \frac{\pi}{4A}(d - \sqrt{A})^2 \tag{10.4}$$

Let N_G be the number of good chips on the wafer, so that N_G/N is the fractional yield.

Most defects from the lithographic process are extremely small and of random nature. Let N_D be the total number of such defects per wafer.

If a random defect is added to a wafer, its probability of destroying a good chip is N_G/N. Thus,

$$dN_G = -\left(\frac{N_G}{N}\right) dN_D \tag{10.5}$$

so that the fractional yield is

$$\frac{N_G}{N} = e^{-N_D/N} \tag{10.6}$$

Minimizing the number of defects on a wafer is thus extremely important for VLSI, where chip area is corresponding large (i.e., where N is small).

A discussion of the different types of defects that can arise during the lithographic process now follows.

10.5.1 Mask Defects

Mask defects of the visual type include pinholes, spots, intrusions, and protrusions. Such defects are a function of the quality control exercised by the manufacturer, as well as of the care taken in using the starting materials. Often these masks are supplied precoated with a strippable lacquer film for surface protection. This film is removed immediately prior to loading in the mask equipment.

Extreme cleanliness is required during the fabrication of reticle masks, and a detailed examination of each exposed mask is essential before it can be replicated. A number of different methods are used for mask inspection today [57]. The simplest approach is to directly compare one die pattern with another on the same mask. This is done by optically overlaying these patterns, followed by visual inspection. A second approach is to compare a die with a defect-free pattern that is stored in digital form in a computer. This approach has the advantage that it can be carried out automatically. A somewhat similar approach is to scan two dies by a dual-beam flying spot scanner and compare the signals by electronic means. Spatial filtering techniques have also been used for this inspection purpose.

Dimensional errors can also occur during mask making. These can only be determined by both line-width and run-out measurements on the mask.

Visual defects on reticle masks at ×10 magnification can be repaired if damaged. Defects in the form of extraneous material in the clear-field regions are usually removed by laser ablation. Voids, and scratches which result in removal of the mask film in the dark-field regions, are considerably harder to repair. Often, the repair operation consists of applying minute droplets of epoxy ink to fill the pinholes. Increasingly, focused ion-beam techniques, involving the deposition of chromium, are being considered for this purpose [48]. Finally, the time taken for mask repair can often exceed that required to make the mask in the first place. Remaking of the mask is a common practice under these circumstances.

10.5.2 Pattern Transfer Defects

A number of problems may be encountered during the opening of windows in the masking film, or in the delineation of fine lines. Some result in defective circuits that are rejected. More serious, however, are those problems that result in an inadvertent spread of device parameters which may impair the reliability of the microcircuit and degrade its performance [58]. Some of these defects are now considered for a few practical situations.

Undercutting of the Resist. If the resist adheres poorly to the masking film, undercutting may occur during the etching process, resulting in an enlarged window (see Fig. 10.19a) and changes in the associated device parameters after diffusion. In extreme cases a defective circuit can result by short-circuiting adjacent regions. Undercutting of the photoresist comes about because of capillary action of the (wet) chemicals that are used during the etching process. This problem is greatly reduced when plasma etching techniques are used.

Dimensional Variations. Dimensional variations may occur in the width of cuts in the masking film (as in Fig. 10.19b) due to lack of control of the exposure time, or limitations of the photographic process. These, too, result in changing the effective area of the masked regions. The problem is especially severe when long, very narrow cuts are required and may lead to breaks in the metallization, as shown in Fig. 10.19c. These variations can also occur because of swelling of the resist during development. Positive resists are less susceptible to this problem than negative ones, so that they are preferred in VLSI applications.

Lack of Registration. This occurs because of mask alignment errors, or because of cumulative errors in the step-and-repeat process. It can lead to

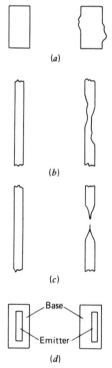

Fig. 10.19 Defects and their results.

altered device configurations, as shown in Fig. 10.19*d* for the emitter and base regions of a transistor. Here, too, defective circuits can result in extreme cases. In some situations, misregistration may result in exposing a protected junction edge, thus degrading its characteristics.

Pinholes. These may be present in the resist and result in the inadvertent opening of a small window in the underlying mask film. Their effect is a function of their specific location. As may be expected, the severity of this problem increases with component packing density in a microcircuit.

Dust Particles. These present a serious problem during the delineation of patterns in mask films. Some dust, such as carbon or metallic particles, is opaque. Its presence on the clear field of a mask will prevent exposure of the underlying resist. This will cause holes in a negative resist after exposure and development, and thus holes in the mask film. The effect with positive resists is to not completely clear an opening in the underlying mask. This latter effect is usually less serious than the former.

Dust particles are often transparent and cause diffraction effects. This can have the same result as opaque particles; often, however, the effect is over a wider area because of the diffraction pattern. In either case, the problems of dust are minimized by the use of good clean-room technique, and by careful visual inspection of the masks prior to installation in the aligner.

Scratches and Tears. The printing of metallization patterns is a particularly difficult task because of the delicate nature of the film. Here scratches are by far the most common type of defect and are caused by rough handling during the fabrication process. Scratches greatly increase the current density in their vicinity, giving rise to (a) the formation of hot spots and (b) the premature onset of electromigration. Thus although the circuit operation is unchanged, its reliability may be greatly impaired in this manner. In extreme cases, scratches result in defective circuits by breaking the metallization path.

Tearing of the metallization may occur during the removal of the photoresist. This is often due to dimensional changes and damage to the photoresist during the alloying step. This defect also results in reducing the effective area of the current path or in an open circuit in extreme cases. The two-level lift-off technique is extremely important in controlling this problem.

Step Coverage. This occurs when an excessively thin layer of resist goes over a step in an oxide. It can give rise to a failure of the resist during the subsequent etching process. The problem is especially severe with negative resists, since they are applied in much thinner layers than are positive resists. Here, too, the use of two (or more) levels of resist can alleviate this problem.

Problems of step coverage can also be reduced by the use of planarization techniques. This important topic is discussed in Section 11.4.

TABLE 10.1 DRAM Production Requirements[a]

DRAM Chip	Time Frame	Chip Size (mm^2)	Feature Size (μm)
1 Mbit	1989–1991	50–40	1–1.2
4 Mbit	1990–1993	80–50	0.65–0.8
16 Mbit	1991–1994	120–80	0.5–0.6
64 Mbit	1992–1995	160–100	0.35–0.45
256 Mbit	1995–1997	240–160	0.25–0.35
1 Gbit	1997–2002	320–240	0.15

TABLE 10.2 Resists for e-Beam and x-Rays

Resist	Type	X-Ray Sensitivity (mJ/cm^2)	20-keV e-Beam Sensitivity ($\mu C/cm^2$)
Naphthoquinonediazine/novolak	AZ145OJ	2100	40
Poly(methylmethacrylate)	PMMA	1800	80
Poly(hexafluorobutylmethacrylate)	FBM-120	35	0.5
Poly(butenesulfone)	PBS	14	2
Poly(glycidylmethacrylate)-co(ethylacrylate)	COP	5	0.6
Chlorinated poly(methylstyrene)	CPMS	8	0.4

REFERENCES

1. S. Wolf and R. N. Tauber, *Silicon Processing for the VLSI Era, Vol. 1—Process Technology*, Lattice Press, Sunset Beach, CA, 1986.

2. G. E. Flores and B. Kirkpatrick, Optical Lithography Stalls X-Rays, *IEEE Spectrum* (Oct. 1991) p. 243.

3. A. Reiser, *Photoreactive Polymers: The Science and Technology of Resists*, John Wiley and Sons, New York, 1989.

4. C. G. Willson, Organic Resist Materials—Theory and Chemistry, in *Introduction to Microlithography: Theory, Materials and Processing*, L. F. Thompson, C. G. Willson and M. J. Bowden, Eds., American Chemical Society Symposium Series 219, Princeton, NJ, 1983, p. 121.

5. *Shipley AZ Photoresists*, The Shipley Company, Newton, MA, 1977.

6. S. Iwamatsu and K. Asanami, Deep UV Projection Systems, *Solid State Technol.* (May 1980) p. 81.

7. M. J. Hatzakis, Electron Resists for Microcircuit and Mask Production, *J. Electrochem. Soc.* **116**, 1033 (1969).

8. N. Taylor, X-Ray Resist Materials, *Solid State Technol.* (May 1980) p. 73.

9. J. Lingnau, R. Dammel, and J. Theis, Recent Trends in X-Ray Resists: Part II, *Solid State Technol.* (Oct. 1989) p. 107.

10. K. Harada, O, Kugure, and K. Murase, Poly(Phenyl Methacrylate-Co-Methyacrylic Acid) as a Dry Etching Durable Positive Electron Resist, *Trans. IEEE Electron Dev.* **ED-29**, 518 (1982).

11. W. R. Sinclair, M. V. Sullivan, and R. A. Fastnacht, Materials for Use in a Durable Selectively Semitransparent Photomask, *J. Electrochem. Soc.* **118**, 341 (1971).

12. E. Alling and C. Stauffer, Image Reversal Photoresist, *Solid State Technol.* (June 1988) p. 37.

13. P. R. Thornton, Electron Physics in Device Microfabrication—I. General Background and Scanning Systems, *Adv. Electron. Electron Phys.* **48**, 271 (1979).

14. P. R. Thornton, Electron Physics in Device Microfabrication—II. Electron Resists, X-Ray Lithography and Electron Beam Lithography Update, *Adv. Electron. Electron Phys.* **54**, 69 (1980).

15. D. E. Davis, R. D. Moore, M. C. Williams, and O. C. Woodard, Automatic Registration in an Electron-Beam Lithographic System, *Solid State Technol.* (Aug. 1978) p. 61.

16. S. Okazaki, K. Mochiji, E. Takeda, and Y. Maruyama, Electron Beam Mask Fabrication for MOSLSI's with 1. 5 μm Design Rule, *Jpn. J. Appl. Phys.* **19**, 51 (1980).

17. D. R. Herriott, R. J. Collier, D. S. Alles, and S. W. Stafford, EBES: A Practical Electron Lithographic System, *IEEE Trans. Electron Dev.* **ED-22**, 385 (1975).

18. G. I. Varnell, D. V. Spicer, and A. C. Rodger, E-Beam Writing Techniques for Semiconductor Device Fabrication, *J. Vac. Sci. Technol.* **10**, 1048 (1973).

19. M. C. King, New Generation of 1 : 1 Optical Project Aligners, in Developments in Semiconductor Microlithography IV, *Proc. SPIE* **174**, 70 (1979).

20. J. M. Bruning, Optical Imaging for Microfabrication, *J. Vac. Sci. Technol.* **17**, 1147 (1980).

21. G. Bouwhuis and S. Wittekoek, Automatic Alignment System for Optical Projecting Printing, *IEEE Trans. Electron Dev.* **ED-26**, 725 (1979).

22. D. W. Widmann and H. Binder, Optical Projection Printing with Step and Repeat Capability, *IEEE Trans. Electron Dev.* **ED-22**, 467 (1975).

23. D. A. Doane, Optical Lithography in the 1 μm Limit, *Solid State Technol.* (Aug. 1980) p. 101.

24. D. L. Flowers, Lubrication in Photolithography, *J. Electrochem. Soc.* **124**, 1608 (1977).

25. F. H. Dill, Optical Lithography, *IEEE Trans. Electron Dev.* **ED-22**, 440 (1975).

26. D. W. Widman and H. Binder, Linewidth Variations in Photoresist Patterns on Profiled Surfaces, *IEEE Trans. Electron Dev.* **ED-22**, 467 (1975).

27. E. J. Walker, Reduction of Standing-Wave Effects by Post-Exposure Bake, *IEEE Trans. Electron Dev.* **ED-22**, 464 (1975).

28. O. Wada, S. Miura, H. Machidas, K. Nakai, and T. Sakurai, A New Fabrication Technique for Optoelectronic Integrated Circuits (OEIC's)—The Graded-Step Process—Applied to the Fabrication of AlGaAs/GaAs PIN/FET and PIN/Amplifier Photoreceivers, *J. Electrochem. Soc.* **132**, 1996 (1985).

29. M. Hatzakis, B. J. Canavello, and J. M. Shaw, Single Step Optical Lift-Off Process, *IBM J. Res. Dev.* **24**, 452 (1980).

30. V. Miller and H. L. Stover, Submicron Optical Lithography: I-Line Wafer Stepper and Photoresist Technology, *Solid State Technol.* (Jan. 1985) p. 127.

31. M. Rothschild and D. J. Erlich, A Review of Excimer Laser Projection Lithography, *J. Vac. Sci. Technol.* **B1**, 1 (1988).

32. L. F. Halle, Trends in Lithography, *IEEE Circuits Dev. Mag*, (Sept. 1988) p. 11.

33. B. J. Lin, Portable Conformable Mask—A Hybrid Near-Ultraviolet and Deep-Ultraviolet Patterning Technique, *Proc. SPIE* **74**, 114 (1979).

34. A. W. McCullough, E. Pavelchek, and H. Windischmann, A Novel Bilevel Resist System, *J. Vac. Sci. Technol.* **B1**, 1241 (1983).

35. M. Hatzakis, Multilayer Resist Systems for Lithography, *Solid State Technol.* (Aug. 1981) p. 74.

36. L. F. Halle, A Water, Soluble Contrast Enhancement Layer, *J. Vac. Sci. Technol.* **B3**, 323 (1985).

37. J. J. Diamond and J. R. Sheats, Simple Algebraic Description of Photoresist Exposure and Contrast Enhancement, *IEEE Electron Dev. Lett.* **EDL-7**, 383 (1986).

38. M. Endo, M. Sasago, H. Nakagawa, Y. Hirai, K. Ogawa, and T. Ishihara, Excimer Laser Lithography Using Contrast Enhancing Material, *J. Vac. Sci. Technol.* **B-6**(2), 559 (1988).

39. M. Born and E. Wolf, *Principles of Optics*, Pergamon Press, Oxford, 1975.

40. M. D. Levenson, N. S. Viswanathan, and R. A. Simpson, Improved Resolution in Photolithography with a Phase Shifting Mask, *IEEE Trans. Electron Dev.* **ED-29**, 1828 (1982).

41. B. J. Lin, The Attenuated Phase-Shifting Mask, *Solid State Technol.* (Jan. 1992) p. 42.

42. K. Kamei, H. Kawasaki, T. Chigua, T. Nakanier, T. Kawabuchi, and M. Yoshimi, Extremely Low-Noise MESFETs Fabricated by Metal-Organic Vapour Deposition, *Electron Lett.* **17**, 450 (1981).

43. S. Okazaki, F. Murai, O. Suga, H. Shirishi and S. Koibuchi, A Practical Electron Beam Direct Writing Process Technology for Submicron Device Fabrication, *J. Vac. Sci. Technol.* **B5**, 402, (1987).

44. M. B. Heritage, Electron-Projection Microfabrication System, *J. Vac. Sci. Technol.* **12**, 1135 (1975).

45. W. L. Brown, T. Venkatesan, and A. Wagner, Ion Beam Lithography, *Solid State Technol.* (Aug. 1981) p. 60.

46. G. Stengl, R. Kaitna, H. Loschner, P. Wolf, and R. Sacher, Ion Projection System for IC Production, *J. Vac. Sci. Technol.* **16**, 1883 (1979).

47. J. Melngalis, Focused Ion Beam Technology and Applications, *J. Vac. Sci. Technol.* **B5**, 459 (1987).

48. I. Banerjee and R. H. Livengood, Applications of Focused Ion Beams, *J. Electrochem. Soc.* **140**, 183 (1993).

49. D. L. Spears and H. I. Smith, High Resolution Pattern Replication Using Soft X-Rays, *Electron. Lett.* **8**, 102 (1972).

50. H. L. Stover, F. L. Hause, and D. McGreevy, X-Ray Lithography for One Micron LSI, *Solid State Technol.* (Aug. 1979) p. 95.

51. B. Yaakobi, H. Kim, J. M. Sources, H. W. Deckman and J. Dunsmuir, Submicron X-Ray Lithography Using Laser-Produced Plasmas as a Source, *Appl. Phys. Lett.* **43**, 686 (1983).

52. E. A. Crawford, A. L. Hoffman, G. F. Albrecht, and M. R. Sogard, Properties of a Laser-Plasma X-Ray Source for X-Ray Lithography, *J. Vac. Sci. Technol.* **B5**, 1575 (1987).

53. D. W. Peters and R. D. Frankel, X-Ray Lithography: The Promise of the Past and the Reality of the Present, *Solid State Technol.* (March 1989) p. 77.

54. R. W. Hill, The Future Cost of Semiconductor Lithography, *J. Vac. Sci. Technol.* **B7**, 1387 (1989).

55. D. Maydan, X-Ray Lithography for Microfabrication, *J. Vac. Sci. Technol.* **17**, 1164 (1980).

56. W. P. Buckley and G. P. Hughes, An X-Ray Lithography System, *J. Electrochem. Soc.* **128**, 1106 (1981).

57. D. B. Novotny and D. R. Ciarlo, Automated Photomask Inspection, Part 1, *Solid State Technol.* (May 1978) p. 51.

58. D. B. Novotny and D. R. Ciarlo, Automated Photomask Inspection, Part 2, *Solid State Technol.* (June 1978) p. 59.

59. J. R. Brauer, Poor QC Yields Bad IC's, *Electron. Eng.* (Aug. 1966), p, 78.

CHAPTER 11

DEVICE AND CIRCUIT FABRICATION

The preceding chapters have described a number of basic processes that are used in semiconductor fabrication technology. The emphasis has been on the theoretical and practical aspects of processes which are in use today, as well as on their limitations. In this chapter, the fabrication of complete microcircuits is treated, with the emphasis on VLSI schemes. This requires a description of some of the ways in which these processes can be combined to form devices and components, which can be interconnected to make a complete microcircuit. These combinations of processes are by no means unique, although a few basic systems have evolved over the last 30 years. However, the requirements of VLSI circuits have resulted in a shift of emphasis to schemes which allow the use of small active device regions, the delineation of fine-line multilevel patterns, and the preservation of relatively planar surfaces during the fabrication process. Technologies such as ion implantation, reactive ion etching, and microlithography are extensively used in these schemes. New sequences of fabrication steps have developed, as each of these technologies has been incorporated into the fabrication process. As a consequence, there is much room for creativity as well as for change, as these different fabrication sequences are evaluated in terms of performance and cost.

Some of the basic integrated circuit fabrication processes are described in this chapter together with their unique features [1, 2]. No attempt has been made to be all-inclusive here; instead, emphasis is placed on well-accepted approaches, and also on those that appear most promising for VLSI circuits. An extensive reference list provides further examples of the variety of fabrication processes that are currently under investigation.

Many different kinds of active and passive components are used in micro-

704

circuits today. In silicon technology, the primary active devices are the metal-oxide-semiconductor field effect transistors (MOSFET, often shortened to MOS) and the bipolar junction transistor (BJT). With GaAs, the most commonly used device is the metal semiconductor field effect transistor (MESFET) [3]. A variation of this device is the high-electron-mobility transistor (HEMT) which provides improved performance over the MESFET. The latest entry into the GaAs arena is the heterojunction bipolar transistor (HBT), which has great promise in both linear and digital circuit applications. Although other components are used, it is convenient to classify microcircuits in terms of their primary active elements. This chapter consequently focuses on the special features of microcircuits based on these devices.

11.1 ISOLATION

A microcircuit can be described as an "ensemble of active and passive components, interconnected within a monolithic block of semiconductor material." Thus the first requirement is to devise schemes for fabricating components that are electrically isolated from each other in order to allow design flexibility. A number of different approaches are in use today. Of these, the most widely used approach is based on the fact that a reverse-biased $p–n$ junction in either silicon or GaAs has an extremely low leakage current (in the picoampere range at room temperature). Thus two regions of a semiconductor are effectively isolated to direct current if they are of the opposite conductivity type and are suitably reverse-biased. The significant coupling between these regions is primarily capacitive in nature, so that its effect on the microcircuit need only be considered at high frequencies.

A second approach, which is conceptually the most straightforward, is to fabricate devices in an active film which is grown on an insulating or semi-insulating (SI) substrate, and to obtain isolation by etching moats around each device. The resulting islands are known as *mesas*, and this process is referred to as *mesa isolation*. Circuits made in silicon films on insulators, as well as those made in epitaxial GaAs on SI GaAs, are examples of this approach. A variation of this technique is to use a semiconductor which is bonded to a substrate by means of an insulating "glue." This *wafer bonding* technique can be used with both silicon and compound semiconductors.

An important variation on this technique is to fabricate devices in pockets of active semiconductor material which are directly formed *in* the insulating substrate. This technique is unique to GaAs, where crystalline SI substrates can be used as the starting material, with ion implantation or diffusion to form the regions of which active devices are made.

Oxide isolation techniques can also be used to form insulating tubs in order to isolate a number of pockets of single-crystal semiconductor. This approach has been widely adopted for the fabrication of microcircuits which must be radiation-hardened, as well as in some special high-voltage applications where the degree of isolation between components must be extremely high.

Finally, isolation can be achieved by a combination of these processes. Thus p–n junction isolation, combined with isolation produced by the localized oxidation of silicon, results in greatly reducing the physical dimensions of active devices, and aids in retaining the planarity of the silicon surface. This combination represents one of the more important approaches to the fabrication of VLSI circuits.

11.1.1 p–n Junction Isolation

The p-n junction is an inherent part of all components and devices. Thus, a resistor formed by making a diffusion into a semiconductor substrate of opposite impurity type is isolated from it as long as a reverse (or zero) bias is maintained across the junction formed in this way. Moreover, any number of resistors, placed side by side, will be isolated from the substrate as well as from each other, provided that this junction bias is maintained.

The condition of self-isolation is also met by silicon and GaAs field effect transistors, since all of their regions (source, drain, and channel) are either reverse- or zero-biased with respect to the substrate. Thus no special arrangement has to be made for these devices.

In silicon-based microcircuits, BJTs can be formed by the successive fabrication of a base and an emitter in a semiconductor substrate. All of these transistors will share a common collector, so that these devices are *not* self-isolating. Consequently, each must be put in an appropriately biased separate tub, of opposite conductivity type to the substrate.* This can be done in a number of different ways.

1. A series of n-type tubs can be diffused into a p-type substrate. A transistor is made in each tub by two additional diffusions, which form its active regions. Again, circuit arrangements must be made to ensure that these tubs are reverse- or zero-biased. This scheme is shown in Fig. 11.1a, together with typical dimensions for a conventional diffused structure in silicon.

Traditionally, triple-diffused transistors of this type have poor performance since their collector series resistance is excessively large. Moreover, they are difficult to make, since the base region is delineated by the placement of three diffusions. Thus relatively wide-base (2–3 μm) structures are necessitated for this process. Recently, however, the exploitation of new technologies involving precise ion implants, oxide isolation, and refractory contacts have allowed the production of devices whose gain-bandwidth product (GBW) is comparable to that of modern transistors [4].

---

*The exception is groups of transistors which share a common collector.

2. A second approach is to form isolated tubs by diffusing completely through the slice from both sides. This has an advantage over the last in that transistors fabricated in them can be of the conventional type, since each tub consists of uniformly doped starting material. However, even with relatively thin substrates, this diffusion step requires at least 40–50 h at 1250°C, and results in excessive contamination and consequently soft junctions.

Figure 11.1b shows a schematic cross section for this process. The first bipolar microcircuits were fabricated in this manner, so that it is of historical significance. Interestingly, this approach was the precursor to the modern BJT, which is fabricated by the double diffused epitaxial (DDE) process.

3. In the DDE process, a thin n-layer is epitaxially grown on a p-type substrate, and isolated tubs are formed in it by p^+-diffusions from the *top* surface. Subsequent base and emitter diffusions result in transistors as shown in Fig. 11.1c. Note the similarity to the configuration resulting from the two-sided diffused process of Fig. 11.1b. The single, but all important, difference is in the thickness of the active n-layer. This greatly reduces the time required to diffuse the isolation tubs, as well as the wasted area due to lateral diffusion.

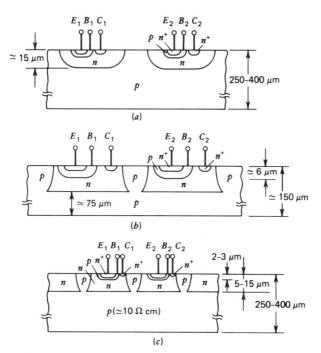

Fig. 11.1 Junction isolation schemes. (a) Triply diffused. (b) Two-sided process. (c) Double-diffused epitaxial process.

Isolated tubs can be used for fabricating both active and passive components, many of which are not self-isolating. Consequently, the DDE process represents the conventional approach to those silicon integrated circuits which are based on the use of BJT devices.

11.1.2   Mesa Isolation

Mesa isolation can be used for circuits which are fabricated on insulating or semi-insulating substrates. These include epitaxial *n*-GaAs on SI GaAs, epitaxial silicon on sapphire (SOS), silicon formed by separation using the implantation of oxygen (SIMOX), and active layers which are bonded to substrates by means of an insulating film. In all cases, isolation of active regions is obtained by masking them with photoresist, and etching the exposed semiconductor.

Wet chemical etching is often employed for this purpose. Here (100) starting material is used, in conjunction with an anisotropic etch of the type described in Chapter 9. Windows are delineated along $\langle 110 \rangle$ directions; next V-grooves are etched deep enough so as to produce electrical separation of the components. HF-based etches are sometimes used with GaAs circuits, because of the rounded etch profile obtained with them. This greatly aids in the step coverage of metal over the mesa edge.

V-groove isolation processes result in nonplanarity of the semiconductor surface, and waste of active chip area. Ion milling, followed by a light plasma etch, has been used to reduce this wasted area. Both approaches create problems along island-edge surfaces, and give rise to enhanced leakage.

With both SOS and SIMOX materials (described in Chapters 5 and 6, respectively), the quality of the starting silicon needs improvement before it is comparable to that of bulk silicon. SOS is considerably poorer than SIMOX, so that its use is restricted to applications such as radiation-hardened digital circuits, where there is no viable alternative.

SIMOX material has the potential for wide use in CMOS and bipolar silicon microcircuits [5]. At the present time, its mobility is not as high as that of bulk or epitaxial silicon, because of dislocations created during the formation of the oxide sublayer. In addition, its cost is high compared to that of epitaxial silicon.

Isolation techniques, which minimize topological variations in surface features, are greatly favored for VLSI fabrication. With silicon microcircuits, one method for achieving this goal involves the oxidation of localized areas of silicon, using Si_3N_4 as a mask. This local oxidation approach is widely used in VLSI fabrication, and is discussed more fully in Section 11.3.

11.1.2.1   Wafer Bonding

The concept of bonding a wafer to an insulating substrate presents the most direct approach for obtaining mesa isolation. The technique has the advantage that widely dissimilar materials can be used. Thus, its applications include silicon on insulators (SOI), GaAs:Si, and a variety of heteroepitaxial combinations of compound semiconductors.

The basic principle underlying this approach is that any two flat, smooth, clean, hydrophilic surfaces can be bonded at room temperature without the use of external forces [6]. This bonding occurs because of the attraction between hydroxyl (OH^-) groups, which is sufficient to cause the spontaneous formation of hydrogen bonds across the gap. Once initiated at a point, bonding spreads across the surface until the entire wafer is contacted.

A number of different approaches, based on this principle, have been demonstrated for contact bonding. In early work [7], bonding of glass-to-metal seals was carried out by establishing a high electric field across the combination, and subjecting it to high temperature and pressure. Glass was successfully bonded to aluminum, germanium, silicon, and GaAs, using this approach.

In later work, it was shown [8] that the success of this method is a critical function of the perfection of surfaces to be joined. Using clean semiconductor technology, flat oxidized silicon wafers could be bonded by the application of a moderate voltage (≈ 20 V) and at high temperature (1100-1200°C).

Wafer-to-wafer bonding can also be achieved without the application of a voltage across the system [9]. This avoids the need for electrical contacting and possible contamination. Moreover, it has the potential for use at the commercial scale in batch processing.

In this approach, two oxidized silicon wafers are pressed together and heated in an oxidizing atmosphere to 700°C. Oxide thicknesses that are used can vary from that of native oxide (≈ 20 Å) to 5000 Å. The mechanism proposed for bonding is the polymerization of Si–OH bonds between the wafer pairs, according to the following reaction

$$Si\text{—}OH + OH\text{—}Si \rightarrow H_2O + Si\text{—}O\text{—}Si \qquad (11.1)$$

Gaseous oxygen between the surfaces is consumed during heat treatment, producing a partial vacuum which assists in the bonding process.

The bond strength, as measured by its surface energy, increases rapidly once viscous flow of the SiO_2 occurs. This is seen in Fig. 11.2 for a number of wet oxides grown on 75 cm Si wafers. Finally, bonding at room temperature has also been demonstrated [10]. Here, a short heat treatment at high temperature is necessary to greatly strengthen this bond.

As seen from this figure, the bonding temperature can be greatly reduced by the use of mixed oxide glasses, in which the onset of viscous flow occurs at a lower temperature. A problem with this approach is that these oxides have considerably more surface roughness than grown oxides. As a result, some form of chemical–mechanical polishing is necessary prior to bonding between these glass surfaces, in order to achieve a void-free contact.

Wafer bonding of compound semiconductors [11] represents a powerful alternative to the heteroepitaxy of lattice-mismatched systems such as GaAs:Si

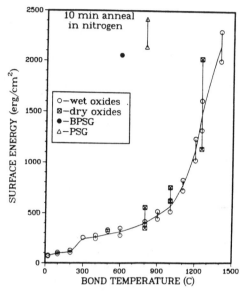

Fig. 11.2 Surface energy versus temperature for wafer bonds. From Maszara [6]. Reprinted with permission of the publisher, The Electrochemical Society, Inc.

and GaAs:InP. As before, the basic approach consists of depositing clean SiO_2 on both materials, which are then contacted and subjected to a heat treatment.

Direct bonding can also be achieved by contacting two substrates which have been freshly dipped in HF, and subjecting the combination to a 650°C heat treatment in hydrogen [12]. It has been proposed that once surface oxides have been removed, the surface becomes strongly reactive. This is followed by surface reconstruction, in order to minimize the surface free energy.

Theories for wafer bonding have not been resolved at the present time. Thus, some workers have questioned the need for having hydrophilic surfaces in order to obtain successful bonding of silicon slices [13]. In fact, some of the strongest bonds have been obtained with HF-treated hydrophobic slices, which were subsequently heat-treated at 800°C.

For all direct bonding schemes, surface defects are a major problem because they lead to void formation. Thus the avoidance of particle contamination is a requirement for successful wafer bonding, in addition to the achievement of surface flatness [14].

An important issue, common to all wafer bonding schemes, is the necessity for thinning the slice, in order that active devices and circuits can be made in it. This is especially true for MOS-based silicon circuits, where active layers of 1–2 μm are required. Here, thinning is accomplished by starting with a heavily doped p^+ substrate, on which the active layer is epitaxially grown. After bonding, this layer can be removed by a combination of mechanical grinding and selective etching. An etchant mixture of $HF : HNO_3 : CH_3COOH$ in a $1 : 3 : 8$

ratio by volume can be used for this purpose. Other etches, described in Chapter 9, have also been used. Thinning is usually not a problem for compound semi-conductors, where the bonded wafer often serves as a substrate for subsequent epitaxial growth.

11.1.3 Oxide Isolation

Isolation in silicon-based microcircuits can also be obtained by the formation of individual tubs of active material, which are lined with an oxide layer [15]. One process for accomplishing this is described here in order to illustrate the approach to this problem. Most practical techniques are based on this process, with various modifications. The process sequence* is as follows:

1. An n-type (100) silicon slice is masked with SiO_2, as shown in Fig. 11.3a, and etched so as to result in the structure of Fig. 11.3b. Anisotropic etching, with windows oriented along $\langle 110 \rangle$ directions, are used to delineate the V-shaped grooves.

2. After removal of the mask, an n^+-layer is diffused across the entire slice (see Fig. 11.3c). This layer is heavily doped so as to provide a low-resistance ohmic contact to this region.

3. A thermal oxide is now grown across the wafer. This oxide becomes the isolation between the single-crystal and the subsequent polycrystalline silicon, which is next deposited (Fig. 11.3d) to a thickness of 250–500 μm. Consider-able stress, accompanied by deformation of the slice, can result from this step. Sometimes the insulating support can be made with a sandwich of alternating layers of polysilicon and silicon dioxide to control this stress [16].

4. The single-crystal side of the slice is now thinned, resulting in the struc-ture shown in Fig. 11.3e. The resulting slice consists of a series of tubs of single-crystal silicon, isolated from each other by a layer of SiO_2. Various active com-ponents may be fabricated within these tubs as desired. Each tub is lined with an n^+-layer which provides a collector connection with a low parasitic resistance, and is essential if high-performance BJTs are fabricated in them. Connection to this layer is also made if the tub is to be used for resistor placement.

This oxide isolation process results in near-perfect isolation between these single-crystal tubs. Its main disadvantage is the same as that for wafer-bonded substrates; that is, considerable thinning of the single-crystal silicon is required to form the active layer. The technique described for wafer-bonded slices; that is, mechanical polishing in the early stages, followed by chemical etching with an etch stop layer, is commonly used. An alternative approach is to use

---

*Details of individual process steps have been described in previous chapters.

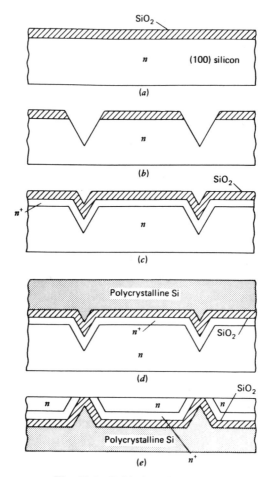

Fig. 11.3 Oxide isolation scheme.

an electrochemical etch which will stop at the appropriately doped epitaxial layer [17].

11.2 SELF-ALIGNMENT

Self-alignment techniques are important in VLSI fabrication technology since they reduce the difficulties of precise alignment, and allow considerable shrinkage of the device size. One example of this technique is the use of a washout emitter in a BJT, to form an ohmic contact to this region. This is shown in Fig. 11.4, where an emitter contact of this type is compared to one of the same contact area, made by a separate alignment step. The advantages of this technique are that it reduces the area of the emitter region by a factor of nine

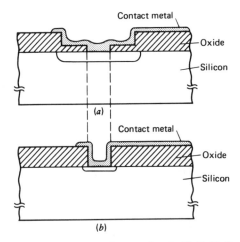

Fig. 11.4 Emitter contact schemes. (*a*) Conventional. (*b*) Self-aligned washout emitter.

to ten, and eliminates an additional mask and alignment step. Thus it is used in high-speed device fabrication technology. For this process, however, considerable care must be taken to prevent exposing the edge of the junction to avoid excessive leakage, or failure due to short circuits caused by diffusion pipes.

Many self-aligned techniques are based on the use of masking materials which can withstand high-temperature processing steps. One such technique, which is widely used in the fabrication of polysilicon gate MOS devices, is outlined in Section 8.1.5. Here a polysilicon mask defines the gate region during the source and drain diffusions. This serves to form these regions so that the gate oxide overlaps them (Fig. 8.3), even when extremely shallow diffusions are used. In addition, the gate becomes heavily doped, thus avoiding an additional process step.

Si_3N_4 can also be used as a mask region in these structures, since it can withstand subsequent high-temperature steps that are required for post-implantation anneals in both silicon and GaAs microcircuits. A unique feature of this material, in addition to its excellent masking quality, is that it can withstand an oxidizing environment without deterioration. It forms the basis for the local oxidation process which is outlined in the following section.

Refractory silicides can be used in a number of self-aligned process applications. Thus, their placement over a polysilicon gate can reduce its sheet resistance by an order of magnitude. They are also used for defining source and drain regions, in order to fully utilize the available contact area.

Doped oxides, which can be used as diffusion sources, have also been used in self-alignment schemes for the fabrication of MOS devices [18]. Using this approach, both phosphorus- and boron-doped oxide patterns can be defined in the same circuit, so that complementary transistors can be fabricated in a single operation.

11.3   LOCAL OXIDATION

This technique is specific to silicon microcircuits. Here the localized oxidation of silicon (LOCOS) serves as the starting point in a technology that shrinks device size and improves device performance [19]. In addition, variations of this approach result in preserving the planarity of the overall microcircuit, thus greatly reducing problems associated with the step coverage of metal. It is understandable, therefore, that LOCOS and its variations are widely used in VLSI fabrication schemes.

The LOCOS process is based on the fact that Si_3N_4 can be used as a mask against thermal oxidation. Typically, its oxidation rate in steam (95°C H_2O) is about 30 times slower than that of silicon. In addition, H_3PO_4 etches are available which will remove Si_3N_4 but not attack SiO_2.

Figure 11.5 shows two basic LOCOS structures. In both, a layer of Si_3N_4 is deposited on the silicon substrate and patterned. The approaches differ in the pre-etch of the silicon that is done prior to the local oxidation (Fig. 11.5b). The resulting structures show that, in each case, the oxide is countersunk in the silicon. As a result, the area of unoxidized silicon is *smaller* than the original nitride mask. This unoxidized silicon is used to form an active device. Its reduced area is a significant advantage over regions formed by diffusion, which are *larger* than the window that is cut in the mask.

The displacement of the plane of the oxide from the silicon surface is controlled by the amount of pre-etch prior to oxidation. (Note that the thickness of the grown silicon dioxide layer is 2.27 times than of the consumed silicon.) Ideally, the most planar situation—namely, the one with the fully recessed oxide (Fig. 11.5b)—is the best for the subsequent placement of metallization patterns.

An important feature of the LOCOS process is that the nitride film can be used as a mask against impurity diffusion, as well as oxidation. This is seen in Fig. 11.6a, where a p^+-diffusion is first made using the nitride film as a diffusion mask, and followed by a deep local oxidation. A variation of this process results in the configuration of Fig. 11.6b. Both of these schemes are useful in the placement of field inversion channel stops between devices in an MOS-based microcircuit. Figure 11.6 illustrates the situation for n-channel devices.

BJTs of equal emitter area, built by conventional processes and by local oxidation, are compared in Fig. 11.7a and 11.7b. The main differences in these structures are seen in the collector–base junction. The device of Fig. 11.7b has a smaller area resulting in smaller size and less parasitic capacitance. Moreover, it has less junction curvature, so that its breakdown voltage is higher. Thus the locally oxidized structure is not only smaller, but also superior in its electrical characteristics.

Variations of this technique are possible. Thus double nitride masking and oxidation steps can be used to form the BJT of Fig. 11.7c, where both the emitter–base and collector–base junctions can be made between the countersunk oxide, with a reduction of the emitter capacitance as well.

An important advantage of local oxidation is that any region, bounded by

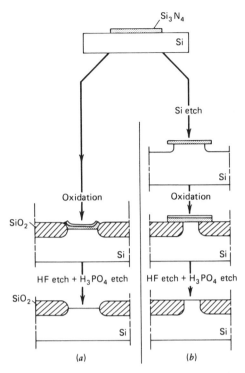

Fig. 11.5 Locally oxidized structures. (*a*) Partially recessed. (*b*) Fully recessed. From Appels et al. [19]. With permission from *Philips Research Reports.*

locally oxidized regions, can have contact made to it without the need for an additional mask and alignment step. This self-alignment feature greatly reduces the minimum size of any exposed region. It provides the same reduction of size as the washout emitter of Fig. 11.4*b*, but is less prone to leakage, since the junction edge is fully protected by the recessed local oxide. Figure 11.8 shows this contact scheme, as well as the washout emitter, by way of comparison. This technique, combined with the use of doped polysilicon for resistors and interconnections, has been exploited in the fabrication of high-speed emitter-coupled logic gates [20]. Integrated Schottky logic has also been developed using LOCOS techniques [21].

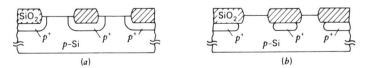

Fig. 11.6 Combination of impurity diffusion with local oxidation.

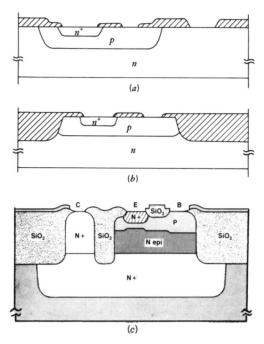

Fig. 11.7 Bipolar transistor. (*a*) Conventional. (*b*) LOCOS. From Appels et al. [19]. With permission from *Philips Research Reports*. (*c*) Doubly oxidized. From Lee and Bass [112], ©1982. The Institute of Electrical and Electronic Engineers.

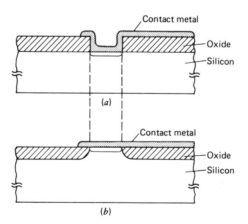

Fig. 11.8 Self-aligned contacting schemes. (*a*) Washout emitter. (*b*) Locally oxidized emitter.

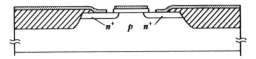

Fig. 11.9 MOS device made by local oxidation. From Appels et al. [19]. With permission from *Philips Research Reports.*

Isolation between MOS devices in a microcircuit is conventionally obtained by the growth of a field oxide, which is usually ≥ 1 μm in thickness. Thus step coverage of the interconnection metal is a particularly severe problem. The use of recessed oxides presents an especially powerful approach in this situation, and is extensively used for this reason [22, 23]. Figure 11.9 shows one version of an MOS device formed by this technique.

There are a number of technological problems which arise during the application of local oxidation technology. The first of these is that the oxidation of silicon proceeds slightly under the nitride as well. A second problem comes about because of the large mismatch in the thermal expansion coefficients of Si_3N_4 and silicon, which results in damage to the semiconductor during local oxidation. It has been shown that this damage can be greatly reduced by growing a thin pad layer of SiO_2 prior to placement of the Si_3N_4 mask [24]. Typically, a 100- to 200-Å thickness has been found sufficient for this purpose. Unfortunately, this greatly enhances the penetration of oxide under the nitride-masked region, resulting in structures of the type shown in Fig. 11.10. These oxide configurations, referred to as "bird beaks," are shown for three different conditions of etching prior to local oxidation [25]. For all cases, a 4500-Å oxide was grown with a pad thickness of 100 Å and a nitride thickness of 4500 Å. Note that the bird's beak becomes more prominent as the oxide is made more recessed.

The thermal oxidation temperature is also an important process parameter, with less beak formation at lower temperatures. Thus there is considerable

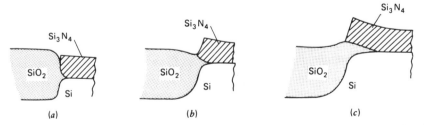

Fig. 11.10 Development of bird beak and crest. From Bassous et al. [25]. Reprinted with permission of the publisher, The Electrochemical Society, Inc. (*a*) No pre-etch. (*b*) 1000-Å pre-etch. (*c*) 2000-Å pre-etch (fully recessed).

emphasis on the use of low-temperature processes for local oxidation. Oxidation in high-pressure steam has been extensively studied [26, 27] for this reason. Here it has been shown that Si_3N_4 can be used as a mask up to at least 90 atm.

The pad oxide is damaged during the local wet oxidation process, and cannot be used for the gate in the subsequent MOS device. It has been postulated that the reaction of Si_3N_4 with water produces NH_3, which penetrates the underlying pad oxide and damages it [28]. It is common practice to etch off this oxide, and replace it with a freshly grown gate oxide.

The preparation of a recessed oxide surface for an MOS transistor is shown in the process sequence of Fig. 11.11. Here, a 100- to 200-Å thick oxide is thermally grown on the silicon slice to serve as the pad oxide. This is covered with a 2000-Å thick layer of Si_3N_4, followed by a 500- to 750 Å-thick SiO_2 layer and a positive photoresist film. At this point, the substrate is as shown in Fig. 11.11*a*.

Next, the photoresist is patterned, and used as a mask for the CVD oxide layer, which is delineated by means of BHF. This oxide serves as a mask to

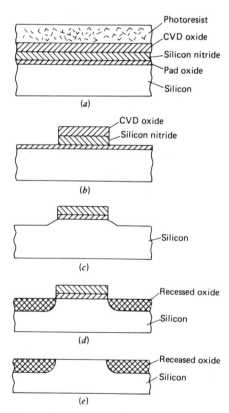

Fig. 11.11 Processing sequence for a fully recessed local oxide.

delineate the underlying nitride, which is etched in hot (180°C) H_3PO_4. Note that photoresist masking cannot be used in this process. The resulting structure is shown in Fig. 11.11b.

Both the CVD oxide and the exposed pad oxide are now removed in BHF. Next, anisotropic etching of the silicon is carried out in an etch of the type described in Chapter 9. Typically, about 2000 Å is removed, as shown in Fig. 11.11c.

A channel stop implant is now placed in the region which will be oxidized. Typically, boron and arsenic implants are used for n- and p-channel devices, respectively. This implant should be deep to prevent its consumption by the local oxidation step. Next follows a 4400-Å field oxide growth, which results in a fully recessed structure of Fig. 11.11d. Steam oxidation is commonly used at 1100°C for this purpose.

Some encroachment of the channel stop implant into the active region occurs during oxidation because of stress-enhanced diffusion effects. Increasingly, high-pressure steam oxidation, at 900°C, is used in order to minimize this encroachment. One alternative approach [29] is to co-implant germanium with the channel stop implant species (boron, in the case of an n-channel device). This alters the point defect population in the silicon, and reduces the diffusivity of boron.

A thin film of oxide ($\simeq$150 Å), which grows on the nitride during this field oxidation step, is removed in BHF. Next, the nitride pattern is removed in hot H_3PO_4. Removal of the pad oxide in BHF results in a wafer (see Fig. 11.11e) which is now ready for MOS device fabrication.

The above process is by no means exclusive, and many modifications can be made by way of improvement. One such uses sidewall oxidation to control the profile of the oxide features [30]. In another scheme a portion of the field oxide is etched back [31], in order to provide a more planar surface. Yet another scheme involves the use of a second nitride layer after substrate etching [32] to result in two thicknesses of local oxide.

The greatest impact of local oxidation technology has been in VLSI applications of MOS-based microcircuits. This is because all of the advantages of this technology can be simultaneously exploited here. These include: recessed oxides to provide a planar surface; field inversion channel stops by diffusion or ion implantation, followed by local oxidation; and extensive use of self-aligned techniques, using Si_3N_4 as a mask for ion implantation, diffusion, and local oxidation.

11.3.1 Trench Techniques

Isolation can also be achieved by the placement of a trench around the active region. This is usually formed by an anisotropic wet chemical etch, or by RIE. Of these approaches, RIE is preferred when deep trenches are required. Once cut, the trenches are filled, and the surface of the filler etched back to provide a planar structure.

The advantages of using a trench over local oxidation is that it consumes less space, and has no bird's beak problem. Moreover, trenches can be used in a number of additional applications. Thus, they can provide isolation between *p*- and *n*-channel devices to prevent latch-up in CMOS circuits. This replaces the conventional guard ring, and greatly improves the packing density of the microcircuit components. Trenches can also be used to form storage capacitors in high-density ($\geq$ 4 Mbit) DRAM applications [33].

There are a number of issues which must be addressed in order to implement this process [34]. Thus, the walls of the trench must be smoothly tapered and nonvertical with a rounded bottom, in order that it can be filled without voids. RIE gas mixtures such as $CHCl_3/O_2/N_2$ are often used since they coat the sidewalls with a "passivating" polymeric film and allow steep etching profiles. Here, the O/N ratio is used to control the trench angle, which is typically held around 85°.

The etching of silicon to form the trench leaves its surface exposed. This presents a serious problem, especially with *p*-type material, which can be inverted at this juncture. Inversion can be prevented by cleaning the trench and thermally oxidizing its surface. The growth of this oxide results in the creation of stress in the silicon, and the formation of dislocations. This is illustrated in Fig. 11.12*a*, where three different trench profiles were etched [35], and subse-

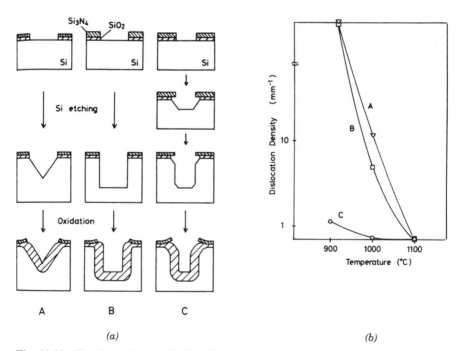

(a)   (b)

Fig. 11.12 Trench profiles and dislocation densities. From Tamaki et al. [35]. Reprinted with permission of the publisher, The Electrochemical Society, Inc.

quently oxidized to form a 1-μm film. Figure 11.12b shows the dislocation density in the silicon, resulting from this oxidation step, as a function of the trench profile and the oxidation temperature. Here, it is seen that a tapered trench, with a flared mouth, is highly desirable to minimize the dislocation density.

An alternative approach is to clean the trench, and deposit an appropriately doped oxide, such as BSG or PSG, so as to form a thin p^+ (or n^+) layer during subsequent heat treatment. In addition to preventing surface inversion, damage to the silicon is reduced because these glasses are closer matched in their TCEs to the silicon. The trench must be filled after thermal oxidation. This is typically done by the LPCVD of polysilicon at 600°C [36]. Next, the surface is planarized after the trench has been filled.

It is necessary to fill the trench with heavily doped material if it is to be used as one plate of a storage capacitor in a memory cell. This creates additional problems, since the dopant affects the growth of polysilicon (see Chapter 8). Selective epitaxy has been used to avoid this problem [37], since it is not sensitive to dopants, and is void-free because of its single-crystal character. However, this requires that the thermal oxide be removed from the bottom of the well, so that seeding of the epitaxial layer can be initiated. Here, a p–n junction must be formed for isolation purposes [38].

11.4 PLANARIZATION

The starting material for a microcircuit is ideally flat. However, the process of circuit fabrication, which includes the growth or deposition of both insulating and conducting films, results in an increasingly nonplanar structure as the wafer proceeds to the metallization stage. For example, the gate oxide of an MOS transistor may be 100–250 Å thick, whereas the thickness of a neighboring field oxide may be as much as 1 μm.

The trend towards as many as four levels of metallization has aggravated this problem. Moreover, as lateral dimensions are further shrunk, it is necessary to increase metallization thickness in order to operate the circuit at usable levels of current. In summary, modern microcircuits are far from planar, and departures from planarity will increase with the thrust toward finer-line geometries and multiple levels of metallization.

This loss of planarity presents two problems. The first is one of maintaining step coverage without breaks in the continuity of fine lines. This is a local issue, with each process tailored to minimize this problem. The second is of a more global character: Loss of planarity eventually results in the inability to image fine-line patterns over the wafer. Table 11.1 at the end of this chapter lists the depth of focus which is necessary for a given resolution using different optical pattern transfer systems, and indicates the magnitude of this problem [39]. As seen from this table, the ability to operate within the depth of focus of a particular lens, over its image field ($\approx$20 mm $\times$ 20 mm), is the overriding factor in the success of all VLSI schemes.

Techniques for flattening steps in a microcircuit are thus increasingly important. These are commonly referred to as *planarization* techniques. Here, both local and global planarization must be considered. Local planarization consists of smoothing steps in topology which occur over short distances. Examples are (a) steps in an MOS transistor oxide and (b) trenches in a dynamic memory. Global planarization, on the other hand, is related to long-range variations in the topology, specifically changes which occur over the entire image field of the stepper.

The local oxidation technique, described in the previous section, results in considerable planarization of itself, especially if the silicon is pre-etched so that the final oxide is fully recessed. Another approach, illustrated in Fig. 11.13a, shows the etch profile obtained with an isotropic etch in an amorphous material such as SiO_2. Here equal etch rates in the lateral and vertical directions result in a circular contour. Taper control, in order to make the step in a cut more gradual, can be achieved by making the etch rate faster at the surface by means of ion implantation, followed by subsequent etching [40]. An inactive ion such as argon is commonly used to damage the surface prior to the photoresist step. The dose is established experimentally so that the resultant cut has a tapered wall as in Fig. 11.13b. This technique has been used successfully for tapering sidewalls in SiO_2, Si_3N_4, and polysilicon windows. One problem here is caused by the fact that photoresist adhesion to an ion-damaged surface is relatively poor, so that an adhesion promoter such as hexamethyldisilazane (HMDS) must be used.

Taper control by ion implantation damage cannot be used with MOS-based microcircuits, since it results in both fixed charge states in the oxide as well as interface traps at the $Si–SiO_2$ interface. An alternative approach involves [41] the use of a two-layer sandwich whose upper layer is more readily etched. This technique is most suited for tapering cuts in a thermally grown oxide. Here a thin layer of CVD silicon dioxide, which etches more rapidly, is used as the taper control layer on which the photoresist is placed. Figure 11.13c shows details of the tapering process using this layer.

Taper control can also be achieved by the use of deposited films of low-melting-point glass, which can be reflowed by heating to elevated temperatures. Details of their flow characteristics are covered in Section 8.3, and will be summarized here. PSG films, grown by the simultaneous oxidation of SiH_4 and PH_3 with about 8 wt.% phosphorus, can be reflowed at 850–900°C to give smooth window edges [42]. However, films grown by the oxidation of TEOS and TMP can be reflowed at about 50°C lower temperature. Finally, films of BPSG, grown by the oxidation of B_2H_6, PH_3 and SiH_4, have been shown to be superior to all of the above, with an additional 50–75°C lowering of the reflow temperature and a reduction in the phosphorus content.

Rapid thermal annealing can be used with both PSG and BPSG films [43]. Typical schedules are from 10 to 60 s, with full planarization occurring in the 1000–1200°C range [44]. This is accomplished by structural changes in the material [45], as evidenced by a continuous modification of its optical properties.

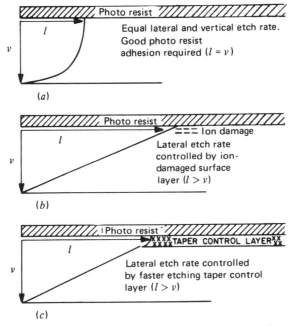

Fig. 11.13 Contours obtained with chemical etch. (*a*) Isotropic. (*b*, *c*) Tapered. From White [41]. Reprinted with permission of the publisher, The Electrochemical Society, Inc.

The reflow characteristics of both PSG and BPSG are greatly influenced by the ambient conditions under which this process is carried out. This is true for conventional furnace anneals as well as for RTA. In both cases, reflow is greatly enhanced [46] when the process is carried out in a steam ambient, as opposed to an inert (N_2 or Ar) ambient. This enhancement is only found in RTA processing if the time is in excess of 20 s, so that it is probably results from the indiffusion of H_2O into the glass network. It has been proposed that this indiffusion causes the formation of hydroxyl groups in the glass, weakening its structure and thus making it susceptible to viscous flow. Interestingly, no water-related problems have been observed in films which have been reflowed in this manner, indicating an absence of molecular H_2O in them.

Planarization can also be accomplished by sputtering SiO_2 with the application of a bias on the target. This promotes the resputtering of the deposited film, and tends to smooth it out [47]. Metal films, such as aluminum and gold, can also be planarized using a pulsed heat source, such as a dye laser [48].

Both photoresists and polyimides can be used for planarization [49]. Polyimide films generally consist of resins such as dimethylformamide in a solvent base, and have excellent thermal and chemical resistance. They have a lower dielectric constant than SiO_2, and can be used for interlayer dielectrics. They are usually cured around 300–400°C, can withstand thermal treatments up to 450–500°C after curing, and can be applied in one or more layers by conventional photoresist spinning techniques. Upon application, they flow around projections and steps on the surface, resulting in the composite structure of Fig. 11.14. The film shape becomes permanent upon curing at around 400–500°C.

A comprehensive study [50] has shown that, in the absence of thermal flow, the viscosity of the material is the most important parameter which determines the effectiveness of the film for planarization, with low-viscosity materials being the best. The materials studied here, ranked in descending order of effectiveness, were novolac resins, PMMA, and polyimides.

With these materials, planarization occurs to a small degree during the spin-on phase, but mostly during the post-spin rest period, followed by a thermal treatment at 100–250°C. This process can be carried out before resist application, or as part of the resist application step itself. However, the thermal stability of polyimides permits them to be used in situations where they are followed by subsequent thermal processing. This study also showed that a second coating followed closely the topology of the first coat [51]. Thus, only a slight improvement resulted from the use of multiple coatings.

The planarizing properties are also a function of the height of the features, as well as the spacing between them [52]. Narrow, close-spaced features are most readily planarized, whereas wide features, spaced apart, are the most difficult. In general, films were conformal when the gap separating features was in excess of 50 μm.

In addition to their lower viscosity, photoresists are also subject to thermal flow if baked to 200–400°C, which is beyond their glass-transition point. This property leads to their wide use in situations where they serve in intermediate process steps. Polyimides, on the other hand, shrink by large amounts during curing. This tends to degrade their planarization properties.

New resins and low-viscosity liquid epoxies are also being considered for planarization. Here, the goal is to develop materials which soften and flow at low temperatures (which improves their leveling properties), and harden at higher temperatures.

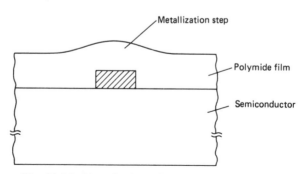

Fig. 11.14 Planarization effect with polyimide film.

Spin-on-glass (SOG) can be used for planarization. Films of this type, upon heat treatment, become SiO_2-like in character, and can be used instead of SiO_2 as interlayer dielectrics in microcircuits. These are also applied by conventional photoresist spinning techniques.

SOG materials usually consist of polysiloxane polymers and can be subdivided into two groups [53]. The first group is relatively stable, and undergoes polymerization during baking. These materials are relatively easy to use in films of thickness greater than 1 μm, and behave much like silicone rubber. Unfortunately, they have a tendency to crack in the harsh environment associated with subsequent RIE processes. Pre-treatments involving low-temperature bake, and low-power plasma treatments, have been found in some cases to inhibit this cracking.

The second group of SOG materials undergoes a reaction during the bake step, converting them to silicates. These are much more tolerant to subsequent plasma treatments; unfortunately, they cannot be applied in thick layers ($\geq 0.25 \mu$m) and are not useful as planarizing agents for modern VLSI applications [54].

Planarization can be carried out by a technique known as *sacrificial etchback*. Steps in this process sequence are illustrated schematically in Fig. 11.15 for a series of metallic projections on an otherwise flat surface [55]. A CVD-grown layer of PSG results in a surface which follows conformally the abrupt contour of the projections. A resist is used for the planarization layer. After planarization, the sandwich is etched in a CF_4–O_2 plasma whose parameters are controlled so as to obtain a 1 : 1 etch ratio for the PSG and the planarizing layer. Figure 11.15c shows the final structure, before contact is made to the tall metallic projections in it.

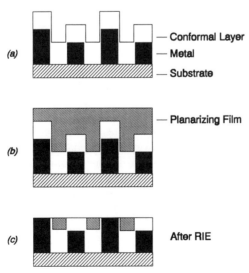

Fig. 11.15 Planarization by plasma etching.

11.4.1 Chemical–Mechanical Polishing

At the present time, chemical–mechanical polishing (CMP) appears to be the only route for achieving truly global planarization over the entire substrate, for microcircuits with minimum feature sizes below 0.35 μm. This approach is conceptually simple; moreover, it requires the removal of only a thin layer of material ($\approx$1 μm), so that problems of maintaining alignment are not as rigid as in the case of the SOI schemes outlined earlier. However, the process requires tight control of both the temperature and the pH of the polishing medium in order to maintain a consistent removal rate during the polishing step, as well as from run to run.

With careful control of these parameters, and choice of polishing solution, it is possible to obtain an etch stop action with CMP [56]. Thus, planarization of structures such as the one shown in Fig. 11.15 is possible on a global basis, when the final RIE step is replaced by CMP.

The CMP of silica and silicaeous materials such as BSG, PSG, and BPSG requires the use [57, 58] of an abrasive slurry in a solution with hydroxyl groups.* Colloidal silica, in a KOH solution, is commonly used for this purpose, although sodium- and potassium-free slurries are also available. A polishing rate of about 750 Å/min, dropping off to 650 Å/min after 3 h of polishing time, is typical during this operation. Elaborate rinsing procedures are necessary for removal of particulate residues from the slurry and the polishing pads, after the CMP operation is completed [59].

Degradation of the microcircuit can be caused by CMP of thin layers which play an active role in device operation. In recent work, the breakdown voltage of 1000 Å gate oxides was reduced from 9 MV/cm to 6 MV/cm as a result of this process. Shorts in the oxide can also be created, probably by local penetration of polishing solutions. Thus, care should be exercised to confine CMP operations to capping layers, and to regions where these effects are of no consequence on subsequent circuit operation.

11.5 METALLIZATION

The process technology for depositing metal films has been described in Section 8.3, and photographic processes for forming interconnection patterns have been covered in Chapter 10. Here, however, we emphasize the special problems of VLSI circuits, and the approaches used for their solution.

First, area utilization of the silicon surface, as well as the ability to lay out a microcircuit, becomes increasingly difficult as circuits become more complex. Both of these problems can be alleviated by the use of multilevel metallization schemes. This not only provides additional surface area, but it also aids in solving many of the topological problems of interconnection, since crossovers

---

*In contrast, the polishing rates for these materials in kerosene or oil-based slurries is negligible.

can be made by connections between layers. Next, planarization techniques must be used to minimize the problems of step coverage. Finally, special photolithographic techniques, which allow the delineation of fine-line patterns, must be used in VLSI circuits. Some of the ways of addressing these problems are now described.

Multilayer metallization is commonplace in VLSI fabrication. Three layers of metal are often used, and some schemes call for even more. All these rely upon the ability to deposit insulating films at low temperature, which serve as barriers between these layers.

The basic requirements of a suitable intermediate layer are many. Thus, it must be capable of deposition with low stress and at low temperature, in order to prevent damage to the underlying metallization. At the same time, it must be sufficiently stable so that additional levels of metallization can be placed upon it, without damage. Next, it should have a reasonably low dielectric constant, and a high electric breakdown electric field strength, in addition to being free from pinholes. Finally, it should be capable of being patterned by conventional photolithography.

Both SiO_2 and BPSG are commonly used for these layers; however, BPSG is preferred since it can be deposited with less stress than undoped films. Recently, spin-on glasses have been considered for this application.

Films of polyimide have also been used. These are laid by spinning, and can be readily etched by conventional photoetching techniques, using hydrazine as an etchant. RIE in an oxygen plasma is also used for this purpose. After curing, these films are stable [60] to temperatures as high as 450–500°C. Electrically, polyimide films are comparable to SiO_2 in their insulating capability. Moreover, they can be made 2–4 μm thick with much lower residual stress, and can provide a considerable degree of planarization.

Sputtered films of SiO_2 and Si_3N_4 have also been used for these intermediate layers. Excellent-quality films can be deposited if careful attention is paid to ensure a clean sputtering system, using oil-free pumping [61]. CVD films, especially those made by the oxidation of TEOS, are particularly suited because of their conformal coverage behavior.

Contacting between aluminum metal layers and silicon is relatively easy, even in the presence of residual oxides. Here the contact system consists of Al–SiO_2–Si. Heat treatment results in breaking down the residual SiO_2 to form an excellent microalloyed contact. With interconnections between layers, however, the problem is considerably more difficult. Here the contact system consists of Al–Al_2O_3–Al, so that great care must be taken to remove the native oxide before the upper layer is deposited. Sputtering techniques are used here, because of the ability to clean the metal area to be contacted by back-sputtering prior to deposition of the upper layer. Here, too, a sputtering system that is oil-free is essential for the formation of these contacts with high repeatability.

The lift-off technique, described in Section 10.3.1.1, is a powerful method for defining high-resolution metallization patterns for VLSI circuits. The use of two-layer photoresist films to provide an overhang is common practice [62], although more complex schemes are sometimes used.

Lift-off techniques rely upon the undercutting effect of wet chemical etches. Consequently, they are difficult to implement with RIE processes which are becoming accepted in VLSI fabrication, unless the process is designed to provide the necessary undercutting. Polyimide films have been successfully used in this application [63]. Here a molybdenum pattern delineation mask is often used, since it allows side etching of the polyimide film during RIE in an oxygen plasma.

11.6   GETTERING

The subject of gettering has been covered in detail in Chapter 1. Here, we shall review this material briefly, and emphasize additional aspects which have become increasingly important with increasing device density.

It goes without saying that successful VLSI fabrication requires the use of starting wafers of optimum quality. Nevertheless, even if fabrication is carried out under completely contamination-free conditions, a number of process-induced defects will eventually limit the circuit yield. The aim of gettering techniques is to remove or reduce these defects (as well as any defects caused by contamination), especially those that are formed in the vicinity of the active devices. Some of these processes remove defects completely; yet others move the defects away from the active region to other areas (the back face of the substrate, for example) where they have no effect.

Metallic impurities are often electronically active in both silicon and GaAs, and reduce device lifetime and increase the leakage current of reverse biased junctions. Often they degrade the reverse breakdown characteristic, resulting in "soft" junctions. Elimination or reduction of these impurities is thus important in VLSI applications [64].

Oxygen and carbon are present in both starting silicon and GaAs. These impurities usually form complexes in the lattice with the host atoms (silicon, gallium, and arsenic) as well as with impurities in the semiconductor. This, in turn, increases the possibility of forming other crystal defects by pinning dislocations at these sites.

Dislocations and stacking faults are also present in the semiconductor. Dislocations are primarily caused by mechanical stress which is induced during high concentration diffusions, and by rapid cool-down of slices after high-temperature processing. Stacking faults, on the other hand, come about because of imperfections during epitaxial layer growth, especially at the layer–substrate interface. In silicon, these can also be created during oxidation (i.e., oxidation-induced stacking faults). The primary effect of stacking faults is that they act as precipitation sites for metallic impurities, which cluster around them in order to relieve the stress in the lattice. This, in turn, results in degraded junction characteristics, as well as in reduced lifetime.

A number of different gettering techniques are used for the removal or reduction of these defects [65]. Some of these are considered here.

Intrinsic Gettering. Silicon wafers, especially those which are grown by the Czochralski process, contain large amounts of dissolved oxygen. During device processing, this oxygen tends to precipitate and create regions of stress, which serve as gettering sites for fast-moving impurities [66].

Halogenic Dopant Sources. The use of these sources during diffusion serves to convert metallic impurities (which are usually fast diffusers) into their volatile halides, which then leave the system by incorporation into the gas stream. These dopants must be used with care, however, since excessive amounts can result in pitting of the semiconductor surface.

Halogenic Oxidation. This process is highly effective in preventing the formation of oxygen-induced stacking faults in silicon. It is also effective in removing metallic impurities from the bulk crystal in much the same way as halogenic dopant sources during diffusion. It results in a significant improvement in the electronic properties of the oxide as well, so that it is routinely used in MOS circuit fabrication.

Mechanical Damage. Sandblasting, physical abrasion, and ion implantation can be used to create a layer of damage on the back surface of the slice. During subsequent high-temperature processing, this damaged region acts as a sink for the removal of metallic impurities as well as stacking faults.

Controlled amounts of damage can also be introduced by depositing a layer of Si_3N_4 on the back surface of the slice [67]. Here the high interfacial stress created by the deposition process results in the formation of a dislocation network which acts as a getter during subsequent processing.

Glassy Layers. Both BSG and PSG, in contact with silicon at elevated temperatures, have been found to be extremely effective in reducing metallic impurities, stacking faults, and dislocations in silicon slices [68]. The effectiveness of these layers comes about because, during heat treatment, they create a diffusion-induced stress which is well ahead of the impurity diffusion front [69]. In addition, they provide a viscous layer in which it is possible for some of the metallic impurities to become entrapped. Often these glassy layers are deposited from a halogenic source, and thus they have some chlorine or bromine incorporated in them. This further provides a gettering role by chemical conversion of the impurity to its halide.

Chemical Cleaning. One of the more powerful methods for reducing process contamination is to use a suitable chemical cleaning process before subjecting the wafer to heat treatment. The "RCA Clean" process and its many variants, described in Chapter 9, are routinely used for this purpose with both silicon and GaAs microcircuits, after each photolithographic sequence of steps. Additionally, a light etch, in a formulation which includes large amounts of deionized water, is often used to flush away surface contaminants prior to thermal processing.

All of the above methods can be used singly during VLSI circuit fabrication, and are used at different points in the process cycle. Often, however, combinations of these techniques are found to give improved results.

11.7 MOS-BASED SILICON MICROCIRCUITS

The MOS transistor is the most promising active component for silicon VLSI circuits at the present time. There are a number of reasons for this choice. First, it is self-isolating, so that devices can be placed side by side on a chip without the need for providing isolation tubs. As a result, it is considerably smaller than its bipolar counterpart, and requires less processing steps. Next, it can be made in bulk silicon, thus avoiding the costly step of epitaxial growth. However, epitaxial structures are increasingly used in high-density applications, to minimize latch-up problems caused by device interactions through a common substrate.

The MOS transistor is a high-impedance device, so that its power dissipation is low. Furthermore, it can be made as an enhancement-mode structure, which dissipates no power in its standby condition. As a result, it is ideally suited for VLSI digital circuits. In addition, its dynamic storage capability can be exploited to minimize the number of devices per memory cell in digital applications. On the other hand, the BJT is faster, and can be made in a variety of special configurations. Recently, the technology has been developed to integrate both bipolar and MOS devices in order to take advantage of their unique characteristics. This technology is described in Section 11.8.8.

Perhaps the most severe requirement on digital integrated circuits using MOS transistors is the necessity of maintaining the threshold voltage of devices to within a few tens of millivolts. This fact, coupled with the need for operating at low voltages in order to minimize the power dissipation, has resulted in an emphasis on self-aligned structures with refractory gates. Polysilicon is the material of choice for this application because of its inherent purity, and its ability to be patterned by conventional microcircuit processes. This is often used in conjunction with silicides, to reduce the sheet resistance of the gate. Silicides are also used for self-aligned connections to the source and drain, to obtain maximum utilization of the available contact area.

A few different basic MOS structures are now described in order to outline their fabrication principles. Many modifications to these processes are continually being developed, so that attempts to be all inclusive would be fruitless. Structures based on DMOS and VMOS devices are not considered here.

The DMOS transistor is ideally suited for discrete power devices, but has highly asymmetrical characteristics which make it less useful in integrated circuit applications. The use of VMOS transistors results in exceptionally good area utilization, but suffers from the disadvantage that all devices must share a common source terminal. Fabrication sequences for these devices have been described elsewhere [2].

11.7.1 The *p*-Channel Transistor

The *p*-channel transistor, based on aluminum-gate technology, was the earliest practical MOS device structure. It is relatively easy to make as an enhancement-mode device, which is preferred choice for digital applications since it minimizes the standby power dissipation. Its process sequence begins with a slice of *n*-type silicon [2]. Typical starting resistivities of 2–5 Ω-cm are commonly used; alternatively, 10-Ω-cm slices, with a substrate-adjust implant, have been used to obtain more precise control of the threshold voltage. The slice is oxidized to about 5000-Å thickness; source-and-drain diffusions are made through windows in this oxide, as shown in Fig. 11.16*a*. Next the slice is oxidized further to build up the field oxide (see Fig. 11.16*b*).

Cuts are made to define the source-and-drain contacts as well as the gate region. A dry thermal oxide, about 500 Å thick, is next grown (Fig. 11.16*c*). The source and drain contact windows are opened for a second time by etching through the thin oxide grown during the thermal oxidation step. Finally an aluminum metal film is deposited on the slice and patterned, resulting in the MOS transistor shown in Fig. 11.16*d*.

The MOS transistor of Fig. 11.16 requires precise definition of the gate so that it overlaps the source and drain. This necessitates that the source and drain regions be deep, to take advantage of the increased lateral diffusion effect. Unfortunately, this leads to devices which are relatively slow because of the increased overlap and drain capacitances. Many of these problems can be avoided by the use of self-aligned refractory gates, as described in Chapter 8. Details of this process will be illustrated in the next section, for an *n*-channel MOS transistor.

11.7.2 The *n*-Channel Transistor

The *n*-channel transistor, fabricated by self-alignment techniques, avoids many of the problems (and disadvantages) of the last device. It is inherently faster than its *p*-channel counterpart, since the majority carrier (electron) mobility is three times the hole mobility. However, it is more difficult to make as an enhancement-mode device, and process controls are generally more stringent than for the aluminum gate, *p*-channel device.

Here the starting silicon is *p*-type, with a resistivity of about 5 Ω-cm. An oxide layer, about 5000 Å thick, is grown over the slice and opened in regions where the source, drain, and gate electrode are to be placed, as shown in Fig.

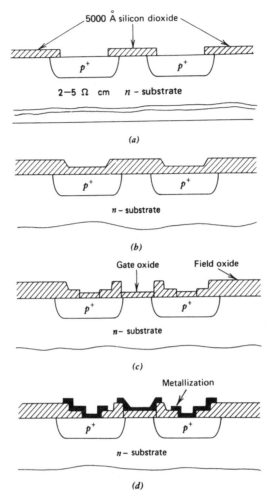

Fig. 11.16 Processing sequence for a *p*-channel aluminum gate MOS transistor. From Colclaser [2].

11.17*a*. A 200- to 500-Å-thick gate oxide is now grown on the slice by dry thermal oxidation. A layer of polycrystalline silicon is next deposited by CVD, and etched in order to define the gate. This polysilicon gate is used as a self-aligned mask to selectively remove the exposed thin oxide, resulting in the structure of Fig. 11.17*b*.

An *n*-type diffusion is made for the source and drain regions. Again, the polysilicon gate serves as a self-aligned mask for this step. The gate is doped heavily *n*-type during the diffusion; this provides the appropriate work function for an *n*-channel transistor (see Fig. 11.17*c*).

Next, a layer of PSG is deposited by CVD and reflowed. This increases the

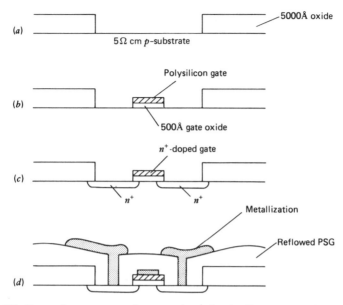

Fig. 11.17 Processing sequence for an *n*-channel polysilicon gate MOS transistor.

thickness of the oxide over the field regions between devices, and thus avoids parasitic MOS action. Moreover, it provides an insulator over the gate metallization, over which a second level of metal can be placed, if required. Contact windows for the gate, source, and drain connections are cut in this PSG layer. The final device, after metallization and pattern definition, is shown schematically in Fig. 11.17*d*.

Variations can be made to this sequence for the purpose of meeting the process requirements, and also to obtain improved device performance. These include: the use of partly or fully recessed local oxidation processes to reduce the parasitic capacitance of the source and drain, and to planarize the structure; the use of ion implantation for substrate adjust or for adjusting the channel resistivity; the use of ion implantation under the field oxide to prevent channel effects, and the extensive use of self-alignment methods which reduce the number of masks that are required, and allow tight physical tolerances to be maintained with less difficulty [70]. A structure, incorporating these features [23], is shown in Fig. 11.18. Here, a deep boron implant is made prior to local oxidation, to provide a channel stop. Selective ion implantation is used to provide both enhancement- and depletion-mode devices, which are shown in this figure.

Further improvements in device performance can be made by the use of silicide ohmic contacts, in order to reduce the parasitic resistance of the source and drain, as well as the sheet resistance of the gate region. These technologies have been described in Chapter 8; their unique advantage results in their lower resistance, as well as in their ability to be surface oxidized [71], so that crossovers

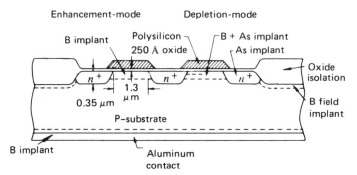

Fig. 11.18 An *n*-channel polysilicon gate MOS with a depletion-type load device. From Dennard et al. [23]. ©1979. The Institute of Electrical and Electronic Engineers.

can be placed on them. Figure 11.19 shows the sequence for a transistor of this type, using a TiSi$_2$ process [72]. Here, the polysilicon gate is oxidized, and selectively etched to preserve its sidewall insulation prior to the deposition of titanium. A heat treatment forms the TiSi$_2$ in the source/drain regions, where encroachment into the junction region is prevented by the presence of this oxide spacer. In this manner, junction leakage due to undercutting effects is prevented.

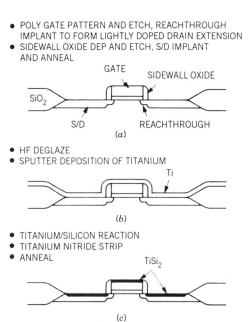

Fig. 11.19 Self-aligned silicide process flow. From Alperin et al. [72]. ©1985. The Institute of Electrical and Electronic Engineers.

11.7.3 Complementary Transistors

Complementary MOS (CMOS) circuits have the advantages of low standby power, high speed, and high noise immunity. Their use in VLSI applications has rapidly advanced for these reasons. Here, self-aligned, silicon gate technology has resulted in low threshold voltages and reduced power dissipation, further increasing their suitability for these applications.

CMOS circuits require a balanced pair of n- and p-channel enhancement-mode devices on the same chip. This is achieved by fabricating one device in the substrate, and the other in a well of opposite impurity type. Both n-well and p-well CMOS circuits can be made, and many arguments have been presented for one structure over the other, depending on the circuit application. It is also possible to use a [73] very lightly doped substrate in which both an n-well and a p-well are made, resulting in a "twin-well" structure. This process provides added device design flexibility, with a small increase in process complexity. At the present time, the single n-well process is commonly used for most microcircuit applications.

Figure 11.20 shows the cross section of a simple CMOS pair, using metal gate technology. Its process sequence is relatively straightforward. Here, the starting material is p-type silicon of about 15-Ω-cm resistivity. The n-channel device is made directly on this substrate, using the process sequence outlined in Fig. 11.16. The p-type device, on the other hand, must be made in an n-type substrate. This is provided by means of an n-type well, which must be diffused into the substrate. This diffusion must have a low surface concentration, and be precisely controlled, if the threshold voltages of the two devices are to match. This combination is extremely difficult to achieve by conventional

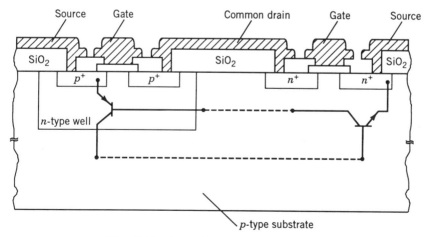

Fig. 11.20 A metal gate CMOS transistor.

diffusion methods. Here, ion implantation is used as a precise predeposition step, followed by drive-in. Subsequent steps in the formation of the p-channel transistor are carried out within this well, following a similar process sequence, except for a change in impurity type.

Parasitic $p–n–p$ and $n–p–n$ transistors are present in CMOS pairs of this type. As shown in Fig. 11.20, a vertical $p–n–p$ is formed in the n-well. At the same time, a lateral $n–p–n$ transistor is present in the p- substrate. Together, these constitute a $p–n–p–n$ transistor which can latch up at some critical current level, and thus present a potential reliability problem. This can occur when the sum of the small signal current gains of the parasitic transistors exceeds unity. Thus, solutions for this problem are aimed at reducing one or both of these current gains. The simplest approach is to increase the separation between the MOS devices; unfortunately, this is contrary to the dictates of obtaining tighter packing densities for VLSI.

A second approach involves the use of guard rings and/or local oxidation around each device. The use of both these methods is shown in the silicon gate structure of Fig. 11.21, together with its associated parasitic transistors. Here, a net space reduction can be achieved [74], notwithstanding the extra space required for these added features.

The gain of the vertical $p–n–p$ can be reduced by placing the p-channel device in a deep well. This, however, results in an increased area because of lateral diffusion. Alternatively, retrograde doping, which provides a retarding field in the base region of this parasitic transistor [75], can be used for this purpose.

Another strategy for increasing the latch-up threshold current is to reduce the lifetime of the substrate region, by which these transistors are interconnected. Gold doping has been attempted for this purpose, but results in leaky junctions.

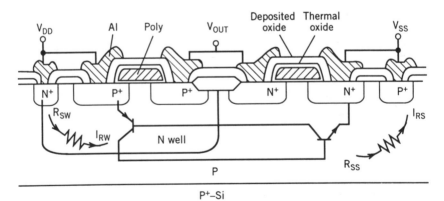

Fig. 11.21 A silicon gate CMOS device. From Swirhun et al. [74]. ©1985. The Institute of Electrical and Electronic Engineers.

Internally gettered silicon slices [76], in which there is a large SiO_2 precipitate concentration in the bulk, is another approach. Finally, it is possible to make the CMOS structure in an epitaxial layer, on a heavily doped substrate. For the structure of Fig. 11.21, this requires the use of a p-epitaxial layer on a p^+-substrate.

Of these methods, the epitaxial substrate, with a retrograde well doping configuration, is perhaps the best approach, and results in an increase in the latch-up threshold current density by a factor of more than 1000. Its disadvantage is the added cost of the epitaxial starting material; however, highly automated reactors have been developed to meet this need, and should minimize this cost penalty. Moreover, the increased packing density resulting from this approach makes it highly suited for VLSI applications.

Trench technology can also be used to improve the packing density in CMOS circuits [77]. Figure 11.22 shows the device structure for a 0.5-μm, polysilicon gate CMOS embodying this approach in addition to many of the improvements outlined above. Here, a 6-μm deep, 2-μm wide trench, filled with SiO_2-polysilicon, is used to provide the isolation function. $TiSi_2$ is used for the self-aligned ohmic contacts, and also to reduce the gate resistance. For this device, threshold currents as large as 200 mA could be sustained without latch-up, while still maintaining a close-spaced configuration for the overall structure.

Because of its many advantages, CMOS is the technology of choice for VLSI microcircuits. As a result, many variations of the basic processes outlined here have been incorporated into modern structures, where gate lengths have shrunk to 0.35 μm and less. These include the use of lightly doped drain (LDD) regions to prevent hot carrier injection into the gate, which is an increasing problem at short gate lengths. Sidewall spacers, which protect the gate oxide during subsequent silicide contact formation, are used in these processes.

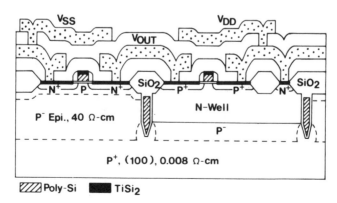

Fig. 11.22 A CMOS device with trench isolation. From Yamaguchi et al. [77], ©1985. The Institute of Electrical and Electronic Engineers.

Figure 11.23 shows a [78] modern CMOS device of this type, with an effective gate length of 0.5 μm, which incorporates many of these advances in technology in a circuit with two levels of metallization.

The flexibility of CMOS circuits can be greatly increased by combining them with other types of active devices [79]. Thus, they are compatible with bipolar junction transistors which can be built on the same substrate. This technology, known as BICMOS, will be described in Section 11.8.8.

SOI structures are also under active investigation for CMOS. Here, a serious problem is the degradation in the channel mobility over that obtained in bulk and epitaxial material, in addition to the higher cost of starting material. Direct wafer bonding has the promise of achieving higher channel mobility. On the other hand, the problems of lapping a 200- to 400-μm thick wafer down to an active layer of 1–2 μm, and maintaining this dimension in a manufacturing environment, have not been solved at the present time.

11.7.4 Memory Devices

Modern computer applications call for ever-increasing amounts of memory. As a result, the thrust of silicon-based integrated circuit technology has been aimed at meeting this need. There are two basic classes of memory devices: (1) volatile memories, which only retain their information as long as the circuit is powered, and (2) nonvolatile memories where information is retained even when the power source is removed. By far the greatest need is for large memories in which read, write, and access functions can be carried out in individual cells, and in a rapid manner. These are usually volatile, and are called *random access memories* (RAMs).

RAMs can be divided into two subsets. The static RAM (SRAM) holds its

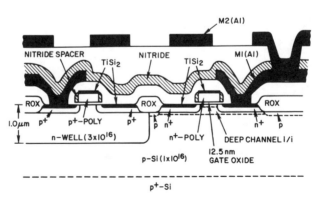

Fig. 11.23 A CMOS device structure using silicide technology. From Chappell et al. [78], ©1989. The Institute of Electrical and Electronic Engineers.

information in a bi-stable circuit, which can be triggered into two different states. Such a memory cell requires a minimum of two transistors, in addition to trigger devices.

An alternative approach is to use a capacitor for dynamic information storage. This requires only one passive component; however, at least one active device is necessary in order to provide read/write/sense functions. A characteristic of this type of memory cell is that its state is altered upon readout, so that it must be reestablished. In addition, there is always loss of charge in the storage capacitor, even when readout does not occur, necessitating that it be refreshed on a periodic basis.

At the present time, all of these functions can be provided by a single MOS transistor/capacitor combination, thus minimizing the area of the silicon per memory bit. As a result, the one-transistor (1-T) dynamic random access memory (DRAM) is the structure of choice for large memories. DRAM chips with 64 Mbits are available commercially today, and 1 Gbit/chip is projected for the turn of the century.

The primary requirement of a *nonvolatile* memory is that it retains its information after the power is removed from the circuit. The simplest form is that of a read only memory (ROM) in which a particular set of information is hardwired at the factory. ROM devices of this type are often used to provide the necessary logic functions for powering up sophisticated equipment in an orderly fashion.

It is also possible to design nonvolatile memory cells which can be programmed by the user. Some of these programmable ROMs (PROMs) can be erased on a bit-by-bit basis; these are referred to as electronically erasable PROMs (E^2PROM or EEPROM) and are extremely useful in system prototyping. In yet others, the information on the entire chip must be erased in one operation, usually by means of ultraviolet (UV) irradiation. These erasable memories are often called EPROMs or UV-EPROMs. Erasure time is typically 20 min or so. Recently, a new class of EPROM has been developed in which block erasure can be accomplished in under 1 ms. These are referred to as "Flash" memories for this reason.

The different basic structures used in this variety of memory devices will now be considered.

11.7.4.1 *Volatile Memories*

The DRAM cell, in its 1-T implementation, represents the smallest configuration for storing one bit of information [80]. It is almost exclusively used in volatile memories today, for this reason. Figure 11.24*a* shows a circuit diagram for a single cell, which consists of an *n*-MOS transistor and a capacitor. The capacitor serves as the information storage element. The MOS transistor is driven by signals on the bit line and word line, and is used to read, write, and sense the storage capacitor. Information is lost during the access

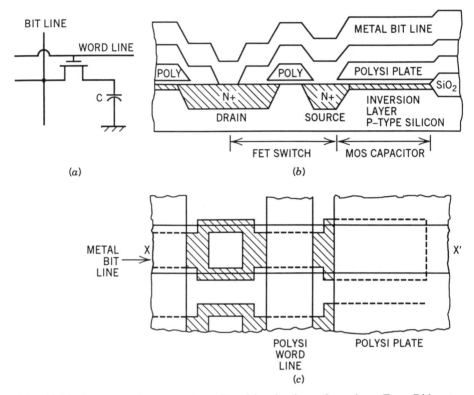

Fig. 11.24 A one-transistor memory cell and its circuit configurations. From Rideout [80], ©1979. The Institute of Electrical and Electronic Engineers.

phase, so that it is necessary for the transistor to provide a refresh function as well.

Many different DRAM cells have been developed, with varying degrees of complexity. In all cases, the aim is to minimize the area of the cell and its interconnections. Thus topological considerations are important, in addition to feature size. Moreover, high-quality oxides are a prime requirement for this application in order to allow the storage element to retain its charge for an appreciable length of time.

The access time of the storage capacitor, as well as the magnitude of the signal, are adversely affected by the parasitic capacitances of the drive lines. Thus, an additional emphasis in DRAM development is to maximize the capacitance of the storage element, while still minimizing the overall size of the cell.

Figure 11.24c shows a simple polysilicon-gate DRAM element [81], together with its cross-sectional view (Fig. 11.24b). Here, a polysilicon plate, in combination with the inversion layer of the transistor, provides the information storage function. The polysilicon gate serves as the word line; this presents a problem

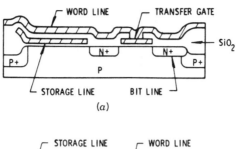

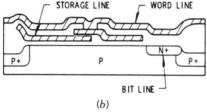

Fig. 11.25 Configurations for one-transistor memory cells. From Tasch et al. [82], ©1978. The Institute of Electrical and Electronic Engineers.

because of its relatively high sheet resistance, even when heavily doped. One approach to reducing this resistance is to cover this line with a refractory silicide such as $TiSi_2$. An alternative approach, shown in Fig. 11.25a, is to connect the polysilicon transfer gate to a metal word line and use the heavily doped drain region as the bit line.

Reduction in cell area can be achieved by the use of a second polysilicon layer to transfer the stored charge from the capacitor to the word line, as shown in Fig. 11.25b. This eliminates the need for the floating n^+-region in the access transistor, and reduces the overall size of the memory element.

The capacitor in cells of this type is about 30 fF. Many approaches have been used to increase this value. One such approach [82], known as the Hi-C concept, consists of altering the doping concentration in the channel region of the storage capacitor by means of selective implantation. A factor of 3 increase in capacitance per unit area can be obtained in this manner.

Further reductions in cell size have been obtained by the use of three-dimensional configurations to increase the capacitance without a penalty in area. These include folded, stacked, and trench capacitors [83, 84]. A comprehensive review of these cell structures is provided elsewhere [85].

Figure 11.26a shows a trench capacitor which is used in one of these advanced cell designs [86]. Here, the starting material is epitaxial p-silicon on a p^+-substrate. The cell structure is placed within an n-well; the trench capacitor extends through the well and into the heavily doped substrate. A considerable increase in storage capacitance is obtained in this manner. Cell capacitance can also be increased by using two or more trenches, as in the design of Fig. 11.26b [87].

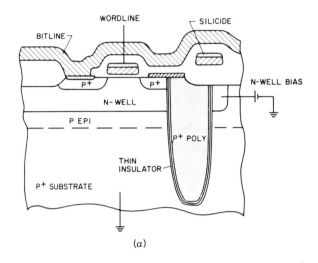

(a)

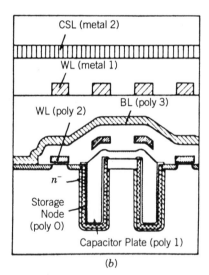

(b)

Fig. 11.26 A memory cell with (a) one trench capacitor. From Lu et al. [86]. ©1986. (b) two trench capacitors. From Fujii et al. [87]. ©1989. The Institute of Electrical and Electronic Engineers.

11.7.4.2 Read-Only Memories

The read-only memory (ROM) is a device which can be programmed only once, and then retains its information permanently. For volume applications, this is accomplished by means of a custom-built mask, which allows individual connections to be made, where required, between the word and bit lines.

The circuit connection of one such ROM element is shown in Fig. 11.27a

[88]. An MOS realization of this circuit is shown in Fig. 11.27*b*. Here, the polysilicon gate is connected to the word line, and the aluminum metal is connected to the bit line. Connection between these lines is made where required, by means of a "through hole" as shown in this figure.

PROMs, which can be programmed one cell at a time, are useful in applications involving new concepts, or for those cases where only a few microcircuits are required. Programming is accomplished by applying an electrical signal to each element in order to alter it permanently. One such element is a polysilicon link, which is necked at one point and which can be fused by subjecting it temporarily to excess current [89], resulting in an open circuit. Another approach is to use a *p–n* diode which is normally in its open circuit condition, and apply a reverse-bias pulse beyond its breakdown voltage [90]. The heat generated during this process produces an Al–Si eutectic alloy,* which effectively shorts the diode on a permanent basis.

11.7.4.3 *Nonvolatile Memories*

A nonvolatile memory device is one that retains its information even when power is removed from the circuit. Thus it is a direct replacement for magnetic core memories, which also have this characteristic. A MOS transistor can be used for this purpose since its gate has a high impedance to ground, and is thus

*This assumes the use of aluminum for one of the lines.

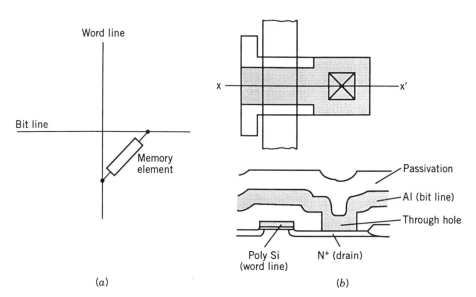

(a)   (b)

Fig. 11.27 A ROM device with a "through hole." From Masuoka et al. [88], ©1984. The Institute of Electrical and Electronic Engineers.

capable of long-term charge retention [91], provided that the oxide is of sufficiently high quality. One such structure consists of a *p*-channel MOS transistor with a floating polysilicon gate. Here, with no charge on the gate, the device is normally in its "off" state. The effect of temporarily biasing the source (or drain) beyond its breakdown voltage results in the avalanche injection of electrons into the oxide, with a resultant accumulation of negative charge on the floating gate. This turns the transistor to its "on" state; moreover, the device remains "on" after removal of the temporary bias signal, since the gate is surrounded by a thermal oxide. The memory retention time for these devices is typically of many years' duration.

Erasure is accomplished by UV irradiation, which increases temporarily the conductivity of the oxide and causes the gate to return to an uncharged state. Cells in a microcircuit array can thus be individually programmed, but must be collectively erased. The erasure time for this process is typically 20–30 min. The schematic configuration for a cell of this type is shown in Fig. 11.28.

A more advanced cell design consists of a stacked polysilicon gate structure [92], where an upper electrode serves as the control gate. This device can use the same photomask for both gates which allows the fabrication of a self-aligned structure, as shown in Fig. 11.29*a*. A modern version of this cell, using local oxidation and a TiSi$_2$ gate, is shown in Fig. 11.29*b* [93].

In this structure, the *n–p–n* transistor is in the "on" state with no charge on the floating gate since the channel is inverted. The application of a temporary positive voltage to the drain and control gate, with the source grounded, results in electron injection from the channel to the gate. This turns the device to its "off" state. Here, the control gate assists in device operation by the application of an aiding electric field across the channel and the floating gate.

Nonvolatile memory operation can also be based [94, 95] on the principle of tunneling through a thin oxide. Typically, this is done by arranging for a small portion of the floating gate to be placed over a thin oxide (100–200 Å) through which tunneling can take place, as shown in Fig. 11.30.

Designs of this type usually result in smaller devices than those based on avalanche injection, and are favored for high-density applications. However, they are more difficult to make because of the need for careful control of both oxide thickness and quality.

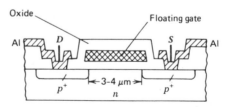

Fig. 11.28 A nonvolatile memory cell with a floating gate. From Scheibe and Schalte [92], ©1977. The Institute of Electrical and Electronic Engineers.

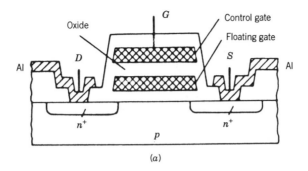

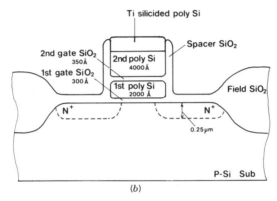

Fig. 11.29 Memory cells with floating and control gates. From Narita et al. [93], ©1985. The Institute of Electrical and Electronic Engineers.

Other stacked configurations have also been developed, with variations in the shape and positioning of the gates over each other. In some cases, the insulator consists of an $SiO_2/Si_3N_4/SiO_2$ sandwich, to obtain a higher breakdown voltage than a single SiO_2 layer. These device structures (and their fabrication steps) are more complex than the structures shown here.

The use of UV erase for EPROMS has the disadvantage that the microcircuit requires an expensive package with a UV-transparent window, and that erasure takes about 20–30 min. Moreover, the entire circuit is erased during this process. These disadvantages have brought about the development of the electrically erasable PROM (E^2PROM).

The cell configuration of the E^2PROM is essentially the same as that of Fig. 11.30, although design optimization is quite different. Setting the cell to the "off" state is done as outlined above. In this structure, the application of a reverse electric field between the control gate and the source causes electrons on the floating gate to tunnel back into the source, and thus turn "on" the device. Oxide thickness must be carefully controlled in the 100- to 200-Å region for the tunneling process to be effective.

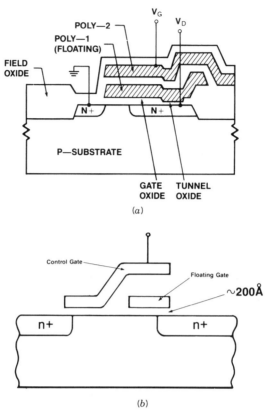

Fig. 11.30 Memory cell with a tunneling oxide. From Oto et al. [94], ©1987. The Institute of Electrical and Electronic Engineers.

It is common practice to use an additional transistor with each control gate in order to provide this second function. This results in added chip size per storage element, so that the E²PROM is considerably more expensive than the EPROM.

A considerable size (and cost) saving can result if storage cells are erased in large blocks. This avoids the use of individual select transistors, and results in a circuit which is cost-competitive with the EPROM. In point of fact, it is an EPROM! However, it can be erased rapidly (≈1 ms) and is called a *flash memory* for this reason. Large memories of this type, with one million or more storage elements on a single chip, can be used as solid-state replacements for floppy discs in computers [96, 97]. Here, the alteration of data in individual cells can be accomplished by block erasure/refresh methods.

In all nonvolatile memory devices, the retention capability is a critical function of the quality of the oxide, and of the ability to seal the floating electrode in an oxide which is free from defects [98]. In addition, since the mechanism for carrier transport is often by tunneling, careful attention should be paid to the thickness of the oxide layer between the floating gate and the silicon, which determines the magnitude of the tunnel current [99]. At the present time, the ability of a memory cell to retain its charge under power-off conditions is, for all practical purposes, indefinite. However, circuit disturbances under powered conditions have been found to affect the stored charge, so that periodic regeneration may be required during operation.

11.7.5 Hard and Soft Errors

The continuing thrust towards reducing the cell size in volatile memories has brought with it problems of reliability. Some of these, such as inter-trench leakage in stacked capacitor cells, result in hard errors which can be solved by eliminating the source of the problem [100]. On the other hand, soft errors, which show up as a transient upset in cells, are much harder to isolate because of their random occurrence. Alpha-particle-induced errors fall in this category.

Alpha particles are doubly charged helium nuclei [101], ejected from the nucleus of high atomic number atoms during radioactive decay. Their energy range is about 8–9 MeV, which corresponds to a penetration depth of about 10–20 μm in silicon. Each alpha particle can produce about 2.5×10^6 electron–hole pairs during its passage. These can result in a transient malfunction of the memory cell if they are created in a region where a potential well is empty, or if the electron–hole pairs can diffuse to such a region.

Alpha particles are generated primarily in the semiconductor packaging, but also in the semiconductor itself. Thus, the use of selected packaging materials with low atomic number can provide a partial solution to this problem.* The use of epitaxial silicon for the active layer can also reduce this problem because of the short lifetime of the substrate. In DRAMs which are made on bulk silicon, a heavily doped buried mesh has been shown to improve the situation. Fabrication of this mesh is accomplished by a high-energy implant, in the 5- to 6-MeV range. Another approach is to use SOI starting material. Here, hole–electron pairs which are generated in the substrate are effectively blocked from penetrating into the active region, because of the intervening SiO_2 barrier.

Device design, to increase the critical charge at which an upset can occur, is yet another approach towards reducing this problem. System design techniques involving error detection and correction are also being investigated for this purpose.

In summary, no complete solution to this problem is available at the present time, so that this type of error represents a fundamental limit to the density of DRAMs [102]. However, a combination of the techniques described here can be used to keep the soft error rate comparable to, or lower than, the hard error rate.

---

*Ceramic packages are especially poor in this regard.

11.7.6 Silicon-on-Insulator Devices

Substrates of silicon on sapphire (SOS) and of oxygen-implanted silicon (SIMOX), as well as those made by wafer bonding, all fall into the general class of silicon on insulator (SOI) materials. Microcircuits made in these materials comprise an important area for VLSI technology, since they are more radiation-resistant than those using bulk silicon, and are also less subject to alpha-particle upsets. Moreover, the insulating substrate greatly reduces the parasitic drain capacitance, so that these circuits are about 50% faster than those made on bulk silicon. An additional advantage is the fact that CMOS devices made on SOI substrates do not suffer from latch-up problems. As a result, they can be made with higher packing densities than conventional CMOS.

Figure 11.31 shows both *p*- and *n*-channel devices, made by a CMOS process in a 4000-Å-thick SOS film [103]. Here a combination of ion milling followed by plasma etching was used in the various delineation steps to minimize undercutting problems, while reducing problems due to ion damage. Glass reflow techniques using a CVD film of PSG, which was flowed at 875°C, served to minimize step problems in the metallization.

A number of special factors must be taken into account in the fabrication of these circuits. In SOI devices, island definition by simple mesa etching leads to exposing the edges of the source and drain junctions, as seen in Fig. 11.31. This results in a parasitic edge transistor which is made on a (111) face* whose threshold voltage is lower than that of the bulk *n*-channel device [104]. This parasitic effect can be reduced if local oxidation techniques are used to form the islands [105]. An alternative process, which results in the formation of rounded islands, can be used to avoid these problems. Although considerably more com-

---

*This assumes that a conventional anisotropic etch process is used for edge definition.

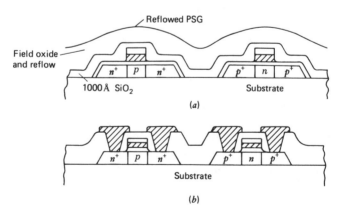

Fig. 11.31 Silicon-on-sapphire devices: (*a*) After reflow and (*b*) after metallization. From Splinter [103], ©1976. The Institute of Electrical and Electronic Engineers.

plex, it results in gate oxides whose breakdown field strength is comparable to that obtained with conventional MOS devices. Details of this process are provided elsewhere [106].

There are some significant differences between the various SOI materials. First, the quality of SIMOX active layers is close to that of bulk silicon, and minority carrier devices can be readily made on them. In contrast, the minority carrier lifetime of SOS material is in the nanosecond range, so that its use is restricted to digital applications. Second, the advantages of LOCOS processes can be exploited in the fabrication of devices with SIMOX material because the substrate bulk is made of silicon. However, the local oxidation process is complicated by the fact that the oxidation kinetics of this layered material are unlike those of bulk silicon, and are not understood at the present time. Finally, the channel mobility in neither of these materials is comparable to that of bulk or epitaxial silicon. Wafer-bonded layers hold out the greatest promise in this regard, and are consequently the subject of considerable research and development at the present time.

11.8 BJT-BASED SILICON MICROCIRCUITS

The bipolar junction transistor (BJT) has been used as the active component in both digital and analog microcircuits since their introduction. It is larger and consumes more power than its MOS counterpart. Consequently, it is being replaced by MOS devices in high-density digital integrated circuits. However, BJT transistors are favored for many analog and digital interface applications, where they are used because of their ultrahigh speed and design flexibility.

Tub-isolated BJTs, of the type shown in Fig. 11.1c, have been the workhorse of this technology. The application of modern VLSI technologies, such as shallow diffusions, oxide and trench isolation, silicide contacts, polysilicon emitters, and self-aligned fabrication schemes, have greatly reduced their size and improved their performance [107, 108] at the same time. Devices with gain-bandwidth (GBW) products in excess of 25 GHz are now possible. Their uses include ultrafast repeater circuits for optical fiber transmission at gigabit rates [109], multiplexers for communication circuits, high-voltage-drive circuits for flat panel displays [110], and circuits which handle both analog and digital signals, to name just a few. Often, they are implemented in a wide variety of designs [111] and in both multiemitter and multicollector configurations.

In BJT-based microcircuits, the active device is often used in conjunction with both active and passive components. Passive elements include many types of diodes, resistors, and capacitors. Active devices include: one or more types of *n–p–n* and *p–n–p* transistors, special-purpose devices such as low-inverse-beta, superbeta, multiemitter, and multicollector transistors, and often junction gate or insulated gate field effect transistors.

In this section, the conventional tub-isolated BJT is described in detail, since many of its concepts carry over to devices which incorporate advanced technologies. This is followed by a description of some of these new structures, and the manner in which they are implemented.

In the double-diffused epitaxial (DDE) process, devices are separated by placing in tubs of the type shown in Fig. 11.1c. For this process, the starting material is a lightly doped slice of p-type silicon. A range of starting resistivity (3–10 Ω-cm) is often specified; in many instances, no upper limit is set on this parameter. A layer of n-type silicon, of the appropriate thickness and resistivity, is epitaxially grown on this slice. Isolation is accomplished by diffusing p^+-moats from the upper surface (the epitaxial layer side); transistors and other components are fabricated in the resulting tubs of n-type material.

The epitaxial layer needs only to be thin enough to fabricate a device. On the other hand, the substrate can be of arbitrary thickness. Substrate thicknesses of 250–400 μm are typical of modern practice for slice diameters of 100 mm or larger. Finally, lateral penetration of the isolation diffusion is approximately equal to its penetration normal to the silicon surface. Consequently, the relatively shallow diffusions required by the DDE process result in minimizing the

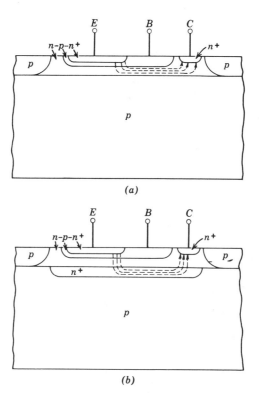

Fig. 11.32 The buried layer.

waste area on the silicon chip that is used for device isolation* and leads to greatly improved process yields.

Devices made in this manner have a high parasitic collector resistance, as is seen from Fig. 11.32a, where the current path from the emitter to the n^+-collector has been delineated. Figure 11.32b shows the same device in which a low-resistivity *buried layer* has been incorporated prior to epitaxial growth. The collector-current flow path is also shown for this transistor. Here much of this path is traversed through the buried layer, resulting in a reduction of the collector parasitic resistance. For a typical high-speed transistor of small geometry, the use of the buried layer can reduce the collector resistance by as much as a factor of 20, with a corresponding improvement in the GBW. In this manner, performance comparable to that of discrete epitaxial transistors may be obtained.

The process sequence for these devices is shown in Fig. 11.33. Also shown in this figure, for illustrative purposes, is a second isolation tub with a diffused resistor. Process variations include (a) the use of implantation for one or more regions and (b) the use of double masking and predeposition for regions such as the buried layer and the isolation wall, which require high conductivity as well as long drive-in times.

Figure 11.34 shows the impurity profile of a conventional transistor made by the DDE process, and illustrates typical dimensions and doping concentrations for devices used in 5-V digital circuits. Numbers in parentheses are for devices used in 15-V analog applications.

Local oxidation techniques described in Section 11.3 are extensively used to fabricate BJT devices of the type illustrated in Fig. 11.7b and 11.7c. These devices are smaller in size than conventional transistors and have lower parasitic capacitance. Moreover, they are ideally suited for self-aligned technology, with a further size reduction.

Figure 11.35 shows an oxide-isolated structure which incorporates many advanced technologies [112]. These include the use of self-aligned silicide pads for contacting the transistor regions. Full utilization of the available contact area is accomplished by this means. In addition, ion implantation is used to form the shallow junctions. This, in turn, allows the use of thin epilayers and results in small-area, high-performance structures which are suitable for high speed computer applications [113].

11.8.1 The Buried Layer

This is the first region to be introduced into the silicon slice, so that it is subjected to considerable movement during subsequent high-temperature processing. The slow diffusers (antimony and arsenic) are used for this application to minimize this movement.

---

*This is not an issue with MOS devices, which are self-isolating.

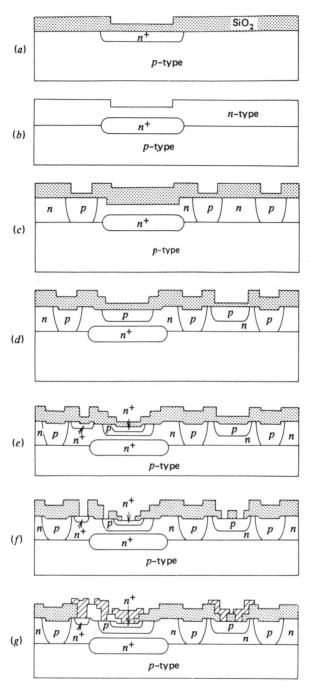

Fig. 11.33 The junction-isolated bipolar integrated circuit process. From Colclaser [2].

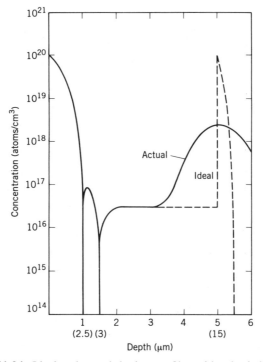

Fig. 11.34 Ideal and actual doping profiles with a buried layer.

The buried layer should be highly doped to minimize its sheet resistance. As a result, considerable out-diffusion takes place during subsequent epitaxial growth and process steps. Partially to offset this problem, this layer is diffused with a long drive-in step to reduce the surface concentration and still main-

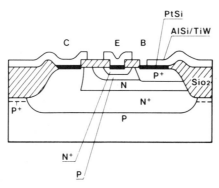

Fig. 11.35 Oxide isolated BJT with silicide contacts. From Tatsuki et al. [113], ©1982. The Institute of Electrical and Electronic Engineers.

tain a low sheet resistance. In addition, the ideal abrupt n^+–n interface is not obtained because of redistribution effects. For thin epitaxial layers (where the collector–base junction is close to the substrate–epitaxial-layer interface) this can significantly alter the doping profile of the collector region, as shown in Fig. 11.34. For comparison, the ideal doping profile of a buried layer transistor is also shown. Typically, the combined effects of out-diffusion and etch-back alter the doping profile of the epitaxial layer for about 2 μm from the epitaxial-layer–substrate interface.* This assumes that the epitaxial layer is grown by the chlorosilane process.

The silane process is commonly used with thin epitaxial (1–2 μm) layers. Here, up-movement of the buried layer is reduced to about 0.2 μm because of the absence of etch-back effects, and also because subsequent thermal processes involve shallow diffusions at lower temperatures and of shorter duration.

Antimony is sometimes used for the buried layer. Its large misfit factor (ϵ = 0.153), and consequent low solid solubility, limits its use to layers with a sheet resistance of about 15–20 Ω/square. In view of the perfect fit of arsenic to the silicon lattice, however, it is possible to incorporate a substantially higher concentration of this dopant without excessive lattice damage. A sheet resistance of less than 5 Ω/square can be achieved with this dopant.

Arsenic has a high vapor pressure, and considerable loss of this dopant occurs during a conventional predeposition and drive-in cycle. This problem can be avoided by predepositing the arsenic by means of high-energy ion implantation. This allows placement of the dopant *below* the silicon surface, and thus greatly reduces this problem.

11.8.2   Choice of Transistor Type

The DDE process results in the possibility of only one type of transistor for a specific type of substrate. Thus only n^+–p–n transistors can be fabricated in the examples shown in Fig. 11.1c. Conversely, for a choice of n-type starting material, only p^+–n–p devices would be possible. In general, the use of n^+–p–n devices is preferred over p^+–n–p devices in microcircuit fabrication for the following reasons:

1. The emitter region of a transistor must be highly doped for efficient operation. This favors arsenic and phosphorus over boron (i.e., an n^+-type emitter over a p^+-type emitter) because of their lower misfit factor.

2. During the drive-in step in an oxidizing ambient, impurity depletion occurs in regions doped with boron but not with phosphorus. In a p^+–n–p transistor, the region that is most lightly doped (the collector) is p-type; hence the depletion effect is most serious for this region. For an n^+–p–n transistor, boron depletion occurs in the more heavily doped base region, so that the problem is easier to handle.

*Note that a long drive-in of the buried layer prior to epitaxy reduces the severity of this problem.

3. The ohmic contact to the collector of an n^+–p–n transistor is made by an n^+-region which is diffused simultaneously with the emitter. For a p^+–n–p device, however, it is necessary to make contact to the base by means of an n^+-region, as seen in Fig. 8.20. Such a region requires an additional masking and diffusion step.

4. The charge state of a thermally grown oxide layer tends to induce an n-type shift in the surface regions beneath it; that is, n-type surface layers become more n-type while p-type layers become less p-type. Thus a lightly doped p-region is most likely to be affected and can actually invert to n-type under certain conditions. The collector of a p^+–n–p transistor is prone to this surface inversion effect because it is the most lightly doped region in this device.

Surface inversion in p^+–n–p transistors leads to short circuits between adjacent base and isolation regions. This can be prevented by diffusing a narrow p^+-guard ring through the middle of the collector surface. Being heavily doped, this remains p-type, and prevents the inversion layer from extending over the entire collector surface. However, its presence necessitates the use of a larger collector tub.

5. The last diffusion step for n^+–p–n based microcircuits involves the n^+-emitter. This is often formed by means of a phosphorus diffusion, resulting in a surface coating of PSG. A glass of this type greatly inhibits sodium ion transport through the oxide, and thus prevents long-term degradation of the underlying device. In contrast, a p^+–n–p based microcircuit has a final layer of BSG. This must be removed and replaced by PSG, in order to stabilize these circuits against ion migration. These extra processing steps are not required for circuits based on n^+–p–n transistors.

As a consequence of the foregoing, almost all bipolar microcircuits are made with n^+–p–n transistors. The starting materials shown in Fig. 11.1 are thus indicative of current practice.

11.8.3 Transistor Properties

The physics of operation of discrete transistors has been described in many texts. Here, however, we consider those device properties that have been altered because of the microcircuit fabrication process [2].

11.8.3.1 Breakdown Voltage

The window-defined microcircuit junction is parallel plane in character, except at the edge of the window cut in the oxide, where it takes on a cylindrical shape. This results in a reduced breakdown voltage in this region.

Computations of the breakdown voltage (BV) of a cylindrical emitter–base junction, formed by an erfc diffusion into a gaussian background, have been made by the use of a semiempirical approach which results in reasonable agreement with those observed in practice [114]. Experimentally, it has been found that, for practical BJTs, the BV is dominated by junction curvature, and is relatively independent of the emitter and base diffusion profiles. Typically, devices with emitter–base junction depths of 2.5 μm , which are commonly used in linear applications, have a breakdown voltage of $BV_{EBO} \simeq 6.5$ V. This voltage falls roughly linearly with junction depth. In high-speed digital transistors, where emitter diffusions are 0.5–1 μm deep, BV_{EBO} is about 3–4 V.

The breakdown voltage of the collector–base junction is more readily determined, since the base diffusion is made into a region of uniform background concentration. Numerical solutions of this problem are available in the literature [115].

BJT transistors, made by local oxidation, have junctions that are essentially parallel plane in character, since the curved region is consumed by this process. Thus, their BV approaches that of the bulk material. For any specific operating voltage, these devices can be designed with more highly doped regions than their tub-isolated counterparts. This results in faster devices which operate at higher current densities, and a further reduction in device dimensions.

11.8.3.2 Gain-Bandwidth Product

Figure 11.36 shows a microcircuit BJT made by the DDE process, together with its discrete counterpart. Differences in both parasitic resistances and capacitances between these devices result in a reduced value of the GBW for the former device, for the following reasons:

1. The collector–base junction of the microcircuit device has a curved sidewall area (and capacitance) in addition to its plane floor area.

2. The collector tub also has a floor and a sidewall which result in a substrate capacitance that is not present in the discrete transistor.

3. The collector contact is made on the same side as the emitter and base

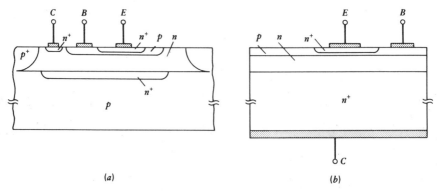

(a)   (b)

Fig. 11.36 Microcircuit and discrete transistors.

contacts. Thus the current flow path in this region is longer than in a discrete transistor, resulting in an increased collector parasitic resistance.

Calculations of the GBW of a transistor have been made, including the effect of these parasitic terms. It has been shown that reduction of the epitaxial layer thickness, and the use of shallow junctions, are important approaches to improving device performance. Concurrently, both techniques result in smaller devices, so that this is an advantage for VLSI circuit fabrication. This is the present trend in advanced high speed digital BJT devices, where the emitter–base junction depth has been reduced from 1.4 μm to 0.2 μm. Simultaneously, collector–base junction depths of 0.35 μm are used, as compared to 1.8 μm in tub-isolated structures. Epilayers of about 1-μm thickness are used, as compared to 5-μm thickness in conventional devices, and curved sidewall capacitances are essentially eliminated by the use of local oxidation.

11.8.3.3 Active Parasitics

In BJT microcircuits, the p-type substrate is tied to the most negative point in the circuit, since this ensures that the collector–substrate junction of the isolation tub is reverse-biased. Under certain conditions of circuit operation, this allows the possibility of obtaining parasitic transistor action in the p–n–p device comprising the base, collector, and substrate regions of Fig. 11.37a. The doping profile of this *vertical* parasitic p–n–p transistor is shown in Fig. 11.37b. Its emitter–base junction has an injection efficiency of about 0.5–0.7, so that the device has a maximum common-emitter current gain of about 2.

In circuit operation the presence of this p–n–p transistor may give rise to a number of different types of effects. Being an active device, it introduces a variable parasitic impedance in the microcircuit of which it is a part. It may provide a "sneak" path by which unwanted coupling occurs between two otherwise isolated components. It may provide a shunt path which diverts part of the current in the circuit. Finally, latch-up of the resulting combination of p–n–p and n–p–n transistors is possible under some circumstances.

In gold-doped devices the diffusion length of minority carriers in the collector region* is short compared to its width. Consequently, the current gain of the p–n–p transistor becomes vanishingly small because of its degraded base transport factor. The presence of a buried layer also serves to reduce its base transport factor to zero by incorporating a retarding field into this region. Thus devices which are gold-doped or have buried layers (or both) are not prone to these effects.

Parasitic p–n–p transistor action can also occur *laterally* to the sidewall, which can serve as the collector. Here the absence of a retarding field can sometimes cause latch-up. However, it can be avoided by widening the effective "base" width, or by placing a p^+-channel stopper in this region.

---

*That is, in the base region of the parasitic p–n–p transistor.

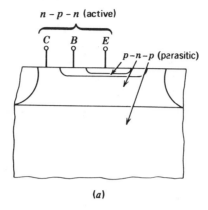

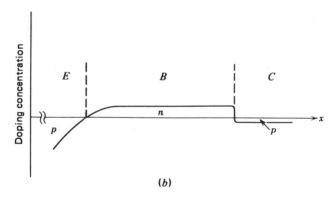

Fig. 11.37 Parasitic *p–n–p* transistor with no buried layer.

11.8.4 *p–n–p* Transistors

There is considerable interest in providing both *p–n–p* and *n–p–n* transistors on the same substrate, with no additional processing steps. Two approaches are reviewed here briefly. Both result in relatively poor transistors, which can be used in circuit schemes where the high-performance demands are made only on the *n–p–n* device. Figure 11.38*a* shows one approach by which such a transistor can be formed at the same time as its *n–p–n* counterpart. This results in a wide-base device with poor frequency performance. Moreover, the substrate of this structure serves as its collector, so that it can only be used in circuit applications where this electrode is at the common ground potential. Transistors of this type are used in the complementary-pair output stage of operational amplifiers.

Figure 11.38*b* shows a lateral *p–n–p* transistor which is tub-isolated from the substrate. A conventional *n–p–n* transistor, with which it is process com-

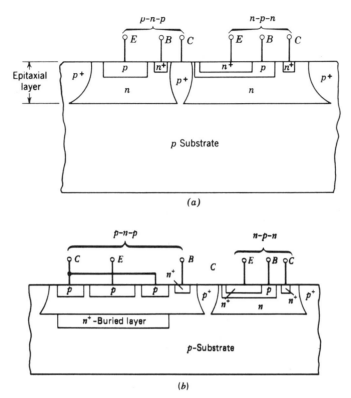

Fig. 11.38 (*a*) Vertical *p–n–p* transistor, (*b*) Lateral *p–n–p* transistor.

patible, is also shown. Here the base width is set by geometrical placement of the emitter and collector so that the device is relatively slow. Moreover, emitter current injection is limited to the top edge, so that its current-carrying capacity is much less than that of a convention vertical *n–p–n* transistor of the same size.

An important problem associated with this transistor is that part of its emitter current is lost by parasitic collection at the isolation wall and at the substrate. This effectively reduces its current gain to ≤5. Lateral injection to the isolation wall can be minimized by making the collector wrap around the emitter. Vertical injection to the substrate is greatly reduced by placement of a buried layer, which provides a retarding field* and suppresses this term. The combined use of these techniques allows common emitter current gains of 50 to be achieved in practice.

Figure 11.39 shows a high-performance lateral *p–n–p* transistor which is oxide-isolated [116]. Here, local oxidation is used to avoid the necessity of the

---

*A minor improvement in base resistance is also accomplished in this manner.

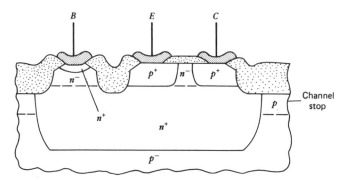

Fig. 11.39   Lateral $p–n–p$ transistor using local oxidation process. From Agraz-Güereña et al. [116], ©1975. The Institute of Electrical and Electronic Engineers.

wrap-around collector, resulting in a small area device compared to the tub-isolated structure of Fig. 11.38b. Here, too, the extensive use of ion implantation allows the complete structure to be made in a 1-μm-thick epilayer.

11.8.5 Special-Purpose Bipolar Transistors

A number of devices, tailored for special applications, are also used in BJT-based microcircuits. A few of these are considered here.

Transistors with multiple emitters are often used in analog–digital conversion operations, and also at the input to T^2L logic gates. These require minor masking changes from the conventional device. Transistors with multiple collectors are also used in current division schemes in analog amplifiers.

Superbeta transistors find use in the input stage of some operational amplifier circuits. These devices, with current gains of 1000 or more, are made by a double masking and diffusion operation, in order to drive the emitter region deeper than that of other transistors in the circuit. They have a very low collector–base punch through* voltage, typically about 2–5 V, because of their extremely narrow base width. This voltage can be increased by using a relatively deep ion implant to position a lightly doped p-base region within the n-pocket, with connection made to it by diffused p^+ regions [117].

An important requirement for satisfactory operation of multi-input logic gates is the need for making an input transistor with an extremely low value of reverse current gain (≤ 0.001). Typically, this device is used in a multiemitter configuration. Figure 11.40 shows the technique by which this may be accomplished in an $n–p–n$ compatible process. Here the base region is shaped so as to have a relatively high lateral resistance. In the forward direction the device

---

*It is worth noting that transistors made by the DDE process usually break down by avalanching of the collector-base junction.

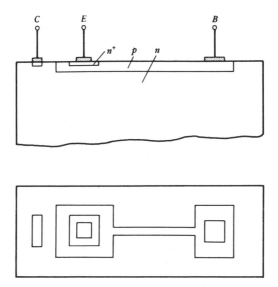

Fig. 11.40 Transistor with low inverse gain.

behaves like a normal transistor with a high extrinsic base resistance. In the reverse direction, however, with the collector acting as an emitter, the voltage drop due to the lateral resistance of the base region causes almost all injection to occur near the base contact. Thus there is negligible transistor action in this direction. Multiemitter devices can be made by a logical extension of this principle.

11.8.6 Field Effect Transistors

Junction gate field effect transistors (JFETs) can also be made on the same chip as bipolar n^+–p–n transistors, by means of a compatible process. Typically, these devices are found in the input stages of linear circuits where an extremely high input impedance is required, as is the case in some operational amplifiers.

Figure 11.41 shows the configuration for a JFET which is compatible with the DDE process. Here the p-base diffusion of the BJT serves as the channel region in which ohmic contacts are made for the source and drain. An n^+-emitter diffusion is used to form the gate on the top side of the resulting p-channel. The isolation tub can also be used as a gate; generally, this region is tied to a fixed reverse potential with respect to the channel.

An important layout requirement of this transistor is that its source region be topologically isolated from the drain, in order to avoid sneak conduction paths. Concentric structures can be used in order to meet this requirement. On the other hand, the layout of Fig. 11.41 is equally satisfactory and takes up less space. Thus it is commonly used for these devices.

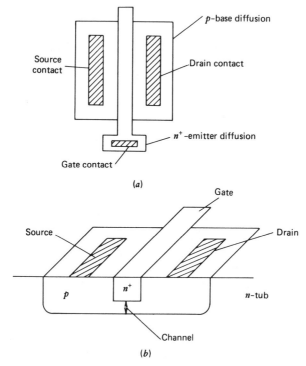

Fig. 11.41 Process-compatible JFET: (*a*) Plan view (*b*) Isometric view.

Compatible processes have also been developed for fabricating CMOS and bipolar transistors on the same chip. This important combination will be described in Section 11.8.8.

11.8.7 Advanced Structures

The tub-isolated transistor has found wide use in digital as well as linear circuits. Until recently, its size and processing complexity have placed it at a competitive disadvantage to MOS devices in VLSI applications. In recent years, a number of new techniques have been developed for removing this disadvantage. All are aimed at reducing the physical dimensions of the device, as well as the number of processing steps. Concurrently, device performance has been improved, so that these approaches are used in very high-speed integrated circuit applications. They share the following characteristics:

1. Use of self-isolated devices to shrink overall area and reduce the number of processing steps.

2. Use of self-aligned techniques to simplify the masking requirements, and thus allow further shrinkage of device dimensions.

3. Use of extremely thin epitaxial layers ($\approx 1 \mu$m), combined with the extensive use of local oxidation, to provide improved high-frequency performance.

4. Use of washout contacts to reduce the device size and improve its high-frequency performance. Silicide technology is often used to reduce their ohmic resistance.

Many of these separate processes have been covered earlier; here, we shall illustrate how they can be integrated into advanced BJT structures for use in microcircuits.

11.8.7.1 *The Triple-Diffused Process*

Transistors made by the triple-diffused process are self-isolating, and thus are smaller than their tub-isolated counterparts. Thus, they have the potential for VLSI applications. As noted earlier, conventional devices of this type are difficult to manufacture, and have extremely poor performance. As a result, they have long been displaced by devices made by the DDE process. Recently, it has been shown that devices with extremely high performance can be fabricated by the triple-diffused process, when all regions are precisely doped by ion implantation [4]. Figure 11.42 shows one such device, which is made in a thin *p*-type epilayer. It uses local oxidation, self-aligned silicide (salicide) contacts and oxide spacers, and makes extensive use of self-aligned technology to shrink device size.

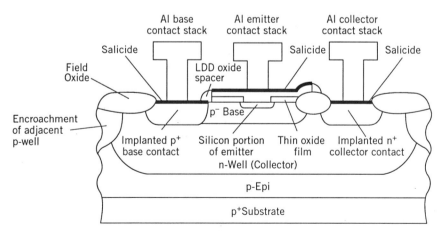

Fig. 11.42 Modern triple diffused bipolar transistor. From Ahmed et al. [4]. With permission from *Solid State Technology*.

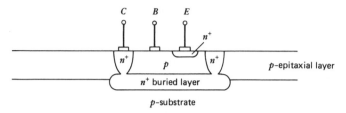

Fig. 11.43 Collector diffusion isolated transistor.

11.8.7.2 Collector Diffusion Isolation

This approach results in a BJT with a conventional doping profile. However, its size is reduced by using the collector region to isolate the device from the substrate [118]. The collector diffusion isolation (CDI) process uses high-resistivity p-type starting silicon in which an n^+-buried layer diffusion is first made. This is followed by a thin ($\approx 1\ \mu$m) p-type epitaxial layer of 0.2 Ω-cm resistivity. Deep n^+-diffusions are next made so as to contact the buried layer; a final n^+-emitter diffusion completes the device, as shown in Fig. 11.43. Here, considerable area saving is achieved since the device is self-isolated. Furthermore, the use of an extremely thin epitaxial layer results in improved high-frequency performance over conventional devices. An additional advantage is that the number of masked diffusion steps is reduced to three, as compared to five for the conventional DDE process.

It is worth noting that the device of Fig. 11.43 has uniform base doping. Consequently, it is prone to surface inversion, resulting from redistribution effects during oxide growth. Moreover, the uniform base doping results in a device with a high inverse beta, so that it is not suitable for many digital circuit applications. These problems can be avoided if the surface doping is enriched by a shallow boron implant, to result in a graded base structure [119].

The CDI process can be combined with local oxidation to result in a greatly improved device structure, as shown in Fig. 11.44. Here local oxidation, together with self-aligned contacts, is used to reduce physical dimensions

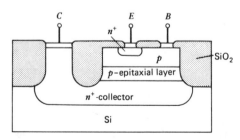

Fig. 11.44 Collector diffusion isolated device with local oxidation.

and improve the breakdown voltage. A second local oxidation within the device serves to further reduce the capacitance of the collector–base junction, with an additional improvement in the its GBW. In addition, boron is implanted in the surface region to result in a graded base structure, with high forward gain and low inverse gain. This approach has proved highly successful and is used in many commercial VLSI circuits.

Ion implantation techniques are used extensively in these devices, and often permit unique device processing steps. For example, it is possible to up-diffuse a base by making a selective implant on top of the buried layer, prior to epitaxy, in order to control its base width.

11.8.7.3 The Polysilicon Emitter Process

The doped polysilicon emitter was developed to allow the formation of extremely shallow junctions which are free from damage. Here, a layer of undoped polysilicon is deposited over an oxide, in which windows are opened for the emitter contact [120]. Next, it is implanted with the appropriate dopant (typically arsenic for an n^+-emitter). A shallow implant is used to avoid damage to the underlying silicon. This polysilicon layer is used as a doped source in the drive-in step. Subsequently, it can be removed; however, it is commonly left in place, and the metallization deposited upon it. This has been shown to result (see Section 8.6.3) in lower ohmic resistance than is obtained when an aluminum contact is made directly to the silicon.

Emitter formation by this approach results in a high yield process because of the absence of anomalous diffusion effects, such as pipe formation. Additionally, the current gain of structures made in this manner exceeds that of conventional transistors by as much as a factor of 5. It has been postulated that this improvement in current gain comes about by a reduction* in the hole current which is back injected into the emitter.

A number of theories have been proposed for this reduction [121, 122]. One such theory is based on a reduction of the hole diffusivity because of the polycrystalline nature of the dopant source. Another is that dopant pile-up at the interface serves as a barrier to hole flow. Finally, it has been proposed that the back-injection of minority carriers is inhibited by the formation of an interfacial oxide at the silicon–polysilicon interface. A detailed comparison of these theories is provided elsewhere [123].

Considerable care must be taken to clean the silicon surface in order to realize this gain improvement. Typically, after a conventional RCA clean step, the substrate is dipped in HF to remove residual oxide traces, just prior to the deposition of the polysilicon layer.

This technique can be adapted to form shallow diffusions with silicided contacts [124]. This is done by depositing consecutive layers of a refractory metal and silicon, and implanting this combination. Subsequent diffusion from this

---

*For an n–p–n transistor.

doped source results in the simultaneous formation of a shallow junction and a silicide contact layer.

11.8.7.4 Self-Aligned Structures

A major problem with conventional BJTs is the relatively large area that is used to provide a low resistance ohmic contact to the base region [125]. This area can be considerably reduced by the use of doped polysilicon to make a wraparound contact, which is self-aligned. A cross-sectional view of a practical device embodying this approach is shown in Fig. 11.45 [126]. Critical dimensions are indicated for a device with a GBW of 12 GHz at a collector–emitter voltage of 1 V.

The base-to-emitter area ratio, which is typically about ten for a tub-isolated BJT and five for an oxide isolated device, can be further reduced to two by means of this approach. This results in a considerable savings in device area. Figure 11.46 compares this structure to a BJT made by local oxidation, and illustrates this reduction in size. This approach is referred to as *super-self-aligned technology* (SST), and is extensively used in the fabrication of BJT-based microcircuits [127]. Further improvements in the technology have been obtained by using a precision implant to effectively reduce the base width by a factor of two, resulting in a GBW of 21.1 GHz at a collector–emitter voltage of 1 V [128].

Trench isolation technology can also be combined with SST, to effect further reductions in device dimensions. Figure 11.47 shows the schematic cross-section of such a structure, using a p^+-polysilicon base contact and an n^+-polysilicon emitter [129]. Here, the trench is filled with undoped single-crystal silicon, grown by selective epitaxy.

In summary, many new combinations of processing sequences are being employed to fabricate BJTs for VLSI applications. Here, ion implantation and doped polysilicon technology, in combination with local oxidation, allow the

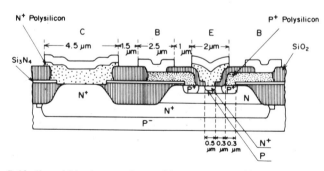

Fig. 11.45 Self-aligned bipolar transistor with a wraparound base contact. From Suzuki et al. [126], ©1985. The Institute of Electrical and Electronic Engineers.

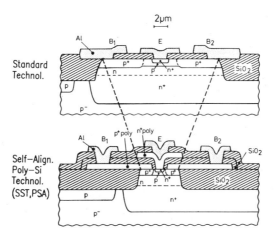

Fig. 11.46 Comparison of standard and self-aligned bipolar technologies. From Rein et al. [127], ©1988. The Institute of Electrical and Electronic Engineers.

use of self-aligned mask technology. Thus they play an important role in simplifying the device process, and also in improving device performance.

11.8.8 Bipolar-CMOS Integrated Circuits (BICMOS)

The advantages of CMOS microcircuits lie in the fact that they consume low power, have high noise immunity, and can be packed into highly dense configurations. However, system speed is limited by their inability to drive highly capacitive loads, such as those associated with word and bit lines in large memories. On the other hand, BJTs can operate faster than CMOS, and have greater current drive capability. Recently, they have been combined with CMOS, with a small increase in process complexity. This combination is commonly known as BICMOS, and is being increasingly used for digital applications which demand more speed than is possible with a purely CMOS circuit. These include sense amplifiers for DRAMs [130] and the integration of memory and peripheral circuits in low-power SRAMs [131].

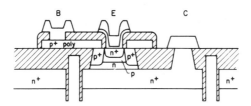

Fig. 11.47 Trench isolated bipolar transistor. From Li et al. [129], ©1987. The Institute of Electrical and Electronic Engineers.

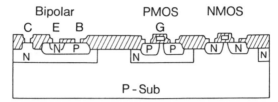

Fig. 11.48 Process-compatible MOS and BJT devices (BICMOS).

CMOS and BJT devices share a number of common process steps, so that the fabrication of a BICMOS circuit presents only a small degree of additional complication. Early circuits, in which the BJT was made by a triple-diffused process, required only one additional masking step over and above those needed for CMOS. Figure 11.48 shows a BICMOS structure of this type, in which the BJT is made by a conventional triple-diffused process outlined in Section 11.8.7.1.

In modern circuits, the performance of the BJT has been greatly improved by the use of buried layer structures. This, however, adds to process complexity, as seen in Fig. 11.49. Here, an n–p–n, a PMOS, and an NMOS transistor are fabricated in a p-type substrate. The starting material for this structure [132] is a lightly doped p-substrate into which the buried layer for the n–p–n is diffused. Next, an n-epilayer is grown. A p^+-diffusion isolation is used for the tub in which the BJT is placed. The p-well is now made by ion implantation. This is followed by local oxidation to isolate devices, and minimize latch-up in the NMOS–PMOS pair.

A shallow n^+-diffusion step is used to form the source/drain regions of the NMOS, the emitter of the BJT, and its collector contact. Shallow p^+-diffusions are used for the source/drain regions of the PMOS device. The base region of the BJT requires considerably lighter doping, necessitating an extra process sequence for this region [133].

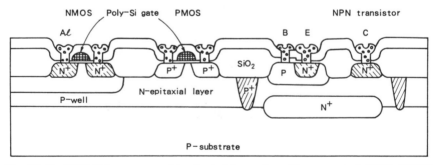

Fig. 11.49 BICMOS structure with a buried layer. From Tsumura et al. [132], ©1988. The Institute of Electrical and Electronic Engineers.

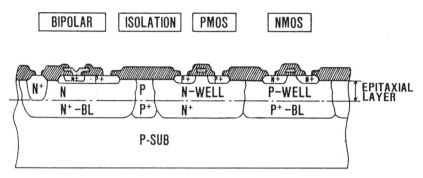

Fig. 11.50 BICMOS structure with multiple wells. From Ogiue et al. [134], ©1986. The Institute of Electrical and Electronic Engineers.

Considerable design flexibility can be obtained by the use of a multiple-well structure, of the type illustrated in Fig. 11.50 [134]. Here, a series of p^+- and n^+-buried layers are diffused into a p-substrate. This is followed by the growth of an n-epilayer which is 1.5 μm thick. Next, a series of n- and p-wells are made in this epilayer, so that they connect to their respective buried layers. A conventional process, with oxide isolation, is used to complete the device structure. Here, local oxidation and a p-well provide isolation between separate devices. A doped polysilicon emitter is used for the BJT.

As in the previous case, the emitter of the BJT and the source/drain regions of the NMOS are made in a single process. The p-region of the BJT is more lightly doped than the source/drain regions of the PMOS, requiring an additional process sequence.

It is also possible to fabricate BICMOS devices [135] which use a vertical p–n–p for the BJT, rather than an n–p–n device. Although not in common use, these devices are useful in the same analog operations. As many as 20 masking operations are necessary for the fabrication of this complex device structure.

11.8.9 Diodes

Diffused junction diodes can be fabricated by means of a single p-type base diffusion into the n-type epitaxial tub. Alternatively, the diode may be obtained by making a transistor and connecting its regions together. Of the many different possible combinations, the device structure shown in Fig. 11.51 is most commonly used, for the following reasons:

1. Under conditions of forward current flow, the collector–base junction is reverse-biased. As a result, the amount of stored charge in the transistor is low, so that it has a rapid recovery characteristic.
2. The forward conductance of the diode is high, because it has a short base width, in addition to a low series resistance.

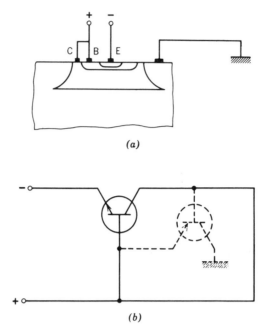

(a)

(b)

Fig. 11.51 A diode configuration with its parasitics.

3. From Fig. 11.51*b* it is seen that the emitter–base diode of the associated parasitic *p–n–p* transistor is short-circuited for this connection. Consequently, there is no possibility of active parasitic effects with this structure.

The single disadvantage of this connection is the fact that it has a breakdown voltage corresponding to that of the emitter–base diode. In the event that a higher breakdown voltage is required, it is necessary to use an alternative configuration, where the emitter and base are tied together.

In some situations, it is necessary to provide multiple diodes with their *p*-regions connected together, as shown in Fig. 11.52*a*. Here, a considerable saving in area is obtained by using a common base tub as in Fig. 11.52*b*. For this configuration, care must be taken to avoid active *n–p–n* transistor action between adjacent emitters. This is done by fabricating the structure with sufficiently large lateral spacing between the various emitter regions, and/or using gold doping to intentionally degrade the minority carrier lifetime.

Other diode structures have been found useful in special applications. Thus diodes, fabricated by making an emitter diffusion into the highly doped *p*-type isolation wall, exhibit a large depletion-layer capacitance per unit area. They suffer from the disadvantage of having a slightly lower breakdown voltage than

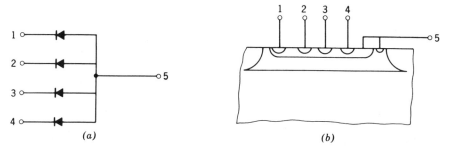

Fig. 11.52 Multiple diodes.

diodes in the form of Fig. 11.51. In addition, they can be only used in circuits where the *p*-side is grounded.

11.8.9.1 Schottky Diodes

Schottky diodes are also used in BJT-based microcircuits. Here their application is primarily confined to high-speed logic stages, where they are placed in shunt with the collector–base regions of transistors to prevent them from going into saturation [139]. An extremely important side benefit is that this approach avoids the necessity for gold doping, which is one of the most difficult processes to control.

The circuit connection for the Schottky diodes, and its integration with a bipolar transistor, has been described in Section 8.5. Commonly used gate materials are molybdenum, tungsten, and the silicides of palladium, platinum, and titanium. Aluminum cannot be used here because its barrier height to silicon is sensitive to subsequent thermal processing.

Schottky diodes can be integrated into the self-aligned schemes for BJTs which have been described earlier. Figure 11.53 shows the schematic cross-section of a Schottky-clamped transistor arrangement, using a Pt–Si diode in the SST process.

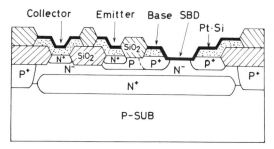

Fig. 11.53 Schottky-clamped transistor with a Pt–Si diode. From Oguie et al. [134], ©1986. The Institute of Electrical and Electronic Engineers.

11.8.10   Resistors

BJT-based microcircuits, especially those used in analog applications, require a number of different resistor types, with widely varying properties. These include diffused resistors which are made during the DDE process, as well as ion-implanted and thin-film types which require additional fabrication steps. Their process technology, as well as their electrical characteristics, are described briefly in the following sections.

11.8.10.1   Diffused Resistors

One approach to making resistors is by means of a p-type diffusion into an n-type background. Resistors of this type are made simultaneously with the base diffusion. Consequently, many of their characteristics are determined by the requirements of the associated active devices which form the basis of the microcircuit design.

The sheet resistance of a base diffusion is typically between 100 and 200 Ω/square. Thus the base diffusion results in reasonably proportioned resistors for the 50-Ω to 10-kΩ range, and is useful for the majority of circuit applications. This sheet resistance is monitored during the diffusion process and can be held to a tolerance of $\pm5\%$ of its room-temperature value. In practice, a tolerance of $\pm10\%$ of the initial value is generally assumed as a design rule for resistors of this type.

The temperature characteristics of a diffused resistor are difficult to compute accurately because of the variation of its doping concentration with diffusion depth [137]. To an approximation, however, they are dominated by the region near the surface where the doping concentration is the highest. Curves of normalized resistance, based on the variation of mobility with temperature, are shown in Fig. 11.54 for p-type material. From these curves it is seen that resistors with a concentration of around 10^{19} cm^{-3} will have a minimum variation with temperature. Fortunately, this is a convenient surface concentration for base diffusions as well, so it is often used in practice. A typical boron-diffused resistor, with this surface concentration, has a temperature variation (from the nominal value at 25°C) of approximately +4% at -55°C to +8% at +125°C.

A diffused resistor is isolated from its background by the contact potential of the associated p–n junction, or by a higher reverse voltage if the tub is appropriately biased. As a consequence, a number of resistors may be placed in the same tub with a physical saving of space on the silicon chip. In practice, it is customary to lump as many resistors as conveniently possible within a single tub.

The breakdown voltage between resistors is equal to the collector–base breakdown voltage of transistors made on the same slice, provided that they are sufficiently far apart so that punch-through does not occur prematurely. In addition, the reverse–leakage current density is the same as that of the associated collector–base junction.

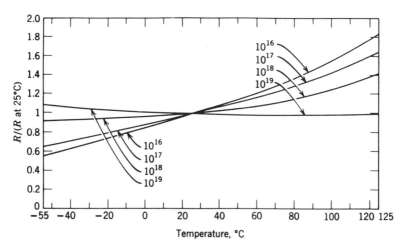

Fig. 11.54 Temperature behavior of doped silicon.

Tubs containing resistors are usually connected to the most positive voltage in the system. This effectively turns off the parasitic p–n–p transistor formed by the p-resistor, the n-tub, and the p-substrate. Consequently, active parasitics, and all their attendant problems, are avoided by this means.

Ohmic contacts are made to the resistor by aluminum, doped polysilicon, or a silicide. The presence of this contact distorts the current flow lines in its vicinity and makes it difficult to calculate its effective resistance. Still, the practical approach is to use a contact scheme that is reasonably uniform in its current flow characteristics and, at the same time, convenient to implement. Two such schemes are shown in Fig. 11.55, together with experimentally measured values of resistance associated with the end effects. Theoretical calculations for these end effects have also been attempted [138].

The effect of a foldover can also been handled empirically. It has been found that a bend with the proportions shown in Fig. 11.56 has an effective resistance

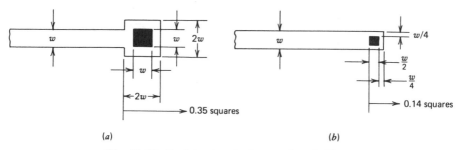

Fig. 11.55 Resistor terminations and end effects.

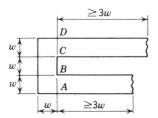

Fig. 11.56 Folded resistors.

of 1 square between *AB* and *CD*, provided that it is followed on either side by a minimum of 3 squares of resistor length [139]. Sharp, right-angle bends of the type shown here lead to a concentration of current at their inside corners. In spite of this fact, however, they are used in low-current situations because of the ease with which they are implemented in the photomask. If necessary, a small fillet can be added to relieve this current concentration.

Low values of resistors may be obtained by using a diffused emitter region, which has a sheet resistance of 2–10 Ω/square. Unfortunately, its value cannot be set with any degree of precision, so that this type of resistor is not commonly used as a discrete circuit element. However, it is often integrated within the emitter of a power output stage where it provides a ballasting function [65].

Large-value resistors, in excess of 10 kΩ/square can be made by utilizing the sheet resistance of the base region *under* the emitter. This is done by making a *p*-diffused resistor and placing an n^+ diffused emitter region (*ABCD*) over it as shown in Fig. 11.57. This emitter region is biased positive with respect to the *p*-diffusion which it covers. Resistor operation in this manner results in pinching the current flow through it, as in the channel of a JFET. The *V–I* characteristic of such a *pinch resistor* is similar to the output curves of a FET, so that it is

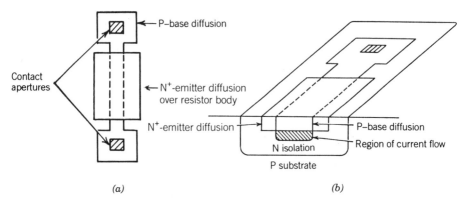

Fig. 11.57 Pinch resistor.

highly nonlinear. Furthermore, resistor operation is limited to voltages below the emitter–base breakdown voltage. Finally, resistors of this type have a wide tolerance (±50%) and are highly temperature-sensitive, changing by as much as 100% over the operating range. Thus, they can only be used in noncritical applications.

11.8.10.2 Ion-Implanted Resistors

Ion-implanted resistors can be made with sheet resistances as high as 5000 Ω/square and with a ±1% tolerance. Thus they allow considerable reduction in chip area, and are especially suited for low-power digital and linear microcircuits. These are fabricated by first forming two base diffusions to which contact can be made, and implanting the resistor between them. A typical process sequence consists of delineating the windows for the contacts, which are made at the same time as the base diffusion. Next, the oxide over the implant region is stripped and a thin oxide, about 1000 Å, is thermally grown in its place. The implant is made through this thermal oxide. Details of this process have been provided in Section 6.5.5.2.

11.8.10.3 Thin-Film Resistors

In some applications, it is necessary to provide ultrastable high-precision resistors, which are preset to a precise value. These are made by vacuum evaporation or sputtering of thin films of resistive materials, such as nichrome, chrome-silicon, or a variety of refractory metal silicides, directly on top of the oxide layer of a microcircuit chip which has in it the various active components that make up the entire circuit [140, 141]. All of these materials adhere firmly to the oxide and are used in film thicknesses of about 100–1000 Å. Typically, these resistors are preset by laser trimming before the chip is encapsulated. Thin-film resistor technology is a subject in itself and will not be covered here.

Polysilicon can also be used for high-value resistors where wide tolerances are permissible. Typically, an undoped film is deposited for this purpose, and subsequently implanted with an impurity (either p- or n-type). This is followed by an anneal step which increases the grain size and lowers the resistivity of the film. Conventional annealing is usually carried out at 600–1000°C, for a period of 30 min. Alternatively, RTA is also used for this purpose [142]. RIE, using a C_2F_6–Cl_2 gas mixture, is often employed for delineating the resistor pattern. Wet etching with mixtures of ethylenediamine, pyrocatecol, and water (EPW) is also used for this purpose.

11.8.11 Capacitors

A capacitor must have a high cutoff frequency and a capacitance value that is independent of the amplitude of the signals that are impressed across it. The only capacitors satisfactorily meeting these requirements are those made by the vacuum evaporation of thin conducting films on either side of an insulating material. These have been treated extensively in the literature [140, 141] and are not considered here.

A reverse-biased diffused *p–n* junction can also be used for this purpose. Figure 11.58 shows some possibilities, each with its equivalent circuit. Note that all of these structures have parasitic resistances which degrade their cutoff frequency (which varies inversely with the *CR* product). In addition, some have parasitic capacitances C_P associated with them. All are voltage dependent and introduce distortion in the circuit in which the capacitor is used. The configuration of Fig. 11.58*c* is restricted to situations where one side of the capacitor

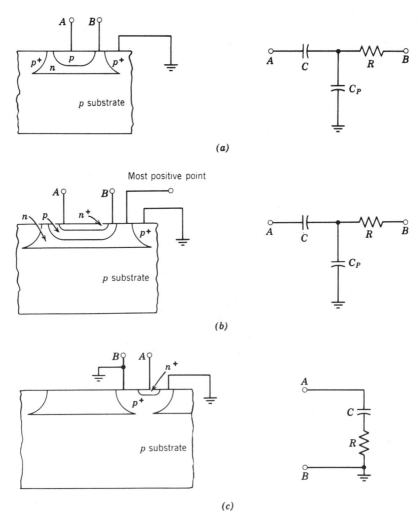

Fig. 11.58 Diffused junction capacitors.

if grounded. Its advantage, however, is that it has the largest capacitance per unit area because the isolation region is heavily doped.

Diffused junction capacitors are used in applications where their nonlinear behavior can be tolerated. Thus they are suitable for use as speedup capacitances in digital circuits, as temporary storage devices in triggering schemes, and as bypass capacitances.

Ion-implanted capacitors can also be used. Here hyperabrupt structures, of the type described in Chapter 6, are widely used in electronic tuning applications.

Oxide insulated capacitors can be fabricated by using the emitter region as one plate, the metallization as the second plate, and the intervening oxide layer as the dielectric. The construction of the capacitor is shown in Fig. 11.59. Its fabrication involves removal of the oxide layer that is normally present over this region (≈ 5000–6000 Å thick) and its replacement by a grown oxide (≈ 500–1000 Å) in order to obtain a higher capacitance per unit area. Although its capacitance per unit area is well below that of a p–n junction, it is superior in all other respects. Thus:

1. The capacitance is essentially independent of voltage, since its lower plate is made of highly doped material.
2. Its dimensions can be held to tighter tolerances, since they are set by the area of the plate and the thickness of the dielectric.
3. Its series parasitic resistance is that of a diffused emitter layer and is considerably lower than that obtained with the p–n junctions of Fig. 11.58.

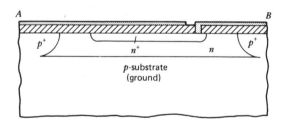

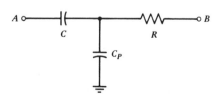

Fig. 11.59   MOS-type capacitor.

The shunt parasitic capacitance is that of the collector–substrate junction and is lower than that of other junctions in the microcircuit.

4. The temperature variation of this capacitance is caused by the change in the dielectric constant of the oxide layer, and is about 20 ppm/°C. This is about 50 times lower than that of a p–n junction.

A parallel-plate capacitor of this type is used in situations where a tolerance requirement is placed on the value of the capacitance, and where it is important that distortion products are not introduced into the circuit. Thus its main uses are in linear integrated circuits. One important application area is its use as a feedback capacitance for phase shift control in operational amplifiers.

Trench capacitors have also been considered for BJT-based microcircuits and have been described in Section 11.3.1. Their characteristics approach those of parallel-plate capacitors. However, their series resistance is higher, so that their frequency response is somewhat degraded. In addition, precise control of the surface area of the wall is difficult, so that the capacitance value cannot be held to tight tolerances. The main advantage of this difficult technology is that it greatly reduces the chip size. This is not of major importance with BJT-based microcircuits, since they are less device-intensive than those made by MOS technology.

11.9   GALLIUM ARSENIDE MICROCIRCUITS

The field effect transistor (FET) is the active component which is used in nearly all of these circuits. Here, a Schottky gate can be used to control the flow of majority carriers between source and drain, resulting in a metal- semiconductor FET (MESFET). Many GaAs microcircuits are based on the use of MESFET devices, and n-type material is used because of its high electron mobility.

There are a number of factors that make the MESFET device ideal for use with GaAs. Thus:

1. The wide energy gap results in a relatively large barrier height (≈ 0.8–0.9 eV) with many metallurigically compatible gate materials, so that the leakage current of the gate to the n-channel is low.
2. The gate is formed by sputtering or vacuum evaporation, which are low-temperature processes. The source and drain regions are usually fabricated by a short (30–60 s) heat treatment at 450°C.
3. The use of photopatterned metal films for the gate as well as for the source/drain regions results in a structure in which extremely tight tolerances can be held, so that these devices are capable of high speed. At the present time, the GaAs MESFET has a speed advantage over silicon devices (of the same physical dimensions) by a factor of 3.

4. Finally, a viable technology for MOS-type enhancement-mode structures in GaAs does not exist at the present time.

Both linear and digital devices and circuits can be fabricated using these devices. These include microwave amplifiers, couplers, mixers, and receivers, and circuits for optical transmission at data rates in excess of 10 Gbit/s [143].

No clearly optimum approach has been developed for GaAs digital integrated circuits at the present time, and there is much room for choice of device type [144, 145] and circuit configurations. Thus, the GaAs MESFET is inherently a depletion-mode device (D-FET) which dissipates power in the absence of a signal on its gate. Relatively complex circuit configurations [146], involving dual power supplies and voltage level shifting, are necessary for satisfactory operation in digital logic circuits. Circuits with enhancement-mode devices (E-FETs) are equally complex; however, these devices are more suited for digital circuits because they do not dissipate power in a standby mode. E-FETs require a narrow channel region which is lightly doped so that it is pinched off by the contact potential of the Schottky junction. They are more difficult to fabricate, and have a lower voltage swing than D-FETs. The use of a junction gate allows an increase in voltage swing as well as in channel width. The junction-gate field effect transistor (JFET) is often used in enhancement mode devices for this reason. Finally, circuits using pairs of enhancement/depletion mode devices (E/D FETs) are highly desirable for reduced complexity and power dissipation while still preserving the advantage of high speed.

The high-electron-mobility transistor (HEMT) is a variant on the FET, in that it uses a multilayer structure to effectively increase the channel mobility by a factor of two or more at room temperature. Moreover, significantly higher mobility values can be obtained by operation of these devices at reduced temperature. HEMT devices are increasingly used as discrete components, and in integrated circuits, for these reasons.

Finally, even further improvements in speed, coupled with high-power handling capability, can be achieved by the use of heterojunction bipolar transistor (HBT) configurations. The technology for these devices is difficult, and much research is being directed along these lines.

This section will discuss the various structures which can be made, and some of their process considerations.

11.9.1 The Depletion-Mode Transistor

The earliest form of this device was fabricated [147] in an n-type layer, epitaxially grown on $\langle 100 \rangle$-oriented SI GaAs. Typically this layer was 0.1–0.2 μm thick, doped to about $(1$–$3) \times 10^{17}$ cm^{-3}. Mesa isolation, using a wet chemical process, was used to delineate devices and passive components. Figure 11.60 shows a cross section through this structure with a D-FET, a Schottky diode, and

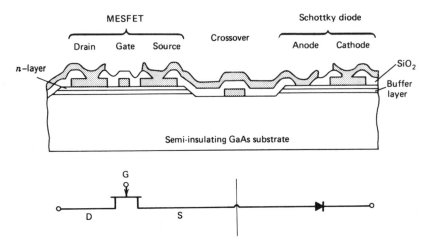

Fig. 11.60 Digital circuit scheme using mesa isolation. From Van Tuyl et al. [148], ©1977. The Institute of Electrical and Electronic Engineers.

a crossover using double-layer metallization [148]. Often, an undoped buffer region, 4–10 μm thick, was grown on the substrate to provide a fresh interface for the active layer.

Variations in the pinch voltage of D-FETs in these microcircuits can be caused by local nonuniformity in epitaxial layer doping and thickness. These variations can be eliminated by selective anodic thinning of the slice so as to obtain a constant doping-thickness product across the wafer [149]. This technique has been outlined in Chapter 7.

The development of high-quality LEC GaAs allows the use of ion implantation to form the active regions in an SI substrate. The doping and penetration depth of this process is sufficiently well controlled so that circuits can be made without the need for anodic thinning techniques [150]. Here, advantage can be taken of the anomalous movement of sulfur implants, described in Chapter 6. Thus, the use of a 350-keV Se-implant and a 300-keV S-implant allows the simultaneous formation of a 0.1-μm-deep channel and 0.4-μm-deep source/drain regions to form the D-FET. After implantation, the layer is capped with Si_3N_4 and annealed. Devices made in this manner are self-isolating, to result in a planar structure.

A wide choice of metals is available for the Schottky gate (see Section 8.7). Of these, the simplest is aluminum, which can be vacuum evaporated, and etched in a fine-line pattern with relative ease. Moreover, it can tolerate the subsequent heat treatment which is required for forming the ohmic contacts for the source and drain. However, it has been replaced by the trimetal system Ti/Pt/Au, described in Chapter 8, which is considerably more temperature-stable. This is usually deposited in a single operation by e-beam evaporation.

The basic approach for gate definition is by the lift-off process. Gate widths of 0.25 μm can be obtained by this method.

The GBW of a MESFET is critically dependent on the parasitic resistance between the gate and the source (including that of the source contact). Contacts with low specific resistance ($\simeq 10^{-5}-10^{-6}$ Ω-cm^2) are thus essential for high-performance devices. The formation of these contacts has been described in detail in Section 8.6. At the present time, the lowest contact resistance values are obtained by the use of evaporated films of Au–Ge eutectic, or Au–Ge–Ni. After deposition and patterning, these films are heat-treated [151] to about 450°C for 30–60 s to form the contact. Rapid thermal annealing has been found to be effective in reducing the contact resistance [152], and is extensively used at the present time. This technique is especially useful for making contacts to AlGaAs layers, which are otherwise difficult because of the surface film of Al$_2$O$_3$ on this semiconductor.

With ohmic contacts of this type, there is deterioration of surface flatness during heat treatment. Often, a cap layer of SiO$_2$ is used to contain the film during its alloying phase. Note, however, that this layer must be grown by an extremely low temperature process ($\leq$150°C) such as sputtering, to prevent excessive interface reaction between the gold and the semiconductor [153].

Yet another approach to reducing the parasitic gate-source resistance is to use etching to form a recessed structure [154]. This is shown in Fig. 11.61a, together with a conventional planar structure in Fig. 11.61b for comparison. Recessed structures of this type are more suited for discrete devices and analog integrated circuits than for high-density digital applications.

A process sequence for a MESFET using a nonrecessed structure is now outlined. The starting material here consists of implanted n-GaAs on an SI substrate, in which active regions have been defined by mesa etching. A typical etch for this purpose is HF:H$_2$O$_2$:H$_2$O in a 1 : 2 : 25 ratio by volume. This gives rounded edges to the mesa, and permits good step coverage for the ohmic contact metal.

Next, a contact mask is used to define the source and drain, which extends over the mesa region. A sequence of e-beam evaporated Au–Ge/Ni/Au in a 1000-Å/200-Å/1000-Å thickness is used for the contact metal, which is partially alloyed at this point. (A typical contact alloying step is 400°C for 45 s; at this point in the process, a 15-s heat treatment at this temperature is used to aid in adhesion of the metal to the GaAs. A lift-off procedure is used, leaving the source/drain metal in place.)

The gate region is now defined, using a chlorobenzene pre-soak for the photoresist so as to provide an overhang after development. The gate metal, consisting of a sequence of Ti/Pt/Au, in a 1500-Å/1500-Å/8000-Å thickness is deposited by e-beam evaporation. More complex, multilayer resists, described in Chapter 10, can also be used for this purpose. This is followed by a second partial alloy step (400°C for 15 s) to promote its adhesion.

At this point, an Si$_3$N$_4$ film, 500–1000 Å thick, is sputtered over the wafer, and holes are cut in it for the gate, source, and drain contacts. The function of

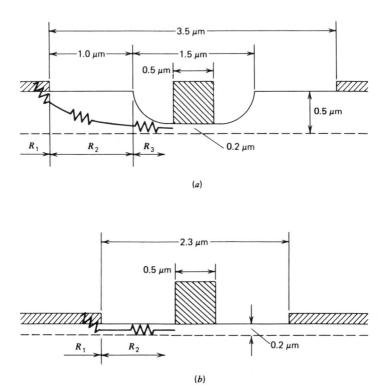

(a)

(b)

Fig. 11.61 MESFET structures. From Ohata et al. [154], ©1980. The Institute of Electrical and Electronic Engineers.

this layer is to provide long-term stabilization of the GaAs surface. A metallization, consisting of Ti/Au in a 1000-Å/8000-Å thickness, is used to connect devices in the integrated circuit. A final heat treatment (400°C for an additional 15 s) serves to complete the ohmic contact formation step, and also improves adhesion of both the gate and the metallization.

There are many possible additions to this process. Often, for example, the wafer is subsequently thinned to reduce its thermal resistance. This is followed by the etching of holes from the back face to expose the source metallization. Next, a thin layer of palladium is sputtered on the back surface, followed by a heavy electroless-plated gold film which serves as the ground plane for the device. In this manner, the source regions of the FETs are directly connected to ground by the shortest path. Details of this process have been provided in Chapter 8.

In microwave circuits, capacitance coupling at crossover points is minimized by the use of a Ti/Au "air bridge" which is made by the process steps shown in Fig. 11.62. Details of the fabrication procedures for air bridges of this type are given elsewhere [155].

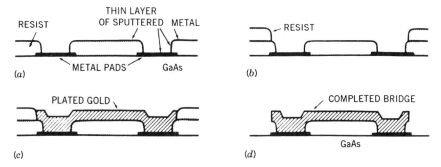

RESIST | THIN LAYER OF SPUTTERED METAL

METAL PADS   GaAs

(a)

RESIST

(b)

PLATED GOLD

(c)

COMPLETED BRIDGE

GaAs

(d)

Fig. 11.62 Process steps in air bridge formation. From Williams [155]. With permission from Artech House, Inc.

11.9.2   The Enhancement-Mode Transistor

As mentioned earlier, circuits based on enhancement-mode devices are highly suited for VLSI applications because of their simplicity and low-power dissipation. One approach to fabricating devices for these circuits [156] follows the sequence outlined in Section 11.8.1 for the depletion-mode MESFET. In this case, however, the n-layer is more lightly doped ($\simeq 10^{17}$ carriers/cm^3), and must be precisely thinned by anodic means to ensure that the channel is pinched off with zero gate bias.

FETs can also be made with a diffused junction gate [157], and are used in both digital and analog circuits. These devices have an advantage over MES-FETs, since their turn-on voltage is related to the energy gap of GaAs, and not to the Schottky barrier height of a gate metal. Typically, the turn-on voltage of JFETs is about 1.1 V as compared to 0.5 V for MESFETs, so that they have greater noise immunity. Figure 11.63 shows the schematic of an E-JFET with a series-connected resistor which serves as a load in a typical inverter configuration [158]. Here multiple silicon ion implants into an SI GaAs substrate were used to form (a) the deep, heavily doped regions over which the contacts were made and (b) the shallower (and less heavily doped) regions for the channel. Next, the implants were annealed for 10 min at 850°C in an arsenic overpres-

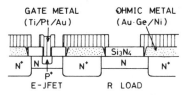

GATE METAL (Ti/Pt/Au)   OHMIC METAL (Au·Ge/Ni)

Si$_3$N$_4$

N$^+$ | N | N$^+$ | N | N$^+$

P$^+$

E - JFET   R LOAD

Fig. 11.63 Schematic of an enhancement-mode JFET inverter circuit. From Gonoi et al. [158], ©1986. The Institute of Electrical and Electronic Engineers.

sure, without any cap film. Next, an Si_3N_4 layer was deposited by PECVD; this layer was used as a mask for the zinc diffusion.

The gate region was formed by an open tube diffusion with diethylzinc as the source, in an overpressure of AsH_3. A junction depth of 1000 Å was required to pinch off the device under conditions of zero bias.

11.9.3 Enhancement/Depletion-Mode Pairs

The use of both enhancement- and depletion-mode devices on the same chip results in considerable circuit simplification, a reduction in device count, and an improved speed–power product in GaAs digital circuits [159]. In addition, these circuits can be operated with low supply voltages, with a concurrent reduction in their power dissipation. Figure 11.64 shows a cross-sectional view of an E/D pair, which can be used in circuits operating with a 1-V supply [160]. Here, a refractory gate of WSi_x was used, with Ti/Au as the second-level metal, and PECVD Si_3N_4 as the insulator. Ion implantation was used to provide different channel depths for the two devices.

11.9.4 The High-Electron-Mobility Transistor

The speed of a MESFET device is directly related to its channel mobility. This, in turn, is limited by extrinsic carrier scattering in the channel which must be quite heavily doped ($\approx 10^{17}$ cm^{-3}) to allow for device operation. One technique for circumventing this limitation is to use undoped GaAs for the channel, and to provide it with electrons from an adjacent heavily doped layer of a wider gap material, such as AlGaAs. Here, the bandgap discontinuity between these materials confines electrons to the GaAs, where they are constrained to be near the interface. High electron mobility, free from scattering effects, can be achieved in this manner. Typically, room-temperature channel mobilities in excess of 9000

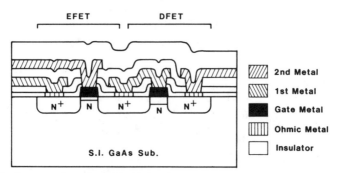

Fig. 11.64 Cross section of an E/D FET pair. From Takano [160], ©1987. The Institute of Electrical and Electronic Engineers.

cm^2 $V^{-1}s^{-1}$ are readily achieved, as compared to 4500 cm^2 V^{-1} s^{-1} for GaAs MESFETs. Moreover, the use of an undoped channel allows mobility improvement by operation at reduced temperature, which is not available in heavily doped GaAs. Channel mobility values as high at 90,000–100,000 cm^2 V^{-1} s^{-1} have been measured in these devices at 77 K.

In a high-electron mobility transistor (HEMT), a heavily doped n^+-AlGaAs layer, with an aluminum fraction around 0.3, is placed on top of an undoped GaAs channel. Electrons in the n^+-AlGaAs will spill over into the GaAs because of the resulting offset in the conduction bands of these materials. These electrons in the GaAs are constrained to stay close to the donor atoms in the n-AlGaAs, and are thus located near the interface, constituting a two-dimensional electron gas (2-DEG). Two important additions are necessary to make this into a practical device. First, in order to minimize interface scattering effects, a thin (50–75 Å) layer of undoped AlGaAs must be placed [161] between the GaAs channel and the n-AlGaAs. Next, in order to make suitable contacts to the n-AlGaAs, a heavily doped layer of n^+-GaAs is grown over it, and subsequently patterned for the source/drain regions. The contact metal is placed on these GaAs regions.

The fabrication technology for the HEMT proceeds along the same lines as that for the MESFET. However, the starting material must be epitaxially grown, and both MBE and OMVPE are used for this purpose. In addition, a selective etch is required to cut through the n^+-GaAs contact layer. Often, the n-AlGaAs layer is etched to some extent, to result [162] in a recessed-gate structure, as shown in Fig. 11.65.

Both enhancement- and depletion-mode HEMTs can be built in the same chip, and used in relatively simple, direct-coupled logic schemes [163]. Here, devices are made in a multilayer structure of the type shown in Fig. 11.66 which is similar to that of Fig. 11.65 except for an additional pair of AlGaAs/GaAs layers. Gate metal is placed on the different AlGaAs layers as shown in this figure [164].

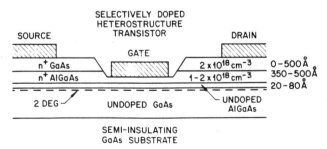

Fig. 11.65 Schematic of an HEMT structure. From Swaminathan and Macrander [162]. ©1991. Reprinted with permission of Prentice-Hall, Englewood Cliffs, NJ.

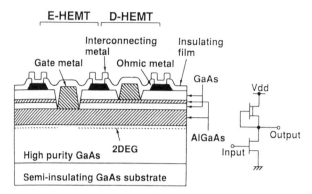

Fig. 11.66 Cross-section of an E/D HEMT pair. From Abe et al. [163], ©1991. The Institute of Electrical and Electronic Engineers.

11.9.5 Self-Aligned Technology

The use of self-aligned techniques is especially advantageous with GaAs-based microcircuits, since it allows the fabrication of small, high-performance devices. These structures are based on the use of one or more refractory materials, which can withstand subsequent thermal processing. In GaAs circuits, the post-implantation anneal presents the most severe problem, since it is often carried out at 900°C.

There are two approaches to self-aligned technology with GaAs. The first exploits refractory materials which become part of the final device. The second uses materials strictly to facilitate processing; these are either removed or left in place after they perform their function, but play no role in device behavior.

Refractory gate materials usually fall into the first category. Of these, WSi$_2$ and Ti-WN are favored since they have a more stable interface than Ti-W with GaAs at high temperatures [165, 166]. They can be deposited by sputtering or CVD, and are readily etched by RIE into the appropriate gate pattern, where they serve as a mask for the channel implant.

Figure 11.67 shows [167] the process sequence for a MESFET with a self-aligned gate made of Ti-WN. The use of an asymmetric placement reduces the parasitic source resistance while still maintaining a high breakdown voltage for the drain. In this process, the Ti-WN layer is grown by sputtering Ti-W in a nitrogen ambient. Next, an Ni pad is used to provide a T-gate mask for the source/drain implants which follow. This pad is removed after implantation, and the slice is capped with silicon oxynitride prior to annealing. Next, RIE is used to planarize the layer which is then delineated for the source/drain contacts. Ti/Au metallization is used for interconnection purposes.

An extremely popular self-aligned technology for GaAs is known as the SAINT (self-aligned implantation for N^+–layer technology) process [168]. The process sequence for fabricating a MESFET using this approach is shown in

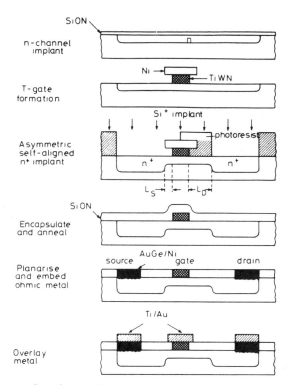

Fig. 11.67 Process flow for a self-aligned gate device. From Geissberger et al. [167]. With kind permission from the Institute of Electrical Engineers, UK.

Fig. 11.68, together with the cross section of the final structure. Here, a PECVD layer of Si_3N_4 is used to stabilize the leakage current of the final device. Next, a trilevel mask, consisting of sequential layers of polytetrafluoropropyl methacrylate (FPM), sputtered SiO_2, and photoresist, is used to provide the necessary overhang for the ion implantation of the source/drain regions. The photoresist and SiO_2 layers are removed prior to the sputter deposition of SiO_2 over the source/drain regions. Eventually, the FPM layer is removed prior to placement of the gate metal. In this manner, it is possible to use a large area gate region with a small gate length and a low parasitic capacitance.

Improvements in the SAINT process have evolved over the years [169]. Thus, integration of the device into a SRAM uses a two-level interconnect scheme as shown in Fig. 11.69a. An improved SAINT structure, using a buried p-layer, is shown in Fig. 11.69b.

11.9.6 The Heterojunction Bipolar Transistor

The heterojunction bipolar transistor (HBT) is a promising candidate for ultra-fast logic circuits, as well as for use in optical transmission systems with bit

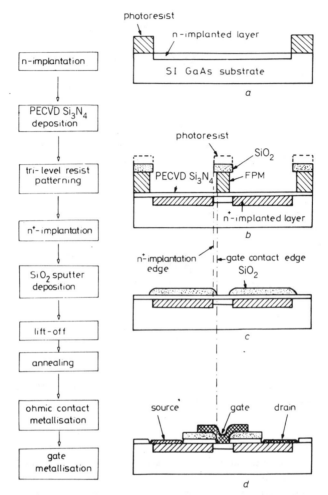

Fig. 11.68   Process flow for a SAINT MESFET. From Yamasaki et al. [168]. With kind permission from the Institute of Electrical Engineers, UK.

rates in excess of 10 Gbit/s [170]. Recently, GBW products in excess of 100 GHz have been reported for these devices. Devices of this type have the high transconductance that is typical of bipolar transistors, and are capable [171] of high-power operation.

HBT devices are of the n–p–n type, with a thin base which is heavily doped in order to obtain a low base-spreading resistance (and hence a high maximum frequency of oscillation). A high injection efficiency into a base of this type requires a heterostructure with an n-AlGaAs emitter.

Discrete devices are usually made on n^+-GaAs substrates which serve as the collector contact. For microcircuit applications, however, SI GaAs is used, and

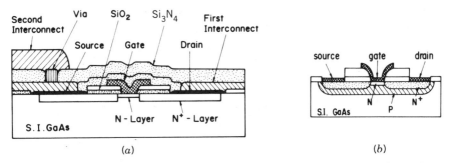

Fig. 11.69 Advanced SAINT structures. From Enoki et al. [169], ©1988. The Institute of Electrical and Electronic Engineers.

contacts are brought out on the top surface. The layer structure of such a device is shown in Fig. 11.70a. Here, the most critical element is the thin base, which is heavily doped. Additionally, the layer should be abrupt, and should not move during subsequent processing. Carbon doping, to a concentration of 10^{20} cm^{-3}, is ideal for this region and is used in OMVPE-grown structures. Beryllium is used in MBE structures, but is somewhat less suitable for this purpose [172].

Figure 11.70b shows a physical realization of a HBT. Here, wraparound base and collector contacts are used to reduce parasitic resistances. Tungsten base and emitter electrodes are used to obtain a self-aligned structure.

Cap				
	n$^+$-GaAs	500A°	2×10^{19}	cm^{-3}
Emitter	N-AlGaAs	1500A°	5×10^{17}	cm^{-3}
Base	P$^+$-GaAs	800A°	7.5×10^{18}	cm^{-3}
Collector	i-GaAs	3000A°		
Buffer	n$^+$-GaAs	5000A°	1×10^{19}	cm^{-3}
S.I. GaAs Substrate				

(a)

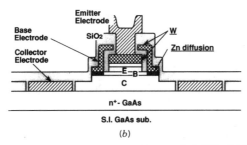

(b)

Fig. 11.70 Cross section of a HBT. From Ichino et al. [172], ©1990. The Institute of Electrical and Electronic Engineers.

TABLE 11.1 Depth of Focus (μm) as a Function of Wave Length and Resolution [39]

Wavelength Å	Resolution (μm)		
	0.8	0.5	0.3
3700	2.2	0.7	0.24
2400	2.7	1.1	0.38
1900	3.4	1.3	0.47

REFERENCES

1. S. Wolf and R. N. Tauber, *Silicon Processing for the VLSI Era, Vol. 2—Process Integration*, Lattice Press, Sunset Beach, CA, 1986.

2. R. A. Colclaser, *Microelectronics: Processing and Device Design*, Wiley, New York, 1980.

3. *An Introduction to Semiconductor Technology: GaAs and Related Compounds*, C. T. Wang, Ed. , Wiley, New York, 1990.

4. S. S. Ahmed, W. W. Asakawa, M. T. Bohr, S. S. Chambers, T. Deeter, M. Denham, J. K. Greason, W. W. Holt, R. R. Taylor, and I. Young, A Triple Diffused Approach for High Performance 0. 8 μm BiCMOS Technology, *Solid State Technol.* (Oct. 1992) p. 33.

5. Y. Yamaguchi, A. Ishibati, M. Shimizu, T. Nishimura, K. Tsukamoto, K. Horie, and Y. Akasaka, A High-Speed 0. 6 μm 16K CMOS Gate Array on a Thin SIMOX Film, *IEEE Trans. Electron Dev.* **40**, 179, (1993).

6. W. P. Maszara, Silicon on Insulator Wafer Bonding: A Review, *J. Electrochem. Soc.* **138**, 341 (1991).

7. G. Wallis, D. I. Pomeranz, Field Assisted Glass–Metal Sealing, *J. Appl. Phys.* **40**, 3496 (1968).

8. R. C. Frye, J. E. Griffith, and Y. H. Wong, A Field Associated Bonding Process for Silicon Dielectric Isolation, *J. Electrochem. Soc.* **133**, 1673 (1986).

9. J. B. Lasky, Wafer Bonding for Silicon-on-Insulator Technologies, *Appl. Phys. Lett.* **48**, 78 (1986).

10. J. Haisma, G. A. C. Spierings, U. K. K. Biermann, and J. A. Pols, *Jpn. J. Appl. Phys.* **28**, 1426 (1989).

11. A. Fathimulla, J. Abrahams, H. Hier, and T. Loughran, Growth and Fabrication of InGaAs/InAlAs HEMTs on Bonded-and-Etch-Back InP on Si, in *Second Inter-*

national Conference on Indium Phosphide and Related Materials, Denver, CO, 1990, p. 57.

12. Y. H. Lo, R. Bhat, D. M. Hwang, M. A. Koza, and T. P. Lee, Bonding by Atomic Rearrangement of InP/InGaAsP 1. 5 μm Wavelength Lasers on GaAs Substrates, *Appl. Phys. Lett.* **58**, 1962 (1991).

13. Y. Bäcklund, K. Hermansson, and L. Smith, Bond-Strength Measurements Related to Silicon Surface Hydrophilicity, *J. Electrochem. Soc.* **139**, 2299 (1992).

14. S. J. Yun, K-Y. Ahn, K-S. Yi, and S-W. Kang, Studies of Microvoids at the Interface of Direct Bonded Silicon Wafers, *J. Electrochem. Soc.* **139**, 2326 (1992).

15. K. E. Bean and W. R. Runyan, Dielectric Isolation: Comprehensive, Current, and Future, *J. Electrochem. Soc.* **124**, 5C (1977).

16. T. Suzuki, A. Mimura, T. Kamei, and T. Ogawa, Deformation in Dielectric Isolated Substrates and Its Control by a Multilayer Polysilicon Support Structure, *J. Electrochem. Soc.* **127**, 1537 (1980).

17. W. R. Runyan, *Semiconductor Measurements and Instrumentation*, McGraw-Hill, New York, 1975.

18. W. M. Cosney and L. H. Hall, The Extension of Self-Registered Gate and Doped-Oxide Diffusion Technology to the Fabrication of Complementary MOS Transistors, *IEEE Trans. Electron Dev.* **ED-20**, 469 (1973).

19. J. A. Appels, E. Kooi, M. M. Paffen, J. J. H. Schatorjé, and W. H. C. G. Verkuylen, Local Oxidation of Silicon and Its Application in Semiconductor Device Technology, *Philips Res. Rep.* **25**, 118 (1970).

20. K. Okada, K. Aomura, T. Nakamura, and H. Shida, A New Polysilicon Process for a Bipolar Device—PSA Technology, *IEEE Trans. Electron Dev.* **ED-26**, 385 (1979).

21. J. Lohstroh, J. D. P. v. d. Crommenacker, and A. J. Linssen, Oxide Isolated ISL Technologies, *IEEE Electron Dev. Lett.* **EDL-2**, 30 (1981).

22. E. Kooi, J. G. van Lierop, W. H. C. G. Verkuijlen, and R. deWerdt, LOCOS Devices *Philips Res. Rep.* **26**, 166 (1971).

23. R. H. Dennard, F. H. Gaensslen, E. J. Walker, and P. W. Cook, 1 μm MOSFET VLSI Technology: Part II. Device Designs and Characteristics for High Performance Logic Applications, *IEEE Trans. Electron Dev.* **ED-26**, 325 (1979).

24. K. Shibata and K. Taniguchi, Generation Mechanism of Dislocations in Local Oxidation of Silicon, *J. Electrochem. Soc.* **127**, 1383 (1980).

25. E. Bassous, H. N. Yu, and V. Maniscalco, Topology of Silicon Structure with Recessed SiO_2, *J. Electrochem. Soc.* **123**, 1729 (1976).

26. R. J. Powell, J. R. Ligenza, and M. S. Schneider, Selective Oxidation of Silicon in Low Temperature High-Pressure Steam, *IEEE Trans. Electron Dev.* **ED-21**, 636 (1974).

27. S. C. Su, Low Temperature Silicon Processing Techniques for VLSI Fabrication, *Solid State Technol.* (March 1981) p. 72.

28. C. A. Goodwin and J. W. Brossman, MOS Gate Oxide Defects Related to Treatment of Silicon Nitride Coated Wafers Prior to Local Oxidation, *J. Electrochem. Soc.* **129**, 1066 (1982).

29. J. R. Pfiester and P. B. Griffin, Anomalous Co-Diffusion of Germanium on Group III and V Dopants in Silicon, *Appl. Phys. Lett.* **54**, 471 (1988).

30. K. Y. Chiu, J. H. Moll, and J. Manoliu, A Bird's Beak Free Local Oxidation Technology Feasible for VLSI Circuit Fabrication, *IEEE Trans. Electron Dev.* **ED-29**, 536 (1982).

31. T. Mizuno, S. Sawada, S. Maeda, and S. Shinozaki, Oxidation Rate Reduction in the Submicrometer LOCOS Process, *IEEE Trans. Electron. Dev.* **ED-34**, 2255 (1987).

32. E. Kooi and J. A. Appels, Selective Oxidation of Silicon and its Device Applications, in *Semiconductor Silicon 1973*, Electrochem. Society, Princeton, NJ, 1973, p. 860.

33. N. C. C. Lu, Advanced Cell Structures for Dynamic RAMS, *IEEE Circuits and Devices Magazine* (Jan. 1989) p. 27.

34. R. N. Carlile, V. C. Liang, O. A. Palusinski, and M. M. Smadi, Trench Etches in Silicon with Controllable Sidewall Angle, *J. Electrochem. Soc.* **135**, 2059 (1988).

35. Y. Tamaki, S. Isomae, K. Sagara, and T. Kure, Evaluation of Dislocation Generation in V-Groove Isolation, *J. Electrochem. Soc.* **135**, 726 (1988).

36. V. J. Silvestri, Growth Kinematics of a Polysilicon Trench Refill Process, *J. Electrochem. Soc.* **133**, 2374 (1986).

37. V. J. Silvestri, Selective Epitaxial Trench (SET), *J. Electrochem. Soc.* **135**, 1808 (1988).

38. T. Morie and J. Murota, Trenches for Isolation, *Jpn. J. Appl. Phys.* **23**, 482 (1984).

39. S. Sivaram, H. Bath, R. Leggett, A. Maury, K. Monnig, and R. Tolles, Planarizing Interlevel Dielectrics by Chemical–Mechanical Polishing, *Solid State Technol.* (May 1992) p. 87.

40. J. Götzlich and H. Rysell, Tapered Windows in SiO_2, Si_3N_4 and Polysilicon Layers by Ion Implantation, *J. Electrochem. Soc.* **128**, 617 (1981).

41. L. K. White, Bilayer Etching of Field Oxides and Passivation Layers, *J. Electrochem. Soc.* **127**, 2687 (1980).

42. W. Kern and R. S. Rosler, Advances in Deposition Processes for Passivation Films, *J. Vac. Sci. Technol.* **14**, 1082 (1977).

43. J. S. Mercier, I. D. Calder, R. P. Beerkens, and H. M. Naguib, Rapid Isothermal Fusion of PSG Films, *J. Electrochem. Soc.* **132**, 2432 (1985).

44. W. Kern, W. A. Kurylo, and C. J. Tino, Optimized Chemical Vapor Deposition of Borophosphosilicate Glass Films, *RCA Rev.* **46**, 117 (1985).

45. I. Barsoni, H. Anzai, and J. Nishizawa, On Structural Impact of Lamp Annealing in Device Back-End Processing, *J. Electrochem. Soc.* **133**, 156 (1986).

46. R. A. Bowling and G. B. Larrabee, Deposition and Reflow of Phosphosilicate Glass, *J. Electrochem. Soc.* **132**, 141 (1985).

47. C. Y. Ting, V. J. Vivalda, and H. G. Schaefer, Study of Planarized Sputter-Deposited SiO_2, *J. Vac. Sci. Technol.* **15**, 1105 (1978).

48. D. B. Tuckerman and A. H. Weisberg, Planarization of Gold and Aluminum Thin Films Using a Pulsed Laser, *IEEE Electron Dev. Lett.* **EDL-7**, 1 (1986).

49. L. B. Rothman, Properties of Thin Polymide Films, *J. Electrochem. Soc.* **127**, 2216 (1980).

50. L. K. White, Planarization Properties of Resist and Polymide Coatings, *J. Electrochem. Soc.* **130**, 1543 (1983).

51. A. G. Emslie, F. T. Bonner, and L. G. Peck, Flow of a Viscous Liquid on a Rotating Disk, *J. Appl. Phys.* **29**, 858 (1958).

52. L. E. Stillwagon, R. G. Larson, and G. N. Taylor, Planarization of Substrate Topology by Spin Coating, *J. Electrochem. Soc.* **134**, 2030 (1987).

53. S. Ito, Y. Homma, E. Sasaki, S. Uchimura, and H. Morishima, Application of Surface Reformed Thick Spin-On Glass to MOS Device Planarization, *J. Electrochem. Soc.* **137**, 1212 (1990).

54. G. Smolinski, N. Lifshiftz, and V. Ryan, Spin-On Dielectrics: Good News and Bad News! in *Electronic Packaging Science IV*, R. Jaccodine, K. A. Jackson, E. D. Little and R. C. Sundahl, Eds., *Mater. Res. Soc. Proc.* **154**, 275 (1989).

55. A. C. Adams and C. D. Capio, Planarization of Phosphorus Doped Silicon Dioxide, *J. Electrochem. Soc.* **128**, 423 (1981).

56. T. H. Daubenspeck, J. K. DeBrosse, C. W. Koburger, M. Armacost, and J. R. Abernathy, Planarization of ULSI Technologies Over Variable Pattern Densities, *J. Electrochem. Soc.* **138**, 506 (1991).

57. L. M. Cook, Chemical Processes in Glass Polishing, *J. Noncryst. Solids* **120**, 152 (1990).

58. W. J. Patrick, W. L. Guthrie, C. L. Standley, and P. M. Schiable, Application of Chemical Mechanical Polishing to the Fabrication of VLSI Circuit Interconnections, *J. Electrochem. Soc.* **138**, 1778 (1991).

59. S. A. Cohen, M. A. Jaso, and A. A. Bright, Electrical Properties of Chemical-Mechanically Polished Tetraethylorthosilicate Films with and without a Capping Layer, *J. Electrochem. Soc.* **139**, 3572 (1992).

60. A. Saiki, S. Harada, T. Okubo, K. Mukai, and T. Kimura, A New Transistor with Two-Level Metal Electrodes, *J. Electrochem. Soc.* **124**, 1619 (1977).

61. H-V. Schreiber and E. Fröschle, High Quality RF-Sputtered Silicon Dioxide Layers, *J. Electrochem. Soc.* **123**, 30 (1976).

62. M. Hatzakis, B. J. Canavello, and J. M. Shaw, Single Step Optical Lift-Off Process, *IBM J. Res. Dev.* **24**, 452 (1980).

63. Y. Homma, H. Nozawa, and S. Harada, Polymide Liftoff Technology for High-Density LSI Metallization, *IEEE Trans. Electron Dev.* **ED-28**, 552 (1981).

64. L. Jastrebski, Origin and Control of Material Defects in Silicon VLSI Technologies: An Overview, *IEEE Trans. Electron Dev.* **ED-29**, 475 (1982).

65. S. K. Ghandhi, *Semiconductor Power Devices*, Wiley, New York, 1977.

66. R. A. Craven and H. W. Korb, Internal Gettering in Silicon, *Solid State Technol.* (July 1981) p. 55.

67. P. M. Petroff, G. A. Rozgonyi, and T. T. Sheng, Elimination of Process-Induced Stacking Faults by Pre-oxidation Gettering of Si Wafers, *J. Electrochem. Soc.* **123**, 565 (1976).

68. M. Nakamura and N. Oi, A Study of Gettering Effect of Metallic Impurities in Silicon, *Jpn. J. Appl. Phys.* **7**, 512 (1968).

69. S. P. Murarka, A Study of the Phosphorus Gettering of Gold in Silicon by the Use of Neutron Activation Analysis, *J. Electrochem. Soc.* **123**, 765 (1976).

70. K. Ohta, K. Yamada, M. Saitoh, K. Shimizu, and Y. Tarui, Quadruply Self-Aligned MOS (QSA-MOS)—A Short-Channel High-Speed High-Density MOSFET for VLSI, *IEEE Trans. Electron Dev.* **ED-27**, 1352 (1980).

71. H. J. Geipel, Jr. , N. Hsieh, M. H. Ishaq, C. W. Koburger, and F. R. White, Composite Silicide Gate Electrodes—Interconnections for VLSI Device Technologies, *IEEE Trans. Electron Dev.* **ED-27**, 1417 (1980).

72. M. E. Alperin, T. C. Hollaway, R. A. Haken, C. D. Gosmeyer, R. V. Karnaugh, and W. D. Parmantie, Development of the Self-Aligned Titanium Silicide Process for VLSI Applications, *IEEE J. Solid-State Circuits* **SC-20**, 61 (1985).

73. T. Furuyama, T. Ohsawa, Y. Watanabe, H. Isiuchi, T. Watanabe, T. Tanaka, K. Natori, and O. Ozawa, An Experimental 4-Mbit CMOS DRAM, *IEEE J. Solid-State Circuits* **SC-21**, 605 (1986).

74. S. E. Swirhun, E. Sangiorgi, A. J. Weeks, R. M. Swanson, K. C. Saraswat, and R. W. Dutton, A VLSI-Suitable Schottky Barrier CMOS Processes, *IEEE J. Solid-State Circuits* **SC-20**, 114 (1985).

75. Y. Taur, G. J. Hu, R. H. Dennard, L. M. Terman, C-Y. Tung, and K. E. Petrillo, A Self-Aligned 1-μm-Channel CMOS Technology with Retrograde n-Well and Thin Epitaxy, *IEEE J. Solid-State Circuits* **SC-20**, 123 (1985).

76. C. N. Anagnostopoulos, E. T. Nelson, J. P. Levine, K. Y. Wong, and D. N. Nichols, Latch-Up and Image Crosstalk Suppression by Internal Gettering, *IEEE J. Solid-State Circuits* **SC-19**, 91 (1984).

77. T. Yamaguchi, S. Morimoto, H. K. Park, and G. C. Eiden, Process and Device Performance of Submicrometer-Channel CMOS Devices Using Deep-Trench Isolation and Self-Aligned $TiSi_2$ Technologies, *IEEE J. Solid-State Circuits* **SC-20**, 104 (1985).

78. T. I. Chappell, S. E. Schuster, B. A. Chappell, J. W. Allan, J. Y. -C. Sun, S. P. Kelpner, R. L. Franch, P. F. Greier, and P. J. Restle, A 3.5-ns/77 K and 6.2-ns/300 K 64K CMOS RAM with ECL Interfaces, *IEEE Trans. Solid-State Circuits* **24**, 859 (1989).

79. G. Zimmer, B. Hoefflinger, and J. Schneider, A Fully Implanted N-MOS, CMOS Bipolar Technology for VLSI of Analog-Digital Systems, *IEEE Trans. Electron Dev.* **ED-26**, 390 (1979).

80. V. L. Rideout, One-Device Cells for Dynamic Random-Access Memories: A Tutorial, *IEEE Trans. Electron Dev.* **ED-26**, 839 (1979).

81. E. Arai and N. Ieda, A 64-kbit Dynamic MOS RAM, *IEEE J. Solid- State Circuits* **SC-13**, 333 (1978).

82. A. F. Tasch, P. K. Chatterjee, H-S. Fu, and T. C. Hollaway, The Hi-C RAM Cell Concept, *IEEE Trans. Electron Dev.* **ED-25**, 33 (1978).

83. T. Kaga, T. Kure, H. Shinriki, Y. Kawamoto, F. Murai, T. Nishida, Y. Nakagome, D. Hisamoto, T. Kisu, E. Takeda, and K. Itoh, Crown-Shaped Stacked-Capacitor

Cell for 1. 5-V Operation 65-Mb DRAM's, *IEEE Trans. Electron Dev.* **38**, 255 (1991).

84. D. Chin, C. Kim, Y. Choi, D-S. Min, H. S. Hwang, H. Choi, S. Cho, T. Y. Chung, C. J. Park, Y. Shin, K. Suh, and Y. E. Park, An Experimental 16-Mbit DRAM with Reduced Peak-Current Noise, *IEEE J. Solid-State Circuits* **24**, 1191 (1989).

85. N. C. C. Lu, Advanced Cell Structures for Dynamic RAMs, *IEEE Circuits and Devices Magazine* (Jan. 1989) p. 27.

86. N. C-C. Lu, P. E. Cottrell, W. J. Craig, S. Dash, D. L. Critchlow, R. L. Mohler, B. J. Machesney, T. H. Ning, W. P. Noble, R. M. Parent, R. E. Scheuerlein, E. J. Sprogis, and L. M. Terman, A Substrate-Plate Trench-Capacitor (SPT) Memory Cell for Dynamic RAMs, *IEEE J. Solid-State Circuits* **SC-21**, 627 (1986).

87. S. Fujii, M. Ogihara, M. Shimizu, M. Yoshida, K. Numata, T. Hara, S. Watanabe, S. Sawada, T. Mizuno, J. Kumagai, S. Yoshikawa, S. Kaki, Y. Saito, H. Aochi, T. Hamamoto, and K. Toita, A 45-ns 16-Mbit DRAM with Triple-Well Structure, *IEEE J. Solid-State Circuits* **24**, 1170 (1989).

88. F. Masuoka, S. Ariizumi, T. Iwase, M. Ono, and N. Endo, An 80 ns 1 Mbit MASK ROM with a New Memory Cell, *IEEE J. Solid-State Circuits* **SC-19**, 651 (1984).

89. L. R. Metzger, A 16K CMOS PROM with Polysilicon Fusible Links, *IEEE J. Solid-State Circuits* **SC-18**, 562 (1983).

90. T. Fukushima, K. Ueno, Y. Matsuzaki, and K. Tanaka, A 40 ns 64 kbit Junction-Shorting PROM, *IEEE J. Solid-State Circuits* **SC-19**, 187 (1984).

91. D. Frohman-Bentchkowsky, FAMOS: A New Semiconductor Charge Storage Device, *Solid State Electron.* **17**, 517 (1974).

92. A. Scheibe and H. Schalte, Technology of a New *n*-Channel One-Transistor EAROM Cell Called SIMOS, *IEEE Trans. Electron Dev.* **ED-24**, 600 (1977).

93. Y. Narita, S. Ohya, Y. Murao, S. Kanauchi, and M. Kikuchi, A High-Speed 1-Mbit EPROM with a Ti-Silicided Gate, *IEEE J. Solid-State Circuits* **SC-20**, 418 (1985).

94. D. H. Oto, V. K. Dham, K. H. Gudget, M. J. Reitsma, G. S. Gongwer, Y. W. Hu, J. F. Olund, H. S. Jones, Jr., and S. T. K. Nieh, High-Voltage Regulation and Process Considerations for High-Density 5 V-Only E^{2}PROM's, *IEEE J. Solid-State Circuits* **SC-18**, 532 (1983).

95. G. Samachisa, C-S. Su, Y-S. Kao, G. Smarandoiu, C-Y. M. Wang, T. Wong, and C. Hu, A 128K Flash EEPROM Using Double-Polysilicon Technology, *IEEE J. Solid-State Circuits* **SC-22**, 676 (1987).

96. V. N. Kynett, N. L. Fandrich, J. Anderson, P. Dix, O. Jungroth, J. A. Kreifels, R. A. Lodenquai, B. Vajdic, S. Wells, M. D. Winston, and L. Yang, A 90-ns One-Million Erase/Program Cycle 1-Mbit Flash Memory, *IEEE J. Solid- State Circuits* **24**, 1259 (1989).

97. M. McConnell, B. Ashmore, R. Bussey, M. Gill, S-W. Lin, D. McElroy, J. F. Schreck, P. Shah, H. Steigler, P. Troung, A. L. Esquivel, J. Paterson, and B. Riemenschneider, An Experimental 4-Mb Flash EEPROM with Sector Erase, *IEEE J. Solid-State Circuits* **26**, 484 (1991).

98. D. K. Brown, S. M. Hu, and J. M. Morrissey, Flaws in Sidewall Oxides Grown on Polysilicon Gates, *J. Electrochem. Soc.* **129**, 1084 (1982).

99. C. Papadas, G. Ghibaudo, G. Pananakakis, G. Gounelle, P. Mortini, and C. Riva, Influence of Tunnel Oxide Thickness Variation on the Programmed Window of Flotox EEPROM Cells, *Solid-State Electronics* **35**, 1195 (1992).

100. T. Hamamoto, S. Yoshikawa, H. Aochi, S. Kaki, and S. Sawada, Trench–Trench Leakage Current Characteristics in the Stacked Trench Capacitor (SST) Cell, *IEEE Trans. Electron Dev.* **38**, 419 (1991).

101. T. C. May and M. H. Woods, Alpha-Particle-Induced Soft Errors in Dynamic Memories, *IEEE Trans. Electron Dev.* **ED-26**, 2 (1979).

102. Y. Idei, N. Homma, H. Nambu, and Y. Sakurai, Soft-Error Characteristics in Bipolar Memory Cells with Small Critical Charge, *IEEE Trans. Electron Dev.* **38**, 2465 (1991).

103. M. R. Splinter, A 2 μm Silicon Gate C-MOS/SOS Technology, *IEEE Trans. Electron Dev.* **ED-25**, 996 (1978).

104. S. N. Lee, R. A. Kjar, and G. Kinoshita, Island Edge Effects in C-MOS/SOS Transistor, *IEEE Trans. Electron Dev.* **ED-25**, 971 (1978).

105. K. Maegushi, M. Okhaski, J. Iwamura, S. Taguchi, E. Sugino, T. Sato, and H. Tango, 4 μm LSI on SIS Using Coplanar-II Process, *IEEE Trans. Electron Dev.* **ED-25**, 945 (1978).

106. M. Haond and D. Le Néel, Lateral Isolation in SOI CMOS Technology, *Solid State Technol.* (July 1991) p. 47.

107. C. Clavin and U. Langmann, Monolithic Multigigabit/s Silicon Bipolar Circuits, *Electron. Lett.* **20**, 471 (1986).

108. R. H. Derkson and H-M. Rein, 7. 3 GHz Dynamic Frequency Dividers Monolithic Integrated in a Standard Bipolar Technology, *IEEE Trans. Microwave Theory Tech.* **MTT-36**, 537 (1988).

109. M. P. Cooke, G. W. Sumerling, T. V. Muoi, and A. C. Carter, Integrated Circuits for a 200-Mbit/s Fiber-Optic Link, *IEEE J. Solid-State Circuits* **SC-21**, 909 (1986).

110. M. Kimura, T. Okabe, I. Shimizu, Y. Nagai, and K. Hoya, A Flat-Panel Display Control IC with 150-V Drivers, *IEEE J. Solid-State Circuits* **SC-21**, 971 (1986).

111. M. Suzuki, M. Hirata, and Y. Ito, An 86K Component Bipolar VLSI Masterslice with a 290-ps Loaded Gate Delay, *IEEE J. Solid-State Circuits* **SC-22**, 1 (1987).

112. S-C. Lee and A. S. Bass, A 2500 Gate Bipolar Macrocell Array with 250 ps Gate Delay, *IEEE J. Solid-State Circuits* **SC-17**, 913 (1982).

113. M. Tatsuki, S. Kato, M. Okabe, H. Yakushiji, and Y. Kuramitsu, An ECL 5000-Gate Gate Array with 190-ps Gate Delay, *IEEE J. Solid-State Circuits* **SC-21**, 234 (1986).

114. P. R. Wilson, The Emitter-Base Breakdown Voltage of Planar Transistors, *Solid State Electron.* **17**, 465 (1974).

115. D. P. Kennedy and R. R. O'Brien, Avalanche Breakdown Calculations for a Planar $p-n$ Junction, *IBM J. Res. Dev.* **10**, 213 (1966).

116. J. Agraz-Güereña, P. T. Panousis, and B. L. Morris, OXIL, A Versatile Bipolar VLSI Technology, *IEEE Trans. Electron Dev.* **ED-27**, 1397 (1980).

117. W. M. Gegg, J. L. Saltich, R. M. Roop, and W. L. George, Ion-Implanted Super-Gain Transistors, *IEEE J. Solid-State Circuits* **SC-11**, 485 (1976).

118. B. T. Murphy, V. J. Glinski, P. A. Gary, and R. A. Pederson, Collector-Diffusion Isolated Integrated Circuits, *Proc. IEEE* **57**, 1523 (1969).

119. B. T. Murphy, S. M. Neville, and R. A. Pedersen, Simplified Bipolar Technology and Its Application to Systems, *IEEE J. Solid-State Circuits* **SC-5**, 7 (1970).

120. J. Graul, A. Glasl and H. Murrmann, High-Performance Transistors with Arsenic-Implanted Polysil Emitters, *IEEE J. Solid-State Circuits* **SC-11**, 491 (1976).

121. T. H. Ning and R. D. Isaac, Effect of Emitter Contact on Current Gain of Silicon Bipolar Devices, *IEEE Trans. Electron Dev.* **ED-27**, 2051 (1980).

122. P. Ashburn, D. Roulston and C. R. Sevakumar, Comparison of Experimental and Computed Results on Arsenic- and Phosphorus-Doped Polysilicon Emitter Bipolar Transistors, *IEEE Trans. Electron Dev.* **ED-34**, 1346 (1987).

123. G. L. Patton, J. C. Bravman and J. D. Plummer, Physics, Technology, and Modeling of Polysilicon Emitter Contracts for VLSI Bipolar Devices, *IEEE Trans. Electron Dev.* **ED-33**, 1754 (1986).

124. M. H. Juang and H. C. Cheng, Formation of Shallow p^+n Junctions by Implanting BF_2^+ Ions Into Thin Cobalt Films on Silicon Substrates, *Solid-State Electronics* **35**, 453 (1992).

125. S. K. Wiedmann, Advancements in Bipolar VLSI Circuits and Technologies, *IEEE J. Solid-State Circuits* **SC-19**, 282 (1984).

126. M. Suzuki, S. Konaka, H. Ichino, T. Sakai, and S. Horiguchi, Design and Application of a 2500-Gate Bipolar Macrocell Array, *IEEE J. Solid- State Circuits* **SC-20**, 1025 (1985).

127. H-M. Rein, Multi-Gigabit-Per-Second Silicon Bipolar IC's for Future Optical-Fiber Transmission Systems, *IEEE J. Solid-State Circuits* **SC-23**, 664 (1988).

128. H. Ichino, N. Isihara, M. Suzuki and S. Konaka, 18-GHz 1/8 Dynamic Frequency Divider Using Si Bipolar Technologies, *IEEE J. Solid-State Circuits* **24**, 1723 (1989).

129. G. P. Li, T. H. Ning, C. T. Chuang, M. B. Ketchen, D. D-L. Tang, and J. Mauer, An Advanced High-Performance Trench-Isolated Self-Aligned Bipolar Technology, *IEEE Trans. Electron Dev.* **ED-34**, 2246 (1987).

130. G. Kitsukawa, K. Yanagisawa, Y. Kobayashi, Y. Kinoshita, T. Ohta, T. Udagawa, H. Miwa, H. Miyazawa, Y. Kawajiri, Y. Ouchi, H. Tsukada, T. Matsumoto, and K. Itoh, A 23-ns 1-Mb BiCMOS Dram, *IEEE J. Solid-State Circuits* **25**, 1102 (1990).

131. H. Van Tran, D. B. Scott, P. K. Fung, R. H. Havemann, R. H. Eklund, T. E. Ham, R. A. Haken, and A. H. Shah, An 8-ns 256K ECL SRAM with CMOS Memory Array and Battery Backup Capability, *IEEE J. Solid-State Circuits* **23**, 1041 (1988).

132. M. Tsumura, R. Takeuchi, and I. Shimizu, BiCMOS Thermal Head Intelligent Driver, *IEEE J. Solid-State Circuits* **23**, 437 (1988).

133. J-I. Miyamoto, S. Saito, H. Momose, H. Shibata, K. Kanzaki, and T. Iizuka, A High-Speed 64K CMOS RAM with Bipolar Sense Amplifiers, *IEEE J. Solid-State Circuits* **SC-19**, 557 (1984).

134. K. Ogiue, M. Odaka, S. Miyaoka, I. Masuda, T. Ikeda, and K. Tonomura, 13-ns,

500-mW, 64-kbit ECL RAM Using HI-BICMOS Technology, *IEEE J. Solid-State Circuits* **SC-21**, 681 (1986).

135. K. Soejima, A. Shida, H. Koga, J. Ukai, H. Sata, and M. Hirata, A BiCMOS Technology with 660-MHz Vertical *p–n–p* Transistor for Analog/Digital ASIC's, *IEEE J. Solid-State Circuits* **25**, 410 (1990).

136. Y. Tarui, Y. Hayashi, H. Teshima, and T. Sekigawa, Transistor Schottky-Barrier-Diode Integrated Logic Circuit, *IEEE J. Solid State Circuits* **SC-4**, 3 (1969).

137. T. Yanagawa, Resistance of Narrow Diffused Layers, *IEEE Trans. Electron Dev.* **ED-19**, 1166 (1972).

138. G. D'Andrea and H. Murrman, Correction Terms for Contacts to Diffused Resistors, *IEEE Trans. Electron. Dev.* **ED-17**, 484 (1970).

139. R. M. Warner, Jr. , and J. N. Fordemwalt, *Integrated Circuits*, McGraw-Hill, New York, 1965.

140. L. I. Maisell and R. Glang, Eds. , *Handbook of Thin Film Technology*, McGraw-Hill, New York, 1970.

141. J. L. Vossen and W. Kern, Eds. , *Thin Film Processes*, Academic Press, New York, 1978.

142. R. A. Powell and R. Chow, Dopant Activation and Redistribution in As⁺-Implanted Polycrystalline Si by Rapid Thermal Processing, *J. Electrochem. Soc.* **132**, 194 (1985).

143. W. R. Wisseman and J. G. Oakes, Eds. , Special Issue on Advances in Monolithic Microwave III–V Devices and Circuits, *IEEE Trans. Electron Dev.* **ED-28**, 133 (1981).

144. R. C. Eden, B. M. Welch, R. Zucca, and S. I. Long, The Prospects for Ultrahigh-Speed VLSI GaAs Digital Logic, *IEEE J. Solid State Circuits* **SC-14**, 221 (1979).

145. K. Lehovec and R. Zuleeg, Analysis of GaAs FET's for Integrated Logic, *IEEE Trans. Electron Dev.* **ED-27**, 1074 (1980).

146. S. L. Long and S. E. Butner, *Gallium Arsenide Integrated Circuit Design*, McGraw-Hill, New York, 1990.

147. R. L. Van Tuyl and C. A. Liechti, High Speed Integrated Logic with GaAs MESFETs, *IEEE J. Solid-State Circuits* **SC-9**, 269 (1974).

148. R. L. Van Tuyl, C. A. Liechti, R. E. Lee, and E. Gowen, GaAs MESFET Logic with 4 GHz Clock Rate, *IEEE J. Solid State Circuits* **SC-12**, 485 (1977).

149. D. L. Rode, B. Schwartz, and J. V. DiLorenzo, Electrolytic Etching and Electron Mobility of GaAs for FET's, *Solid State Electron.* **17**, 1119 (1974).

150. B. M. Welch, Y-D. Shen, R. Zucca, R. C. Eden, and S. I. Long, LSI Processing Technology for Planar GaAs Integrated Circuits, *IEEE Trans. Electron Dev.* **ED-27**, 1116 (1980).

151. M. N. Yoder, Ohmic Contacts in GaAs, *Solid State Electron.* **23**, 117 (1980).

152. Laser and Electron-Beam Solid Interactions and Materials Processing, J. F. Gibbons, L. D. Hess, and T. W. Sigmon, Eds. , *Proceedings of the Materials Research Society Annual Meeting*, Boston, MA, 1980.

153. D. C. Miller, The Alloying of Gold and Gold Alloy Ohmic Contact Metallizations with Gallium Arsenide, *J. Electrochem. Soc.* **127**, 467 (1980).

154. K. Ohata, H. Itoh, F. Hasegawa, and Y. Fujiki, Super Low-Noise GaAs MESFET's with a Deep-Recess Structure, *IEEE Trans. Electron Dev.* **ED-27**, 1029 (1980).

155. R. Williams, *Modern GaAs Processing Methods*, Artech House, Boston, 1990.

156. M. Ida, T. Mizutani, K. Asai, M. Uchida, K. Shimada, and S. Ishida, Fabrication Technology for an 80-ps Normally-Off GaAs MESFET Logic, *IEEE Trans. Electron Dev.* **ED-28**, 489 (1981).

157. R. Zuleeg, J. K. Notthoff, and K. Lehovec, Femtojoule High-Speed Planar GaAs E-JFET Logic, *IEEE Trans. Electron Dev.* **ED-25**, 628 (1978).

158. K. Gonoi, I. Honbori, M. Wada, K. Togashi, and Y. Kato, A GaAs 8 × 8-bit Multiplier/Accumulator Using JFET DCFL, *IEEE J. Solid-State Circuits* **SC-21**, 523 (1986).

159. R. S. Hinds, S. R. Canaga, G. M. Lee, and A. K. Choudhury, A 20K GaAs Array With 10K of Embedded SRAM, *IEEE J. Solid-State Circuits* **26**, 245 (1991).

160. S. Takano, H. Makino, N. Tanino, M. Noda, K. Nishitani, and S. Kayano, A GaAs 16K SRAM with a Single 1-V Supply, *IEEE J. Solid-State Circuits* **SC-22**, 699 (1987).

161. T. Nimura, S. Hiyamizu, K. Joshin, and K. Hikosaka, Enhancement Mode High Electron Mobility Transistor for Logic Applications, *Jpn. J. Appl. Phys.* **20**, L317 (1981).

162. V. Swaminathan and A. T. Macrander, *Materials Aspects of GaAs and InP Based Structures*, Prentice-Hall, Englewood Cliffs, NJ, 1991.

163. M. Abe and T. Nimura, Ultrahigh-Speed HEMT LSI Technology for Supercomputer, *IEEE J. Solid-State Circuits* **26**, 1337 (1991).

164. K. Kajii, Y. Watanabe, M. Suzuki, I. Hanyu, M. Kosugi, K. Odani, T. Mimura, and M. Abe, A 40-ps High Electron Mobility Transistor 4.1 K Gate Array, *IEEE J. Solid-State Circuits* **23**, 485 (1988).

165. Y. Katayama, M. Morioka, Y. Sawada, K. Ueyanagi, T. Mishima, Y. Ono, Y. Usogawa, and Y. Shiraki, A New Two-Dimensional Electron Gas Field-Effect Transistor Fabricated on Undoped AlGaAs–GaAs Heterostructure, *Jpn. J. Appl. Phys.* **23** L150 (1984).

166. G. Nuzillat, E. H. Perea, G. Bert, F. Damay-Kavala, M. Gloanec, M. Peltier, T. P. Ngu, and C. Arnodo, GaAs MESFET IC's for Gigabit Logic Applications, *IEEE J. Solid-State Circuits* **SC-17**, 569 (1982).

167. A. E. Geissberger, R. A. Sadler, E. L. Griffin, I. J. Bahl, and M. L. Balzan, Refractory Self-Aligned Gate Technology for GaAs Microwave FETs and MMICs, *Electron. Lett.* **23**, 1073 (1987).

168. K. Yamasaki, K. Asai, T. Mizutani, and K. Kurumada, Self-Align Implantation for n^+-Layer Technology (SAINT) for High-Speed GaAs ICs, *Electronics Lett.* **18**, 119 (1982).

169. T. Enoki, K. Yamasaki, K. Osafune, and K. Ohwada, 0. 3-μm Advanced SAINT FET's Having Asymmetric n^+-Layers for Ultra-High-Frequency GaAs MMIC's, *IEEE Trans. Electron. Dev.* **ED-35**, 18 (1988).

170. H. Kroemer, Heterostructure Bipolar Transistors and Integrated Circuits, *Proc. IEEE* **70**, 13 (1982).

171. W. Liu and B. Bayraktaroglu, Theoretical Calculations of Temperature and Current Profiles in Multi-finger Heterojunction Bipolar Transistors, *Solid-State Electron.* **36**, 125 (1993).

172. H. Ichino, N. Ishiharo, Y. Yamauchi, O. Nakajima, K. Nagata, and T. Nittono, 12-Gb/s Decision Circuit IC Using AlGaAs/GaAs HBT Technology, *IEEE J. Solid-State Circuits* **25**, 1538 (1990).

THE MATHEMATICS OF DIFFUSION

This appendix considers some important mathematical techniques which can be used for solving problems involving diffusion in seminconductors. Exhaustive treatments of this subject are given elsewhere [1–3].

A.1 SOLUTIONS FOR A CONSTANT DIFFUSION COEFFICIENT

Many diffusion situations can be handled if the diffusion coefficient is assumed to be concentration independent. For this case, Fick's second law, in one dimension, takes the form

$$\frac{\partial N}{\partial t} = D\frac{\partial^2 N}{\partial x^2} \tag{A.1}$$

where N is the volume concentration of the diffusing impurity. It can be shown by direct differentiation that

$$N = \frac{A}{t^{1/2}} e^{-x^2/4Dt} \tag{A.2}$$

is a solution of this equation, where A is an arbitrary constant. This expression is symmetric in x, and tends to zero as x tends to infinity. The total amount of impurity involved in this diffusional process, at any given time, is

$$Q = \int_{-\infty}^{\infty} N\,dx \tag{A.3}$$

Substituting in Eq. (A.2), gives

$$Q = 2A\sqrt{\pi D} \tag{A.4}$$

This is also the initial amount of material at $t = 0$ and $x = 0$. Solving for A, the volume concentration is obtained as

$$N(x, t) = \frac{Q}{2\sqrt{\pi Dt}} e^{-(x/2\sqrt{Dt})^2} \tag{A.5}$$

A.1.1 Reflection and Superposition

The symmetric nature of the solution of Eq. (A.5) allows us to build up solutions for other diffusion situations by using the concept of reflection at a boundary, combined with superposition. This is possible because of the linear character of the diffusion equation. Consider, for example, the above situation for diffusion in the positive direction alone, i.e., for diffusion from a sheet of impurities Q (cm^{-2}) into a semi-infinite body ($0 \le x \le \infty$). A solution can be obtained from Eq. (A.5) by assuming that the solution for negative values of x is reflected in the plane of $x = 0$, and superposed on the distribution of $x \ge 0$. Thus for this case,

$$N = \frac{Q}{\sqrt{\pi D t}} e^{-(x/2\sqrt{Dt})^2} \tag{A.6}$$

This is the gaussian distribution given in Eq. (4.51); it is useful for describing base diffusion profiles in a transistor (see Fig. 4.17).

A.1.2 Extended Initial Conditions

It is also possible to use Eq. (A.5) to determine the solution for other initial conditions. Consider, for example, diffusion from an *infinite* source into an infinite region ($-\infty < x < \infty$). Assume that this source has a concentration N_0, extending over $-\infty \le x \le 0$. Any finite element of this source, of thickness $d\xi$, contains a quantity of impurity $N_0 \, d\xi$, at a distance ξ from any given plane P.

The concentration of P due to this sheet source is

$$\frac{N_0 \, d\xi}{2\sqrt{\pi D t}} e^{-\xi^2/4Dt} \tag{A.7}$$

For an infinite source it follows that

$$N(x, t) = \frac{N_0}{2\sqrt{\pi D t}} \int_x^\infty e^{-\xi^2/4Dt} \, d\xi \tag{A.8}$$

The error function erf y is defined by

$$\text{erf } y = \frac{2}{\sqrt{\pi}} \int_0^y e^{-z^2} \, dz \tag{A.9}$$

Values of erf y for different values of y are given in Table A.1. Combining with Eq. (A.8), gives

$$N(x, t) = \frac{N_0}{2} \left(1 - \text{erf} \frac{x}{2\sqrt{Dt}} \right) \tag{A.10}$$

The function $(1 - \text{erf } y)$ is often written as the complementary error function, erfc y.

Again, the concept of reflection from a boundary can be used to determine the solution for diffusion into a semi-infinite region. It follows that, for this case ($\infty < x \leq 0$),

$$N(x, t) = N_0 \, \mathrm{erfc} \, \frac{x}{2\sqrt{Dt}} \qquad (A.11)$$

This is the complementary error function distribution given in Eq. (4.49); it is useful for describing emitter diffusion profiles in a transistor. A plot of this function is also shown in Fig. 4.17.

Both Eqs. (A.6) and (A.11) show that a single dimensionless parameter $x/2\sqrt{Dt}$ is involved in describing these impurity distributions. It can be readily shown that this is true for any diffusion situation involving a semi-infinite medium with zero background concentration, whose surface is fixed. It follows that the penetration depth x is proportional to the square root of time in these situations. In practice, this relationship is found to hold for the vast majority of diffusion situations in silicon as well as in gallium arsenide. This is true for cases where D is concentration dependent, as well as for situations where D is constant.

A.1.3 Diffusion through a Narrow Slot

This serves as an example of direct solution for a three-dimensional situation. It is of practical importance in high-speed, shallow-diffused devices. The diffusion profile for this case is derived by first considering a limited point source. Then the equation describing diffusion from this source into an infinite volume is

$$\frac{\partial N}{\partial t} = D \left(\frac{\partial^2 N}{\partial x^2} + \frac{\partial^2 N}{\partial y^2} + \frac{\partial^2 N}{\partial z^2} \right) \qquad (A.12)$$

The solution of this equation takes the form

$$N = \frac{A}{\sqrt{t}} e^{-(x^2 + y^2 + z^2)/4Dt} \qquad (A.13)$$

If M is the amount of impurity at this point, then

$$M = \int_{-\infty}^{\infty} \int_{-\infty}^{\infty} \int_{-\infty}^{\infty} N \, dx \, dy \, dz = 8\pi DA \qquad (A.14)$$

The concentration at a distance r from a point source in an infinite volume is obtained by combining Eqs. (A.13) and (A.14), to give

$$N(r, t) = \frac{M}{8\sqrt{\pi Dt}} e^{-r^2/4Dt} \qquad (A.15)$$

Diffusion from a line source into an infinite volume can be determined by integrating this solution with respect to a space variable. Thus if diffusion is considered into an infinite volume from a line source along the y axis, then

$$N = \int_{-\infty}^{\infty} \frac{\delta}{4\pi Dt} e^{-(x^2+y^2+z^2)/4Dt} \, dy \tag{A.16}$$

where δ is the amount of diffusing impurity deposited initially, per unit length of the line source. Solving, the concentration at a distance r is obtained as

$$N(r, t) = \frac{\delta}{4\pi Dt} e^{-r^2/4Dt} \tag{A.17}$$

Solution for diffusion into a semi-infinite plane follows from the reflecting boundary method. For this situation,

$$N(r, t) = \frac{\delta}{2\pi Dt} e^{-r^2/4Dt} \tag{A.18}$$

Diffusion from a narrow slot is approximated by this distribution. In this case, $\delta \simeq QW$, where W is the width of the slot and Q is the surface density of dopant in atoms/cm^2.

A.1.4 Miscellaneous Useful Solutions

A number of solutions of the diffusion equation are available in the published literature. Many of these are of practical importance in integrated circuit fabrication technology. Some of these are now given, without their derivation.

1. *Diffusion from a limited source of finite width.*   Consider an impurity of concentration N_0, deposited in a finite width of x_0. Then the resulting impurity profile in the semiconductor after time t is given by

$$N(x, t) = \frac{N_0}{2} \left[\text{erf} \left(\frac{x - x_0}{2\sqrt{Dt}} \right) - \text{erf} \left(\frac{x + x_0}{2\sqrt{Dt}} \right) \right] \tag{A.19}$$

The surface concentration resulting from this diffusion is

$$N(0, t) = N_0 \left[1 - \text{erfc} \left(\frac{x_0}{2\sqrt{Dt}} \right) \right] \tag{A.20}$$

2. *Out-diffusion from a uniformly doped buried layer.*   Consider a uniformly doped buried layer, of doping concentration N^+, penetrating

from $x = 0$ to $x = x^+$ into a substrate. An undoped epitaxial layer, from $x = 0$ to $x = -x_{epi}$, is grown on this substrate. This is subsequently processed by a series of high-temperature steps, resulting in an effective value of Dt. Then the out-diffusion of the buried layer into the epitaxial region $(-x_{epi} \le x \le 0)$ is given by

$$N(x, t) = \frac{N^+}{2} \left[\text{erfc} \left(\frac{-x}{2\sqrt{Dt}} \right) + \text{erfc} \left(\frac{2x_{epi} + x}{2\sqrt{Dt}} \right) \right.$$

$$\left. - \text{erfc} \left(\frac{x^+ - x}{2\sqrt{Dt}} \right) - \text{erfc} \left(\frac{2x_{epi} - x^+ + x}{2\sqrt{Dt}} \right) \right] \quad \text{(A.21)}$$

3. *Diffusion from a doped oxide source.* Consider an oxide, of thickness x_{ox}, having a dopant concentration of N_{ox}. Let D_{ox} and D be the diffusion coefficients of the impurity in the oxide and the semiconductor, respectively, and let m be the impurity segregation coefficient at the oxide–semiconductor interface (see Section 4.8.1). Then, for short diffusion times, such that $x_{ox} > 4\sqrt{Dt}$, the impurity concentration in the semiconductor is given by

$$N(x, t) \simeq \frac{N_{ox} (D_{ox}/D)^{1/2}}{1 + k} \text{erfc} \left(\frac{x}{2\sqrt{Dt}} \right) \quad \text{(A.22)}$$

where

$$k = \frac{1}{m} \left(\frac{D_{ox}}{D} \right)^{1/2} \quad \text{(A.23)}$$

4. *Bilateral diffusion into a finite body.* Consider a slice of thickness x_0, into which diffusions are made from both sides. For all t let

$$N(0, t) = N(x_0, t) = N_0 \quad \text{(A.24)}$$

Then

$$N(x, t) = N_0 \left[1 - \frac{4}{\pi} e^{-\pi^2 Dt/x_0^2} \sin \left(\frac{\pi x}{x_0} \right) \right] \quad \text{(A.25)}$$

Out-diffusion from two sides of a uniformly doped slice of thickness x_0 is given by

$$N(x, t) = \frac{4N_0}{\pi} \left[e^{-\pi^2 Dt/x_0^2} \sin \left(\frac{\pi x}{x_0} \right) \right] \quad \text{(A.26)}$$

where N_0 is the initial doping concentration.

A.1.5 Some Useful Error Function Relations

The following relations are useful in the solution of many diffusion problems:

1. $\displaystyle \int_0^z e^{-y^2}\, dy = \frac{\sqrt{\pi}}{2}\, \text{erf } z$ (A.27)

2. $\displaystyle \int_0^z e^{-y^2}\, dy = z - \frac{z^3}{3 \times 1!} + \frac{z^5}{5 \times 2!} - \cdots$ (A.28)

3. $\displaystyle \int_0^\infty e^{-y^2}\, dy = \frac{\sqrt{\pi}}{2}$ (A.29)

4. $\displaystyle \frac{d}{dz}\,(\text{erf } z) = \frac{2}{\sqrt{\pi}}\, e^{-z^2}$ (A.30)

5. $\displaystyle \frac{d^2}{dz^2}\,(\text{erf } z) = -\frac{4}{\sqrt{\pi}}\, z\, e^{-z^2}$ (A.31)

6. $\text{erfc } z = 1 - \text{erf } z$ (A.32a)

For large z, $\text{erfc } z = \dfrac{1}{2\sqrt{\pi z}}\, e^{-z^2}$ (A.32b)

7. $\displaystyle i \text{ erfc } z = \int_z^\infty \text{erfc } z\, dz$ (A.33)

8. $\displaystyle i^2 \text{ erfc } z = \int_z^\infty \int_z^\infty \text{erf } z\, dz\, dz$ (A.34)

9. $\text{erf}\,(-z) = -\text{erf } z$ (A.35)

10. $\text{erf } 0 = 0$ (A.36)

11. $\text{erf } \infty = 1$ (A.37)

A short list of the function erfc z is given in Table A.1.

A.1.6 General Solution for Diffusion into a Semi-Infinite Body

We begin this solution with Fick's second law,

$$\frac{\partial N}{\partial t} = D\frac{\partial^2 N}{\partial x^2}$$ (A.38)

The initial distribution of the impurity is given by

$$N(x, 0) = f(x)$$ (A.39)

One solution of this equation may be written as the product of two functions, one of time and one of space. Thus let

$$N(x, t) = X(x)T(t)$$ (A.40)

TABLE A.1   Error Function erf z^a

z	erf(z)	z	erf(z)	z	erf(z)	z	erf(z)
0.00	0.000 000	0.50	0.520 500	1.00	0.842 701	1.50	0.966 105
0.01	0.011 283	0.51	0.529 244	1.01	0.846 810	1.51	0.967 277
0.02	0.022 565	0.52	0.537 899	1.02	0.850 838	1.52	0.968 413
0.03	0.033 841	0.53	0.546 464	1.03	0.854 784	1.53	0.969 516
0.04	0.045 111	0.54	0.554 939	1.04	0.858 650	1.54	0.970 586
0.05	0.056 372	0.55	0.563 323	1.05	0.862 436	1.55	0.971 623
0.06	0.067 622	0.56	0.571 616	1.06	0.866 144	1.56	0.972 628
0.07	0.078 858	0.57	0.579 816	1.07	0.869 773	1.57	0.973 603
0.08	0.090 078	0.58	0.587 923	1.08	0.873 326	1.58	0.974 547
0.09	0.101 281	0.59	0.595 936	1.09	0.876 803	1.59	0.975 462
0.10	0.112 463	0.60	0.603 856	1.10	0.880 205	1.60	0.976 348
0.11	0.123 623	0.61	0.611 681	1.11	0.883 533	1.61	0.977 207
0.12	0.134 758	0.62	0.619 411	1.12	0.886 788	1.62	0.978 038
0.13	0.145 867	0.63	0.627 046	1.13	0.889 971	1.63	0.978 843
0.14	0.156 947	0.64	0.634 586	1.14	0.893 082	1.64	0.979 622
0.15	0.167 996	0.65	0.642 029	1.15	0.896 124	1.65	0.980 376
0.16	0.179 012	0.66	0.649 377	1.16	0.899 096	1.66	0.981 105
0.17	0.189 992	0.67	0.656 628	1.17	0.902 000	1.67	0.981 810
0.18	0.200 936	0.68	0.663 782	1.18	0.904 837	1.68	0.982 493
0.19	0.211 840	0.69	0.670 840	1.19	0.907 608	1.69	0.983 153
0.20	0.222 703	0.70	0.677 801	1.20	0.910 314	1.70	0.983 790
0.21	0.233 522	0.71	0.684 666	1.21	0.912 956	1.71	0.984 407
0.22	0.244 296	0.72	0.691 433	1.22	0.915 534	1.72	0.985 003
0.23	0.255 023	0.73	0.698 104	1.23	0.918 050	1.73	0.985 578
0.24	0.265 700	0.74	0.704 678	1.24	0.920 505	1.74	0.986 135
0.25	0.276 326	0.75	0.711 156	1.25	0.922 900	1.75	0.986 672
0.26	0.286 900	0.76	0.717 537	1.26	0.925 236	1.76	0.987 190
0.27	0.297 418	0.77	0.723 822	1.27	0.927 514	1.77	0.987 691
0.28	0.307 880	0.78	0.730 010	1.28	0.929 734	1.78	0.988 174
0.29	0.318 283	0.79	0.736 103	1.29	0.931 899	1.79	0.988 641
0.30	0.328 627	0.80	0.742 101	1.30	0.934 008	1.80	0.989 091
0.31	0.338 908	0.81	0.748 003	1.31	0.936 063	1.81	0.989 525
0.32	0.349 126	0.82	0.753 811	1.32	0.938 065	1.82	0.989 943
0.33	0.359 279	0.83	0.759 524	1.33	0.940 015	1.83	0.990 347
0.34	0.369 365	0.84	0.765 143	1.34	0.941 914	1.84	0.990 736
0.35	0.379 382	0.85	0.770 668	1.35	0.943 762	1.85	0.991 111
0.36	0.389 330	0.86	0.776 100	1.36	0.945 561	1.86	0.991 472
0.37	0.399 206	0.87	0.781 440	1.37	0.947 312	1.87	0.991 821
0.38	0.409 009	0.88	0.786 687	1.38	0.949 016	1.88	0.992 156
0.39	0.418 739	0.89	0.791 843	1.39	0.950 673	1.89	0.992 479
0.40	0.428 392	0.90	0.796 908	1.40	0.952 285	1.90	0.992 790
0.41	0.437 969	0.91	0.801 883	1.41	0.953 852	1.91	0.993 090
0.42	0.447 468	0.92	0.806 768	1.42	0.955 376	1.92	0.993 378
0.43	0.456 887	0.93	0.811 564	1.43	0.956 857	1.93	0.993 656
0.44	0.466 225	0.94	0.816 271	1.44	0.958 297	1.94	0.993 923
0.45	0.475 482	0.95	0.820 891	1.45	0.959 695	1.95	0.994 179
0.46	0.484 655	0.96	0.825 424	1.46	0.961 054	1.96	0.994 426
0.47	0.493 745	0.97	0.829 870	1.47	0.962 373	1.97	0.994 664
0.48	0.502 750	0.98	0.834 232	1.48	0.963 654	1.98	0.994 892
0.49	0.511 668	0.99	0.838 508	1.49	0.964 898	1.99	0.995 111

TABLE A.1 *(Continued)*

2.00	0.995 322	2.50	0.999 593	3.00	0.999 977 91	3.50	0.999 999 257
2.01	0.995 525	2.51	0.999 614	3.01	0.999 979 26	3.51	0.999 999 309
2.02	0.995 719	2.52	9.999 634	3.02	0.999 980 53	3.52	0.999 999 358
2.03	0.995 906	2.53	0.999 654	3.03	0.999 981 73	3.53	0.999 999 403
2.04	0.996 086	2.54	0.999 672	3.04	0.999 982 86	3.54	0.999 999 445
2.05	0.996 258	2.55	0.999 689	3.05	0.999 983 92	3.55	0.999 999 485
2.06	0.996 423	2.56	0.999 706	3.06	0.999 984 92	3.56	0.999 999 521
2.07	0.996 582	2.57	0.999 722	3.07	0.999 985 86	3.57	0.999 999 555
2.08	0.996 734	2.58	0.999 736	3.08	0.999 986 74	3.58	0.999 999 587
2.09	0.996 880	2.59	0.999 751	3.09	0.999 987 57	3.59	0.999 999 617
2.10	0.997 021	2.60	0.999 764	3.10	0.999 988 35	3.60	0.999 999 644
2.11	0.997 155	2.61	0.999 777	3.11	0.999 989 08	3.61	0.999 999 670
2.12	0.997 284	2.62	0.999 789	3.12	0.999 989 77	3.62	0.999 999 694
2.13	0.997 407	2.63	0.999 800	3.13	0.999 990 42	3.63	0.999 999 716
2.14	0.997 525	2.64	0.999 811	3.14	0.999 991 03	3.64	0.999 999 736
2.15	0.997 639	2.65	0.999 822	3.15	0.999 991 60	3.65	0.999 999 756
2.16	0.997 747	2.66	0.999 831	3.16	0.999 992 14	3.66	0.999 999 773
2.17	0.997 851	2.67	0.999 841	3.17	0.999 992 64	3.67	0.999 999 790
2.18	0.997 951	2.68	0.999 849	3.18	0.999 993 11	3.68	0.999 999 805
2.19	0.998 046	2.69	0.999 858	3.19	0.999 993 56	3.69	0.999 999 820
2.20	0.998 137	2.70	0.999 866	3.20	0.999 993 97	3.70	0.999 999 833
2.21	0.998 224	2.71	0.999 873	3.21	0.999 994 36	3.71	0.999 999 845
2.22	0.998 308	2.72	0.999 880	3.22	0.999 994 73	3.72	0.999 999 857
2.23	0.998 388	2.73	0.999 887	3.23	0.999 995 07	3.73	0.999 999 867
2.24	0.998 464	2.74	0.999 893	3.24	0.999 995 40	3.74	0.999 999 877
2.25	0.998 537	2.75	0.999 899	3.25	0.999 995 70	3.75	0.999 999 886
2.26	0.998 607	2.76	0.999 905	3.26	0.999 995 98	3.76	0.999 999 895
2.27	0.998 674	2.77	0.999 910	3.27	0.999 996 24	3.77	0.999 999 903
2.28	0.998 738	2.78	0.999 916	3.28	0.999 996 49	3.78	0.999 999 910
2.29	0.998 799	2.79	0.999 920	3.29	0.999 996 72	3.79	0.999 999 917
2.30	0.998 857	2.80	0.999 925	3.30	0.999 996 94	3.80	0.999 999 923
2.31	0.998 912	2.81	0.999 929	3.31	0.999 997 15	3.81	0.999 999 929
2.32	0.998 966	2.82	0.999 933	3.32	0.999 997 34	3.82	0.999 999 934
2.33	0.999 016	2.83	0.999 937	3.33	0.999 997 51	3.83	0.999 999 939
2.34	0.999 065	2.84	0.999 941	3.34	0.999 997 68	3.84	0.999 999 944
2.35	0.999 111	2.85	0.999 944	3.35	0.999 997 838	3.85	0.999 999 948
2.36	0.999 155	2.86	0.999 948	3.36	0.999 997 983	3.86	0.999 999 952
2.37	0.999 197	2.87	0.999 951	3.37	0.999 998 120	3.87	0.999 999 956
2.38	0.999 237	2.88	0.999 954	3.38	0.999 998 247	3.88	0.999 999 959
2.39	0.999 275	2.89	0.999 956	3.39	0.999 998 367	3.89	0.999 999 962
2.40	0.999 311	2.90	0.999 959	3.40	0.999 998 478	3.90	0.999 999 965
2.41	0.999 346	2.91	0.999 961	3.41	0.999 998 582	3.91	0.999 999 968
2.42	0.999 379	2.92	0.999 964	3.42	0.999 998 679	3.92	0.999 999 970
2.43	0.999 411	2.93	0.999 966	3.43	0.999 998 770	3.93	0.999 999 973
2.44	0.999 441	2.94	0.999 968	3.44	0.999 998 855	3.94	0.999 999 975
2.45	0.999 469	2.95	0.999 970	3.45	0.999 998 934	3.95	0.999 999 977
2.46	0.999 497	2.96	0.999 972	3.46	0.999 999 008	3.96	0.999 999 979
2.47	0.999 523	2.97	0.999 973	3.47	0.999 999 077	3.97	0.999 999 980
2.48	0.999 547	2.98	0.999 975	3.48	0.999 999 141	3.98	0.999 999 982
2.49	0.999 571	2.99	0.999 976	3.49	0.999 999 201	3.99	0.999 999 983

[a] For a more complete table, see L. J. Comrie, *Chambers Six Figure Mathematical Tables*, Vol. 2, W. & R. Chambers, Edinburgh, 1949.

Substituting in Eq. (A.38) and rearranging terms,

$$\frac{1}{DT}\frac{dT}{dt} = \frac{1}{X}\frac{d^2X}{dx^2} \tag{A.41}$$

Since the left-hand side of this equation is only a function of t and the right-hand side only a function of x, each side must be equal to some constant that is independent of both t or x. Writing this constant as $-\lambda^2$, Eq. (A.40) can be broken into

$$\frac{dT}{T} = -\lambda^2 D\ dt \tag{A.42}$$

and

$$\frac{d^2X}{dx^2} = -\lambda^2 X \tag{A.43}$$

Solving,

$$T(t) = \gamma e^{-\lambda^2 Dt} \tag{A.44}$$

$$X(x) = \alpha \cos \lambda x + \beta \sin \lambda x \tag{A.45}$$

and

$$N(x, t) = \lambda e^{-\lambda^2 Dt}(A \cos \lambda x + B \sin \lambda x) \tag{A.46}$$

where γ, α, β are constants of integration, $A = \alpha\gamma$, and $B = \beta\gamma$.

The general solution of Eq. (A.38) can be written as a sum of partial solutions of this type because it is a linear equation. In addition, since the body has infinite dimensions, the choice of λ is quite arbitrary, and the summation over discrete values of λ can be replaced by an integral. Hence this solution may be written as

$$N(x, t) = \int_{-\infty}^{+\infty} e^{-\lambda^2 Dt}(A \cos \lambda x + B \sin \lambda x)\ d\lambda \tag{A.47}$$

It is now necessary to solve for the various constants of integration. Fourier's integral theorem states that

$$f(x) = \frac{1}{2\pi}\int_{-\infty}^{+\infty}\int_{-\infty}^{+\infty} f(\xi) \cos [\lambda(\xi - x)]\ d\xi\ d\lambda \tag{A.48}$$

$$= \frac{1}{2\pi}\int_{-\infty}^{+\infty}\left\{\left[\int_{-\infty}^{+\infty} f(\xi) \cos \lambda\xi\ d\xi\right] \cos \lambda x\ dx\right.$$

$$\left. + \left[\int_{-\infty}^{+\infty} f(\xi) \sin \lambda\xi\ d\xi\right] \sin \lambda x\ dx\right\}\ d\lambda \tag{A.49}$$

At time $t = 0$ the initial concentration is given by $f(x)$.

Substituting in Eq. (A.47),

$$f(x) = N(x, 0) = \int_{-\infty}^{\infty} (A \cos \lambda x + B \sin \lambda x) \, d\lambda \qquad (A.50)$$

Comparing Eqs. (A.49) and (A.50),

$$A = \frac{1}{2\pi} \int_{-\infty}^{\infty} f(\xi) \cos \lambda \xi \, d\xi \qquad (A.51)$$

$$B = \frac{1}{2\pi} \int_{-\infty}^{\infty} f(\xi) \sin \lambda \xi \, d\xi \qquad (A.52)$$

Substituting these values of A and B in Eq. (A.47), gives

$$N(x, t) = \frac{1}{2\pi} \int_{-\infty}^{\infty} f(\xi) \left[\int_{-\infty}^{\infty} e^{-\lambda^2 Dt} \cos \lambda(\xi - x) \, d\lambda \right] d\xi \qquad (A.53)$$

But

$$\int_{-\infty}^{\infty} e^{-\lambda^2 Dt} \cos \lambda(\xi - x) \, dx = \sqrt{\pi/Dt} \, e^{-(\xi - x)^2/4Dt} \qquad (A.54)$$

Hence

$$N(x, t) = \frac{1}{2\sqrt{\pi Dt}} \int_{-\infty}^{\infty} f(\xi) \, e^{-(\xi - x)^2/4Dt} \, d\xi \qquad (A.55)$$

This is the general solution of the diffusion equation for an infinite body. Here

$$f(\xi) = N(\xi, 0) \qquad (A.56)$$

A.2   SOLUTION FOR A TIME-DEPENDENT DIFFUSION COEFFICIENT

Consider the situation where the diffusivity is given by

$$D = D(t) \qquad (A.57)$$

Introducing a new time variable T, such that

$$\partial T = D(t) \, \partial t \qquad (A.58)$$

the diffusion equation now becomes

$$\frac{\partial N}{\partial T} = \frac{\partial^2 N}{\partial x^2} \qquad (A.59)$$

so that it can be solved as before. Note that this new time variable can be written as

$$T = (Dt)_{\text{eff}} = \int_0^{t_0} D(t)\, dt \qquad (A.60)$$

where $(Dt)_{\text{eff}}$ is the effective Dt product, and t_0 is the time interval over which the diffusivity varies.

A.2.1 Ramping of a Diffusion Furnace

An important situation which occurs in practice is the ramping of a diffusion furnace to prevent warpage of slices. The effect of ramping can be readily estimated for a linear ramp rate C. Assuming a ramp-down situation, the furnace temperature is given by

$$T = T_0 - Ct \qquad (A.61)$$

where T_0 is the initial temperature. Assume that ramp-down occurs for a time t_0. Then the effective Dt product during this time is given by

$$(Dt)_{\text{eff}} = \int_0^{t_0} D(t)\, dt \qquad (A.62)$$

During a typical diffusion, ramping is carried out until the diffusivity is negligibly small. Thus this integration can be simplified with little error if the upper limit is taken as infinity. Since

$$\frac{1}{T} = \frac{1}{T_0 - Ct} \simeq \frac{1}{T_0}\left(1 + \frac{Ct}{T_0} + \cdots\right) \qquad (A.63)$$

and

$$D = D_0\, e^{-E_0/kT} \qquad (A.64)$$

we can write

$$D(t) \simeq D_0 \exp\left[-\frac{E_0}{kT_0}\left(1 + \frac{Ct}{T_0} + \cdots\right)\right]$$

$$= D(T_0)\, e^{-(CE_0/kT_0^2)t} \qquad (A.65)$$

where $D(T_0)$ is the diffusivity at T_0, and is given by

$$D(T_0) = D_0\, e^{-E_0/kT} \qquad (A.66)$$

Substituting into Eq. (A.62) and using an upper integration limit of infinity, gives

$$(Dt)_{eff} \simeq D(T_0) \left(\frac{kT_0^2}{CE_0} \right) \tag{A.67}$$

Thus the ramp-down process results in an effective additional time equal to kT_0^2/CE_0 at the initial diffusion temperature T_0. A similar relationship can be derived for the ramp-up situation.

A.3 SOLUTION FOR CONCENTRATION-DEPENDENT DIFFUSION COEFFICIENTS

This situation is encountered often in practice, as shown in Chapter 4. The equation for diffusion is then

$$\frac{\partial N}{\partial t} = \frac{\partial}{\partial x} \left(D \frac{\partial N}{\partial x} \right) \tag{A.68}$$

where D is a function of N. This equation can be converted into an ordinary differential equation by means of Boltzmann's transformation. Let

$$\eta = x/2\sqrt{t} \tag{A.69}$$

Then

$$\frac{\partial N}{\partial x} = \frac{1}{2t^{1/2}} \frac{dN}{d\eta} \tag{A.70a}$$

$$\frac{\partial N}{\partial t} = -\frac{x}{4t^{3/2}} \frac{dN}{d\eta} \tag{A.70b}$$

so that Fick's second law reduces to

$$-2\frac{dN}{d\eta} = \frac{d}{d\eta} \left(D \frac{dN}{d\eta} \right) \tag{A.71}$$

This equation can be solved for those situations where the boundary conditions can be expressed in terms of η alone. An important case is that of diffusion in a semi-infinite medium with a constant surface concentration, such that

$$N = N_{sur}, \qquad x = 0, t > 0 \tag{A.72a}$$

$$N = N_1, \qquad x > 0, t = 0 \tag{A.72b}$$

TABLE A.2   Numerical Solutions to the Concentration Dependent Diffusion Equation[a,b]

y	Case A	Case B	Case C
		N	
0.001	0.999	0.999	0.999
0.005	0.996	0.996	0.997
0.01	0.991	0.992	0.993
0.02	0.982	0.985	0.986
0.05	0.955	0.961	0.964
0.10	0.907	0.918	0.925
0.15	0.858	0.872	0.881
0.20	0.807	0.822	0.830
0.25	0.753	0.766	0.769
0.30	0.697	0.703	0.697
0.32	0.674	0.677	0.663
0.34	0.650	0.648	0.624
0.36	0.626	0.617	0.577
0.38	0.602	0.584	0.521
0.40	0.578	0.550	0.449
0.42	0.553	0.512	0.340
0.43			0.237
0.435			0.058
0.436			0.000
0.44	0.528	0.471	
0.46	0.503	0.425	
0.48	0.477	0.373	
0.50	0.451	0.312	
0.52	0.425	0.234	
0.54	0.398	0.107	
0.545		0.022	
0.546		0.000	
0.56	0.370		
0.58	0.343		
0.60	0.315		
0.62	0.287		
0.64	0.258		
0.66	0.229		
0.68	0.199		
0.70	0.169		
0.72	0.139		
0.74	0.108		
0.76	0.077		
0.78	0.045		
0.80	0.013		
0.808	0.000		

[a] Here $N = N(x, t)/N_{sur}$ and $y = x/2 \sqrt{D_{sur}t}$.
[b] See reference 4.

For this situation the boundary conditions become

$$N = N_{sur}, \qquad \eta = 0 \qquad (A.73a)$$

$$N = N_1, \qquad \eta = \infty \qquad (A.73b)$$

and Eq. (A.71) can be used for this case.

The diffusion of many impurities in silicon and gallium arsenide was described in Section 4.5.2 by a diffusion coefficient D, such that

$$\text{Case } A: \qquad D = D_{sur} \left(\frac{N}{N_{sur}} \right) \qquad (A.74a)$$

$$\text{Case } B: \qquad D = D_{sur} \left(\frac{N}{N_{sur}} \right)^2 \qquad (A.74b)$$

$$\text{Case } C: \qquad D = D_{sur} \left(\frac{N}{N_{sur}} \right)^3 \qquad (A.74c)$$

Numerical solutions for these cases have been made [4]. Tabulated values for N/N_{sur} versus $y \ (= x/2\sqrt{D_{sur} \, t})$ are given in Table A.2 for these solutions.

A.4   DETERMINATION OF THE DIFFUSION CONSTANT

The diffusion of impurities in semiconductors is characterized at any given temperature by a diffusion constant D, which may be a function of impurity concentration. Many diffusions in silicon, particularly those with low surface concentrations (base diffusions, for example), can be characterized by a concentration independent value of D. In general, this is not true for emitter diffusions, particularly for shallow emitters. Nevertheless, the assumption of concentration independence is often made for simplicity. Diffusions in gallium arsenide are usually described by a concentration dependent value of D.

A.4.1   The p–n Junction Method

This technique assumes that D is concentration independent. It is extremely popular because the experimental technique involves the formation of junctions. As a result, the answers obtained in this way are especially useful in device fabrication. The starting point for this technique is to make one or more diffusions into semiconductors of known background concentration and opposite impurity type. Assuming that diffusion takes place from a constant concentration impurity source into a background concentration

N_B, we obtain

$$N_B = N_0 \operatorname{erfc}\left(\frac{x_j}{2\sqrt{Dt}}\right) \tag{A.75}$$

where x_j is the junction depth and N_0 is the surface concentration. Both x_j and N_0 can be obtained by the methods outlined in Section 4.10. N_B is given from measurement of the specific resistivity of the original material.

An alternate method, which avoids determination of N_0, is to conduct simultaneous diffusions into two slices of background concentration N_{B1} and N_{B2} under identical conditions. If the junction depths are x_1 and x_2, respectively, then

$$\frac{N_{B1}}{N_{B2}} = \frac{\operatorname{erfc}(x_1/2\sqrt{Dt})}{\operatorname{erfc}(x_2/2\sqrt{Dt})} \tag{A.76}$$

This equation can be solved approximately with

$$D \simeq \frac{1}{4t}\left[\frac{x_1^2 - x_2^2}{\ln(N_{B2}x_1/N_{B1}x_2)}\right] \tag{A.77}$$

A more general method, for an arbitrary impurity profile, now follows.

A.4.2   The Boltzmann–Matano Method [5]

The starting point for this method is a knowledge of the impurity concentration as a function of diffusion depth. This can be obtained by conducting a diffusion with the impurity, followed by direct measurement by secondary ion mass spectrometry. An alternative technique is diffusion with a radioactive isotope of the impurity, and measurement of the radiation intensity as successive layers are removed. A third technique consists of measuring the sheet resistance upon successive removals of layers. Details of these techniques are beyond the scope of this book.

Once the concentration profile is known, it is necessary to establish that the diffusion process can be described by means of the Boltzmann transformation. This requires first that a linear relation exist between the impurity penetration depth and the square root of the diffusion time. This is true for the majority of diffusion situations in silicon as well as in gallium arsenide.

Consider the situation of a diffusing impurity into an undoped background. Here

$$N = N_0, \qquad x < 0, t = 0 \tag{A.78a}$$

$$N = 0, \qquad x > 0, t = 0 \tag{A.78b}$$

Using Boltzmann's transformation, Fick's second law is given by

$$-2\eta \frac{dN}{d\eta} = \frac{d}{d\eta}\left(D\frac{dN}{d\eta}\right) \tag{A.79}$$

so that

$$-2\int_0^{N_1} \eta\, dN = \left[D\frac{dN}{d\eta}\right]_{N=0}^{N=N_1} \tag{A.80}$$

But $D(dN/d\eta) = 0$ when $N = 0$. Making this substitution and setting $\eta = x/2\sqrt{t}$, gives

$$D_{N=N_1} = -\frac{1}{2t}\frac{dx}{dN}\int_0^{N_1} x\, dN \tag{A.81}$$

Eq. (A.81) shows that the diffusion constant can be determined at any given concentration N_1 from knowledge of the slope of the N versus x profile at $N = N_1$, and the total number of diffused impurities between $N = 0$ and $N = N_1$.

REFERENCES

1. B. I. Boltaks, *Diffusion in Semiconductors*, Academic, New York, 1963.
2. J. Crank, *The Mathematics of Diffusion*, Clarendon, Oxford, England, 1975.
3. W. Jost, *Diffusion in Solids, Liquids, and Gases*, Academic, New York, 1962.
4. L. R. Weisberg and J. Blanc, Diffusion with Interstitial-Substitutional Equilibrium. Zinc in GaAs, *Phys. Rev.* **131**, 1548 (15 Aug. 1963).
5. P. G. Shewmon, *Diffusion in Solids*, McGraw Hill, New York, 1963.

INDEX

Italic page numbers indicate pertinent data concerning the item.